Aquatic Ecotoxicology

Aquatic Ecotoxicology

Advancing Tools for Dealing with Emerging Risks

Edited by

Claude Amiard-Triquet, PhD, DSc

Honorary Research Director, Centre National de la Recherche Scientifique (CNRS), University of Nantes, France; Invited Professeur at Ocean University of China, Qingdao

Jean-Claude Amiard, PhD, DSc

Emeritus Research Director, CNRS, University of Nantes, France; Invited Professeur at Ocean University of China, Qingdao

Catherine Mouneyrac, PhD, MSc

Professor of Aquatic Ecotoxicology and Dean of the Faculty of Sciences, Universite Catholique de L'Ouest, Angers, France

AMSTERDAM • BOSTON • HEIDELBERG • LONDON
NEW YORK • OXFORD • PARIS • SAN DIEGO
SAN FRANCISCO • SINGAPORE • SYDNEY • TOKYO

Academic Press is an imprint of Elsevier

Academic Press is an imprint of Elsevier
125 London Wall, London EC2Y 5AS, UK
525 B Street, Suite 1800, San Diego, CA 92101-4495, USA
225 Wyman Street, Waltham, MA 02451, USA
The Boulevard, Langford Lane, Kidlington, Oxford OX5 1GB, UK

ISBN: 978-0-12-800949-9

British Library Cataloguing-in-Publication Data
A catalogue record for this book is available from the British Library

Library of Congress Cataloging-in-Publication Data
A catalog record for this book is available from the Library of Congress

For information on all Academic Press publications
visit our website at http://store.elsevier.com/

Publisher: Mica Haley
Acquisition Editor: Erin Hill-Parks
Editorial Project Manager: Molly McLaughlin
Production Project Manager: Julia Haynes
Designer: Mark Rogers

Typeset by TNQ Books and Journals
www.tnq.co.in

Printed and bound in the United States of America

Contents

Contributors

Jean-Claude Amiard, PhD, DSc French National Research Center (CNRS), LUNAM University, Nantes, France; and Université de Nantes, Nantes, France

Rachid Amara, PhD Laboratoire d'Océanologie et Géosciences, Université du Littoral, Wimereux, France

Claude Amiard-Triquet, PhD, DSc French National Research Center (CNRS), France

Jean Armengaud, PhD, HDR CEA-Marcoule, DSV/IBICTEC-S/SPI/Li2D, Laboratory Innovative Technologies for Detection and Diagnostic, Bagnols-sur-Cèze, France

M.J. Bebianno, PhD CIMA–Centre of Marine Environmental Research, University of Algarve, Faro, Portugal

Brigitte Berthet, PhD, ICES La Roche sur Yon, France; LUNAM University, Nantes, France; and Université de Nantes, Nantes, France

Angel Borja, PhD Marine Research Division, AZTI-Tecnalia, Herrera Kaia, Pasaia (Gipuzkoa), Spain

Julie Bremner, BSc, MRes, PhD Cefas, Suffolk, England, UK

Arnaud Chaumot, PhD Irstea, Unité de Recherche MALY, Laboratoire d'écotoxicologie, Villeurbanne, France

Tracy K. Collier National Marine Fisheries Service, National Oceanic and Atmospheric Administration, Washington, DC, USA

Gael Dur Université Lille 1 Sciences et Technologies, CNRS UMR 8187 LOG, Wimereux, France

Olivier Geffard, PhD Irstea, Unité de Recherche MALY, Laboratoire d'écotoxicologie, Villeurbanne, France

Patrice Gonzalez, PhD Researcher, French National Research Center (CNRS), University of Bordeaux, Arcachon, France

M. Gonzalez-Rey CIMA–Centre of Marine Environmental Research, University of Algarve, Faro, Portugal

Scott A. Hecht National Marine Fisheries Service, National Oceanic and Atmospheric Administration, Washington, DC, USA

John P. Incardona National Marine Fisheries Service, National Oceanic and Atmospheric Administration, Washington, DC, USA

Kevin W.H. Kwok, BSc, PhD Food Safety and Technology Research Centre, Department of Applied Biology and Chemical Technology, The Hong Kong Polytechnic University, Hung Hom, Hong Kong

Cathy A. Laetz, M.S. Marine Environmental Science, National Marine Fisheries Service, National Oceanic and Atmospheric Administration, Washington, DC, USA

Jae-Seong Lee Department of Biological Science, College of Science, Sungkyunkwan University, Suwon, South Korea

Mario Lepage Irstea, UR EABX, Cestas, France

Lorraine Maltby, BSc, PhD Department of Animal and Plant Sciences, The University of Sheffield, Sheffield, UK

James C. McGeer Biology, Wilfrid Laurier University, Waterloo, ON, Canada

Christophe Minier, PhD ONEMA, Vincennes, France

Catherine Mouneyrac, PhD, DSc LUNAM University, Nantes, France; and Université Catholique de l'Ouest, Angers, France

Iñigo Muxika Marine Research Division, AZTI-Tecnalia, Herrera Kaia, Pasaia (Gipuzkoa), Spain

Fabien Pierron French National Research Center (CNRS), University of Bordeaux, Arcachon, France

J. Germán Rodríguez Marine Research Division, AZTI-Tecnalia, Herrera Kaia, Pasaia (Gipuzkoa), Spain

Nathaniel L. Scholz, PhD National Marine Fisheries Service, National Oceanic and Atmospheric Administration, Washington, DC, USA

Henriette Selck, MSc, PhD Department of Environmental, Social and Spatial Change, Roskilde University, Roskilde, Denmark

Sami Souissi, PhD Université Lille 1 Sciences et Technologies, CNRS UMR 8187 LOG, Wimereux, France

Kristian Syberg, MSc, PhD Department of Environmental, Social and Spatial Change, Roskilde University, Roskilde, Denmark

Katrin Vorkamp Department of Environmental Science, Aarhus University, Roskilde, Denmark

Eun-Ji Won Department of Biological Science, College of Science, Sungkyunkwan University, Suwon, South Korea

CHAPTER 1

Introduction

Claude Amiard-Triquet

Abstract

The aquatic environment appears as the final destination for most of anthropogenic contaminants released from industry, agriculture, urbanization, transport, tourism, and everyday life. On the other hand, inland, coastal, and marine waters provide many services important for human well-being. The conservation of ecosystems and human health is based on a sound assessment of the risks associated with the presence of contaminants in the aquatic environment. The aim of this book is to use cross-analyses of procedures, biological models, and contaminants to design ecotoxicological tools suitable for better environmental assessments, particularly in the case of emerging contaminants and emerging concern with legacy pollutants.

Keywords: Aquatic environment; Bioaccumulation; Bioassays; Bioindicators; Biomarkers; Ecotoxicological tools; Emerging contaminants; Emerging risks; Exposure; Risk assessment.

Chapter Outline

Many different classes of contaminants enter the environment as a consequence of human activities, including industry, agriculture, urbanization, transport, tourism, and everyday life. Initially, air pollutants are atmospheric contaminants and solid wastes are terrestrial contaminants, whereas liquid effluents are aquatic contaminants. Processes involved in the fate of contaminants in each compartment—air, soil, water—lead to many intercompartment exchanges, governed by advection (e.g., deposition, run-off, erosion), diffusion (e.g., gas absorption, volatilization), and degradation, both biotic and abiotic (Figures 2.2 and 2.3, this book), and by large-scale transport from atmospheric and marine currents. The aquatic environment appears as the final destination for most of anthropogenic contaminants, and aquatic sediments, either deposited or in suspension, as the major sink for their storage with only few exceptions (e.g., perfluorooctanoic acid [PFOA], Post et al., 2012; water-soluble pesticides such as alachlor, atrazine, and diuron).

Aquatic Ecotoxicology. http://dx.doi.org/10.1016/B978-0-12-800949-9.00001-2

It is generally admitted that about 100,000 molecules are introduced in aquatic media intentionally (e.g., pesticides, antifouling paints) and more often unintentionally as incompletely treated sewages or because of accidents. In most cases, several classes of contaminants are present concomitantly, thus being able to act in addition, synergy, or antagonism (Chapter 18).

The Millennium Ecosystem Assessment (MEA, 2005) highlights how ecosystem services are important as determinants and constituents of human well-being. Inland, coastal, and marine waters are important contributors of core services (nutrient cycling, primary production). They are also part contributors in providing services (water, food, biochemicals, genetic resources) and cultural services (such as recreation and ecotourism, aesthetic and educational benefits). As emphasized by Maltby (2013), applying approaches based on the ecosystem service concept to the protection, restoration, and management of ecosystems requires the development of new understanding, tools, and frameworks.

Legislation has been adopted on a worldwide scale to improve the status of aquatic ecosystems (e.g., United States' Clean Water Act, 1972; European Community Water Framework Directive, ECWFD, 2000). Environmental management aiming at the improvement of chemical and ecological quality in aquatic media must be based on robust risk assessments. Retrospective risk assessments are performed when sites have potentially been impacted in the past. When they show a degradation of environmental quality, the restoration of degraded habitats and ecosystems must be addressed. Prospective, or predictive, risk assessments aim at assessing the future risks of anthropogenic pressure such as climate change or releases of new chemicals into the environment. Strategies to limit the risks of both new and existing chemicals include the federal Toxic Substances Control Act (1976) in the United States, and a new chemical policy, Registration, Evaluation, and Authorization of CHemicals in Europe (2006).

1.1 Ecotoxicological Tools Currently Used for Risk Assessment in Aquatic Media

Conventional risk assessment (Chapter 2) in different environments aims at establishing a comparison between the degree of exposure expected or measured in the field and the effects induced by a contaminant or a class of contaminants. It is mainly based on the determination of predicted environmental concentrations (PECs) and predicted no effect concentrations (PNECs). The procedure has been described in the Technical Guidance Document on Risk Assessment in support of European Commission regulations (TGD, 2003). PECs and PNECS are then used in a risk quotient approach: very simplistically, if the PEC/PNEC ratio is lower than 1, the substance is not considered to be of concern; if the PEC/PNEC ratio is higher than 1, further testing must be carried out to improve the determination of PEC or PNEC with subsequent revision of PEC/PNEC ratio, or risk reduction measures must be envisaged (TGD, 2003).

Environmental quality standards ([EQS] concentration in water, sediment, or biota that must not be exceeded) are a major tool to protect the aquatic environment and human health

(Chapter 3). An overshoot of EQS at a given site triggers management actions (e.g., research for contamination sources, reduction of contaminant discharges). EQS for sediment and biota are needed to ensure protection against indirect effects and secondary poisoning. To date, no EQSs are available for sediments under the ECWFD (2000), partly because the total dose of a pollutant in sediment has a low ecotoxicological significance and the bioavailable fraction must be determined using specific methods (Chapter 3). In addition, different sediment quality guidelines are commonly used by official organisms in the US (National Oceanic and Atmospheric Administration) (Long et al., 1995; MacDonald et al., 1996), Canada (http://ceqg-rcqe.ccme.ca/), Australia (McCready et al., 2006), etc. An overshoot of these guidelines at a given site triggers additional investigations on the impacts and their extent.

Environmental monitoring is then indispensable to assess if environmental concentrations meet standards/guidelines (Chapter 3). An excellent example is provided by the Coordinated Environmental Monitoring Program undertaken under the OSPAR Commission that aims at protecting and conserving the Northeast Atlantic and its resources. Guidelines for monitoring of hazardous substances in sediment and biota are available at http://www.ospar.org/content/content.asp?menu=00900301400135_000000_000000. OSPAR monitoring guidance is regularly reviewed in collaboration with the International Council for the Exploration of the Sea and, where necessary, updated to take account of new developments such as the inclusion of new monitoring parameters.

However, chemical measurements of contaminants in environmental matrices pose a number of problems in many monitoring programs:

1. Analytical efforts focus on chemicals that are perceived to be relatively easy to analyze (heavy metals, DDT and its metabolites, γHCH, αHCH, some congeners of polychlorobiphenyls [PCBs], some individual polycyclic aromatic hydrocarbons [PAHs], etc.);
2. Complex mixtures present in multipolluted environments include many classes of compounds that are not yet accessible to analysis or are extremely expensive to analyze, particularly emerging contaminants (nanomaterials) or known contaminants of emerging concern (pharmaceuticals, personal care products) and their metabolites;
3. As previously mentioned for sediments, the total dose of a pollutant in any compartment of the environment (water, sediment, biota) has a low ecotoxicological significance since their physicochemical forms govern their bioaccessibility and biological effects.

1.2 How Can We Improve Risk Assessment?

To improve exposure assessment, it is indispensable to take into account the physicochemical characteristics of different classes of contaminants (Chapter 4). In the case of metals, a number of chemical speciation models allows a good characterization of the metal chemical species in a solution containing inorganic ligands and well-characterized organic ligands, particularly natural organic matter that is one of the most dominating processes in freshwater and salinity (chlorinity) in seawater (Paquin et al., 2002; VanBriesen et al., 2010). Different procedures have been described to take into account bioavailability concepts in the risk assessment process or environmental quality criteria setting. The Free Ion Activity Model (FIAM) has been designed to take into account the central role of the activity of the free metal ion as a regulator of interactions (both uptake and toxicity) between metals and aquatic organisms (Campbell, 1995). As the FIAM, the Biological Ligand Model is a chemical equilibrium-based model but at the center of this model is the site of action of toxicity in the organism that corresponds to the biotic ligand. The Biological Ligand Model can be used to predict the degree of metal binding at this site of action, and this level of accumulation is in turn related to a toxicological response (Paquin et al., 2002).

Passive samplers are devices that rely on diffusion and sorption to accumulate analytes in the sampler (Mills et al., 2010). Among these techniques, diffusive equilibration in thin films and diffusive gradients in thin films allow a better understanding of the speciation of metals in the environment, differentiating between free-, inorganic-, and organic-bound metal species and organometallic compounds. Other passive samplers can be used for different classes of organic chemicals, also providing a partial determination of the physicochemical characteristics that govern fate and effects of contaminants. For instance, semipermeable membrane devices are relevant for nonpolar contaminants such as PAHs, whereas polar organic chemical integrative samplers are relevant for polar compounds such as detergents including alkylphenols, pharmaceuticals, and pesticides (Mills et al., 2010).

Chemical monitoring in different environmental matrices (seawater, freshwater, underground water, and effluents; sediment and leachates; organismal tissue and fluids) may be carried out by using either a priori or "global" approach. In the first case, the analyses focus on main classes of known contaminants, particularly the priority hazardous substances listed in European legislation (http://eur-lex.europa.eu/LexUriServ/LexUriServ.do?uri=OJ:L:2013:226:0001:0017:EN:PDF) and USEPA (http://water.epa.gov/scitech/methods/cwa/pollutants.cfm). For instance, the Joint Assessment and Monitoring Program guidelines for monitoring contaminants in biota (https://www.google.fr/#q=JAMP+Guidelines+for+Monitoring+Contaminants+in+Biota) and sediments (https://www.google.fr/#q=jamp+guidelines+for+monitoring+contaminants+in+sediments) provide procedures for metals (including organotin compounds), parent and alkylated PAHs, hexabromocyclododecane, perfluorinated compounds (PFCs), polybromodiphenyl ethers, PCBs, dioxins, furans, and dioxin-like PCBs. However, this a priori approach is not adapted for all the unknown contaminants present in mixtures in most of the aquatic media. The global approach combining biotesting, fractionation, and chemical analysis, helps to identify hazardous compounds in complex environmental mixtures (Burgess et al., 2013). Toxicity identification evaluation (TIE), which was mainly developed in North America in support to the US Clean Water Act and effects-directed analysis (EDA) that originates from both Europe and North America, differed primarily by the biological endpoints used to reveal toxicity (whole organism toxicity tests vs cellular toxicity tests able to reveal mutagenicity, genotoxicity, and endocrine disruption) (Figure 15.2, this book).

In the case of emerging contaminants such as nanomaterials (NMs), the situation is even more challenging because efficient methods and techniques for their detection and quantification in the complex environmental media are not yet available—not to mention difficulties currently insurmountable for investigating the transport and fate of NMs in water systems (Wong et al., 2013). However, advanced nuclear analytical and related techniques recently reviewed by Chen et al. (2013) are powerful tools that can be applied (1) to study their transformation in vitro; (2) to analyze the bio–nano interactions at the molecular level; and (3) for the study of in vivo biodistribution and quantification of nanoparticles in animals. But to date, many of these analytical resources located in large-scale facilities are not available for routine applications in nanotoxicology (Chen et al., 2013).

Processes leading to bioaccumulation of environmental contaminants in aquatic organisms are reviewed in Chapter 5. They include direct uptake of compounds from water (bioconcentration) as well as dietary uptake and incorporation of sediment-bound contaminants. Even when considering only waterborne exposure, bioconcentration factors (concentration in biota/concentration in water) indicate that nearly all the contaminants are incorporated at a level higher than encountered in water. As mentioned previously, the chemical characteristics of contaminants in water and other sources (preys, sediment) are a major driver of bioaccumulation but biological factors also influence bioaccumulation (Eggleton and Thomas, 2004; Abarnou, in Amiard-Triquet and Rainbow, 2009; Rainbow et al., 2011). The concepts of bioaccessibility and trophic bioavailability are often used concurrently. Release of a chemical from ingested food is a prerequisite for uptake and assimilation. Bioaccessibility of a food-bound contaminant can be measured by its extractability from food (or sediment frequently ingested with food by deposit-feeding invertebrates and flatfish). Trophic bioavailability should be used in the strict sense to describe the proportion of a chemical ingested with food which enters the systemic circulation (Versantvoort et al., 2005). Similarly, only a fraction of contaminants present in water is readily available for organisms (FIAM, Chapter 4).

Thus the Tissue Residue Approach has been developed to link toxicity to incorporated doses of contaminants rather than external doses (Chapter 5). However, the relationship between global concentrations in organisms and noxious effects is not simple. The limitation of uptake—responsible for the gap between bioaccessibility and bioavailability—has been described as contributing to the ability of organisms to cope with the presence of contaminants in their medium as well as increased elimination, or storage in nontoxic forms (Amiard-Triquet and Rainbow, in Amiard-Triquet et al., 2011). For instance, when an organism has high metal concentrations in its tissues, it does not necessarily exhibit toxicity effects (Luoma and Rainbow, 2008).

However, to date the results of bioassays generally remained expressed as an external concentration–effect relationship (Figure 1.1). In addition, these classical bioassays exhibit a number of weaknesses. Considering the conditions of exposure (*X* axis), acute concentrations are most often tested, whereas in the real world, low concentrations are present except

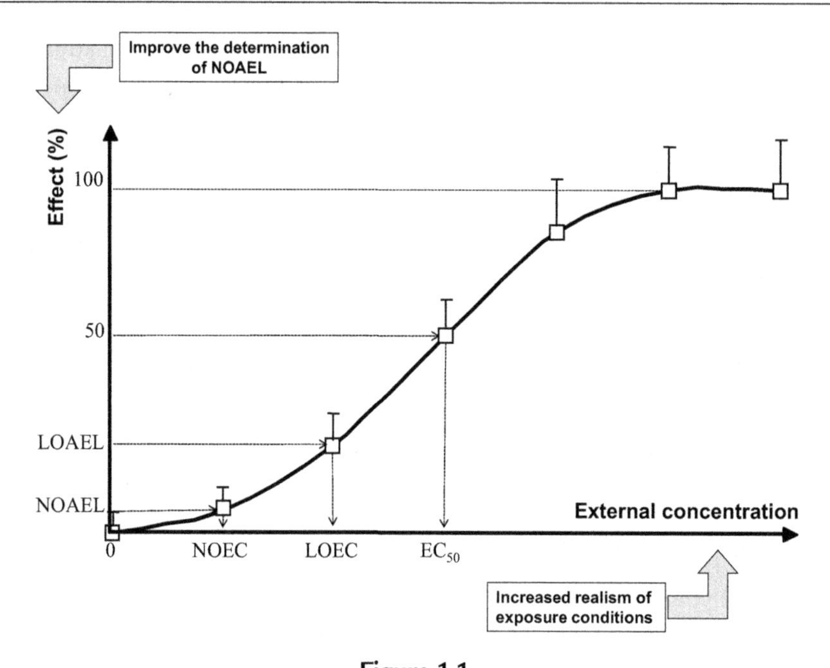

Figure 1.1
How to improve the assessment of the concentration–effect relationship in classical bioassays.

in the case of accidents. The contaminant under examination is generally added in water, whereas dietary and sediment exposures are neglected. Interactions between different classes of contaminants are neglected. Considering the observed effects (*Y* axis), test organisms are most often the equivalent for aquatic media of laboratory rats. Mortality tests and short-term tests are predominantly used. Proposals for improving bioassays (Chapter 6) include three pillars:

1. Improving the realism of exposure (low concentrations, chronic exposures, mesocosms, experiments in the field including transplantations);
2. Improving the determination of the no observed adverse effect level (for individual effects, prefer growth, behavior; develop subindividual effects such as biochemical markers; focus on those impacting reproduction since the success of reproduction is key for population fate);
3. Improve the statistical determination of toxicological parameters for instance by using the benchmark dose method as advocated by the European Food Safety Authority (EFSA, 2009) and the US Environmental Protection Agency (USEPA, 2012a).

Moreover, improving the extrapolation of experimental toxicity data to field situations needs a relevant choice of reference species (Chapter 9) by using organisms from the wild, representative of their environment as well as sensitive organisms or life stages (Galloway et al., 2004; Berthet et al., in Amiard-Triquet et al., 2011; Berthet, in Amiard-Triquet et al., 2013)

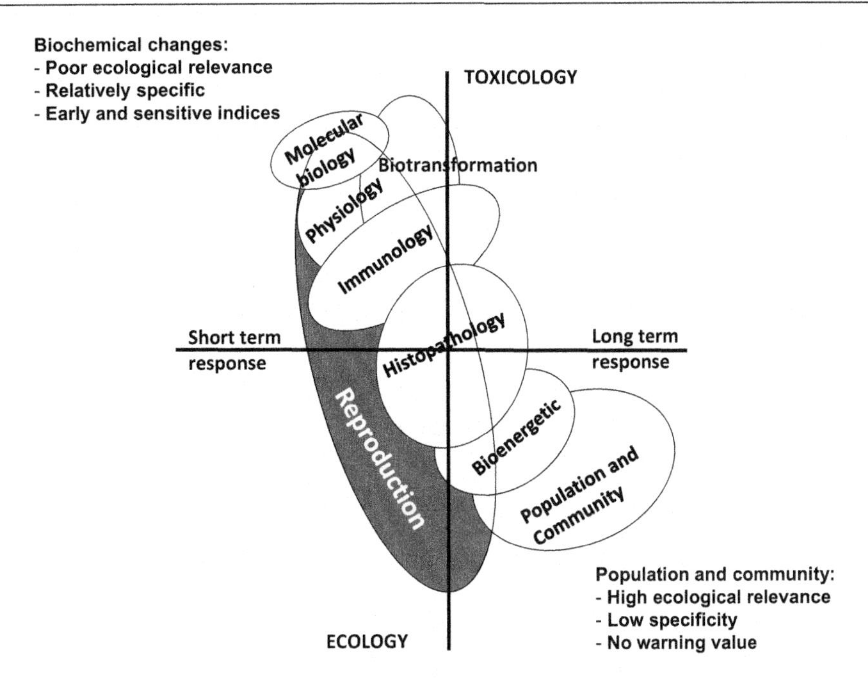

Figure 1.2
Pros and cons of biological responses at different levels of organization as biomarkers/bioindicators of the presence and/or effects of environmental stress including chemical contaminants. *Modified after Adams et al. (1989); Amiard-Triquet and Amiard, in Amiard-Triquet et al. (2013).*

In addition to chemistry and bioassays, the triad of analyses classically used for environmental assessment also includes the analysis of assemblages and communities (Chapman, 1990; ECWFD, 2000). Other tools are recommended including the use of biological responses at different levels of biological organization (Chapters 7 and 8) as biomarkers of the presence/effects of contaminants in the environment (Allan et al., 2006; Chapman and Hollert, 2006; Amiard-Triquet et al., 2013).

Because biomarkers were defined by Depledge in 1994 ("A biochemical, cellular, physiological or behavioral variation that can be measured in tissue or body fluid samples or at the level of whole organisms that provides evidence of exposure to and/or effects of, one or more chemical pollutants (and/or radiations)"), they are more and more frequently used despite recurrent criticisms about their responsiveness to confounding factors, their insufficient specificity of response toward a given class of chemicals, and their lack of ecological relevance. These weaknesses are analyzed in Chapter 7 and strategies to overcome these limits and take advantage of the potential of biomarker tools are recommended. The main achievement expected from the methodology of biomarkers is to provide an early signal of environmental degradation, well before effects at the community level become significant (Figure 1.2), a sign that severe environmental degradation has already occurred, thus leading to expensive remediation processes.

A comprehensive methodology initially proposed to assess the health status of estuarine ecosystems (Amiard-Triquet and Rainbow, 2009) may possibly be generalized to other aquatic ecosystems. The first step is based on the detection of abnormalities revealed by high level biomarkers, linking alterations at molecular, biochemical, and individual levels of organization to adverse outcomes in populations and communities (Mouneyrac and Amiard-Triquet, 2013), in keystone species or functional groups important for the ecosystem. When such impairments are revealed, the end-users need to know the nature of pollutant exposure, indispensable for any risk reduction measure or remediation decision. Core biomarkers validated in international intercalibration exercises and more specific of the main classes of contaminants are to be used for this second step such as those recommended in the Joint Assessment and Monitoring Program Guidelines for Contaminant-Specific Biological Effects (OSPAR Agreement, 2008-09). They include specific biological effects for monitoring metals, PAHs, tributyltin, and estrogenic chemicals. Used in battery, they are able to reveal the presence/effects of environmental mixtures (e.g., metals, PAHs, PCBs, pesticides, endocrine disruptors). And finally, analytical chemistry will be used to validate the hypotheses provided by biomarkers.

During the past decade, "omics" technologies (Chapter 8) covering genomics, proteomics, and metabolomics have emerged and their potential for the risk assessment of chemicals has been addressed (Garcia-Reyero and Perkins, 2011; Connon et al., 2012; Van Straalen and Feder, 2012; SCHER, SCENIHR, SCCS, 2013). Transcriptomics corresponds to a global analysis of gene expression; proteomics focuses on the functional responses of gene expression (proteins and peptides); and metabolomics measures the concentrations of endogenous metabolites (end products of cellular processes) or xenometabolites that represent enzymatic activity upon foreign substances such as environmental contaminants. Molecular approaches are clearly suitable as early warning systems and provide a powerful tool for high-throughput screening of substances/mixtures. Gene expression is also expected to be specific to the type of stress, and to respond quickly (hours to days), compared with tests based upon growth and reproduction that can last several weeks. Thus several regulatory authorities are considering how genomics tools could contribute to environmental pollution assessment (SCHER, SCENIHR, SCCS, 2013). According to Connon et al. (2012), these technologies have proven to be useful in elucidating modes of action of toxicants (a key point for the risk assessment of chemical mixtures, see Chapter 18). Omics have certainly an interest for ecotoxicology of mixtures because "the few ecotoxicogenomics studies that have considered mixtures suggest they may induce other genes than either of the constituent chemicals. On the gene expression level, a mixture appears like a new chemical" (SCHER, SCENIHR, SCCS, 2013). From 2002 to 2011, 41 studies were published with a focus on mixture toxicity assessment (for a review, see Altenburger et al., 2012).

Today, the relationship between molecular effects and responses at higher hierarchical levels (population, community) is largely unknown, despite several examples that suggest the existence of mechanistic links between omics responses and effects at other levels of biological organization such as behavior, growth, predation risk, fitness, and mortality (Vandenbrouck et al., 2009; Connon et al., 2012; SCHER, SCENIHR, SCCS, 2013).

As underlined in a recent book devoted to ecotoxicology modeling (Devillers, 2009a), "the fate and effects of chemicals in the environment are governed by complex phenomena and modeling approaches have proved to be particularly suited not only to better understand these phenomena but also to simulate them in the frame of predictive hazard and risk assessment schemes." Modeling may be used in each field of the ecotoxicology triad: exposure, bioaccumulation, and effects (Chapters 11 and 12). For exposure, preference should be given to adequately measured, representative exposure data where these are available. For existing substances, monitoring programs often include only spatiotemporal spot check of environmental concentrations that have limited interest, whereas no measured environmental concentrations will normally be available for new substances as already mentioned for nanomaterials. Therefore, PECs must often be calculated (TGD, 2003). Measured data can then be used to revise the calculated concentrations. This exercise increased the confidence in the modeling of contaminant release into the environment as exemplified for radionuclides emitted by a nuclear reprocessing plant in northwest France since the model was modified using the long series of measurements that were available for some radionuclides (Nord-Cotentin Radioecology Group, 2000).

Bioaccumulation results from various interacting mechanisms that depend on the characteristics of the compounds and on biological factors. Various attempts have been made to model bioaccumulation in order to describe and possibly to predict the fate of organic contaminants in food webs (Abarnou, in Amiard-Triquet and Rainbow, 2009). From a practical standpoint, the Canadian Center for Environmental Modeling and Chemistry has launched a Bioaccumulation Fish Model software (http://www.trentu.ca/academic/aminss/envmodel/models/Fish.html) that requires input data concerning chemical properties of the organic contaminant under assessment and properties of the fish and its environment. The correlation between the effects of molecules and their physicochemical properties is at the basis of the Quantitative Structure–Activity Relationship (QSAR) discipline. The QSAR models represent key tools in the development of drugs as well as in the hazard assessment of chemicals. They are an alternative to in vivo animal testing and are recommended in a number of legislations/regulations (e.g., Registration, Evaluation, and Authorization of Chemicals). Their potential is well-documented for endocrine disruption modeling (Devillers, 2009b) or for grouping of mixture components based on structural similarities (SCCS, SCHER, SCENIHR, 2011).

We have already mentioned that the responsiveness of biomarkers to confounding factors, and their lack of ecological relevance have partly hampered their use. Modeling the influence of confounding factors (for details, see Chapter 11) and the use of population dynamics models provide a significant improvement for a sound interpretation of biomarker data (Chapters 11 and 12).

1.3 The Choice of Biological Models for Bioassays, Biomarkers, and Chemical Monitoring

The pros and cons of using different species to support the ecotoxicological methodologies in biota are reviewed in Chapter 9. Standardized biological test methods validated by official bodies (International Organization for Standardization, Organization for Economic Co-operation and Development, ASTM International, Environnement Canada, etc.) are often based on laboratory reared organisms such as microalgae, zooplankton (e.g., daphnids) or fish (e.g., *Danio rerio*). Standard bioassay organisms can be relevant considering issues of comparability and consistency when the relative toxicity of different compounds is to be determined. However, the loss of genetic variation resulting from maintaining populations in the laboratory must be taken into consideration (Athrey et al., 2007). It is also needed to be clearly aware of this potential change of genetic pattern when extrapolating from laboratory to natural populations. Using organisms from the wild is certainly attractive to improve the environmental realism of bioassays in the framework of prospective risk assessment. In this case, test organisms must be obtained from relatively uncontaminated field sites to avoid the risk of undervaluation because of the tolerance acquired by organisms chronically exposed to contaminants in their environment (Amiard-Triquet et al., 2011). The species used in bioassays should be determined using an appropriate taxonomic key. All organisms should be as uniform as possible in age and size class (ASTM, 2012) to avoid any influence of these potential confounding factors (Chapter 7).

Confounding factors must also be avoided in the case of biomarkers and chemical monitoring in biota in support of retrospective assessment. Wild organisms collected in the field are used for the so-called "passive" biomonitoring whereas "active" biomonitoring is based on caged organisms that may be obtained from aquaculture or natural populations from clean areas. A clear advantage of active biomonitoring is the possibility of selecting organisms with the same history, age, and size.

For each of the major ecotoxicological tools (bioassays, biomarkers, and chemical monitoring in biota), a multispecies approach is recommended. The calculation of PNECs using statistical extrapolation techniques are based on the species sensitivity distribution (Dowse et al., 2013) and may be used only when many NOECs, determined in different taxa (e.g., algae, invertebrates, fish, amphibians) are available (Chapter 2). In the ECOMAN project (Galloway et al., 2004, 2006), various biomarkers were determined in common coastal organisms showing different feeding types (filter-feeding, grazing, and predation) and habitat requirements (estuary and rocky shore). The authors highlighted how this holistic integrated approach is essential to identify the full impact of chemical contamination for ecosystem management. The variability of biological responses (either in terms of bioaccumulation or effects) between different taxa and different feeding habits is well documented but even within more restricted groups, dramatic differences may be expressed as exemplified in the case of silver in filter-feeding bivalves (Berthet et al., 1992).

Chapters 10 to 13 focus on some reference species that have allowed important achievements in ecotoxicological studies of the aquatic environment such as endobenthic invertebrates in the sediment compartment, gammarid crustaceans in freshwater, copepod crustaceans in estuarine and marine waters, and fish in different water masses. Because they belong to vertebrates, ecotoxicology of fish can take advantage of the more advanced research of mammal toxicology. For one decade, the fathead minnow *Pimephales promelas* and the mummichog *Fundulus heteroclitus* have been recognized as relevant models (Ankley and Villeneuve, 2006; Burnett et al., 2007).

Primary producers have a crucial role in the functioning of aquatic ecosystems particularly for their role in nutrient biogeochemical cycles and as the first step in food webs. Eutrophication (Hudon, in Férard and Blaise, 2013) including "green tides" of macroalgae as well as algal blooms involving harmful phytoplankton (Watson and Molot, in Férard and Blaise, 2013) is a clear sign of trophic disequilibrium. Changes in communities of both macrophytes and microalgae may be used to assess the ecological status of aquatic environment. Microalgae and macrophytes may also be used in toxicity tests including a number of standardized bioassays (Hanson; Debenest et al., both in Férard and Blaise, 2013) and as matrices for the determination of biomarkers of photosynthesis (Eullaffroy, in Férard and Blaise, 2013). The range of applications for microalgae in ecotoxicology include their potential for toxicogenomic studies, their use in flow cytometry (Stauber and Adams, in Férard and Blaise, 2013) and the determination of various biomarkers (metal chelators, stress proteins, defenses against oxidative stress, xenobiotic detoxification systems, reviewed by Torres et al., 2008) or the use of diatoms as indicators of metal pollution (Morin et al., 2012). Recent studies on phytotoxicity of engineered nanomaterials have revealed the toxic potential of these emerging contaminants toward both higher plants and algae (Petit, in Férard and Blaise, 2013; Chapter 17, this book). Ecotoxicological research with respect to phytoremediation has also been reviewed recently (Dosnon-Olette and Eullafroy, in Férard and Blaise, 2013). Thus a specific chapter in the present book would be largely redundant with these recent papers.

Rotifers are also a group that is not reviewed in this book. The use of rotifers in ecotoxicology has been documented in 1995 by Snell and Janssen. This review has been quoted 185 times until now, an eloquent testimony of the interest of the scientific community. It has been updated recently by Dahms et al. (2011) and Rico-Martínez (in Férard and Blaise, 2013).

There is no distinct chapter on bivalves living in the water column despite—or because—mussels and oysters are the among the most commonly used species in fundamental and applied ecotoxicology. Their role in biomonitoring programs will be evoked in Chapter 5 and their contribution to the study of emerging contaminants in Chapter 16 dedicated to pharmaceuticals and care products. Canesi et al. (2012) claim that bivalve mollusks, in particular *Mytilus* spp., represent a unique target group for nanoparticle toxicity. Matranga and Corsi (2012) underscore the existence of "Mytibase," an interactive catalog of 7112 transcripts of the mussel *M. galloprovincialis* that can help using the "omic" tools for marine organisms. Very recently, Binelli et al. (2015) have reviewed the ecotoxicological studies carried out with the zebra mussel *Dreissena polymorpha* to suggest this bivalve species as possible reference organism for inland waters.

In many contaminated environments where living beings have been historically exposed to high anthropogenic pressures, ecotoxicologists are not in the role of providing tools to prevent further aggravation of the already existing problems. In this case, it is needed to qualify the ecological status of water masses in agreement with the different legislation aiming at the conservation/improvement of environmental quality, including the conservation of biodiversity with reference to a nearly undisturbed situation (Chapter 14). The ecological status can be determined by using biological indicators (bioindicators) as surrogates to indicate the quality of the environment in which they are present. They have been designed either at the level of species or communities. Sentinel species may be considered as any species providing a warning of a dysfunction or an imbalance of the environment, or, more restrictively, a warning of the dangers of substances to human and environmental health. In addition to bioaccumulative species (Chapter 5) and those used for the determination of infraindividual and individual biomarkers (Chapter 7), the sentinel species can be bioindicator species, providing information by their absence (or presence) and/or the abundance of individuals in the environment under study (Berthet, in Amiard-Triquet et al., 2013). Among bioindicators at the community/assemblage level, there are five biological compartments retained in the ECWFD (2000): phytoplankton, macroalgae, angiosperms, macrozoobenthos, and fish. Additional groups may be recommended such as meiobenthic groups (e.g., foraminifera, copepods, nematodes) and zooplankton (Dauvin et al., 2010).

The assessment methods used to classify the ecological status of rivers, lakes, coastal, and transitional waters in Member States of the European Community according to the ECWFD are available at http://www.wiser.eu/results/method-database/. Tools and methodologies used in assessing ecological integrity in estuarine and coastal systems have been reviewed by Borja et al. (2008) considering the situation in North America, Africa, Asia, Australia, and Europe. Many of the biotic indices in current use may not be specific enough in terms of the different kinds of stress. However, the AZTI's Marine Biotic Index (software freely available at http://www.azti.es) despite being designed to assess the response of soft-bottom macrobenthic communities to the introduction of organic matter in ecosystems, has been validated in relation to other environmental impact sources (e.g., drilling cuts with ester-based mud, submarine outfalls, industrial and mining wastes, jetties, sewerage works) (Borja et al., 2003). Positive correlations were particularly indicated between AZTI's Marine Biotic Index and metals or PCBs (Borja et al., 2000). In certain cases, more specific indices may be used such as the nematode/copepod ratio (Carman et al., 2000) and the polychaete/amphipod ratio to identify petroleum hydrocarbon exposure (Gómez Gesteira and Dauvin, 2000).

The "static" look at structural ecosystem properties must be complemented using an approach toward the ecosystem function and dynamics (Borja et al., 2008). Monitoring and assessment tools for the management of water resources are generally more effective if they are based on a clear understanding of the mechanisms that lead to the presence or absence of species groups in the environment (Usseglio-Polatera et al., 2000). The theory of traits (life history, ecological and biological traits) states that a species' characteristics might enable its persistence and development in given environmental conditions (Logez et al., 2013). Biological Traits Analysis (BTA) could reveal which environmental factors may be responsible for a given observed impairment, thus providing causal insight into the interaction between species and stressors (Culp et al., 2010; Van den Brink et al., 2011). This is well illustrated by a trait-based indicator system that was developed to identify species at risk of being affected by pesticides, with reference to life history and physiological traits (Liess et al., 2008). In an experiment with outdoor stream mesocosms, long-term community effects of the insecticide thiacloprid were detected at concentrations 1000 times below those detected by the principal response curve approach (Liess and Beketov, 2011). There is now a widespread conviction that biological traits should be used for environmental risk assessment (Artigas et al., 2012). However, the BTA is not always more powerful than the traditional taxonomic approach as observed by Alvesa et al. (2014), studying the subtidal nematode assemblages from a temperate estuary (Mondego estuary, Portugal), anyway providing useful knowledge of the functional structure and characterization of nematode communities in the estuary. Thus it is necessary to analyze carefully the strengths, weaknesses, opportunities and threats of using BTA (Van den Brink et al., 2011). Improved data analysis and the development of relevant traits are key for a sound ecological risk assessment (Bremner et al., 2006; for details, see Chapter 14).

1.4 Emerging Concern with Legacy Pollutants and Emerging Contaminants

Environmental contaminants may be assigned to two categories: legacy pollutants that have been present in the environment for decades and emerging chemicals that have only recently been detected and appreciated as possible environmental threats. Because effective analytical procedures have existed since the 1970s, the ecotoxicology of metals has been particularly well-developed (Luoma and Rainbow, 2008). Other legacy pollutants (PAHs, PCBs, dioxins and furans, and chlorinated pesticides such as DDT) became accessible to analysis more recently and because of the variety of environmental levels and biological effects among their different compounds/congeners/metabolites, analytical developments are still needed to improve their ecotoxicological assessment (Chapter 4). The ecotoxicological knowledge about PCBs, fire-retardants, cadmium, etc., was already important (Eisler, 2007) when more recently their endocrine disrupting potential was discovered (Amiard et al., in Amiard-Triquet et al., 2013). To date, close to 800 chemicals are known or suspected to be endocrine disruptor compounds (EDCs), among which only a small fraction have been investigated with procedures that allows the identification of endocrine effects at the level of the whole organism (Bergman et al., 2013). Aquatic organisms are simultaneously exposed to many EDCs that can interact depending on their mode of action (estrogenic, antiestrogenic, androgenic, antiandrogenic and thyroid effects). It is impossible, for both technical and cost-effective reasons, to determine concentrations of all such compounds. Against this background, TIE or EDA-based strategies (Chapter 4) are particularly useful to characterize more accurately the environmental exposure to EDCs. Bioanalytical tools are developed using in vitro and in vivo models (OECD, 2012), including the generation of transgenic models such as tadpole fluorescent screens and fluorescent zebrafish transgenic embryos, for sexual and thyroid hormone disruptors (Brack et al., 2013). Low-dose effects, nonmonotonic dose responses, and the changes of biological susceptibility depending on life stage pose problems that cannot be solved by using classical strategies of risk assessment (Bergman et al., 2013). Chapter 15 will be dedicated to EDCs, a category of contaminants that are not defined as usually by their chemical characteristics (e.g., metals, PAHs, PCBs) but by the hazard associated to their presence in the aquatic environment. It will explore the more recent ecoepidemiological studies that try to explore the links between effects on the development, growth, and reproduction in individuals and the effects at the population and community levels.

Pharmaceuticals (Chapter 16) are submitted to precise regulations concerning their therapeutic value and potential secondary negative effects on human health but their environmental impact was not initially envisaged. Drugs are not totally assimilated in human organisms. Residues (urine) are released in the environment (with or without treatment in a waste water treatment plant) and numerous persistent molecules may be detected in natural waters. It should be noted that ecological footprints of active pharmaceuticals depend on risk factors that can differ substantially in low-, middle-, and high-income countries (Kookana et al., 2014).

Until recently, effects of pharmaceuticals on aquatic organisms were observed at very high doses "typically at least 1 order of magnitude higher than concentrations normally found in surface waters," suggesting that environmental doses were not deleterious (Corcoran et al., 2010) with the exception of antibiotic compounds in the environment and their potential for selection of resistant microbial strains. However, the lack of consideration given to the chronic nature of the exposures, the absence of knowledge on the significance of metabolites and transformation products resulting from the parent active pharmaceutical ingredients, or the potential for mixture effects were recognized (Corcoran et al., 2010; Kümmerer, 2010). More recent studies have demonstrated that at much lower doses and even realistic doses able to be found in the environment, deleterious effects may be observed. Pharmaceuticals can act as endocrine disruptors, the most potent being the synthetic estrogen 17-α-ethynylestradiol used in birth control pills (Kidd et al., 2007), but strong presumptions of effects in the field have been recently published (Sanchez et al., 2011; Vieira Madureira et al., 2011). Also at low doses, behavioral effects were induced by antidepressant (fluoxetine), analgesic ibuprofen, and antiepileptic carbamazepine (De Lange et al., 2006; Gaworecki and Klaine, 2008; Painter et al., 2009; Di Poi et al., 2013). Recent research on the aquatic toxicity of human pharmaceuticals to aquatic organisms has been critically reviewed by Brausch et al. (2012) and the most critical questions to aid in development of future research programs on the topic has been extended to veterinary pharmaceuticals and personal care products (moisturizers, lipsticks, shampoos, hair colors, deodorants, and toothpastes) (Boxall et al., 2012). In 2010, the European Environment Agency held a specialized workshop that drew up proposals to reduce the environmental footprint of pharmaceuticals including (1) the eco-classification of all pharmaceuticals according to their environmental hazardousness and (2) the definition of environmental quality standards for pharmaceuticals; both of these approaches needing more data to describe the fate and long-term effects of pharmaceuticals in the aquatic environment (EEA, 2010).

Chapter 17 is dedicated to another category of emerging contaminants of environmental concern, the NMs. An NM is defined as any material that has unique or novel properties, because of the nanoscale (nanometer-scale) structuring. At this scale, physical and chemical properties of materials differ significantly from those at a larger scale. Nanomaterials can have one (e.g., nanosheet), two (e.g., nanotube), or three dimensions (e.g., nanoparticle) in the nanoscale. Engineered nanomaterials have multiple uses in nanometrology, electronics, optoelectronics, information and communication technology, bionanotechnology and nanomedicine (Royal Society, 2004). An inventory of nanotechnology-based consumer products introduced on the market (http://www.nanotechproject.org/cpi/) reveals that the number of products exploded from 54 in 2005 to 1628 in October 2013, the main product categories being in the field of health and fitness (personal care, clothing, cosmetics, sporting goods, filtration), home and garden, automotive, food, and beverages. The highest consumers were mainly in the United States (741 products), Europe (440), and East Asia (276). The major materials are silver, titanium, carbon, silicon/silica, zinc, and gold. The economic developments of nanotechnologies should not compromise the safety for human health and the environment. Products have already come to market, so first attention should be paid to postmarketing risks. Safety research should contribute to the sustainable development of nanotechnologies (Royal Society, 2004). The ecotoxicological history of nanomaterials parallels that of pharmaceuticals: nanotoxicity was first examined in tests carried out in the short term with very high doses, generally higher than mg/L^{-1} in water (Buffet, 2012; Gottschalk et al., 2013; Wong et al., 2013). In contrast, predicted concentrations of ENPs arising from use in consumer products are generally lower than the $\mu g/L^{-1}$ in water (Tiede et al., 2009; Sun et al., 2014) but with a severe degree of uncertainty (Gottschalk et al., 2011). However, recent studies carried out at much lower doses were able to reveal that all the nanoparticles tested were incorporated in the whole organisms but also enter the cells (Chapter 17) and even the nucleus (e.g., Joubert et al., 2013).

Procedures for improving the assessment of the risk of nanomaterials are reviewed/proposed in Chapter 17. They include a better characterization of exposure, not only by measuring environmental concentrations (Chapter 4) but also by considering the many physicochemical parameters that govern the fate and effects of NMs (Card and Magnuson, 2010; Chen et al., 2013). Bioaccumulation studies must evolve, taking into account not only the concentration in the whole organism/organ but the intracellular uptake and localization. Concerning both the evaluation of uptake and effects of NMs, different ways of incorporation must be explored (water, preys, sediment). Harmonization of test protocols can enable screening of different NMs and meaningful comparison between studies (Wong et al., 2013). However, innovating methods and techniques must be encouraged to improve the realism of test procedures.

PFCs are a large group of manufactured compounds that are widely used in everyday products (cookware, sofas and carpets, clothes and mattresses, food packaging, firefighting materials) and in a variety of industries (aerospace, automotive, building and construction, electronics) (NIEHS, 2012). Because of these widespread applications and their environmental persistence (OECD, 2002), PCFs are commonly detected in the environment and their presence in sediment and aquatic biota even in remote sites (Arctic biota) is well-documented. PFOA and perfluorooctane sulfonate (PFOS) are the PFCs that generally show the highest environmental concentrations (Wang et al., 2011 and literature cited therein). PFOS does not only bioconcentrate in fish tissues but it depurates slowly (50% clearance times of up to 116 days in the bluegill sunfish) (OECD, 2002). Using the limited information available, fish and fishery products seem to be one of the primary sources of human exposure to PFOS (USEPA, 2012b). PFOS and PFOA have a long half-life in humans (approximately 4 years), increasing the risk of adverse outcomes since different health effects are suspected in humans (USEPA, 2012b). Based on the assumption that consumption of fish by humans is the most critical route, Moermond et al. (2010) have proposed water quality standards in accordance with the ECWFD. The reader will not find a distinct chapter about PFCs in this book because it would be redundant with the recent reviews by Giesy et al. (2010) and Ding and Peijnenburg (2013).

Exposure of aquatic organisms to hazardous compounds is primarily through complex environmental mixtures, those that occur in water, sediment, and preys (Chapter 18). Interactions of chemical factors with physical and/or biological stressors in the environment are beyond the scope of this chapter (see the subsection on confounding factors in Chapter 7). Kortenkamp et al. (2009) have distinguished four categories of mixtures: (1) substances that are mixtures themselves (e.g., metallic alloys); (2) products that contain more than one chemical (e.g., cosmetics, biocidal products); (3) chemicals jointly emitted at any step of their lifecycle; and (4) mixtures of several chemicals emitted from various sources, via multiple pathways that might occur together in environmental media. Guidance for conducting cumulative risk assessments has been published by regulatory bodies in United States, United Kingdom, Norway, and Germany but except for the two first categories, risk assessments in the European Union deal mainly with individual substances (SCCS, SCHER, SCENIHR, 2011).

The main effort must be directed toward ecosystems where significant exposure is likely or at least plausible. Individual components in a mixture have specific and different physicochemical properties that govern their fate (and consequently effects) in the environment. Theoretically, it would be possible to identify each individual component of a mixture and then to determine a PEC for each of them but, in practice, this approach requires an unrealistic and extremely expensive analytical investment. Recently, strategies based upon the similarity of physicochemical properties (e.g., log K_{ow}, water solubility), and environmental-degradation potentials (e.g., photodegradation and hydrolysis rates), have been proposed for the identification of "blocks" of components that may be considered together with the help of QSARs (SCCS, SCHER, SCENIHR, 2011). However, to date, it seems much more difficult to take into account biological degradation that is a key process for the elimination of chemicals in the aquatic environment.

Mixture studies have been mainly conducted (1) to evaluate and quantify the overall toxicity of complex environmental samples (whole mixture approach) or (2) to reveal the joint action of individual molecules (component-based approach) (Kortenkamp et al., 2009). Recent studies try to fulfill the gap between these two approaches with promising results expected from the use of TIE and EDA (ECETOC, 2011; EC STAR, 2012).

Regulatory risk assessment of chemical mixtures needs at the minimum a sound knowledge of the different modes of action (MoA) of individual contaminants. It is generally admitted that the effects of a mixture composed of individual molecules with similar MoA can be estimated by summing the doses/concentrations, scaled for relative toxicity to take into account the different potency of each substance. This has been illustrated in the case of EDCs by Jin et al. (2012), who have examined the biological traits of the fish *Gobiocypris rarus* submitted to a coexposure to three estrogenic compounds (17β-estradiol, diethylstilbestrol, and nonylphenol) and Pottinger et al. (2013) in the case of a coexposure to four antiandrogenic compounds of the fish *Gasterosteus aculeatus*.

In the case of a mixture composed of molecules with dissimilar MoA, it may be proposed to assess the effects using models of response addition (based on the probability of responses to the individual components) or effect addition (by summing of biological responses) (Chapter 18). Consequently, it is expected that mixtures composed of dissimilarly acting chemicals at levels below NOECs will not induce significant effects. However, at such low doses, the interpretation of data is tricky and controversial, as illustrated by the conclusions derived by different groups of experts (Kortenkamp et al., 2009; SCCS, SCHER, SCENIHR, 2011) from a study with fish (Hermens et al., 1985), two studies with algae (Walter et al., 2002; Faust et al., 2003), and one study using an in vitro cellular test (Payne et al., 2001). At higher doses, interactions either synergistic (supra-additive) or antagonistic (infra-additive) are more easy to identify. They include toxicokinetic, metabolic, and toxicodynamic interactions. Toxicokinetic interactions occur when a contaminant modifies the absorption of others (e.g., Tan et al., 2012; Su et al., 2013). Toxicodynamic interactions occur when the different constituents of a mixture have a similar target, a situation encountered in the case of ligand–receptor interactions and well-documented for EDCs (Kortenkamp et al., 2009). Synergistic interactions have been documented for pesticides considering both biocidal products (Bjergager et al., 2011; Backhaus et al., 2013) and mixtures commonly detected in aquatic habitats (Laetz et al., 2009). The overall toxicity of a pharmaceutical mixture is in general substantially higher than the toxicity of each individual substance at its concentration present in the mixture (EEA, 2010). Antagonistic effects are reported in response to co-exposure to EDCs with known MoA, namely estrogens and antiestrogens (Sun et al., 2011; Wu et al., 2014). In such cases, the models are of no help and direct experimentation remains the only available tool (Chapter 18).

The aim of this book is to use cross-analyses of procedures, models, and contaminants to design ecotoxicological tools suitable for better environmental assessments, particularly in the case of emerging contaminants and emerging concern with legacy pollutants.

References

Adams, S.M., Shepard, K.L., Greeley Jr., M.S., et al., 1989. The use of bioindicators for assessing the effects of pollutant stress on fish. Mar. Environ. Res. 28, 459–464.

Allan, I.J., Vrana, B., Greenwood, R., et al., 2006. A "toolbox" for biological and chemical monitoring requirements for the European Union's Water Framework Directive. Talanta 69, 302–322.

Altenburger, R., Scholz, S., Schmitt-Jansen, M., et al., 2012. Mixture toxicity revisited from a toxicogenomic perspective. Environ. Sci. Technol. 46, 2508–2522.

Alves, A.S., Veríssimo, H., Costa, M.J., et al., 2014. Taxonomic resolution and Biological Traits Analysis (BTA) approaches in estuarine free-living nematodes. Estuar. Coast. Shelf Sci. 138, 69–78.

Amiard-Triquet, C., Amiard, J.C., Rainbow, P.S., 2013. Ecological Biomarkers: Indicators of Ecotoxicological Effects. CRC Press, Boca Raton.

Amiard-Triquet, C., Rainbow, P.S., 2009. Environmental Assessment of Estuarine Ecosystems. A Case Study. CRC Press, Boca Raton.

Amiard-Triquet, C., Rainbow, P.S., Roméo, M., 2011. Tolerance to Environmental Contaminants. CRC Press, Boca Raton.

Ankley, G.T., Villeneuve, D.L., 2006. The fathead minnow in aquatic toxicology: past, present and future. Aquat. Toxicol. 78, 91–102.

Artigas, J., Arts, G., Babut, M., et al., 2012. Towards a renewed research agenda in ecotoxicology. Environ. Pollut. 160, 201–206.

ASTM, 2012. Standard Guide for Behavioral Testing in Aquatic Toxicology. http://enterprise1.astm.org/DOWNLOAD/E1604.1186927-1.pdf.

Athrey, N.R.G., Leberg, P.L., Klerks, P.L., 2007. Laboratory culturing and selection for increased resistance to cadmium reduce genetic variation in the least killifish, *Heterandria formosa*. Environ. Toxicol. Chem. 26, 1916–1921.

Backhaus, T., Altenburger, R., Faust, M., et al., 2013. Proposal for environmental mixture risk assessment in the context of the biocidal product authorization in the EU. Environ. Sci. Eur. 25 (4)http://www.enveurope.com/content/25/1/4 (last accessed 11.12.13).

Bergman, A., Heindel, J.J., Jobling, S., et al., 2013. State of the Science of Endocrine Disrupting Chemicals 2012. United Nations Environment Programme and the World Health Organization, Genevahttp://unep.org/pdf/9789241505031_eng.pdf.

Berthet, B., Amiard, J.C., Amiard-Triquet, C., et al., 1992. Bioaccumulation, toxicity and physico-chemical speciation of silver in Bivalve Molluscs: ecotoxicological and health consequences. Sci. Total Environ. 125, 97–122.

Binelli, A., Della Torre, C., Magni, S., et al., 2015. Does zebra mussel (*Dreissena polymorpha*) represent the freshwater counterpart of *Mytilus* in ecotoxicological studies? A critical review. Environ. Pollut. 196, 386–403.

Bjergager, M.B.A., Hanson, M.L., Lissemorec, L., et al., 2011. Synergy in microcosms with environmentally realistic concentrations of prochloraz and esfenvalerate. Aquat. Toxicol. 101, 412–422.

Borja, A., Muxika, I., Franco, J., 2003. The application of a marine biotic index to different impact sources affecting soft-bottom benthic communities along European coasts. Mar. Pollut. Bull. 46, 835–845.

Borja, A., Franco, J., Pérez, V., 2000. A Marine Biotic Index to establish the ecological quality of soft-bottom benthos within European estuarine and coastal environments. Mar. Pollut. Bull. 40, 1100–1114.

Borja, A., Bricker, S.A., Daniel, M., Dauer, D.M., et al., 2008. Overview of integrative tools and methods in assessing ecological integrity in estuarine and coastal systems worldwide. Mar. Pollut. Bull. 56, 1519–1537.

Boxall, A.B.A., Rudd, M.A., Brooks, B.W., et al., 2012. Pharmaceuticals and personal care products in the environment: what are the big questions? Environ. Health Perspect. 120, 1221–1229.

Brack, W., Govender, S., Schulze, T., et al., 2013. EDA-EMERGE: an FP7 initial training network to equip the next generation of young scientists with the skills to address the complexity of environmental contamination with emerging pollutants. Environ. Sci. Eur. 25, 18.

Brausch, J.M., Connors, K.A., Brooks, B.W., et al., 2012. Human pharmaceuticals in the aquatic environment: a review of recent toxicological studies and considerations for toxicity testing. Rev. Environ. Contam. Toxicol. 218, 1–99.

Bremner, J., Rogers, S.I., Frida, C.L.J., 2006. Methods for describing ecological functioning of marine benthic assemblages using Biological Traits Analysis (BTA). Ecol. Indic 6, 609–622.

Buffet, P.E., 2012. Évaluation du risque environnemental des nanoparticules métalliques: biodisponibilité et risque potentiel pour deux espèces clés des écosystèmes estuariens (Ph.D. thesis) University of Nantes,France. http://archive.bu.univ-nantes.fr/pollux/show.action?id=4222b036-2cc9-45e7-a146-038b3361bae3.

Burgess, R.M., Ho, K.T., Brack, W., et al., 2013. Effects-directed analysis (EDA) and toxicity identification evaluation (TIE): complementary but different approaches for diagnosing causes of environmental toxicity. Environ. Toxicol. Chem. 32, 1935–1945.

Burnett, K.G., Bain, L.J., Baldwin, W.S., et al., 2007. *Fundulus* as the premier teleost model in environmental biology: opportunities for new insights using genomics. Comp. Biochem. Physiol. 2D, 257–286.

Van den Brink, P.J., Alexander, A.C., Desrosiers, D., et al., 2011. traits-based approaches in bioassessment and ecological risk assessment: strengths, weaknesses, opportunities and threats. Integrated Environ. Assess. Manage. 7, 198–208.

Campbell, P.G.C., 1995. Interactions between trace metals and aquatic organisms: a critique of the free-ion activity model. In: Tessier, A., Turner, D.R. (Eds.), Metal Speciation in Aquatic Systems. John Wiley & Sons, New York, USA, pp. 45–102.

Canesi, L., Ciacci, C., Fabbri, R., et al., 2012. Bivalve molluscs as a unique target group for nanoparticle toxicity. Mar. Environ. Res. 76, 16–21.

Card, J.W., Magnuson, B.A., 2010. A method to assess the quality of studies that examine the toxicity of engineered nanomaterials. Int. J. Toxicol. 29, 402–410.

Carman, K.R., Fleeger, J.W., Pomarico, S.M., 2000. Does historical exposure to hydrocarbon contamination alter the response of benthic communities to diesel contamination? Mar. Environ. Res. 49, 255–278.

Chapman, P.M., 1990. The sediment quality triad approach to determining pollution-induced degradation. Sci. Total Environ. 97-98, 815–825.

Chapman, P.M., Hollert, H., 2006. Should the sediment quality triad become a tetrad, a pentad, or possibly even a hexad? J. Soils Sediments 6, 4–8.

Chen, C.Y., Li, Y.F., Qu, Y., et al., 2013. Advanced nuclear analytical and related techniques for the growing challenges in nanotoxicology. Chem. Soc. Rev. 42, 8266–8303.

Connon, R.E., Geist, J., Werner, I., 2012. Effect-based tools for monitoring and predicting the ecotoxicological effects of chemicals in the aquatic environment. Sensors 12, 12741–12771.

Corcoran, J., Winter, M.J., Tyler, C.R., 2010. Pharmaceuticals in the aquatic environment: a critical review of the evidence for health effects in fish. Crit. Rev. Toxicol. 40, 287–304.

Culp, J.M., Armanini, D.G., Dunbar, M.J., et al., 2010. Incorporating traits in aquatic biomonitoring to enhance causal diagnosis and prediction. Integr. Environ. Ass. Manag. 7, 187–197.

Dahms, H.U., Hagiwara, A., Lee, J.S., 2011. Ecotoxicology, ecophysiology, and mechanistic studies with rotifers. Aquat. Toxicol. 101, 1–12.

Dauvin, J.C., Bellan, G., Bellan-Santini, D., 2010. Benthic indicators: from subjectivity to objectivity – where is the line? Mar. Pollut. Bull. 60, 947–953.

Depledge, M.H., 1994. The rational basis for the use of biomarkers as ecotoxicological tools. In: Fossi, M.C., Leonzio, C. (Eds.), Nondestructive Biomarkers in Vertebrates. Lewis Publishers, Boca Raton, pp. 261–285.

Devillers, J., 2009a. Ecotoxicology Modeling. Springer, Dordrecht.

Devillers, J., 2009b. Endocrine Disruption Modeling. CRC Press, Taylor & Francis Group, Boca Raton, FL.

Ding, G., Peijnenburg, W.J.G.M., 2013. Physicochemical properties and aquatic toxicity of poly- and perfluorinated compounds. Crit. Rev. Environ. Sci. Technol. 43, 598–678.

Dowse, R., Tang, D., Palmer, C.G., et al., 2013. Risk assessment using the species sensitivity distribution method: data quality versus data quantity. Environ. Toxicol. Chem. 32, 1360–1369.

EC STAR, 2012. Critical Review of Existing Approaches, Methods and Tools for Mixed Contaminant Exposure, Effect and Risk Assessment in Ecotoxicology and Evaluation of Their Usefulness for Radioecology. Contract Number: Fission20103.5.1269672. https://wiki.ceh.ac.uk/download/attachments/148996380/STAR+deliverable+4.1+Final.pdf (last accessed 12.12.13).

ECETOC, 2011. Development of Guidance for Assessing the Impact of Mixtures of Chemicals in the Aquatic Environment. Technical Report 211.

ECWFD, 2000. Directive 2000/60/EC of the European Parliament and Council of 23 October 2000 Establishing a Framework for Policy in the Field of Water (JO L 327 of 22 December 2000). European Commission, Brussels.

EEA, 2010. Pharmaceuticals in the Environment. EEA. Technical Report No 1/2010.

EFSA, 2009. Use of the benchmark dose approach in risk assessment 1. Guidance of the Scientific Committee (Question No EFSA-Q-2005-232). EFSA J. 1150, 1–72.

Eggleton, J., Thomas, K.V., 2004. A review of factors affecting the release and bioavailability of contaminants during sediment disturbance events. Environ. Intern. 30, 973–980.

Eisler, R., 2007. Eisler's Encyclopedia of Environmentally Hazardous Priority Chemicals. Elsevier, Amsterdam.

Faust, M., Altenburger, R., Backhaus, T., et al., 2003. Joint algal toxicity of 16 dissimilarly acting chemicals is predictable by the concept of independent action. Aquat. Toxicol. 63, 43–63.

Férard, J.F., Blaise, C., 2013. Encyclopedia of Aquatic Ecotoxicology. Springer, Dordrecht.

Galloway, T.S., Brown, R.J., Browne, M.A., et al., 2004. Ecosystem management bioindicators: the ECOMAN project—A multibiomarker approach to ecosystem management. Mar. Environ. Res. 58, 233–237.

Galloway, T.S., Brown, R.J., Browne, M.A., et al., 2006. The ECOMAN project: a novel approach to defining sustainable ecosystem function. Mar. Pollut. Bull. 53, 186–194.

Garcia-Reyero, N., Perkins, E.J., 2011. Systems biology: leading the revolution in ecotoxicology. Omics and Environmental Science, Critical Review. Environ. Toxicol. Chem. 30, 265–273.

Gaworecki, K.M., Klaine, S.J., 2008. Behavioral and biochemical responses of hybrid striped bass during and after fluoxetine exposure. Aquat. Toxicol. 88, 207–213.

Giesy, J.P., Naile, J.E., Khim, J.S., et al., 2010. Aquatic toxicology of perfluorinated chemicals. In: Whitacre, D.M. (Ed.), Rev. Environ. Contam. Toxicol, vol. 202, pp. 1–52.

Gómez Gesteira, J.L., Dauvin, J.C., 2000. Amphipods are good bioindicators of the impact of oil spills on soft-bottom macrobenthic communities. Mar. Pollut. Bull. 40, 1017–1027.

Gottschalk, F., Ort, C., Scholz, R.W., et al., 2011. Engineered nanomaterials in rivers – exposure scenarios for Switzerland at high spatial and temporal resolution. Environ. Pollut. 159, 3439–3445.

Gottschalk, F., Sun, T.Y., Nowack, B., 2013. Environmental concentrations of engineered nanomaterials: review of modeling and analytical studies. Environ. Pollut. 181, 287–330.

Hermens, J., Leeuwangh, P., Musch, A., 1985. Joint toxicity of mixtures of groups of organic aquatic pollutants to the guppy (*Poecilia reticulata*). Ecotoxicol. Environ. Saf. 9, 321–326.

Jin, S., Yang, F., Liao, T., et al., 2012. Enhanced effects by mixtures of three estrogenic compounds at environmentally relevant levels on development of Chinese rare minnow (*Gobiocypris rarus*). Environ. Toxicol. Pharmacol. 33, 277–283.

Joubert, Y., Pan, J.F., Buffet, P.E., et al., 2013. Subcellular localization of gold nanoparticles in the estuarine bivalve *Scrobicularia plana* after exposure through the water. Gold Bull. 46, 47–56.

Kidd, K.A., Blanchfield, P.J., Mills, K.H., et al., 2007. Collapse of a fish population after exposure to a synthetic estrogen. Proc. Nat. Acad. Sci. U.S.A. 104, 8897–8901.

Kookana, R.S., Williams, M., Boxall, A.B.A., et al., 2014. Potential ecological footprints of active pharmaceutical ingredients: an examination of risk factors in low-, middle- and high-income countries. Phil. Trans. R. Soc. 369B 20130586.

Kortenkamp, A., Backhaus, T., Faust, M., 2009. State of the Art Report on Mixture Toxicity. Study Contract Number 070307/2007/485103/ETU/D.1, Final Report, 391 pp. http://ec.europa.eu/environment/chemicals/effects/pdf/report_mixture_toxicity.pdf (last accessed 11.12.13).

Kümmerer, K., 2010. Pharmaceuticals in the Environment. Ann. Rev. Environ. Resour. 35, 57–75.

De Lange, H.J., Noordoven, W., Murk, A.J., et al., 2006. Behavioural responses of *Gammarus pulex* (Crustacea, Amphipoda) to low concentrations of pharmaceuticals. Aquat. Toxicol. 78, 209–216.

Laetz, C.A., Baldwin, D.H., Collier, T.K., et al., 2009. The synergistic toxicity of pesticide mixtures: implications for risk assessment and the conservation of endangered Pacific Salmon. Environ. Health Perspect. 117, 348–353.

Liess, M., Beketov, M., 2011. Traits and stress: keys to identify community effects of low levels of toxicants in test systems. Ecotoxicology 20, 1328–1340.

Liess, M., Schäfer, R.B., Schriever, C.A., 2008. The footprint of pesticide stress in communities—species traits reveal community effects of toxicants. Sci. Total Environ. 406, 484–490.

Logez, M., Bady, P., Melcher, A., et al., 2013. A continental-scale analysis of fish assemblage functional structure in European rivers. Ecography 36, 80–91.

Long, E.R., MacDonald, D.D., Smith, S.L., et al., 1995. Incidence of adverse biological effects within ranges of chemical concentrations in marine and estuarine sediments. Environ. Manag. 19, 81–97.

Luoma, S., Rainbow, P.S., 2008. Metal Contamination in Aquatic Environments: Science and Lateral Management. Cambridge University Press, Cambridge.

MacDonald, D.D., Carr, R.S., Calder, F.D., et al., 1996. Development and evaluation of sediment quality guidelines for Florida coastal waters. Ecotoxicology 5, 253–278.

Maltby, L., 2013. Ecosystem services and the protection, restoration, and management of ecosystems exposed to chemical stressors. Environ. Toxicol. Chem. 32, 974–983.

Matranga, V., Corsi, I., 2012. Toxic effects of engineered nanoparticles in the marine environment: model organisms and molecular approaches. Mar. Environ. Res. 76, 32–40.

McCready, S., Birch, G.F., Long, E.R., et al., 2006. An evaluation of Australian sediment quality guidelines. Arch. Environ. Contam. Toxicol. 50, 306–315.

MEA (Millenium Ecosystem Assessment), 2005. Ecosystems and Human Well-being: Synthesis. Island Press, Washington, DC. World Resources Institute. http://pdf.wri.org/ecosystems_human_wellbeing.pdf.

Mills, G.A., Greenwood, R., Allan, I.J., et al., 2010. Application of passive sampling techniques for monitoring the aquatic environment. In: Namieski, J., Szefer, P. (Eds.), Analytical Measurements in Aquatic Environments. CRC Press, Boca Raton, pp. 41–68.

Moermond, C., Verbruggem, E., Smit, C., 2010. Environmental Risk Limits for PFOS: A Proposal for Water Quality Standards in Accordance with the Water Framework Directive. www.rivm.nl/bibliotheek/rapporten/601714013.pdf.

Morin, S., Cordonier, A., Lavoie, I., et al., 2012. Consistency in diatom response to metal-contaminated environments. In: Guasch, H., Ginebreda, A., Geiszinger, A. (Eds.), Emerging and Priority Pollutants in Rivers. Springer-Verlag, Berlin Heidelberg, pp. 117–146.

Mouneyrac, C., Amiard-Triquet, C., 2013. Biomarkers of ecological relevance. In: Férard, J.F., Blaise, C. (Eds.), Comprehensive Handbook of Ecotoxicological Terms. Springer, Dordrecht, pp. 221–236.

NIEHS (National Institute of Environmental Health Sciences), 2012. Perfluorinated Chemicals (PFCs). http://www.niehs.nih.gov/health/materials/perflourinated_chemicals_508.pdf (last accessed 12.12.12).

Nord-Cotentin Radioecology Group, 2000. Estimation of Exposure Levels to Ionizing Radiation and Associated Risks of Leukemia for Populations in the Nord-Cotentin: Summary Report. 357 pp.

OECD, 2002. Co-operation on Existing Chemicals Hazard Assessment of Perfluorooctane Sulfonate (Pfos) and its Salts. Report ENV/JM/RD(2002)17/FINAL. http://www.oecd.org/chemicalsafety/risk-assessment/2382880.pdf (last accessed 30.12.13).

OECD, 2012. Conceptual Framework for Testing and Assessment of Endocrine Disrupters. http://www.oecd.org/env/ehs/testing/OECD%20Conceptual%20Framework%20for%20Testing%20and%20Assessment%20of%20Endocrine%20Disrupters%20for%20the%20public%20website.pdf.

OSPAR Agreement 2008–09. https://www.google.fr/#q=JAMP+Guidelines+for+Contaminant-Specific+Biological+Effects+%28OSPAR+Agreement+2008-09%29.

Di Poi, C., Darmaillacq, A.S., Dickel, L., et al., 2013. Effects of perinatal exposure to waterborne fluoxetine on memory processing in the cuttlefish *Sepia officinalis*. Aquat. Toxicol. 132–133, 84–91.

Painter, M.M., Buerkley, M.A., Julius, M.L., et al., 2009. Antidepressants at environmentally relevant concentrations affect predator avoidance behavior of larval fathead minnows (*Pimephales promelas*). Environ. Toxicol. Chem. 28, 2677–2684.

Paquin, P.R., Gorsuch, J.W., Apte, S., et al., 2002. The biotic ligand model: a historical overview. Comp. Biochem. Physiol. 133C, 3–35.

Payne, J., Scholze, M., Kortenkamp, A., 2001. Mixtures of four organochlorines enhance human breast cancer cell proliferation. Environ. Health Perspect. 109, 391–397.

Post, G.B., Cohn, P.D., Cooper, K.R., 2012. Perfluorooctanoic acid (PFOA), an emerging drinking water contaminant: a critical review of recent literature. Environ. Res. 116, 93–117.

Pottinger, T.G., Katsiadaki, I., Jolly, C., et al., 2013. Anti-androgens act jointly in suppressing spiggin concentrations in androgen-primed female three-spined sticklebacks – prediction of combined effects by concentration addition. Aquat. Toxicol. 140–141, 145–156.

Rainbow, P.S., Luoma, S.N., Wang, W.X., 2011. Trophically available metal – a variable feast. Environ. Pollut. 159, 2347–2349.

Royal Society, 2004. Nanoscience and Nanotechnologies: Opportunities and Uncertainties. RS Policy Document 19/04. Available 12.12.13. at http://www.raeng.org.uk/news/publications/list/reports/nanoscience_nanotechnologies.pdf.

Sanchez, W., Sremski, W., Piccini, B., et al., 2011. Adverse effects in wild fish living downstream from pharmaceutical manufacture discharges. Environ. Int. 37, 1342–1348.

SCCS, SCHER, SCENIHR, 2011. Toxicity and Assessment of Chemical Mixtures. http://ec.europa.eu/health/scientific_committees/environmental_risks/docs/scher_o_150.pdf (last accessed 11.12.13).

SCHER, SCENIHR, SCCS, 2013. Addressing the New Challenges for Risk Assessment. http://ec.europa.eu/health/scientific_committees/emerging/docs/scenihr_o_037.pdf.

Snell, T.W., Janssen, C.R., 1995. Rotifers in ecotoxicology: a review. Hydrobiologia 314–314, 231–247.

Su, Y., Yan, X., Pu, Y., et al., 2013. Risks of single-walled carbon nanotubes acting as contaminants-carriers: potential release of Phenanthrene in Japanese Medaka (*Oryzias latipes*). Environ. Sci. Technol. 47, 4704–4710.

Sun, L., Shao, X., Hua, X., et al., 2011. Transcriptional responses in Japanese Medaka (*Oryzias latipes*) exposed to binary mixtures of an estrogen and anti-estrogens. Aquat. Toxicol. 105, 629–639.

Sun, T.Y., Gottschalk, F., Hungerbühler, K., Nowack, B., 2014. Comprehensive probabilistic modelling of environmental emissions of engineered nanomaterials. Environ. Pollut. 185, 69–76.

Tan, C., Fan, W.H., Wang, W.X., 2012. Role of titanium dioxide nanoparticles in the elevated uptake and retention of cadmium and zinc in Daphnia magna. Environ. Sci. Technol. 46, 469–476.

TGD, 2003. Technical Guidance Document on Risk Assessment in Support of Commission Directive 93/67/EEC on Risk Assessment for New Notified Substances, Commission Regulation (EC) N° 1488/94 on Risk Assessment for Existing Substances and Directive 98/8/EC of the European Parliament and of the Council Concerning the Placing of Biocidal Products on the Market. European Commission, Joint Research Centre, EUR 20418 EN/2.

Tiede, K., Hassellöv, M., Breitbarth, E., et al., 2009. Considerations for environmental fate and ecotoxicity testing to support environmental risk assessments for engineered nanoparticles. J. Chromatogr. 1216A, 503–509.

Torres, M.A., Barros, M.P., Campos, S.C.G., et al., 2008. Biochemical biomarkers in algae and marine pollution: a review. Ecotoxicol. Environ. Saf. 71, 1–15.

USEPA, 2012a. Benchmark Dose Technical Guidance and Other Relevant Risk Assessment Documents. http://www.epa.gov/raf/publications/benchmarkdose.htm.

USEPA, 2012b. Emerging Contaminants – Perfluorooctane Sulfonate (PFOS) and Perfluorooctanoic Acid (PFOA). EPA 505-F-11–002 http://www.epa.gov/fedfac/pdf/emerging_contaminants_pfos_pfoa.pdf (last accessed 30.12.13).

Usseglio-Polatera, P., Bournaud, M., Richoux, P., et al., 2000. Biological and ecological traits of benthic freshwater macroinvertebrates: relationships and definition of groups with similar traits. Freshwater Biol. 43, 175–205.

Van Straalen, N.M., Feder, M.E., 2012. ecological and evolutionary functional genomics—how can it contribute to the risk assessment of chemicals? Environ. Sci. Technol. 46, 3–9.

VanBriesen, J.M., Small, M., Weber, C., et al., 2010. Modelling chemical speciation: thermodynamics, kinetics and uncertainty. In: Hanrahan, G. (Ed.), Modelling of Pollutants in Complex Environmental Systems, vol. II. ILM Publications, a Trading Division of International Labmate Limited, pp. 133–149.

Vandenbrouck, T., Soetaert, A., van der Ven, K., et al., 2009. Nickel and binary metal mixture responses in *Daphnia magna*: molecular fingerprints and (sub)organismal effects. Aquat. Toxicol. 92, 18–29.

Versantvoort, C.H.M., Oomen, A.G., Van de Kamp, E., et al., 2005. Applicability of an *in vitro* digestion model in assessing the bioaccessibility of mycotoxins from food. Food Chem. Toxicol. 43, 31–40.

Vieira Madureira, T., Rocha, M.J., Cruzeiro, C., et al., 2011. The toxicity potential of pharmaceuticals found in the Douro River estuary (Portugal). Aquat. Toxicol. 105, 292–299.

Walter, H., Consolaro, F., Gramatica, P., et al., 2002. Mixture toxicity of priority pollutants at no observed effect concentrations (NOECs). Ecotoxicology 11, 299–310.

Wang, T., Lu, Y., Chen, C., et al., 2011. Perfluorinated compounds in estuarine and coastal areas of north Bohai Sea, China. Mar. Pollut. Bull. 62, 1905–1914.

Wong, S.W.Y., Leung, K.M.Y., Djurišі, A.B., 2013. A comprehensive review on the aquatic toxicity of engineered nanomaterials. Rev. Nanosci. Nanotechnol 2, 79–105.

Wu, F., Lin, L., Qiu, J.W., et al., 2014. Complex effects of two presumably antagonistic endocrine disrupting compounds on the goldfish *Carassius aumtus*: a comprehensive study with multiple toxicological endpoints. Aquat. Toxicol. 155, 43–51.

CHAPTER 2

Conventional Risk Assessment of Environmental Contaminants

Jean-Claude Amiard, Claude Amiard-Triquet

Abstract

Conventional risk is assessed by using a four-tier approach including hazard identification, assessment of exposure (predicted environmental concentrations), hazard characterization (predicted no effect concentrations) and risk characterization based upon the risk quotient (predicted environmental concentration/predicted no effect concentration for water, sediment, and biota). This core procedure is used worldwide with some differences in the details and is described here using the European Union recommendations. Ecotoxicity databases have been compiled in many countries. The current procedures suffer many limitations, mainly because of the dominant use of standardized bioassays involving single species and unique substances, generally neglecting the dietary route of exposure and the effects of mixtures. The most important weaknesses include large uncertainties in extrapolating data across doses, species, and life stages, poor assessment of contaminants with nonmonotonic dose–response relationship, the lack of data (e.g., environmental degradation) for emerging contaminants, and the noninclusion of adaptation in polluted environments.

Keywords: Dose–effect relationship; Emission assessment; Environmental fate; Hazard identification; PECs; PNECs; Risk quotient approach; SSDs.

Chapter Outline

Aquatic Ecotoxicology. http://dx.doi.org/10.1016/B978-0-12-800949-9.00002-4

Introduction

Environmental influences on health, particularly the role of water quality, have been recognized as anciently as in the treatise called "Airs, Waters, Places" by the Greek physician Hippocrates in the second half of the fifth century BC. However, environmental concern has developed more recently, particularly with the use of synthetic organic chemicals after the Second World War. "Silent Spring" by Rachel Carson (1962) has had a key role for environmental science and the society, documenting the detrimental effects on the environment—particularly on birds—of the unreasonable use of pesticides. The term "ecotoxicology" was coined by René Truhaut in 1969 who defined it as "the branch of toxicology concerned with the study of toxic effects, caused by natural or synthetic pollutants, to the constituents of ecosystems, animal (including human), vegetable and microbial, in an integral context" (published in Truhaut, 1977). The development of ecotoxicology has allowed the implementation of retrospective risk assessment, considering the effects of the dispersion of chemical compounds into the environment and possibly mitigating them, as also prospective risk assessment that aims at assessing the future risks from releases of new and existing chemicals into the environment.

Environmental risk assessment aims at the protection of ecosystems, considering their structure, functioning, and services. Predictive risk assessment aims at assessing the future risks from releases of chemicals into the environment. In the United States, the federal Toxic Substances Control Act gives the US Environmental Protection Agency (USEPA) the authority to regulate, and even ban, the manufacture, use, and distribution of both new and existing chemicals (Schierow, 2009). In Europe, a significant improvement occurred recently with a new chemical policy, REACH, for Registration, Evaluation, and Authorization of Chemicals (CEC, 2003). Procedures needed to reach this aim have been adopted in many countries (USEPA, 1998; ANZECC and ARMCANZ, 2000; CCME, 2007; NITE, 2010; Gormley et al., 2011) and supranational organizations such as the Organisation for Economic Co-operation and Development (OECD, 2012b) and European Environment Agency (EEA, 1998; TGD, 2003). These procedures are regularly updated using scientific enhancements. A systematic literature review conducted in the Elsevier database (ScienceDirect) using the terms "environmental risk assessment" AND "aquatic" have shown that about 200 papers were published from 2010 until May 13, 2014, including several reviews regrouped in the third edition of the *Encyclopedia of Toxicology*. The Society of Environmental Toxicity and Chemistry is also very active in this field, as shown by the workshop "Closing the gap between academic research and regulatory risk assessment of chemicals" held in 2013 in Glasgow, Scotland (http://www.setac.org/members/group_content_view.asp?group=90708&id=189652&hhSearchTerms=%22Environmental+and+risk+and+assessment%22, accessed 13.05.14). The European Centre for Ecotoxicology and Toxicology of Chemicals (ECETOC) is an independent association that cooperates in a scientific context with intergovernmental agencies, governments, health authorities, and other public and professional institutions with interests in ecotoxicological and toxicological issues relating to chemicals. ECETOC's Targeted Risk Assessment tool calculates the risk of exposure from chemicals to workers, consumers, and the environment (http://www.ecetoc.org/research?q=Targeted%20Risk%20Assessment%20%28TRA%29%20tool).

Despite different guidelines having been launched in many national and supranational regulatory bodies, the procedures follow the same general scheme.

2.1 Principles for Environmental Risk Assessment

The meaning of the words *hazard* and *risk* is not totally clear for everybody, even in dictionaries. For example, one dictionary defines hazard as "a danger or risk," which helps explain why many people use the terms interchangeably. Among specialists of (eco)toxicology, hazard is defined as any source of potential damage, harm, or adverse health effects on something or someone under certain environmental conditions. However, it is clear that damage can occur only if organisms are exposed to hazard. Risk is the chance or probability that an organism will be harmed or experience an adverse health effect if exposed to a hazard.

The procedures currently in use for conventional risk assessment are depicted in Figure 2.1. The first step consists in the identification of hazard based on physicochemical properties, ecotoxicity, and intended use (EEA, 1998). Criteria for the selection of priority substances include their degree of persistency, toxicity, and bioaccumulation (for details, see

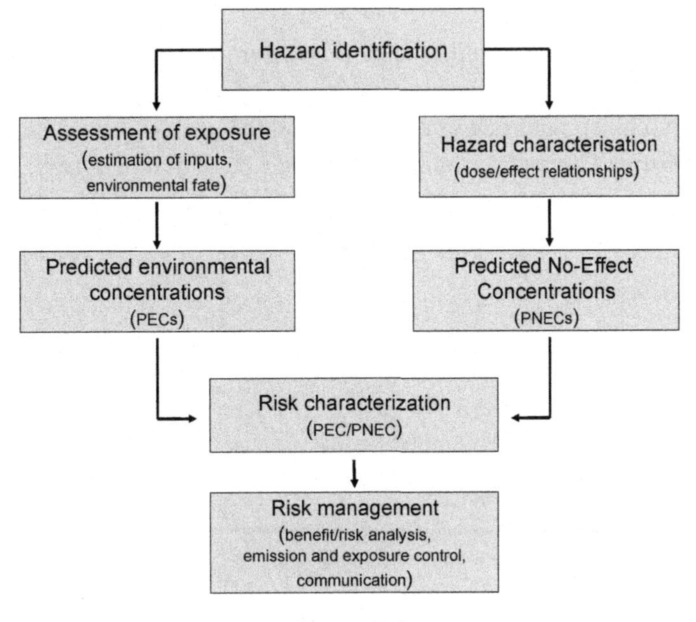

Figure 2.1
Principles for environmental risk assessment.

http://www.miljostatus.no/en/Topics/Hazardous-chemicals/Hazardous-chemical-lists/List-of-Priority-Substances/Criteria-for-the-selection-of-Priority-Substances/, accessed 13.05.14). In Europe, 45 substances or groups of substances are on the list of priority substances for which environmental quality standards were set in 2008 (amended in 2013), including selected existing chemicals, plant protection products, biocides, metals, and other groups such as polyaromatic hydrocarbons and polybrominated biphenyl ethers (PBDEs). The complete list is available at http://eur-lex.europa.eu/legal-content/EN/ALL/?uri=CELEX:32013L0039 (accessed 4.12.14). Two lists have special significance to water quality regulatory programs in the US Clean Water Act: a list of 65 toxic pollutants and a list of 129 priority pollutants (http://water.epa.gov/scitech/methods/cwa/pollutants-background.cfm#pp, accessed 13.05.14).

When a hazard has been identified, there is the need to assess the fate and effects of pollutants. Studying the fate and the biogeochemical cycle of pollutants in ecosystems is crucial in determining the environmental risk assessment because knowledge is needed on the release of pollutants in water, air, and sediment/soil because many exchanges occur between them (Figure 2.2). It is also important to take into account the trophic transfer of pollutants from sediment (suspended in the water column or deposited) and microorganisms to invertebrates (filter-feeders or deposit feeders) then predatory fish (omnivorous, carnivorous, supercarnivorous). When these data are available, they can be used in models for predicted environmental concentrations (PECs). Modeling is particularly useful for certain emerging contaminants such as nanoparticles that cannot be directly measured in environmental matrices (Chapter 17).

When a substance has been recognized as hazardous, there is the need to carry out hazard characterization (Figure 2.1). This step mainly aims at determining the relationship between the concentration of a given pollutant in a medium and the noxious effects that this substance can induce in organisms. The main parameters that may be determined from experimental

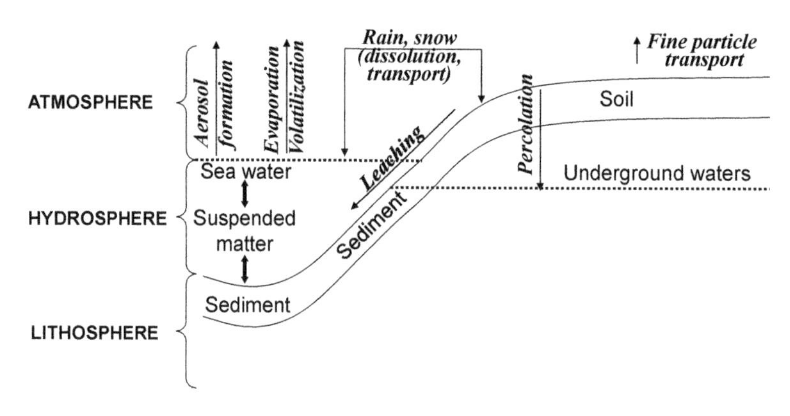

Figure 2.2
Contaminant dispersion in the physical environment.

tests include the no observed adverse effect level (NOAEL) corresponding to the no observed effect concentration (NOEC); the lowest observed adverse effect level corresponding to the lowest observed effect concentration (LOEC); and the median effect concentration (EC_{50}) producing a deleterious effect in 50% of the experimental population. When NOEC is available for a sufficient number of species belonging to different taxa/trophic level, modeling allows the determination of a predicted no effect concentration (PNEC) for the studied contaminant (Figure 2.1). Risk characterization is based on the risk quotient approach using the ratio PEC/PNEC. Very simplistically, if the PEC/PNEC ratio is greater than 1, the substance is considered to be of concern and risk reduction measures must be envisaged.

Environmental risk assessment has three main functions: (1) it allows the identification of environmental compartments and organisms at risk resulting from the use of a given substance; (2) it is the basis of risk management decisions (reduction of environmental inputs); and (3) it leads to restrictions or bans of certain substances including both new and existing chemicals.

2.2 Exposure: Determination of Predicted Environmental Concentrations

Chemical monitoring is labor-intensive and chemical analyses are expensive. In addition, until recently and until the use of passive samplers, traditional spot sampling procedures only reflected a short-lived situation. It is a serious problem because at a given site, water quality fluctuates greatly. Low concentrations of micropollutants are difficult to detect and the risk of secondary contamination when handling the samples is important. Thus, the available measured environmental concentrations have to be validated, taking into account the quality of the applied measuring techniques (OECD, 2000). Then representative data for the environmental compartment of concern will be selected. Another possibility is modeling. On the other hand, no measured environmental concentrations will normally be available for new substances (e.g., nanoparticles; see Chapter 17). Therefore, concentrations of these substances in the environment must be estimated.

2.2.1 Emission Assessment

Modeling PECs can use the measured values of a substance under examination in real releases. When such values are not available, the release rate of each given substance will be estimated based upon its use pattern. Emission scenario documents have been published under the auspices of the European Chemicals Bureau (TGD, 2003; part 4) for 12 categories of anthropogenic activities (chemical industry, metal extraction industry, refining and processing industry, biocides used in various applications, personal, domestic, and public domains, etc.) processing and producing chemicals able to enter the environment. The release estimation is based on the emission factors at different steps of the industrial process (production, formulation, processing, etc.), the production volume per time unit, the elimination in on-site treatment facilities (industrial activities), and the elimination in wastewater treatment facilities (domestic and public domains). All the releases and the receiving environmental compartment(s)—air, soil/sediment, water—must be identified.

When detailed information on the use patterns, release into the environment, and elimination are missing, generic exposure scenarios are applied. The basic assumption is that substances are emitted into a model environment characterized by environmental parameters that can be average values or reasonable (excluding accidental situations) worst-case values. When more specific data on the emission of a substance may be obtained, the generic assessment may be improved for a more realistic result, including, for instance, topographical and climatological variability.

Local emissions ($Elocal_{i,j}$ in $kg{\cdot}d^{-1}$) can be calculated for each life-cycle stage i of the life-cycle and each compartment j according to the following formula:

$$Elocal_{i,j} = Fmainsource_i * 1000/Temission_i * RELEASE_{i,j}$$

with

> $Fmainsource_i$ = fraction of release at the local main source at stage i,
> $Temission_i$ = number of days per year for the emission in stage i ($d{\cdot}yr^{-1}$),
> $RELEASE_{i,j}$ = release during stage i to compartment j ($t{\cdot}yr^{-1}$).

For the regional scale assessments, the emissions are assumed to be a constant and continuous flux during the year, thus the $Eregional_j$ is the total emission to compartment j (annual average) in $kg{\cdot}d^{-1}$. Regional emissions can be calculated by summing the release fractions for each stage of the life-cycle according to the following formula:

$$Eregional_j = 1000/365 * \sum_{i=1\ to\ n} * RELEASE_{i,j}$$

with $RELEASE_{i,j}$ = release during life-cycle stage i to compartment j ($t{\cdot}yr^{-1}$).

2.2.2 Behavior and Fate in the Environment

The fate of a given substance once released into the environment needs to be estimated by considering exchanges between physical compartments of the environment (Figure 2.2) and biotic and abiotic transformation processes. The quantification of distribution and degradation of the substance (as a function of time and space) leads to an estimate of PEClocal and PECregional. The PEC calculation is described for water (ground- and freshwaters, transitional and marine waters), soil and sediment, and air (TGD, 2003; part 2). Transport of the substance between the compartments must be taken into account to determine the full biogeochemical cycle of substances.

2.2.2.1 Abiotic and biotic degradation

In each physical compartment, a given substance is submitted to physicochemical and biological processes leading to its degradation. This includes:

- hydrolysis and oxidation in water;
- photolysis in surface water and in the atmosphere;

- microbial degradation in surface water, soil and sediment (as also in sewage treatment plants); and
- metabolization in macroorganisms.

The study of degradation does not need to be conducted if the substance is inorganic. For organic substances, test guidelines that may be used to conclude on ready biodegradability for organic substances have been recently reviewed (Kapanen et al., 2013) under the auspices of the European Chemicals Agency. Substance properties influence the applicability of specific test guidelines. Information on physicochemical properties enables the identification of the most appropriate test guideline. In addition, marine screening tests are available (OECD 306 "Biodegradability in Seawater").

According to the US Geological Survey (http://toxics.usgs.gov/definitions/biodegradation.html), biodegradation may be characterized for the purpose of hazard assessment as:

- primary, corresponding to the alteration of the chemical structure of a substance resulting in loss of a specific property of that substance;
- environmentally acceptable, meaning that biodegradation occurs to such an extent that undesirable properties of the compound are removed;
- ultimate, corresponding to the complete breakdown of a compound to either fully oxidized or reduced simple molecules (such as carbon dioxide/methane, nitrate/ammonium, and water). However, in some cases, the products of biodegradation can be more harmful than the substance degraded.

Degradation can occur under both aerobic and anaerobic conditions. A substance may be considered as easily degradable provided that the degradation rate is >60% over 28 days. In addition to the degradation rate, other useful parameters are the half-life (denoted DT_{50}, which is the time required for the disappearance of 50% of the applied substance) and the specific degradation rate constant k. By definition, the specific degradation rate constant is equal to the relative change in concentration per time:

$$k = (1/C) \cdot (dC/dt)$$

First-order kinetics implies that the rate of degradation ($mg \cdot L^{-1} \cdot day^{-1}$) is proportional to the concentration of substrate, which declines over time. With true first-order kinetics, the specific degradation rate constant, k, is independent of time and concentration. First-order kinetics are normally expected under the conditions prescribed for standardized tests (e.g., OECD 309 "Aerobic Mineralisation in Surface Water – Simulation Biodegradation Test"). However, deviations from first-order kinetics may be observed, for instance, if the diffusion rate, rather than the biological reaction rate, limits the biotransformation rate.

2.2.2.2 Distribution

Generally, hydrophilic compounds are mainly present in water, whereas hydrophobic compounds are present in air, soil/sediment, and biota. Multimedia models have been developed to examine the multimedia environmental fate of organic chemicals that are discharged to the environment. One of the models currently in use in the European regulatory framework (TGD, 2003; part 2) employs the fugacity concept and treats four bulk compartments: air, water, soil, and bottom sediment, which consist of subcompartments of varying proportions of air, water, and mineral and organic matter (Mackay and Peterson, 1991). These authors assume that equilibrium (equifugacity) applies within each compartment (i.e., between subcompartments), but not between compartments. Within the same compartment, partition coefficients may be useful to examine the distribution of chemicals.

The transfer of a substance from the water to the air subcompartment (e.g., volatilization from surface water, Figure 2.2) may be assessed by using its Henry's law constant. The air–water partitioning coefficient ($K_{air\text{-}water}$) can be estimated according to the following equation:

$$K_{air-water} = H/(R * TEMP)$$

with

H = Henry's law constant ($Pa \cdot m^3 \cdot mol^{-1}$),
R = gas constant ($Pa \cdot m^3 \cdot mol^{-1} \cdot k^{-1}$),
TEMP = temperature at the air–water interface (K).

Partition coefficients solid-water in suspended matter (Kp_{susp}), in sediment (Kp_{sed}), and soil (Kp_{soil}) expressed in ($L \cdot kg^{-1}$) are calculated according to the same approach (TGD, 2003; part 2).

$$Kp_{comp} = Foc_{comp} \cdot K_{oc} \text{ with comp} \in \{soil, sed, susp\}$$

with

K_{oc} = partition coefficient organic carbon-water ($L \cdot kg^{-1}$),
Foc_{comp} = weight fraction of organic carbon in compartment *comp* ($kg \cdot kg^{-1}$).

K_{oc} may be measured by adsorption studies (EC C18; OECD 106, 2000) or by the high-performance liquid chromatography method (EC C19; OECD 121, 2001). Kp may be expressed as the concentration of the substance sorbed to solids (in $mg_{chem} \cdot kg_{solid}^{-1}$) divided by the concentration dissolved in porewater ($mg_{chem} \cdot L_{water}^{-1}$). This formulation is similar to the distribution coefficient Kd used for inorganic substances (metals, metalloids, radionuclides) (OECD 106 "Adsorption - Desorption Using a Batch Equilibrium Method").

$$Kd = Cs(eq)/Caq(eq)$$

with

Cs(eq) = concentration of the inorganic substance adsorbed onto the solid phase ($mg \cdot kg^{-1}$) at equilibrium,
Caq(eq) = concentration of the inorganic substance dissolved in the aqueous phase ($mg \cdot L^{-1}$) at equilibrium.

2.2.2.3 Predicted Environmental Concentrations

Emissions can be assessed on a local scale (PEClocal) when only one source of emission is recognized; for instance, a sewage treatment plant (for a detailed example, see TGD, 2003; part 2). However, point source releases can also contribute to the environmental concentrations on a larger scale thus leading to the assessment of a PECregional. The concentrations of substances released from diffuse sources over a wider area are assessed on a regional scale (PECregional). Concerning the aquatic environment, procedures for the calculation of PEClocal and PECregional are described for surface waters, marine waters, and sediments. In addition, the PECoral is calculated to assess bioaccumulation and secondary poisoning of predators.

2.2.2.3.1 Calculation of PECs for the aquatic compartment

Figure 2.3 shows the most important fate processes in the aquatic compartment. For local PEC, the procedure recommended in the TGD (2003, part 2) is based on the assumption of complete mixing of the effluent in surface water whereas volatilization, degradation, and sedimentation are ignored because of the short distance between the point of effluent discharge and the exposure location. The calculation of the PEClocal for the aquatic compartment includes the calculation of the discharge concentration to a given water body, dilution effects, and removal from the aqueous medium by adsorption to suspended matter.

For PECregional, it is also important to take into account the general movement of the contaminated plume and the long-range transport of suspended particles by the river flow or drift and marine currents (Figure 2.3). In this case, volatilization, degradation, and sedimentation must be taken into account, should all the different processes of exchange exist between compartments (Figure 2.2). For regional computations, the TGD (2003, part two) recommend the use of multimedia fate models described by Mackay et al. (1992), Van de Meent (1993),

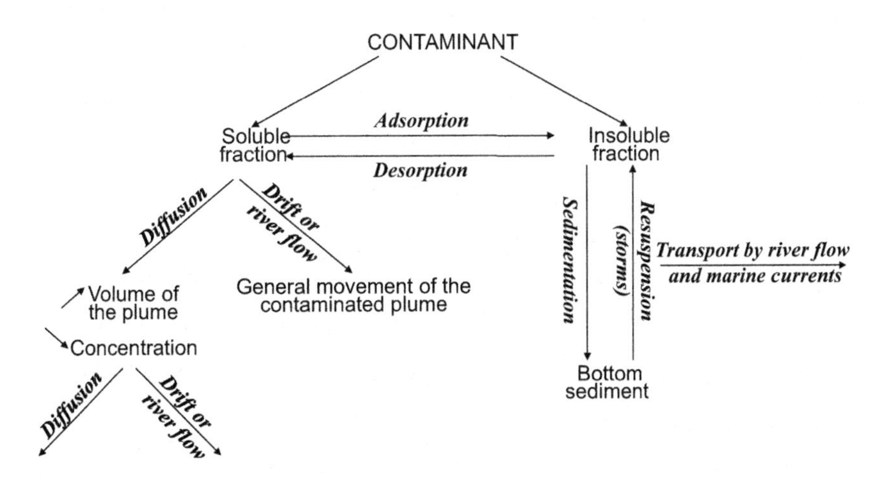

Figure 2.3
Circulation of contaminants in water masses.

and Brandes et al. (1996). In these models, each compartment (air, water, soil, and bottom sediment) is considered homogeneous and well mixed.

2.2.2.3.2 Calculation of PECs for sediment

The concentration in freshly deposited sediment is taken as the PEClocal for sediment and according to (Di Toro et al., 1991) may be determined by using the equation:

$$PEClocal_{sed} = K_{susp-water}/RHO_{susp} * PEClocal_{water} * 1000$$

with

$PEClocal_{water}$ concentration in surface water during emission episode ($mg \cdot L^{-1}$),
$K_{susp\text{-}water}$ suspended matter–water partitioning coefficient ($m^3 \cdot m^{-3}$),
RHO_{susp} bulk density of suspended matter ($kg \cdot m^{-3}$),
$PEClocal_{sed}$ predicted environmental concentration in sediment ($mg \cdot kg^{-1}$).

In this equation, the $PEClocal_{sed}$ is derived from the corresponding water body concentration, assuming a thermodynamic partitioning equilibrium. This model has certain limits underlined in the TGD (2003, part 2), such as the frequent lack of equilibrium distribution between water and suspended matter is also an underestimation of sediment concentration when release to the surface water predominately occurs as particles.

The multimedia fate model will be used again to calculate PECregional for sediment. The scenario used for freshwater PEC calculations must be modified to allow dispersive exchange between the coastal zone to the continental seawater (Figure 2.3). The results from regional models should be interpreted with caution because PECs (water, sediment) are averaged for all the regional compartments and are considered homogeneous and well mixed. However, there is a considerable uncertainty in the determination of input parameters (e.g., degradation rates, partitioning coefficients) and much higher concentrations can be encountered locally (TGD, 2003; part 2).

2.2.2.3.3 Calculation of PECs for biota

Contaminant concentrations in fish and their predators is a result of uptake from the aqueous phase and intake of contaminated food (prey species, sediment). Direct uptake from the water phase is predominant for hydrophilic substances (octanol–water partition coefficient $\log K_{ow} < 4.5$), whereas intake from food becomes increasingly important for lipophilic substances ($\log K_{ow} \geq 4.5$). The calculation of PECoral uses the bioconcentration factor (BCF) and the biomagnification factor (BMF). The TGD (2003, part 2) provides equations allowing the calculation of a given substance's BCF from the value of the K_{ow} for this substance. However, experimentally determined BCF values are often preferable and standardized procedures have been recently updated (OECD 305 "Bioaccumulation in Fish: Aqueous and Dietary Exposure"). The BMF is defined (TGD, 2003; part 2) as the relative concentration in a predatory animal compared to the concentration in its prey (BMF=Cpredator/Cprey). It is recommended to use lipid normalized Cpredator and Cprey.

$$PECoral_{predator} = PEC_{water} * BCF_{fish} * BMF$$

with

$PECoral_{predator}$ = PEC in food ($mg \cdot kg^{-1}_{wet\ fish}$),
PEC_{water} = PEC in water ($mg \cdot L^{-1}$),
BCF_{fish} = bioconcentration factor for fish on wet weight basis ($L \cdot kg^{-1}_{wet\ fish}$),
BMF = biomagnification factor in fish (–).

When carrying risk assessment, it is important to validate the models used for the calculation of PECs considering measured data from monitoring programs and from high-quality literature.

2.3 Ecotoxicity: Determination of Predicted No Effect Concentrations

Once the hazardous effects of concern are identified, it is then needed to assess the relationship between dose (concentration) and response (effect) in different species. When results are available for different trophic levels and taxa, it is then possible to determine PNECs (Figure 2.1).

2.3.1 Hazard Characterization

Hazard characterization is mainly based on toxicity data obtained by using experimental toxicity tests. In both the European REACH and US Toxic Substances Control Act regulations, the level of ecotoxicity assessment is linked to the tonnage of new or existing substances produced by chemical industries. Standard information requirements for aquatic toxicity data under REACH includes short-term toxicity testing on invertebrates (preferred species *Daphnia*) and growth inhibition of aquatic plants (algae preferred) for a tonnage of 1–10 tons $year^{-1}$. Short-term toxicity testing on fish is needed for a tonnage of 10–100 tons $year^{-1}$. In addition, long-term toxicity testing on invertebrates (preferred species *Daphnia*) and fish are needed for tonnages >100 tons $year^{-1}$ (Tarazona et al., 2014).

Many standardized bioassays have been published by many regulatory bodies (OECD, International Organization for Standardization, ASTM, USEPA) for sediment and water toxicity testing (listed in Cesnaitis et al., 2014; Tarazona et al., 2014). The most commonly used are single-test species, carried out in dramatically simplified laboratory "ecosystems," most often without food and substratum. Tests can be carried out under static, semistatic, or flow-through conditions (EC, 2005). Static describes aquatic toxicity tests in which test solutions are not renewed during the test. Flow-through described tests in which solutions in test vessels are renewed continuously by the constant inflow of a fresh solution or by a frequent intermittent inflow. Semistatic describes aquatic tests in which test solutions are replaced periodically during the test. Depending on the strategy adopted, the level of exposure integrated over the whole duration of the test may be highly variable. However, many reports provide only the nominal concentration at the beginning of the test because monitoring the real concentrations in experimental units is costly and labor-intensive.

In the oldest acute tests, the observed endpoint was most often lethality, whereas more recently, growth, reproduction parameters, and even behavior are chosen as endpoints in long-term toxicity tests (for examples, see Cesnaitis et al., 2014; Tarazona et al., 2014). Short-term toxicity tests allow the determination of EC_{50}s when the endpoint is a sublethal effect or median lethal concentrations (LC_{50}) when the endpoint is lethality. Long-term toxicity tests are relevant for the determination of NOECs and LOECs. The determination of these parameters is shown in Figure 2.4. In this example, six doses have been tested. No difference was observed between the response of controls and the response of specimens exposed to D1, whereas a significant change was observed at D2. Thus D1 corresponds to the observed NOEC and D2 to the observed LOEC. In fact the true values of NOEC and LOEC termed "biological" are between D1 and D2. The major disadvantage of this procedure is that it assimilates the NOEC and LOEC to the corresponding experimental doses. Thus the results are deeply influenced by the dose spacing and also by the number of tested organisms at each dose (influencing the statistical significance). Less "rustic" derivations of LC_{50}/EC_{50} and NOEC values from raw values (probit analysis, analysis of variance, and post hoc statistical tests) are given in the TGD (2003, part 2).

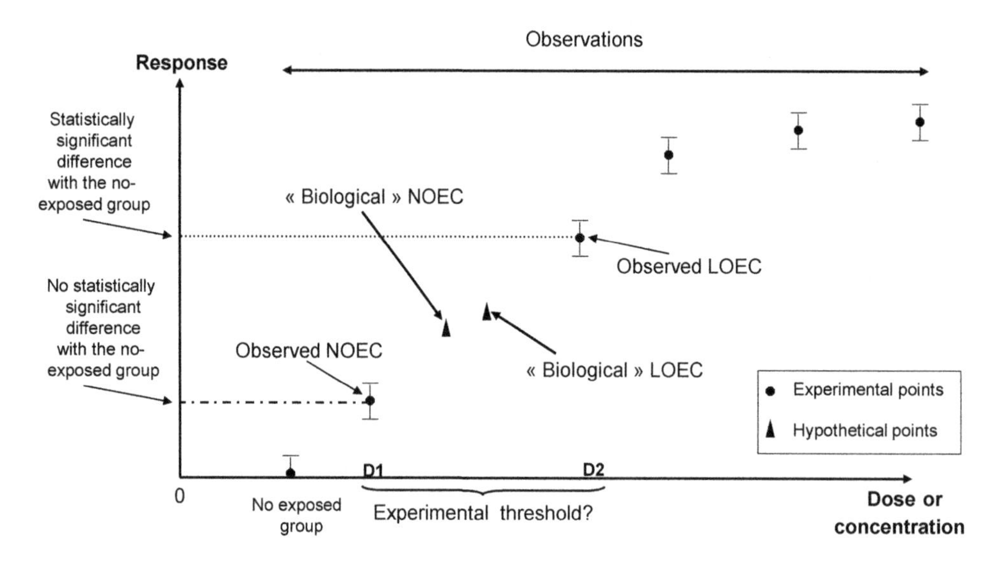

Figure 2.4
Experimental determination of the dose-response relationship: biological versus observed NOEC and LOEC.

However, it remains necessary to improve the determination of the NOECs that will be used at the next step (Figure 2.1) for the calculation of PNECs. Thus, another approach has been proposed by both European and US agencies (EFSA, 2009; USEPA, 2009) (Figure 2.5). First, a software freely available online allows the modeling of the dose–response curve (USEPA software) by choosing the model that fits well with the experimental points. Then, it is needed to choose a level of effects (benchmark response)—e.g., BMR_{10} corresponding to the benchmark dose (BMD_{10}) that induces a response in 10% of experimental specimens. Last, it is possible to choose the limit of the confidence interval (90 or 95%). The lower benchmark dose (BMDL) is the lower limit of the confidence interval at 90 or 95% of the BMD.

2.3.2 Calculation of PNECs

It is very important to evaluate available data with regard to their completeness and their reliability and relevance for environmental hazard and risk assessment (adequacy). The TGD (2003, part 2) puts forward general guidelines on the evaluation of ecotoxicity data. Two main points must be examined: the procedures used to carry out the study and the way the results have been interpreted. Greater weight should normally be attached to studies carried out according to standardized bioassays, conducted following good laboratory practices (GLP; European Directive 2004/10/EC; OECD, 1998; CFR, 2011a,b). The adequacy of a bioassay also lies in the way that the performance and results are described (critical pieces of information are missing; the design of the test is insufficiently detailed, etc.).

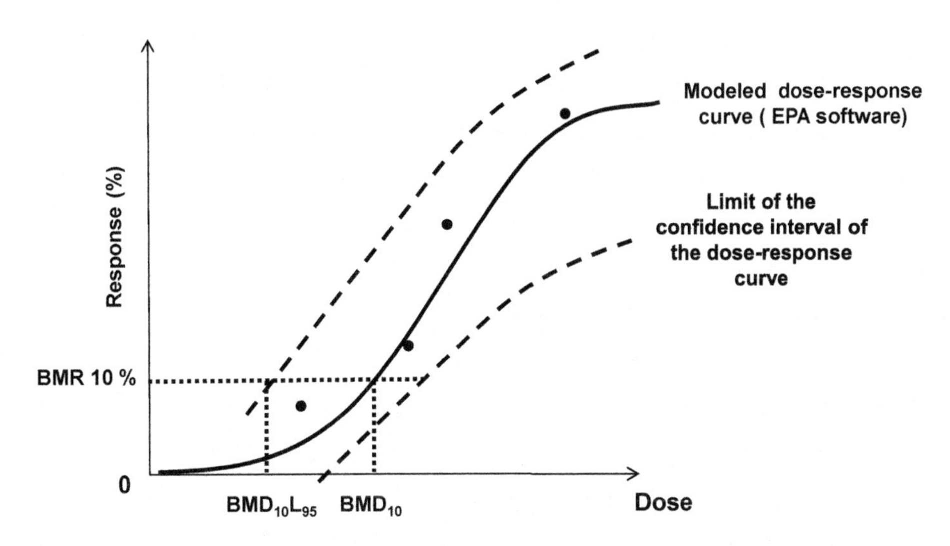

Figure 2.5
Assessment of the different parameters of the benchmark dose (BMD). BMR_{10}, 10% of experimental specimens are affected; $BMD_{10}L_{95}$, lower limit of the confidence interval at 95% of the BMD.

Determination of the PNECs takes into account the number of bioassay results available for different species representative of different trophic levels and the availability of long-term toxicity data. PNECs in current use for risk assessment in the aquatic environment are $PNEC_{aquatic}$ (fresh- and seawater), $PNEC_{sediments}$, and PNECoral (because of secondary poisoning).

2.3.2.1 Calculation of $PNEC_{aquatic}$

PNEC calculations are carried out using either assessment factors or statistical extrapolation techniques.

The use of assessment factors is needed to take into account the uncertainties resulting of (1) intra- and interlaboratory variation of toxicity data, (2) intra- and interspecies variations (biological variance), (3) short-term to long-term toxicity extrapolation, and (4) laboratory data to impact field extrapolation. The uncertainty decreases with larger and more relevant datasets, thus allowing the use of lower assessment factors (AFs). The AFs applicable in the framework of the European regulations are shown in Table 2.1. Seawater AFs are consistently higher than freshwater AF because fewer marine data are available. It also remains questionable if marine species are more sensitive than freshwater species. A recent literature review (Klok et al., 2012) based on 3627 references concludes that there is no systematic difference in sensitivity to pesticides between fresh- and saltwater species.

The calculation of $PNEC_{aquatic}$ using statistical extrapolation techniques is based upon the species sensitivity distribution (SSD). SSDs are cumulative distributions of measures of species sensitivity to a stressor or toxicant. In the case of triclosan depicted in Figure 2.6, the normal model provided the best fit of the models tested among the possible range of distributions (European Commission, 2011; Health Canada, Environment Canada, 2012). SSDs are used to estimate concentrations that will protect a certain percentage of a community. It is frequently accepted that 5% of sensitive species may be neglected. For instance, in the case depicted in Figure 2.6, the provisional hazardous concentration 5% (HC_5) is $115\,ng{\cdot}L^{-1}$. Only a substantial amount of toxicity data from several taxonomic groups can result in a robust HC_5. In Europe, to meet the requirements of the REACH guidance, the database must contain preferably more than 15, but at least 10 NOECs/EC_{10}s (effect concentration producing a deleterious effect in 10% of the experimental population), from different species covering at least eight taxonomic groups. For estimating a quality standard for the freshwater community, fish and a second family in the phylum Chordata (e.g., fish, amphibian), a crustacean (e.g., cladoceran, copepod, ostracod, isopod, amphipod, crayfish), an insect (e.g., mayfly, dragonfly, damselfly, stonefly, caddisfly, mosquito, midge), a family in a phylum other than Arthropoda or Chordata (e.g., Rotifera, Annelida, Mollusca), a family in any order of insect or any phylum not already represented, algae and higher plants would normally need to be represented (European Commission, 2011). Similar taxa requirements have also been adopted in the water quality guidelines of other countries (CCME, 2007).

Table 2.1: Assessment factors to derive a $PNEC_{aquatic}$ in fresh- or seawater

Available Data	Assessment Factor
Freshwater	
At least one short-term LC_{50}/EC_{50} from each of three trophic levels of the base set (fish, *Daphnia*, and algae)	1000
One long-term NOEC (either fish or *Daphnia*)	500
Two long-term NOECs from species representing two trophic levels (fish and/or *Daphnia* and/or algae)	50
Long-term NOECs from at least three species (normally fish, *Daphnia*, and algae) representing three trophic levels	10
Species sensitivity distribution method	1–5
Field data or model ecosystems	Reviewed on a case-by-case basis
Seawater	
Lowest short-term LC_{50}/EC_{50} from freshwater or saltwater representatives of three taxonomic groups (algae, crustaceans, and fish) of three trophic levels	10,000
Lowest short-term LC_{50}/EC_{50} from freshwater or saltwater representatives of three taxonomic groups (algae, crustaceans, and fish) of three trophic levels + two additional marine taxonomic groups (e.g., echinoderms, mollusks)	1000
One long-term NOEC (from freshwater or saltwater crustacean reproduction or fish growth studies)	1000
Two long-term NOECs from freshwater or saltwater species representing two trophic levels (algae and/or crustaceans and/or fish)	500
Lowest long-term NOECs from three freshwater or saltwater species (normally algae and/or crustaceans and/or fish) representing three trophic levels	100
Two long-term NOECs from freshwater or saltwater species representing two trophic levels (algae and/or crustaceans and/or fish) + one long-term NOEC from an additional marine taxonomic group (e.g., echinoderms, mollusks)	50
Lowest long-term NOECs from three freshwater or saltwater species (normally algae and/or crustaceans and/or fish) representing three trophic levels + two long-term NOECs from additional marine taxonomic groups (e.g., echinoderms, mollusks)	10

Modified after TGD (2003), part 2.

The calculation of PNECs again includes an assessment factor, but much lower than normal as few toxicological data are available (Table 2.1):

$$PNEC = HC_5/AF \quad (\text{with } 1 < AF < 5)$$

The exact value of the AF depends on the overall quality of the database concerning the mode of exposure (chronic vs acute studies, mesocosm/field studies), the biological models tested (representative of different taxonomic groups, feeding strategies and trophic levels, sensitive life stages), the goodness of fit of the SSD curve, or the size of confidence interval around the

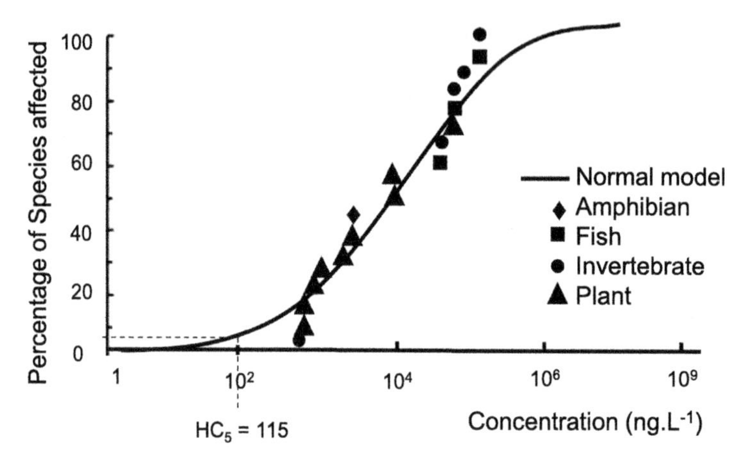

Figure 2.6

Species-sensitivity distribution for the preliminary assessment of triclosan using chronic toxicity data for freshwater organisms. HC_5, hazardous concentration 5%. *Modified after Health Canada, Environment Canada (2012).*

fifth percentile. Information on the mode of action of chemicals is also important to judge of the relevance of taxonomic groups tested, realizing that the mode of action may differ between short-and long-term effects and between taxonomic groups (e.g., special sensitivity of algae to copper; of insects and also crustaceans to insecticides).

Many procedures for environmental risk assessment are based upon single-species laboratory acute toxicity data (Raimondo et al., 2013). They seem relatively robust for estimating environmental quality standards because the geographical distribution of the species used to construct SSDs generally do not have a significant influence (Maltby et al., 2005; Feng et al., 2013; Wang et al., 2014). However, it remains questionable if SSDs derived from single-species laboratory acute toxicity data can be used to protect species assemblages in aquatic ecosystems. This question has been addressed for insecticides by collating single-species acute toxicity data and (micro)mesocosm data (Maltby et al., 2005). These authors concluded that "the corresponding median HC_5 (95% protection level with 50% confidence) was generally protective of single applications of insecticide but not of continuous or multiple applications. In the latter cases, a safety factor of at least five should be applied to the median HC_5".

In the case of an intermittent release (less than once per month and for no more than 24 h), the major risk is due to acute toxic effects. Thus, it is recommended (TGD, 2003; part 2) to calculate a $PNEC_{water}$, applying an assessment factor of 100 to the lowest LC_{50}/EC_{50} of at least three short-term tests from three trophic levels.

2.3.2.2 Calculation of $PNEC_{sediment}$

Sediments in both freshwater and marine environments are the final sink for most of the contaminants entering the aquatic environment as a consequence of the high sorption capacity of particulate matter. Substances with a low partitioning coefficient (Kd for inorganic substances, K_{oc} or K_{ow} for organic substances) have a low capacity to bind to sediments. Thus to avoid extensive testing of chemicals a log K_{oc} or log K_{ow} of ≥3 can be used as a trigger value for sediment effects assessment (TGD, 2003; part 2).

The calculation of the $PNEC_{sed}$ according to the equilibrium partitioning method has been described for both fresh- and seawater. The following equation is applied:

$$PNEC_{sed} = \left(K_{susp-water}/RHO_{susp}\right) * PNEC_{water} * 1,000$$

with

$PNEC_{water}$ = PNEC in water ($mg{\cdot}L^{-1}$),
RHO_{susp} = bulk density of wet suspended matter ($kg{\cdot}m^{-3}$),
$K_{susp\text{-}water}$ = partition coefficient suspended matter–water ($m^3{\cdot}m^{-3}$)
and $PNEC_{sed}$ = PNEC in sediment ($mg{\cdot}kg^{-1}$).

Sediment toxicity testing with endobenthic species is widely developed (Chapter 10), and these bioassays are extensively used in the framework of REACH (Cesnaitis et al., 2014) and in other countries, based on test guidelines published by OECD, USEPA, and ASTM. A $PNEC_{sed}$ can be derived from these tests using assessment factors (Table 2.2) according to a strategy similar to the one already described for the determination of $PNEC_{aquatic}$ (for detail, see TGD, 2003; part 2).

Table 2.2: Assessment factors to derive a $PNEC_{sediment}$ in freshwater or marine environment

Available Test Result	Assessment Factor
Freshwater	
One long-term test (NOEC or EC_{10})	100
Two long-term tests (NOEC or EC_{10}) with species representing different living and feeding conditions	50
Three long-term tests (NOEC or EC_{10}) with species representing different living and feeding conditions	10
Seawater	
Short-term freshwater or marine test	10,000
Two short-term tests including a minimum of one marine test with an organism of a sensitive taxa	1000
One long-term freshwater sediment test	1000
Two long-term freshwater sediment tests with species representing different living and feeding conditions	500
One long-term freshwater and one saltwater sediment test representing different living and feeding conditions	100
Three long-term sediment tests with species representing different living and feeding conditions	50
Three long-term tests with species representing different living and feeding conditions including a minimum of two tests with marine species	10

Modified after TGD (2003), part 2.

2.3.2.3 Calculation of PNECoral

Secondary poisoning can arise from the uptake of contaminants from prey and sediment (bioaccumulation, biomagnification) in addition to uptake from water (bioconcentration). Bioaccumulation is termed biomagnification when the transfer of chemicals via the food chain results in an increase of body concentrations at higher levels in the trophic chain. Most of the commonly studied compounds that possess K_{ow} in the 10^4–10^9 range exhibit a capacity to be bioconcentrated and thus, depending on their persistency, to be bioaccumulated and biomagnified in food webs. The maximum bioaccumulation potential has been observed for compounds with log K_{ow} between 5.5 and 7.5 (Abarnou, in Amiard-Triquet and Rainbow, 2009).

The risk to the freshwater fish-eating predators is calculated as the ratio between the concentration in their food ($PECoral_{predator}$) and the no-effect-concentration for oral intake ($PNECoral_{predator}$). The calculation must be preferably based on toxicity data provided by long-term studies, including NOECs established considering not only survival but also growth or reproduction. The following equations are applied respectively for fish-eating birds and mammals:

$$NOEC_{bird} = NOAEL_{bird} * CONV_{bird}$$

with

$NOEC_{bird}$ = NOEC for birds ($kg \cdot kg_{food}^{-1}$),
$NOAEL_{bird}$ = NOAEL for birds ($kg \cdot kg^{-1} bw \cdot d^{-1}$),
$CONV_{bird}$ = conversion factor from NOAEL to NOEC ($kg\,bw \cdot d^{-1} \cdot kg_{food}^{-1}$).

$$NOEC_{mammal,\ food\_chr} = NOAEL_{mammal,\ oral\_chr} * CONV_{mammal}$$

with

$NOEC_{mammal,food\_chr}$ = NOEC for mammals ($kg \cdot kg_{food}^{-1}$),
$NOAEL_{mammal,oral\_chr}$ = NOAEL for mammals ($kg \cdot kg^{-1} bw \cdot d^{-1}$),
$CONV_{mammal}$ = conversion factor from NOAEL to NOEC ($kg\,bw \cdot d^{-1} \cdot kg_{food}^{-1}$).

The PNECoral is ultimately derived from the toxicity data (food basis) applying an assessment factor. The following equation is applied:

$$PNECoral = TOX_{oral}/AF_{oral}$$

with

PNECoral = PNEC for secondary poisoning of birds and mammals (in $kg \cdot kg_{food}^{-1}$),
AF_{oral} = assessment factor applied in extrapolation of PNEC
and $TOX_{oral} = LC50_{bird}$, $NOEC_{bird}$, or $NOEC_{mammal,food,chronic}$ (in $kg \cdot kg_{food}^{-1}$).

The assessment factors for extrapolation differ from 30 to 3000 when the TOX_{oral} is a LC_{50} or an NOEC established by using either a short-term or a chronic test. In the marine environment, this strategy must be adapted to take into account the fact that the marine food chains are very long and complex, including marine fish, fish predators, and top predators.

2.4 Risk Characterization

The risk quotient approach is based on the determination of PECs and PNECs for different environmental compartments. Very simplistically, if the PEC/PNEC ratio is greater than 1, the substance is of concern. It is clear that modeling PECs and PNECs includes a vast amount of simplifications and uncertainties, and strategies have been proposed to improve uncertainty analysis in risk assessment of chemicals (Verdonck et al., 2007). The decision by the competent authority to request additional data must take into account strategies implying lowest cost and effort, highest gain of information, and the avoidance of unnecessary testing on animals. Improvements of the determination of the PEC/PNEC ratio rely on additional information about release and fate of contaminants, better assessment of environmental concentrations based on monitoring programs, additional bioassays (bioaccumulation, chronic toxicity), or alternative methodologies (in silico and in vitro tests). If information/further testing allows a lowering of the PEC/PNEC ratio, it may be concluded that there is no need for risk reduction measures. This iterative approach has precautionary aspects because data gaps are filled by worst-case assumptions or high assessment factors. If the PEC/PNEC ratio remains greater than 1, risk reduction measures must be decided. Variations on this general procedure (applied for existing substances) are offered for new substances and biocides (TGD, 2003; part 2; EC Directive 98/8 concerning the placing of biocidal products on the market).

Several ecotoxicity databases are available in France (INERIS, 2014), Europe (IUCLID, 2014; EUSES, 2014), the United States (ECOTOX (AQUIRE); PubChem Compound Database), and Japan (Japanese Ministry of Environment, 2014).

2.5 Conclusions

It is generally accepted that about 100,000 chemical substances are able to enter the aquatic environment. The information on the dangerous properties of chemicals is incomplete or even lacking completely for most of them. This lack of data has stimulated a new chemical policy for industrial chemical control in the European Union called REACH. The REACH legislation, enacted in June 2007, has allowed the documentation of 12,349 unique substances (updated April 17, 2014) marketed in quantities above 1 ton $year^{-1}$, stimulating the collection of existing data but also the production of new data to fill the gaps. However, risk assessment under REACH is carried out mainly by applying existing methodologies, particularly standardized bioassays involving single species and unique substances (Cesnaitis et al., 2014; Tarazona et al., 2014). Because the traditional approaches do not provide mechanistic information on how the chemicals exert toxicity, there remain large uncertainties (Verdonck et al., 2007) in extrapolating data across doses, species, and life stages. In the field of human toxicology, the criticism of current toxicity testing systems has led to a global call to action to bring toxicology into the twenty-first century (NRC, 2007; Seidle and Stephens, 2009; Ritter et al., 2012). Among new technologies that have been recently developed, gene transcripts (transcriptomics), proteins (proteomics), and small cellular molecules (metabolomics) have been tentatively applied in the field of ecotoxicology. However, according to Fent and Sumpter (2011), "the use of these techniques in aquatic toxicology is still in its infancy, and data cannot yet be applied to environmental risk assessment or regulation until more emphasis is placed on interpreting the data within their physiological and toxicological contexts."

Hormesis and tolerance are biological responses that are important in the life of a majority of species (in fact, in all the species that have been investigated from this point of view). Hormesis is a biphasic dose–response phenomenon characterized by a low-dose stimulation and a high-dose inhibition. Hormetic responses reflect both direct stimulatory or overcompensation responses to damage induced by relatively low doses of chemical or physical agents (Calabrese and Blain, 2011). To be able to take into account hormesis, conventional bioassay designs must be modified to include lower doses. In addition, improved statistical methods for testing nonlinear dose–response relationships are needed (Qin et al., 2010). Tolerance may be defined as the ability of organisms to cope with stress, either natural (heat shock, salinity variations, hypoxia, plant toxins) or chemical, depending on anthropogenic inputs of many different classes of contaminants into the environment. If differences in tolerance ratios are relatively small, then they may be encompassed by conventional safety factors used to establish protective guidelines (Berthet et al., in Amiard-Triquet et al., 2011). However, some examples suggest that safety margins may not be adequate for all contaminant classes. For instance, the lethal concentration for 20% of the individuals tested for PCB126 determined in killifish embryos originating from 24 estuarine sites ranged over four orders of magnitude, depending on their chronic exposure to sediment-bound total PCBs (Nacci et al., 2009). For a more realistic in situ risk assessment, the computation of more relevant PNECs could be obtained by using ecotoxicological data on species endemic to the considered area, as recommended for pesticides in an estuarine environment (Guérit et al., 2008).

In their safety evaluations of chemicals, public health agencies grant special prominence to studies based on standard bioassays carried out by using GLP. Much less weight in risk assessments is given to a large number of non-GLP studies conducted in academic research units by the leading experts in various fields of science from around the world. In a review dealing with the case of bisphenol A, Myers et al. (2009) concluded that "public health decisions should be based on studies using appropriate protocols with appropriate controls and the most sensitive assays, not GLP [since] research using state-of-the-art techniques should play a prominent role in safety evaluations of chemicals." This conclusion can be extended to many other chemical substances, particularly emerging contaminants or contaminants of emerging concern, among which those that exhibit nonlinear dose–response relationships are especially difficult.

Because of the uncertainties in environmental risk assessment according to the conventional scheme, the environmental quality standards can differ consistently between different countries. In addition, environmental quality standards must be reviewed periodically to take into account the progress of scientific knowledge. A typical case is that of legacy contaminants with emerging concern, such as endocrine disruptors. That endocrine disruptors induce a nonmonotonic dose–response relationship and are efficient only at certain stages of the life-cycle make their noxiousness inaccessible in standardized bioassays.

In the case of manufactured nanomaterials, OECD test guidelines or other internationally agreed-on methods used for "traditional chemicals" have been examined for their suitability toward a list of endpoints including physicochemical properties, degradation and accumulation, and effects on biotic systems (OECD, 2010, 2014). Sample preparation and dosimetry for the safety testing of manufactured nanomaterials are particularly important and difficult to master, whereas they are crucial to characterize precisely the real exposure of test organisms (OECD, 2012a).

Data about degradation of chemical substances in the environment are crucial for the determination of the PECs but largely unknown for emerging contaminants. Thus US Geological Survey scientists are conducting research to document the potential for biodegradation of caffeine, hormones, antibiotics, antimicrobial disinfectants, and other chemicals found in the effluent discharged from wastewater treatment plants (http://toxics.usgs.gov/regional/emc/ec_biodegradation2.html).

Another specific issue is the question of risk assessment of fluctuating concentrations of chemicals (e.g., leaching of pesticides by surface run-off in agricultural catchments). The question of pulses in the aquatic environment has been recently reviewed (Chèvre and Valloton, 2013) and showed that the main characteristics governing ecotoxicological effects are not only the level and the duration of the exposure because of the pulse, but also the time between two successive pulses, allowing (or not allowing) biological recovery.

Because most of the bioassays in current use are single-species tests, it is a complicated task to extrapolate to the level of the ecosystem, even if it is recommended to carry out toxicity testing with at least three different taxa (algae, *Daphnia*, fish). Even when more data are available, allowing the use of statistical extrapolation techniques, based on SSDs, the specificity of the mode of action of the substance should be a priority factor in calculating the ecological risk (Guérit et al., 2008). According to these authors, "PNECs for herbicides should therefore be calculated from NOEC from toxicity tests on vegetal species mainly or PNECs for insecticides should give priority to NOEC from tests on crustaceans" (phylogenetically linked to insects). In addition, intraspecific and interspecific competition is rarely considered when extrapolating from lower to higher biological organization levels (Kattwinkel and Liess, 2014 and literature quoted therein). According to these authors, "ignoring these potential negative effects of competitors on population recovery in risk assessment might lead to overvaluing the recovery potential of a species and an under-protective estimated safe toxicant concentration."

Although regulations are crucial for chemical risk management, the current procedures are primarily based on individual substances for a single toxicity assessment approach, without taking into account mixture of all toxic chemicals found in the environment. Several studies have documented the intuitive approach that the effect of whole mixtures will be higher than the effects of a single individual compound with different species and different classes of chemicals, in the laboratory, in mesocosms, and in the field (Gregorio and Chèvre, 2014). The challenge for ecotoxicology is therefore to develop robust methods to assess the environmental risk of chemicals in mixtures. From a review on the use of mixture toxicity assessment in different regulatory frameworks, Syberg et al. (2009) recommended the construction of a database that includes data on chemicals in the European environment that could be used for mixture toxicity assessment of the chemicals with individual PECs/PNECs >0.1. Guidance for assessing the impact of mixtures of chemicals in the aquatic environment has recently been developed (ECETOC, 2011; SCCS, SCHER and SCENIHR, 2011).

The use of animals in standardized toxicity tests widely used in environmental risk assessment has raised ethical concern, thus leading to the launching of the European directive on the protection of animals used for scientific purposes (EU, 2010). However, it deals exclusively with amphibians, fish, and, among invertebrates, only cephalopods. In addition, the large gap that exists between assessment capacities and assessment needs raises economic concerns particularly when long-term toxicity tests are considered. Both of these reasons have led to the development of alternative methodologies (in silico and in vitro methods, weight-of-evidence approach, etc.) that have already been used in the framework of REACH (ECHA, 2011). The strategies and methods aim to meet the 3Rs principle of Refinement, Reduction, and Replacement. They have been reviewed by a panel of 40 experts from academic, industrial, and regulatory origins (Scholz et al., 2013), who conclude that it is difficult to replace an animal test by a single experimental or nontesting approach and that the availability of high-quality in vivo reference data are important for developing alternatives. In any case, the development of alternative testing approaches requires future action and research.

References

Amiard-Triquet, C., Rainbow, P.S., 2009. Environmental Assessment of Estuarine Ecosystems. A Case Study. CRC Press, Boca Raton.

Amiard-Triquet, C., Rainbow, P.S., Roméo, M., 2011. Tolerance to Environmental Contaminants. CRC Press, Boca Raton.

ANZECC, ARMCANZ, 2000. Australian and New Zealand Guidelines for Fresh and Marine Water Quality. National Water Quality Management Strategy Paper No. 4. ANZECC and ARMCANZ, Canberra.

Brandes, L.J., den Hollander, H., van de Meent, D., 1996. SimpleBox 2.0: A Nested Multimedia Fate Model for Evaluating the Environmental Fate of Chemicals. National Institute of Public Health and Environmental Protection (RIVM), Bilthoven, The Netherlands. RIVM Report 719101 029.

Calabrese, E.J., Blain, R.B., 2011. The hormesis database: the occurrence of hormetic dose responses in the toxicological literature. Reg. Toxicol. Pharmacol. 61, 73–81.

Carson, R., 1962. Silent Spring. Houghton Mifflin, Boston, re-published by Mariner Books (2002).

CCME, 2007. A Protocol for the Derivation of Water Quality Guidelines for the Protection of Aquatic Life. Canadian Council of Ministers of the Environment, Winnipeg.

CEC (Commission of the European Communities), 2003. Proposal for a Regulation of the European Parliament and of the Council Concerning the Registration, Evaluation, Authorisation and Restrictions of Chemicals (REACH) Establishing a European Chemical Agency and Amending Directive 1999/45/EC and Regulation (EC) (On Persistent Organic Pollutants). COM 644.

Cesnaitis, R., Sobanska, M.A., Versonnen, B., et al., 2014. Analysis of the ecotoxicity data submitted within the framework of the REACH regulation. Part 3. Experimental sediment toxicity assays. Sci. Total Environ. 475, 116–122.

CFR (Code of Federal Regulations), 2011a. Title 40-Protection of Environment. Chapter I - Environmental Protection Agency (Continued). Subchapter E - Pesticide programs. Part 160 Good laboratory practice standards http://www.gpo.gov/fdsys/pkg/CFR-2011-title40-vol24/xml/CFR-2011-title40-vol24-part160.xml.

CFR (Code of Federal Regulations), 2011b. Title 40-Protection of Environment. Chapter I - Environmental Protection Agency (Continued). Subchapter R - Toxic Substances Control Act (continued). Part 792 Good laboratory practice standards http://www.gpo.gov/fdsys/pkg/CFR-2011-title40-vol32/xml/CFR-2011-title40-vol32-part792.xml.

Chèvre, N., Valloton, N., 2013. Pulse exposure in ecotoxicology. In: Férard, J.F., Blaise, C. (Eds.), Encyclopedia of Aquatic Ecotoxicology. Springer Reference, Dordrecht, pp. 917–925.

Di Toro, D.M., Zarba, C.S., Hansen, D.J., et al., 1991. Technical basis of establishing sediment quality criteria for nonionic organic chemicals using equilibrium partitioning. Environ. Toxicol. Chem. 10, 1541–1583.

EC (Environment Canada), 2005. Guidance Document on Statistical Methods for Environmental Toxicity Tests. EPS 1/RM/46 with June 2007 amendments http://publications.gc.ca/collections/collection_2012/ec/En49-7-1-46-eng.pdf (accessed 10.01.14).

ECETOC, 2011. Development of Guidance for Assessing the Impact of Mixtures of Chemicals in the Aquatic Environment. Technical report 211.

ECHA, 2011. The Use of Alternatives to Testing on Animals for the REACH Regulation. ECHA-11-R-004.2-EN.

ECOTOX (AQUIRE). http://www.epa.gov/ecotox/ (accessed 3.06.14).

EEA, 1998. Environmental Risk Assessment - Approaches, Experiences and Information Sources. Environmental issue report No 4 http://www.eea.europa.eu/publications/GH-07-97-595-EN-C2 (accessed 13.05.14).

EFSA (European Food Security Authority), 2009. Use of the benchmark dose approach in risk assessment. 1. Guidance of the Scientific Committee (Question No EFSA-Q-2005-232). The EFSA Journal 1150, 1–72.

EU, 20.10.2010. Directive 2010/63/EU of the European Parliament and of the council of September 22, 2010 on the protection of animals used for scientific purposes. Off. J. Eur. Union10, 33–79.

European Commission, 2011. WFD-CIS Guidance Document N° 27 Technical Guidance for Deriving Environmental Quality Standards. Office for Official Publications of the European Communities, Luxembourg. 204 p. https://circabc.europa.eu/sd/a/0cc3581b-5f65-4b6f-91c6-433a1e947838/TGD-EQS%20CIS-WFD%2027%20EC%202011.pdf.

EUSES, 2014. http://ihcp.jrc.ec.europa.eu/our_activities/public-health/risk_assessment_of_Biocides/new-version-of-euses-2.1.2/?searchterm=None (accessed 3.06.14).

Feng, C.L., Wua, F.C., Dyer, S.D., et al., 2013. Derivation of freshwater quality criteria for zinc using interspecies correlation estimation models to protect aquatic life in China. Chemosphere 90, 1177–1183.

Fent, K., Sumpter, J.P., 2011. Progress and promises in toxicogenomics in aquatic toxicology: is technical innovation driving scientific innovation? Aquat. Toxicol. 105, 25–39.

Gormley, A., Pollard, S., Rocks, S., Black, E., 2011. UK Guidance. Guidelines for Environmental Risk Assessment and Management: Green Leaves III. 80 p. https://www.gov.uk/government/publications/guidelines-for-environmental-risk-assessment-and-management-green-leaves-iii.

Gregorio, V., Chèvre, N., 2014. Assessing the risks posed by mixtures of chemicals in freshwater environments: case study of Lake Geneva, Switzerland. WIREs Water 1, 229–247.

Guérit, I., Bocquené, G., James, A., Thybaud, E., Minier, C., 2008. Environmental risk assessment: a critical approach of the European TGD in an in situ application. Ecotoxicol. Environ. Saf. 71, 291–300.

Health and the Environment (RIVM), RIVM Report No. 672720 001, Bilthoven, The Netherlands Health Canada, Environment Canada, 2012. Preliminary Assessment. Triclosan; Chemical Abstracts Service Registry Number 3380-34-5 http://www.ec.gc.ca/ese-ees/6EF68BEC-5620-4435-8729-9B91C57A9FD2/Triclosan_EN.pdf.

INERIS, 2014. http://www.ineris.fr/substances/fr/page/21, (accessed 21.05.14).

IUCLID, 2014. http://www.oecd.org/fr/env/ess/risques/electronictoolsfordatasubmissionevaluationandexchangeintheoecdcooperativechemicalsassessmentprogramme.htm (accessed 3.06.14).

Japanese Ministry of Environment, March 2014. Japan ecotoxicity Tests Data. http://www.env.go.jp/chemi/sesaku/02e.pdf.

Kapanen, A., Heinonen, J., Dilhac, B., 2013. How to Bring Your Registration Dossier in Compliance with REACH – Tips and Hints Part 4 Biodegradation II – How to Choose the Appropriate Method for Ready Biodegradability Testing? http://echa.europa.eu/documents/10162/13628/compliance_pt4_webinar_05_biodegradationII_en.pdf.

Kattwinkel, M., Liess, M., 2014. Competition matters: species interactions prolong the long-term effects of pulsed toxicant stress on populations. Environ. Toxicol. Chem. 33, 1458–1465.

Klok, C., de Vries, P., Jongbloed, R., et al., 2012. Literature Review on the Sensitivity and Exposure of Marine and Estuarine Organisms to Pesticides in Comparison to Corresponding Fresh Water Species. http://www.efsa.europa.eu/fr/supporting/doc/357e.pdf (accessed 21.05.14).

Mackay, D., Paterson, S., 1991. Evaluating the multimedia fate of organic chemicals: a level III fugacity model. Environ. Sci. Technol. 25, 427–436.

Mackay, D., Paterson, S., Shiu, W.Y., 1992. Generic models for evaluating the regional fate of chemicals. Chemosphere 24, 695–717.

Maltby, L., Blake, N., Brock, T.C.M., Van den Brink, P.J., 2005. Insecticide species sensitivity distributions: importance of test species selection and relevance to aquatic ecosystems. Environ. Toxicol. Chem. 24, 379–388.

Myers, J.P., vom Saal, F.S., Akingbemi, B.T., et al., 2009. Why public health agencies cannot depend on good laboratory practices as a criterion for selecting data: the case of bisphenol A. Environ. Health Perspect. 117, 309–315.

Nacci, D., Huber, M., Champlin, D., et al., 2009. Evolution of tolerance to PCBs and susceptibility to a bacterial pathogen (*Vibrio harveyi*) in Atlantic killifish (*Fundulus heteroclitus*) from New Bedford Harbor (MA, USA) harbor. Environ. Pollut. 157, 857–864.

NITE, 2010. National Institute of Technology and Evaluation (Japan). Guidelines for initial risk assessment of chemicals substances (Summary), 15 p.

NRC (National Research Council), 2007. Toxicity Testing in the 21st Century: A Vision and a Strategy. National Academies Press, Washington D.C.

OECD, 1998. Series on Principles of Good Laboratory Practice and Compliance Monitoring. Number 1. OECD Principles on Good Laboratory Practice (as revised in 1997). ENV/MC/CHEM(98)17.

OECD, 2000. Report of the OECD workshop on improving the use of monitoring data in the exposure assessment of industrial chemicals. Series on Testing and Assessment, vol. 18, Organisation for Economic Cooperation and Development (OECD), OECD Environmental Health and Safety Publications, Paris.

OECD, 2010. Preliminary Guidance Notes on Sample Preparation and Dosimetry for the Safety Testing of Manufactured Nanomaterials. ENV/JM/MONO(2010)25.

OECD, 2012a. Guidance on sample preparation and dosimetry for the safety testing of manufactured nanomaterials. Series on the Safety of Manufactured Nanomaterials, vol. 36, ENV/JM/MONO(2012)40.

OECD, 2012b. The OECD Environmental Risk Assessment Toolkit: Tools for Environmental Risk Assessment and Management. http://www.oecd.org/chemicalsafety/risk-assessment/theoecdenvironmentalriskassessmenttoolkittoolsforenvironmentalriskassessmentandmanagement.htm.

OECD, 2014. Ecotoxicology and environmental fate of manufactured nanomaterials: test guidelines expert meeting report. Series on the Safety of Manufactured Nanomaterials, vol. 40, ENV/JM/MONO(2014)1.

PubChem Compound Database. http://www.ncbi.nlm.nih.gov/pccompound (accessed 19.03.15).

Qin, L.T., Liu, S.S., Liu, H.L., et al., 2010. Support vector regression and least squares support vector regression for hermetic dose–response curves fitting. Chemosphere 78, 327–334.

Raimondo, S., Jackson, C.R., Barron, M.G., 2013. Web-based Interspecies Correlation Estimation (Web-ice) for Acute Toxicity: User Manual. Version 3. 2, EPA/600/R-12/603 U. S. Environmental Protection Agency, Office of Research and Development, Gulf Ecology Division, Gulf Breeze, FLhttps://www.google.fr/#q=%28 Web-ICE%29+for+Acute+Toxicity+raimondo.

Ritter, L., Austin, C.P., Bend, J.R., et al., 2012. Integrating Emerging Technologies into Chemical Safety Assessment. The Expert Panel on the Integrated Testing of Pesticides. Council of Canadian Academies, Ottawa (Ontario), Canada. 291 p.

SCCS, SCHER, SCENIHR, 2011. Toxicity and Assessment of Chemical Mixtures. http://ec.europa.eu/health/scientific_committees/environmental_risks/docs/scher_o_150.pdf (accessed 11.12.13).

Schierow, L.J., 2009. The Toxic Substances Control Act (TSCA): Implementation and New Challenges. https://www.acs.org/content/dam/acsorg/policy/acsonthehill/briefings/toxicitytesting/crs-rl34118.pdf.

Scholz, S., Sela, E., Blaha, L., et al., 2013. A European perspective on alternatives to animal testing for environmental hazard identification and risk assessment. Reg. Toxicol. Pharmacol. 67, 506–530.

Seidle, T., Stephens, M.L., 2009. Bringing toxicology into the 21st century: a global call to action. Toxicol. Vitro 23, 1576–1579.

Syberg, K., Jensen, T.S., Cedergreen, N., et al., 2009. On the use of mixture toxicity assessment in REACH and the Water Framework Directive: a review. Hum. Ecol. Risk Assess. 15, 1257–1272.

Tarazona, J.V., Sobanska, M.A., Cesnaitis, R., et al., 2014. Analysis of the ecotoxicity data submitted within the framework of the REACH Regulation. Part 2. Experimental aquatic toxicity assays. Sci. Total Environ. 472, 137–145.

TGD, 2003. Technical Guidance Document on Risk Assessment in Support of Commission Directive 93/67/EEC on Risk Assessment for New Notified Substances, Commission Regulation (EC) No 1488/94 on Risk Assessment for Existing Substances and Directive 98/8/EC of the European Parliament and of the Council Concerning the Placing of Biocidal Products on the Market. European Commission. Joint Research Centre. EUR 20418 EN/2.

Truhaut, R., 1977. Eco-Toxicology - objectives, principles and perspectives. Ecotoxicol. Environ. Saf. 1, 151–173.

USEPA, April 1998. Guidelines for Ecological Risk Assessment. EPA/630/R-95/002F. U.S. Environmental Protection Agency, Washington, DC. 124 p. and Annexes.

USEPA, 2009. Benchmark Dose Software (BMDS). http://www.epa.gov/ncea/bmds/ (accessed 18.03.2015).

Van de Meent, D., 1993. Simplebox: A Generic Multimedia Fate Evaluation Model. National Institute of Public.

Verdonck, F.A.M., Souren, A., Van Asselt, M.B.A., et al., 2007. Improving uncertainty analysis in European Union risk assessment of chemicals. Integr. Environ. Assess. Manage. 3, 333–343.

Wang, X., Yan, Z., Liu, Z., et al., 2014. Comparison of species sensitivity distributions for species from China and the USA. Environ. Sci. Pollut. Res. 21, 168–176.

CHAPTER 3

Quality Standard Setting and Environmental Monitoring

Jean-Claude Amiard, Claude Amiard-Triquet

Abstract

Environmental quality standards (EQSs in water, sediment, or biota) are briefly described because the derivation of predicted no effect concentrations that form the basis of EQS values have been delineated previously (Chapter 2). Because EQSs are a major tool in protecting the aquatic environment and human health, it is crucial to verify that they are respected, carrying out chemical monitoring in water, sediment, and biota. Complementarily, monitoring with biomarkers allows "preventive" risk assessment by detecting biological changes before population effects may be evident. Monitoring with bioindicators and biotic indices allows "retrospective" risk assessment. Integrated monitoring (chemical, biological, ecological) is a prominent method of facing the scientific, economic, and health challenges arising from anthropogenic impacts. Environmental sample banking has the potential to conduct retrospective monitoring of emerging contaminants.

Keywords: Biomonitoring; Chemical monitoring; Ecological tools; Environmental quality standards; Environmental specimen banking; Integrated monitoring; Quality assurance/quality control.

Chapter Outline

Introduction

Environmental quality standards (EQSs, concentrations in water, sediment, or biota that must not be exceeded) are a major tool to protect the aquatic environment and human health. "EQSs should protect freshwater and marine ecosystems from possible adverse effects of

Aquatic Ecotoxicology. http://dx.doi.org/10.1016/B978-0-12-800949-9.00003-6

chemicals as well as human health via drinking water or ingestion of food originating from aquatic environments. Several different types of receptor therefore need to be considered, i.e., the pelagic and benthic communities in freshwater, brackish or saltwater ecosystems, the top predators of these ecosystems and human health" (TGD, 2011).

Predicted no effect concentrations are traditionally used in risk assessment (Chapter 2) and form the basis of most EQS values (Matthiessen et al., 2010). Procedures used for quality standard setting or risk assessment are very similar, including the application of assessment (i.e., safety) factors (AFs) depending on the quality and quantity of available toxicity data, the use of statistical extrapolation methods (i.e., species sensitivity distributions), and the use of mesocosm studies to assess impacts of chemicals under more realistic conditions than laboratory single species tests allow for (Lepper, 2005; Matthiessen et al., 2010, Chapter 2).

The number of aquatic EQS derivations completed to date is 111 for Australia, 200 for Canada, 46 for the United States, and 33 for the European Union (EU). The number of sediment EQS completed to date is 34 for Australia and approximately 20 for the EU (Matthiessen et al., 2010). However, the authors state that "the overwhelming majority of priority chemicals have as yet no [water and sediment] EQS at all." Much fewer EQSs are available for biota, as exemplified in the European Community that has launched 12 EQS_{biota} only among 45 priority substances in the field of water policy (EU, 2013). In addition to these regulatory values of EQSs, tentative values are also available such as sediment quality guidelines proposed for the assessment of contaminated sediments (Wenning et al., 2005). Even regulatory EQSs deserve to be validated in the field (e.g., EQSs for polyaromatic hydrocarbons [PAHs] and metals validated by the US Environmental Protection Agency (USEPA, 2003, 2005) or by using mesocosm data (Matthiessen et al., 2010). Reviewing established EQSs may be fruitful to derive more robust standards when new data or new derivation techniques become available. Thus, improving the protection of the aquatic environment and human health clearly need important research efforts.

Environmental monitoring is then indispensable to verify if EQSs are honored. Guidelines for chemical monitoring of hazardous substances in water, sediment, and biota are available under the auspices of different official bodies (Lauenstein and Cantillo, 1998; Schmitt et al., 1999; EC, 2010; OSPAR Commission, 2011, 2012b, 2013). However, to validate the EQS values, it is important to examine not only the level of contaminants in the physical media (water, sediment) and in organisms, but also the biological effects at different levels of biological organization by using biomarkers (at infraindividual and individual levels, Chapters 7 and 8) and ecological indicators (at population and community levels, Chapter 14).

In this chapter, we will review how the implementation of EQSs and monitoring tools can benefit from advances of the knowledge and which future research is needed.

3.1 Environmental Quality Standards and Guidelines

To meet the objectives of protection as defined previously (TGD, 2011), receptors and compartments at risk must be identified for each substance under examination to take into account its physicochemistry and fate in the environment (preferential partitioning in water or sediment, potential for bioaccumulation and biomagnification). More precisely, EQS assessment sediments are relevant only if there is an evidence of sorption potential (Log $K_{oc} \geq 3$ or Log $K_{ow} \geq 3$) or of high toxicity to benthic organisms. Biota assessment of organic substances is undertaken on the basis of the evidence of bioaccumulation potential (biomagnification factor >1 or bioconcentration factor ≥100, or Log $K_{ow} \geq 3$), provided that there is no mitigating property (for instance, rapid biodegradation), or the evidence of high intrinsic toxicity to mammals and birds (TGD, 2011). The applicability of these procedures of prioritization remains questionable in the case of emerging contaminants.

The procedures recommended in European countries for the derivation of EQSs are summarized in Figure 3.1, but those adopted in other countries are very similar (Matthiessen et al., 2010). The first step consists in the calculation of temporary quality standards (QS) in fresh- and sea-water, in the compartments at risk for the substance under examination. For more detail, the mode of calculation is based on the methodologies described in Chapter 2 for predicted no effect concentrations. Several assessments are generally carried out for the same compartment. For instance, QSs are derived in surface waters considering the protection of pelagic species (direct ecotoxicity based on bioassays), the protection of human health from drinking water, the protection of biota from secondary poisoning, and the protection of human health from consuming fisheries products. In the two latter cases, QS_{biota} are converted into equivalent water concentrations by using bioconcentration and biomagnification factors (Figure 3.1, after TGD, 2011). Then, the lowest standard calculated for these different objectives of protection will be adopted as the overall EQS for that compartment.

It is needed to protect human and environmental health in case of chronic contamination as well as in case of accidents such as oil spills. To fulfill this aim, two water column EQSs will be required: (1) a long-term standard, expressed as an annual average concentration (AA-EQS) and normally based on chronic toxicity data and (2) a short-term standard, referred to as a maximum acceptable concentration EQS (MAC-EQS), which is based on acute toxicity data. In the first case, the use of the annual mean induces a definite underestimation of risk, particularly when concentrations are highly dispersed. Another possibility is to use the 90th percentile concentration of the samples. Because chronic toxicity data are largely missing for many substances , the calculation of AA-EQSs may be based on data of acute toxicity tests by using AFs, the value of which depends on the number and quality of the toxicological data (Tables 2.1 and 2.2). Because sediment and biota are integrative matrices, representative of exposure over long periods, it is not appropriate to derive MAC-EQSs for these compartments.

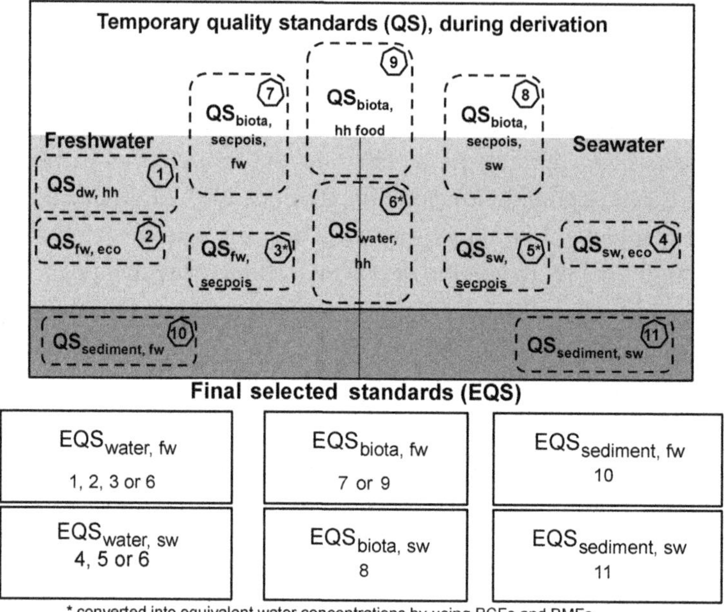

Figure 3.1
Determination of environmental quality standards for the protection of the environment and human health. Temporary quality standards (1) for drinking water ($QS_{dw, hh}$); considering direct ecotoxicity for (2) freshwater ($QS_{fw, eco}$) and (4) seawater ($QS_{sw, eco}$); considering secondary poisoning expressed in (3) freshwater ($QS_{fw, secpois}$) and (5) seawater ($QS_{sw, secpois}$) after conversion; (6) for food human consumption of fishery products, expressed in water ($QS_{water, hh}$) after conversion; secondary poisoning based on concentrations in (7) freshwater biota ($QS_{biota, secpois, fw}$) and (8) marine biota ($QS_{biota, secpois, sw}$); (9) food human consumption of fishery products, based on concentrations in biota ($QS_{biota, hh}$); for sediment, based on sediment toxicity data for (10) freshwater ($QS_{sediment, fw}$) and (11) seawater ($QS_{sediment, sw}$). Selected overall environmental quality standards for water compartments ($EQS_{fw - EQSsw}$), biota ($EQS_{biota, fw}$ $EQS_{biota, sw}$), and sediment ($EQS_{sediment, fw}$ $EQS_{sediment, sw}$). BCFs: bioconcentration factors; BMFs: biomagnification factors. *After TGD (2011).*

The use of AFs introduces an inevitable degree of uncertainty in the determination of EQSs. Thus, the relevance of using of a single EQS value for each protection goal toward a given substance is questionable. Consequently, it is recommended that each EQS should be framed as a range of at least two values representing higher and lower levels of protection (Matthiessen et al., 2010). However, this procedure has not yet been applied in the framework of regulatory practices.

Because the number of tested chemicals with reliable test data remains small compared with the number of substances potentially at risk for the environment and human health, the use of nontesting approaches have been discussed (Matthiessen et al., 2010; TGD, 2011). Available methods include, for instance, the category approach—grouping chemicals whose physicochemical properties, fate and behavior, and toxicological properties are likely to be similar—and the quantitative structure–activity relationships, possible calculation methods for "reading across" between media (e.g., equilibrium partitioning for PAH mixtures, Hansen et al., 2003; biotic ligand model, Chapter 4). Other methods such as acute-to-chronic ratios and toxicity correlations between taxa (e.g., fish and daphnids) are considered by many as unacceptable in EQS derivation (Matthiessen et al., 2010), as exemplified by the heterogeneity of responses between fish and arthropods to exposure to chlorpyrifos (Figure 10.3, http://www.epa.gov/oppefed1/ecorisk/ecofram/sra.htm).

The need for separate marine and freshwater standards remain a topic of discussion (Matthiessen et al., 2010; TGD, 2011), considering the possibility that salinity can influence both the physicochemistry of contaminants and organisms as exemplified in the case of cadmium by Rainbow and Black (2005); also, there is the possibility that sensitivity differs between marine and freshwater organisms. Examples of substances (ammonia, pyrethroid insecticides, organometallics, metals) for which toxicity is different are provided in Matthiessen et al. (2010). On the other hand, a recent literature review (Klok et al., 2012) based on 3627 references concludes that there is no systematic difference in sensitivity to pesticides between fresh- and salt-water species. However, at least in the European strategy, seawater AFs are consistently higher than freshwater AFs because fewer marine data are available (Tables 2.1 and 2.2).

No consensus exists concerning the way to take into account the background concentrations by naturally occurring substances when higher than EQS. The added risk approach used in the EU is described with some detail in Matthiessen et al. (2010). It is mainly recommended for metals because for other substances (e.g., natural steroidal estrogens, PAHs) the philosophy is even more unclear. It is accepted that organisms are not disturbed by metals that have been present in the environment over geological time. More scientifically, depending on their origin, naturally occurring versus anthropogenic metals do not show the same distribution in the sediment matrix. The former are preferentially present in the mineral matrix (residual phase of operationally defined geochemical fractions), whereas the latter are mainly present in more labile phases as exemplified in the metal-rich Gironde estuary, France (Amiard et al., 2007). In Australia, Canada, and some other jurisdictions, the added risk procedure (that allows the incorporation of the background concentration of naturally occurring substances such as metals in environmental risk limits; for details, see Crommentuijn et al., 2000) is not followed but it is now recognized worldwide that bioavailability must be included when implementing EQSs (Matthiessen et al., 2010).

Chemicals that are carcinogenic, mutagenic, and toxic to reproduction are particularly at risk. Cancer is a disease that affects the majority of metazoan species (Vittecoq et al., 2013). It has been widely studied in fish (Wirgin and Waldman, 1998; Nacci et al., 2002;

Collier et al., in Amiard-Triquet et al., 2013) and is well documented in certain marine mammals (Martineau et al., 2002). Because in most cases, cancer is postreproductive (Romeo and Wirgin, 2011), many scientists consider that it is not a critical factor for the long-term viability of most populations (Matthiessen et al., 2010). However, Wirgin and Waldman (2004) have reported the simultaneity of a reduced abundance, a high prevalence of cancer, and truncated age structure in the Atlantic tomcod *Microgadus tomcod* from the highly contaminated Hudson River. Another reason not to neglect the potential effect of carcinogenic compounds is the diversity of ways in which oncogenic phenomena may influence the competitive abilities of individuals, their susceptibility to pathogens, their vulnerability to predators, and their ability to disperse, thus affecting ecological processes that govern biotic interactions (Vittecoq et al., 2013). Genotoxic assessment methods are exemplified in Matthiessen et al. (2010).

Reproductive success is a key for maintaining healthy fish and wildlife populations. A few papers have strongly suggested (Miller and Ankley, 2004; Mills and Chichester, 2005; Miller et al., 2007; Sanchez et al., 2011) and even demonstrated (Kidd et al., 2007) that the effect of endocrine disruptors had clear effects on fish abundance. That they induce a nonmonotonic dose–response relationship (Vandenberg et al., 2012; Bergman et al., 2013) and are efficient only at certain stages of the life-cycle make their noxiousness inaccessible in standardized bioassays. Accurate description of nonmonotonic dose–response curves is a key step for the characterization of hazards associated to the presence of pollutants in the medium and thus need the development of dedicated models.

The success and accuracy with which MAC-EQSs and AA-EQSs are used will depend crucially on the design of the aquatic monitoring program that provides the data on chemical concentrations in water, sediment, and biota (Matthiessen et al., 2010; TGD, 2011).

To date, EQSs are operational for water, whereas in the case of sediment, it remains preferable to use sediment quality guidelines. A critical review of published studies from a wide range of laboratory and field studies in freshwater, estuarine, and marine environments (encompassing more than 8000 sediment samples) has been launched under the auspices of the Society of Environmental Toxicology and Chemistry that summarizes discussions among scientists, environmental regulators, and environmental managers (Wenning et al., 2005). Effects-based sediment quality guidelines can be used to assess the probability of observing adverse biological effects to benthic organisms with known levels of statistical confidence, a practice that is well-developed for the interpretation of monitoring data as exemplified recently by Zhuang and Gao (2014).

3.2 Chemical Monitoring Strategies

Monitoring hazardous substances in the environment targets two main objectives: show the spatial distribution of contaminants to determine which sites are at risk and examine temporal trends at different sites to determine if the situation is improving or worsening. Three classes of environmental matrices may be used for monitoring aquatic environments: water, sediment, and biota.

3.2.1 Chemical Monitoring in the Water Column

It is often deliberated that water is not a suitable matrix to assess trace contamination at environmentally pertinent levels. Indeed, low concentrations of micropollutants are difficult to detect, thus inducing risks of secondary contamination when handling the samples. Depending on the chemicals of interest, handling can include water filtration (analysis of either filtered or total water in different monitoring programs), flow-through centrifugation, acidification, storage, pretreatment and preconcentration steps for metals, extraction, and enrichment steps for organic compounds (OSPAR Commission, 2013). In seawater, where concentrations of many contaminants are extremely low, combined with too high quantification limits of the laboratories, monitoring contaminants in water under the EU Water Framework Directive "resulted in thousands of non significant results, acquired at great expense" (Knoery and Claisse, 2010).

Water masses are highly fluctuating, particularly in transitional waters that are among the most exposed to anthropogenic impacts worldwide. Depending on hydrological conditions (low tide vs high tide, spring tide vs neap tide, seasonal variations of river flow rate), water collected at a given site by using the traditional spot sampling procedures does not belong to the same water mass (as indicated by different salinities), depending on the precise moment of the day/month/year, and on the sampling depth. Thus it may be recommended to standardize sampling procedures by selecting a time during the tidal cycle (ebb tide or flow tide), a position in the water column (surface, middle, or bottom water column), and constant physical and chemical parameters, especially salinity as a marker of water masses (Cailleaud et al., 2009). In addition, hydrological conditions associated with wind-generated erosion events interfere with the erosion–sedimentation equilibrium, modifying the characteristics of the water masses in term of turbidity, thus influencing the partitioning between components of the aqueous phase. As a result, spatial and temporal representativeness of water samples collected at insufficient frequencies do not allow the proper assessment of chronic contamination levels (Knoery and Claisse, 2010). However, in the case of contaminants that partition strongly into the water rather than the sediment or biota, water can be the preferred matrix for monitoring (OSPAR Commission, 2013).

Time-weighted average concentrations can be obtained by using continuous water intake over a prolonged period, followed by filtration and extraction, but the operationality is limited (OSPAR Commission, 2013). Passive sampler devices (PSDs) developed for chemical substance monitoring in water have been reviewed by Mills et al. (2010). They may be deployed in the medium over 15 days for as long 1 month. Consider that (1) semipermeable membrane devices are relevant for nonpolar contaminants such as PAHs; (2) polar organic chemical integrative samplers for polar compounds (detergents including alkylphenol, pharmaceuticals, pesticides); (3) diffusive equilibration in thin films; and diffusive gradients in thin films for trace metals (free ions, labile organic or inorganic complexes). This variety of PSDs has a potential for providing partial characterization of the physicochemical forms that govern fate and effects of contaminants.

PSDs are devices that rely on diffusion and sorption to accumulate analytes in the sampler. This ensures the preconcentration of the contaminants, allowing a lowering of the detection and quantification limits. On the other hand, this will limit their applicability for monitoring programs requiring chemical analyses of total water (OSPAR Commission, 2013).

However, considering the availability of the necessary calibration parameters for the target analytes, PSDs for polar contaminants were considered insufficiently mature for quantitative spatial and temporal trend monitoring under the OSPAR Commission, but may be useful in initial surveys (OSPAR Commission, 2013). This document underlines that diffusive gradients in thin films are a mature PSD technique for trace metals, but that its application remains limited in the marine environment. It is also pointed out that PSDs are deployed on sites such as jetties and buoys, at which some losses may be expected (currents, wind events, malevolence).

3.2.2 Chemical Monitoring in Sediment

Contrary to water, sediment is an integrative matrix for most of the contaminants entering the aquatic environment. High concentrations of micropollutants tremendously reduce the risks of sample contamination and allow easy detection. However, sediments are not a homogeneous medium but a complex matrix, including both mineral and organic fractions; spatial heterogeneity is also important, in relation with hydrodynamism.

Sediments represent a record of past contamination but at different time scales according to whether they are surface or subsurface samples. Surface sediments are representative of recently deposited particles because they are submitted to changes on time scales dictated by deposition, particularly in areas with strong hydrodynamic conditions (estuaries, swollen rivers). In estuaries, changes in concentrations of both metals and organic contaminants have been shown as a consequence of successive events of deposition and resuspension (Delofre and Lafite; Amiard et al., both in Amiard-Triquet and Rainbow, 2009). Furthermore, sampling the sedimentary column allows access to retrospective assessment of temporal changes, particularly the identification of "background" or "preindustrial" conditions (OSPAR Commission, 2011), provided that the age of the sediment at a given depth can be established by using sedimentation rates usually determined with radiometric analyses in both freshwater (e.g., Klaminder et al., 2012) and coastal environments (e.g., Alvarez-Iglesias et al., 2007).

The mineral fraction of sediment itself is complex, with different particle size and different mineral composition. Sediment textures are classified by comparison with the fractions present in a soil (sand, silt, and clay). The classification generally accepted is based on the equivalent spherical diameter as follows: gravel 2 mm–2 cm, sand 0.05–2 mm, silt 0.05–0.002 mm, clay <0.002 mm (for more details, see soil texture triangle at http://www.nrcs.usda.gov/wps/portal/nrcs/detail/soils/edu/kthru6/?cid=nrcs142p2_054311). According to the mineral composition (granite, shale, deep-sea clays, sandstone, carbonate rocks), the background concentrations of many metals differ greatly with the highest ones observed in clays and the lowest ones in sandstone and carbonate rocks (Forstner and Wittmann, 1979). The organic fraction of sediment includes both living (bacteria) and inert (postmortem decomposition of organisms, products of excretion) organic matter. Fine-grained sediments have generally a higher content of organic matter. Because carbon is a major component of organic matter, the level of total organic carbon (TOC) is often used as a proxy for organic matter.

Metal concentrations in sediments are generally linearly correlated to grain size, whereas the organic content does not play a major role in metal sorption. Consequently, it is absolutely necessary to normalize the contaminant data to improve comparability of sediment analyses for metals. Normalization of metal data to conservative elements (Al, Li) is one of the recommended options as also the selection of grain size (Figure 3.2).

At the opposite end of the spectrum, organic contaminants generally do not correlate linearly with the grain size distribution, whereas they generally correlate more strongly with TOC. Despite counterexamples existing, normalization of organic contaminant data to TOC is a procedure frequently adopted, as exemplified in Figure 3.3. Considering PAHs, the pollution gradient appears relatively similar when results are expressed in $\mu g/g^{-1}$ dry weight of sediment or in $\mu g/g^{-1}$ TOC, with the exception of the Seine estuary. Using raw data, this estuary, that is one of the most contaminated in Europe, does not stand out above others (Amiard et al., in Amiard-Triquet and Rainbow, 2009). For polychlorinated biphenyls (PCBs) and organochlorine pesticides, the well-established pollution of the Seine estuary appears more clearly when normalized values are used, whereas using raw data suggests that several sites have a degree of contamination nearing the one of the Seine.

Normalization of the different procedures for relevant sampling, storage, treatment of sediments before chemical analysis and instrumental determination have been described in detail for different classes of contaminants (PAHs, organotins, metals, polybrominated diphenyl ethers, hexabromocyclododecane, perfluorinated compounds, dioxins, furans, and dioxin-like PCBs) in marine sediments (OSPAR Commission, 2011). However, this document does not evoke the problem of bioaccessibility/bioavailability that has a crucial role to establish a sound link between total bulk concentrations in sediments and their potential toxic effects on living organisms (Chapter 5). More recently, the utility of the Passive Sampling Methods for Contaminated Sediments has been assessed in the framework of a SETAC Technical Workshop "Guidance on Passive Sampling Methods (PSMs) to Improve Management of Contaminated Sediments," (Integrated

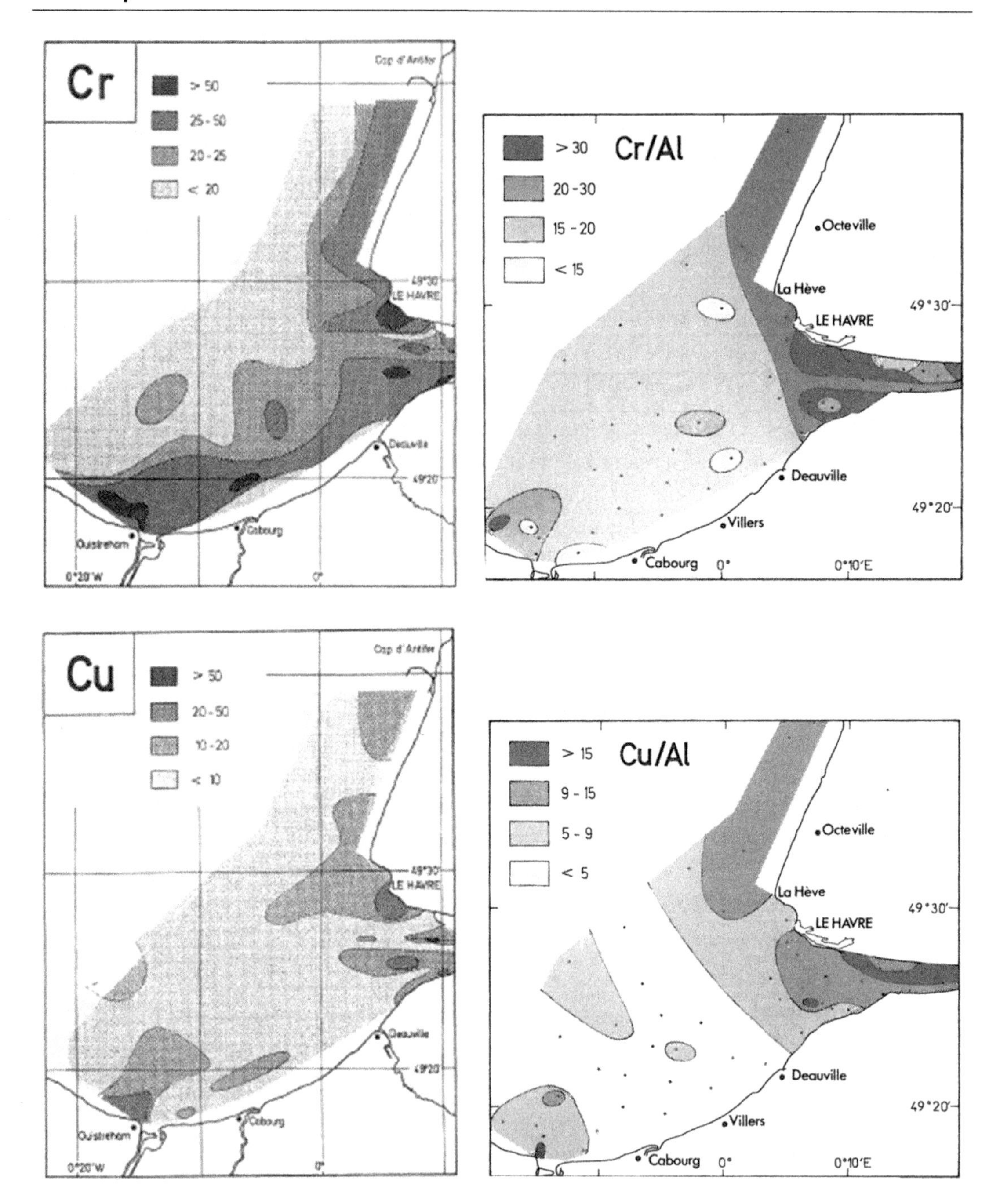

Figure 3.2

Comparison of metal concentrations in sediments from the Bay of Seine, France, expressed either on the basis of sediment dry weight or as a ratio to a conservative element, Al. *After Boust (1981), with permission.*

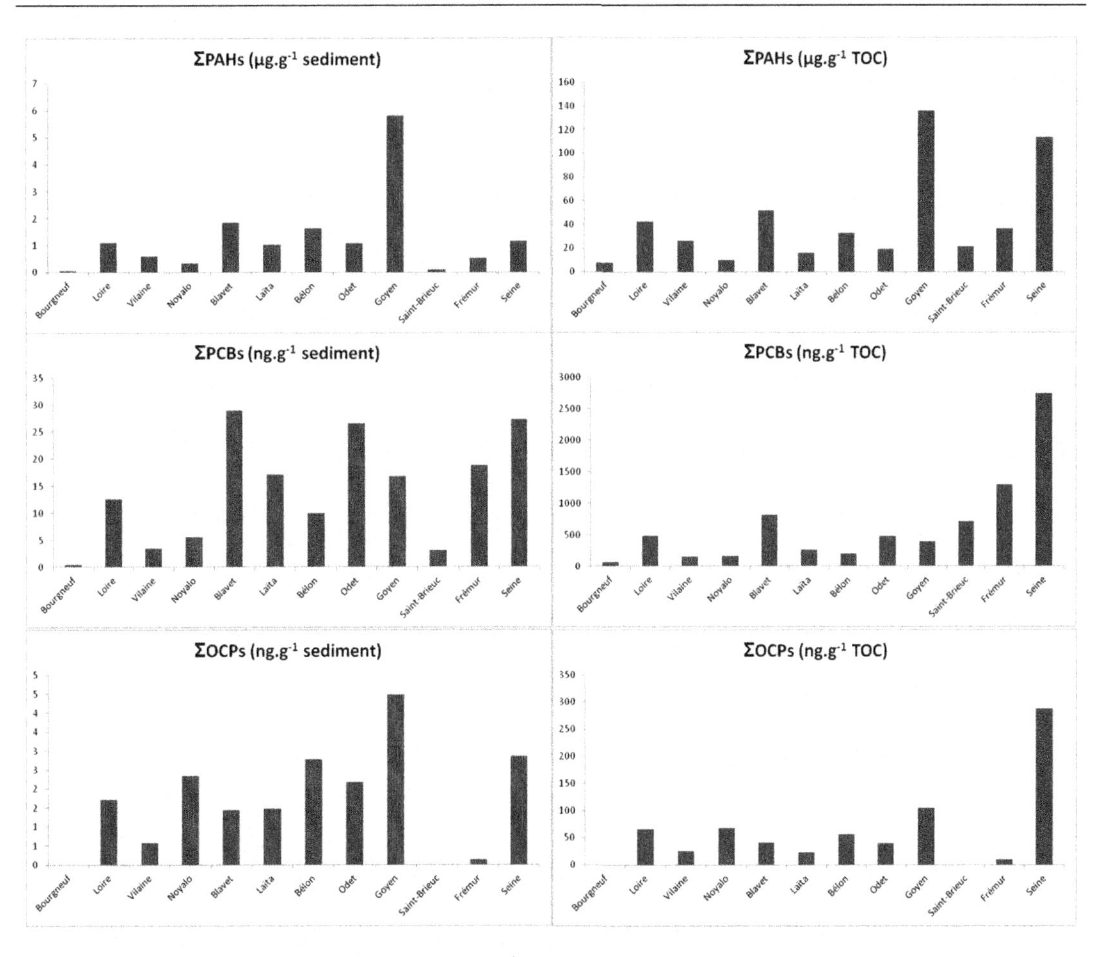

Figure 3.3

Comparison of organic contaminant concentrations in sediments from estuaries of northwest France expressed either on the basis of sediment dry weight or on TOC. *After Fossi Tankoua (2011), with permission.*

Environmental Assessment and Management 10 (2), 2014). Lydy et al. (2014) have reviewed the literature on PSMs, considering the various polymers that have been used as the sorbing phase (e.g., polydimethylsiloxane, polyethylene, polyoxymethylene) and their potential for both laboratory and field studies. Practical guidance on the use of PSMs for improved exposure assessment of hydrophobic organic chemicals in sediments is provided by Ghosh et al. (2014), whereas the state of the art is reviewed by Peijnenburg et al. (2014) in the case of metals.

Because PSMs allow the measurement of the freely dissolved contaminant concentration in porewater (C_{free}), they give access to the chemical activity of contaminants. Thus they have a potential to decrease uncertainty in site investigation and management, offering a better estimation than total bulk concentrations in sediments for crucial endpoints: benthic organism toxicity, bioaccumulation, sediment flux, and water column exposures (Greenberg et al., 2014; Lydy et al., 2014). As previously mentioned for PSDs in the water column, recent developments in PSMs have significantly improved our ability to reliably measure even very low levels of C_{free} (Mayer et al., 2014). Programmatic applications are provided by Greenberg et al. (2014); for instance, USEPA Superfund and Great Lakes Legacy Act sites or sediments from the Grenlandsfjord (Norway) submitted to emissions of different persistent organic pollutants by a smelter. A brief overview of these five papers has been proposed by Parkenson and Maruya (2014), who emphasize the needs for future research and communication for future consensus among the science, management, and practitioners on the appropriate use of PSMs in supporting contaminated sediment management (Mayer et al., 2014).

3.2.3 Chemical Monitoring in Biota

As previously mentioned for sediment, many organisms represent an integrative matrix for most of the contaminants entering the aquatic environment. Again, high concentrations of micropollutants tremendously reduce the risks of sample contamination and allow easy detection. In addition, contaminants incorporated in tissues of organisms have a high biological significance that has been leveraged by the Tissue Residue Approach for toxicity assessment (Chapter 5).

The most popular biomonitoring program is certainly the International Mussel Watch, developed after Goldberg (1975) and improved through a workshop sponsored by the environment studies board of the US National Academy of Sciences (NAS, 1980). Regional, national, and international applications include the French National Monitoring Network for the Quality of the Marine Environment (RNO, 1974–2007), renamed the Monitoring Network for the Chemical Contamination since 2008 (Knoery and Claisse, 2010), the monitoring of the marine environment of the northeast Atlantic under the OSPAR Convention (OSPAR Commission, 2012a). The National Oceanic and Atmospheric Administration's (NOAA) Center for Coastal Monitoring and Assessment has maintained the National Status and Trends in coastal and estuarine areas since 1984. In addition, since 1992, the Mussel Watch Program uses invasive zebra mussels (*Dreissena* spp.) to monitor for contaminants in the Great Lakes (Kimbrough et al., 2008). In Europe, *Dreissena polymorpha* is also used in the framework of the Joint Commission for the Protection of Italian-Swiss Waters against Pollution (www.cipais.org) and the German Environmental Specimen Programme (Environmental Sample Banking [ESB], http://www.umweltprobenbank.de/en/documents).

In this choice of filter-feeder bivalves (mussels and oysters) as a sentinel species, many criteria have been considered:

- ability to concentrate numerous pollutants;
- species with a long life (integrative indicators);
- worldwide distribution when considering different species within each genus;

- tolerance to brackish waters (thus present in estuaries);
- nonmigrant and representative of the pollution at the site of collection;
- abundance, reasonable size, and easy sampling (ensuring practicability).

In addition to chemical monitoring based on bivalves, contaminant concentrations were measured in piscivorous and benthivorous fish collected in 1995–2004 from large US river basins as part of the Biomonitoring of Environmental Status and Trends (BEST) Program (Hinck et al., 2008a). More recently, updated guidelines have been launched under the auspices of the OSPAR commission (2012b) for the sampling and analysis of contaminants in fish, shellfish, and seabird eggs that are considered as suitable for hazardous substances including trace metals and many classes of organic compounds (chlorinated compounds, parent and alkylated PAHs, brominated flame retardants, perfluorinated compounds, tributyltin and its breakdown products, dioxins, furans, and dioxin-like PCBs). A new European Commission guidance document for the implementation of EQS_{biota}(in progress at http://www.wca-environment.com/drafting-supplementary-guidance-for-the-implementation-of-eqsbiota/) provides an inventory of species/tissues currently used in European biota monitoring programs that includes many different freshwater fish species.

Compiling data about residue of total DDT and total HCH in fish from lakes and rivers in the world, Tsuda (2012) has been able to establish temporal trends over a 30-year period, and in differentiated uses in western, eastern, and developing countries. In addition to show spatio-temporal variations of contamination, chemical monitoring in biota can be interpreted in terms of risk to wildlife that forage at the studied sites. From the data obtained in the framework of the BEST Program, Hinck et al. (2009) concluded that dieldrin, total DDT, total PCBs, toxaphene, TCDD-EQ, and several elements exceeded toxicity thresholds to protect fish and piscivorous wildlife in samples from at least one of the monitored sites with most exceedances for total PCBs, mercury, and zinc. Data collected in bivalves and fish are also of interest for comparison with maximum permissible levels in food in different countries or regions.

However, many factors can influence bioaccumulation, among which:

- belonging to a given taxonomic group (e.g., Cu concentrations are especially high in crustaceans, the respiratory pigment of which is hemocyanin that contains two copper atoms responsible for binding oxygen)
- biological role of essential (Cu, Fe, Zn, etc.) versus nonessential (Cd, Pb, Hg, etc.) metals
- interspecific differences (RNO, 2006; Fletcher et al., 2014)
- reproductive status ("biological dilution" of metals)
- age (cumulative pollutants)
- biological specificities (molting in crustaceans; biomagnification of persistent organic pollutants in lipid-rich fish)

For a long time, different strategies have been recommended to cope with confounding factors (NAS, 1980). More recently, active biomonitoring by translocation of organisms between clean and contaminated sites has been proposed for bivalves (De Kock and Kramer, 1994) then expanded to other species (e.g., Besse et al., 2013). This allows the use of specimens from the same age and size classes, the same genetic pattern, and the same history of previous exposure to contaminants. To go beyond interspecific differences between mussels and oysters, a procedure has been recommended in the framework of the French Mussel Watch. Metal concentrations [M] have been compared in specimens from each species collected from the same site then a mean ratio [M] in oysters/[M] in mussels has been calculated. This ratio is the highest for silver (50) followed by zinc (15), copper (10), and cadmium (2.5), whereas interspecific differences were not significant for organic contaminants (RNO, 2006).

The reproductive status is a well-known driver of contaminant concentrations in bivalves. For metals (except organometals) that do not bind to lipids, the increased body weight because of lipid-rich genital products at a constant level of metal quantity in the whole body represents a kind of "biological dilution." Of course, this is not true for lipophilic organic contaminants. In addition, food availability at different sites influences the condition, the total body weight and the relative importance of lipids, with possible consequences on contaminant concentrations. An adjustment method for bioaccumulation data has been proposed for metals in mussels (Andral et al., 2004) then expended to other bioaccumulative species (Poirier et al., 2006). More generally, for lipophilic compounds, it is recommended to expressed the concentrations on both a wet weight and lipid weight basis (OSPAR Commission, 2012b).

3.3 Biomarkers, Bioindicators, and Biotic Indices in Monitoring Programs

To date, biomarker-based monitoring programs remain limited (Collier et al., in Amiard-Triquet et al., 2013). However, particularly in Western countries, a number of biomarkers are already recommended for regulatory purposes. In the framework of the BEST Program's Large River Monitoring Network (http://www.cerc.usgs.gov/data/best/search/), in addition to chemical monitoring, fish health was examined by using a suite of fish health indicators (condition factor, somatic indices, health assessment index, macrophage aggregate parameters) and reproductive biomarkers (steroid hormones, vitellogenin, gonadal histology) with the objective of assessing less persistent chemicals in the environment and detecting molecular-level changes before population effects may be evident. Hinck et al. (2007, 2008b) have compiled the results for the Colorado River Basin and river basins in the Southeastern United States, showing that a number of sites are at risk for different contaminants.

In the framework of integrated monitoring and assessment of contaminants under the auspices of the International Council for the Exploration of the Sea (ICES), biomarkers of defense and damage have been recommended in invertebrates and fish (for details, see Table 7.1.). They are particularly useful when background assessment criteria (estimated from data typical of remote areas) and environmental assessment criteria (derived from toxicological data and indicating a sublethal effect) are available (Davies and Vethaak, 2012).

Despite the fact that biomarkers can respond to confounding factors, both extrinsic (temperature, salinity, dissolved oxygen, etc.) and intrinsic (size, weight, age, sex, etc.) in addition to specific stress, the so-called "core biomarkers" remain useful to target the classes of contaminants responsible for observed impairments, provided that they are used adequately (Chapter 7). On the other hand, biomarkers of "ecological relevance" provide an opportunity to make the link between responses at infraindividual, individual, and supraindividual levels (Amiard-Triquet et al., 2013; Mouneyrac and Amiard-Triquet, 2013). However, despite the concept of the Adverse Outcome Pathway (Ankley et al., 2010) is particularly attractive to establish such links, the cases documented in the field remain limited to populational effects of endocrine disruptors in gastropods (Gubbins et al., in Amiard-Triquet et al., 2013) and fish (Kidd et al., 2007). Cascading effects have also been observed in multipolluted estuaries such as those reported in worms (Durou et al., 2007; Gillet et al., 2008), fish, and crustaceans (Weis et al., 2001, 2011) or in freshwater fish exposed to pharmaceutical manufacture discharges (Sanchez et al., 2011). In other cases, the possibility of infraindividual or individual biomarker responses having an interpretative capacity in terms of supraindividual effects is still tentative but is supported by exciting results (Coulaud, 2012; Brousseau et al., Moore et al., Vasseur et al., all in Amiard-Triquet et al., 2013).

Determining ecological integrity to answer the needs of legislations (US Clear Water Act and Oceans Policy, EU Water Framework Directive and Marine Strategy Framework Directive, Oceans Act in Canada and Australia) is mainly based on five biological compartments: phytoplankton, macroalgae, angiosperms, benthic macrofauna, and ichthyofauna. Additional groups may be recommended such as meiobenthos including foraminifera, copepods, and nematodes (Debenay, Ferrero, both in Amiard-Triquet and Rainbow, 2009; Dauvin et al., 2010 and literature quoted therein) and zooplankton, particularly useful in lakes (Dauvin et al., 2010; literature quoted therein; Jeppesen et al., 2011). In Europe, the WISER Project (Water bodies in Europe – Integrative Systems to assess Ecological status and Recovery, http://www.wiser.eu/programme/) has mobilized the scientific community of ecologists from most of the European countries, existing data were revisited, whereas new data were obtained and the results were summarized in a special issue of the journal *Hydrobiologia* (2013;704(1)).

Compared with all the other procedures previously described for monitoring, this ecological approach provides a direct assessment of the health status of water bodies, even if the question of uncertainties must be carefully taken into account (Clarke, 2013). However, a deterioration revealed through ecological approaches generally has its roots in stresses that have begun to impact the ecosystems years ago and establishing a cause-and-effect relationship is nearly an impossible task. The question of confounding factors must be evoked here again because the degradation of an ecosystem is generally multifactorial, rarely only due to chemical stress. This is particularly obvious in highly heterogeneous environments such as estuaries (Dauvin, 2007; Elliott and Quintino, 2007).

Ecological tools such as the biotic index were initially designed to assess the impact of increasing levels of organic matter in ecosystems (Chapter 14). Thus its relevance for chemical impacts remains a topic of discussion. However, Borja et al. (2000) have shown that the biotic index used to investigate the status of benthic communities is consistent with the percentage of samples that goes beyond the effects range-low, representing concentrations in sediments above which adverse effects are expected to occur rarely (Long et al., 1995) for contaminants such as arsenic, mercury, nickel, lead, copper, chromium, PCB, and DDT.

Using ecological tools, despite their high degree of relevance for the assessment of the health status of ecosystems, have certain limitations both scientific and practical. In addition to their lack of specificity already mentioned, they cannot be applied consistently across all the biogeographic regions. Their use can be labor-intensive, needing the recognition of many different species at a moment when the number of taxonomists has decreased dramatically in most countries. The taxonomic identification may be done at levels higher than the species (e.g., at the genus or family levels), according to the taxonomic sufficiency concept described by Ellis (1985). In freshwater, according to Mueller et al. (2013), "TS was universally applicable within taxonomic groups for different habitats in one biogeographic region. Aggregation to family or order was suitable for quantifying biodiversity and environmental gradients, but multivariate community analyses required finer resolution in fishes and macrophytes than in periphyton and macro-invertebrates." For marine macrobenthos, authors agree that no substantial loss of information occurs when identification is carried out at the family level (e.g., Warwick, 1988; Ferraro and Cole, 1995; Dauvin et al., 2003; Mendes et al., 2007; Bacci et al., 2009). Mendes et al. (2007) underline that faunal patterns at different taxonomic levels tend to become similar with increased pollution. Thus it is attractive to invest most of the sampling effort in quantitative data and number of spatial and temporal replicates rather than in taxonomic detail (Mueller et al., 2013). However, Dauvin et al. (2003) underline that in certain cases it may be of interest to identify particular species whose presence or abundance is indicative of particular pollutant impacts. More recently, Borja and Elliott (2013) have warned against a fast, cost-effective, and superficially attractive approach that in certain cases is not fit for purpose. The complementarity of ecosystem assessments based on structure and functioning is fully recognized (Chapter 14) and both strategies are already incorporated in the legislation. However, functioning analysis using biological traits requires information on species, not on families.

3.4 Integrated Monitoring and Assessment of Contaminants and Their Effects

The complementarity of different monitoring strategies has been initially recognized in the case of sediment. The sediment quality triad, as originally developed 30 years ago, involved three lines of evidence: sediment chemistry, sediment toxicity, and benthic community structure. It has widely evolved with the inclusion of new lines of evidence considering bioaccumulation and biomagnification (to assess secondary poisoning), toxicity identification evaluation, effect directed analysis, and the use of biomarkers, etc. (Chapman and Hollert, 2006). This strategy has inspired many researchers involved in environmental assessment (quoted 129 times in the international literature to date), such as Beketov and Liess (2012) calling for integration between the bottom-up approach of ecotoxicology (use of small-scale experiments to predict effects on the entire ecosystems) and the top-down macroecological approach (focusing directly on ecological effects at large spatial scales). We have already mentioned that the BEST Program was based on the integration of chemical monitoring in fish tissues, fish health indicators, and reproductive biomarkers. An even more integrated approach is recommended for the assessment of contaminants and their effects in the marine environment of the North-East Atlantic (OSPAR Commission, 2012c). In the key environmental matrices (water, sediment, and biota), the chemical concentrations, the biological effects data (biomarkers at different levels of biological organization, bioassays, benthic community indices), and other supporting parameters (e.g., hydrography, sediment characteristics) are combined for assessment.

Because of their high heterogeneity, transitional waters pose specific problems. Chapman et al. (2013) have recently reviewed methods to assess sediment contamination in estuaries, particularly chemical assessments, biomarkers, bioindicators, and biological surveys. A vast case study of the Seine estuary (France)—recognized as one of the most polluted estuaries in Europe—has been permitted to launch recommendations for the environmental assessment of estuarine ecosystems, by using chemical and biological tools in the sediment compartment and the water column (Amiard-Triquet and Rainbow, 2009).

Many field studies were carried out in this area to test the effectiveness of many different biomarkers and bioindicators (Poisson et al., 2011). In situ biomonitoring was mainly based on molluscs (49% of biological models). The polychaete worm *Nereis diversicolor* and fish (roach *Rutilus rutilus*, flounder *Platichthys flesus*, sole *Solea solea*, dab *Limanda limanda*) were also widely used whereas the copepod crustacean *Eurytemora affinis* accounted for only 0.5%. Studies devoted to populations and communities represented 7% of the whole. The Figure 3.4 depicted the effects observed at different levels of biological organization including the results obtained either for all the zoological groups (Figure 3.4(A)) or for fish only (Figure 3.4(B)). In the first case, the highest values of the effect index [(very important effects + important effects)/(sum of the whole categories) in percentage] were observed for individual biomarkers and bioindicators at population or community levels. For fish, effect indices were high for all the categories of biomarkers and bioindicators, highlighting the higher sensitivity of fish compared with invertebrates.

Among the advantages of using macroinvertebrates to assess ecological quality is that these organisms are relatively sedentary and thus cannot avoid deteriorating water/sediment quality, conditions are often emphasized. The sensitivity of fish and the sedentary lifestyle of benthic invertebrates are clearly complementary in environmental monitoring. It is also relevant to use biomarkers and bioindicators at different levels of biological organization because at the

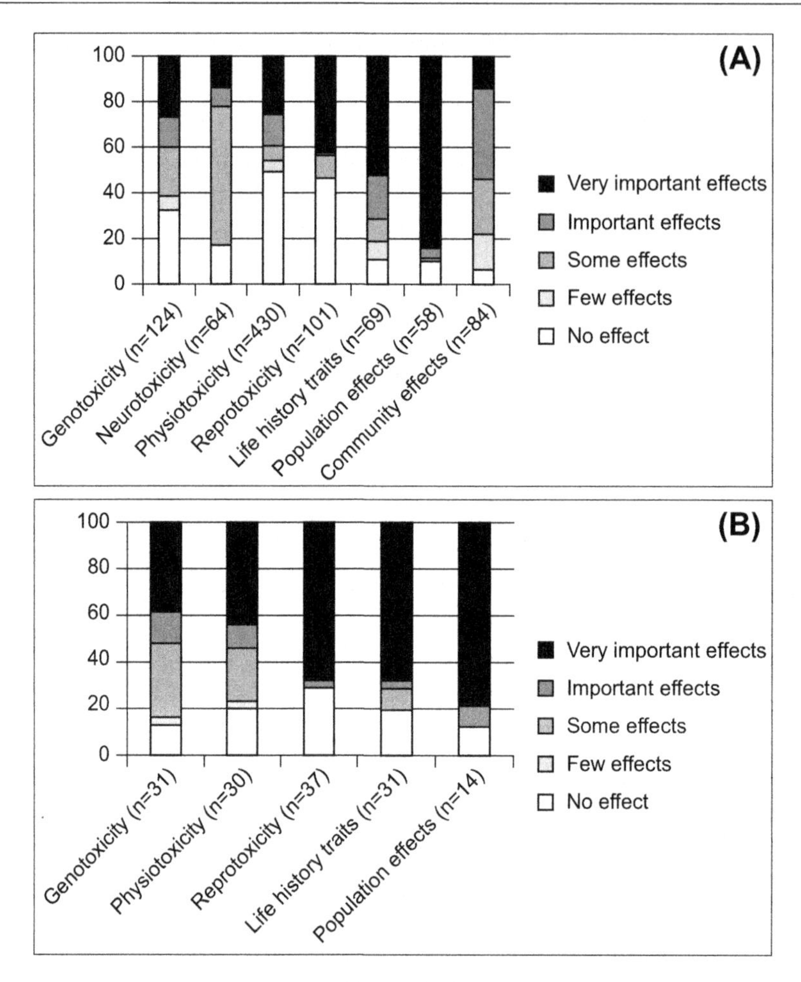

Figure 3.4

Responsiveness of different biomarkers tested in the multipolluted Seine estuary (France). Effect index = 100×(very important effects + important effects)/(sum of the whole categories). (A) Whole of the zoological groups; (B) fish. *Modified after Poisson et al. (2011) and GIP Seine-Aval (2011).*

increasing levels, more and more external factors can influence the biological response that is consequently less and less specific of a given stress.

3.5 Conclusions

As analyzed with some details in the case of predicted no effect concentrations (Chapter 2), the current procedures for the determination of EQSs suffer many limitations, thus leading to major uncertainties concerning the selected values for EQS_{water}, $EQS_{sediment}$, and EQS_{biota}. Despite a quantity of work carried out for decades, only a few EQSs have been established for water and sediment and even less for biota, thus leading the European

Commission to launched a new guidance document for the implementation of EQS_{biota}. The procedures are costly and most countries have similar priority substances and fairly similar EQS derivation procedures. As a result, Matthiessen et al. (2010) pleaded for international harmonization.

Monitoring in the physical media (water, sediment) provides an assessment of exposure to a number of contaminants. Mandatory requirements in many regulations are still based on total concentrations of contaminants in these matrices. However, scientific advancements have been incorporated by several regulatory bodies that highlight the potential of tools such as passive samplers (OSPAR commission, 2012c) that give access to physico-chemical characteristics of contaminants and the importance of bioavailability (ITRC, 2011; TGD, 2011).

Monitoring in bioaccumulators (bivalves, fish) is a well-established procedure widely used in many countries. In Western countries, more or less centralized programs, funded by regional, national or supranational authorities are running for decades (CIPAIS; ESB; Hinck et al., 2008a; Kimbrough et al., 2008; Knoery and Claisse, 2010; OSPAR Commission, 2012a). Similarly to data obtained in physical matrices, chemical monitoring in biota show spatiotemporal variations of contamination, but in addition it provides direct measurements of bioavailable contaminants, the fraction that is the most able to exert a toxic effect. Thus strategies (Critical Body Residue approach, the Tissue Residue Approach) have been developed to link toxicity to incorporated doses of contaminants rather than external doses (Chapter 5). Last, chemical monitoring in biota can be interpreted in terms of risk to wildlife that forage at the studied sites (Hinck et al., 2009) and these data are also of interest for comparison with maximum permissible levels in food used by human consumers.

Assessing compliance between the chemical residue data measured in environmental matrices at a given site and corresponding EQSs is not a simple task. The procedures are described in detail in International Organization for Standardization guidance on the use of sampling data for decision making, based on the compliance with thresholds and classification systems (ISO, 2008).

Monitoring with biomarkers aims at assessing less persistent chemicals in the environment and detecting molecular-level changes before population effects may be evident, an objective clearly announced in the BEST Program's Large River Monitoring Network (http://www.cerc.usgs.gov/data/best/search/). Carrying out such a "preventive risk assessment" is certainly ambitious and it has been argued by many scientists that biomarkers at infraindividual or individual levels have no predictive value because an event at a low level of biological organization does not predict consequences at supra-individual levels (population, community, ecosystem). However, an extensive review of the literature has shown that in some cases, impairments of biochemical, physiological, or behavioral biomarkers can lead to ecological disturbances through a series of cascading events (Amiard-Triquet et al., 2013). Thus a tier approach has been proposed (Figure 7.5), using complementarily core biomarkers such as those recommended by the ICES (Davies and Vethaak, 2012) or the BEST Program and biomarkers of ecological relevance.

When sites are under stress for a long duration (years), it is expected that effects will be conspicuous at supraindividual levels of biological organization (population, communities). In this case, monitoring with bioindicators and biotic indices (Chapter 14) will be very efficient for assessing the ecological status of the studied sites but in terms of retrospective risk assessment, it will most probably not help for identifying the origin of observed disturbances.

Analytical quality assurance/quality control is crucial for fruitful comparisons between sites, sampling periods (covering decades in some programs), and comparison between measurements and EQSs. Although it has long been accepted that analytical quality assurance/quality control is required in chemistry laboratories (e.g., Quality Assurance of Information for Marine Environmental Monitoring in Europe), then for biomarkers (e.g., Biological Effects Quality Assurance in Monitoring Programmes; Mediterranean Action Plan for the Barcelona Convention; Working Group of Biological Effect of Contaminants in the framework of ICES), there has been resistance to adopting this in ecological analyses (Elliott, 1993; Gray and Elliott, 2009) but the EU Water Framework Directive requires the intercalibration of assessment methods to set consistent and comparable standards for rivers, lakes, and coastal waters across Europe. However, as recently as 2013, Borja and Elliott still stated that there is insufficient training of taxonomists so that even if analyses are performed they may not be to a sufficient standard.

According to Hutchinson et al. (2013), analytical chemistry integrated with the application of biomarkers and bioassays has strengthened the evidence base to support an ecosystem approach to manage marine pollution problems because of legacy contaminants. This strategy is also applicable for freshwater environments. New developments may be recommended to support policy needs considering the complexity of environmental pollution with emerging contaminants and biotoxins in addition to legacy contaminants. Integrated monitoring appears as a prominent way to face the scientific, economic, and health challenges arising from anthropogenic impacts.

A huge number of samples have been collected as part of the various monitoring programs conducted worldwide. Therefore, it has been possible to establish sample banks in several countries (http://www.inter-esb.org/). These specimens representing a wide range of environmental matrices can be used (1) for evaluating governmental environmental policy-making and regulations; (2) as a resource for animal health evaluation; (3) to investigate time trends in ecosystems; (4) for the detection of newly emerging chemicals in the time trends; (5) for validations of computer models for environmental phenomena; (6) for source identification of contaminants; (7) as a tool for food safety; and (8) for the evaluation of genetic selection pressure because of environmental changes (Koizumi et al., 2009). The potential of ESB to conduct retrospective monitoring is illustrated for organotin (Rüdel et al., 2007), brominated flame retardants (Sawal et al., 2011), and perfluorinated compounds (Theobald et al., 2011). The basic concept and requirements of ESB have been recently reviewed (UNEP, 2011). Until now, ESB has been mainly used for environmental chemistry purposes but in the field of human health, biobanks are essential resources for the development of biomarkers (e.g., Holland et al., 2005; Jackson and Banks, 2010). In environmental studies, using archived environmental samples for the determination of biomarkers is not a common practice despite old examples existing (e.g., stress proteins, Sanders and Martin, 1993). More recently, Koizumi et al. (2009) consider that ESB have an interesting future for DNA profiling. In the framework of one of the most developed ESB program, blood samples for biomarker studies (e.g., immunological, hormonal, genotoxic) are already integrated (Paulus et al., 2005; Klein et al., 2012). Biometric data (physiological condition indices) and timeline changes in populations and biocenoses registered in parallel make possible to establish direct correlations to the data on chemical media.

Acknowledgments

Thanks are due to Dr Marc Babut (IRSTEA, Lyon, France) for his kind revision.

References

Alvarez-Iglesias, P., Quintana, B., Rubio, B., et al., 2007. Sedimentation rates and trace metal input history in intertidal sediments from San Simón Bay (Ría de Vigo, NW Spain) derived from ^{210}Pb and ^{137}Cs chronology. J. Environ. Radioact. 98, 229–250.

Amiard, J.C., Geffard, A., Amiard-Triquet, C., et al., 2007. Relationship between lability of sediment-bound metals (Cd, Cu, Zn) and their bioaccumulation in benthic invertebrates. Estuar. Coastal Shelf Sci. 72, 511–521.

Amiard-Triquet, C., Amiard, J.C., Rainbow, P.S., 2013. Ecological Biomarkers: Indicators of Ecotoxicological Effects. CRC Press, Boca Raton.

Amiard-Triquet, C., Rainbow, P.S., 2009. Environmental Assessment of Estuarine Ecosystems. A Case Study. CRC Press, Boca Raton. 355 pp.

Andral, B., Stanisière, J.Y., Sauzade, D., et al., 2004. Monitoring chemical contamination levels in the Mediterranean based on the use of mussel caging. Mar. Pollut. Bull. 49, 704–712.

Ankley, G.T., Bennett, R.S., Erickson, R.J., et al., 2010. Adverse outcome pathways: a conceptual framework to support ecotoxicology research and risk assessment. Environ. Toxicol. Chem. 29, 730–741.

Bacci, T., Trabucco, B., Marzialetti, S., 2009. Taxonomic sufficiency in two case studies: where does it work better? Mar. Ecol. 30 (Suppl. 1), 13–19.

Beketov, M.A., Liess, M., 2012. Ecotoxicology and macroecology–time for integration. Environ. Pollut. 162, 247–254.

Bergman, A., Heindel, J.J., Jobling, S., et al. (Eds.), 2013. State of the Science of Endocrine Disrupting Chemicals – 2012. WHO/UNEPhttp://unep.org/pdf/9789241505031_eng.pdf.

Besse, J.P., Coquery, M., Lopes, C., et al., 2013. Caged *Gammarus fossarum* (crustacea) as a robust tool for the characterization of bioavailable contamination levels in continental waters. Toward the determination of threshold values. Water Res. 47, 650–660.

Borja, A., Franco, J., Pérez, V., 2000. A marine biotic index to establish the ecological quality of soft-bottom benthos within European estuarine and coastal environments. Mar. Pollut. Bull. 40, 1100–1114.

Borja, A., Elliott, M., 2013. Marine monitoring during an economic crisis: the cure is worse than the disease. Mar. Pollut. Bull. 68, 1–3.

Boust, D., 1981. Les métaux-traces dans l'estuaire de la Seine et ses abords. Thèse de spécialité de l'université de Caen, France. 186 pp.

Cailleaud, K., Forget-Leray, J., Peluhet, L., et al., 2009. Tidal influence on the distribution of hydrophobic organic contaminants in the Seine estuary and biomarker responses on the copepod *Eurytemora affinis*. Environ. Pollut. 157, 64–71.

Chapman, P.M., Hollert, H., 2006. Should the sediment quality triad become a tetrad, a pentad, or possibly even a hexad? J. Soils Sed. 6, 4–8.

Chapman, P.M., Wang, F., Caeiro, S.S., 2013. Assessing and managing sediment contamination in transitional waters. Environ. Int. 55, 71–91.

Clarke, R.T., 2013. Estimating confidence of European WFD ecological status class and WISER bioassessment uncertainty guidance software (WISERBUGS). Hydrobiologia. 704, 39–56.

Coulaud, R., 2012. Modélisation et changement d'échelles pour l'évaluation écotoxicologique: application à deux macroinvertébrés aquatiques, *Gammarus fossarum* (Crustacé amphipode)et *Potamopyrgus antipodarum* (Mollusque gastéropode). Thesis Lyon University, France.

Crommentuijn, T., Polder, M., Sijm, D., et al., 2000. Evaluation of the dutch environmental risk limits for metals by application of the added risk approach. Environ. Toxicol. Chem. 19, 1692–1701.

Dauvin, J.C., 2007. Paradox of estuarine quality: benthic indicators and indices in estuarine environments, consensus or debate for the future. Mar. Pollut. Bull. 55, 271–281.

Dauvin, J.C., Gomez Gesteira, J.L., Salvande Fraga, M., 2003. Taxonomic sufficiency: an overview of its use in the monitoring of sublittoral benthic communities after oil spills. Mar. Pollut. Bull. 46, 552–555.

Davies, I.M., Vethaak, A.D., 2012. Integrated Marine Environmental Monitoring of Chemicals and Their Effects. ICES Cooperative Research. Report No. 315, 277 pp.

Durou, C., Smith, B.D., Roméo, M., et al., 2007. From biomarkers to population responses in *Nereis diversicolor*: assessment of stress in estuarine ecosystems. Ecotoxicol. Environ. Saf. 66, 402–411.

Dauvin, J.C., Bellan, G., Bellan-Santini, D., 2010. Benthic indicators: from subjectivity to objectivity – where is the line? Mar. Pollut. Bull. 60, 947–953.

EC, 2010. Guidance Document No. 25 on Chemical Monitoring of Sediment and Biota under the Water Framework Directive. Common Implementation Strategy for the Water Framework Directive. Technical Report-2010-041 https://circabc.europa.eu/sd/a/7f47ccd9-ce47-4f4a-b4f0-cc61db518b1c/Guidance%20No%2025%20-%20Chemical%20Monitoring%20of%20Sediment%20and%20Biota.pdf (last accessed 29.08.14).

Elliott, M., 1993. The quality of macrobiological data. Mar. Pollut. Bull. 25, 6–7.

Elliott, M., Quintino, V., 2007. The estuarine quality paradox, environmental homeostasis and the difficulty of detecting anthropogenic stress in naturally stressed areas. Mar. Pollut. Bull. 54, 640–645.

Ellis, D., 1985. Toxonomic sufficiency in pollution assessment. Mar. Pollut. Bull. 16, 459.

EU, 2013. Directive 2013/39/EU of the European parliament and of the council of 12 August 2013. Off. J. Eur. Union L 226, 1–17. http://repath.ivl.se/download/18.372c2b801403903d2758c14/1385131891692/EU+vattendir+PFOS.pdf (last accessed 29.08.14).

Ferraro, S.P., Cole, F.A., 1995. Taxonomic level sufficient for assessing pollution impacts on the Southern California Bight macrobenthos – revisited. Environ. Toxicol. Chem. 14, 1031–1040.

Fletcher, D.E., Lindell, A.H., Stillings, G.K., et al., 2014. Variation in trace-element accumulation in predatory fishes from a stream contaminated by coal combustion waste. Arch. Environ. Contam. Toxicol. 66, 341–360.

Forstner, U., Wittmann, G.T.W., 1979. Metal Pollution in the Aquatic Environment. Springer-Verlag, Berlin; New York. 486 pp.

Fossi Tankoua, O., 2011. Perturbations du comportement et de la reproduction: des outils pour l'évaluation précoce de la dégradation de la qualite de l'environnement estuarien et côtier (Ph.D. thesis). University of Nantes, France. 308 pp.

Ghosh, U., Driscoll, S.K., Burgess, R.M., et al., 2014. Passive sampling methods for contaminated sediments: practical guidance for selection, calibration, and implementation. Integr. Environ. Assess. Manag. 10, 210–223.

Gillet, P., Mouloud, M., Durou, C., et al., 2008. Response of *Nereis diversicolor* population (Polychaeta, Nereididae) to the pollution impact – Authie and Seine estuaries (France). Estuar. Coast. Shelf Sci. 76, 201–210.

GIP Seine-Aval, 2011. Etat des ressources biologiques: Effets de la contamination chimique sur les poissons de l'estuaire de la Seine. http://seine-aval.crihan.fr/web/attached_file/componentId/kmelia218/attachmentId/26730/lang/fr/name/Effets%20contamination%20sur%20poissons.pdf (last accessed 20.08.14).

Goldberg, E.D., 1975. The mussel watch. A first step in global marine monitoring. Mar. Pollut. Bull. 6, 111–113.

Gray, J.S., Elliott, M., 2009. Ecology of Marine Sediments: Science to Management. OUP, Oxford. p. 260.

Greenberg, M.S., Chapman, P.M., Allan, I.J., et al., 2014. Passive sampling methods for contaminated sediments: risk assessment and management. Integr. Environ. Assess. Manag. 10, 224–236.

Hansen, D.J., Di Toro, D.M., McGrath, J.A., et al., 2003. Procedures for the Derivation of Equilibrium Partitioning Sediment Benchmarks (ESBs) for the Protection of Benthic Organisms: PAH Mixtures. EPA/600/R-02/013. Office of Research and Development, U.S. Environmental Protection Agency, Narragansett, RI.

Hinck, J.E., Schmitt, C.J., Chojnacki, Kimberly A., et al., 2009. Environmental contaminants in freshwater fish and their risk to piscivorous wildlife based on a national monitoring program. Environ. Monit. Assess. 152, 469–494.

Hinck, J.O., Blazer, V.S., Denslow, N.D., et al., 2007. Chemical contaminants, health indicators, and reproductive biomarker responses in fish from the Colorado river and its tributaries. Sci. Total Environ. 378, 376–402.

Hinck, J.O., Blazer, V.S., Denslow, N.D., et al., 2008b. Chemical contaminants, health indicators, and reproductive biomarker responses in fish from rivers in the Southeastern United States. Sci. Total Environ. 390, 538–557.

Hinck, J.O., Schmitt, C.J., Ellersieck, M.R., et al., 2008a. Relations between and among contaminant concentrations and biomarkers in black bass (*Micropterus* spp.) and common carp (*Cyprinus carpio*) from large U.S. rivers, 1995–2004. J. Environ. Monit. 10, 1499–1518.

Holland, N.T., Pfleger, T.L., Berger, E., et al., 2005. Molecular epidemiology biomarkers – sample collection and processing considerations. Toxicol. Appl. Pharmacol. 206, 261–268.

Hutchinson, T.H., Lyons, B.P., Thain, J.E., et al., 2013. Evaluating legacy contaminants and emerging chemicals in marine environments using adverse outcome pathways and biological effects-directed analysis. Mar. Pollut. Bull. 74, 517–525.

ISO 5667-20, 2008. Water Quality—Sampling—Part 20: Guidance on the Use of Sampling Data for Decision Making—Compliance with Thresholds and Classification Systems. http://wwwwfdukorg/sites/default/files/Media/ISO_5667-20_2008%28E%29-Character_PDF_documentpdf.

ITRC (Interstate Technology & Regulatory Council), 2011. Incorporating Bioavailability Considerations into the Evaluation of Contaminated Sediment Sites. CS-1. Interstate Technology & Regulatory Council, Contaminated Sediments Team, Washington, D.C.http://www.itrcweb.org/contseds-bioavailability/cs_1.pdf.

Jackson, D.H., Banks, R.E., 2010. Banking of clinical samples for proteomic biomarker studies: a consideration of logistical issues with a focus on pre-analytical variation. Proteomics – Clin. Appl. 4, 250–270.

Jeppesen, E., Nõges, P., Davidson, T., et al., 2011. Zooplankton as indicators in lakes: a scientific-based plea for including zooplankton in the ecological quality assessment of lakes according to the European Water Framework Directive (WFD). Hydrobiologia 676, 279–297.

De Kock, W.C., Kramer, K.J.M., 1994. Active biomonitoring (ABM) by translocation of bivalve mollusks. In: Kramer, K.J.M. (Ed.), Biomonitoring of Coastal Waters and Estuaries. CRC Press, Boca Raton (FL), pp. 51–84.

Kidd, K.A., Blanchfield, P.J., Mills, K.H., et al., 2007. Collapse of a fish population after exposure to a synthetic estrogen. Proc. Nat. Acad. Sci. 104, 8897–8901.

Kimbrough, K.L., Johnson, W.E., Lauenstein, G.G., et al., 2008. An assessment of two decades of contaminant monitoring in the nation's coastal zone. Silver Spring, MD. In: NOAA Technical Memorandum NOS NCCOS 74. , p. 105. http://ccma.nos.noaa.gov/publications/MWTwoDecades.pdf.

Klaminder, J., Appleby, P., Crook, P., et al., 2012. Post-deposition diffusion of ^{137}Cs in lake sediment: implications for radiocaesium dating. Sedimentology 59, 2259–2267.

Klein, R., Paulus, M., Tarricone, K., et al., 2012. Guideline for Sampling and Sample Treatment – Bream (*Abramis brama*). http://www.umweltprobenbank.de/upb_static/fck/download/SOP_Bream.pdf.

Klok, C., de Vries, P., Jongbloed, R., et al., 2012. Literature Review on the Sensitivity and Exposure of Marine and Estuarine Organisms to Pesticides in Comparison to Corresponding Fresh Water Species. http://www.efsa.europa.eu/fr/supporting/doc/357e.pdf (last accessed 21.05.14).

Knoery, J., Claisse, D., 2010. Insights from 30 years of experience in running the French marine chemical monitoring network. ICES-CM 2010/F08. In: ICES Annual Science Conference, 20e24 September 2010, Nantes, France. www.ices.dk/products/CMdocs/CM-2010/F/F0810.pdf.

Koizumi, A., Harada, K.H., Inoue, K., et al., 2009. Past, present, and future of environmental specimen banks. Environ. Health Prev. Med. 14, 307–318.

Lauenstein, G.G., Cantillo, A.Y. (Eds.), 1998. Sampling and Analytical Methods of the National Status and Trends Program Mussel Watch Project: 1993–1996 Update. NOAA/National Ocean Service/Office of Ocean Resources Conservation and Assessment, Silver Spring, MD (NOAA Technical Memorandum NOS ORCA, 130). http://www.ccma.nos.noaa.gov/publications/tm130.pdf.

Lepper, P., 2005. Manual on the Methodological Framework to Derive Environmental Quality Standards for Priority Substances in Accordance with Article 16 of the Water Framework Directive (2000/60/EC). http://www.helpdeskwater.nl/algemene-onderdelen/structuur-pagina%27/zoeken-site/@6240/manual_environmental/.

Long, E.R., MacDonald, D.D., Smith, S.L., et al., 1995. Incidence of adverse biological effects within ranges of chemical concentrations in marine and estuarine sediments. Environ. Manag. 19, 81–97.

Lydy, M.J., Landrum, P.F., Oen, A.M.P., et al., 2014. Passive sampling methods for contaminated sediments: state of the science for organic contaminants. Integr. Environ. Assess. Manag. 10, 167–178.

Martineau, D., Lemberger, K., Dallaire, A., et al., 2002. Cancer in wildlife, a case study: Beluga from the St Lawrence estuary, Québec, Canada. Environ. Health Perspect. 110, 285–292.

Matthiessen, P., Babut, M., Batley, G., et al., 2010. Water and sediment EQS derivation and application. In: Crane, M., Matthiessen, P., Maycock, D.S., et al. (Eds.), Derivation and Use of Environmental Quality and Human Health Standards for Chemical Substances in Water and Soil. SETAC Publications, Pensacola, pp. 47–103.

Mayer, P., Parkerton, T.F., Adams, R.G., et al., 2014. Passive sampling methods for contaminated sediments: scientific rationale supporting use of freely dissolved concentrations. Integr. Environ. Assess. Manag. 10, 197–209.

Mendes, C.L.T., Tavares, M., Soares-Gomes, A., 2007. Taxonomic sufficiency for soft-bottom sublittoral mollusks assemblages in a tropical estuary, Guanabara Bay, Southeast Brazil. Mar. Pollut. Bull. 54, 377–384.

Miller, D.H., Ankley, G.T., 2004. Modeling impacts on populations: fathead minnow (*Pimephales promelas*) exposure to the endocrine disruptor 17[beta]-trenbolone as a case study. Ecotoxicol. Environ. Saf. 59, 1–9.

Miller, D.H., Jensen, K.M., Villeneuve, D.L., et al., 2007. Linkage of biochemical responses to population-level effects: a case study with vitellogenine in the fathead minnow (*Pimephales promelas*). Environ. Toxicol. Chem. 26, 521–527.

Mills, G.A., Greenwood, R., Allan, I.J., et al., 2010. Application of passive sampling techniques for monitoring the aquatic environment. In: Namieski, Szefer (Eds.), Analytical Measurements in Aquatic Environments. CRC Press, pp. 41–68.

Mills, L.J., Chichester, C., 2005. Review of evidence: are endocrine-disrupting chemicals in the aquatic environment impacting fish populations? Sci. Tot Environ. 343, 1–34.

Mouneyrac, C., Amiard-Triquet, C., 2013. Biomarkers of ecological relevance. In: Férard, J.F., Blaise, C. (Eds.), Encyclopedia of Aquatic Ecotoxicology. Springer, Dordrecht, pp. 221–236.

Mueller, M., Pander, J., Geist, J., 2013. Taxonomic sufficiency in freshwater ecosystems: effects of taxonomic resolution, functional traits, and data transformation. Freshwater Sci. 32, 762–778.

Nacci, D.E., Kohan, M., Pelletier, M., et al., 2002. Effects of benzo[a]pyrene exposure on a fish population resistant to the toxic effects of dioxin-like compounds. Aquat. Toxicol. 57, 203–215.

NAS, 1980. The international mussel watch. In: Report of a Workshop Sponsored by the Environment Studies Board. Natural Resources Commission, National Research Council, National Academy of Sciences, Washington, DC.

OSPAR Commission, 2011. JAMP Guidelines for Monitoring Contaminants in Sediments (last revision) Available at: http://www.ospar.org/v_measures/browse.asp?menu=01290301120125_000002_000000 (last accessed 07.08.14).

OSPAR Commission, 2012a. CEMP 2011 Assessment Report. Monitoring and Assessment Series. http://www.ospar.org/documents/dbase/publications/p00563/p00563_cemp_2011_assessment_report.pdf (last accessed 18.08.14).

OSPAR Commission, 2012b. JAMP Guidelines for Monitoring Contaminants in Biota. http://www.ospar.org/v_measures/browse.asp?menu=01290301120125_000002_000000.

OSPAR Commission, 2012c. JAMP Guidelines for the Integrated Monitoring and Assessment of Contaminants and Their Effects. Available at: http://www.ospar.org/v_measures/browse.asp?menu=01290301120125_000002_000000 (last accessed 07.08.14).

OSPAR Commission, 2013. JAMP Guidelines for Monitoring of Contaminants in Seawater (Agreement 2013-03). Available at: http://www.ospar.org/v_measures/browse.asp?menu=01290301120125_000002_000000 (last accessed 07.08.14).

Parkenson, T.F., Maruya, K.A., 2014. Passive sampling in contaminated sediment assessment: building consensus to improve decision making. Integr. Environ. Assess. Manag. 10, 163–166.

Paulus, M., Bartel, M., Klein, R., et al., 2005. Effect of monitoring strategies and reference data of the German environmental specimen banking program. Nukleonika. 50 (Suppl. 1), S53–S58. http://www.ichtj.waw.pl/ichtj/nukleon/back/full/vol50_2005/v50s1p53f.pdf.

Peijnenburg, W.J.G.M., Teasdale, P.R., Reible, D., et al., 2014. Passive sampling methods for contaminated sediments: state of the science for metals. Integr. Environ. Assess. Manag. 10, 179–196.

Poirier, L., Berthet, B., Amiard, J.C., et al., 2006. A suitable model for the biomonitoring of trace metals bioavailabilities in estuarine sediment: the annelid polychaete *Nereis diversicolor*. J. Mar. Biol. Ass. UK 86, 71–82.

Poisson, E., Fisson, C., Amiard-Triquet, C., et al., 2011. Effets de la contamination chimique. Des organismes en danger ? Fascicule Seine-Aval 2.7. http://seine-aval.crihan.fr/web/attached_file/componentId/kmelia106/attachmentId/24907/lang/fr/name/maquetteVF_reduite.pdf (last accessed 20.08.14).

Rainbow, P.S., Black, W.H., 2005. Physicochemistry or physiology: cadmium uptake and effects of salinity and osmolality in three crabs of different ecologies. Mar. Ecol. Prog. Ser. 286, 217–229.

RNO, 2006. Surveillance de la qualité du milieu marin. Ministère de l'Environnement and Institut français de recherche pour l'exploitation de la mer (Ifremer), Paris and Nantes, France. http://envlit.ifremer.fr/content/download/27640/224803/version/2/file/rno06.pdf (last accessed 27.08.08).

Romeo, M., Wirgin, I., 2011. Biotransformation of organic contaminants and the acquisition of resistance. In: Amiard-Triquet, C., Rainbow, P.S., Roméo, M. (Eds.), Tolerance to Environmental Contaminants. CRC Press, Boca Raton, pp. 175–208 (Chapter 8).

Rüdel, H., Müller, J., Steinhanses, J., et al., 2007. Retrospective monitoring of organotin compounds in freshwater fish from 1988 to 2003: results from the German environmental specimen bank. Chemosphere 66, 1884–1894.

Sanchez, W., Sremski, W., Piccini, B., et al., 2011. Adverse effects in wild fish living downstream from pharmaceutical manufacture discharges. Environ. Int. 37, 1342–1348.

Sanders, B.M., Martin, L.S., 1993. Stress proteins as biomarkers of contaminant exposure in archived environmental samples. Sci. Total Environ. 139–140, 459–470.

Sawal, G., Windmüller, L., Würtz, A., et al., 2011. Brominated flame retardants in bream (*Abramis brama*) from six rivers and a lake in Germany. Organohalogen Comp. 73, 515–518.

Schmitt, C.J., Blazer, V.S., Dethloff, G.M., et al., 1999. Biomonitoring of Environmental Status and Trends (BEST) Program: Field Procedures for Assessing the Exposure of Fish to Environmental Contaminants. Information and Technology Report, USGS/BDR/TTR-1999-0007 www.cerc.usgs.gov/pubs/center/pdfDocs/91116.pdf (last accessed 02.09.14).

TGD, 2011. Common Implementation Strategy for the Water Framework Directive (2000/60/EC). Technical Guidance for Deriving Environmental Quality Standards. Guidance Document No. 27, Technical Report -2011–055.203pp.https://circabc.europa.eu/sd/a/0cc3581b-5f65-4b6f-91c6-433a1e947838/TGD-EQS%20CIS-WFD%2027%20EC%202011.pdf (last accessed 29.08.14).

Theobald, N., Schäfer, S., Baass, A.C., et al., 2011. Retrospective monitoring of perfluorinated compounds in fish from German rivers and coastal marine ecosystems. Organohalogen Comp. 73, 440–443.

Tsuda, T., 2012. Residue of DDT and HCH in fish from lakes and rivers in the world. In: Pesticides – Recent Trends in Pesticide Residue Assay, Chapter 2. http://cdn.intechopen.com/pdfs-wm/38057.pdf.

UNEP, 2011. Draft Revised Guidance on the Global Monitoring Plan for Persistent Organic Pollutants. Rep. UNEP/POPS/COP.5/INF/27. http://www.chem.unep.ch/Pops/GMP/New_POPS_monitoring_guidance-COP.5-INF-27.English.pdf.

USEPA (US Environmental Protection Agency), November 2003. Procedures for the Derivation of Equilibrium Partitioning Sediment Benchmarks (ESBs) for the Protection of Benthic Organisms: PAH Mixtures. Washington; USEPA-600-R-02–013.

USEPA (US Environmental Protection Agency), January 2005. Procedures for the Derivation of Equilibrium Partitioning Sediment Benchmarks (ESBs) for the Protection of Benthic Organisms: Metal Mixtures (Cadmium, Copper, Lead, Nickel, Silver, and Zinc). Washington; USEPA-600-R-02–011.

Vandenberg, L.N., Colborn, T., Hayes, T.B., et al., 2012. Hormones and endocrine-disrupting chemicals: low-dose effects and nonmonotonic dose responses. Endocr. Rev. 33, 378–455.

Vittecoq, M., Roche, B., Daoust, S.P., et al., 2013. Cancer: a missing link in ecosystem functioning? Trends Ecol. Evol. 28, 628–635.

Warwick, R.M., 1988. The level of taxonomic discrimination required to detect pollution effects on marine benthic communities. Mar. Pollut. Bull. 19, 259–268.

Weis, J.S., Bergey, L., Reichmuth, J., et al., 2011. Living in a contaminated estuary: behavioral changes and ecological consequences for five species. Biosciences 61, 375–385.

Weis, J.S., Smith, G., Zhou, T., et al., 2001. Effects of contaminants on behavior. Biochemical mechanisms and ecological consequences. Biosciences 51, 209–217.

Wenning, R.J., Batley, G.E., Ingersoll, C.G., et al. (Eds.), 2005. Use of Sediment Quality Guidelines and Related Tools for the Assessment of Contaminated Sediments. SETAC Publications, Pensacola.

Wirgin, I., Waldman, J.R., 1998. Altered gene expression and genetic damage in North American fish populations. Mutat. Res. 399, 193–219.

Wirgin, I., Waldman, J.R., 2004. Resistance to contaminants in North American fish populations. Mutat. Res. 552, 73–100.

Zhuang, W., Gao, X., 2014. Integrated assessment of heavy metal pollution in the surface sediments of the Laizhou bay and the coastal waters of the Zhangzi island, China: comparison among typical marine sediment quality indices. PloS One 9 (4), e94145.

CHAPTER 4

How to Improve Exposure Assessment

Katrin Vorkamp, James C. McGeer

Abstract

This chapter describes approaches to environmental exposure assessments of metals and organic contaminants. The free ion activity model is an equilibrium-based geochemical speciation model to predict free metal ion concentrations, which are mainly associated with toxic effects. Biotic ligand models expand on this concept and predict the short-term binding of free ions to a biotic ligand, also based on equilibrium processes. Working either at equilibrium or in the kinetic phase, passive samplers are increasingly used to determine hydrophobic organic contaminants in the environment. Passive sampling techniques also exist for hydrophilic compounds and metals. The reserved concept of passive dosing can provide stable exposure concentrations in toxicity tests. Chemical monitoring in the context of the European Union Water Framework Directive or the Oslo–Paris Commission generates concentrations to be compared to toxicity-based environmental quality standards or assessment criteria. New analytical methods have been developed for the high quality trace analysis of organic compounds of emerging concern.

Keywords: Bioaccumulation; Bioavailability; BLM; Diagnostic methods; Emerging contaminants; Environmental monitoring; FIAM; Impact assessment; Metals; Passive sampling.

Chapter Outline

Aquatic Ecotoxicology. http://dx.doi.org/10.1016/B978-0-12-800949-9.00004-8

Introduction

The key to assessing contaminant exposure in relation to potential impacts to aquatic biota depends on an understanding of its physicochemical characteristics and interactions in relation to bioavailability and uptake. Although total contaminant concentration in the medium of interest has traditionally been used to characterize exposure, there is now a suite of tools that help to refine assessments. Tools for assessing exposure in this context can take two different approaches: modeling of environmental behavior to predict toxic forms and their biological interactions and use of devices and methods designed to characterize accessible and/or toxic forms in situ. In relation to the former, geochemical modeling software that estimates the speciation of metals (defined as the distribution of total metal into its different forms, physical and chemical) can be used to account for the influence of local environments on the relative abundance of toxic forms at equilibrium. These approaches can also be expanded to include biotic components that link uptake and bioaccumulation to effect thresholds. For example, natural organic matter can complex the free ion forms of metals thereby reducing toxicity as the metal–organic matter complex has very low availability. Similarly, as estuarine waters increase in salinity, the toxicity of metals generally decreases; this is due to the combined effects of increase chloride complexations and competition with Na at the site of uptake into the organism.

The second approach to assessing exposure to contaminants in relation to biotic impacts involves direct rather than modeled estimates. The use of specialized sampling devices and analytical techniques to detect contaminants and characterize their bioavailable/toxic forms of a contaminant provides for an assessment of a wide range of substances in the environment. These methodologies can be applied to both organics as well as inorganics. Passive sampling devices rely on the ability to distinguish forms based on diffusion and ultimately, chemical activity, which is linked with the potential for toxic impact. Passive dosing is an increasingly applied technique in exposure studies. Improvements in the analytical chemistry of emerging contaminants or those already included in monitoring programs also support the developments in environmental and exposure studies. This chapter discusses some of the different procedures applied in the exposure phase of risk assessment. It begins with modeling approaches for metals, namely the free ion activity model (FIAM) and the biotic ligand model (BLM) and then discusses the current state of the science on passive sampling techniques as well as approaches for improving the understanding of exposure to organic contaminants and substances of emerging concern.

4.1 Free Ion Activity Model

Although metals must bioaccumulate to induce impairment, not all forms of metal are bioavailable. The concentration of total or dissolved metal associated with toxicity can vary dramatically depending on chemistry of the exposure medium. Research as far back as the

1970s demonstrated that free ion forms of metals are associated with toxicity (Sunda et al., 1978). Metals bound to particulate matter and complexed forms in the dissolved phase (species) have a reduced bioavailability (or are not available) and therefore contribute much less (or not at all) to toxic responses. This understanding of the relationship between geochemical speciation and impacts to aquatic biota is the basis for the FIAM. FIAM states that the biotic response to exposure is due to the interaction of metal cations with negatively charged ligands on the cell surface. Metal cations can also form complexes with negatively charged ligands in solution and the extent to which this occurs defines the degree to which free metal cations will be available to interact at the cell surface. Therefore the composition of the aquatic medium is an important factor to consider in exposure assessment because it leads to an understanding of the species of metals that will exist under those conditions, particularly the free ion.

Geochemical factors that influence aquatic speciation and the formation of free ions will also influence toxicity. Inorganic ligands such as F^-, $C1^-$, SO_4^{-2}, HCO_3^-, CO_3^-, and HPO_3^- are among the potentially important anions that may reduce free metal ion concentrations via complexation. This explains, at least in part, how aquatic alkalinity, pH, and salinity can modulate toxicity: decreases in pH and alkalinity favors reduced complexation and higher amounts of free metal ions and therefore toxicity is increased. Increased salinity on the other hand results in greater Cl^- complexation of free metal ions and therefore mitigation of toxicity. Dissolved natural organic matter (NOM) and other organic compounds can also complex free metal ions. NOMs are complex and heterogeneous compounds that arise from the degradation of terrestrial or aquatic biota and can present a number of different negatively charged ligands. For example, metal binding can be via carboxyl, phenolic, sulphidic, and protein groups on the NOM (Paquin et al., 2002). NOM complexation capacity can vary with both the concentration of NOM (measured as dissolved organic carbon) as well as the relative composition of binding groups (i.e., its quality).

Complexation reactions are at the heart of the FIAM, both in terms of reactions that define the free ion concentration of metal as well as their subsequent interaction with negatively charged ligands on the biotic surface. The FIAM applies equilibrium-based geochemical speciation modeling principles to predict free metal ion concentrations. The most common equilibrium speciation models are software packages such as the Windemere Humic Aqueous Model (the current version is VII, Tipping, 1994; Tipping et al., 2011), Chemical Equilibria in Soils and Solutions (Santore and Driscoll, 1995), MINEQL+ (Schecher and McAvoy, 2001), and MINTEQA2 (Allison et al., 1991). All of these models estimate free ion concentrations at equilibrium using Log K values (equilibrium constants). The ability to accurately predict free ion concentrations is therefore dependent on the database of Log K values and these tend to be relatively similar for inorganic complexation across the different models. They differ, however, in the approach to NOM complexation. For example, MINEQL+ does not include NOM complexation, whereas the Windemere Humic Aqueous Model applies a sophisticated approach that includes both chemical and electrostatic interactions across a family of binding sites with different characteristics, represented as fulvic acid and humic acid (Tipping, 1994; Tipping et al., 2011).

The FIAM is underpinned by rigorous science and, because of the wide availability of geochemical equilibrium software, is relatively simple to apply in the context of exposure assessment. However, these strengths undercut some important considerations. One is that although FIAM recognizes that other cations may complete with free metal ions for binding and uptake into cells, it does not incorporate these competitive effects in terms of toxicity reduction. In fact, a description of how the FIAM can be applied to estimate toxicity is lacking because it only provides a conceptual framework for metal speciation and bioavailability in relation to toxic forms. As explained by Wang (2014), the FIAM is a conceptual model. As such, it provides a useful way to understand the dynamics of exposure conditions within assessments but it does not predict threshold for toxic effects.

4.2 Biotic Ligand Model

The BLM approach predicts the response to metal exposure by accounting for the variation induced by water chemistry and in this way it applies the concepts of the FIAM (Di Toro et al., 2001; Niyogi and Wood, 2004). The success of the BLM is evidenced by the fact that it has expanded beyond its initial suite of metals (Cd, Cu, Zn, Pb, and Ag: Di Toro et al., 2001) and is increasingly accepted as a tool for setting thresholds on a water chemistry-specific basis. Similar to the FIAM, the BLM applies equilibrium-based geochemical modeling to estimate metal speciation with a focus on the bioavailable forms/species. The BLM takes the additional step of predicting uptake into an organism and then bioaccumulation at the site of toxicity, the biotic ligand (BL, Figure 4.1).

The original description of the conceptual framework for the BLM is credited to Pagenkopf (1983). His gill surface interaction model considered geochemical speciation with the addition of a biological ligand into the model to represent toxic impacts. The approach was also developed by Morel (1983) and Campbell (1995) with the focus on toxicity being associated with the free ion form of metal (i.e., the FIAM). The biotic ligand approach considers metal interactions at the site of toxicity as a predictable formation of complexes between the metal cation with a negatively charged BL. The BLM therefore integrates predictions by modeling interactions at three levels: aquatic speciation, tissue bioaccumulation, and toxicological impacts. This integrated model provides for prediction of impact on a site specific basis and therefore offers a powerful tool for exposure assessment.

The interactions between metal species and the BL are incorporated into the equilibrium modeling framework via description of additional ligands that account for biological interactions. These additional ligands are incorporated as equilibrium constants; for example, for the binding of free metal ions to the BL. If other species of metal are associated with toxicity; for example, hydroxide or carbonate forms—these can also be incorporated into the model.

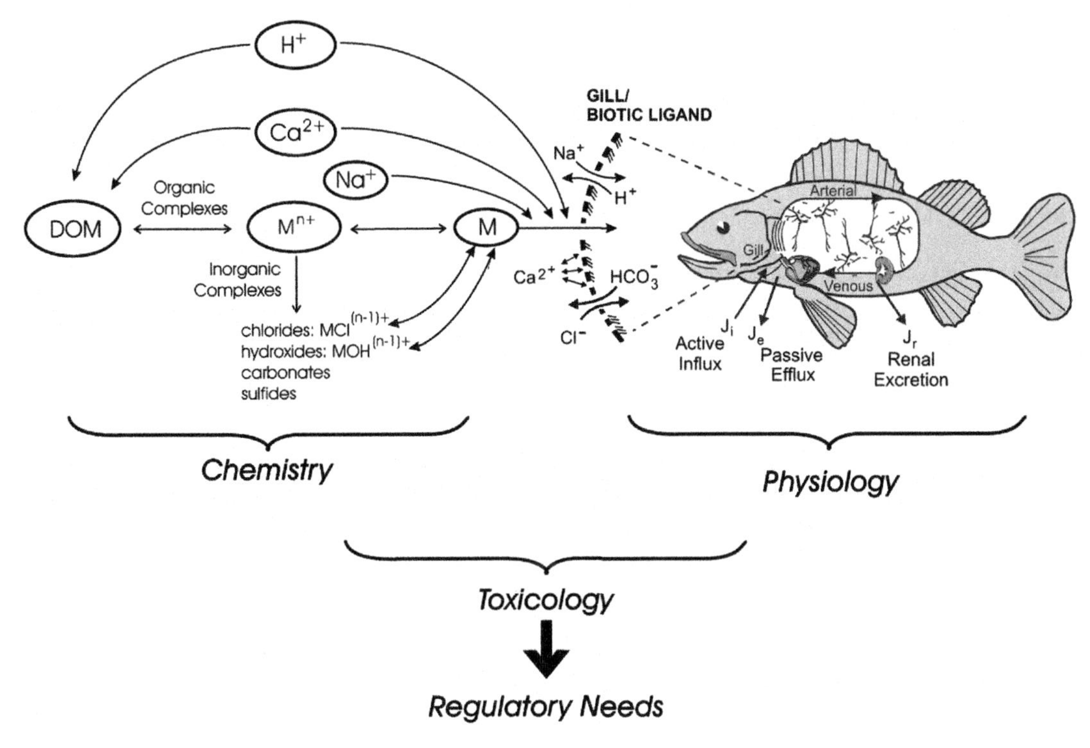

Figure 4.1
Schematic diagram from Figure 1 of Paquin et al. (2002), describing the integration of chemistry, physiology, bioaccumulation and toxicology within the BLM approach. BLMs incorporate chemistry (geochemical equilibrium speciation), physiology (metal uptake and cationic competition), and toxicology (accumulation thresholds associated with impacts).

In addition to the binding of toxic species, cations that bind to the BL and reduce toxicity are also accounted for. Examples of cations that competitively bind to the BL and therefore reduce bioaccumulation of toxic forms of the metal are Ca^{2+}, Na^{+}, Mg^{2+}, and H^{+}. The exposure, uptake, and accumulation interactions accounted for in BLMs are illustrated through schematic diagram shown in Figure 4.1 (also see Figure 4.1 of Di Toro et al., 2001).

The strength of BLM approach is that it simultaneously accounts for the geochemical speciation as well as the relative binding of metal species to the BL (the site of toxicity). Thresholds for effects are established based on the concentration of metal bound to the biotic ligand. The concentration of metal bound to the biotic ligand is established through equilibrium constants and effects are assumed to follow an accumulation–response model. Within any given water chemistry, increased exposure results in greater accumulation and this is assumed to result in increased severity of impact. Thresholds for accumulation are linked with measured toxicity endpoints. For example in the case of acute lethality, the measured (dissolved) $L(E)C_{50}$ (the concentration associated with 50% lethality (or effect)) is associated with accumulation on the BL and is given as the LA_{50} (the accumulation associated with 50% lethality). Although $L(E)C_{50}$ values can vary with water chemistry and associated changes in bioavailability and uptake, the LA_{50} value is constant. Therefore the BLM is a tissue bioaccumulation based model in that it is the accumulation of metal that is used to predicted toxicity. By linking accumulation at the site of toxicity (i.e., the BL) and measured effects, any endpoint, whether acute or chronic, can be used to establish a BLM.

Using equilibrium-based modeling approaches to predict toxicity would appear to be conceptually flawed because toxicity is a disequilibrium event. However, this approach is supported physiologically under the theoretical assumption that short-term accumulation (24 h or less) is predictive of subsequent impacts (e.g., 48 or 96 h $L(C)E_{50}$). During exposures, the initial bioaccumulation of metal occurs via nutritionally required ion (e.g., Ca, Na, Mg) uptake mechanisms and follows a rapid rise to a plateau (equilibrium) that can be characterized via Michaelis Menton kinetics. The assumption within the model is that if the threshold concentration for short-term accumulation is breached, this will set in motion a series of physiological disruptions that will subsequently result in toxicity and this has been demonstrated by Morgan et al. (1997) as well as MacRae et al. (1999). Playle et al. (1992, 1993a,b) reported Log K values for Cd and Cu accumulation to trout gills using 3-h accumulations, whereas MacRae et al. (1999) characterized Cu binding to rainbow trout and brook trout gills based on 24-h exposures. It is likely that the accumulation of metal on the BL will increase beyond the short term, particularly if the case of lethality where deterioration may be extreme. However it is the understanding of the characteristics of short-term uptake and accumulation that is used in the model and this is described as an equilibrium-based event that induces disequilibrium (i.e., toxicity).

The linkage between accumulations of bioavailable forms via ion uptake mechanisms and leading to the induction of toxicity also explains the protective effect that cations can have on metal toxicity. Cationic competition with metal free ion occurs at the site of uptake and is specific to the mechanism through which the metal is taken up. As described in Niyogi and Wood (2004), different metals disrupt different ionic uptake processes (see Figure 4.2). Both Cu^{2+} and Ag^{+} are taken up through a Na^{+} channel whereas Cd^{2+}, Zn^{2+}, Pb^{2+}, and Co^{2+} are associated with Ca^{2+} uptake (Niyogi and Wood, 2004). These metal ion interactions explain not only how cations protect against toxicity but also the mechanisms of toxicity itself (disruption of ion regulation and internal ionic balance).

In summary, the BLM applies FIAM principles and then extends them using equilibrium based approaches to predict the short-term binding of toxic metal species to the BL as well as the competitive interactions with cations that reduce toxicity. Thresholds for accumulation of metal on the BL are linked to standard toxicity endpoints that can be acute or chronic. Estimating toxicity is done by applying measured water chemistry with a geochemical speciation modeling software with the addition of equilibrium constants for the uptake of forms of metal that are associated with toxicity as well as competitive interactions with cations that inhibit metal

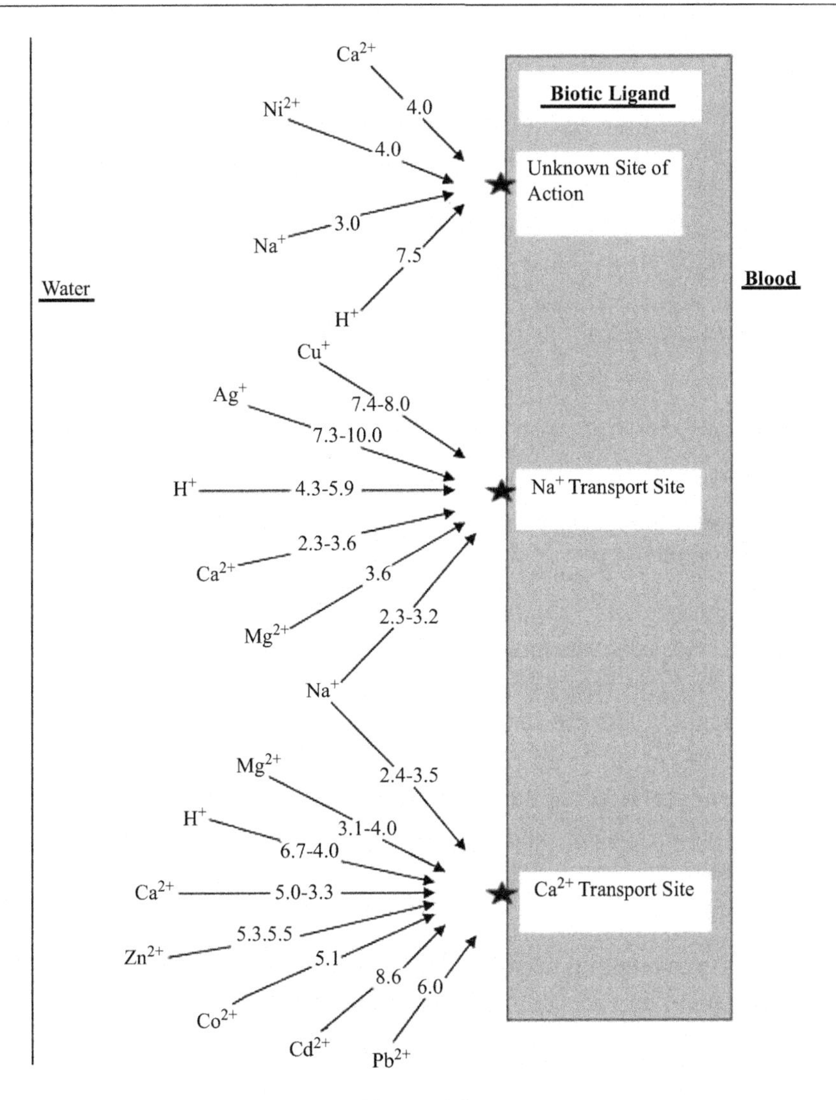

Figure 4.2
Schematic diagram showing the binding affinities free ion forms of Zn, Co, Cd, Pb, Ag, Cu, and Ni as well as the ranges for those for competitive cations that reduce the binding and accumulation at the site of toxic action, the biotic ligand. *This figure is taken from Figure 1 of the BLM review of Niyogi and Wood (2004).*

uptake. Application of the BLM within exposure assessment can be done in different ways. Known concentrations of metal can be provided as a model input (in addition to the site water chemistry) to determine if the resulting predicted bioaccumulation on the BL reaches the threshold in the model (i.e., LA_{50}). Alternatively, the model can also be applied to determine the dissolved concentration of metal that would produce the LA_{50} in that specific water chemistry. These two approaches offer flexibility in estimating the potential for effects as well as how changing conditions (metal concentrations or site chemistry) can influence toxic responses.

4.3 Passive Samplers

Passive sampling is based on the molecular diffusion of the target analytes into or onto a sorbent, with adsorption processes mainly used for polar compounds (Smedes et al., 2010). The sampling process begins with the exposure of the sorbent to the sampling medium and continues until the chemical potential in both media is equal—i.e., equilibrium is reached (Seethapathy et al., 2008). The uptake process proceeds in a linear (kinetic) phase of time-integrating sampling, an intermediate phase and the equilibrium phase (Figure 4.3) following first-order kinetics as given in Eqn (4.1) (Mayer et al., 2003; Vrana et al., 2005).

$$C_{Sampler} = C_{Medium} \cdot \frac{k_1}{k_2} \cdot \left(1 - e^{-k_2 \cdot t}\right) \tag{4.1}$$

$C_{Sampler}$: concentration on the sampler, C_{Medium}: concentration in the sampling medium, k_1: uptake rate, k_2: elimination rate, t: time.

Passive sampling techniques are valuable tools in the context of exposure and risk assessment as the equilibrium analyte concentration of the sampler can be converted to fugacity, chemical potential/activity, or freely dissolved concentrations (Mayer et al., 2003). The freely dissolved concentration C_{free} is often used synonymously to a bioaccessible concentration, but conceptual differences exist, as detailed by Reichenberg and Mayer (2006): C_{free} is the result of a phase distribution (partitioning) that is driven by differences in chemical potential, thus an expression of free energy rather than of an accessible pool of molecules. The authors also give several examples of equilibrium sampling in the context of baseline toxicity and bioconcentration studies.

In the following, passive sampling theory and techniques will be described for different groups of analytes, taking into account that concepts and practical aspects of passive sampling

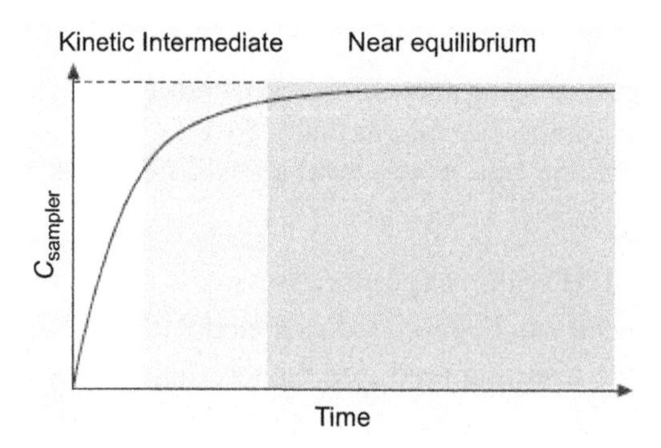

Figure 4.3
Kinetic and equilibrium phases of passive sampling. *Printed with permission from Mayer et al. (2003). Copyright 2003, American Chemical Society.*

are largely determined by the target analytes' physicochemical properties. In addition, some information will be given on passive dosing which reverses equilibrium passive sampling with promising opportunities for toxicity and risk assessment studies (Smith et al., 2010).

4.3.1 Hydrophobic Compounds

As outlined in Eqn (4.1), the conversion of a passive sampler concentration to a concentration in the medium (e.g., water) requires a sampling rate or, at equilibrium, an equilibrium partition coefficient (Eqn (4.2)). Sampling rates depend on several factors, such as sampler geometry and environmental conditions. Therefore, passive sampling in water often uses performance reference compounds (PRCs) for in situ calibration, which are similar to the target analytes, but do not occur in the environment and provide dissipation rates under the same environmental conditions as the sampling rates to be calculated (Booij et al., 2002; Huckins et al., 2002).

$$C_{Sampler} = C_{Medium} \cdot \frac{k_1}{k_2} = C_{Medium} \cdot K \tag{4.2}$$

K: sampler-medium equilibrium partition coefficient.

Partition coefficients between the passive sampler and the sampling medium (e.g., water) are primarily temperature-dependent and usually determined under controlled laboratory conditions (Yates et al., 2007; Smedes et al., 2009). Because the equilibrium concentrations of hydrophobic compounds in the water phase usually are low and experimental artefacts might complicate their measurements, these authors suggest a cosolvent method: Mixing water with a solvent increases the partitioning of hydrophobic molecules into this phase where they can be determined more accurately. From a series of experiments with varying water:solvent ratios, the analyte concentration at 0% solvent can be extrapolated (Yates et al., 2007; Smedes et al., 2009).

There is general agreement in the scientific community that passive sampling techniques are most mature for hydrophobic compound, such as most polychlorinated biphenyls (PCBs), polycyclic aromatic hydrocarbons (PAHs) and polybrominated diphenyl ethers (PBDEs) (ICES, 2013; Mills et al., 2014). However, for compounds with very high $\log K_{OW}$ values, such as BDE-209 (with an estimated $\log K_{OW} > 10$), partition coefficients are difficult to determine, mainly because of long equilibration times. Some compounds might also be subject to photolytic degradation during passive sampling deployment, whereas effects of biofouling can be accounted for to some extent by PRCs (Vrana et al., 2005).

Semipermeable membrane devices (SPMDs) have been in use for 25 years to determine hydrophobic compounds in water and other media (Södergren, 1987; Huckins et al., 1990; Petty et al., 2000). They consist of a tube of low-density polyethylene (LDPE) filled with triolein or another receiving organic phase. Only truly dissolved and nonionized compounds diffuse through the LDPE and can subsequently be extracted with organic solvents

(Vrana et al., 2005). It is a well-studied and widely used technique for which standard operating procedures exist, including the use of PRCs, whereas the main disadvantage is the complex procedure to recover and analyses of the compounds from the triolein, in addition to the sampler's susceptibility to biofouling (Górecki and Namniesnik, 2002; Vrana et al., 2005). Recent examples of the use of SPMDs for the monitoring of hydrophobic compounds in water include the publications by Booij et al. (2014a) and Kim et al. (2014).

A simplified version of the SPMDs is the use of LDPE without triolein, for which an efficient uptake of hydrophobic compounds has been shown as well (Carls et al., 2004; Adams et al., 2007). This technique also allows the use of PRCs for its calibration (Booij et al., 2002). LDPE overcomes some of the analytical difficulties described for SPMDs. Applications of LDPE for the sampling of hydrophobic compounds in the aquatic environment have recently been published by Jacquet et al. (2014) and Bruemmer et al. (2015).

Silicone rubber, also known as polydimethylsiloxane, is an increasingly popular material for use as a passive sampler of hydrophobic compounds, beyond its use in solid phase microextraction (Arthur and Pawliszyn, 1990; Cornelissen et al., 2008; Smedes and Booij, 2012). It offers large flexibility in terms of polymer thickness, sampler geometry, and thus sampling rates and has shown a range of application possibilities as an equilibrium sampler (Mayer et al., 2003). Experience exists with the use of PRCs and the determination of silicone–water partition coefficients (Booij et al., 2002; Smedes et al., 2009). Recently, silicone has been used for the passive sampling of hydrophobic compounds from water by e.g., Prokes et al. (2012) and Jahnke et al. (2014).

Several other polymers and passive sampling formats are available and have been applied for the sampling of hydrophobic compounds in aquatic systems (Vrana et al., 2005; Rusina et al., 2007; Allan et al., 2009).

4.3.2 Polar Organic Compounds

As many so-called emerging contaminants such as pharmaceuticals, personal care products, phosphorous flame retardants and polar pesticides are more polar than traditionally monitored halogenated persistent organic pollutants, i.e., with $\log K_{OW}$ values <4, there is growing interest in possibilities of their passive sampling, both from a monitoring and a risk assessment perspective (Alvarez et al., 2004; ICES, 2013). Samplers for more hydrophilic compounds generally consist of an adsorption phase covered by a thin diffusion-limiting membrane (Mills et al., 2011). The uptake mechanism is thus conceptually different from the partitioning process described for hydrophobic compounds and not fully understood, in particular the transfer kinetics across the membrane, consequently lacking theoretical models to describe and predict uptake parameters (Mills et al., 2014). As a result, the passive sampling of polar compounds is generally considered less mature than that of hydrophobic compounds, but able to provide qualitative and semiquantitative information (ICES, 2013).

The strong binding capacity for the target analytes (and potentially, competing molecules) generally excludes the use of PRCs for calibration purposes although some promising exceptions have been reported (Mazzella et al., 2010; Belles et al., 2014). The transfer of a laboratory-based calibration to field situations increases the uncertainty connected with this type of sampling, but research into in situ calibration is ongoing (Ibrahim et al., 2013; Lissalde et al., 2014; Mills et al., 2014).

The polar organic chemical integrative sampler (POCIS) is commonly applied for the passive sampling of polar compounds. It consists of a solid phase between two microporous polyethersulfone diffusion-limiting membranes (Alvarez et al., 2004; Morin et al., 2012). The solid phase can be a single (e.g., Oasis HLB) or multiphasic system, for example combining Isolute ENV+ and Ambersorb 1500 dispersed on SX-3 Bio Beads (Morin et al., 2012). POCIS has mainly been used for screening purposes, but also for the semiquantitative determination of time-weighted average concentrations and in combination with bioassays, producing toxicity information (Morin et al., 2012).

The Chemcatcher has the same principal design, in this case an immobilized disk as the receiving phase combined with a thin diffusion-limiting membrane, which extends the kinetic regime (Kingston et al., 2000). Different disk and membrane materials enable the use of this format for a wide range of analytes with different physicochemical properties (Vrana et al., 2005). For polar compounds, styrene-divinylbenzene polymers are commonly applied (Mills et al., 2011). For some hydrophilic compounds, a considerable release from the styrene-divinylbenzene disks and nonisotropic kinetics for the uptake and desorption processes have been shown (Shaw et al., 2009; Vermeirssen et al., 2013). Consequently, PRCs have limited suitability. However, calibration data are available for many chemicals sampled with the Chemcatcher (Vrana et al., 2005; Smedes et al., 2010).

Recent studies using the POCIS and Chemcatcher designs for polar compounds include those of Vermeirssen et al. (2012) and Kaserzon et al. (2014).

4.3.3 Metals

The measurement and risk assessment of metals in the environment require considerations about their speciation. Metals can be sorbed to particles and colloids, form complexes with dissolved organic matter or smaller ligands, and dissolve in water as free ions. Toxicity is generally related to the free ions and labile metal fractions (Allan et al., 2007). It has been recognized that metal concentrations determined by passive sampling have toxicological relevance (ICES, 2013).

The passive sampling of metals often relies on the diffusive gradients in thin films (DGT) technique originally developed by Zhang et al. (1995), which uses a defined diffusion layer consisting of a porous gel film and a layer of complexing cation exchange beads (e.g., Chelex 100). Concentration gradients are established in the gel layer as the metals pass through it and accumulate in the resin (van Leeuwen et al., 2005). From the amount in the resin and the time of deployment, a time-weighted average concentration in the bulk water can be determined if diffusion coefficients are known (Garmo et al., 2003; Mills et al., 2011). However, some authors highlight that effects of ambient water quality and deployment conditions should be given attention in the interpretation of results as factors like flow rates, temperature, pH, nutrients etc. will have an influence on the process (Mills et al., 2014). Besides the DGT principle, the Chemcatcher format described for polar organics previously also is applicable to metals if used with a chelating disk and an appropriate membrane (e.g., cellulose acetate) (Allan et al., 2007). It requires metal-specific uptake rates depending on temperature and water turbulences and usually obtained under laboratory conditions (Mills et al., 2011).

Regarding its function in a risk assessment context, the DGT technique has also been applied to assess metal bioavailability in soil and sediments (e.g., Zhao et al., 2006; Zhang et al., 2014). Several publications are available on the use of DGT for metal determination in water, a recent one being that by Turner et al. (2015).

4.3.4 Passive Dosing

Common aquatic toxicity tests of hydrophobic compounds face the challenge of producing exposure concentrations close to aqueous solubility and keeping these constant for the duration of the test. Originally introduced as partitioning driven administering, passive dosing can overcome these difficulties by loading a sorbent with hydrophobic compounds and equilibrating it with water (Mayer et al., 1999). Furthermore, it avoids coexposure to solvents commonly used for spike solutions (Mayer and Holmstrup, 2008). In tests with PAH toxicity on *Daphnia magna*, silicone was loaded with PAHs, which subsequently partitioned into water, reached equilibrium within a few hours, and could be kept constant over the test period of 10 weeks (Smith et al., 2010). The authors emphasized the reproducibility of this approach as another advantage. It has also been successfully applied in mixture toxicity tests, with the advantage of keeping mixture compositions stable (Schmidt et al., 2013).

With a particular interest in mixture toxicity, extracts of silicone applied as passive samplers of hydrophobic compounds in water have been used as exposure media in in vitro toxicity tests (Emelogu et al., 2013). Potential direct effects from the extraction solvent were controlled and excluded. Although the use of sampled—rather than loaded—compounds appears closer to the real environmental situation and includes unidentified compounds, hydrophobic compounds do not necessarily reach equilibrium during the field sampling, which will lead to differences between their exposure and environmental compositions and concentrations (ICES, 2013).

The passive dosing technique also has potential in the assessment of sediment and soil toxicity (Mayer and Holmstrup, 2008; ICES, 2013).

4.4 Improving Techniques for the Determination of Organic Contaminants

The environmental monitoring of organic contaminants traditionally strives to determine the total concentration of a given compound in an environmental compartment, for instance sediment or water. Decisions about the environmental compartment as well as sampling times, locations, and frequencies are largely determined by the objectives of the monitoring program (e.g., compliance checks, temporal trend monitoring) and the physicochemical properties of the compound, among these its partitioning behavior and environmental stability. As described in Chapter 1, environmental monitoring programs usually follow "a priori approaches,"—i.e., the determination of selected chemicals, which will be described here for the European Union (EU) Water Framework Directive (WFD) and the Coordinated Environmental Monitoring Programme of OSPAR for the protection of the northeast Atlantic. The "global approaches" described in Chapter 1 try to link observed effects to chemicals identified in a certain fraction of the environmental sample. The diagnostic methods Toxicity Identification Evaluation (TIE) and Effects-Directed Analysis (EDA) will be introduced here. In addition, this chapter will briefly present the "integrated monitoring" concept as developed by OSPAR and the International Council for Exploration of the Sea (ICES) in an attempt to build a bridge between conventional chemical monitoring and biological effects observed in the environment (Davies and Vethaak, 2012).

4.4.1 A Priori Approaches in European Environmental Monitoring Programs

Environmental monitoring programs have been established at the national level (e.g., the US Clean Water Act) and in relation to certain water bodies or regions (i.e., the regional sea conventions), among these the OSPAR coordinated environmental monitoring program (CEMP) covering the northeast Atlantic. In Europe, the monitoring of inland surface waters and coastal waters is described by the WFD at the level of the EU (2000). One purpose of these monitoring programs is to assess trends of selected chemicals to evaluate the effect of regulatory actions. Furthermore, the monitoring data form the basis of environmental assessments, which aim to characterize the status of the environment with regard to chemical pollution, by comparisons with toxicology-based effect concentrations.

Table 4.1 summarizes the organic compounds currently included as "priority hazardous substances" in the EU WFD with their associated environmental quality standards (EQSs) in water or biota (EU, 2013) as well as the mandatory compounds of the OSPAR CEMP with their environmental assessment criteria (EAC) (OSPAR, 2009). The priority hazardous substances are a subset of the total list of priority substances in the EU WFD (EU, 2013). In contrast to the EU WFD and its EQS values, the OSPAR EACs are not legally binding limits. To achieve high and comparable analytical quality, technical specifications and guidance documents have been issued. The EU WFD requires that all analytical methods are validated and documented in accordance with EN ISO/IEC-17025 or an equivalent standard (EU, 2009a). Quality assurance and quality control (QA/QC) requirements also include the laboratory's participation in relevant proficiency testing schemes and the analyses of suitable reference materials. Minimum performance criteria have been defined for the analytical methods in relation to EU WFD monitoring, including measurement uncertainties ≤50% at the level of the EQS and limits of quantification equal or below a value of 30% of the EQS (EU, 2009a).

For a parameter to be included in the OSPAR CEMP, certain conditions have to be fulfilled: Assessment criteria have to be developed, including EACs (Table 4.1) and background assessment concentrations, which help to assess whether or not a concentration significantly exceeds a natural or methodology-based background. Furthermore, relevant proficiency testing schemes have to be in place and guidelines have to be available. These guidelines cover the matrices biota, sediment, and water and include general guidance on sampling strategies, equipment, storage, and analytical procedures (OSPAR, 2012a,b, 2013). Compound-specific technical annexes to these guidelines provide detailed information on analytical methods including QA/QC. As part of a common implementation strategy of the EU WFD, several technical guidance documents have been published, among these guidelines on the chemical monitoring in surface water (EU, 2009b) and sediment and biota (EU, 2010). These guidelines are comparable to the level of the general OSPAR guidelines and do not provide compound-specific guidance on analytical methods. Recent reports on WFD monitoring, including methodological challenges have been published by e.g., Miège et al. (2012), Lava et al. (2014), and Vorkamp et al. (2014).

4.4.2 Toxicity Identification Evaluation and Effects-Directed Analysis

The motivation to include a chemical in an environmental monitoring program, such as those of Table 4.1, is usually the knowledge—or suspicion—of adverse effects on environmental or human health. However, the monitoring programs usually do not address whether or not an effect is present, but whether or not a contaminant is present at a level potentially causing effects. The TIE and EDA approaches start with the observation of an effect and aim at identifying the compound or compounds causing this effect. Although identical in their objectives, the approaches are different in their assumptions, strategies, methodologies and often endpoints (Burgess et al., 2013).

The TIE was originally developed in the United States in the 1980s to monitor and characterize the toxicity of effluents under the US Clean Water Act and to ultimately avoid the discharge of toxic chemicals to the environment (Ho and Burgess, 2013). As described by these authors, the TIE uses as three-phase approach, in which phase I characterizes the toxicants as cationic metals or nonionic organic chemicals, phase II identifies the specific toxicant, and phase III confirms these findings. Unlike EDA, TIE considers toxicity of the original

Table 4.1: Organic Priority Hazardous Substances and Their Associated Annual Average Environmental Quality Standards (AA-EQSs) of the EU Water Framework Directive (WFD) (EU, 2013) and Organic Compounds Mandatory for Monitoring under the OSPAR Coordinated Environmental Monitoring Program (CEMP) with Their Associated Environmental Assessment Criteria (EACs) (OSPAR, 2009)

EU WFD		OSPAR CEMP	
Priority hazardous substances	AA-EQS for inland surface water (ISW), coastal waters (CW), or biota (B)	Compounds	EAC for mussels (M), fish (F), or sediment (S)
Anthracene	0.1 μg/L (ISW, CW)	Anthracene	78 μg/kg dw (S), 290 μg/kg dw (M)
Benzo[a]pyrene	1.7×10^{-4} μg/L (ISW, CW), 5 μg/kg ww (B)	Benzo[a]pyrene	325 μg/kg dw (S), 600 μg/kg dw (M)
Brominated diphenyl ethers (PBDEs)	0.0085 μg/kg ww (B)	Naphthalene	43 μg/kg dw (S), 340 μg/kg dw (M)
Chloroalkanes, $C_{10\text{-}13}$	0.4 μg/L (ISW, CW)	Phenanthrene	1250 μg/kg dw (S), 1700 μg/kg dw (M)
Dicofol	1.3×10^{-3} μg/L (ISW), 3.2×10^{-5} μg/L (CW), 33 μg/kg ww (B)	Fluoranthene	250 μg/kg dw (S), 110 μg/kg dw (M)
Di(2-ethylhexyl)-phthalate (DEHP)	1.3 (ISW, CW)	Pyrene	350 μg/kg dw (S), 100 μg/kg dw (M)
Dioxins and dioxin-like compounds	0.0065 μg/kg TEQ (B)	Benzo[a]-anthracene	1.5 μg/kg dw (S), 80 μg/kg dw (M)
Endosulfan	0.005 μg/L (ISW), 0.0005 μg/L (CW)	Indeno[1,2,3,c,d]-pyrene	1.5 μg/kg dw (S)
Hexabromocyclododecane (HBCD)	0.0016 μg/L (ISW), 0.0008 μg/L (CW), 167 μg/kg ww (B)	Benzo[g,h,i]-perylene	2.1 μg/kg dw (S), 110 μg/kg dw (M)
Heptachlor/heptachlor epoxide	2×10^{-7} μg/L (ISW), 1×10^{-8} μg/L (CW), 0.0067 μg/kg ww (B)	PCB-28	1.7 μg/kg dw (S), 13.5 μg/kg dw (M)
Hexachlorobenzene (HCB)	10 μg/kg ww (B)	PCB-52	2.7 μg/kg dw (S), 80 μg/kg dw (M)
Hexachlorobutadiene (HCBD)	55 μg/kg ww (B)	PCB-101	3.0 μg/kg dw (S), 5 μg/kg dw (M)
Hexachlorocyclohexane (HCH)	0.02 μg/L (ISW), 0.002 μg/L (CW)	PCB-118	0.6 μg/kg dw (S), 1 μg/kg dw (M),
Nonylphenols	0.3 μg/L (ISW, CW)	PCB-138	7.9 μg/kg dw (S), 100 μg/kg dw (M)
Pentachlorobenzene	0.007 μg/L (ISW), 0.0007 μg/L (CW)	PCB-153	40 μg/kg dw (S), 1790 μg/kg dw (M)
Perfluorooctane sulfonate (PFOS)	0.00013 μg/L (CW), 9.1 μg/kg ww (B)	PCB-180	12 μg/kg dw (S), 26.5 μg/kg dw (M)
Quinoxyfen	0.15 μg/L (ISW), 0.015 μg/L (CW)	Brominated diphenyl ethers (PBDEs)	NA
Tributyltin compounds	0.0002 μg/L (ISW, CW)	Hexabromocyclododecane (HBCD)	NA
Trifluralin	0.03 μg/L (ISW, CW)	-	-

Dw, dry weight; NA, not available; TEQ, toxicity equivalents; ww: wet weight. Sediment values are normalized to 2.5% organic carbon.

medium, as opposed to a solvent extract, because the matrix can have implications for the compound's availability (Burgess et al., 2013). Following an initial whole organism toxicity test, phase I tests characterize the physicochemical properties of the toxicant through sample manipulation combined with toxicity tests (USEPA, 1991). Successive manipulations render certain groups of potential toxicants unavailable for the toxicity test and thus limit the spectrum of toxicity-causing agents. Phase II focuses on structure elucidation or other identification of the toxicant, which according to Burgess et al. (2013) is less complicated for cationic metals, but would benefit from more research into robust methods for organic chemicals. Originally developed for effluents, TIE was later developed further to be applicable to marine species as well as sediments (USEPA, 1996; Golding et al., 2006; USEPA, 2007). Further research has focused on including emerging contaminants in phase II tests. Recent publications on TIE include those by Greenstein et al. (2014) and Matos et al. (2014).

Likewise developed in the 1980s, EDA is based on an iterative process combining physical-chemical fractionation of a sample with biotesting (Brack, 2003). Unlike TIE, EDA uses solvent extracts for the identification of toxicity inducing agents and usually chooses endpoints that primarily respond to organic compounds (Burgess et al., 2013). However, the authors highlight that EDAs do have the potential to use whole-organism endpoints and thus a broader range of toxicants. Typical endpoints in EDA include, but are not limited to, endocrine disruption (Chapter 15), fish embryo toxicity (Zielke et al., 2011) mutagenicity, and aryl hydrocarbon receptor effects (Brack et al., 2005).

Brack (2003) outlined the following steps of EDA: (1) extraction, recognizing that this step is necessarily selective and usually focuses on organic toxicants; (2) fractionation (i.e., reducing the complexity of the sample by removing nontoxic components); (3) biotesting; and (4) identification and confirmation. EDA has also been applied in the context of the EU WFD (Brack et al., 2007, 2015). Recent examples of EDA applications have been published by Booij et al. (2014b) and Fetter et al. (2014).

4.4.3 Integrated Chemical and Biological Effects Monitoring According to OSPAR and International Council for Exploration of the Sea

OSPAR and ICES developed a framework to link chemical monitoring with that of biological effects (Davies and Vethaak, 2012). The idea was to use the same samples of water, sediment, and biota for both purposes (i.e., the simultaneous measurement of chemical concentrations, biomarkers, and various supporting parameters). Biological effect results could then be assessed in the same way as described for chemicals in Section 4.4.1 (i.e., by comparisons with environmental and background assessment criteria). In a next step, the overall assessment of environmental quality would be based on results of chemical measurements as well as those of biological effects. For this purpose, a list of biomarkers has been suggested, including effects at the cellular, tissue, and organism level for mussels and fish (Davies and Vethaak, 2012).

4.5 Improving Techniques for the Determination of Emerging Contaminants

Research into new contaminants points to many potentially problematic compounds in the environment, by means of the diagnostic TIE or EDA approaches described in Section 4.4.2 (Brack et al., 2007), of modeling exercises (Howard and Muir, 2010; McLachlan et al., 2014), structure elucidation methods (Hoh et al., 2006), or screening studies (Langford et al., 2012). The growing knowledge of specific new contaminants has been summarized in a number of comprehensive review articles (Covaci et al., 2011; Sverko et al., 2011; Vorkamp and Rigét, 2014; Van Doorslaer et al., 2014; Wei et al., 2015). The term "emerging compounds" is not clearly defined. Some authors also consider the priority hazardous substances of the EU WFD (Table 4.1) as emerging contaminants, by contrast to legacy persistent organic pollutants traditionally included in environmental monitoring programs (e.g., Miège et al., 2012). Furthermore, some authors prefer the term "contaminants of emerging concern" to specify that these compounds are not newly introduced in the environment, but are receiving growing attention (e.g., Masoner et al., 2014). Analytical methods are often based on existing methods that are extended to accommodate these new contaminants (Covaci et al., 2011). This has the advantage of existing QA/QC methods that are transferred and extended as well, but might have the disadvantage of lacking specificity. Furthermore, new hydrophobic contaminants might require detection limits clearly below those of legacy contaminants (Vorkamp and Rigét, 2014), whereas new hydrophilic contaminants often require liquid rather than gas chromatographic instrumentation (Vorkamp et al., 2014). Perfluorinated alkylated substances do not fit into either of these categories and have also shown previously unfamiliar environmental behavior (Butt et al., 2010). In the following, some new hydrophobic and hydrophilic contaminant groups will be presented and a brief introduction to nanomaterials will be given.

4.5.1 Examples of Hydrophobic Contaminants

Without a clear definition of "emerging contaminants," broad approaches to this category include all nonlegacy organic contaminants (i.e., those not included in the Stockholm Convention on persistent organic pollutants). An example are the short-chained chlorinated paraffins, which have been on the market for decades, such as plasticizers, flame retardants, and general successor to PCBs (Stiehl et al., 2008). They are high-production-volume chemicals with an annual production of about 300,000 tons (Reth et al., 2006), candidates for the Stockholm Convention, on OSPAR's list of "Chemicals for Priority Action" and among the priority hazardous substances of the EU WFD (Table 4.1). Still, environmental information is surprisingly sparse, probably owing to the analytical challenges of these complex mixtures that cannot be separated by conventional gas chromatographic methods and for which no individual analytical standards are available (Bayen et al., 2006; Sverko et al., 2012). The most common analytical procedures include attempts to increase selectivity by removing potentially interfering compounds as part of the clean-up process and apply gas chromatography (GC) in combination with high-resolution mass spectrometry(MS), or low-resolution MS (LRMS) (Tomy et al., 1997; Reth and Oehme, 2004). Because of its higher selectivity, high-resolution MS is generally considered the superior technique, but LRMS is chosen increasingly for practical reasons (Sverko et al., 2012). An interlaboratory test gave coefficients of variations >200% for short-chained chlorinated paraffin determination, confirming the need for further method development and consolidation (Pellizzato et al., 2009). Alternative techniques are those of two-dimensional GC (Korýtar et al., 2005) and carbon-skeleton analysis (Hussy et al., 2012).

Another compound group of growing interest is that of novel brominated flame retardants (NBFRs) (Covaci et al., 2011). Following the ban of polybrominated diphenyl ethers (PBDEs), replacement products might have been introduced or nonregulated flame retardants might be used increasingly. Table 4.2 gives some examples of these NBFRs, but it should be noted that many other compounds exist as detailed by Bergman et al. (2012) and Law et al. (2013). The analytical methods are usually similar to those of PBDE determinations—i.e., using established extraction and clean-up methods and instrumental analysis by GC-LRMS with electron capture negative ionization, which usually allows low detection limits. However, the brominated phthalate bis(2-ethylhexyl)-tetrabromophthalate (Table 4.2) is not stable to the acid treatment commonly applied in the removal of lipids in biota samples (Ali et al., 2011; Vorkamp et al., 2015). Therefore, the clean-up of lipid-rich samples for the analysis of bis(2-ethylhexyl)-tetrabromophthalate usually includes gel permeation chromatography (Vorkamp et al., 2015). Along with the NBFR, dechlorane plus, a chlorinated flame retardant, has received increasing interest (Möller et al., 2010; Vorkamp et al., 2015).

4.5.2 Examples of Hydrophilic Contaminants

Being a research area of growing interest, pharmaceuticals and personal care products cover a large spectrum of different purpose drugs, such as antibiotics, anti-inflammatory and antipyretic analgesics and local anaesthetic drugs, antidepressants, antidiabetics, cardiovascular drugs, hormones, hypnotics, and many others (Chapter 16). Each pharmaceutical usually consists of one or several active pharmaceutical ingredients as well as a number of excipients and additives (Kümmerer, 2010). In addition, transformation of the parent compounds can occur inside and outside the human body, increasing the number of substances of potential environmental relevance. As outlined by Kümmerer (2010), besides their use to improve human health, drugs in the environment can also originate from veterinary use or from the unofficial transfer as growth promoters. Whereas their discharge to the environment is mainly related to wastewater, various other sources have been identified, including pharmaceutical production and use of sewage sludge as a fertilizer (Cooper et al., 2008). Although the compounds are not persistent in the environment, their continuous release into the environment might exceed their degradation, which has led to their characterization as "pseudo-persistent" environmental pollutants.

Table 4.2: Example of Novel Brominated Flame Retardants of Current Interest

Acronym	Full Name	$\log K_{OW}$	Structure
EH-TBB	2-Ethylhexyl-2,3,4,5-Tetrabromobenzoate	8.75	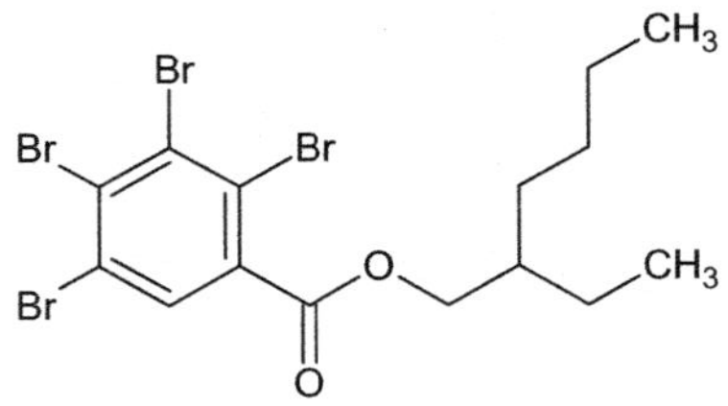
BEH-TBEP	Bis(2-ethylhexyl)-tetrabromophthalate	10.08; 11.95	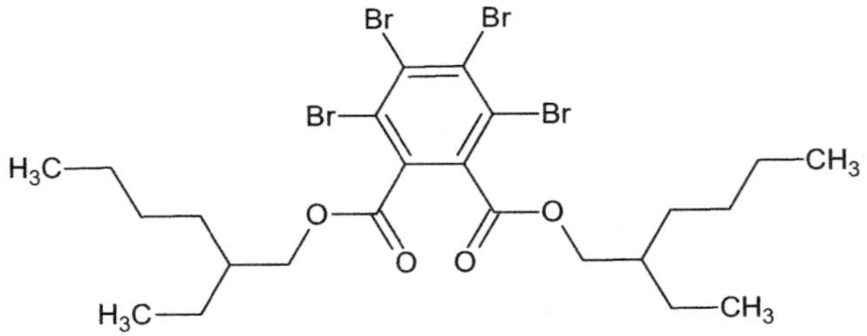
BTBPE	1,2-bis(2,4,6-tribromophenoxy)-ethane	7.88; 9.15	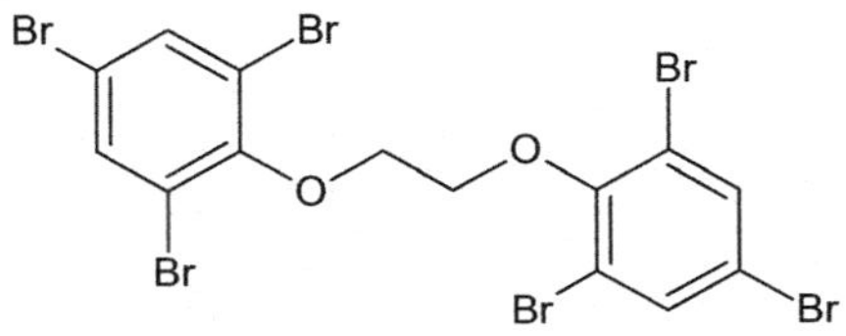
DBDPE	Decabromodiphenyl ethane	7–10; 11.1	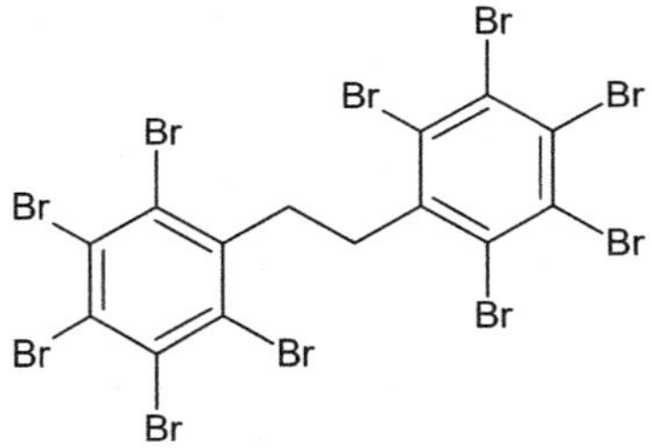
TBP-DBPE	2,3-Dibromopropyl-2,4,6-tribromophenyl ether	5.9; 6.34	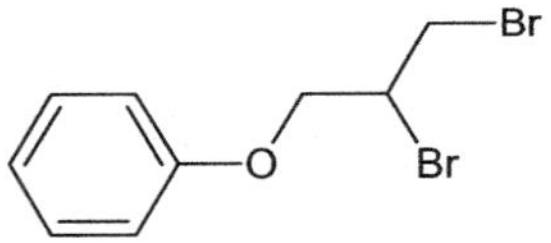

Acronyms according to Bergman et al. (2012); $\log K_{OW}$ values according to Vorkamp and Rigét (2014).

Also resulting from wastewater-related sources and being largely similar in their physicochemical characteristics, ingredients of personal care products are often addressed together with pharmaceuticals in environmental studies. These ingredients can include surfactants, bactericides, ultraviolet filters, and antioxidants, each of which consists of several individual chemicals (Huber et al., 2013). As with pharmaceuticals, many of these compounds are designed to exhibit a certain effect on organisms, which causes concern about their presence in the environment (Kolpin et al., 2002).

Analytical methods are versatile, taking into account differences in physicochemical properties of the target analytes, expected concentrations, and the sample medium (Kolpin et al., 2002). Based on water samples, these can be analyzed directly after appropriate preconcentration if no matrix interference occurs or subjected to extraction, for example using solid-phase extraction. Whereas liquid chromatography, often in combination with MS/MS, is the method of choice for direct analysis of water samples and many extracts, GC-MS has also been applied for the determination of some pharmaceuticals (Masoner et al., 2014). Recent publications on pharmaceuticals and personal care products include those of Thomaidi et al. (2015) and Zenobio et al. (2015).

4.5.3 Nanomaterials

Nanomaterials cover a large and diverse group of materials of high commercial significance. Their environmental detection and risk assessment have received growing interest, but procedures established for chemicals might not be meaningful or applicable for nanomaterials (Hansen et al., 2013, Chapter 17). Analytical methods commonly applied in chemical analysis might be adapted to some extent, but include the challenge that nanomaterials exist in colloidal systems and therefore need to be addressed in terms of chemical and physical form (Klaine et al., 2012; von der Kammer et al., 2012). Klaine et al. (2012) further detailed the challenges of analyses of complex environmental systems in terms of transferring environmental conditions to the laboratory, distinguishing from backgrounds and consequently, producing sufficiently low detection limits, recognizing nanomaterial alteration and, on top of all this, covering a large range of diverse materials. Because of the multitude of analytical challenges, predicted environmental concentrations and predicted no-effect concentrations, central parameters in risk characterization, are difficult to determine for nanomaterials (Quik et al., 2011). These authors suggest including nanomaterial specific processes in exposure assessments, whereas Hansen et al. (2013) discussed using information on novelty, persistence, ready dispersion, bioaccumulation, and potentially irreversible actions as early warning signs, following considerations of the European Environment Agency on the precautionary principle (EEA, 2001).

4.6 Conclusions

Originally based on determinations of total concentrations of metals or organic contaminants, developments within exposure analyses and assessments have moved toward more differentiated, refined, and, ultimately, relevant methods. At the same time, the analysis of emerging contaminants as well as the trace analysis of well-known contaminants provides new and increasingly precise and accurate environmental data, often in a combination of research and monitoring. Future work within exposure assessments will benefit from the collaboration of different schools and disciplines as well as researchers and regulators.

References

Adams, R.G., Lohmann, R., Fernandez, L.A., et al., 2007. Polyethylene devices: passive samplers for measuring dissolved hydrophobic organic compounds in aquatic environments. Environ. Sci. Technol. 41, 1317–1323.

Ali, N., Harrad, S., Muenhor, D., et al., 2011. Analytical characteristics and determination of major novel brominated flame retardants (NBFRs) in indoor dust. Anal. Bioanal. Chem. 400, 3073–3083.

Allan, I.J., Booij, K., Paschke, A., et al., 2009. Field performance of seven passive sampling devices for monitoring of hydrophobic substances. Environ. Sci. Technol. 43, 5383–5390.

Allan, I.J., Knutsson, J., Guigues, N., et al., 2007. Evaluation of the Chemcatcher and DGT passive samplers for monitoring metals with highly fluctuating water concentrations. J. Environ. Monit. 9, 672–681.

Allison, J.D., Brown, D.S., Novo-Gradac, K.J., 1991. MINTEQA2/PRODEFA2, a Geochemical Assessment Model for Environmental Systems, Version 3.0. User's Manual; U.S. Environmental Protection Agency, Washington, DC.

Alvarez, D.A., Petty, J.D., Huckins, J.N., et al., 2004. Development of a passive, in situ, integrative sampler for hydrophilic organic contaminants in aquatic environments. Environ. Toxicol. Chem. 23, 1640–1648.

Arthur, C.L., Pawliszyn, J., 1990. Solid phase microextraction with thermal desorption using fused silica optical fibers. Anal. Chem. 62, 2145–2148.

Bayen, S., Obbard, J.P., Thomas, G.O., 2006. Chlorinated paraffins: a review of analysis and environmental occurrence. Environ. Int. 32, 915–929.

Belles, A., Tapie, N., Pardon, P., et al., 2014. Development of the performance reference compound approach for the calibration of "polar organic chemical integrative sampler" (POCIS). Anal. Bioanal. Chem. 406, 1131–1140.

Bergman, Å., Rydén, A., Law, R.J., et al., 2012. A novel abbreviation standard for organobromine, organochlorine and organophosphorus flame retardants and some characteristics of the chemicals. Environ. Int. 49, 57–82.

Booij, K., Smedes, F., van Weerlee, E.M., 2002. Spiking of performance reference compounds in low density polyethylene and silicone passive water samplers. Chemosphere 46, 1157–1161.

Booij, K., van Bommel, R., van Aken, H.M., et al., 2014a. Passive sampling of nonpolar contaminants at three deep-ocean sites. Environ. Pollut. 195, 101–108.

Booij, P., Vethaak, D., Leonards, P.E.G., et al., 2014b. Identification of photosynthesis inhibitors of pelagic marine algae using 96-well plate microfractionation for enhanced throughput in effect-directed analysis. Environ. Sci. Technol. 48, 8003–8011.

Brack, W., 2003. Effect-directed analysis: a promising tool for the identification of organic toxicants in complex mixtures. Anal. Bioanal. Chem. 377, 397–407.

Brack, W., Altenburger, R., Schüürmann, G., et al., 2015. The SOLUTIONS project: challenges and responses for present and future emerging pollutants in land and water resources management. Sci. Total Environ. 503–504, 22–31.

Brack, W., Klamer, H.J.C., de Alda, M.L., et al., 2007. Effect-directed analysis of key toxicants in European River Basins—a review. Environ. Sci. Pollut. Res. 14, 30–38.

Brack, W., Schirmer, K., Erdinger, L., et al., 2005. Effect-directed analysis of mutagens and ethoxyresorufin-o-deethylase inducers in aquatic sediments. Environ. Toxicol. Chem. 24, 2445–2458.

Bruemmer, J., Falcon, R., Greenwood, R., et al., 2015. Measurement of cyclic volatice methylsiloxanes in the aquatic environment using low-density polyethylene passive sampling devices using and in-field calibration study—challenges and guidance. Chemosphere 122, 38–44.

Burgess, R.M., Ho, K.T., Brack, W., et al., 2013. Effects-directed analysis (EDA) and toxicity identification evaluation (TIE): complementary but different approaches for diagnosing causes of environmental toxicity. Environ. Toxicol. Chem. 32, 1935–1945.

Butt, C.M., Berger, U., Bossi, R., et al., 2010. Levels and trends of poly- and perfluorinated compounds in the arctic environment. Sci. Total Environ. 408, 2936–2965.
Campbell, P.G.C., 1995. Interactions between trace metals and aquatic organisms: a critique of the free-ion activity model in metal speciation and bioavailability in aquatic systems. In: Tessier, A., Turner, D.R. (Eds.), Trace Metal Speciation and Bioavailability in Aquatic Systems. John Wiley and Sons, Chichester, pp. 45–102.
Carls, M.G., Holland, L.G., Short, J.W., et al., 2004. Monitoring polynuclear aromatic hydrocarbons in aqueous environments with passive low-density polyethylene membrane devices. Environ. Toxicol. Chem. 23, 1416–1424.
Cooper, E.R., Siewicki, T.C., Phillips, K., 2008. Preliminary risk assessment database and risk ranking of pharmaceuticals in the environment. Sci. Total Environ. 398, 26–33.
Cornelissen, G., Pettersen, A., Broman, D., et al., 2008. Field testing of equilibrium passive samplers to determine freely dissolved native polycyclic aromatic hydrocarbon concentrations. Environ. Toxicol. Chem. 27, 499–508.
Covaci, A., Harrad, S., Abdallah, M.A.E., et al., 2011. Novel brominated flame retardants: a review of their analysis, environmental fate and behaviour. Environ. Int. 37, 532–556.
Davies, I.M., Vethaak, A.D., 2012. Integrated Marine Environmental Monitoring of Chemicals and Their Effects. ICES Cooperative Research Report No. 315, 277 pp. www.ices.dk.
Di Toro, D.M., Allen, H.E., Bergman, H., et al., 2001. Biotic ligand model of the acute toxicity of metals. 1 Technical basis. Environ. Toxicol. Chem. 20, 2383–2396.
EEA, 2001. Late Lessons from Early Warning: the Precautionary Principle 1896–2000. European Environmental Agency, Copenhagen, Denmark.
Emelogu, E.S., Pollard, P., Robinson, C.D., et al., 2013. Investigating the significance of dissolved organic contaminants in aquatic environments: coupling passive sampling with in vitro bioassays. Chemosphere 90, 210–219.
EU, 2000. Directive 2000/60/EC of the European Parliament and of the Council of 23 October 2000 establishing of framework for Community action in the field of water policy. Off. J. Eur. Communities L 327, 22.12.2000, pp. 1–73.
EU, 2009a. Commission Directive 2009/90/EC of 31 July 2009 laying down, pursuant to Directive 2000/60/EC of the European Parliament and of the Council, technical specifications for chemical analysis and monitoring of water status. Off. J. Eur. Communities L 201, 1.8.2009, pp. 36–38.
EU, 2009b. Guidance on Surface Water Chemical Monitoring under the Water Framework Directive. Guidance Document No. 19. Common Implementation Strategy for the Water Framework Directive (2000/60/EC). Technical Report 2009–025.
EU, 2010. Guidance Document No. 25 on Chemical Monitoring of Sediment and Biota under the Water Framework Directive. Common Implementation Strategy for the Water Framework Directive (2000/60/EC). Technical Report 2010–041.
EU, 2013. Directive 2013/39/EU of the European Parliament and of the Council of 12 August 2013 amending Directives 2000/60/EC and 2008/105/EC as regards priority substances in the field of water policy. Off. J. Eur. Communities L 226, 24.8.2013, pp. 1–17.
Fetter, E., Krauss, M., Brion, F., et al., 2014. Effect-directed analysis for estrogenic compounds in a fluvial sediment sample using transgenic *cyp19a1b*-GFP zebrafish embryos. Aquat. Toxicol. 154, 221–229.
Garmo, Ø.A., Røyset, O., Stelnnes, E., et al., 2003. Performance study of diffusive gradients in thin films for 55 elements. Anal. Chem. 75, 3573–3580.
Golding, C., Krassoi, R., Baker, E., 2006. The development and application of a marine toxicity identification evaluation (TIE) protocol for use with an Australian bivalve. Australas. J. Ecotoxicol. 12, 37–44.
Górecki, T., Namiesnik, J., 2002. Passive sampling. Tr. Anal. Chem. 21, 276–291.
Greenstein, D.J., Bay, S.M., Young, D.L., et al., 2014. The use of sediment toxicity identification evaluation methods to evaluate clean up targets in an urban estuary. Integr. Environ. Assess. Manag. 10, 260–268.
Hansen, S.F., Nielsen, K.N., Knudsen, N., et al., 2013. Operationalization and application of "early warning signs" to screen nanomaterials for harmful properties. Environ. Sci. Process. Impacts 15, 190–203.
Ho, K.T., Burgess, R.B., 2013. What's causing toxicity in sediments? Results of 20 years of toxicity identification and evaluations. Environ. Toxicol. Chem. 32, 2424–2432.
Hoh, E., Zhu, L., Hites, R.A., 2006. Dechlorane plus, a chlorinated flame retardant, in the Great Lakes. Environ. Sci. Technol. 40, 1184–1189.

Howard, P.H., Muir, D.C.G., 2010. Identifying new persistent and bioaccumulative organics among chemicals in commerce. Environ. Sci. Technol. 44, 2277–2285.

Huber, S., Remberger, M., Goetsch, A., et al., 2013. Pharmaceuticals and Additives in Personal Care Products as Environmental Pollutants—Faroe Islands, Iceland and Greenland. Nordic Council of Ministers. TemaNord 2013, 541.

Huckins, J.N., Petty, J.D., Lebo, J.A., et al., 2002. Development of the permeability/performance reference compound approach for in situ calibration of semipermeable membrane devices. Environ. Sci. Technol. 36, 85–91.

Huckins, J.N., Tubergen, M.W., Manuweera, G.K., 1990. Semipermeable membrane devices containing model lipid: a new approach to monitoring the bioavailability of lipophilic contaminants and estimating their bioconcentration potential. Chemosphere 20, 533–552.

Hussy, I., Webster, L., Russell, M., et al., 2012. Determination of chlorinated paraffins in sediments of the Firth of Clyde by gas chromatography with electron capture negative ionisation mass spectrometry and carbon skeleton analysis by gas chromatography with flame ionisation detection. Chemosphere 88, 292–299.

Ibrahim, I., Togola, A., Conzales, C., 2013. In-situ calibration of POCIS for the sampling of polar pesticides and metabolites in surface water. Talanta 116, 495–500.

ICES, January 29–31, 2013. Report of the Workshop on the Application of Passive Sampling and Passive Dosing to Contaminants in Marine Media (WKPSPD). www.ices.dk.

Jacquet, R., Miège, C., Smedes, F., et al., 2014. Comparison of five integrative samplers in laboratory for the monitoring of indicator and dioxin-like polychlorinated biphenyls in water. Chemosphere 98, 18–27.

Jahnke, A., Mayer, P., McLachlan, M., et al., 2014. Silicone passive equilibrium samplers as "chemometers" in eels and sediments of a Swedish lake. Environ. Sci. Process. Impacts 16, 464–472.

Kaserzon, S.L., Hawker, D.W., Kennedy, K., et al., 2014. Characterisation and comparison of the uptake of ionizable and polar pesticides, pharmaceuticals and personal care products by POCIS and Chemcatchers. Environ. Sci. Process. Impacts 16, 2517–2526.

Kim, U.J., Kim, H.Z., Alvarez, D., et al., 2014. Using SPMDs for monitoring hydrophobic organic compounds in urban river water in Korea compared with using conventional water grab samples. Sci. Total Environ. 470–471, 1537–1544.

Kingston, J.K., Greenwood, R., Mills, G.A., et al., 2000. Development of a novel passive sampling system for the time-averaged measurement of a range of organic pollutants in aquatic environments. J. Environ. Monit. 2, 487–495.

Klaine, S.J., Koelmans, A.A., Horne, N., et al., 2012. Paradigms to assess the environmental impact of manufactured nanomaterials. Environ. Toxicol. Chem. 31, 3–14.

Kolpin, D.A., Furlong, E.T., Meyer, M.T., 2002. Pharmaceuticals, hormones, and other organic wastewater contaminants in U.S. streams, 1999–2000: a national reconnaissance. Environ. Sci. Technol. 36, 1202–1211.

Korýtar, P., Parera, J., Leonards, P.E.G., 2005. Quadrupole mass spectrometer operating in the electron-capture negative ion mode as detector for comprehensive two-dimensional gas chromatography. J. Chromatogr. A. 1067, 255–264.

Kümmerer, K., 2010. Pharmaceuticals in the environment. Annu. Rev. Environ. Resour. 35, 57–75.

von der Kammer, F., Ferguson, P.L., Holden, P.A., et al., 2012. Analysis of engineered nanomaterials in complex matrices (environment and biota): general considerations and conceptual case studies. Environ. Toxicol. Chem. 31, 32–49.

Langford, K.H., Beylich, B.A., Bæk, K., et al., 2012. Screening of Selected Alkylphenolic Compounds, Biocidees, Rodenticides and Current Use Pesticides. NIVA-report 6343/2012, SPFO-report: 1116/2012, TA-2899/2012.

Lava, R., Majoros, L.I., Dosis, I., et al., 2014. A practical example of the challenges of biota monitoring under the Water Framework Directive. Tr. Anal. Chem. 59, 103–111.

Law, R.J., Losada, S., Barber, J.L., et al., 2013. Alternative flame retardants, dechlorane plus and BDEs in the blubber of harbor porpoises (*Phocoena phocoena*) stranded or bycaught in the UK during 2008. Environ. Int. 60, 81–88.

Lissalde, S., Mazzella, N., Mazellier, P., 2014. Polar organic chemical integrative samplers for pesticides monitoring: impacts of field exposure conditions. Sci. Total Environ. 488–489, 188–196.

van Leeuwen, H.P., Town, R.M., Buffle, J., et al., 2005. Dynamic speciation analysis and bioavailability of metals in aquatic systems. Environ. Sci. Technol. 39, 8546–8556.

MacRae, R.K., Smith, D.E., Swoboda-Colberg, N., et al., 1999. Copper binding affinity of rainbow trout (*Oncorhynchus mykiss*) and brook trout (*Salvelinus fontinalis*) gills: Implications for assessing bioavailable metal. Environ. Toxicol. Chem. 18, 1180–1189.

Masoner, J.R., Kolpin, D.W., Furlong, E.T., 2014. Contaminants of emerging concern in fresh leachate from landfills in the conterminous United States. Environ. Sci. Process. Impacts 16, 2335–2354.

Matos, M.F., Botta, C.M.R., Fonseca, A.L., 2014. Toxicity identification evaluation (phase I) of water and sediment samples from a tropical reservoir contaminated with industrial and domestic effluents. Environ. Monit. Assess. 186, 7999–8006.

Mayer, P., Holmstrup, M., 2008. Passive dosing of soil invertebrates with polycyclic aromatic hydrocarons: limited chemical activity explains toxicity cutoff. Environ. Sci. Technol. 42, 7516–7521.

Mayer, P., Tolls, J., Hermens, J.L.M., et al., 2003. Equilibrium sampling devices. Environ. Sci. Technol. 185A–191A.

Mayer, P., Wernsing, J., Tolls, J., 1999. Establishing and controlling dissolved concentrations of hydrophobic organics by partitioning from a solid phase. Environ. Sci. Technol. 33, 2284–2290.

Mazzella, N., Lissalde, S., Moreira, S., et al., 2010. Evaluation of the use of performance reference compounds in an Oasis-HLB adsorbent based passive sampler for improving water concentration estimates of polar herbicides in freshwater. Environ. Sci. Technol. 44, 1713–1719.

McLachlan, M.S., Kierkegaard, A., Radke, M., et al., 2014. Using model-based screening to help discover unknown environmental contaminants. Environ. Sci. Technol. 48, 7264–7271.

Miège, C., Peretti, A., Labadie, P., et al., 2012. Occurrence of priority and emerging organic compounds in fishes from the Rhone River (France). Anal. Bioanal. Chem. 404, 2721–2735.

Mills, G.A., Fones, G.R., Booij, K., et al., 2011. Passive sampling technologies. In: Quevauviller, P., Roose, P., Verreet, G. (Eds.), Chemical Marine Monitoring: Policy Framework and Analytical Trends. John Wiley & Sons, Ltd, pp. 397–432.

Mills, G.A., Gravell, A., Vrana, B., et al., 2014. Measurement of environmental pollutants using passive sampling devices—and updated commentary on the current state of the art. Environ. Sci. Process. Impacts 16, 369–373.

Morel, F.M.M., 1983. Principles of Aquatic Chemistry. Wiley Interscience, New York.

Morgan, I.J., Henry, R.P., Wood, C.M., 1997. The mechanism of acute silver nitrate toxicity in freshwater rainbow trout (*Oncorhynchus mykiss*) is inhibition of gill Na^+ and Cl^- transport. Aquat. Toxicol. 8, 145–163.

Morin, N., Miège, C., Randon, J., et al., 2012. Chemical calibration, performance, validation and applications of the polar organic chemical integrative sampler (POCIS) in aquatic environments. Tr. Anal. Chem. 36, 144–175.

Möller, A., Xie, Z., Sturm, R., et al., 2010. Large-scale distribution of dechlorane plus in air and seawater from the Arctic to Antarctica. Environ. Sci. Technol. 44, 8977–8982.

Niyogi, S., Wood, C.M., 2004. Biotic Ligand Model, a flexible tool for developing sitespecific water quality guidelines for metals. Environ. Sci. Technol. 38, 6177–6192.

OSPAR, 2009. Background Document on CEMP Assessment Criteria for QSR 2010. Monitoring and Assessment Series 461/2009. ISBN: 978-1-907390-08-1.

OSPAR, 2012a. JAMP Guidelines for Monitoring Contaminants in Biota. Agreement 1999-2 www.ospar.org.

OSPAR, 2012b. JAMP Guidelines for Monitoring Contaminants in Sediment. Agreement 2002-16 www.ospar.org.

OSPAR, 2013. JAMP Guidelines for Monitoring Contaminants in Seawater. Agreement 2013-03 www.ospar.org.

Pagenkopf, G.K., 1983. Gill surface interaction model for trace-metal toxicity to fishes: role of complexation, pH, and water chemistry. Environ. Sci. Technol. 17, 342–346.

Paquin, P.R., Gorsuch, J.W., Apte, S., et al., 2002. The biotic ligand model: a historic overview. Comp. Biochem. Physiol. 133C, 3–35.

Pellizzato, F., Ricci, M., Held, A., et al., 2009. Laboratory intercomparison study on the analysis of short-chain chlorinated paraffins in an extract of industrial soil. Tr. Anal. Chem. 28, 1029–1035.

Petty, J.D., Orazio, C.E., Huckins, J.N., et al., 2000. Considerations involved with the use of semipermeable membrane devices for monitoring environmental contaminants. J. Chromatogr. A 879, 83–95.

Playle, R.C., Dixon, D.G., Burnison, K., 1993a. Copper and cadmium-binding to fish gills—modification by dissolved organic-carbon and synthetic ligands. Can. J. Fish. Aquat. Sci. 50, 2667–2677.
Playle, R.C., Dixon, D.G., Burnison, K., 1993b. Copper and cadmium-binding to fish gills—estimates of metal gill stability-constants and modeling of metal accumulation. Can. J. Fish. Aquat. Sci. 50, 2678–2687.
Playle, R.C., Gensemer, R.W., Dixon, D.G., 1992. Copper accumulation on gills of fathead minnows—influence of water hardness, complexation and pH of the gill microenvironment. Environ. Toxicol. Chem. 11, 381–391.
Prokes, R., Vrana, B., Klánová, J., 2012. Levels and distribution of dissolved hydrophobic organic contaminants in the Morava river in Zlin district, Czech Republic as derived from their accumulation in silicone rubber passive samplers. Environ. Pollut. 166, 157–166.
Quik, J.T.K., Vonk, J.A., Hansen, S.F., et al., 2011. How to assess exposure of aquatic organisms to manufactured nanoparticles? Environ. Int. 37, 1068–1077.
Reichenberg, F., Mayer, P., 2006. Two complementary sides of bioavailability: accessibility and chemical activity of organic contaminants in sediments and soils. Environ. Toxicol. Chem. 25, 1239–1245.
Reth, M., Ciric, A., Christensen, G.N., et al., 2006. Short- and medium-chain chlorinated paraffins in biota from the European Arctic—differences in homologue group patterns. Sci. Total Environ. 367, 252–260.
Reth, M., Oehme, M., 2004. Limitations of low resolution mass spectrometry in the electron capture negative ionization mode for the analysis of short- and medium-chain chlorinated paraffins. Anal. Bioanal. Chem. 378, 1741–1747.
Rusina, T.P., Smedes, F., Klanova, J., et al., 2007. Polymer selection for passive sampling: a comparison of critical properties. Chemosphere 68, 1344–1351.
Santore, R.C., Driscoll, C., 1995. The CHESS Model for Calculating Chemical Equilibria in Soils and Solutions. Chemical Equilibrium and Reaction Models. Soil Science Society of America, American Society of Agronomy, Madison, WI.
Schecher, W.D., McAvoy, D.C., 2001. MINEQL+, Version 4.5. Environmental Research Software, Hallowell, ME.
Schmidt, S.N., Holmstrup, M., Smith, K.E.C., et al., 2013. Passive dosing of polycyclic aromatic hydrocarbon (PAH) mixtures to terrestrial springtails: linking mixture toxicity to chemical activities, equilibrium lipid concentrations, and toxic units. Environ. Sci. Technol. 47, 7020–7027.
Seethapathy, S., Górecki, T., Li, X., 2008. Passive sampling in environmental analysis. J. Chromatogr. A 1184, 234–253.
Shaw, M., Eaglesham, G., Mueller, J.F., 2009. Uptake and release of polar compounds in SDB-RPS Empore™ disks: implications for their use as passive samplers. Chemosphere 75, 1–7.
Smedes, F., Bakker, D., de Weert, J., 2010. The Use of Passive Sampling in WFD Monitoring. Deltares report 1202337-004-BCS-0027, 59 pp.
Smedes, F., Booij, K., 2012. Guidelines for Passive Sampling of Hydrophobic Contaminants in Water Using Silicone Rubber Samplers. ICES Techniques in Marine Environmental Sciences No. 52.
Smedes, F., Geertsma, R.W., van der Zande, T., et al., 2009. Polymer-water partition coefficients of hydrophobic compounds for passive sampling: application of cosolvent models for validation. Environ. Sci. Technol. 43, 7047–7054.
Smith, K.E.C., Dom, N., Blust, R., et al., 2010. Controlling and maintaining exposure of hydrophobic organic compounds in aquatic toxicity tests by passive dosing. Aquat. Toxicol. 98, 15–24.
Stiehl, T., Pfordt, J., Ende, M., 2008. Globale Destillation. I. Evaluierung von Schadsubstanzen aufgrund ihrer Persistenz, ihres Bioakkumulationspotentials und ihrer Toxizität im Hinblick auf ihren potentiellen Eintrag in das arktische Ökosystem. J. Consumer Prot. Food Saf. 3, 61–81.
Sunda, W.G., Engel, D.W., Thuotte, R.M., 1978. Effect of chemical speciation on toxicity of cadmium to grass shrimp, *Palaemonetes pugio*: importance of free cadmium ion. Environ. Sci. Technol. 12, 409–413.
Sverko, E., Tomy, G.T., Märvin, C.H., et al., 2012. Improving the quality of environmental measurements on short chain chlorinated paraffins to support global regulatory efforts. Environ. Sci. Technol. 46, 4697–4698.
Sverko, E., Tomy, G.T., Reiner, E.J., et al., 2011. Dechlorane plus and related compounds in the environment: a review. Environ. Sci. Technol. 45, 5088–5098.
Södergren, A., 1987. Solvent-filled dialysis membranes simulate uptake of pollutants by aquatic organisms. Environ. Sci. Technol. 21, 855–859.

Thomaidi, V.S., Stasinakis, A.S., Borova, V.L., et al., 2015. Is there a risk for the aquatic environment due to the existence of emerging organic contaminants in treated domestic wastewater? Greece as a case-study. J. Hazard. Mater. 283, 740–747.

Tipping, E., 1994. WHAM A computer equilibrium model and computer code for waters, sediments, and soils incorporating a discrete site/electrostatic model of ion-binding by humic substances. Comput. Geosci. 20, 973–1023.

Tipping, E., Lofts, S., Sonke, J.E., 2011. Humic Ion-Binding Model VII: a revised parameterisation of cation-binding by humic substances. Environ. Chem. 8, 225–235.

Tomy, G.T., Stern, G.A., Muir, D.C.G., et al., 1997. Quantifying C10-C13 polychloroalkanes in environmental samples by high-resolution gas chromatography/electron capture negative ion high-resolution mass spectrometry. Anal. Chem. 69, 2762–2771.

Turner, G.S.C., Mills, G.A., Burnett, J.L., et al., 2015. Evaluation of diffusive gradients in thin-films using a Diphonix® resin for monitoring dissolved uranium in natural waters. Anal. Chim. Acta 854, 78–85.

USEPA, 1991. Methods for Aquatic Toxicity Identification Evaluations. Phase I Toxicity Characterization Procedures, second ed. EPA/600/6–91/003.

USEPA, 1996. Marine toxicity Identification Evaluation (TIE). Phase I Guidance Document. EPA/600/R-96/054.

USEPA, 2007. Sediment Toxicity Identification Evaluation (TIE). Phases I, II, and III Guidance Document. EPA/600/R-07/060.

Van Doorslaer, X., Dewulf, J., van Langenhove, H., et al., 2014. Fluoroquinolone antibiotics: an emerging class of environmental micropollutants. Sci. Total Environ. 500–501, 250–269.

Vermeirssen, E.L.M., Dietschweiler, C., Escher, B.I., 2012. Transfer kinetics of polar organic compounds over polyethersulfone membranes in the passive samplers POCIS and Chemcatcher. Environ. Sci. Technol. 46, 6759–6766.

Vermeirssen, E.L.M., Dietschweiler, C., Escher, B.I., et al., 2013. Uptake and release kinetics of 22 polar organic chemicals in the Chemcatcher passive sampler. Anal. Bioanal. Chem. 405, 5225–5236.

Vorkamp, K., Bossi, R., Bester, K., et al., 2014. New priority substances of the European Water Framework Directive: biocides, pesticides and brominated flame retardants in the aquatic environment of Denmark. Sci. Total Environ. 470–471, 459–468.

Vorkamp, K., Bossi, R., Rigét, F.F., et al., 2015. Novel brominated flame retardants and dechlorane plus in Greenland air and biota. Environ. Pollut. 196, 284–291.

Vorkamp, K., Rigét, F.F., 2014. A review of new and current-use contaminants in the Arctic environment: evidence of long-range transport and indications of bioaccumulation. Chemosphere 111, 379–395.

Vrana, B., Mills, G.A., Allan, I.J., et al., 2005. Passive sampling techniques for monitoring pollutants in water. Tr. Anal. Chem. 24, 845–868.

Wang, W.-X., 2014. Prediction of metal toxicity in aquatic organisms. Chin. Sci. Bull. 58, 194–202.

Wei, G.L., Li, D.Q., Zhuo, M.N., 2015. Organophosphorus flame retardants and plasticizers: sources, occurrence, toxicity and human exposure. Environ. Pollut. 196, 29–46.

Yates, K., Davies, I., Webster, L., et al., 2007. Passive sampling: partition coefficients for a silicone rubber reference phase. J. Environ. Monit. 9, 1116–1121.

Zenobio, J.E., Sanchez, B.C., Leet, J.K., et al., 2015. Presence and effects of pharmaceutical and personal care products on the Baca National Wildlife Refuge, Colorado. Chemosphere 120, 750–755.

Zhang, C., Ding, S., Xu, D., et al., 2014. Bioavailability assessment of phosphorous and metals in soils and sediments: a review of diffusive gradients in thin films (DGT). Environ. Monit. Assess. 186, 7367–7378.

Zhang, H., Davison, W., Miller, S., et al., 1995. In situ high resolution measurements of fluxes of Ni, Cu, Fe and Mn and concentrations of Zn and Cd in porewaters by DGT. Geochim. Cosmochim. Acta 59, 4181–4192.

Zhao, F.J., Rooney, C.P., Zhang, H., et al., 2006. Comparison of soil solution speciation and diffusive gradients in thin-films measurements as an indicator of copper bioavailability to plants. Environ. Toxicol. Chem. 25, 733–742.

Zielke, H., Seiler, T.-B., Niebergall, S., et al., 2011. The impacts of extraction methodologies on the toxicity of sediments in the zebrafish (*Danio rerio*) embryo test. J. Soils Sediments 11, 352–363.

CHAPTER 5

From Incorporation to Toxicity

Jean-Claude Amiard, Claude Amiard-Triquet

Abstract

Many different mechanisms interfere with environmental contaminants before they can reach their targets (biological membranes, specific enzymes and receptors, DNA, RNA). The characteristics of chemicals that govern their bioaccessibility and their ability to reach the systemic circulation of receptor organisms (bioavailability) and their subcellular sites of action are reviewed. The physico-chemical characteristics of contaminants are key parameters controlling both their bioaccumulation and fate within the body/organ/cell. Strategies to assess bioavailability are discussed, among which passive sampling techniques, partial extraction methods, and in vitro digestion appear as the most promising tools. Using the tissue residue approach for toxicity assessment allows less dependence on confounding factors influencing exposure and bioavailability. This methodology is well-developed for organics and organometallics, whereas its application is much more complicated for metals. An important application of tissue residue approach relates to evaluating the potential hazards of bioaccumulative chemicals in field biota that are measured in biomonitoring programs worldwide.

Keywords: Bioaccessibility; Bioavailability; Equilibrium partitioning models; In vitro digestion; Passive sampling techniques; Sediment resuspension; Tissue quality guidelines; Tissue residue approach.

Chapter Outline

Aquatic Ecotoxicology. http://dx.doi.org/10.1016/B978-0-12-800949-9.00005-X

Aquatic organisms are exposed to a large number of chemical contaminants in the water column, sediments, and their food preys. These contaminants can cross the biological membranes and interfere with biological macromolecules, leading to deleterious effects. However, a series of events can happen between the initial exposure and the final effect, often depending on the physicochemical characteristics of contaminants and their distribution between the different compartments of the environment (water, sediment, biota). The total dose of a given pollutant in any of these compartments has a poor ecotoxicological significance. Many examples exist of discrepancies between the levels of contaminants in the medium and the quantities incorporated in organisms. In a number of cases, this situation results from the lack of bioavailability of environmental contaminants and it may be explained by their physicochemical characteristics. At the next step, similar discrepancies may be observed between high levels of incorporated contaminants and the lack of severe impacts, at least in certain biota. The explanation is the ability of living organisms to cope with contaminants through different processes (Amiard-Triquet et al., 2011). In the case of metals, detoxification occurs mainly through biomineralization or binding to specific detoxificatory proteins (Amiard et al., 2006; Luoma and Rainbow, 2008), whereas for organic contaminants the major processes include biotransformation favoring excretion. These processes govern the distribution of contaminants in organisms between different organs and at the subcellular level. The physicochemical characteristics of contaminants within prey species strongly influence the subsequent transfer to predators (Amiard-Triquet and Rainbow, in Amiard-Triquet et al., 2011).

In this chapter, we will first examine how the physicochemical characteristics of contaminants in sediment and food influence their uptake in organisms and the extrinsic and intrinsic factors that can modify bioavailability. Then we will illustrate that the global concentrations in sources is far from being always significantly correlated with toxic effects in organisms. To offset this difficulty for risk assessment, we will explore strategies developed to link toxicity to incorporated doses of contaminants rather than external doses. This includes the critical body residue approach and the tissue residue approach (TRA) that have been reviewed in a special issue of *Integrated Environmental Assessment and Management* (2011) after a Pellston Workshop under the auspices of the Society of Environmental Toxicity and Chemistry.

5.1 From Bioaccessibility to Bioavailability

Bioavailability is a term generally applied by environmental scientists (such as many of those quoted in the present chapter) to describe how contaminants are released from the sediment matrix, transported, and taken up by a target organism. However, it is somewhat confusing because it covers processes that take place both outside and inside the organism, the digestive content itself being clearly "outside" from a physiological point of view (Figure 5.1). More clearly, the release of a chemical from the solid to the dissolved phase is termed bioaccessibility

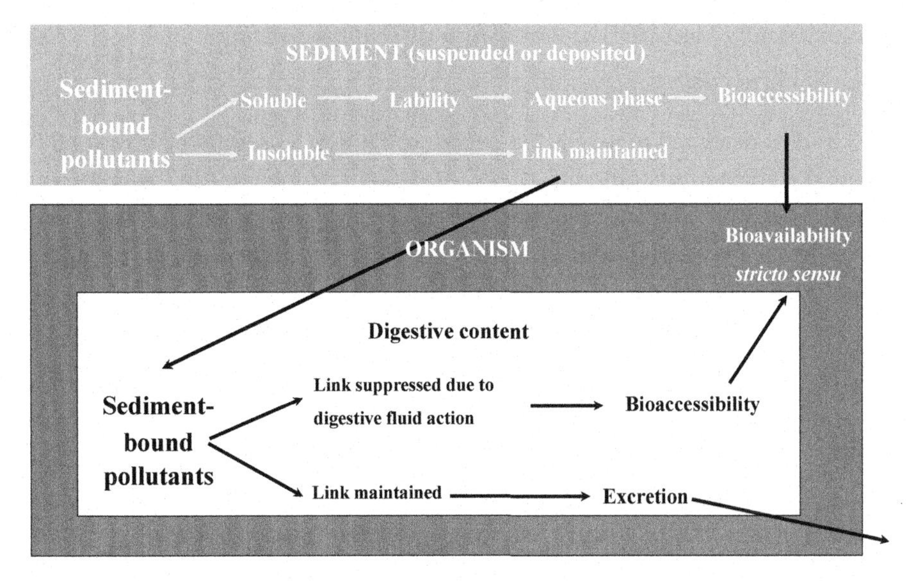

Figure 5.1
Uptake of contaminants from sediments: bioaccessibility and bioavailability.

(ITRC, 2011). It is a prerequisite for uptake and assimilation into the systemic circulation of receptor organisms, which are termed bioavailability stricto sensu (ITRC, 2011).

These concepts are also applicable when prey species constitute the source of contaminants (Versantvoort et al., 2005). Once ingested, food (as with sediments, particularly for endogenous organisms) is submitted to digestion processes (enzymes, pH) that induce the release of a fraction of food-bound contaminants. The maximum quantity of soluble contaminants in the digestive tract constitutes the bioaccessible fraction. A part of these bioaccessible contaminants will cross the gut wall, contaminating the biological fluids, and finally reaching the site(s) of toxic action: this constitutes the bioavailable fraction (Lanno et al., 2004).

For both sediment and food matrices, bioavailability is mainly dependent upon the physicochemical characteristics of contaminants at the site of absorption. Besides chemical processes (speciation, complexation), attention should also be given to physiological aspects (transport sites, biotic ligands) for predicting toxicity of chemicals (Janssen et al., 2003).

5.1.1 Bioavailability from Sediment

The importance of sediment as a major source of chemical contaminants for endobenthic species, including deposit- and filter-feeders, is well-documented in Chapter 10. In addition, the fish feeding on the intra-sediment fauna ingest significant quantities of sediment at the same time (up to 65% of the digestive contents for flounder *Platichthys flesus*, 55% for sole *Solea solea* in the Loire Estuary, according to Marchand (1978), in Amiard (1992)).

5.1.1.1 Methods Used to Measure Bioaccessibility and/or Bioavailability

Different procedures have been proposed:

- chemical studies designed to assess the degree of lability of sediment-bound pollutants;
- experimental studies mimicking the effects of digestive processes in the gut tract on the fate of particle-bound contaminants;
- direct assessment of bioavailability based on the determination of bioaccumulation from sediments (see Chapter 10);
- indirect assessment through effects observed in exposed organisms or communities.

Extractability can be measured using chemical extractants such as those recommended for metal partitioning among geochemical fractions (Amiard et al., 2007; literature cited therein; Community Bureau of Reference sequential extraction scheme BCR EUR 1476EN) or those tested to develop surrogates to estimate the bioavailability of polychlorinated biphenyls (PCBs) (e.g., You et al., 2007). Other examples of extraction methods are provided in the review by Hunter et al. (2011). Biological extractants such as digestive fluids of endobenthic invertebrates (polychaetes, sipunculans) and fish have been also used for the extraction of trace elements, polyaromatic hydrocarbons (PAHs), or PCBs (Mayer et al., 1996, 2001; Weston and Maruya, 2002; Goto and Wallace, 2009; Baumann et al., 2012; Wang et al., 2012). Artificial gut fluids consisting of chemical extractants able to recreate conditions relatively similar to those encountered in guts of invertebrates or fish were also used, such as acid solutions at physiological pH levels for metal extraction (Amiard et al., 1995, 2007; Ettajani et al., 1996; Peña-Icart et al., 2014) or sodium dodecyl sulfate for PAHs (Ahrens et al., 2005).

Solid-phase samplers work according to the principle of equilibrium partitioning theory (Di Toro et al., 1991; Hansen et al., 2005). Examples of equilibrium samplers include solid-phase microextraction (SPME), dialysis and polyethylene membranes, and polyoxymethylene (Hunter et al., 2011 and literature cited therein). The use of passive samplers has been developed and include polyoxymethylene for PCBs (Janssen et al., 2011), passive samplers with low-density polyethylene as the sorbent phase for degradation products of the organochlorine pesticide DDT (Liu et al., 2013), polyethylene samplers for polychlorinated dibenzo-*p*-dioxin/dibenzofuran (PCDD/F) concentrations (Friedman and Lohmann, 2014) or diffusive gradients in thin films for metals (Yin et al., 2014). The manipulation of contaminated sediments with Tenax resin and matrix solid-phase microextraction provide direct evidence of a causal relationship between sequestration and bioavailability of sorbed organic contaminants (PAHs, PCBs, permethrin) to deposit feeders evaluated by using the freshwater oligochaete tests (Kraaij et al., 2002; You et al., 2007; Mackenbach et al., 2014), the determination of BSAFs (biota to sediment accumulation factors) in *Chironomus tentans* larvae (Cui et al., 2010) or the determination of biolipid PCB concentrations in the marine and estuarine worm *Neanthes arenaceodentata* and *Leptocheirus plumulosus* (Werner et al., 2010). Biomimetic extractions (seawater or Tenax extractions, solubilization with the anionic surfactant sodium dodecyl sulfate) over several weeks consistently underestimated PAH bioaccumulation in bivalve samples (the deposit-feeder *Macomona liliana* and the filter-feeder *Austrovenus stutchburyi*), whereas PAH bioavailability was considerably lower than predictions based on equilibrium partitioning theory (Ahrens et al., 2005).

Sediment equilibrium partitioning models and passive sampling techniques have been used to estimate PCDD/F concentrations in the deposit-feeding worm *Pectorina gouldii* and the filter-feeding clam *Mya arenaria* (Friedman and Lohmann, 2014). Water partitioning models failed to estimate correctly biota concentrations, whereas polyethylene-based porewater samplers provided the best estimates for biota, particularly for deposit feeders (within 1.6x). Similarly, for metals, Yin et al. (2014) compared passive samplers (diffusive gradients in thin films) and conventional methods (including simultaneously extracted metals–acid volatile sulfide models, Community Bureau of Reference sequential extraction, and total metal concentrations) validated the use of diffusive gradients in thin films to predict metal accumulation in snails *Bellamya aeruginosa* from freshwater lake sediments. More generally, Akkanen et al. (2012), who have reviewed passive samplers and their applicability in sediments for organic contaminants, concluded that these methods have proven to be feasible for bioavailability estimations instead of more laborious methods.

Indirect determination of bioavailability through the assessment of toxicological parameters (median lethal concentrations [LC_{50}], median effect concentrations) has been proposed for pyrethroid insecticides (Xu et al., 2007; Harwood et al., 2013; Li et al., 2013). Expressing the LC_{50}s for *C. tentans* in different sediments with regard to insecticide concentrations on a bulk sediment basis, organic carbon–normalized sediment basis, porewater basis, dissolved organic carbon–normalized porewater basis, and freely dissolved porewater basis has shown that the latter was the most representative of the bioaccessible fraction (Xu et al., 2007). SPME fibers and Tenax beads allow a more accurate assessment of toxicity across sediment types than those based on total sediment concentrations of pyrethroid insecticides (Harwood et al., 2013; Li et al., 2013). Similarly, free aqueous PCB concentrations or surrogate measures such as the PCB concentration assimilated by passive samplers (for example, single-phase polyoxymethylene or SPME), or the Tenax extractable PCB concentration, correlate better with biolipid concentrations than relationships based only on total organic carbon normalized solid-phase concentrations (Werner et al., 2010).

The assessment of bioavailability stricto sensu may be based considering biological responses in terms of bioaccumulation from sediments (in endobenthic organisms, see Chapter 10; or in fish models, e.g., Gaillard et al., 2014) or effects observed in exposed organisms or communities. Sediment toxicity tests that have been widely developed (Chapters 6 and 10) provides an indirect evidence of bioavailability as exemplified in Figure 5.2. Oyster larvae were exposed to elutriates of sediments collected monthly from the Gironde estuary (France), which is

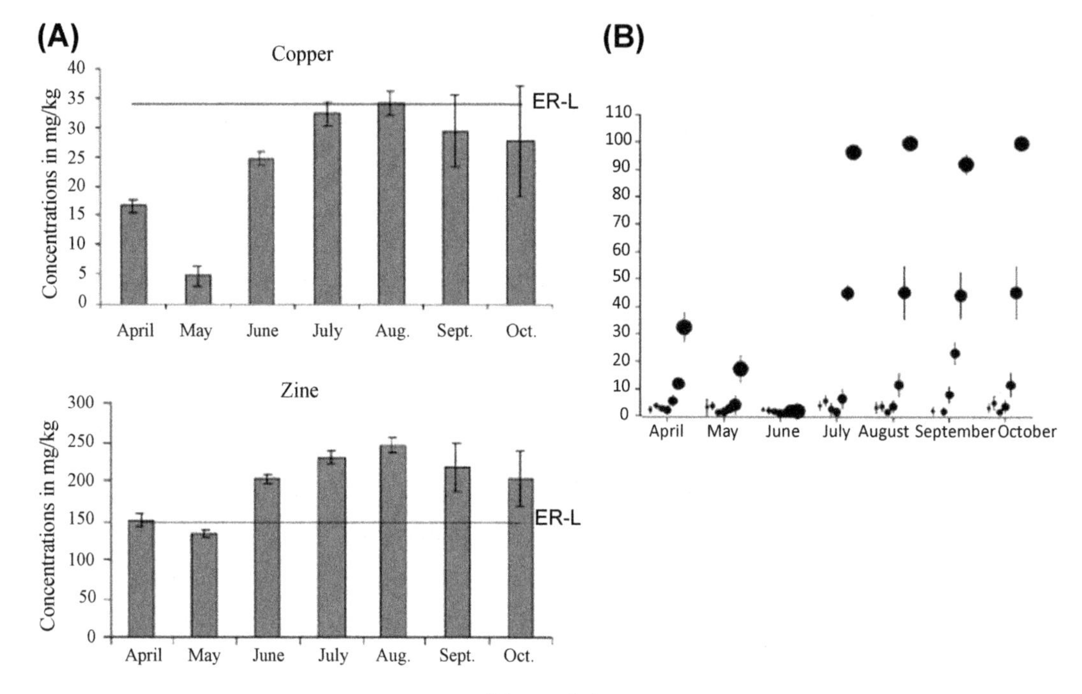

Figure 5.2

Response of oyster larvae to sediment contamination in the Gironde estuary, France. *(Modified after Geffard et al. (2005).)* (A) Metal concentrations in sediments; ER-L: effect range-low value as defined by Long et al. (1995). (B) Oysters: percent of abnormal larvae after exposure to sediment elutriate (0%, 1%, 5%, 10%, 25%, 50%, and 100%); symbols increase in size with increasing concentrations.

highly contaminated by metals. In several samples, copper and zinc concentrations were nearing or even higher than the effect range low values (Long et al., 1998), which are the concentrations above which biological effects are expected. Such high values were observed particularly from June to October, corresponding to the highest percentages of larval abnormalities revealed by ecotoxicity tests.

At the community level, Lindgren et al. (2014) have shown that PAH effects strongly depend on contaminant bioavailability. They underline the implications for risk analysis studies of petroleum-contaminated marine sediments, stating that sediment characteristics and their effects on bioavailability are important to include.

5.1.1.2 Factors Affecting Organic Contaminant Bioavailability

Physicochemical properties of the contaminant, such as hydrophobicity, aromaticity, size (e.g., aliphatic compounds, Meggo and Schnoor, 2011), and planarity are important factors controlling the partitioning and bioavailability of hydrophobic organic contaminants (HOCs) (Hunter et al., 2011). The route of uptake into an organism can influence the bioavailability of the HOC. It is generally thought that for a HOC to pass through the cellular membrane, it must first be freely dissolved in the aqueous phase, regardless of whether the HOC is in the gut or in the environmental matrix in contact with the organism (Hunter et al., 2011).

Biotic and abiotic factors affecting HOC bioavailability have been recently reviewed (Hunter et al., 2011; Meggo and Schnoor, 2011). The role of organic carbon content is underscored as the main property controlling the bioavailability of HOCs in soils and sediments. Unless the organic contaminant content is extremely low (<0.2%), adsorption is generally insignificant. Many studies have shown a correlation between the rapid desorption HOC pool and bioavailability to benthic organisms. However, Cui et al. (2010) have shown that the slow desorption pool of phenanthrene or permethrin was readily available to *C. tentans*, with BSAF (slow) values ranging from 25.3% to 73.9% of BSAF (rapid). Therefore, they conclude that the kinetically slow desorption fraction should not be ignored in sediment toxicity assessment. As contaminants persist in soils or sediments, they become increasingly recalcitrant to desorption to the aqueous phase, a phenomenon generally termed sequestration. Because the resistant fraction (slowly desorbing fraction) increases with time, the process is termed aging or weathering (Ruus et al., 2005; Hunter et al., 2011). Minor differences in the quality of relatively simple organic molecules (extracellular polymeric and humic substances) can affect contaminant behavior in ecotoxicological studies, as shown in the case of chlorpyrifos bioavailability determined by measuring bioaccumulation in *Chironomus riparius* larvae (Lundqvist et al., 2010). Because of their strong sorption capacities, interactions between granulated activated carbon or carbon nanotubes in sediments are able respectively to reduce the bioavailability of PCBs to freshwater oligochaete (Sun and Ghosh, 2007) and suppressed the BSAF of PAHs in *Chironomus plumosus* larvae (Shen et al., 2014). Likewise, petroleum residues in sediments are thought to provide a sorptive phase that reduces the bioavailability of PCBs (Zwiernik et al., 1999).

Redox potential and pH must also be taken into account to understand the fate (and consequently bioavailability) of organic pollutants in soils and sediments. Temperature affects chemical equilibrium partition coefficients as does the biochemistry and physiological response of organisms (Meggo and Schnoor, 2011).

5.1.1.3 Factors Affecting Metal Bioavailability

Several factors affecting HOC bioaccessibility and bioavailability are also important in the case of metals. Many factors can influence the lability and bioavailability of trace metals in sediments, including salinity, pH, redox, dissolved oxygen, sediment composition (mineralogy, grain size, organic contaminant, acid volatile sulfide), bonding strength between the metals and various solid phases, uptake routes, metal speciation, and characteristics of target organisms (Wu et al., 2013; Hamzeh et al., 2014). Among these factors, the speciation and concentration of metals have the most significant impact on the assimilation of metals by benthic animals. In the case of metal-based nanoparticles, both environmental and biological factors govern metal assimilation, but only some information is already available (Chapter 17).

Strong interacting effects of a sewage treatment plant through decreased oxygenation and increased level of organic matter have been highlighted (Teuchies et al., 2011; Thevenon et al., 2011) because under anoxic conditions, the formation of insoluble metal sulfides is known to reduce metal availability. Recent water quality improvements in freshwater ecosystems of certain countries are generally accompanied with an increase in oxygen concentration. This may result in the leaching of sediment-bound metals to overlying surface water, even in undisturbed watercourses (De Jonge et al., 2012).

5.1.1.4 Changes of Contaminant Bioavailability During Sediment Disturbance Events

Remobilization of sediment-associated contaminants can occur during natural events, such as tidal movement, floods, and storms, or during human activities such as boat traffic (Superville et al., 2014), dredging, dredge disposal (OSPAR, 2004), and trawling (Allan et al., 2012; Bradshaw et al., 2012). Metal remobilization experiments have been designed for assessing the effects of resuspension (Acquavita et al., 2012; Bataillard et al., 2014; Hamzeh et al., 2014). Field studies using passive sampling devices have also been carried out (Allan et al., 2012; Bradshaw et al., 2012; Superville et al., 2014). The results indicated that metals (in decreasing order: Cd, Ni, Pb, Cu, Zn, Cr) are released rapidly from freshwater sediment with a sharp peak at the beginning of the experiment, followed by a fast coprecipitation and/or adsorption processes on the suspended particles (Hamzeh et al., 2014). Similarly, the release of Hg species from the solid to the dissolved phase of sediments from lagoons along the Adriatic Sea (Italy) became negligible quickly after a resuspension event (Acquavita et al., 2012).

In unperturbed, anoxic sediment from a marina on the south coast of France, arsenic was almost fully under its trivalent As(III) form (Bataillard et al., 2014). After oxidation during the resuspension event, and in abiotic conditions, As was fully pentavalent As(V) in the oxidized zone of the resettled sediment. On the contrary, in the presence of a bacterial mat that consumed oxygen for respiration processes, the sediment was preserved from total oxidation. Consequently, As was present under both As(III) and As(V) forms (Bataillard et al., 2014). In the Deûle River (northern France), the resuspension of sedimentary particles into the overlying water significantly increased the dissolved electrolabile Pb and Zn contents. This increase does not persist for long because when the boat traffic slows down, Pb and Zn concentrations quickly drop again (Superville et al., 2014). Toxicity studies have shown that the concentrations of metals released during resuspension are not sufficient to be acutely toxic, although some chronic effects have been observed (Eggleton and Thomas, 2004).

Substantial amounts of PCDD/Fs and non-ortho PCBs are released from the sediments as a consequence of commercial bottom trawling (Bradshaw et al., 2012). On average, a one order of magnitude increase in freely dissolved PCDD/F concentrations was seen within minutes of the sediment being resuspended (Allan et al., 2012). Contrary to the fate described for metals, the phenomenon seems relatively persistent. Indeed, in a field study by Bradshaw et al. (2012), PCDD/Fs from the sediments were taken up by mussels that, during one month, accumulated them to levels above the European Union maximum advised concentration for human consumption. An increase in the bioavailability of lipophilic organic contaminants has also been shown to occur during the dredging of contaminated sediments (Eggleton and Thomas, 2004).

5.1.2 Bioavailability from Food

As already mentioned in the case of sediment, the bioaccessibility of contaminants present in the food matrix is broadly governed by their physicochemical characteristics. Prey species living in contaminated environments are able to cope with the presence of chemicals, thanks to different defense mechanisms. Very simplistically, biotransformation, favoring excretion, is the main strategy allowing a defense against organic contaminants. For metals, storage under detoxified forms either soluble (metallothioneins, MTs) or insoluble (granules resulting from biomineralization) constitutes the major defense system. Because endobenthic species are important prey species in many aquatic food chains, their role in the transfer of contaminants and associated toxicity have been reviewed with some details in Chapter 10.

When elimination is preferred, a limited risk of transfer toward predators could be expected. This is generally the case for chemicals such as PAHs that are prone to biotransformation and elimination in most species, including invertebrates. Consequently, PAH concentrations generally decrease in organisms at increasing levels in the food chain. The capacity of organisms to metabolize PCBs is minimal or absent in mollusks, and increases in decapod crustaceans, fish, sea birds, and marine mammals. However, metabolites of PCBs such as the organic contaminant pesticide DDT and brominated flame retardants (such as polybrominated biphenyl ethers) are found in the tissues of organisms, favoring biomagnification in food chains (Abarnou, 2009). The capacity to store HOC in lipids is also an important factor on which bioaccumulation obviously depends, thus leading to a particular concern in the case of lipid-rich aquatic foodstuffs for humans. Because PCB concentrations in fish from many water bodies exceed the regulation threshold for human consumption, fishing and/or fish consumption are determined by local limitations.

Depending on the categories of lipids contributing to the storage of HOC, health effects can be contrasted. When HOCs are concentrated in storage lipids, they are temporarily kept away from the general metabolism and do not exert noxious effects. However, when storage lipids are remobilized, HOCs could recover their noxiousness—for instance, during the reproductive migrations of eels (Robinet and Feunteun, 2002) or in migrating humpback whales, an iconic mammal at the top of the Antarctic food chain (Cropp et al., 2014).

On the other hand, contaminants such as PCBs or methylmercury may be associated to lipids in the brain. Relationships between environmentally relevant concentrations of Hg and biomarkers of neurochemical function or neurobehavior in fish, piscivorous mammals, and marine mammals have been documented (Basu et al., 2005a,b; Weis, 2009; Dietz et al., 2013), and further studies are required to explore the physiological and ecological significance of these findings. Because 1 billion people rely on fish and other seafood as their main source of animal proteins, the consequences for human populations is also a topic of high concern (Dewailly and Knap, 2006; Donaldson et al., 2010; Grandjean et al., 2012 and literature quoted therein), leading the European Union to launch a pioneering project on track to assess level of contaminants in seafood (http://eurizon.es/pagina/noticias/?lang=en).

Chemical properties that strongly influence biomagnification potential are hydrophobicity, polarity, and metabolizability (Norström and Letcher, 1997). The use of bioconcentration factors and their prediction from a bioconcentration factor-K_{ow} (octanol–water partition coefficient) correlation is a basic but very efficient tool to determine the bioaccumulation of a substance (Abarnou, 2009). However, mechanistic models must be developed to take into account the ecological and physiological factors that control bioaccumulation. For instance, modeling bioaccumulation of PCBs in the common sole *Solea solea* or the hake *Merluccius merluccius* include contaminant uptake from food, contaminant losses from spawning, and the influence of physiological variables, such as sex, body size, and lipid content on PCB kinetics (Bodiguel et al., 2009; Eichinger et al., 2010). These authors have neglected biotransformation because in the case of PCBs, this phenomenon is negligible from of the high stability of these compounds. Including biotransformation of organic xenobiotics that can affect elimination kinetics increases the complexity of the analysis. In vitro biotransformation assays are being explored to improve estimates of bioconcentration factors, but differences in dosing techniques can influence the determination of biotransformation rates as illustrated for PAHs (Lee et al., 2014).

The physicochemical form of metals in their prey clearly influences subsequent trophic transfer with a high accessibility of soluble forms (including MT-bound metals) and varying accessibility of insoluble forms (Figure 5.3), depending on the strength of the digestive processes in a particular predator that governs what metal-binding fractions can be assimilated and to what extent (Rainbow et al., 2011a).

The measurement of bioaccessibility can lead to a more accurate risk assessment compared with the measurements of total concentrations of pollutant in food when establishing maximum acceptable concentrations of metals in food. A simulated in vitro digestion method has been developed to assess bioaccessibility of metals in mollusks and fish to human consumers (Amiard et al., 2008; Metian et al., 2009; He and Wang, 2011, 2013). It appears to be a powerful tool for seafood safety assessment.

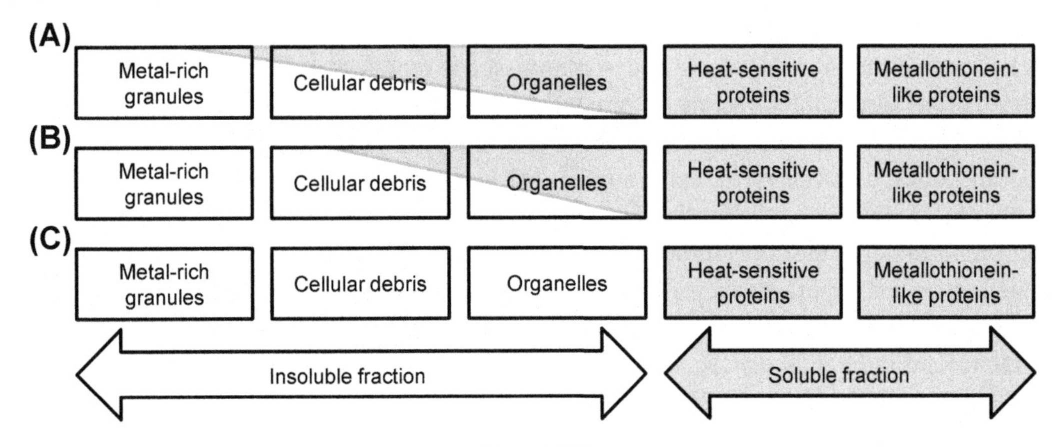

Figure 5.3

Fractionation of metal accumulated in prey into five components (after Wallace et al., 2003). (A) Highlighted areas covering all five fractions to some degree represent metal accumulated in prey that is trophically available to a neogastropod mollusc (Cheung and Wang, 2005; Rainbow et al., 2007). (B) Highlighted areas (from four fractions) represent metal accumulated in prey that is trophically available to a predator with weaker digestive powers than a neogastropod mollusc. (C) Highlighted areas (from two fractions) represent metal accumulated in prey that is trophically available to a planktonic copepod filtering phytoplankton (Reinfelder and Fisher, 1991). *In Rainbow et al., 2011(a) with permission.*

5.2 Relationship Between Bioaccumulated Concentrations and Noxious Effects

Conventional assessment of chemical hazards to aquatic life are mainly based on the relationship between external exposure (contaminant concentrations in water, much less frequently sediment or food) and biological effects in organisms (Chapter 2). Tissue residues, measured in either whole-body or organ-specific concentrations, have been proposed as a better metric to assess the toxicity of environmental contaminants (McCarty and Mackay, 1993). Guidelines for the protection of wildlife that consume aquatic biota have been established using this approach in Canadian regulations as early as 1999 (CCME, 1999), whereas the US Environmental Protection Agency (USEPA, 2005) is working on incorporating tissue residue into standards for bioaccumulative chemicals. On the contrary, there is a lack of formal use of tissue residue in Europe. In particular, the European Union Water Framework Directive is oriented primarily toward water-based quality standards.

The utility of the TRA for toxicity assessment has been assessed in the framework of a Pellston Workshop (*Integrated Environmental Assessment and Management* 7 (1), 2011), showing significant advances in theory and application. The major findings are summarized in Meador et al. (2011). The TRA relies on the toxicology principle that a toxic effect is not observed unless the chemical reaches the site of action. Therefore, for most compounds in which a concentration in the whole body/organ/tissue is representative of the concentration at the level of the target, the direct measurement of bioaccumulated concentrations can be preferred to measurements in the exposure media. This is the case for baseline toxicants, organometallics, chlorophenols, and a number of chemicals that are weakly metabolized. On the other hand, TRA is rarely applied for compounds in which bioaccumulated concentration is not proportional to the concentration at the level of the target (e.g., highly metabolized compounds, metals, mutagens, cyanide). The most important difference between metals and organics is that bioavailability, uptake processes, and factors that determine toxicity are more specific toward individual metals, species, environmental conditions, and exposure routes than for organic chemicals (McCarty et al., 2011).

5.2.1 Metals

In the endobenthic bivalve *Scrobicularia plana* originating from contaminated sites in Cornwall (UK), zinc concentrations as high as 4810 mg kg^{-1} were determined in the years 1977–1979. Despite these contaminations, bivalves were able to survive and reproduce and were still present in 2006 with tissue concentrations still ≥1000 mg kg^{-1} (Rainbow et al., 2011b). On the other hand, acute toxicity tests were carried out in our laboratory with clams collected from a reference site with background levels of zinc. This metal was analyzed in specimens that had survived the LC_{50} for zinc, showing mean concentrations of 650 mg kg^{-1} (Amiard, 1991). Paradoxically, the lowest dose showed the worst effect. In fact, this discrepancy may be explained as a result of the tolerance acquired in organisms chronically exposed to contaminants in their environment. Only a fraction of the global concentration is reactive and can induce noxious effects.

As highlighted previously in the case of trophic transfer, the different fractions in which incorporated metals are stored have different toxic potentials. Those that are important for toxic pressure are heat-sensitive proteins, organelles and microsomes, and cell debris (Vijver et al., 2004; Adams et al., 2011). Many studies have shown that the distribution of metals is highly variable depending on the metal, the degree of contamination, the organs, and the source of contamination, even if very simplistically they are often limited to the relative importance of the soluble versus insoluble fractions (Figure 5.3). The ability to distinguish between heat-resistant (MT-like proteins) and heat-sensitive proteins (i.e., enzymes) according to the popular method described by Wallace et al. (2003) is hampered, at least in certain species, because the possibility of metal exchanges between MTs and others compounds has been shown as a consequence of a lack of stability of metal–SH links during heating (Bragigand and Berthet, 2003). Convincing relationships between physicochemical forms of metals and their toxic effects have been demonstrated in grass shrimp *Palaemonetes pugio* fed cadmium-contaminated prey by using a Sephadex G-75 elution profile of the cytosol (Wallace et al., 2000). These authors showed that successful prey (live *Artemia salina*) capture decreased with increased Cd concentration bound to high-molecular-weight proteins (i.e., enzymes). Similarly, size-exclusion chromatography was applied to specimens of the freshwater bivalve *Pyganodon grandis* collected from 10 lakes located along a known metal gradient in a mining area in northwestern Québec (Wang et al., 1999). Symptoms of toxic effects at different levels of biological organization were associated with a biochemical anomaly in bivalves exposed to concentrations of dissolved free Cd^{2+} higher than ~1 nM in the external medium. Sound research about the subcellular partitioning of metals is needed to identify a surrogate measure for the metabolically active pool of accumulated metal that would be both reliable and more practical than the time-consuming methodologies based on cytosol chromatography.

From a review by Adams et al. (2011), the use of a tissue residue-based approach for deriving a threshold for effects in aquatic organisms is not universally applicable with exceptions for trace metals that are not detoxified (Luoma and Rainbow, 2008) such as Se and methylmercury and possibly other organometallics (Meador et al., 2002a). Despite the complexity of applying the TRA to metals, it represents at least the advantage of avoiding the influence of many confounding factors influencing bioavailability. Strategies have been developed to take advantages of the TRA such as the kinetic bioenergetic-based approach (Luoma and Rainbow, 2005) and the biologically inactive metal and biologically active metal model (Steen-Redeker and Blust, 2004). Complementarily, the biotic ligand model has been proposed as a tool to evaluate quantitatively the manner in which water chemistry affects the speciation and biological availability of metals in aquatic systems (Paquin et al., 2002; Adams et al., 2011). Recent developments of the biotic ligand model are reviewed in Chapter 4.

5.2.2 Organic Chemicals

As previously mentioned in the case of metals, using tissue residue concentrations allows less dependence on confounding factors influencing exposure and bioavailability, as does the integration of multiple routes of exposure that contribute to chemical bioaccumulation and toxicity (McCarty et al., 2011). The vast majority of data available on TRA is derived from laboratory studies of acute lethal responses to aquatic contaminants (McElroy et al., 2011) in which the dietary route is generally neglected, but on the other hand, throughout most of the twentieth century, researchers measured various organic compounds and metals in myriad organisms (Meador et al., 2011). In a number of cases, exposure and effect assessment of contaminants were carried out complementarily, for instance, in wildlife (particularly marine mammals) and fish (Jepson et al., 2005; Letcher et al., 2010; Evans, 2011; Brousseau et al., 2013).

Recently, comprehensive datasets of tissue residue vs effect concentrations became available (USACE/USEPA, 2009; USEPA, 2009). The majority of the data originates from acute toxicity studies thus the most frequent endpoint is lethality and the use of sublethal endpoints is highly desirable for a better link to population-level effects (Chapters 6 and 7). Desirable attributes that should be considered when developing databases of critical body residues have been summarized in McElroy et al. (2011). Concentrations of hydrophobic organic compounds are usually normalized to lipid content to minimize intra- and interspecific variability. Lipid normalization using the octanol–water partition coefficient K_{ow} is usually accepted as a reasonable surrogate for total lipid in general, but it may fail to explain toxicodynamic processes that depend on specific lipid classes (McElroy et al., 2011). According to these authors, exposure to contaminants can alter lipid metabolism, storage, use, and lipid class composition. For instance, in the clam *S. plana*, a significant relationship between an increase of triacylglycerol percentages and an increase of dioxin-like compounds (Eq-TCDD) concentrations was shown by principal component analysis and supported by correlation tests, whereas an increase of glycolipid percentages was correlated to an increase of estrogenic pollutant concentrations and concentrations of pollutants representative of PAHs in sediments (Figure 5.4(A)). Plot of scores for PC1 and PC2 clearly allow the separation of three sampling sites differing by their pollution pattern (Figure 5.4(B)) (Perrat et al., 2013). In addition, McElroy et al. (2011) emphasize that the importance of analytical techniques used to measure lipids is an often overlooked and important problem associated with interpreting reported lipid-normalized critical body residue residues; the same is true for environmental

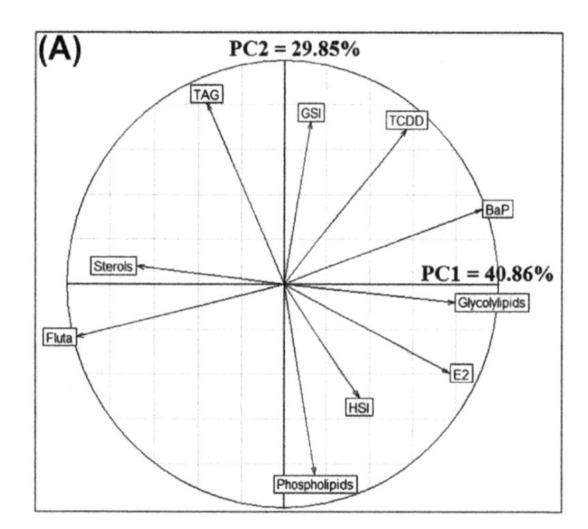

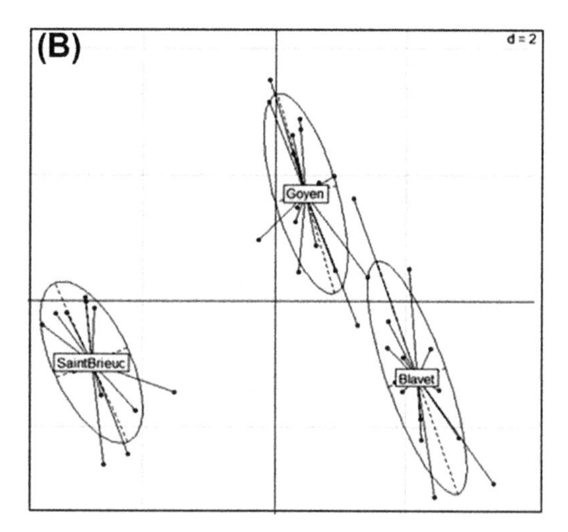

Figure 5.4

Principal component analysis of lipid classes and biometrical indices in the clam *Scrobicularia plana* from three estuaries (Blavet, Goyen, Bay of St Brieuc, France) exhibiting different contamination patterns. (A) Correlation circle; (B) plot of scores for PC1 and PC2 with the barycenters of the clouds separating clearly the results of the three estuaries. BaP: PAH-like activity (all four determined with cellular tests on sediment extracts); E2: estrogenic activity; fluta: antiandrogenic activity; GSI: gonadosomatic index; HIS: hepatosomatic index; TAG: triacylglycerol; TCDD: dioxin-like activity. *After Perrat et al., 2013 with permission.*

contaminants. This problem is taken into account in the framework of environmental monitoring, and the solution lies in the organization of training workshops and intercalibration exercises (Chapter 3).

Among the factors governing toxicokinetics and toxicodynamics, the most frequently quoted are environmental factors (temperature, pH), even if their influence is considerably less than for exposure-media concentrations, time dependency (Escher et al., 2011), and biotransformation. Fish and many aquatic invertebrates are able to metabolize organic contaminants at different degrees, thus reducing concentrations of the parent compounds in organisms, but the presence of metabolites that may contribute to toxicity must not be ignored. At least in the case of PAHs, using biliary metabolites as the concentration measurement to establish a relationship with biological effects seems a promising tool as shown in natural populations of haddock (*Melanogrammus aeglefinus*) and Atlantic cod (*Gadus morhua*) in two North Sea areas with extensive oil production (Balk et al., 2011).

Meador (2006) has described a set of procedures for developing tissue, water, and sediment quality guidelines for the protection of aquatic life by using the TRA for toxicity assessment. Examples of tissue quality guidelines that can help evaluating organismal health in the field, include tributyltin, mercury, selenium, chlorophenols, DDT, PCBs, and dioxin (Meador et al., 2002a,b; USEPA, 2004; Beckvar et al., 2005; Steevens et al., 2005; Meador, 2006). "Although guidelines based on tissue residues are more ecologically relevant and preferred, the development of water or sediment values may be desirable for many situations and statutory requirements primarily because these media are considered easier to manage and cleanup" (Meador, 2006). Applications are provided for narcotic chemicals and PAHs (Di Toro et al., 2000; Di Toro and McGrath, 2000), nonspecific acting compounds, and tributyltin (Meador, 2006).

Another important application of TRA relates to evaluating the potential hazards posed by those chemicals that bioaccumulate in field biota (McElroy et al., 2011; Sappington et al., 2011). Passive and active biomonitoring using bivalves may be used as a basis for in situ applications of the TRA (Sappington et al., 2011) but discrepancy can occur between predicted effects using accumulation of studied toxicants and observed toxicity. This is likely because of other (nonmeasured) contaminant stressors or confounding factors (McElroy et al., 2011).

5.2.3 Mixtures

Because "mixtures are the norm rather than the exception in environmental studies," recent papers have explored the interest of internal residue concentrations in the assessment of mixtures of similarly and dissimilarly acting organic and metal compounds, despite estimating the toxicity of metal mixtures is highly controversial (Dyer et al., 2011; Escher et al., 2011; McCarty et al., 2011). Dyer et al. (2011) provide a framework that integrates the TRA and mixture toxicology in a three-tier approach. Very simplistically, the first tier consists in an estimation of the toxicity of mixtures by using concentration addition regardless of the mechanism of action of the components. However, because of the differences between metals and organic chemicals underlined in the previous subsections, the tier I methodology has been designed in two different ways. The general scheme used for organics must be modified for metals, taking into account the background concentration of elements that are natural constituents of the body burden and differential tolerance according to taxa (Luoma and Rainbow, 2008; Adams et al., 2011). If the combination of measured concentrations in the mixture exceeds that predicted to produce adverse effects or previously referenced levels, Dyer et al. (2011) consider that it is necessary to proceed to tier II. To obtain a more accurate estimate of the toxicity of the mixture, it is recommended to use a model that integrates both concentration addition and independent action approaches. Then, incorporating species sensitivity distribution methods would allow the extrapolation of the mixed model for single species (tiers I and II) to communities (tier III).

5.3 Conclusions

Many different mechanisms can interfere with a contaminant present in the environment before it can reach its target (biological membranes, specific enzymes and receptors, DNA, RNA; Escher et al., 2011), where it will exert its toxicity. This is a reason for many discrepancies observed between concentrations in the media that can sometimes be very high with only few consequences in terms of toxic effect. It is generally observed that only a fraction of contaminants present in the source can be incorporated in aquatic organisms. Release of a chemical from ingested food or sediment is a prerequisite for uptake and assimilation, governing bioaccessibility. The next step, termed bioavailability stricto sensu corresponds to the uptake of the fraction of the chemical ingested with food or sediment that enters the systemic circulation (Versantvoort et al., 2005; ITRC, 2011).

Many abiotic and biotic factors are known as influencing both bioaccessibility and bioavailability. They include the physicochemical characteristics of contaminants (metal speciation, dissolution of metals from nanoparticles and their agglomeration/aggregation, hydrophobicity, aromaticity, molecular size, and planarity of organics) and the physiological characteristics of organisms (strength of the digestive processes, processes devoted to defense against organics such as biotransformation and excretion, defense against metals mainly based upon storage under detoxified forms, soluble, or insoluble). However, for emerging contaminants, the state of knowledge is very much limited and techniques for the measurement of physicochemical parameters are not easily implemented. Both the chemistry of contaminants and the physiology of organisms may be modified by environmental factors such as temperature, pH and salinity. In addition, each of them can respond to specific factors, such as the importance of organic carbon controlling the bioavailability of HOCs in sediments (Hunter et al., 2011; Meggo and Schnoor, 2011), whereas redox, dissolved oxygen, and sediment composition influence the lability and bioavailability of trace metals (Wu et al., 2013; Hamzeh et al., 2014).

In this chapter, we have reviewed many different strategies that have been proposed to assess bioavailability. Among chemical measures, the use of passive sampling techniques is one of the most promising tools, being already included in the strategy recommended by ITRC (2011) to help incorporate bioavailability considerations into the evaluation of contaminated sediment sites. Operationality of solid-phase microextraction for the assessment of bioavailability and contaminant mobility has also been tested by ESTCP (2012) in the laboratory and field. Partial extraction methods and in vitro digestion are also advanced approaches that provide a relevant assessment of the fraction of a contaminant in sediment or in food that is the most bioaccessible and thus more prone to bioavailability. From an operational viewpoint, chemical assessments may be complemented by biological measures of bioaccumulation and toxicity, also incorporated in the strategy of the ITRC (2011).

In addition to be used for a direct estimation of bioavailability, bioaccumulation is at the basis of the tissue residue approach for toxicity assessment (*Integrated Environmental Assessment and Management* 7 (1), 2011). Instead of linking toxicological effects to exposure concentrations determined in the environment (Chapter 2), this methodology is based on the relationship between concentrations in organisms and toxicity. This allows less dependence on confounding factors influencing exposure and bioavailability. However, this methodology has been well-developed for organics and organometallics, whereas its application is much more complicated in the case of metals. An important application of TRA relates to evaluating the potential hazards posed by chemicals that bioaccumulate in field biota and are well-documented in biomonitoring programs worldwide (Chapter 3). The TRA has also been proposed as a promising tool for chemical mixtures and a framework by which the toxicity of mixtures can be estimated for single species as well as multiple species and communities (Dyer et al., 2011). However, ongoing research efforts are needed to reach an operational stage.

As already mentioned, for many situations and statutory requirements, water or sediment quality guidelines are needed. It is possible to keep the advantages of the TRA (particularly that tissue residue toxicity metrics are less variable among species and environmental conditions compared to responses expressed as a function of an ambient exposure concentration) by using methods for translating between environmental compartments (biota, water, sediment, and soil) with bioaccumulation factors. These methods are well-established for organic chemicals, but are somewhat more limited for most metals (Meador, 2006; Sappington et al., 2011).

The TRA often associates data issued from laboratory tests and field monitoring. Meador (2006) states that in many cases tissue concentrations considered adverse may not be found in organisms collected from the field because sensitive individuals and species would likely have been eliminated or severely impacted. On the other hand (exemplified in Section 5.2.1 for the clam *S. plana*), tolerance acquired by organisms chronically exposed to contaminants in their environment can blur the relationship between bioaccumulation and toxicity. Intraspecific differences of tolerance toward chemicals must also be taken into account (Escher et al., 2011) whereas interspecific differences can be processed using the classical species sensitivity distribution (Sappington et al., 2011).

As already mentioned for toxicity assessment based on concentrations in the external media (Chapter 2), the vast majority of data available on TRA is derived from laboratory studies of acute lethal responses to aquatic contaminants, in which the dietary route is generally neglected. "Additional residue-effects data on sublethal endpoints, early life stages, and a wider range of legacy and emergent contaminants will be needed to improve the ability to use TRA for organic and organometallic compounds" (McElroy et al., 2011).

References

Abarnou, A., 2009. Organic contaminants in coastal and estuarine food webs. In: Amiard-Triquet, C., Rainbow, P.S. (Eds.), Environmental Assessment of Estuarine Ecosystems. A Case Study. CRC Press, Bocaton, pp. 107–134.

Acquavita, A., Emili, A., Covelli, S., et al., 2012. The effects of resuspension on the fate of Hg in contaminated sediments (Marano and Grado Lagoon, Italy): Short-term simulation experiments. Estuarine Coastal Shelf Sci. 113, 32–40.

Adams, W.J., Blust, R., Borgmann, U., et al., 2011. Utility of tissue residues for predicting effects of metals on aquatic organisms. Integr. Environ. Assess. Manag. 7, 75–98.

Ahrens, M., Depree, C.V., Golding, L., 2005. How bioavailable are sediment-bound contaminants in New Zealand harbours? In: 4th South Pacific Conference on Stormwater and Aquatic Resource Protection: Conference Proceedings 4–6 May 2005, Auckland, N.Z. http://www.researchgate.net/publication/236954905.

Akkanen, J., Slootweg, T., Mäenpää, K., et al., 2012. Bioavailability of organic contaminants in freshwater environments. In: Emerging and Priority Pollutants in RiversThe Handbook of Environmental Chemistry, pp. 25–53.

Allan, I.J., Nilsson, H.C., Tjensvoll, I., et al., 2012. PCDD/F release during benthic trawler-induced sediment resuspension. Environ. Toxicol. Chem. 31, 2780–2787.

Amiard, J.C., 1991. Réponses des organismes marins aux pollutions métalliques. In: Réactions des êtres vivants aux changements de l'environnement, Actes des Journées de l'Environnement du C.N.R.S., pp. 197–205 Paris.

Amiard, J.-C., 1992. Bioavailability of sediment-bound metals for benthic aquatic organisms. In: Vernet, J.-P. (Ed.), Impact of Heavy Metals on the Environment, vol. 2. Elsevier, Amsterdam, pp. 183–202.

Amiard, J.-C., Ettajani, H., Jeantet, A.Y., Ballan-Dufrançais, C., Amiard-Triquet, C., 1995. Bioavailability and toxicity of sediment-bound lead to a filter-feeder bivalve *Crassostrea gigas* (Thunberg). BioMetals 8, 280–289.

Amiard, J.C., Amiard-Triquet, C., Barka, S., et al., 2006. Metallothioneins in aquatic invertebrates: their role in metal detoxification and their use as biomarkers. Aquat. Toxicol. 76, 160–202.

Amiard, J.C., Geffard, A., Amiard-Triquet, C., et al., 2007. Relationship between lability of sediment-bound metals (Cd, Cu, Zn) and their bioaccumulation in benthic invertebrates. Estuarine Coastal Shelf Sci. 72, 511–521.

Amiard, J.-C., Amiard-Triquet, C., Charbonnier, L., et al., 2008. Bioaccessibility of essential and non-essential metals in commercial shellfish from Western Europe and Asia. Food Chem. Toxicol. 46, 2010–2022.

Amiard-Triquet, C., Rainbow, P.S., Roméo, M., 2011. Tolerance to Environmental Contaminants. CRC Press, Boca Raton.

Balk, L., Hylland, K., Hansson, T., et al., 2011. Biomarkers in natural fish populations indicate adverse biological effects of offshore oil production. PLoS ONE 6 (5), e19735. http://dx.doi.org/10.1371/journal.pone.0019735.

Basu, N., Scheuhammer, A., Grochowina, N., et al., 2005a. Effects of mercury on neurochemical receptors in wild river otters (*Lontra canadensis*). Environ. Sci. Technol. 39, 3585–3591.

Basu, N., Klenavic, K., Gamberg, M., et al., 2005b. Effects of mercury on neurochemical receptor-binding characteristics in wild mink. Environ. Toxicol. Chem. 24, 1444–1450.

Bataillard, P., Grangeon, S., Quinn, P., et al., 2014. Iron and arsenic speciation in marine sediments undergoing a resuspension event: the impact of biotic activity. J. Soils Sediments 14, 615–629.

Baumann, Z., Koller, A., Fisher, N., 2012. Factors influencing the assimilation of arsenic in a deposit-feeding polychaete. Comp. Biochem. Physiol. 156C, 42–50.

Beckvar, N., Dillon, T., Reed, L., 2005. Approaches for linking whole body fish residues of mercury or DDT to biological effects threshold. Environ. Toxicol. Chem. 24, 2094–2105.

Bodiguel, X., Maury, O., Mellon-Duval, C., et al., 2009. A dynamic and mechanistic model of PCB bioaccumulation in the European hake (*Merluccius merluccius*). J. Sea Res. 62, 124–134.

Bradshaw, C., Tjensvoll, I., Sköld, M., et al., 2012. Bottom trawling resuspends sediment and releases bioavailable contaminants in a polluted fjord. Environ. Pollut. 170, 232–241.

Bragigand, V., Berthet, B., 2003. Some methodological aspects of metallothionein evaluation. Comp. Biochem. Physiol. 134A, 55–61.

Brousseau, P., Pillet, S., Frouin, H., et al., 2013. Linking immunotoxicity and ecotoxicological effects at higher biological levels. In: Amiard-Triquet, C., Amiard, J.-C., Rainbow, P.S. (Eds.), Ecological Biomarkers: Indicators of Ecotoxicological Effects. CRC Press, Boca Raton, pp. 131–154.

CCME (Canadian Council of Ministers of the Environment), 1999. Protocol for the Derivation of Canadian Tissue Residue Guidelines for the Protection of Wildlife that Consume Aquatic Biota. http://ceqg-rcqe.ccme.ca/download/en/290 (last accessed 30.07.14).

Cheung, M.S., Wang, W.X., 2005. Influences of subcellular compartmentalization in different prey on the transfer of metals to a predatory gastropod from different prey. Mar. Ecol. Prog. Ser. 286, 155–166.

Cropp, R., Bengtson Nash, S., Hawker, D., 2014. A model to resolve organochlorine pharmacokinetics in migrating humpback whales. Environ. Toxicol. Chem. 33, 1638–1649.

Cui, X., Hunter, W., Yang, Y., et al., 2010. Bioavailability of sorbed phenanthrene and permethrin in sediments to *Chironomus tentans*. Aquat Toxicol. 98, 83–90.

Dietz, R., Sonne, C., Basu, N., et al., 2013. What are the toxicological effects of mercury in Arctic biota? Sci. Total Environ. 443, 775–790.

Di Toro, D.M., Zarba, C.S., Hansen, D.J., et al., 1991. Technical basis for establishing sediment quality criteria for nonionic chemicals using equilibrium partitioning. Environ. Sci. Technol. 10, 1541–1583.

Di Toro, D.M., McGrath, J.A., Hansen, D.J., 2000. Technical basis for narcotic chemicals and polycyclic aromatic hydrocarbon criteria. I. Water and tissue. Environ. Toxicol. Chem. 19, 1951–1970.

Di Toro, D.M., McGrath, J.A., 2000. Technical basis for narcotic chemicals and polycyclic aromatic hydrocarbon criteria. II. Mixtures and sediments. Environ. Toxicol. Chem. 19, 1971–1982.

De Jonge, M., Teuchies, J., Meire, P., et al., 2012. The impact of increased oxygen conditions on metal contaminated sediments part I: effects on redox status, sediment geochemistry and metal bioavailability. Water Res. 46, 2205–2214.

Dewailly, É., Knap, A., 2006. Food from the oceans and human health; balancing risks and benefits. Oceanography 19, 84–93.

Donaldson, S.G., Van Oostdam, J., Tikhonov, C., et al., 2010. Environmental contaminants and human health in the Canadian Arctic. Sci. Total Environ. 408, 5165–5234.

Dyer, S., St.J.Warne, M., Meyer, J.S., et al., 2011. Tissue residue approach for chemical mixtures. Integr. Environ. Assess. Manag. 7, 99–115.

Eggleton, J., Thomas, K.V., 2004. A review of factors affecting the release and bioavailability of contaminants during sediment disturbance events. Environ. Intern. 30, 973–980.

Eichinger, M., Loizeau, V., Roupsard, F., et al., 2010. Modelling growth and bioaccumulation of polychlorinated biphenyls in common sole (*Solea solea*). J. Sea Res. 64, 373–385.

Escher, B.I., Ashauer, R., Dyer, S., et al., 2011. Crucial role of mechanisms and modes of toxic action for understanding tissue residue toxicity and internal effect concentrations of organic chemicals. Integr. Environ. Assess. Manag. 7, 28–49.

ESTCP, 2012. Demonstration and Evaluation of Solid Phase Microextraction for the Assessment of Bioavailability and Contaminant Mobility Cost and Performance Report. Environmental security technology certification program U.S. Department of Defense (ER-200624).

Ettajani, H., Amiard-Triquet, C., Jeantet, A.Y., et al., 1996. Fate and effects of soluble or sediment-bound arsenic in oysters (*Crassostrea gigas* Thun). Arch. Environ. Contam. Toxicol. 31, 38–46.

Evans, P.G.H. (Ed.), 2011. Chemical Pollution and Marine Mammals. ECS special publication series no. 55.

Li, H., Sun, B., Xin Chen, X., et al., 2013. Addition of contaminant bioavailability and species susceptibility to a sediment toxicity assessment: application in an urban stream in China. Environ. Pollut. 178, 135–141.

Friedman, C.L., Lohmann, R., 2014. Comparing sediment equilibrium partitioning and passive sampling techniques to estimate benthic biota PCDD/F concentrations in Newark Bay, New Jersey (U.S.A.). Environ. Pollut. 186, 172–179.

Gaillard, J., Banas, D., Thomas, M., et al., 2014. Bioavailability and bioaccumulation of sediment-bound polychlorinated biphenyls to carp. Environ. Toxicol. Chem. 33, 1324–1330.

Geffard, O., Geffard, A., Budzinski, H., et al., 2005. Mobility and potential toxicity of sediment-bound metals in a tidal estuary. Environ. Toxicol. 20, 407–417.

Goto, D., Wallace, W.G., 2009. Influences of prey- and predator-dependent processes on cadmium and methylmercury trophic transfer to mummichogs (*Fundulus heteroclitus*). Can. J. Fish. Aquat. Sci. 66, 836–846.

Grandjean, P., Weihe, P., Nielsen, F., et al., 2012. Neurobehavioral deficits at age 7 years associated with prenatal exposure to toxicants from maternal seafood diet. Neurotoxicol. Teratol 34, 466–472.

Hamzeh, M., Ouddane, B., Daye, M., et al., 2014. Trace metal mobilization from surficial sediments of the Seine River Estuary. Water Air Soil Pollut. 225, 1878.

Hansen, D.J., Di Toro, D.M., Berry, W.J., et al., 2005. Procedures for the Derivation of Equilibrium Partitioning Sediment Benchmarks (Esbs) for the Protection of Benthic Organisms: Metal Mixtures (Cadmium, Copper, Lead, Nickel, Silver and Zinc). EPA/600/R-02/011.

Harwood, A.D., Landrum, P.F., Lydy, M.J., 2013. Bioavailability-based toxicity endpoints of bifenthrin for *Hyalella azteca* and *Chironomus dilutes*. Chemosphere 90, 1117–1122.

He, M., Wang, W.-X., 2011. Factors affecting the bioaccessibility of methylmercury in several marine fish species. J. Agric. Food Chem. 59, 7155–7162.

He, M., Wang, W.-X., 2013. Bioaccessibility of 12 trace elements in marine molluscs. Food Chem. Toxicol. 55, 627–636.

Hunter, W.H., Gan, J., Kookana, R.S., 2011. Bioavailability of hydrophobic organic contaminants in soils and sediments. In: Xing, B., Senesi, N., Huang, P.M. (Eds.), Biophysico-chemical Processes of Anthropogenic Organic Compounds in Environmental Systems. John Wiley & Sons, Hoboken (NJ, USA), pp. 517–534.

ITRC (Interstate Technology & Regulatory Council), 2011. Incorporating Bioavailability Considerations into the Evaluation of Contaminated Sediment Sites. CS-1. Interstate Technology & Regulatory Council, Contaminated Sediments Team, Washington, D.C.http://www.itrcweb.org/contseds-bioavailability/cs_1.pdf (last accessed 31.07.14).

Janssen, C.R., Heijerick, D.G., De Schamphelaere, K.A.C., et al., 2003. Environmental risk assessment of metals: tools for incorporating bioavailability. Environ. Int. 28, 793–800.

Janssen, E.M., Oen, A.M., Luoma, S.N., et al., 2011. Assessment of field-related influences on polychlorinated biphenyl exposures and sorbent amendment using polychaete bioassays and passive sampler measurements. Environ. Toxicol. Chem. 30, 173–180.

Jepson, P.D., Bennett, P.M., Deaville, R., Allchin, C.R., Baker, J.R., Law, R.J., 2005. Relationships between polychlorinated biphenyls and health status in harbour porpoises (*Phocoena phocoena*) stranded in the United Kingdom. Environ. Toxicol. Chem. 24, 238–248.

Kraaij, R., Seinen, W., Tolls, J., et al., 2002. Direct evidence of sequestration in sediments affecting the bioavailability of hydrophobic organic chemicals to benthic deposit-feeders. Environ. Sci. Technol. 36, 3525–3529.

Lanno, R., Wells, J., Conder, J., et al., 2004. The bioavailability of chemicals in soils for earthworms. Ecotoxicol. Environ. Saf. 57, 39–47.

Lee, Y.S., Lee, D.H.Y., Delafoulhouze, M., et al., 2014. In vitro biotransformation rates in fish liver S9: effect of dosing techniques. Environ. Toxicol. Chem. 33, 1885–1893.

Letcher, R.J., Bustnes, J.O., Dietz, R., 2010. Exposure and effects assessment of persistent organohalogen contaminants in Arctic wildlife and fish. Sci. Total Environ. 408, 2995–3043.

Lindgren, J.F., Hassellöva, I.M., Dahllöf, I., 2014. PAH effects on meio- and microbial benthic communities stronglydepend on bioavailability. Aquat. Toxicol. 146, 230–238.

Liu, H.H., Bao, L.J., Feng, W.H., et al., 2013. A multisection passive sampler for measuring sediment porewater profile of dichlorodiphenyltrichloroethane and its metabolites. Anal. Chem. 85, 7117–7124.

Long, E.R., Field, L.J., Macdonald, D.D., 1998. Predicting toxicity in marine sediments with numerical sediment quality guidelines. Environ. Toxicol. Chem. 17, 714–727.

Long, E.R., McDonald, D.D., Smith, S.L., et al., 1995. Incidence of adverse biological effects within ranges of chemical concentrations in marine and estuarine sediments. Environ. Manag. 19, 81–97.

Lundqvist, A., Bertilsson, S., Goedkoop, W., 2010. Effects of extracellular polymeric and humic substances on chlorpyrifos bioavailability to *Chironomus riparius*. Ecotoxicology 19, 614–622.

Luoma, S.N., Rainbow, P.S., 2005. Why is metal bioaccumulation so variable? Biodynamics as a unifying concept. Environ. Sci. Technol. 39, 1921–1931.

Luoma, S.N., Rainbow, P.S., 2008. Metal Contamination in Aquatic Environments. Science and Lateral Management. Cambridge University Press, Cambridge.

Mackenbach, E.M., Harwood, A.M., Mills, M.A., et al., 2014. Application of a tenax model to assess bioavailability of polychlorinated biphenyls in field sediments. Environ. Toxicol. Chem. 33, 286–292.

Marchand, J., 1978. Villes et ports, développement portuaire, croissance spatiale des villes, environnement littoral, 587, CNRS, Paris, pp. 445–462.

Mayer, L.M., Chen, Z., Findlay, R., et al., 1996. Bioavailability of sedimentary contaminants subject to deposit-feeder digestion. Environ. Sci. Technol. 30, 2641–2645.

Mayer, L.M., Weston, D.P., Bock, M.J., 2001. Benzo[*a*]pyrene and zinc solubilization by digestive fluids of benthic invertebrates—a cross-phyletic study. Environ. Toxicol. Chem. 20, 1890–1900.

McCarty, L.S., Mackay, D., 1993. Enhancing ecotoxicological modeling and assessment: body residues and modes of toxic action. Environ. Sci. Technol. 27, 1719–1728.

McCarty, L.S., Landrum, P.F., Luoma, S.N., et al., 2011. Advancing environmental toxicology through chemical dosimetry: external exposures versus tissue residues. Integr. Environ. Assess. Manag. 7, 7–27.

McElroy, A.E., Barron, M.G., Beckvar, N., et al., 2011. A review of the tissue residue approach for organic and organometallic compounds in aquatic organisms. Integr. Environ. Assess. Manag. 7, 50–74.

Meador, J.P., 2006. Rationale and procedures for using the tissue-residue approach for toxicity assessment and determination of tissue, water, and sediment quality guidelines for aquatic organisms. Hum. Ecol. Risk Assess. 12, 1018–1073.

Meador, J.P., Collier, T.K., Stein, J.E., 2002a. Determination of a tissue and sediment threshold for tributyltin to protect prey species of juvenile salmonids listed under the US Endangered Species Act. Aquat. Conserv. Mar. Freshwat. Ecosyst. 12, 539–551.

Meador, J.P., Collier, T.K., Stein, J.E., 2002b. Determination of a tissue and sediment threshold for tributyltin to protect prey species of juvenile salmonids listed under the US Endangered Species Act. Aquat. Conserv. Mar. Freshwat. Ecosyst. 12, 493–516.

Meador, J.P., Adams, W.J., Escher, B.I., et al., 2011. The tissue residue approach for toxicity assessment: findings and critical reviews from a Society of Environmental Toxicology and Chemistry Pellston Workshop. Integr. Environ. Assess. Manag. 7, 2–6.

Meggo, R.E., Schnoor, J.L., 2011. Abiotic and biotic factors affecting the fate of organic pollutants in soils and sediments. In: Xing, B., Senesi, N., Huang, P.M. (Eds.), Biophysico-chemical Processes of Anthropogenic Organic Compounds in Environmental Systems. John Wiley & Sons, Hoboken (NJ, USA), pp. 535–558.

Metian, M., Charbonnier, L., Oberhaensli, F., et al., 2009. Assessment of metal, metalloid and radionuclide bioaccessibility from mussels to human consumers, using centrifugation and simulated digestion methods coupled with radiotracer techniques. Ecotox. Environ. Saf. 72, 1499–1502.

Norstrom, R.J., Letcher, R.J., 1997. Role of biotransformation in bioconcentration and bioaccumulation. In: Sijm, D., de Bruin, J., de Voogt, P., de Wolf, P. (Eds.), Biotransformation in Environmental Risk Assessment. Society of Environmental Toxicology and Chemistry (SETAC), Bruxelles, Europe Worshop, p. 103–113 Europe Worshop.

OSPAR Commission, 2004. Revised OSPAR Guidelines for the Management of Dredged Material. Reference 2004-08. Convention for the Protection of the Marine Environment of the North-East Atlantic. 30 pp. http://www.dredging.org/documents/ceda/downloads/environ-ospar-reviseddredged-material-guidelines.pdf.

Paquin, P.R., Gorsuch, J.W., Apte, S., et al., 2002. The biotic ligand model: a historical review. Comp. Biochem. Physiol. 133C, 3–35.

Peña-Icart, M., Mendiguchia, C., Villanueva-Tagle, M.E., et al., 2014. Revisiting methods for the determination of bioavailable metals in coastal sediments. Mar. Pollut. Bull. 89, 67–74.

Perrat, E., Couzinet-Mossion, A., Fossi Tankoua, O., et al., 2013. Variation of content of lipid classes, sterols and fatty acids in gonads and digestive glands of *Scrobicularia plana* in relation to environment pollution levels. Ecotox. Environ. Saf. 90, 112–120.

Rainbow, P.S., Amiard, J.C., Amiard-Triquet, C., et al., 2007. Trophic transfer of trace metals: subcellular compartmentalizationin bivalve prey, assimilation by a gastropod predator and in vitro digestion simulations. Mar. Ecol. Prog. Ser. 348, 125–138.

Rainbow, P.S., Luoma, S.N., Wang, W.X., 2011a. Trophically available metal – a variable feast. Environ. Pollut. 159, 2347–2349.

Rainbow, P.S., Kriefman, S., Smith, B.D., et al., 2011b. Have the bioavailabilities of trace metals to a suite of biomonitors changed over three decades in SW England estuaries historically affected by mining? Sci. Total Environ. 409, 1589–1602.

Reinfelder, J.R., Fisher, N.S., 1991. The assimilation of elements ingested by marine copepods. Science 251, 794–796.

Robinet, T., Feunteun, E., 2002. Sublethal effects of exposure to chemical compounds: a cause for the decline in Atlantic eels? Ecotoxicology 11, 265–277.

Ruus, A., Schaanning, M., Øxnevad, S., et al., 2005. Experimental results on bioaccumulation of metals and organic contaminants from marine sediments. Aquat. Toxicol. 72, 273–292.

Sappington, K.G., Bridges, T.S., Bradbury, S.P., et al., 2011. Application of the tissue residue approach in ecological risk assessment. Integr. Environ. Assess. Manag. 7, 116–140.

Shen, M.H., Xia, X.H., Zhai, Y.W., et al., 2014. Influence of carbon nanotubes with preloaded and coexisting dissolved organic matter on the bioaccumulation of polycyclic aromatic hydrocarbons to *Chironomus plumosus* larvae in sediment. Environ. Toxicol. Chem. 33, 182–189.

Steen-Redeker, E., Blust, R., 2004. Accumulation and toxicity of cadmium in the aquatic oligochaete *Tubifex tubifex*: a kinetic modeling approach. Environ. Sci. Technol. 38, 537–543.

Steevens, J.A., Reiss, M.R., Pawlisz, A.V., 2005. A methodology for deriving tissue residue benchmarks for aquatic biota: a case study for fish exposed to 2,3,7,8-tetrachlorodibenzo- *p*-dioxin and equivalents. Integr. Environ. Assess. Manag. 1, 142–151.

Sun, X.L., Ghosh, U., 2007. PCB bioavailability control in Lumbriculus variegatus through different modes of activated carbon addition to sediments. Environ. Sci. Technol. 41, 4774–4780.

Superville, P.J., Prygiel, E., Magnier, A., et al., 2014. Daily variations of Zn and Pb concentrations in the Deûle River in relation to the resuspension of heavily polluted sediments. Sci. Total Environ. 470-471, 600–607.

Teuchies, J., Bervoets, L., Cox, T.J.S., et al., 2011. The effect of waste water treatment on river metal concentrations: removal or enrichment? J. Soils Sediments 11, 364–372.

Thevenon, F., Graham, N.D., Herbez, A., et al., 2011. Spatio-temporal distribution of organic and inorganic pollutants from Lake Geneva (Switzerland) reveals strong interacting effects of sewage treatment plant and eutrophication on microbial abundance. Chemosphere 84, 609–617.

USACE/USEPA (US Army Corps of Engineers/US Environmental Protection Agency), Environmental Residue-Effects Database (ERED). Technical Points of Contact Bridges, T.S., Lutz, C. http://el.erdc.usace.army.mil/ered/ (last accessed 31.07.14).

USEPA, 2004. Aquatic Life Ambient Freshwater Quality Criteria for Selenium. United Stated Environmental Protection Agency Office of Water, Washington, D.C. EPA-822-D-04–001.

USEPA (US Environmental Protection Agency), 2009. US Environmental Protection Agency Comprehensive Toxicity/Tissue Residue Database. Technical point of contact C. Russom. Available from: http://www.epa.gov /med/Prods_Pubs/tox_residue.htm (accessed 31.07.14).

USEPA, 2005. Tissue-based Criteria for "bioaccumulative" Chemicals. http://www.epa.gov/scipoly/sap/meetings/2008/october/aquatic_life_criteria_guidelines_tissue_08_26_05.pdf (last accessed 30.07.14).

Versantvoort, C.H.M., Oomen, A.G., Van de Kamp, E., Rompelberg, C.J.M., Sips, A.J.A.M., 2005. Applicability of an in vitro digestion model in assessing the bioaccessibility of mycotoxins from food. Food Chem. Toxicol. 43, 31–40.

Vijver, M.G., van Gestel, C.A.M., Lanno, R.P., et al., 2004. Internal metal sequestration and its ecotoxicological relevance – a review. Environ. Sci. Technol. 38, 4705–4712.

Wallace, W.G., Hoexum Brouwer, T.M., Brouwer, M., 2000. Alterations in prey capture and induction of metallothioneins in grass shrimp fed cadmium-contaminated prey. Environ. Toxicol. Chem. 19, 962–971.

Wallace, W.G., Lee, B.G., Luoma, S.N., 2003. Subcellular compartmentalization of Cd and Zn in two bivalves. I. Significance of metal-sensitive fractions (MSF) and biologically detoxified metal (BDM). Mar. Ecol. Prog. Ser. 249, 183–197.

Wang, D., Couillard, Y., Campbell, P.G.C., et al., 1999. Changes in subcellular metal partitioning in the gills of freshwater bivalves (*Pyganodon grandis*) living along an environmental cadmium gradient. Can. J. Fish. Aquat. Sci. 56, 774–784.

Wang, F., Wang, W.X., Huang, X.P., 2012. Spatial distribution of gut juice extractable Cu, Pb and Zn in sediments from the Pearl River Estuary, Southern China. Mar. Environ. Res. 77, 112–119.

Weis, J., 2009. Reproductive, developmental, and neurobehavioral effects of methylmercury in fishes. J. Environ. Sci. Health 27C, 212–225.

Werner, D., Hale, S.E., Ghosh, U., Luthy, R.G., 2010. Polychlorinated biphenyl sorption and availability in field-contaminated sediments. Environ. Sci. Technol. 44, 2809–2815.

Weston, D.P., Maruya, K.A., 2002. Predicting bioavailability and bioaccumulation with in vitro digestive fluid extraction. Environ. Toxicol. Chem. 21, 962–971.

Wu, X., Xie, L., Xu, L., et al., 2013. Effects of sediment composition on cadmium bioaccumulation in the clam *Meretrix meretrix* Linnaeus. Environ. Toxicol. Chem. 32, 841–847.

Xu, Y., Spurlock, F., Wang, Z., et al., 2007. Comparison of five methods for measuring sediment toxicity of hydrophobic contaminants. Environ. Sci. Technol. 41, 8394–8399.

Yin, H., Cai, Y., Hongtao Duan, H., et al., 2014. Use of DGT and conventional methods to predict sediment metal bioavailability to a field inhabitant freshwater snail (*Bellamya aeruginosa*) from Chinese eutrophic lakes. J. Hazard. Mat 264, 184–194.

You, J., Landrum, P.F., Trimble, T.A., et al., 2007. Availability of polychlorinated biphenyls in field-contaminated sediment. Environ. Toxicol. Chem. 26, 1940–1948.

Zwernick, M.J., Quensen III, J.F., Boyd, S.A., 1999. Residual petroleum in sediments reduces the bioavailability and rate of reductive dechlorination of Aroclor 1242. Environ. Sci. Technol. 33, 3574–3578.

CHAPTER 6

How to Improve Toxicity Assessment? From Single-Species Tests to Mesocosms and Field Studies

Claude Amiard-Triquet

Abstract

The relationship between the concentration of a chemical substance in the medium and its effects on organisms must be quantified as well as possible because it is at the basis of the determination of the predicted no effect concentration that is a key parameter for environmental risk assessment. Thus, designing the procedures aiming at the determination of toxicological parameters must mobilize the most recent scientific knowledge. The present chapter provides a literature review allowing advances in biotesting practices, considering both the improvement of exposure scenarios (increasing the realism of the doses tested and of the ways of exposure), and the choice of more sensitive endpoints (growth, reproduction, behavior, and also subindividual effects linked to the success of reproduction). The quality of bioassays lies also on the relevant choice of biological models (Chapter 10) and the use of the best available statistic treatments.

Keywords: Behavior; Caging; Ecological biomarkers; Endpoints; Exposure scenarios; Field bioassays; Growth; Low-dose effects; Reproduction.

Chapter Outline

Aquatic Ecotoxicology. http://dx.doi.org/10.1016/B978-0-12-800949-9.00006-1

Bioassays carried out in the framework of conventional risk assessment (Chapter 2) suffer a number of weaknesses. Very often, the doses tested are much higher than those encountered in the field even in the most polluted areas of the world and the duration of exposure is very short, mimicking at best situations observed in the case of accidents but in no way chronic contamination. Waterborne exposure is the most commonly used source of contamination, neglecting contamination through food or sediments that may be important (Wang, 2011): the former particularly in the case of many organic lipophilic contaminants (as exemplified for polychlorinated biphenyls by Péan et al., 2013); the latter for many species living in close contact and even ingesting sediment (endobenthic invertebrates, flatfish and also filter-feeding bivalves). Although exposure of aquatic organisms to hazardous compounds is primarily through complex environmental mixtures, bioassays are classically carried out with individual substances and interactions between different classes of contaminants are scarcely taken into account.

Among the effects observed in classical bioassays, mortality is the most commonly used endpoint in short-term bioassays (Chapter 2). Again, with a view to the extrapolation of bioassay results to field situations, the relevance of mortality seems questionable because, except in accidents, many organisms are able to cope at least partly with the presence of pollutants in their environment. Consequently, mortality occurs only after prolonged exposure, sublethal effects (e.g., on growth, reproduction) being more likely to occur.

Organisms used in many standardized bioassays are equivalent for the aquatic environment to laboratory rats (Berthet et al., in Amiard-Triquet et al., 2011). It is again a problem of environmental realism; in addition, it is suspected that a loss of genetic variation could result from maintaining populations in the laboratory (Chapter 10). In *Daphnia*, Messiaen et al. (2013) report that genetic variation complicates predictions of response of natural populations to chemical stressors from results of standard single-clone laboratory ecotoxicity tests.

Therefore the objective of this chapter is to propose improvements of bioassays considering the mode of exposure and the choice of the observed endpoints, the selection of the biological models (see also Chapter 10) and, in agreement with the recommendations of regulatory bodies (EFSA, 2009; USEPA, 2012), a better determination of toxicological parameters (see also Chapter 2).

6.1 Improve the Realism of Conditions of Exposure

The main experimental approaches in ecotoxicology are presented in Figure 6.1 along a gradient of scale. Experiments with single or few species rely on the reference species concept developed in Chapter 10. They may be carried out in batch or chemostat for unicellular species such as microalgae or in small laboratory units or microcosms for small invertebrates and fish (Figure 6.1). However, small-scale experiments are biased against the detection of processes requiring a slow speed and space (Sommer, 2012). Mesocosms are sections of a natural ecosystem or reconstructed ecosystems that are separated from the environment by

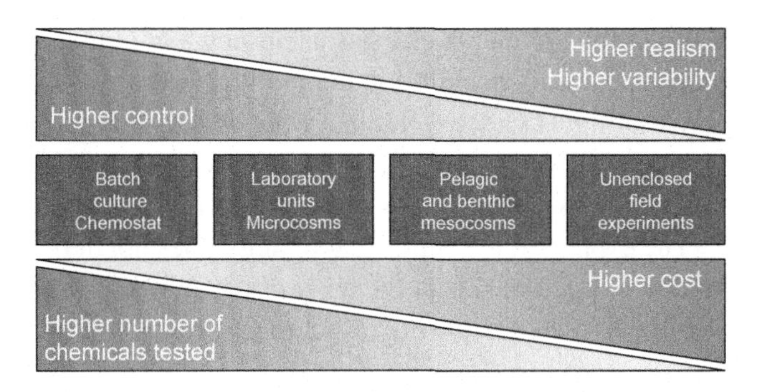

Figure 6.1
The main experimental approaches in ecotoxicology presented along a gradient of scale.

physical barriers. The literature provides many examples of mesocosms representative of benthic or pelagic systems, and in the case of freshwater environments of both lotic and lentic conditions. Mesocosms can operate with natural species assemblages, but experimental manipulations including removal or addition of organisms are also well-represented in the literature (e.g., Buffet et al., 2013, 2014). The last step consists of field experiments. They can be carried out in spatially well-confined ecosystems such as the small lakes used by Kidd et al. (2007) to study the fate of a fish population after exposure to a synthetic estrogen. More frequently, they are based on active biomonitoring procedures (De Kock and Kramer, 1994) such as caging of organisms in contaminated areas exposed to a well-defined source of contamination compared to a relatively clean area (e.g., Sanchez et al., 2011).

6.1.1 Choice of Experimental Concentrations and Exposure Scenario

Considering both laboratory units and experimental ecosystems, the choice of the chemical concentrations tested is crucial. Typically, short-term toxicity tests frequently used in conventional risk assessment were based on a large range of concentrations including the median lethal concentration (LC_{50}). Even in academic studies, high concentrations have also been used to obtain "clearer" responses. Such high concentrations, even when they are sublethal (e.g., one-tenth of the LC_{50}) remain environmentally unrealistic, never being encountered in the environment, even in aquatic systems strongly impacted by anthropogenic activities. Despite being clearer, results may be artifactual, as shown for many proteins, enzymatic or not, used as biomarkers: increasing chemical concentrations first induce an increase of the biomarker response then when the degree of exposure increases again, the biomarker response is dramatically depleted. Such bell-shaped relationships are known for glutathione S-transferase and catalase (Dagnino et al., 2007), ethoxyresorufin-O-deethylase (Ortiz-Delgado et al., 2008), and metallothioneins (Barka et al., 2001; Amiard-Triquet and Roméo, in Amiard-Triquet et al., 2011). The decrease of biomarker responses at very high toxicant concentrations is most probably from severe pathological impact upon target organs—abolishing any adaptive potential to cope with pollutants—as explained by Köhler et al. (2000) in the case of hsp 70.

In the clam *Scrobicularia plana*, lethal toxicity has been shown in individuals exposed to zinc in acute toxicity tests that had zinc concentration in their body lower than those of individuals chronically exposed in Restronguet Creek (UK) that were able to survive and reproduce (Amiard, 1991). Such a counterintuitive pattern—a higher incorporated dose of metals having a lower effect—has also been observed in the polychaete worm *Arenicola marina* (Casado-Martinez et al., 2010). The explanation again is that toxicity does not depend only on the accumulated concentration of toxic metals in the body of the animal, but also on the rate at which the toxic metal is accumulated, allowing or not the expression of adaptive potential.

The initial physicochemical characteristics and the fate of contaminants added in the experimental medium can deeply influence the results of bioassays as reviewed by Bejarano et al. (2014) and Martin et al. (2014) for fuel oils and dispersants and for nanoparticles in Chapter 17. Regardless of the conditions (static, semistatic or flow-through; see Chapter 2), it is generally useful to provide measured concentrations in addition to nominal concentrations. The total concentration may be insufficient when certain constituents are mainly responsible for the observed toxicity. For instance, Martin et al. (2014) indicate that polyaromatic hydrocarbon monitoring is key to accurate risk assessment of heavy fuel oils.

A more specific technical problem is the use of solvents for chemicals that are relatively water-insoluble. In this case, water-miscible solvents are currently used for testing them and Organization for Economic Co-operation and Development (OECD) recommends a maximum concentration of $100\,\mu L^{-1}$. However, several studies highlighted effects of lower concentrations of solvents (e.g., Turner et al., 2012; for a review, see Hutchinson et al., 2006) leading Lecomte et al. (2013) to recommend avoiding their use or, where this use is inevitable, the concentrations and type of solvents should be investigated for suitability for the measured endpoints before use in chemical testing strategies.

6.1.2 Single-Species Bioassay

Conventional risk assessment is mainly based on this category of tests (Chapter 2). Analyses of the ecotoxicity data submitted within the framework of the Registration, Evaluation, Authorization and Restriction of Chemicals (REACH) regulation in Europe have been recently published. They report both experimental sediment toxicity assays (Cesnaitis et al., 2014) and experimental aquatic toxicity assays (Tarazona et al., 2014). In this context, single-species bioassays were predominantly used, whereas multispecies studies appeared only at the margin.

6.1.3 Experimental Ecosystems

6.1.3.1 Strategies Used for Mimicking a Natural Environment

Experimental ecosystems are man-made constructions intended to simulate a natural environment. They are characterized by the presence of sediment, rebuilt food webs and interspecies relationships, mimic of hydrodynamics (lotic vs lentic mesocosms, devices reproducing the tidal cycle in marine mesocosms). They are most often used outdoors or, in some manner, incorporated intimately with the ecosystem that they are designed to reflect. They include microcosms, mesocosms, limnocorrals, and littoral enclosures (Nordberg et al., 2009). Lotic mesocosms or microcosms consist in streams of various sizes used to evaluate the effects of substances. Lentic systems aim at mimicking nonflowing water such as lakes and ponds. According to the International Union of Pure and Applied Chemistry, a mesocosm is an enclosed and essentially self-sufficient experimental environment or ecosystem that is on a larger scale than a laboratory microcosm. However, it is difficult to distinguish between microcosms and mesocosms based on spatial scale (size dimensions or volume) because there is some overlap and differences of opinion (OECD, 2004). Other definitions highlight the main characteristics that confer a major interest to mesocosms: a limited body of water, close to natural conditions, while allowing the manipulation of environmental factors (http://www.mesocosm.eu/node/16; FAO, 2009). Worldwide mesocosm facilities are many, particularly in Europe and North America, as exemplified by the experimental stream facility developed by the US Environmental Protection Agency (USEPA) in Cincinnati, Ohio (USEPA, 2011). The scientific production based on their use is listed at http://www.mesocosm.eu/ and http://www.science.gov/topicpages/y/year+mesocosm+approach.html#. Three main objectives have been assigned to mesocosm approaches in this field of ecotoxicological research: (1) determine the physicochemical forms, behavior, and fate of contaminants; (2) examine the bioaccumulation and/or the effects of contaminants on an individual species, with the distinctive feature that it lives in its reconstructed ecosystem; and (3) examine the effects of contaminants on populations, food webs, and communities.

The characteristics that allow the assessment of exposure of living organisms to contaminants have been highlighted for many different chemicals such as metals and metalloids (Muller et al., 2003; Riedel and Sanders, 2003; Smolyakov et al., 2010a,b; Bennett et al., 2012; Correia et al., 2013a; Lagauzère et al., 2013), metal-based nanoparticles (Cleveland et al., 2012; Lowry et al., 2012; Buffet et al., 2013, 2014), veterinary drug ivermectin (Sanderson et al., 2007), algicide Irgarol (Sapozhnikova et al., 2009), many different pesticides (Bromilow et al., 2006; Pablo et al., 2008), brominated flame retardants (de Jourdan et al., 2013), and petroleum hydrocarbons and polyaromatic hydrocarbons (Yamada et al., 2003; Coulon et al., 2012). Studies of behavior and fate of chemicals and oils, in particular in the event of real spills at sea, have been carried out using flume tanks, the most recent of which allow the reproduction of the height and frequency of waves, the exposure of surface slicks to the recreated spectrum of natural light, and air circulation at the water surface (Centre of Documentation, Research and Experimentation on Accidental Water Pollution, http://www.cedre.fr/en/cedre/flume-tank.php).

When studying the effects of different contaminants toward aquatic organisms, the presence of food and sediment provides better conditions and therefore time to complete the test may be longer than in laboratory experiments with water alone. The recovery of affected taxa after the application of a test toxic substance may be examined over important durations (99 days–4 years in the works of Sanderson et al., 2007; Caquet et al., 2007; Hanson et al., 2007; Mohr et al., 2007, 2008; López-Mancisidor et al., 2008; Liess and Beketov, 2011).

When considering individual species, mesocosms—including all the possible sources of contamination (water, sediment, diet)—permit an improvement of the realism of exposure conditions that has been used, for instance, for the assessment of environmental hazards associated with metal-based nanoparticles (Ferry et al., 2009; Buffet et al., 2013, 2014). Other examples include responses of molecular indicators in the common carp (*Cyprinus carpio*) exposed to the herbicides alachlor and atrazine (Chang et al., 2005) and vitellogenin gene expression in male fathead minnows (*Pimephales promelas*) exposed to 17α-ethynylestradiol (Gordon et al., 2006). In addition to improving realism, under these favorable conditions, organisms can live, develop, and reproduce, allowing the determination of life traits (development, growth, fecundity) as exemplified with the freshwater gastropod *Lymnaea* spp. exposed to pesticides and/or nonylphenol polyethoxylate as adjuvant (Baturo et al., 1995; Jumel et al., 2002) or with two amphibian species after chronic cadmium exposure in outdoor aquatic mesocosms (James et al., 2005).

Mesocosms are of particular interest for multispecies tests because they allow us to take into account both direct effects of contaminants and indirect effects from interaction between species in experimental ecosystems. Many studies have been devoted to assessing the effects of pesticides, but over the past decade, this experimental approach has been applied to other classes of contaminants (metal, petroleum hydrocarbons), particularly in coastal and estuarine mesoscosms (Table 6.1).

6.1.3.2 Replicability, Repeatability, Ecological Realism

Kraufvelin (1999) working with Baltic Sea hard bottom littoral mesocosms, has examined differences in community structure between controls run during the same year (replicability), differences among mesocosms run during different years (repeatability), and differences between mesocosms and the field mother system (ecological realism). He concluded that "the degrees of replicability, repeatability and ecological realism are too low for straightforward use of these and probably most other mesocosms in predictive risk assessment or in extrapolation of results to natural ecosystems." A limitation to mesocosm studies is the problem of replication, particularly with increasing size of the experimental system. In articles published from 1983 to 1993, 37% of mesocosms studies were not replicated (Petersen et al., 1999). Among about 100 papers reporting marine phytoplankton studies from 2000 to 2010, Watka (2012) found 32% lacking replication. In outdoor freshwater lentic mesocosms, a very low variability was observed for water temperature, pH, oxygen, and suspended solids, whereas nutrient concentrations were highly variable. Biological parameters measured at the individual level were less variable than those measured at the community level and functional descriptors frequently showed a lower variability than structural ones (Caquet et al., 2001).

Table 6.1: An Overview of the Use of Mesocosms to Reveal Direct and Indirect Effects of Contaminants in Multispecies Tests

Mesocosm	No. of Replicates	Contaminant	Duration	Organisms	Endpoint	Reference
Freshwater Mesocosms						
Outdoor tanks, 5000 L	2	Atrazine, herbicide	25 d	Freshwater phyto pK community	Chlorophyll fluorescence Tolerance Taxonomic composition	Seguin et al. (2002).
Outdoor artificial ponds, 1480 L	3	Carbaryl, insecticide Atrazine, herbicide	56–88 d	Amphibian community	Survival, body size, development, and time to metamorphosis	Boone and James (2003).
Outdoor artificial ponds, 12,000 L	3	Ivermectin veterinary drug	265 d	Zoo pK	Abundance Species richness	Sanderson et al. (2007).
Indoor ponds 15 m^3 Indoor streams 40 m^3	3 (control) 1 (treatment)	Metazachlor herbicide	140–170 d	Macrophytes	Biomass of different species	Mohr et al. (2007).
Indoor ponds 15 m^3 Indoor streams 40 m^3	3 (control) 1 (treatment)	Metazachlor herbicide	140–170 d	pK community	Abundance, dynamics Community structure	Mohr et al. (2008).
Outdoor streams, 20 m long × 1 m wide × 30 cm deep	3	Copper	18 mo	Primary producers Macroinvertebrates Fish	Abundance, richness, biomass Leaf litter decomposition Individual and population responses of fish	Roussel et al. (2007a,b, 2008).
Outdoor tanks 11 m^3	5	Chlorpyrifos OP insecticide	99 d	Zoo pK	Population and community effects	López-Mancisidor et al. (2008).
Outdoor streams (4 m long, 0.1 m wide)	4	Lambda-cyhalothrin Pyrethroid	10 d	Macroinvertebrate assemblage	Drift, algal biomass Decomposition of leaf litter	Rasmussen et al. (2008).
Outdoor lentic microcosm (Ø: 1.05 m; height: 0.9 m, 780 L)	2	Gamma-cyhalothrin Pyrethroid	73 d	Lentic invertebrates	Population dynamics Species composition Chlorophyll-*a* (phyto pK) Macrophyte biomass Leaf decomposition	Wijngaarden et al. (2009).
Outdoor ponds 9 m^3	4 (control) 3 (treatment)	Different pesticides used in crop protection programs	10 mo	Benthic macroinvertebrate communities	Drift, abundance of different taxa Decomposition of leaf litter	Auber et al. (2011).

Continued

Table 6.1: An Overview of the Use of Mesocosms to Reveal Direct and Indirect Effects of Contaminants in Multispecies Tests—cont'd

Mesocosm	No. of Replicates	Contaminant	Duration	Organisms	Endpoint	Reference
Outdoor stream mesocosms, 5000 L	4–10 (control) 2 (treatment)	Thiacloprid neonicotinoid insecticide	4 y	Community transplanted from an uncontaminated stream	Abundance Species richness SPEAR index	Liess and Beketov (2011).
Indoor experimental stream (12 m long, 30–50 cm wide)	Pseudoreplicates	Triclosan Antimicrobial	56 d	Micro-, meio-, macrobenthos	Periphyton chlorophyll *a*, composition Periphytic bacteria cell density, resistance Macrobenthos composition	Nietch et al. (2013).
Lotic mesocosm, 20 m long × 1 m wide	3	Bisphenol A Plastics and resin constituent	165 d	Different trophic levels	Watercress volume Macroinvertebrates' community structure Abundance of fish groups, size, gonad morphology	de Kermoysan et al. (2013).
Estuarine and Coastal Mesocosms						
Indoor tanks, 5000 L	None	Water-soluble fraction of heavy oil	6 d	Coastal bacteria, virus, HNF	FDC of bacteria Population densities	Ohwada et al. (2003).
Outdoor, 1 m^3 Flow-through with natural water	4–8	Trace elements + nutrients	28 d	Estuarine phyto pK community	Primary production Size class frequency Biomass of different species	Reidel et al. (2003) and Reidel and Sanders (2003).

In situ estuarine mesocosms, 25 L	2	Diesel fuel	6 d	Bacteria, phyto pK	Production, densities, Chlorophyll *a*	Nayar et al., 2005.
Pelagic enclosure in a fjord, 3 m^3	3	Zn pyrithione, antifungal, antibacterial	6 d	pK community	Functional responses PICT	Hjorth et al. (2006).
Pelagic enclosure in a fjord, 3 m^3	3	Pyrene	12 d	pK food web	Function and structure	Hjorth et al. (2007).
Coastal stream mesocosms, 4 m long, 30 cm diameter	2 × 2 (control) 3 (treatments)	Chlorpyrifos OP insecticide	124 d	Macroinvertebrate communities	Abundance Richness	Colville et al. (2008).
Tidal mesocosms 42 cm long, 32 cm wide, 25.5 cm deep	3	Simulated oil spill	21d	Bacteria Benthic phototrophs	rRNA sequences Chlorophyll *a* fluorescence Photosystem II	Coulon et al. (2012) and Chronopoulou et al. (2013).
Tidal mesocosms 25 × 15 × 6 cm deep	5	Metal contaminated sediments	6 wk	Benthic community	Number of individuals, number of taxa	O'Brien and Keough (2013).

FDC, frequency of dividing cells; HNF, heterotrophic nano-flagellates; PICT, pollution-induced community tolerance; pK, plankton; SPEAR, species at risk.

Similarly, in stream mesocosms, a good replicability of physicochemistry was shown. Both benthic and drifting assemblages were comparable among mesocosms, whereas small differences in taxonomic composition were evident among particular mesocosms, with higher or lower abundances observed for a minority (5%) of taxa. The authors concluded that "large ($4\,m^2$) outdoor flow-through mesocosms can be replicable when located near to the source system and allowed to establish naturally" (Harris et al., 2007). The synchronous use of pond and stream mesocosms has been used to compare the reaction of same species to the pesticide metazachlor under different environmental conditions (Mohr et al., 2008). Comparing the principal response curves obtained by ordination technique and similarity index data obtained for phyto- and zooplankton as well as macrophytes, variability was lower in ponds than in streams, maybe because these groups played a minor role in stream mesocosms. The authors suggest that "the resulting lower abundance of macrophytes and plankton led to a higher variability and less clear results as compared to the ponds." In this context, the statistical treatment of data must be carefully chosen. Univariate statistical methods may be used for parameters showing moderate intermesocosm variability (physicochemical or species-level biometric parameters), whereas multivariate techniques seems more appropriate for other levels of investigation (Caquet et al., 2001).

In addition to replicability, repeatability should be considered when results of such studies are used in risk assessment (Wong et al., 2004). In outdoor stream mesocosm studies conducted over 4 years at the same facility, these authors have shown that invertebrate assemblages were structurally distinct at the start of each study and the taxa responsible for differences in the assemblages were also different each year.

More recently, Beaudouin et al. (2012) have developed a tool able to precisely predict control mesocosm outputs, to which endpoints in mesocosms exposed to chemicals could be compared. Monitoring mosquitofish (*Gambusia holbrooki*) populations in lentic mesocosms, they applied a stochastic modeling of the control fish populations to assess the probabilistic distributions of population endpoints. These authors concluded that this strategy allows the detection of more significant and biologically relevant perturbations than classical methods.

6.1.4 In Situ Bioassays

Chemical monitoring in biota is well-developed for a number of anthropogenic contaminants (https://www.google.fr/#q=JAMP+Guidelines+for+Monitoring+Contaminants+in+Biota). However, not all contaminants or synergistic and antagonistic toxic effects can be determined by solely analyzing the physicochemical factors. Thus, various bioassays may be built on the ability of aquatic organisms to sense a wide range of pollutants (special series on in situ–based effects measures linking responses to ecological consequences in aquatic ecosystems, *Integr. Environ. Assess. Manag.*, 2007, Volume 3, Number 2).

A first approach consists in the collection of environmental samples (effluents, sediments) in the ecosystem under examination for quality assessment. Then the test organisms may be exposed to (1) dilution series of the effluent as recommended in the USEPA's whole effluent toxicity testing program (http://water.epa.gov/scitech/methods/cwa/wet/), (2) different concentrations of total decanted sediment (reflecting whole-sediment toxicity), or (3) sediment elutriates (a treatment reflecting the toxicity of the contaminant fraction which can be released from sediment during resuspension, for instance on the occasion of a dredging operation) (Geffard et al., 2005). The biological models may be standard test species, such as *Daphnia magna* (e.g., Maltby et al., 2000), *Chironomus* spp. (e.g., Li et al., 2013), invertebrate larvae (His et al., 1999; OEHHA, 2004), amphipod *Hyalella azteca* (Bundschuh et al., 2011), and fish *Pimephales promelas* (recommended by USEPA; e.g., Garcia-Reyero et al., 2011). The toxicological endpoints may be classical parameters such as the immobilization of *D. magna* (Straus) in a 48-h test (OECD, 1984; Maltby et al., 2000), oyster embryo-larval abnormalities (Leverett and Thain, 2013), or more innovative endpoints such as competitive nest holding behavior, changes in vitellogenin, correlated to changes in gene expression in *P. promelas* (Garcia-Reyero et al., 2011).

Very often, the so-called in situ toxicity assays are not totally carried out in the field. The first step consists in the exposure of caged test organisms in the ecosystem under investigation or in water diverted from the natural medium. Then the organisms are brought back to the laboratory to determine behavioral, physiological, or biochemical impairments. The biological models may be again standard test species, such as *D. magna* (e.g., Maltby et al., 2000), oyster larvae (Quiniou et al., 2007), or microalgae *Scenedesmus acutus* grown in dialysis membrane bags (Bauer et al., 2012). In most cases, indigenous organisms are preferred in these in situ bioassays, such as endobenthic invertebrates (Chapter 10), freshwater gastropods (e.g., Correia et al., 2013b; De Castro-Català et al., 2013), gammarids (Chapter 11), other crustaceans (e.g., Perez and Wallace, 2004; Moreira et al., 2006a; Steward et al., 2010), and fish (e.g., Triebskorn et al., 1997; Weis et al., 2011).

Biological early warning systems, often based on behavioral monitoring, involve the use of many organisms representative of various trophic levels (Gerhardt et al., 2006; Bae and Park, 2014). On-line biomonitoring systems are used to examine the responses of caged test organisms exposed in situ either in a bypass system or directly instream, as widely exemplified with the Multispecies Freshwater Biomonitor (http://www.limco-int.com/#!scientific-publications/c15cb). This device allows quantitative records of behavioral responses of vertebrates and invertebrates to environmental changes. It has been successfully employed with freshwater invertebrates and fish before being adapted to the marine context (Stewart et al., 2010; Buffet et al., 2013).

These in situ tests are sometimes associated with environmental chemistry, more or less explicitly in the framework of a toxicity identification evaluation procedure (described in

Chapter 4). The global approach combining biotesting, fractionation, and chemical analysis helps to identify hazardous compounds in complex environmental mixtures (Burgess et al., 2013). Information combining results of toxicity tests and the body burden of animals could offer the ability to identify the chemicals that should be reduced in amount to improve the state of a site (Hellou, 2011).

6.1.5 Comparing the Results Obtained with Laboratory, In Situ Bioassays and Mesocosms

Many factors can bias the extrapolation of experimental situations to natural environments. For instance, sediment toxicity assays using juvenile *Mercenaria mercenaria* have shown that growth rates of field-deployed clams tended to be higher than growth rates determined in laboratory assays, especially at the reference sites. Field studies indicated a higher potential for toxicity than did the laboratory studies at degraded sites (Ringwood and Keppler, 2002). Comparing laboratory and in situ assays with the polychaete *Hediste diversicolor*, Moreira et al. (2006b) have shown differences for postexposure feeding rates and for the activity of a biotransformation enzyme (glutathione *S*-transferase) under control conditions. The presence of potential stress factors, to which the organisms may be exposed in the field (handling, transport, caging) has been proposed as explanatory hypothesis. Sediments that were demonstrated to reduce fecundity of the amphipod *Melita plumulosa* in a laboratory reproduction test were not similarly toxic when amphipods were exposed to the same sediments in the field. The regular tidal renewal of overlying water experienced by these amphipods caged in a large saltwater lagoon likely prevented the accretion of dissolved zinc in the overlying water, also providing additional or alternative nutrition that limited the need for amphipods to interact with the contaminated sediments (Mann et al., 2010).

Pablo et al. (2008) designed a study allowing the comparison of the fate of chlorpyrifos and its toxicity to common freshwater invertebrates in the laboratory and in stream mesocosms. Despite the differences in the dynamics of chlorpyrifos in the laboratory and mesocosm systems—likely from the mass transport of chlorpyrifos from the mesocosm via stream flow—the sensitivities of the mayfly *Atalophlebia australis* and the cladoceran *Simocephalus vetulus* were similar in both systems. Wijngaarden et al. (2009) have also reported consistent results when they compared the effects of the pyrethroid insecticide gamma-cyhalothrin on aquatic invertebrates in laboratory and outdoor microcosm tests (with size shown in Table 6.1, they would be considered as mesocosms by other authors). Studying behavioral changes in the freshwater amphipod *Gammarus pulex*, the stonefly *Leuctra nigra*, and the mayfly *Heptagenia sulphurea* exposed to the pyrethroid lambda-cyhalothrin, Nørum et al. (2010) concluded that it was possible to extrapolate directly from pyrethroid-induced behavioral changes observed in the laboratory to drift under more realistic conditions in stream microcosms. Comparing biomarker responses of ragworms *Hediste diversicolor* and clams *Scrobicularia plana* exposed to copper or silver either soluble or as nanoparticles in laboratory microcosms and in outdoor mesocosms, Mouneyrac et al. (2014) have shown a certain degree of consistency. However, greater responses were observed in both species exposed in mesocosms for certain parameters. This could be due to the fact that in laboratory microcosms, the only source of contamination was seawater whereas in outdoor mesocosms organisms were exposed in addition to dietary and sediment-bound contaminants.

6.2 Improve the Determination of No Observed Effect Concentrations

As underlined by Alonso and Camargo (2012), "short-term lethal bioassays are not suited for assessing the real effects of pollutants in natural ecosystems, as their concentrations are usually unrealistic under an ecological point of view." Chronic bioassays are more realistic for assessing effects on aquatic animals and, since the beginning of the 2000s, several sub-acute and chronic tests have been recommended as exemplified in Figure 6.2 for benthic marine organisms. In this case, as also when in situ bioassays are carried out or when mesocosms are run, mortality is generally limited and the endpoints of interest are more often growth and reproduction.

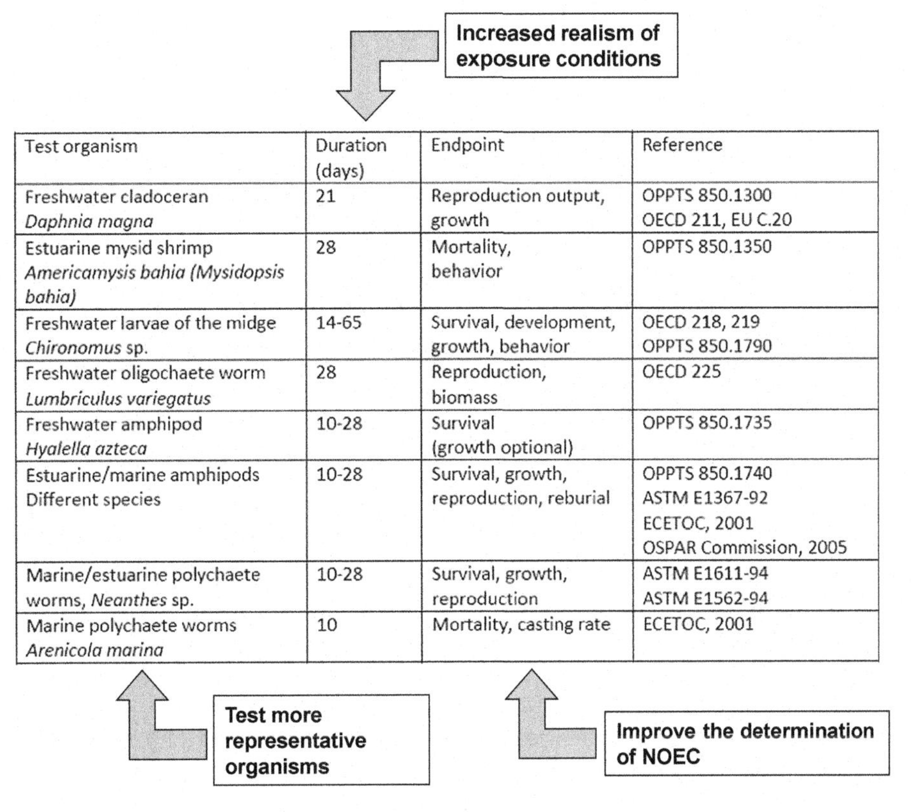

Test organism	Duration (days)	Endpoint	Reference
Freshwater cladoceran *Daphnia magna*	21	Reproduction output, growth	OPPTS 850.1300 OECD 211, EU C.20
Estuarine mysid shrimp *Americamysis bahia (Mysidopsis bahia)*	28	Mortality, behavior	OPPTS 850.1350
Freshwater larvae of the midge *Chironomus* sp.	14-65	Survival, development, growth, behavior	OECD 218, 219 OPPTS 850.1790
Freshwater oligochaete worm *Lumbriculus variegatus*	28	Reproduction, biomass	OECD 225
Freshwater amphipod *Hyalella azteca*	10-28	Survival (growth optional)	OPPTS 850.1735
Estuarine/marine amphipods Different species	10-28	Survival, growth, reproduction, reburial	OPPTS 850.1740 ASTM E1367-92 ECETOC, 2001 OSPAR Commission, 2005
Marine/estuarine polychaete worms, *Neanthes* sp.	10-28	Survival, growth, reproduction	ASTM E1611-94 ASTM E1562-94
Marine polychaete worms *Arenicola marina*	10	Mortality, casting rate	ECETOC, 2001

Figure 6.2
Toxicity tests with benthic marine organisms.

Focusing on responses linked to the success of reproduction is ecologically relevant because it is crucial for the conservation of populations. Cascading effects of chemical stress through the different levels of biological organization can lead to local extinction of populations. The use of subindividual effects may be very interesting to improve the sensitivity of no observed effect concentrations because they respond precociously and at low doses. However, among subindividual effects, it is needed to give priority to those that are a good link between physiological and supraindividual effects such as the so-called "ecological biomarkers," including those linked to energy metabolism, biomarkers of neurotoxicity, genotoxicity, or endocrine disruption (Amiard-Triquet et al., 2013; Mouneyrac and Amiard-Triquet, 2013). At the individual level, behavioral biomarkers are particularly relevant (Hellou, 2011; Weis et al., 2011; Alonso and Camargo, 2012; Amiard-Triquet and Amiard, in Amiard-Triquet et al., 2013) because they are sensitive (responding at concentrations existing in the environment as exemplified for polychlorinated biphenyls by Péan et al., 2013), noninvasive, cost-effective, and rely on diverse intimate mechanisms of life (neurotransmitters, hormones, energy) that can be used as biomarkers of effects and result in disturbances that can lead until "ecological death" (Roast et al., 2000; Scott and Sloman, 2004). In addition, the development and application of a biological early warning system based on behavioral responses and the computational methods used to process the behavioral monitoring data have been recently reviewed (Bae and Park, 2014).

The large gap that exists between assessment capacities and assessment needs raises economic concerns, favoring the development of high throughput decision support tools for chemical risk management, an approach that have been developed first in the field of human toxicology (Kavlock et al., 2012). However, ecotoxicogenomics are also developing into a key tool for the assessment of environmental impacts and environmental risk assessment for aquatic ecosystems (Piña and Barata, 2011). Changes detected at the most intimate level of interaction between biological systems and environmental contaminants are usually considered at the most sensitive ones. However, for aquatic species exposed to cadmium, responses in gene expression were on average four times above the no observed effect concentrations and 11 times below the LC_{50} values (Fedorenkova et al., 2010). Thus these authors concluded that "to confirm the sensitivity claimed by ecotoxicogenomics, more testing at low concentrations is needed." It has also been claimed that ecotoxicogenomics would be able to bridge the gap between genes and populations (Fedorenkova et al., 2010). Similarly, Vandenbrouck et al. (2009) suggest that genes linked to the energy budget parameters could become molecular biomarkers with ecological relevance. To reach such ambitious objectives, there is a need to go many steps further.

Omics techniques can clearly help learning how chemical causes toxicity, provided that functional aspects are soundly taken into account, relating the mRNA and protein levels, and protein activity of individual genes (Chapter 8). The next step is to become able to integrate the expression of mRNA and protein profiles within their physiological and toxicological contexts (Fent and Sumpter, 2011). Here we join the concept of cascading effects evoked above as well as that of adverse outcome pathway defined as the linkage between an initial molecular event and an adverse outcome at the individual or population level (Ankley et al., 2010; Patlewicz et al., 2013). Application of adverse outcome pathways at higher levels of biological organization (community, ecosystem functioning, and ecosystem services) is more challenging, particularly because it does not include the potential for homeostasis or other compensatory feedbacks, such as those that might occur from interactions among individuals or species (Connon et al., 2012; Forbes and Calow, 2012; Kattwinkel and Liess, 2014).

6.3 Improve the Selection of Test Species

The criteria that make a species relevant for experimental aquatic testing are reviewed in Chapter 10. Ecotoxicity data remain mainly based on single-species bioassays (Cesnaitis et al., 2014; Tarazona et al., 2014), the most frequently used species being available from in-house or commercial sources as recommended for many standardized bioassays (USEPA, 2002). To improve the realism of experimental aquatic tests, the use of field-collected organisms, representative of the real world, has strong appeal (Chapter 10).

Using sensitive species or life stages may be a basis for risk assessments insuring more protective strategies for the environment. A review on the use of embryos and larvae of mollusks in toxicity tests confirmed the hypothesis of the relative sensitivity of early life stages, indicating, however, some considerable exceptions such as resistance to cadmium or insecticides (His et al., 1999). Despite some other counterexamples may be found in the literature (Berthet, in Amiard-Triquet et al., 2013), early life stages are already in current use for ecotoxicological bioassays, including many standardized ones (for an overview, see OSPAR Commission, 2000; USEPA, 2002). Short-term toxicity test on embryo and larval stages with freshwater or marine fish are also available from OECD (OECD 212) and ISO (ISO 12890) as also early-life stage toxicity test (OECD 210) and juvenile fish growth test (OECD 215). Invertebrate embryos (sea urchin, bivalves) and fish larvae are proposed for monitoring contaminant effects under the auspices of the International Council for the Exploration of the Sea, a strategy that can be applied for the implementation of the European Union Marine Strategy Framework Directive (Lyons et al., 2010). The design (e.g., Leverett and Thain, 2013) and interpretation (Wheeler et al., 2014) of tests based on early and sensitive life stages are still in progress.

6.4 Improve Mathematical and Computational Data Analyses

Even in a basic approach such as the determination of a classical sigmoid dose–effect relationship, the statistical determination of toxicological parameters needs improvements such as the use of the benchmark dose approach (for details, see Chapter 2). Nonmonotonic dose–response relationships may be observed in the case of hormesis, characterized by a low dose stimulation and a high dose inhibition for many environmental contaminants, as reviewed by Calabrese and Blain (2011). It is also well-documented in the case of endocrine disrupting chemicals (Vandenberg et al., 2012; Bergman et al., 2013). Accurate description of nonmonotonic dose–response curves is a key step for the characterization of hazards associated to the presence of pollutants in the medium, thus needing the development of dedicated models (Qin et al., 2010; Zhu et al., 2013).

The importance of relevant statistical treatments have been underlined in complementarity with several methodologies described previously, such as the replicability of mesocosm studies. Appropriate mathematical and computational data analyses are also necessary to manage and interpret the large datasets generated by biotests based on long-term and automatic observation studies (Bae and Park, 2014). The vast quantity of data produced by high-throughput methods such as toxicogenomics has needed the development in parallel of powerful analytical tools. Afshari et al. (2011) have highlighted the impact that bioinformatics and statistical analysis have had on the field of toxicogenomics over the past 5–10 years and proposed their vision of advancing toxicogenomics in future years in the field of human toxicology from which lessons can be learnt for ecotoxicology. With increasing realism of exposure conditions from laboratory units to field studies (Figure 6.1), a number of confounding factors (natural habitat differences, seasonal variations) can interfere with the biological responses to chemical stress. Modeling the influence of confounding factors can allow an improvement of the interpretation of in situ assays (Chapters 7 and 11).

6.5 Conclusion

"Are we in the dark ages of environmental toxicology?" asked McCarty in 2013. In fact, standardized environmental toxicity test methods that were designed decades ago are still in current use, despite their limitations, and have been underlined in the scientific literature for many years. However, regulatory objectives have substantially increased in breadth and depth (McCarty, 2013), thus the guidelines for toxicity testing launched by regulatory bodies have evolved, providing a better basis for environmental risk assessment (Chapter 2). But certainly "updated testing methods could provide some improvements in toxicological data quality" (McCarty, 2013).

If the use of experimental ecosystems or in situ bioassays can generate substantial additional cost, favoring the use of realistic concentrations for toxicity testing has no financial impact. Realistic concentrations can be chosen taking into account the highest concentrations of a given contaminant reported in the literature, for instance using Eisler's encyclopedia of environmentally hazardous priority chemicals (Eisler, 2007). When environmental data are not available as it may be the case for new chemicals or nanoparticles, predicted environmental concentrations can be used (Chapter 2).

The results of comparative in situ and laboratory sediment bioassays highlight that distinguishing contaminant effects from effects associated with natural habitat differences is a major challenge for in situ bioassays. However, normalization procedures have already been proposed to effectively account for natural differences in habitat conditions (Chapters 7 and 13). Papers by Burton et al. (2012) and Rosen et al. (2012) that have been accepted for inclusion in U.S. Navy Research, support the operational value of an integrated deployment system, the Sediment Ecotoxicity Assessment Ring, which incorporates rapid in situ hydrological, chemical, bioaccumulation, and toxicological lines of evidence for assessing sediment and overlying water contamination. The authors even consider that this approach is rapid, and should prove cost-effective, with many of the measurements being made on-site in less than one week.

Because of the different responses of organisms from the culture and the field to contaminants, it may be concluded that test species obtained under laboratory conditions are not totally relevant for risk assessment. On the other hand, considering the seasonal variability in the sensitivity of aquatic organisms (Chapter 7), the influence of confounding factors must be clearly assessed when using field-collected organisms. No univocal recommendation can be made as implicitly acknowledged by the USEPA (2002) stating that the use of test organisms taken from the receiving water is impractical but providing guidelines for collection (USEPA, 1973; USEPA, 1990; in USEPA, 2002).

Natural phenomena (quality and quantity of available food; washout effect of tidal renewal of overlying water in estuarine and coastal areas) can limit the effects of contaminants in the wild. Laboratory tests as well as field tests with caged animals, are likely to overestimate toxicity for species, also because organisms would avoid contamination in heterogeneous field settings (Ward et al., 2013). Avoidance is a common phenomenon, well documented in many zoological taxa toward different chemical classes and contaminated environmental matrices (Amiard-Triquet and Amiard, in Amiard-Triquet et al., 2013). However, despite contributing to the protection of individuals, avoidance can led to the local extinction of diverse species in contaminated medium and this risk should not been neglected.

Because the biological mode of action of contaminants and species' homologies are largely unknown, extrapolation from the toxic responses of laboratory test species to all species representing that group in the environment is an important source of uncertainty. Understanding the mechanistic basis for toxicological responses and identifying molecular response pathways can allow more accurate extrapolation of species responses as conceptualized by Celander et al. (2011).

The use of mesocosms can appear as the most complete procedure for a sound and realistic assessment of the fate and effects of contaminants in ecosystems. However, the counterparts are important, including scientific aspects (replicability, repeatability) and operational aspects (only a few contaminants can be assessed due to the high cost and the duration of experiments). It is not certain that the society is ready to pay for a better integration of scientific knowledge at a higher cost for the conservation of environment because the importance of ecosystem services are not very clear for many people in the general population. In addition, tests with living beings raise ethical and economic concerns and are considered as inappropriate for assessing all of the substances and effluents that require regulatory testing (Scholz et al., 2013). Hence, alternative methods (in vitro and in silico methods) have been proposed and are already accepted for integration in REACH registration dossiers (Grindon et al., 2006; ECHA, 2011).

Assessing the ecotoxicity of mixtures of contaminants is clearly a challenge for the future of ecotoxicology. The most direct way forward consists in the exposure of test organisms to mixtures belonging to the same class (e.g., different metals) or to different classes (pesticides, metals, oil, surfactants). However, this approach seems unrealistic because the inventory of chemical substances that have been submitted under the Toxic Substances Control Act includes the identities of more than 83,000 substances (http://www.epa.gov/oppt/existingchemicals/pubs/tscainventory/howto.html). New strategies are needed (Chapter 18).

In the case of emerging contaminants, the use of conventional risk assessment is highly questionable. For instance, nonmonotonic dose-response relationships are frequent for endocrine disrupting chemicals, as are the variations of their impact at different life stages, and need a special approach. In the case of manufactured nanomaterials, their physicochemical characterization is crucial for a sound assessment of exposure but it is particularly difficult to master (OECD, 2012). These issues will be documented in the chapters of this book devoted to different classes of emerging contaminants.

References

Afshari, C.A., Hamadeh, H.K., Bushel, P.R., 2011. The evolution of bioinformatics in toxicology: advancing toxicogenomics. Toxicol. Sci. 120 (S1), S225–S237.

Alonso, Á., Camargo, J.A., 2012. A video-based tracking analysis to assess the chronic toxic effects of fluoride ion on the aquatic snail *Potamopyrgus antipodarum* (Hydrobiidae, Mollusca). Ecotoxicol. Environ. Saf. 81, 70–75.

Amiard, J.C., 1991. Réponses des organismes marins aux pollutions métalliques. In: Réactions des êtres vivants aux changements de l'environnement, Actes des Journées de l'Environnement du C.N.R.S. CNRS, Paris, pp. 197–205.

Amiard-Triquet, C., Amiard, J.C., Rainbow, P.S., 2013. Ecological Biomarkers: Indicators of Ecotoxicological Effects. CRC Press, Boca Raton.

Amiard-Triquet, C., Rainbow, P.S., Romeo, M., 2011. Tolerance to Environmental Contaminants. CRC Press, Boca Raton.

Ankley, G.T., Bennett, R.S., Erickson, R.J., et al., 2010. Adverse outcome pathways: a conceptual framework to support ecotoxicology research and risk assessment. Environ. Toxicol. Chem. 29, 730–741.

Auber, A., Roucaute, M., Togola, A., et al., 2011. Structural and functional effects of conventional and low pesticide input crop-protection programs on benthic macroinvertebrate communities in outdoor pond mesocosms. Ecotoxicology 20, 2042–2055.

Bae, M.J., Park, Y.S., 2014. Biological early warning system based on the responses of aquatic organisms to disturbances: a review. Sci. Total Environ. 466–467, 635–649.

Barka, S., Pavillon, J.F., Amiard, J.C., 2001. Influence of different essential and non-essential metals on MTLP levels in the copepod *Tigriopus brevicornis*. Comp. Biochem. Physiol. 128C, 479–493.

Baturo, W., Lagadic, L., Caquet, T., 1995. Growth, fecundity and glycogen utilization in *Lymnaea palustris* exposed to atrazine and hexachlorobenzene in freshwater mesocosms. Environ. Toxicol. Chem. 14, 503–511.

Bauer, D.E., Conforti, V., Ruiz, L., et al., 2012. An *in situ* test to explore the responses of *Scenedesmus acutus* and *Lepocinclis acus* as indicators of the changes in water quality in lowland streams. Ecotox. Environ. Saf. 77, 71–78.

Beaudouin, R., Ginot, V., Monod, G., 2012. Improving mesocosm data analysis through individual-based modelling of control population dynamics: a case study with mosquitofish (*Gambusia holbrooki*). Ecotoxicology 21, 155–164.

Bejarano, A.C., Clark, J.R., Coelho, G.M., 2014. Issues and challenges with oil toxicity data and implications for their use in decision making: a quantitative review. Environ. Toxicol. Chem. 33, 732–742.

Bennett, W.W., Teasdale, P.R., Panther, J.G., et al., 2012. Investigating arsenic speciation and mobilization in sediments with DGT and DET: a mesocosm evaluation of oxic-anoxic transitions. Environ. Sci. Technol. 46, 3981–3989.

Bergman, A., Heindel, J.J., Jobling, S., et al. (Eds.), 2013. State of the Science of Endocrine Disrupting Chemicals – 2012. WHO/UNEP. http://unep.org/pdf/9789241505031_eng.pdf.

Boone, M.D., James, S.M., 2003. Interactions of an insecticide, herbicide, and natural stressors in amphibian community mesocosms. Ecol. Appl. 13, 829–841.

Bromilow, R.H., de Carvalho, R.F., Evans, A.A., et al., 2006. Behavior of pesticides in sediment/water systems in outdoor mesocosms. J. Environ. Sci. Health B 41, 1–16.

Buffet, P.E., Richard, M., Caupos, F., et al., 2013. A mesocosm study of fate and effects of CuO nanoparticles on endobenthic species (*Scrobicularia plana, Hediste diversicolor*). Environ. Sci. Technol. 47, 1620–1628.

Buffet, P.E., Zalouk-Vergnoux, A., Châtel, A., et al., 2014. A marine mesocosm study on the environmental fate of silver nanoparticles and toxicity effects on two endobenthic species: the ragworm *Hediste diversicolor* and the bivalve mollusc *Scrobicularia plana*. Sci. Tot. Environ. 470–471, 1151–1159.

Bundschuh, M., Zubrod, J.P., Seitz, F., et al., 2011. Mercury-contaminated sediments affect amphipod feeding. Arch. Environ. Contam. Toxicol. 60, 437–443.

Burgess, R.M., Ho, K.T., Brack, W., et al., 2013. Effects-directed analysis (EDA) and toxicity identification evaluation (TIE): complementary but different approaches for diagnosing causes of environmental toxicity. Environ. Toxicol. Chem. 32, 1935–1945.

Burton Jr., G.A., Rosen, G., Bart Chadwick, D., et al., 2012. A sediment ecotoxicity assessment platform for in situ measures of chemistry, bioaccumulation and toxicity. Part 1: system description and proof of concept. Environ. Pollut. 162, 449–456.

Calabrese, E.J., Blain, R.B., 2011. The hormesis database: the occurrence of hormetic dose responses in the toxicological literature. Reg. Toxicol. Pharmacol. 61, 73–81.

Caquet, T., Hanson, M.L., Roucaute, M., et al., 2007. Influence of isolation on the recovery of pond mesocosms from the application of an insecticide. II. Benthic macroinvertebrate responses. Environ. Toxicol. Chem. 26, 1280–1290.

Caquet, T., Lagadic, L., Monod, G., et al., 2001. Variability of physicochemical and biological parameters between replicated outdoor freshwater lentic mesocosms. Ecotoxicology 10, 51–66.

Casado-Martinez, M.C., Smith, B.D., Luoma, S.N., et al., 2010. Metal toxicity in a sediment-dwelling polychaete: threshold body concentrations or overwhelming accumulation rates? Environ. Pollut. 158, 3071–3076.

Celander, M.C., Goldstone, J.V., Denslow, N.D., 2011. Species extrapolation for the 21st century. Environ. Toxicol. Chem. 30, 52–63.

Cesnaitis, R., Sobanska, M.A., Versonnen, B., et al., 2014. Analysis of the ecotoxicity data submitted within the framework of the REACH Regulation. Part 3. Experimental sediment toxicity assays. Sci. Total Environ. 475, 116–122.

Chang, L.W., Toth, G.P., Gordon, D.A., et al., 2005. Responses of molecular indicators of exposure in mesocosms: common carp (*Cyprinus carpio*) exposed to the herbicides alachlor and atrazine. Environ. Toxicol. Chem. 24, 190–197.

Chronopoulou, P.M., Fahy, A., Coulon, F., et al., 2013. Impact of a simulated oil spill on benthic phototrophs and nitrogen-fixing bacteria in mudflat mesocosms. Environ. Microbiol. 15, 242–252.

Cleveland, D., Long, S.E., Pennington, P.L., et al., 2012. Pilot estuarine mesocosm study on the environmental fate of silver nanomaterials leached from consumer products. Sci. Total Environ. 421–422, 267–272.

Colville, A., Jones, P., Pablo, F., et al., 2008. Effects of chlorpyrifos on macroinvertebrate communities in coastal stream mesocosms. Ecotoxicology 17, 173–180.

Connon, R.E., Geist, J., Werner, I., 2012. Effect-based tools for monitoring and predicting the ecotoxicological effects of chemicals in the aquatic environment. Sensors 12, 12741–12771.

Correia, R.R.S., Martins de Oliveira, D.C., Guimarães, J.R.D., 2013a. Mercury methylation in mesocosms with and without the aquatic macrophyte *Eichhornia crassipes (mart.) Solms*. Ecotox. Environ. Saf. 96, 124–130.

Correia, V., Ribeiro, R., Moreira-Santos, M., 2013b. A laboratory and *in situ* postexposure feeding assay with a freshwater snail. Environ. Toxicol. Chem. 32, 2144–2152.

Coulon, F., Chronopoulou, P.M., Fahy, A., et al., 2012. Central role of dynamic tidal biofilms dominated by aerobic hydrocarbonoclastic bacteria and diatoms in the biodegradation of hydrocarbons in coastal mudflats. Appl. Environ. Microbiol. 78, 3638–3648.

Dagnino, A., Allen, J.I., Moore, M.N., et al., 2007. Development of an expert system for the integration of biomarker responses in mussels into an animal health index. Biomarkers 12, 155–172.

De Castro-Català, N., López-Doval, J., Gorga, M., et al., 2013. Is reproduction of the snail *Physella acuta* affected by endocrine disrupting compounds? An *in situ* bioassay in three Iberian basins. J. Hazard. Mater. 263, 248–255.

De Kock, W.C., Kramer, K.J.M., 1994. Active biomonitoring (ABM) by translocation of bivalve *Molluscs*. In: Kramer, K.J.M. (Ed.), Biomonitoring of Coastal Waters and Estuaries. CRC Press, Boca-Raton, pp. 51–84.

ECETOC, 2001. Risk Assessment in Marine Environments. Technical Report No. 82, 141 pp.

ECHA (European Chemicals Agency), 2011. The Use of Alternatives to Testing on Animals for the REACH Regulation. http://echa.europa.eu/documents/10162/13639/alternatives_test_animals_2011_en.pdf.

EFSA (European Food Security Authority), 2009. Use of the benchmark dose approach in risk assessment. 1. Guidance of the Scientific Committee (Question No EFSA-Q-2005-232). EFSA J. 1150, 1–72.

Eisler, R., 2007. Eisler's Encyclopedia of Environmentally Hazardous Priority Chemicals. Elsevier Science Ltd, Oxford.

FAO (Food and Agriculture Organization of the United Nations), 2009. Biosafety of Genetically Modified Organisms: Basic Concepts, Methods and Issues. FAO, Rome.

Fedorenkova, A., Arie Vonk, J., Rob Lenders, H.J., et al., 2010. Ecotoxicogenomics: bridging the gap between genes and populations. Environ. Sci. Technol. 44, 4328–4333.

Fent, K., Sumpter, J.P., 2011. Progress and promises in toxicogenomics in aquatic toxicology: is technical innovation driving scientific innovation? Aquat Toxicol. 105 (3–4 Suppl.), 25–39.

Ferry, J.L., Craig, P., Hexel, C., et al., 2009. Transfer of gold nanoparticles from the water column to the estuarine food web. Nat. Nano. 4, 441–444.

Forbes, V.E., Calow, P., 2012. Promises and problems for the new paradigm for risk assessment and an alternative approach involving predictive systems models. Environ. Toxicol. Chem. 31, 2663–2671.

Garcia-Reyero, N., Lavelle, C.M., Escalon, B.L., et al., 2011. Behavioral and genomic impacts of a wastewater effluent on the fathead minnow. Aquat. Toxicol. 101, 38–48.

Geffard, O., Geffard, A., Budzinski, H., et al., 2005. Mobility and potential toxicity of sediment-bound metals in a tidal estuary. Environ. Toxicol. 20, 407–417.

Gerhardt, A., Ingram, M.K., Kang, I.J., et al., 2006. *In situ* on-line toxicity biomonitoring in water: recent developments. Environ. Toxicol. Chem. 25, 2263–2271.

Gordon, D.A., Toth, G.P., Graham, D.W., et al., 2006. Effects of eutrophication on vitellogenin gene expression in male fathead minnows (*Pimephales promelas*) exposed to 17α-ethynylestradiol in field mesocosms. Environ. Pollut. 142, 559–566.

Grindon, C., Combes, R., Cronin, M.T.D., et al., 2006. Integrated decision-tree testing strategies for environmental toxicity with respect to the requirements of the EU REACH legislation. Altern Lab Anim. 34, 651–664.

Hanson, M.L., Graham, D.W., Babin, E., et al., 2007. Influence of isolation on the recovery of pond mesocosms from the application of an insecticide. I. Study design and planktonic community responses. Environ. Toxicol. Chem. 26, 1265–1279.

Harris, R.M.L., Armitage, P.D., Milner, A.M., et al., 2007. Replicability of physicochemistry and macroinvertebrate assemblages in stream mesocosms: implications for experimental research. Freshwater Biol. 52, 2434–2443.

Hellou, J., 2011. Behavioral ecotoxicology, an "early warning" signal to assess environmental quality. Environ. Sci. Pollut. Res. 18, 1–11.

His, E., Beiras, R., Seaman, M.N.L., 1999. The assessment of marine pollution—bioassays with bivalve embryos and larvae. Adv. Mar. Biol. 37, 1–178.

Hjorth, M., Dahllöf, I., Forbes, V.E., 2006. Effects on the function of three trophic levels in marine plankton communities under stress from the antifouling compound zinc pyrithione. Aquat. Toxicol. 77, 105–115.

Hjorth, M., Vester, J., Henriksen, P., et al., 2007. Functional and structural responses of marine plankton food web to pyrene contamination. Mar. Ecol. Prog. Ser. 338, 21–31.

Hutchinson, T.H., Shillabeer, N., Winter, M.J., et al., 2006. Acute and chronic effects of carrier solvents in aquatic organisms: a critical review. Aquat. Toxicol. 76, 69–92.

James, S.T., Little, E.E., Semlitsch, R.D., 2005. Metamorphosis of two amphibian species after chronic cadmium exposure in outdoor aquatic mesocosms. Environ. Toxicol. Chem. 24, 1994–2001.

de Jourdan, B.P., Hanson, M.L., Muir, D.C.G., et al., 2013. Environmental fate of three novel brominated flame retardants in aquatic mesocosms. Environ. Toxicol. Chem. 32, 1060–1068.

Jumel, A., Coutellec, M.A., Cravedi, J.P., et al., 2002. Nonylphenol polyethoxylate adjuvant mitigates the reproductive toxicity of fomesafen on the freshwater snail *Lymnaea stagnalis* in outdoor experimental ponds. Environ. Toxicol. Chem. 21, 1876–1888.

Kattwinkel, M., Liess, M., 2014. Competition matters: species interactions prolong the long-term effects of pulsed toxicant stress on populations. Environ. Toxicol. Chem. 33, 1458–1465.

Kavlock, R., Chandler, K., Houck, K., et al., 2012. Update on EPA's ToxCast program: providing high throughput decision support tools for chemical risk management. Chem. Res. Toxicol. 25, 1287–1302.

de Kermoysan, G., Joachim, S., Baudoin, P., et al., 2013. Effects of bisphenol A on different trophic levels in a lotic experimental ecosystem. Aquat. Toxicol. 144–145, 186–198.

Kidd, K.A., Blanchfield, P.J., Mills, K.H., et al., 2007. Collapse of a fish population after exposure to a synthetic estrogen. Proc. Natl. Acad. Sci. U.S.A. 104, 8897–8901.

Köhler, H.R., Zanger, M., Eckwert, H., et al., 2000. Selection favours low hsp70 levels in chronically metal-stressed soil arthropods. J. Evol. Biol. 13, 569–582.

Kraufvelin, P., 1999. Baltic hard bottom mesocosms unplugged: replicability, repeatability and ecological realism examined by non-parametric multivariate techniques. J. Exp. Mar. Biol. Ecol. 240, 229–258.

Lagauzère, S., Motelica-Heino, M., Viollier, E., et al., 2013. Remobilisation of uranium from contaminated freshwater sediments by bioturbation. Biogeosciences Discuss. 10, 17001–17041.

Lecomte, V., Nourya, P., Tutundjian, R., 2013. Organic solvents impair life-traits and biomarkers in the New Zealand mudsnail *Potamopyrgus antipodarum* (Gray) at concentrations below OECD recommendations. Aquat. Toxicol. 140–141, 196–203.

Leverett, D., Thain, J., 2013. Oyster Embryo-larval Bioassay (Revised). ICES Techniques in Marine Environmental Sciences, No. 54. http://www.ices.dk/sites/pub/Publication%20Reports/Techniques%20in%20Marine%20Environmental%20Sciences%20%28TIMES%29/times54/TIMES%2054%20web.pdf (last accessed 10.12.14).

Li, H.Z., Sun, B.Q., Lydy, M.J., et al., 2013. Sediment-associated pesticides in an urban stream in Guangzhou, China: implication of a shift in pesticide use patterns. Environ. Toxicol. Chem. 32, 1040–1047.

Liess, M., Beketov, M., 2011. Traits and stress: keys to identify community effects of low levels of toxicants in test systems. Ecotoxicology 20, 1328–1340.

López-Mancisidor, P., Carbonell, G., Marina, A., et al., 2008. Zooplankton community responses to chlorpyrifos in mesocosms under Mediterranean conditions. Ecotox. Environ. Saf. 71, 16–25.

Lowry, G.V., Espinasse, B.P., Badireddy, A.R., et al., 2012. Long-term transformation and fate of manufactured Ag nanoparticles in a simulated large scale freshwater emergent wetland. Environ. Sci. Technol. 46, 7027–7036.

Lyons, B.P., Thain, J.E., Stentiford, G.D., et al., 2010. Using biological effects tools to define Good Environmental Status under the European Union Marine Strategy Framework Directive. Mar. Pollut. Bull. 60, 1647–1651.

Maltby, L., Clayton, S., Yu, H., et al., 2000. Using single-species toxicity tests, community-level responses, and toxicity identification evaluations to investigate effluent impacts. Environ. Toxicol. Chem. 19, 151–157.

Mann, R.M., Ross, Y., Hyne, V., et al., 2010. A rapid amphipod reproduction test for sediment quality assessment: *in situ* bioassays do not replicate laboratory bioassays. Environ. Toxicol. Chem. 29, 2566–2574.

Martin, J.D., Adams, J., Hollebone, B., et al., 2014. Chronic toxicity of heavy fuel oils to fish embryos using multiple exposure scenarios. Environ. Toxicol. Chem. 33, 677–687.

McCarty, L.S., 2013. Are we in the dark ages of environmental toxicology? Regul. Toxicol. Pharmacol. 67, 321–324.

Messiaen, M., Janssen, C.R., De Meester, L., et al., 2013. The initial tolerance to sub-lethal Cd exposure is the same among ten naïve pond populations of *Daphnia magna*, but their micro-evolutionary potential to develop resistance is very different. Aquat. Toxicol. 144–145, 322–331.

Mohr, S., Berghahn, R., Feibicke, M., et al., 2007. Effects of the herbicide metazachlor on macrophytes and ecosystem function in freshwater pond and stream mesocosms. Aquat. Toxicol. 82, 73–84.

Mohr, S., Feibicke, M., Berghahn, R., et al., 2008. Response of plankton communities in freshwater pond and stream mesocosms to the herbicide metazachlor. Environ. Pollut. 152, 530–542.

Moreira, S.M., Lima, I., Ribeiro, R., et al., 2006b. Effects of estuarine sediment contamination on feeding and on key physiological functions of the polychaete *Hediste diversicolor*: laboratory and *in situ* assays. Aquat. Toxicol. 78, 186–201.

Moreira, S.M., Moreira-Santos, M., Guilhermino, L., et al., 2006a. An *in situ* postexposure feeding assay with *Carcinus maenas* for estuarine sediment-overlying water toxicity evaluations. Environ. Pollut. 139, 318–329.

Mouneyrac, C., Amiard-Triquet, C., 2013. Biomarkers of ecological relevance. In: Férard, J.F., Blaise, C. (Eds.), Comprehensive Handbook of Ecotoxicological Terms. Springer, pp. 221–236.

Mouneyrac, C., Buffet, P.E., Poirier, L., et al., 2014. Fate and effects of metal-based nanoparticles in two marine invertebrates, the bivalve mollusc Scrobicularia plana and the annelid polychaete Hediste diversicolor. Environ. Sci. Pollut. Res. 21, 7899–7912.

Muller, F.L.L., Jaquet, S., Wilson, W.H., 2003. Biological factors regulating the chemical speciation of Cu, Zn and Mn under different nutrient regimes in a marine mesocosm experiment. Limnol. Oceanogr. 48, 2289–2302.

Nayar, S., Goh, B.P., Chou, L.M., 2005. Environmental impacts of diesel fuel on bacteria and phytoplankton in a tropical estuary assessed using in situ mesocosms. Ecotoxicology 14, 397–412.

Nietch, C.T., Quinlan, E.L., Lazorchak, J., et al., 2013. Effects of a chronic lower range of triclosan exposure to a stream mesocosm community. Environ. Toxicol. Chem. 32, 2874–2887.

Nordberg, M., Templeton, D.M., Andersen, O., et al., 2009. Glossary of terms used in ecotoxicology (IUPAC Recommendations 2009). Pure Appl. Chem. 81, 829–970.

Nørum, U., Friberg, N., Jensen, M.R., et al., 2010. Behavioural changes in three species of freshwater macroinvertebrates exposed to the pyrethroid lambda-cyhalothrin: laboratory and stream microcosm studies. Aquat. Toxicol. 98, 328–335.

O'Brien, A.L., Keough, M.J., 2013. Detecting benthic community responses to pollution in estuaries: a field mesocosm approach. Environ. Pollut. 175, 45–55.

OECD (Organization for Economic Co-operation and Development), 1984. *Daphnia* sp. Acute Immobilization Test and Reproduction Test. Guideline 202. OECD Guidelines for Testing Chemicals. OECD, Paris.

OECD, 2004. Draft Guidance Document on Simulated Freshwater Lentic Field Tests (Outdoor Microcosms and Mesocosms). http://www.oecd.org/fr/securitechimique/essais/32612239.pdf (last accessed 13.01.14).

OECD, 2012. The OECD Environmental Risk Assessment Toolkit: Tools for Environmental Risk Assessment and Management. http://www.oecd.org/chemicalsafety/risk-assessment/theoecdenvironmentalriskassessmenttoolkittoolsforenvironmentalriskassessmentandmanagement.htm.

OEHHA (Office of Environmental Health Hazard Assessment), 2004. Overview of Freshwater and Marine Toxicity Tests: A Technical Tool for Ecological Risk Assessment. http://oehha.ca.gov/ecotox/pdf/marinetox3.pdf (last accessed 09.01.14).

Ohwada, K., Nishimura, M., Wada, M., et al., 2003. Study of the effect of water-soluble fractions of heavy-oil on coastal marine organisms using enclosed ecosystems, mesocosms. Mar. Pollut. Bull. 47, 78–84.

Ortiz-Delgado, J.B., Behrens, A., Segner, H., et al., 2008. Tissue-specific induction of EROD activity and CYP1A protein in *Sparus aurata* exposed to B(a)P and TCDD. Ecotoxicol. Environ. Saf. 69, 80–88.

OSPAR Commission, 2000. OSPAR Background Document Concerning the Elaboration of Programmes and Measures Relating to Whole Effluent Assessment. http://www.ospar.org/documents/dbase/publications/p00117/p00117_wea%20elaboration%20of%20programmes%20and%20measures.pdf (last accessed 09.12.14).

OSPAR, 2005. Protocols on Methods for the Testing of Chemicals Used in the Offshore Oil Industry. OSPAR Commission 2005–11.

Pablo, F., Krassoi, F.R., Jones, P.R.F., et al., 2008. Comparison of the fate and toxicity of chlorpyrifos–laboratory versus a coastal mesocosm system. Ecotoxicol. Environ. Saf. 71, 219–229.

Patlewicz, G., Simon, T., Goyak, K., 2013. Use and validation of HT/HC assays to support 21st century toxicity evaluations. Regul. Toxicol. Pharm. 65, 259–268.

Péan, S., Daouk, T., Vignet, C., et al., 2013. Long-term dietary-exposure to non-coplanar PCBs induces behavioral disruptions in adult zebrafish and their offspring. Neurotoxicol. Teratol. 39, 45–56.

Perez, M.H., Wallace, W.G., 2004. Differences in prey capture in grass shrimp, *Palaemonetes pugio*, collected along an environment impact gradient. Arch. Environ. Contam. Toxicol. 46, 81–89.

Petersen, J.E., Cornwell, J.C., Kemp, W.M., 1999. Implicit scaling in the design of experimental aquatic ecosystems. Oikos. 85, 3–18. http://www.jstor.org/stable/3546786.

Piña, B., Barata, C., 2011. A genomic and ecotoxicological perspective of DNA array studies in aquatic environmental risk assessment. Aquat Toxicol. 105, 40–49.

Qin, L.T., Liu, S.S., Liu, H.L., et al., 2010. Support vector regression and least squares support vector regression for hormetic dose–response curves fitting. Chemosphere 78, 327–334.

Quiniou, F., Damiens, G., Gnassia-Barelli, M., et al., 2007. Marine water quality assessment using transplanted oyster larvae. Environ. Intern. 32, 27–33.

Rasmussen, J.J., Friberg, N., Larsen, S.E., 2008. Impact of lambda-cyhalothrin on a macroinvertebrate assemblage in outdoor experimental channels: implications for ecosystem functioning. Aquat. Toxicol. 90, 228–234.

Riedel, G.F., Sanders, J.G., 2003. The interrelationships among trace element cycling, nutrient loading, and system complexity in estuaries: a mesocosm study. Estuaries 26, 339–351.

Riedel, G.F., Sanders, J.G., Breitburg, D.L., 2003. Seasonal variability in response of estuarine phytoplankton communities to stress: linkages between toxic trace elements and nutrient enrichment. Estuaries 26, 323–338.

Ringwood, A.H., Keppler, C.J., 2002. Comparative *in situ* and laboratory sediment bioassays with juvenile *Mercenaria mercenaria*. Environ. Toxicol. Chem. 21, 1651–1657.

Roast, S.D., Widdows, J., Jones, M.B., 2000. Disruption of swimming in the hyperbenthic mysid *Neomysis integer* (Peracarida: Mysidacea) by the organophosphate pesticide chlorpyrifos. Aquat. Toxicol. 47, 227–241.

Rosen, G., Bart Chadwick, D., Burton, G.A., et al., 2012. A sediment ecotoxicity assessment platform for *in situ* measures of chemistry, bioaccumulation and toxicity. Part 2: Integrated application to a shallow estuary. Environ. Pollut. 162, 457–465.

Roussel, H., Chauvet, E., Bonzom, J.M., 2008. Alteration of leaf decomposition in copper-contaminated freshwater mesocosms. Environ. Toxicol. Chem. 27, 637–644.

Roussel, H., Joachim, S., Lamothe, S., et al., 2007b. A long-term copper exposure on freshwater ecosystem using lotic mesocosms: individual and population responses of three-spined sticklebacks (*Gasterosteus aculeatus*). Aquat. Toxicol. 82, 272–280.

Roussel, H., Ten-Hage, L., Joachim, S., et al., 2007a. A long-term copper exposure on freshwater ecosystem using lotic mesocosms: primary producer community responses. Aquat. Toxicol. 81, 168–182.

Sanchez, W., Sremski, W., Piccini, B., et al., 2011. Adverse effects in wild fish living downstream from pharmaceutical manufacture discharges. Environ. Intern. 37, 1342–1348.

Sanderson, H., Laird, B., Pope, L., et al., 2007. Assessment of the environmental fate and effects of ivermectin in aquatic mesocosms. Aquat. Toxicol. 85, 229–240.

Sapozhnikova, Y., Pennington, P., Wirth, E., et al., 2009. Fate and transport of Irgarol 1051 in a modular estuarine mesocosm. J. Environ. Monit. 11, 808–814.

Scholz, S., Sela, E., Blaha, L., et al., 2013. A European perspective on alternatives to animal testing for environmental hazard identification and risk assessment. Regul. Toxicol. Pharm. 67, 506–530.

Scott, G.R., Sloman, K.A., 2004. The effects of environmental pollutants on complex fish behavior: integrating behavioural and physiological indicators of toxicity. Aquat. Toxicol. 68, 369–392.

Seguin, F., Le Bihan, F., Leboulanger, C., et al., 2002. A risk assessment of pollution: induction of atrazine tolerance in phytoplankton communities in freshwater outdoor mesocosms, using chlorophyll fluorescence as an endpoint. Water Res. 36, 3227–3236.

Smolyakov, B.S., Ryzhikh, A.P., Bortnikova, S.B., et al., 2010a. Behavior of metals (Cu, Zn and Cd) in the initial stage of water system contamination: effect of pH and suspended particles. Appl. Geochem. 25, 1153–1161.

Smolyakov, B.S., Ryzhikh, A.P., Romanov, R.E., 2010b. The fate of Cu, Zn, and Cd in the initial stage of water system contamination: the effect of phytoplankton activity. J. Hazard. Mater. 184, 819–825.

Sommer, U., 2012. Experimental Systems in Aquatic Ecology. eLS, John Wiley & Sons Ltd, Chichester. http://dx.doi.org/10.1002/9780470015902.a0003180.pub2http://www.els.net.

Stewart, S.C., Dick, J.T., Laming, P.R., et al., 2010. Assessment of the Multispecies Freshwater Biomonitor (MFB) in a marine context: the green crab (*Carcinus maenas*) as an early warning indicator. J. Environ. Monit. 12, 1566–1574.

Tarazona, J.V., Sobanska, M.A., Cesnaitis, R., et al., 2014. Analysis of the ecotoxicity data submitted within the framework of the REACH Regulation. Part 2. Experimental aquatic toxicity assays. Sci. Tot. Environ. 472, 137–145.

Triebskorn, R., Köhler, H.R., Honnen, W., et al., 1997. Induction of heat shock proteins, changes in liver ultrastructure, and alterations of fish behavior: are these biomarkers related and are they useful to reflect the state of pollution in the field? J. Aquat. Ecosyst. Stress Recov. 6, 57–73.

Turner, C., Sawle, A., Fenske, M., et al., 2012. Implications of the solvent vehicles dimethylformamide and dimethylsulfoxide for establishing transcriptomic endpoints in the zebrafish embryo toxicity test. Environ. Toxicol. Chem. 31, 593–604.

USEPA, 1973. In: Weber, C.I. (Ed.), Biological Field and Laboratory Methods for Measuring the Quality of Surface Waters and Effluents. U.S. Environmental Protection Agency, Methods Development and Quality Assurance Research Laboratory, Cincinnati, Ohio. EPA 670/4-73-001. 200 pp.

USEPA, 1990. In: Klemm, D.J., Lewis, P.A., Kessler, F., Lazorchak, J.M. (Eds.), Macroinvertebrate Field and Laboratory Methods for Evaluating the Biological Integrity of Surface Waters. Environmental Monitoring Systems Laboratory, Office of Modeling, Monitoring Systems, and Quality Assurance, Office of Research and Development, U.S. Environmental Protection Agency, Cincinnati, Ohio. 45268 EPA/600/4–90/030.

USEPA, 2002. Methods for Measuring the Acute Toxicity of Effluents and Receiving Waters to Freshwater and Marine Organisms. http://www.epa.gov/region6/water/npdes/wet/wet_methods_manuals/atx.pdf (last accessed 09.12.14).

USEPA, 2011. Experimental Stream Facility: Design and Research. EPA 600/F-11/004. Technical Report, Cincinnati, OH. http://nepis.epa.gov/Adobe/PDF/P100E4QH.pdf (last accessed 23.01.14).

USEPA, 2012. Benchmark Dose Technical Guidance and Other Relevant Risk Assessment Documents. http://www.epa.gov/raf/publications/benchmarkdose.htm.

Vandenberg, L.N., Colborn, T., Hayes, T.B., et al., 2012. Hormones and endocrine-disrupting chemicals: low-dose effects and nonmonotonic dose responses. Endocr. Rev. 33, 378–455.

Vandenbrouck, T., Soetaert, A., van der Ven, K., 2009. Nickel and binary metal mixture responses in *Daphnia magna*: molecular fingerprints and (sub)organismal effects. Aquat. Toxicol. 92, 18–29.

Wang, W.X., 2011. Incorporating exposure into aquatic toxicological studies: an imperative. Aquat. Toxicol. 105 (Suppl.), 9–15.

Ward, D.J., Simpson, S.L., Jolley, D.F., 2013. Avoidance of contaminated sediments by an amphipod (*Melita plumulosa*), a harpacticoid copepod (*Nitocra spinipes*), and a snail (*Phallomedusa solida*). Environ. Toxicol. Chem. 32, 644–652.

Watka, L.A., 2012. Using Mesocosms in Marine Ecology: A Laboratory Experiment in the Rocky Intertidal Zone with Extended Commentary on Current Uses. M.A. thesis Brown University, Providence, Rhode Island. http://envstudies.brown.edu/theses/archive20112012/LaurenWatkaThesis.pdf (last accessed 06.02.14).

Weis, J.S., Bergey, L., Reichmuth, J., et al., 2011. Living in a contaminated estuary: behavioral changes and ecological consequences for five species. Biosciences 61, 375–385.

Wheeler, J.R., Maynard, S.K., Crane, M., 2014. An evaluation of fish early life stage tests for predicting reproductive and longer-term toxicity from plant protection product active substances. Environ. Toxicol. Chem. 33, 1874–1878.

Wijngaarden, R.P.A., Barber, I., Brock, T.C.M., 2009. Effects of the pyrethroid insecticide gamma-cyhalothrin on aquatic invertebrates in laboratory and outdoor microcosm tests. Ecotoxicology 18, 211–224.

Wong, D.C., Maltby, L., Whittle, D., et al., 2004. Spatial and temporal variability in the structure of invertebrate assemblages in control stream mesocosms. Water Res. 38, 128–138.

Yamada, M., Takada, H., Toyoda, K., et al., 2003. Study on the fate of petroleum-derived polycyclic aromatic hydrocarbons (PAHs) and the effect of the chemical dispersant using enclosed ecosystems, mesocosms. Mar. Pollut. Bull. 47, 78–84.

Zhu, X.W., Liu, S.S., Qin, L.T., et al., 2013. Modeling non-monotonic dose-response relationships: model evaluation and hormetic quantities exploration. Ecotoxicol. Environ. Saf. 89, 130–136.

CHAPTER 7

Individual Biomarkers

Claude Amiard-Triquet and Brigitte Berthet

Abstract

The application of biomarkers is well-advanced for the marine environment in Western countries, whereas additional efforts are needed for the freshwater environment and developing countries. Criticisms that have mainly hampered a more general use of biomarkers are their lack of specificity and ecological relevance. It is important to cope with these problems because biomarkers have a special interest as a cost-effective, sensitive, and early warning system. Thus strategies are proposed to mitigate the lack of specificity resulting from the influence of confounding factors and a three-step integrative methodology is recommended including (1) the detection of abnormalities based on ecological biomarkers, linking infraindividual to individual and population levels; (2) measurements of lower level (core) biomarkers in battery because the end-users need to know the nature of pollutant exposure with a view to remediation; and (3) use of analytical chemistry to validate the hypotheses provided by biomarkers.

Keywords: Confounding factors; Core biomarkers; Ecological biomarkers; Integrated biomarker response (IBR).

Chapter Outline

A biomarker was defined by Depledge (1994) as "a biochemical, cellular, physiological or behavioral variation that can be measured in tissue or body fluid samples or at the level of whole organisms that provides evidence of exposure to and/or effects of, one or more chemical pollutants (and/or radiations)." Biomarkers, particularly those detected at low levels of biological organization, are generally early and sensitive indices of chemical stress (see for instance Smit et al., 2009). These sublethal biomarkers are recommended in operational procedures for assessing and managing chemical contamination in aquatic media, in complementarity with chemical assessments, bioindicators, and biological surveys (e.g., Chapman et al., 2013; Dagnino and Variengo, 2014). Guidelines for the integrated aquatic environmental monitoring

Aquatic Ecotoxicology. http://dx.doi.org/10.1016/B978-0-12-800949-9.00007-3

of chemicals and their effects have been launched under the auspices of regulatory bodies in the United States and Europe, including many different biomarkers. They are in current use for the Biomonitoring of Environmental Status and Trends Program (Schmitt and Dethloff, 2000), the OSPAR Hazardous Substances Strategy (OSPAR Agreement 2003–2021), and the European Union (EU) Marine Strategy Framework Directive (Davies and Vethaak, 2012). From a recent review of biomarkers currently used in environmental monitoring (Collier et al., in Amiard-Triquet et al., 2013) in Europe, the United States, and Asia, the application of biomarkers is well-advanced for the marine environment in Western countries, whereas additional efforts are needed for the freshwater environment and developing countries. In the latter, the choice of the suitable biomarkers to be used is crucial to make the biomonitoring tool a cost-effective strategy (UNEP/MAP/MED POL, 2007).

Criticisms that have mainly hampered a more general use of biomarkers are their lack of specificity (Cairns, 1992) and ecological relevance (Forbes et al., 2006). Comparing the relative importance of natural fluctuations of a biomarker response with stress-induced response, Cairns (1992) has shown that a highly variable background can conceal stress response, whereas a relatively stable background allows the detection of significant variations because of stress. In 2006, the ecological relevance of many biomarkers was highly questionable because the studies highlighting the links between infra- and supraindividual biological responses were scarce (Weis et al., 2001; De Coen and Janssen, 2003).

In the present chapter, we will review successively "core biomarkers," the effectiveness of which has been recognized in both laboratory and field studies and biomarkers of ecological relevance (Davies and Vethaak, 2012; Amiard-Triquet et al., 2013; Mouneyrac and Amiard-Triquet, 2013; Gagné, 2014). Among core biomarkers we will distinguish between those that allow the organisms to cope with the presence of contaminants in their medium and those that reveal deleterious effects, termed, respectively, biomarkers of defense and damage (De Lafontaine et al., 2000). The operational modes for the application of the methodology of biomarkers will be described, including active biomonitoring by translocation of organisms between clean and contaminated sites initially proposed for bivalves (De Kock and Kramer, 1994) and field studies. Of course, it is impossible to provide an exhaustive report of all the papers dealing with confounding factors. However, the analysis of about 100 papers enable the building of a comprehensive picture of the problem. Then strategies will be proposed to mitigate the lack of specificity resulting from the influence of confounding factors. This chapter also includes a section about the use of batteries of biomarkers and the different indices that have been proposed for the classification of sites according to their degree of pollution will be discussed.

7.1 Core Biomarkers

Among core biomarkers, several are recommended for regulatory purposes (Table 7.1). For all these biomarkers, background assessment criteria and environmental assessment criteria are

Table 7.1: Biomarkers of Defense and Damage Recommended in the Framework of Integrated Monitoring and Assessment of Contaminants under the Auspices of ICES

	Matrix	Quality Assurance
	Biomarkers of Defense	
Cytochrome P450 1A activity (EROD)	Fish liver	BEQUALM, MEDPOL
Metallothionein	Fish	BEQUALM
	Mussels	MEDPOL
	Biomarkers of Damage	
Acetylcholinesterase	Fish Invertebrates	BEEP project
Lysosomal stability (cytochemical and neutral red)	Mussels	MEDPOL
Comet assay	Fish Invertebrates	WGBEC
Micronucleus formation	Fish Bivalves	MED POL
DNA adducts	Fish Invertebrates	BEQUALM
Vitellogenin (messenger RNA transcripts or protein)	Male fish plasma	WGBEC
Externally visible diseases Liver histopathology Macroscopic liver neoplasms	Fish	BEQUALM
Intersex		No formal QA
Histopathology (gametogenesis)	Mussels	Currently not available
Imposex/intersex	Gastropods	QUASIMEME
Stress on stress	Bivalves	Not required
Scope for growth	Bivalves	IOC
Reproductive success	Fish (eelpout)	BEQUALM

BEEP project, Biological Effects of Environmental Pollution—in marine coastal ecosystems, EU project EVK3-2000-00543); BEQUALM, Biological Effects Quality Assurance in Monitoring Programmes; IOC, Intergovernmental Oceanographic Commission biological effects workshops; MEDPOL, Mediterranean Action Plan for the Barcelona Convention; QUASIMEME, Quality Assurance of Information for Marine Environmental Monitoring in Europe; WGBEC, Working Group of Biological Effect of Contaminants (ICES).
Modified after Davies and Vethaak (2012).

available (Davies and Vethaak, 2012). Briefly, background assessment criteria are estimated from data typical of remote areas, whereas environmental assessment criteria are usually derived from toxicological data and indicate a sublethal effect.

Mechanisms of tolerance in aquatic organisms include uptake limitation, detoxification of bioaccumulated contaminants and improved excretion (Amiard-Triquet et al., 2011). The detoxification dimension of tolerance provides core biomarkers (Table 7.1) such as cytochrome P450 1A activity (EROD) and metallothioneins (MTs). In addition to those recommended by regulatory bodies, scientists have proposed a number of other useful biomarkers of defense, including (1) biotransformation enzymes of organic chemicals such as glutathione *S*-transferase (GST) (Roméo and Wirgin, in Amiard-Triquet et al., 2011); (2) multixenobiotic resistance that exerts a protective role toward fuel oil, polyaromatic hydrocarbons (PAHs), sediment extracts, and pesticides in fish (Chapter 13); (3) antioxidant defenses against environmental prooxidants including both enzymatic (superoxide dismutase [SOD], catalase [CAT], peroxidases [e.g., GST]) and nonenzymatic defenses (Regoli et al., in Amiard-Triquet et al., 2011; Abele et al., 2012); and (4) stress proteins also called chaperones because they protect other proteins from denaturation and aggregation and refold damage proteins to a functional conformation (Frydman, 2001; Mouneyrac and Roméo, in Amiard-Triquet et al., 2011).

Saturation of defense mechanisms can occur for several core biomarkers (MT, EROD, heat shock proteins, CAT, GST, SOD, multixenobiotic resistance) along a pollution gradient. In this case, the relationship between the level of exposure and the response intensity fits well a bell-shaped curve (Dagnino et al., 2007). This is clearly a practical problem for biomarkers used in biomonitoring programs since the same response intensity may be observed at different exposures, either before or after the peak (Amiard-Triquet and Roméo, in Amiard-Triquet et al., 2011). This is also a problem considering environmental conservation because it means that in highly contaminated environments, defenses are overwhelmed and that noxious effects can occur, that can be used as biomarkers of damage, several of which are recognized as core biomarkers (Table 7.1).

In addition to lysosomal membrane stability, certain other lysosomal biomarkers (lysosomal swelling, lipid accumulation, lipofuscin) can be used in protistans, coelenterates, annelid worms, mollusks, crustaceans, and fish (Moore et al., in Amiard-Triquet et al., 2013). Recent results illustrate the potential use of lysosomal membrane stability as a pragmatic indicator of biotic injury in environmental monitoring programs because in 10 estuaries in Australia, there was a strong association between lysosomal membrane stability in oysters *Saccostrea glomerata* and metal exposure (Edge et al., 2014).

When they are not neutralized by antioxidant defenses, reactive oxygen species could induce different types of cellular damage. Assessing lipid and protein oxidation is classically used in environmental studies (Lushchak, 2011). Reactive oxygen species may also cause oxidative DNA damages in aquatic organisms, providing biomarkers of genotoxicity such as DNA adducts, micronucleus, and DNA breaks revealed by Comet assay tests (Abele et al., 2012; Vasseur et al., in Amiard-Triquet et al., 2013) currently recommended under the auspices of the International Council for the Exploration of the Sea (ICES) (Table 7.1). Particular attention has been devoted to DNA damage in germ cells (Table 7.2).

Vitellogenins (VTGs) in fish have been proposed as a useful biomarker of estrogenic contamination (Davies and Vethaak, 2012). Recent researches in invertebrates have suggested that the levels of vitellogenin-like protein or VTG gene expression in bivalves may be also useful biomarkers (De los Ríos et al., 2013; Andrew et al., 2010), whereas in freshwater gastropods (Gust et al., 2014) and arthropods (Ara and Damrongphol, 2014; Short et al., 2014; literature

Table 7.2: DNA Damage to Sperm Induced by Paternal Genotoxin Exposure in Invertebrates and Fish

Zoological Taxa	Contaminant	References
	Polychaete Worms	
Arenicola marina[b]	MMS, B[*a*]P	Lewis and Galloway (2008, 2009)
Nereis diversicolor[b]	MMS	Lewis and Galloway (2008)
Nereis virens[b]	Cu, PUA	Caldwell et al. (2011)
	Bivalves	
Mytilus edulis[b]	MMS, B[*a*]P	Lewis and Galloway (2009)
Mytilus galloprovincialis[b]	Nano-iron	Kadar et al. (2011)
	Amphipod Crustacean	
Gammarus fossarum[a,b]	MMS	Lacaze et al. (2010)
	Caging in a river	Lacaze et al. (2011a)
	$K_2Cr_2O_7$, MMS, caging	Lacaze et al. (2011b)
	Fish	
Salmo trutta[b]	MMS	Devaux et al. (2011)
Salvelinus alpinus[b]	MMS	

B[*a*]P, benzo[*a*]pyrene; MMS, methyl methane sulfonate; PUA, polyunsaturated aldehyde (4*E*-decadhienal).
[a]Field study.
[b]Laboratory study.

quoted therein; see also Chapter 11) studies investigating VTG induction have resulted in negative or inconclusive results.

In complementarity, histopathological biomarkers linked to reproductive impairments are recommended as core biomarkers (Table 7.1). Externally visible diseases such as imposex in gastropods are easy to observe (Gubbins et al., in Amiard-Triquet et al., 2013), whereas other impairments such as intersex in fish and bivalves, needing microscopic observations, are much more time-consuming. Histocytological biomarkers respond to organic contaminants, metal contamination, nanoparticles, mixed contamination in marine, and brackish and freshwater environments (Au, 2004; Amiard and Amiard-Triquet, in Amiard-Triquet et al., 2013).

Behavioral impairments occur in many different taxa as a consequence of exposure to many different legacy or emerging contaminants at low doses in experiment or in the field in contaminated areas (Amiard-Triquet and Amiard, in Amiard-Triquet et al., 2013). More recent papers confirm behavior as a responsive biomarker for pharmaceuticals (Di Poi et al., 2013; Hedgespeth et al., 2014) or nanoparticles (Gambardella et al., 2014; Mouneyrac et al., 2014). The development of advanced techniques, allowing quantitative behavioral monitoring in real time has allowed the implementation of biological early warning systems to continuously detect a wide range of pollutants for effective water quality monitoring and management (Bae and Park, 2014).

Chemical stress can affect energy metabolism either directly such as the impairment of feeding behavior and digestive enzyme activities or indirectly such as the energy cost of defense mechanisms (Mouneyrac et al., in Amiard-Triquet et al., 2011) or the decrease of food availability when prey species are susceptible to contaminants (Dedourge-Geffard et al., in Amiard-Triquet et al., 2013). Energy metabolism also greatly depends on the reproductive cycle and then biomarkers are operational only when the natural fluctuations of energy are well-known in the species used as biological models. However, the scope for growth is recognized as a core biomarker (Table 7.1) and adenylate energy charge, cellular energy allocation, and condition indices may be also relevant (Mouneyrac et al., in Amiard-Triquet et al., 2013). The utilization of energy reserves as an accurate biomarker of pollution is still being debated but at least in fish, the ratio of lipids involved in storage (triacylglycerols, TAG) to structural lipids such as sterols seems a good predictor of the animal's physiological status (Chapter 13).

7.2 Ecological Biomarkers

The core biomarkers (Table 7.1), lysosomal stability, and scope for growth are positively correlated (Moore et al., in Amiard-Triquet et al., 2013). The possibility of lysosomal biomarker reactions having a predictive capacity in terms of ecosystem level effects has been explored in marine areas, showing a statistical link between the responses at the cellular and community levels (Crowe et al., 2004; Moore et al., in Amiard-Triquet et al., 2013).

To date, biomarkers of immunotoxicity have not been accepted as core biomarkers under the auspices of the ICES, mainly because the problem of confounding factors is particularly crucial. However, many studies are available for bivalves, fish, and marine mammals (Brousseau et al., in Amiard-Triquet et al., 2013; Dupuy et al., 2014 and literature quoted therein). Immunotoxicological approaches seem pertinent, the hypothesis of a chemically induced immunodeficiency being strongly supported by the high occurrence of opportunistic infections and parasitic infestations in fish, thus leading to a lower fitness (Dautremepuits et al., 2004; Danion et al., 2012), and in marine mammals (Martineau et al., 2002).

Behavioral impairments result from a number of causes at the subindividual level (Figure 7.1) and may have consequences at the supraindividual levels of population and community. Among causes, the relationships between the core biomarker acetylcholinesterase (AChE) inhibition (Table 7.1) and behavioral impairments are well-documented for aquatic biota because many studies have investigated the ecotoxicological effects of pesticides (Amiard-Triquet and Amiard, in Amiard-Triquet et al., 2013) and it remains a topic of interest (Peltzer et al., 2013). Linkages between behavioral biomarker responses of invertebrates (gammarids, bivalves) and aspects of community structure and functioning have been statistically established (Maltby et al., 2002; Seabra Pereira et al., 2014). Reviewing the effects of pollution on reproductive behavior of fishes, Jones and Reynolds (1997) concluded potential consequences for reproductive success and population effects in *Tilapia rendalli* living in areas treated with

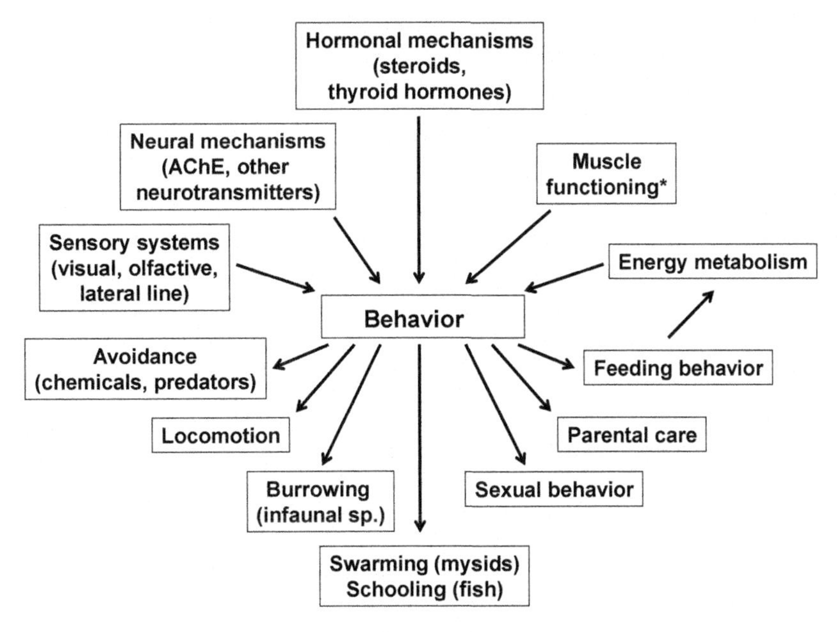

Figure 7.1
Linking biological responses at different levels of biological organization: behavioral ecotoxicology. *After Weis et al. (2001, 2011); Amiard-Triquet and Amiard, in Amiard-Triquet et al. (2013); de Carvalho (2013). *Fritsch et al. (2013).*

pesticides. "Foraging processes may be used as behavioral endpoints that link effects on individuals to the population and community levels, enabling risk assessment of environmental contaminants at larger ecological scales" (Hedgespeth et al., 2014). This concept has been strikingly illustrated in a comprehensive study with two fish and three crustacean species chronically exposed in contaminated estuaries compared to conspecifics from a cleaner estuary (Weis et al., 2001, 2011), using laboratory tests and field observations to decipher the mechanisms linking subindividual effects to impairments of behavior and ultimately effects at the community level.

Impairments of energy metabolism also result from a number of causes at the subindividual level and may have consequences at the supraindividual levels of population and community as illustrated with a case study with the endobenthic worm *Nereis diversicolor* (Figure 7.2). In this species living in multipolluted estuaries, biochemical markers have revealed disturbances (impairment of AChE and digestive enzyme activities) able to affect behavior through neurotoxicity and decreased available energy linked to impaired assimilation. Induction of defenses can have induced an energy cost of tolerance, reinforced by the depletion of food sources in a contaminated estuary. Both burrowing and feeding behaviors were disturbed. The former has direct implications for survival, favoring predation. The later has indirect implications in energy metabolism, adding a decrease of energy acquired from food to decreased assimilation and cost of tolerance. This is most probably the origin of reduced growth and

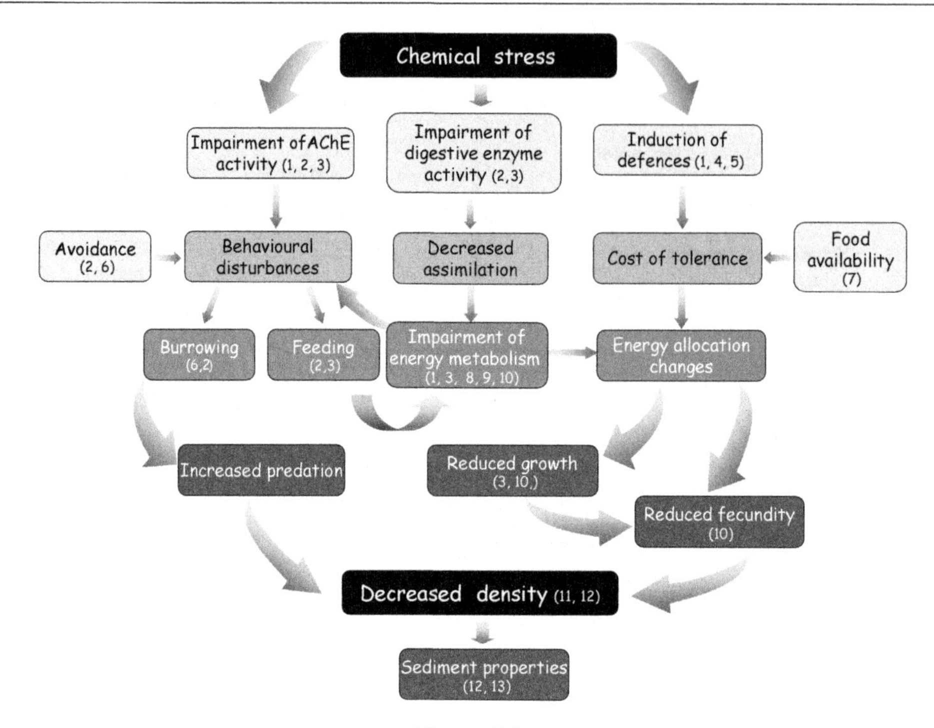

Figure 7.2

Linking biological responses at different levels of biological organization: a case study with the endobenthic worm *Nereis diversicolor*. [1]Durou et al. (2007a); [2]Kalman et al. (2009); [3]Fossi Tankoua et al. (2012a); [4]Berthet et al. (2003); [5]Mouneyrac et al. (2003); [6]Mouneyrac et al. (2010); [7]Debenay (2009); Ferrero (2009); Sylvestre (2009), all in Amiard-Triquet and Rainbow (2009); [8]Durou et al. (2005); [9]Durou et al. (2007b); [10]Durou et al. (2008); [11]Mouneyrac et al., in Amiard-Triquet and Rainbow (2009); [12]Gillet et al. (2008); [13]Bessineton, in Amiard-Triquet and Rainbow (2009).

reduced fecundity observed in worms from a severely contaminated estuary (Seine estuary, France). Reduced fecundity and increased predation has led to a dramatic depletion of population density that may be worsened by nonchemical anthropogenic impacts. Because these endobenthic worms are "ecological engineers" (Chapter 10), reduced bioturbation in populations with low densities could lead to ecosystemic effects.

DNA damages (currently recommended as core biomarkers, Table 7.1) have been observed in fish or invertebrates exposed to chemicals, particularly PAHs, in rivers and estuaries, as a consequence of oil spills or production by anthropogenic activities (Davies and Vethaak, 2012; Vasseur et al., in Amiard-Triquet et al., 2013). Linkages between oxidative stress and DNA damage in invertebrates (bivalve *Scrobicularia plana*) and benthic community structure have been statistically established (Silva et al., 2012). Because the consequences may be limited by DNA repair (Peterson and Côté, 2004), it is only recently that the transfer of toxicity to the progeny has been investigated in full, including genotoxic effects on sperm (Table 7.2) and consequences such as severe developmental abnormalities and prolonged effects (increased mortality and malformations) in fish *Salmo trutta* (Devaux et al., 2011). Despite being not absolutely conclusive, the studies about the Exxon Valdez oil spill suggest a link between genotoxic, physiological, and populational effects in the Pacific herring *Clupea pallasi* (Vasseur et al., in Amiard-Triquet et al., 2013). In gammarid crustaceans, reproduction defects were observed for a level of DNA damage that could have significant consequences for population dynamics (Lacaze et al., 2011b). The genotoxic effects of environmentally relevant diuron exposure on oyster genitors can be transmitted to offspring, likely impairing development (decrease in hatching rate, higher level of larval abnormalities, delay in metamorphosis) and growth (Barranger et al., 2014). Thus these authors underline the risk associated to chemical contamination for oyster recruitment and fitness.

The ecotoxicological risks of endocrine disrupting compounds are reviewed in Chapter 17. Some of the mechanisms involved in endocrine disruption have already been incorporated as biomarkers in regulatory strategies, such as VTG levels and the degree of imposex/intersex (Table 7.1). The development of imposex has been documented in many different species of gastropods, practically leading to the local extinction of the dogwhelk *Nucella lapillus* in several areas impacted by **tributyltin**-based antifouling paints (Bryan et al., 1987; Gubbins et al., in Amiard-Triquet et al., 2013). There is no clear link to date between intersex in bivalves and consequences on population fate as shown for instance in *S. plana* (Fossi Tankoua et al., 2012b). Intersex was also shown in fish. In gudgeons (*Gobio gobio*) exposed in the wild to pharmaceutical manufacture discharges, intersex was associated with VTG induction and male-biased sex ratio as also to a decrease of population density and disturbances of fish assemblage (Sanchez et al., 2011). In another fish species, the fathead minnow *Pimephales promelas*, experiments in the field (Kidd et al., 2007) and modeling (Miller et al., 2007) have shown that endocrine disrupting compounds impair the normal levels of VTG with population effects which can nearly lead to local extinction.

7.3 Confounding Factors

Confounding factors have been documented for both core biomarkers (Davies and Vethaak, 2012; Farcy et al., 2013) and ecological biomarkers (Amiard-Triquet et al., 2013). The influence of the most studied confounding factors (temperature, salinity, and the complex size/weight/age) on many different biomarkers are shown in Table 7.3.

General patterns of pollutant toxicity as a function of ambient temperature in aquatic ectotherms have been reviewed by Sokolova and Lannig (2008). The major types of responses able to generate variations in biomarkers are depicted in Figure 7.3. For toxic metals, toxicity generally increases with increasing temperature, without threshold (case A) or only above a threshold (case B), whereas decreasing toxicity with increasing temperature (case C) has been described for some organic pollutants. Case D, termed optimum response, is well-documented

Table 7.3: Influence of the Most Studied Confounding Factors on Biomarker Responses at Different Levels of Biological Organization in Aquatic Organisms

Biomarker	Temperature	Salinity	Size/Weight/Age
	Defense		
MT	Oligochaete worm (Gillis et al., 2004[b]), gastropod (Leung et al., 2000[b]), bivalve (Serafim et al., 2002[b]; Kamel et al., 2014[a])	Gastropod (Leung et al., 2002[b]), bivalve (Fossi Tankoua et al., 2011[a])	Gastropod (Leung et al., 2000[b]), bivalves (Serafim et al., 2002[b]; Mouneyrac et al., 1998[a])
Antioxydant defense (GST, CAT, SOD)	Bivalve (Kamel et al., 2014[a]; Robillard et al., 2003[a]; López-Galindo et al., 2014[b]; Lesser et al., 2010[a]; Greco et al., 2011[b]; Verlecar et al., 2007[b]; Giarratano et al., 2011[a]; Damiens et al., 2004[b]), crustacean (Cailleaud et al., 2007[b]; Tu et al., 2012[b])	Bivalves (Fossi Tankoua et al., 2011[a]; Zaccaron da Silva et al., 2005[b]), crustacean (Cailleaud et al., 2007[b]; Tu et al., 2012[b]), fish (Kopecka and Pempkowiak, 2008[a]; Martinez-Alvarez et al., 2002[b])	Bivalves (Fossi Tankoua et al., 2011[a]; Robillard et al., 2003[a]; Ahmad et al., 2011[a])
EROD	Bivalve (Ricciardi et al., 2006[b]), fish (Lange et al., 1998[a]; Sleiderink et al., 1995a[a]; Sleiderink et al., 1995b[b]; Russo et al., 2009[b])	Fish (Livingstone et al., 2000[a])	
Heat shock proteins	Bivalves (Lesser et al., 2010[a]; Dimitriadis et al., 2012[b]; Liu et al., 2012[b]), fish (Yang et al., 2012[b]; Triebskorn et al., 1997[b])	Bivalve (Werner, 2004[b])	
	Damage		
AChE	Bivalves (Kamel et al., 2014[a]; Robillard et al., 2003[a]; López-Galindo et al., 2014[b]; Ricciardi et al., 2006[b]; Dimitriadis et al., 2012[b]; Pfeifer et al., 2005[a]; Burgeot et al., 2010[a]), crustaceans (Cailleaud et al., 2007[b]; Menezes et al., 2006[b])	Bivalves (Fossi Tankoua et al., 2011[a]; Pfeifer et al., 2005[a]; Damiens et al., 2004[b]), crustacean (Cailleaud et al., 2007[b])	Bivalve (Fossi Tankoua et al., 2011[a]), crustacean (Xuereb et al., 2009[b])
Digestive enzymes	Invertebrates (Charron et al., 2013[b]; Dedourge-Geffard et al. (review in Amiard-Triquet et al., 2013), fish (Dedourge-Geffard et al. (review in Amiard-Triquet et al., 2013))	Polychaete worm (Kalman et al., 2010[a]), bivalve (Fossi Tankoua et al., 2011[a])	Crustacean (Gamboa-Delgado et al., 2003[a]), fish (Kuz'mina, 1996[a])
Immune parameters	Bivalves (Gagnaire et al., 2006[b]; Monari et al., 2007[b]; Gagné et al., 2008[a]; Greco et al., 2011[b])	Bivalves (Gagnaire et al., 2006[b]; Gagné et al., 2008[a])	Bivalve (Pichaud et al., 2009[a])
Lysosomal fragility	Gastropod (Russo et al., 2009[b]), bivalve (Dimitriadis et al., 2012[b])		
DNA damage	Bivalves (Buschini et al., 2003[b]; Almeida et al., 2011[a]; Kolarević et al., 2013[a,b])		Fish (Cachot et al., 2013[a])
Oxidative damage	Bivalves (Kamel et al., 2014[a]; Greco et al., 2011[b]; Abele et al., 2002[b]; Verlecar et al., 2007[b]; Giarratano et al., 2011[a])	Bivalve (Damiens et al., 2004[b]), crustaceans (Paital and Chainy, 2012[b]), fish (Martinez-Alvarez et al., 2002[b])	Bivalve (Ahmad et al., 2011[a])

Histocytopathology	Bivalve (López-Galindo et al., 2014[b]), fish (Triebskorn et al., 1997[b])		Fish (Cachot et al., 2013[a]; Lang et al., 2006[a]; Basmadjian et al., 2008[a]; Myers et al., 2008[a]; Stentiford et al., 2010[a])
Behavior			
Locomotion, burrowing	Bivalve (Monari et al., 2007[b]), crustaceans (Wijnhoven et al., 2003[b]), fish (Triebskorn et al., 1997[b])	Polychaete worm (Kalman et al., 2010[a]), bivalve (Fossi Tankoua et al., 2011[a]; Bonnard et al., 2009[b])	Bivalves (Fossi Tankoua et al., 2011[a]; Townsend et al. 2009[a,b]), amphibian (Denoël et al., 2010[b])
Feeding, egestion rates	Gastropods (Hylleberg, 1975[b]; Pascal et al., 2008[b]; Krell et al., 2011[b]), crustaceans (Maltby et al., 2002[a]; Lozano et al., 2003[b]; Moreira et al., 2006[b])	Gastropods (Hylleberg, 1975[b]; Krell et al., 2011[b]), crustacean (Moreira et al., 2006[b])	Polychaete worm (Kalman et al., 2010[a]), bivalve (Townsend et al. 2009[a,b]), crustaceans (Lozano et al., 2003[b]; Nilsson, 1974[b])
Reproductive behavior	Crustacean (Dick et al., 1998[b])		
Energy			
O_2 consumption	Gastropod (Leung et al., 2000[b])	Crustacean (Paital and Chainy, 2012[b])	Gastropod (Leung et al., 2000[b])
Lactate dehydrogenase		Polychaete worm (Kalman et al., 2010[a]), bivalve (Fossi Tankoua et al., 2011[a]), crustacean (Menezes et al., 2006[b])	
Energy reserves	Bivalve (López-Galindo et al., 2014[b])		Polychaete worm (Kalman et al., 2010[a]), bivalve (Fossi Tankoua et al., 2011[a])
Growth	Oligochaete worm (Gillis et al., 2004[b]), gastropod (Selck et al., 2006[b]), crustaceans (Pöckl and Humpesch, 1990[b]; Neuparth et al., 2002[b]; Winkler and Greve, 2002[b]; Prato et al., 2008[b]), fish (Sandström et al., 2005[a])	Crustacean (Delgado et al., 2011[b])	Bivalve (Fossi Tankoua et al., 2011[a])
Reproduction	Oligochaete worm (Gillis et al., 2004[b]), gastropods (Wayne, 2001; (review); Gust et al., 2011[b]; Sieratowicz et al., 2011[b]), bivalves (MacInnes and Calabrese, 1979[b]), crustaceans (Neuparth et al., 2002[b]; Prato et al., 2008[b]; Pöckl, 1992[b]; Sutcliffe, 1992[a]; Maranhão and Marques, 2003[b]; Eriksson Wiklund and Sundelin, 2004[a]), fish (Körner et al., 2008[b])	Bivalves (MacInnes and Calabrese, 1979[b]), Crustaceans (Prato et al., 2008[b]), sea urchins (Carballeira et al., 2011[b])	Polychaete worm (Durou et al., 2008[a]), crustaceans (Prato et al., 2008[b]; Maranhão and Marques, 2003[b]; Pöckl, 1993[a]; Thatje et al., 2004[b]; Castellani and Altunbas, 2006[a]; Pöckl, 2007[a]; Buchholz and Buchholz, 2010) (review)

[a]Field study.
[b]Laboratory study.

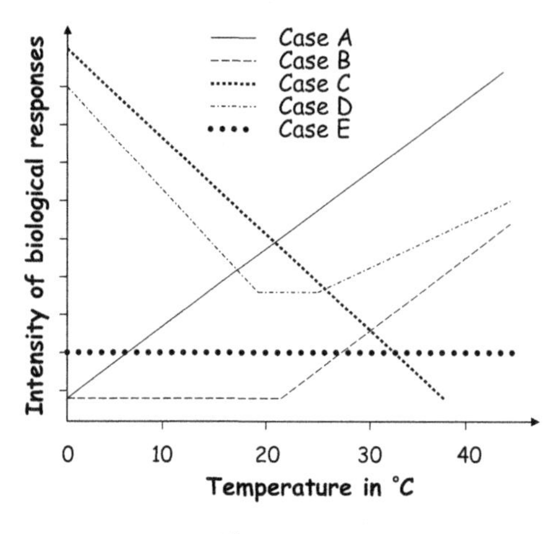

Figure 7.3

General patterns of biological responses to pollutant toxicity as a function of ambient temperature in aquatic ectotherms. For details, see Section 7.3. *Modified after Sokolova and Lannig (2008).*

in mammals; this minimum toxicity corresponding to the thermal preferendum has also been observed in invertebrates (MacInnes and Calabrese, 1979; Attig et al., 2014). Case E corresponds to the absence of any effect of temperature on biomarkers. It is well-established that temperature influences enzyme activities, many of which are used as subindividual markers. In addition to this direct effect, temperature can influence biomarker responses as a consequence of its effects on contaminant kinetics.

Salinity can exert direct effects on biota, depending on the euryhalinity of species and also indirect effects through its influence on the physicochemical characteristics of contaminants that govern their bioavailability and toxicity (Rainbow, 1997; Cailleaud et al., 2009).

The influence of the size/weight/age complex on biomarker responses (Table 7.3) is at least partly the consequence of differences in the uptake of contaminants, which are well-documented after many previous studies in the bivalves used in Mussel Watch programs. Several biomarkers are influenced by the duration of exposure and thus the age of organisms.

Many other factors can also influence many different biomarkers as exemplified in Table 7.4. In the present framework, it is impossible to quote all the papers devoted to other confounding factors (light and circadian cycle, diet, sexual maturity, life stage).

For seasonal factors, many reports exist on variations of a large range of biomarkers in many different species. In bivalves frequently used as model species, seasonal variations have been recognized considering a large set of individual and infraindividual biomarkers as reviewed by Sheehan and Power (1999). It remains an area of concern and many data are also available for fish (Table 7.5).

Table 7.4: Influence of Less Studied Confounding Factors on Biomarker Responses at Different Levels of Biological Organization in Aquatic Organisms

	Extrinsic Parameters				Intrinsic Parameter
Biomarker	**pH**	**Oxygen**	**Tidal Height**	**Parasitism**	**Sex**
			Defense		
MT				Bivalve (Canesi et al., 2010)	
Antioxidant defense (e.g., GST, CAT, SOD)	Bivalves (Vidal et al., 2002; Robillard et al., 2003)	Bivalves (Vidal et al., 2002; Robillard et al., 2003); crustacean (Gorokhova et al., 2013)	Bivalve (Letendre et al., 2009)	Bivalve (Canesi et al., 2010)	Fish (Kopecka and Pempkowiak, 2008)
EROD					Fish (Kopecka and Pempkowiak, 2008)
Heat shock proteins	Bivalve (Liu et al., 2012)		Bivalve (Halpin et al., 2004); gastropods (Pöhlmannet al., 2011)		
			Damage		
AChE	Bivalve (Robillard et al., 2003)				
Immune parameters			Bivalve (Alix et al., 2013)		Bivalves (Duchemin et al., 2007); crustacean (Matozzo et al., 2013)
VTG					Bivalve (Matozzo et al., 2012)
Lysosomal biomarkers				Bivalves (Canesi et al., 2010; Minguez et al., 2009)	
Oxidative damage	Bivalve (Vidal et al., 2002)	Bivalve (Vidal et al., 2002)	Bivalve (Schmidt et al., 2012)		
Genotoxicity			Bivalve (Schmidt et al., 2012)		Bivalves (Almeida et al., 2013)

Continued

Table 7.4: Influence of Less Studied Confounding Factors on Biomarker Responses at Different Levels of Biological Organization in Aquatic Organisms—cont'd

	Extrinsic Parameters				Intrinsic Parameter
Biomarker	**pH**	**Oxygen**	**Tidal Height**	**Parasitism**	**Sex**
Histopathology					Fish (Koehler, 2004)
Behavior	Crustacean (Naylor et al., 1989; Maltby et al., 2002)	Crustacean (Maltby et al., 1990)		Crustacean (McCahon et al., 1988; Pascoe et al., 1995; Fielding et al., 2003)	
Energy				Crustacean (Lettini and Sukhdeo, 2010)	
Reproduction		Crustaceans (Eriksson Wiklund and Sundelin, 2004; van den Heuvel-greve et al., 2007), fish (Wu et al., 2003; Shang et al., 2006; Landry et al., 2007; Thomas et al., 2007; Cheek et al., 2009; Wu, 2009; Thomas and Rahman, 2012)		Gastropod (Wayne, 2001)	

Table 7.5: Insight into Seasonal Variations of Biomarkers in Invertebrates and Fish

Species	Biological Responses	References
	Invertebrates	
Mytilus sp.	MT	Ivanković et al. (2005)
	MT, AChE, CAT, GST	Leiniö and Lehtonen (2005)
	mRNA abundance	Banni et al. (2011)
	SOD, GST, LPO	Giarratano et al. (2011)
	Energy metabolism, antioxidant defenses	Nahrgang et al. (2013)
	GST, VTG-like proteins, LPO, DNA damage	Schmidt et al. (2013); Almeida et al. (2013)
	Lysosomal biomarkers	Lekube et al. (2014)
Crassostrea gigas	Immune parameters	Duchemin et al. (2007)
Ruditapes decussatus	Biotransformation and antioxidant enzymes, LPO, MT, ALAD	Cravo et al. (2012)
	Genotoxicity	Almeida et al. (2013)
	Histopathological indices	Costa et al. (2013)
Tapes semidecussatus	Genotoxicity	Hartl et al. (2004)
Ruditapes philippinarum	Genotoxicity, behavior	Sacchi et al. (2013)
Macoma balthica	MT, AChE, CAT, GST	Leiniö and Lehtonen (2005)
	GST, CAT, GR	Barda et al. (2014)
Chlamys islandica	Energy metabolism, antioxidant defenses	Nahrgang et al., 2013
Placopecten magellicanus	Immunocompetence, digestive enzyme (lipase)	Pichaud et al. (2009)
Carcinus aestuarii	Haemolymph parameters	Matozzo et al. (2013)
	Fish	
Platichthys flesus	Liver histopathology	Lang et al. (2006); Cachot et al. (2013)
	AChE, EROD, CAT, GST, GSI	Kopecka and Pempkowiak (2008)
Mugil cephalus	Oxidative stress	Padmini et al. (2008)
Gadus morhua	CAT, TBARS, energy metabolism	Nahrgang et al. (2013)

AChE, acetylcholinesterase; ALAD, δ-aminolevulinic acid dehydratase; CAT, catalase; GR, glutathione reductase; GSI, gonadosomatic index; GST, glutathione *S*-tranferase; LPO, lipid peroxidation; MT, metallothionein; SOD, superoxide dismutase; TBARS, thiobarbituric acid reactive substances; VTG, vitellogenin.

Seasonal fluctuations depend on both extrinsic and intrinsic factors, among which the most important are certainly temperature, food availability, and reproductive status, but also the pollutant inputs associated with the intensity of hydrologic events (Hallberg et al., 2007; Sänkiaho, 2009; Barletta et al., 2012), the efficiency of sewage treatment plants (seasonality of microbial activity, temperature, rainfall) (Kumar et al., 2011), and applications of pesticide treatments.

7.4 Multibiomarker Approach in Field Studies

7.4.1 Active Biomonitoring

Active biomonitoring provides an opportunity for selecting standard organisms (homogeneous size, sex, reproductive status) with similar previous histories (specimens produced by aquaculture or collected from a reference site). This methodology has been widely developed for marine (e.g., Tsangaris et al., 2011; Dabrowska et al., 2013; De los Rios et al., 2013; Marigomez et al., 2013a; Lekube et al., 2014; Turja et al., 2014; Vidal-Liñan et al., 2014) and freshwater (Lacaze et al., 2013) mussels. In addition, it has been applied to other invertebrates such as polychaetes (Ramos-Gómez et al., 2011), mudsnails (Gust et al., 2010, 2014), and amphipods (see Chapter 11). Since Oikari (2006) has reviewed caging techniques for field exposures of fish, this methodology has been well-developed (e.g., Brammell et al., 2010; Klobučar et al., 2010; Kerambrun et al., 2011; Wang et al., 2011).

7.4.2 Integrative Biomarker Indices

To summarize biomarker responses and simplify their interpretation in biomonitoring programs, several integrative biomarker indices have been proposed (Wosniok et al., 2005; Broeg and Lehtonen, 2006; Marigomez et al., 2013b). Although the original information is expressed by several parameters, the index can provide a monitoring criterion by itself, on the basis of which spatiotemporal comparisons can be performed. For instance, Fossi Tankoua et al. (2013) have combined infra-individual biomarkers relatively "specific" such as MTs, GST, AChE, and individual biomarkers, more representative of general health (condition index, burrowing speed) to calculate an integrated biomarker response (IBR), initially developed by Beliaeff and Burgeot (2002) with only biochemical markers. They have shown (Figure 7.4) that this index allows the discrimination of stress levels between a reference (Bay of Bourgneuf, France) and a multipolluted site (Loire estuary, France).

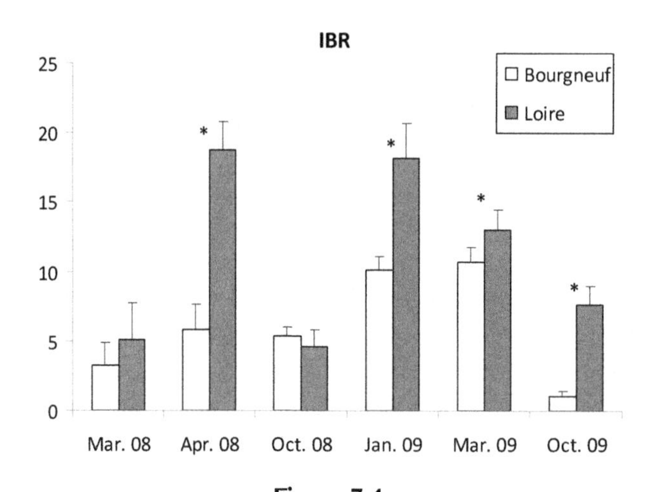

Figure 7.4
IBR indices in clams *Scrobicularia plana* collected from the Bay of Bourgneuf (reference) and the multipolluted Loire estuary. *Significant intersite differences. *After Fossi Tankoua et al. (2013) (with permission).*

Despite different sensitivity, resolution, and informative output, different integrative biomarker indices provide coherent information, showing accordance with the known contamination levels in the different study areas (Broeg and Lehtonen, 2006; Tsangaris et al., 2011; literature cited therein; Wang et al., 2011; Fossi Tankoua et al., 2013).

The Aquatic Ecosystem Health Index (Yeom and Adams, 2007) and the Estuarine Fish Health Index (Richardson et al., 2011) integrate biomarkers at the subindividual and individual levels as also bioindicators at the population and community levels. Despite the widespread appeal for using such integrative procedures to put more "eco" in ecotoxicology, such indices have been rarely used in practice. Refinements have been recently proposed, adding toxicity test responses and environmental variables in the index (Kim et al., 2014).

7.5 Conclusions

Assessing the ecological status of aquatic environments can be based upon the determination of biological quality elements (taxonomic composition, abundance and diversity of phytoplankton, macrophytes and phytobenthos, benthic invertebrate, and fish fauna) at supraindividual levels of biological organization by comparison with "totally—that no longer exist—or nearly totally undisturbed conditions" as recommended, for instance, in the European Community Water Framework Directive (Chapter 14). However, the highest level of biological organization, is the latency between exposure to pollutants and the occurrence of effects, increasing dramatically the magnitude of remediation problems. At the other end of the scale, infraindividual biomarkers provide a cost-effective and early and sensitive warning system. A comprehensive methodology may be proposed to reconcile the benefits of biochemical markers and ecological responses (Figure 7.5).

However, the success of this strategy depends on our ability to cope with confounding factors that affect most of the biomarkers commonly used for ecotoxicological assessment. In fact, certain biomarkers are robust, being not influenced or only marginally so, at least in certain species (e.g., *Nereis diversicolor*, Berthet et al., in Amiard-Triquet et al., 2011; *Platichthys flesus* studied by Laroche et al., 2013). Thus, when developing a new sentinel species, it is indispensable to assess carefully its sensitivity to confounding factors. From an operational point of view, the determination of biochemical markers in a whole organism or in the whole soft tissue seems attractive, being more cost-effective than a careful dissection. However, to obtain sound information from biomarkers, it is generally required to choose a peculiar organ (e.g., Ahmad et al., 2011, 2013) or even a peculiar cell type in the case of genotoxic damages (Lewis and Galloway, 2008). The interpretation of biomarker data (both core and ecological) is only possible if the baseline values have been established (Davies and Vethaak, 2012) or can be compared with those determined in reference sites with again "totally or nearly totally undisturbed conditions." Active biomonitoring can help to avoid the influence of intrinsic factors (size/weight/age, sometimes sex, sexual status) but it is must be kept in mind that

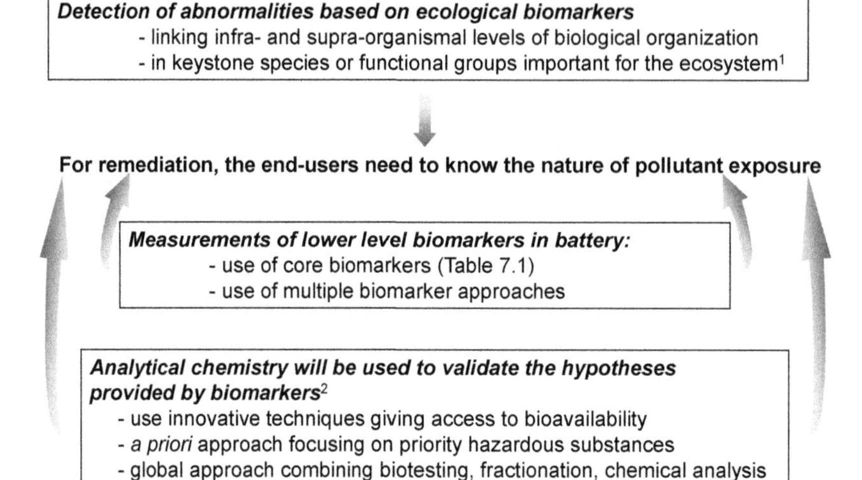

Figure 7.5

A three-step comprehensive methodology to assess the health status of aquatic ecosystems. [1]See Chapter 9; [2]Chapter 4.

"naïve" individuals used in caging experiments will not be perfectly representative of the response of local individuals that may have developed tolerance to chronic exposure (Berthet et al., in Amiard-Triquet et al., 2011).

Modeling may be used to establish correction factors, taking into account the physiological condition of the organisms as initially developed to normalize body concentrations of contaminants (Andral et al., 2004) then extended to biomarkers (Mouneyrac et al., 2008). Models may be also established for extrinsic factors, combining laboratory experiments, statistical treatments, and field tests as exemplified in gammarids (Chapter 11).

Certain biomarkers can respond to general stress associated with handling, capture, time since collection, anoxia, oxidative burst experienced by living organisms in laboratory, transplantation, and field studies. Examples include total oxyradical scavenging capacity responses and DNA damage in mussels (Wilson et al., 1998; Camus et al., 2004), responses of enzymes (AChE, GST, lactate dehydrogenase) in shrimps (Menezes et al., 2006), the induction of plasma cortisol in fish (Hontela, 2000), and immune functions (Brousseau et al., in Amiard-Triquet et al., 2013). When designing a sampling strategy, it is also important to take into account the fact that shore location may have a significant effect on biomarker expression (Gagné et al., 2009; Schmidt et al., 2012).

When developing a sampling plan, it is generally poorly reliable to limit the effort to spot sampling, except if the seasonal variability of the parameters of interest is well-established. Depending on the objectives and selected biomarkers, it may be possible to focus (e.g., biomarkers linked to reproduction) or on the contrary avoid key periods in the life of the sentinel species. By integrating the individual biomarker responses into a biomarker response index, Hagger et al. (2010) have been able to identify times of the year when environmental impact was highest and hence when the timing of monitoring programs using biomarkers should be carried out. The IBR generally shows accordance with the known contamination levels in studied sites and is useful to simplify the interpretation of biological effects of pollution in biomonitoring. However, some misuses and bias were recorded in certain studies and improvements based on a new calculation method have been recently proposed (Devin et al., 2014). It is generally calculated from values of infraindividual biomarkers, but individual biomarkers may be easily added in the index (Fossi Tankoua et al., 2013; Tlili et al., 2013) as also bioindicators at the population and community level (Aquatic Ecosystem Health Index by Yeom and Adams, 2007; Estuarine Fish Health Index by Richardson et al., 2011), toxicity test responses and environmental variables (integrated health responses by Kim et al., 2014). However, as the ecological relevance increases, the practicability clearly decreases, a reason why the IBR is one of the most used among integrative indices.

In addition to their use for the assessment of health status in areas submitted to anthropogenic inputs, biomarkers may be useful tools for improving toxicity tests used for instance in toxicity assessment of sediments collected in contaminated areas or for predictive assessment of new chemicals (Chapter 6).

References

Abele, D., Heise, K., Pörtner, H.O., et al., 2002. Temperature-dependence of mitochondrial function and production of reactive oxygen species in the intertidal mud clam *Mya arenaria*. J. Exp. Biol. 205, 1831–1841.

Abele, D., Vázquez-Medina, J.P., Zenteno-Savín, T., 2012. Oxidative Stress in Aquatic Ecosystems. Wiley-Blackwell, Chichester, UK.

Ahmad, I., Mohmood, I., Pacheco, M., et al., 2013. Mercury's mitochondrial targeting with increasing age in *Scrobicularia plana* inhabiting a contaminated lagoon: damage-protection dichotomy and organ specificities. Chemosphere 92, 1231–1237.

Ahmad, I., Mohmood, I., Mieiro, C.L., et al., 2011. Lipid peroxidation vs. antioxidant modulation in the bivalve *Scrobicularia plana* in response to environmental mercury–organ specificities and age effect. Aquat. Toxicol. 103, 150–158.

Alix, G., Beaudry, A., Brousseau-Fournier, C., et al., 2013. Increase sensitivity to metals of hemocytes obtained from *Mya arenaria* collected at different distances from the shore. J. Xenobiotics 3 (s1), 29–30 e11.

Almeida, C., Pereira, C., Gomes, T., et al., 2011. DNA damage as a biomarker of genotoxic contamination in *Mytilus galloprovincialis* from the south coast of Portugal. J. Environ. Monit. 13, 2559–2567.

Almeida, C., Pereira, C.G., Gomes, T., et al., 2013. Genotoxicity in two bivalve species from a coastal lagoon in the south of Portugal. Mar. Environ. Res. 89, 29–38.

Amiard-Triquet, C., Rainbow, P.S., 2009. Environmental Assessment of Estuarine Ecosystems. A Case Study. CRC Press, Boca Raton.

Amiard-Triquet, C., Amiard, J.C., Rainbow, P.S., 2013. Ecological Biomarkers: Indicators of Ecotoxicological Effects. CRC Press, Boca Raton.

Amiard-Triquet, C., Rainbow, P.S., Romeo, M., 2011. Tolerance to Environmental Contaminants. CRC Press, Boca Raton.

Andral, B., Stanisière, J.Y., Sauzade, D., et al., 2004. Monitoring chemical contamination levels in the Mediterranean based on the use of mussel caging. Mar. Pollut. Bull. 49, 704–712.

Andrew, M.N., O'Connor, W.A., Dunstan, R.H., et al., 2010. Exposure to 17α-ethynylestradiol causes dose and temporally dependent changes in intersex, females and vitellogenin production in the Sydney rock oyster. Ecotoxicology 19, 1440–1451.

Ara, F., Damrongphol, P., 2014. Vitellogenin gene expression at different ovarian stages in the giant freshwater prawn, *Macrobrachium rosenbergii*, and stimulation by 4-nonylphenol. Aquac. Res. 45, 320–326.

Attig, H., Kamel, N., Sforzini, S., et al., 2014. Effects of thermal stress and nickel exposure on biomarkers responses in *Mytilus galloprovincialis* (Lam). Mar. Environ. Res. 94, 65–71.

Au, D.W.T., 2004. The application of histo-cytopathological biomarkers in marine pollution monitoring: a review. Mar. Pollut. Bull. 48, 817–834.

Bae, M.J., Park, Y.S., 2014. Biological early warning system based on the responses of aquatic organisms to disturbances: a review. Sci. Total Environ. 466–467, 635–649.

Banni, M., Negri, A., Mignone, F., et al., 2011. Gene expression rhythms in the mussel *Mytilus galloprovincialis* (Lam.) across an annual cycle. PLoS One 6 (5), e18904.

Barda, I., Purina, I., Rimsa, E., et al., 2014. Seasonal dynamics of biomarkers in infaunal clam *Macoma balthica* from the Gulf of Riga (Baltic Sea). J. Marine Syst. 129, 150–156.

Barletta, M., Lucena, L.R.R., Costa, M.F., et al., 2012. The interaction rainfall vs. weight as determinant of total mercury concentration in fish from a tropical estuary. Environ. Pollut. 167, 1–6.

Barranger, A., Akcha, F., Rouxel, J., et al., 2014. Study of genetic damage in the Japanese oyster induced by an environmentally-relevant exposure to diuron: evidence of vertical transmission of DNA damage. Aquat. Toxicol. 146, 93–104.

Basmadjian, E., Perkins, E.M., Phillips, C.R., et al., 2008. Liver lesions in demersal fishes near a large ocean outfall on the San Pedro Shelf, California. Environ. Monit. Assess. 138, 239–253.

Beliaeff, B., Burgeot, T., 2002. Integrated biomarker response, a useful tool for ecological risk assessment. Environ. Toxicol. Chem. 21, 1316–1322.

Berthet, B., Mouneyrac, C., Amiard, J.C., et al., 2003. Accumulation and soluble binding of Cd, Cu and Zn in the polychaete *Hediste diversicolor* from coastal sites with different trace metal bioavailabilities. Arch. Environ. Contam. Toxicol. 45, 468–478.

Bonnard, M., Romeo, M., Amiard-Triquet, C., 2009. Effects of copper on the burrowing behavior of estuarine and coastal invertebrates, the polychaete *Nereis diversicolor* and the bivalve *Scrobicularia plana*. Hum. Ecol. Risk Assess. 15, 11–26.

Brammell, B.F., McClain, J.S., Oris, J.T., et al., 2010. CYP1A expression in caged rainbow trout discriminates among sites with various degrees of polychlorinated biphenyl contamination. Arch. Environ. Contam. Toxicol. 58, 772–782.

Broeg, K., Lehtonen, K.K., 2006. Indices for the assessment of environmental pollution of the Baltic Sea coasts: integrated assessment of a multibiomarker approach. Mar. Pollut. Bull. 53, 508–522.

Bryan, G.W., Gibbs, P.E., Hummerstone, L.G., et al., 1987. Copper, zinc, and organotin as long-term factors governing the distribution of organisms in the Fal Estuary in southwest England. Estuaries 10, 208–219.

Buchholz, F., Buchholz, C., 2010. Growth and moulting in Northern krill (*Meganyctiphanes norvegica* Sars). Adv. Mar. Biol. 57, 173–197.

Burgeot, T., Gagné, F., Forget-Leray, J., et al., 2010. Acetylcholinesterase: methodology development of a biomarker and challenges of its application for biomonitoring. ICES CM Code: F-25 In: International Council for the Exploration of the Sea, Annual Science Conference, September 20–24, 2010, Nantes, France e-paper: http://www.ices.dk/products/CMdocs/CM-2010/F/F2510.pdf.

Buschini, A., Carboni, P., Martino, A., et al., 2003. Effects of temperature on baseline and genotoxicant-induced DNA damage in haemocytes of *Dreissena polymorpha*. Mutat. Res. Gen. Tox. En. 537, 81–92.

Cachot, J., Cherel, Y., Larcher, T., et al., 2013. Histopathological lesions and DNA adducts in the liver of European flounder (*Platichthys flesus*) collected in the Seine estuary versus two reference estuarine systems on the French Atlantic coast. Environ. Sci. Pollut. Res. 20, 723–737.

Cailleaud, K., Forget-Leray, J., Peluhet, L., et al., 2009. Tidal influence on the distribution of hydrophobic organic contaminants in the Seine Estuary and biomarker responses on the copepod *Eurytemora affinis*. Environ. Pollut. 157, 64–71.

Cailleaud, K., Maillet, G., Budzinski, H., et al., 2007. Effects of salinity and temperature on the expression of enzymatic biomarkers in *Eurytemora affinis* (Calanoida, Copepoda). Comp. Biochem. Physiol. A147, 841–849.

Cairns, J.J., 1992. The threshold problem in ecotoxicology. Ecotoxicology 1, 3–16.

Caldwell, G.S., Lewis, C., Pickavance, G., et al., 2011. Exposure to copper and a cytotoxic polyunsaturated aldehyde induces reproductive failure in the marine polychaete *Nereis virens* (Sars). Aquat. Toxicol. 104, 126–134.

Camus, L., Pampanin, D.M., Volpato, E., et al., 2004. Total oxyradical scavenging capacity responses in *Mytilus galloprovincialis* transplanted into the Venice lagoon (Italy) to measure the biological impact of anthropogenic activities. Mar. Pollut. Bull. 49, 801–808.

Canesi, L., Barmo, C., Fabbri, R., et al., 2010. Effects of vibrio challenge on digestive gland biomarkers and antioxidant gene expression in *Mytilus galloprovincialis*. Comp. Biochem. Physiol. 152C, 399–406.

Carballeira, C., Martín-Díaz, L., DelValls, T.A., 2011. Influence of salinity on fertilization and larval development toxicity tests with two species of sea urchin. Mar. Environ. Res. 72, 196–203.

de Carvalho, P.S.M., 2013. Behavioral biomarkers and pollution risks to fish health and biodiversity. In: Alves de Almeida, E., Alberto de Oliveira Ribeiro, C. (Eds.), Pollution and Fish Health in Tropical Ecosystems. CRC Press, Boca Raton, pp. 350–377.

Castellani, C., Altunbaş, Y., 2006. Factors controlling the temporal dynamics of egg production in the copepod *Temora longicornis*. Mar. Ecol. Prog. Ser. 308, 143–153.

Chapman, P.M., Wang, F.Y., Caeiro, S.S., 2013. Assessing and managing sediment contamination in transitional waters. Environ. Int. 55, 71–91.

Charron, L., Geffard, O., Chaumot, A., et al., 2013. Effect of water quality and confounding factors on digestive enzyme activities in *Gammarus fossarum*. Environ. Sci. Pollut. Res. 20, 9044–9056.

Cheek, A.O., Landry, C.A., Steele, S.L., et al., 2009. Diel hypoxia in marsh creeks impairs the reproductive capacity of estuarine fish populations. Mar. Ecol. Prog. Ser. 392, 211–221.

Costa, P.M., Carreira, S., Costa, M.H., et al., 2013. Development of histopathological indices in a commercial marine bivalve (*Ruditapes decussatus*) to determine environmental quality. Aquat. Toxicol. 126, 442–454.

Cravo, A., Lopes, B., Serafim, A., et al., 2012. Spatial and seasonal biomarker responses in the clam *Ruditapes decussatus*. Biomarkers 18, 30–43.

Crowe, T.P., Smith, E.L., Donkin, P., et al., 2004. Measurement of sublethal stress effects on individual organisms indicate community-level impacts of pollution. J. Appl. Ecol. 41, 114–123.

Dabrowska, H., Kopko, O., Turja, R., et al., 2013. Sediment contaminants and contaminant levels and biomarkers in caged mussels (*Mytilus trossulus*) in the southern Baltic Sea. Mar. Environ. Res. 84, 1–9.

Dagnino, A., Viarengo, A., 2014. Development of a decision support system to manage contamination in marine ecosystems. Sci. Total Environ. 466–467, 119–126.

Dagnino, A., Allen, J.I., Moore, M.N., et al., 2007. Development of an expert system for the integration of biomarker responses in mussels into an animal health index. Biomarkers 12, 155–172.

Damiens, G., His, E., Gnassia-Barelli, M., et al., 2004. Evaluation of biomarkers in oyster larvae in natural and polluted conditions. Comp. Biochem. Physiol. C138, 121–128.

Danion, M., Le Floch, S., Castric, J., et al., 2012. Effect of chronic exposure to pendimethalin on the susceptibility of rainbow trout, *Oncorhynchus mykiss* L., to viral hemorrhagic septicemia virus (VHSV). Ecotoxicol. Environ. Saf. 79, 28–34.

Dautremepuits, C., Betoulle, S., Paris-Palacios, S., et al., 2004. Humoral immune factors modulated by copper and chitosan in healthy or parasitised carp (*Cyprinus carpio* L.) by *Ptychobothrium* sp. (Cestoda). Aquat. Toxicol. 68, 325–338.

Davies, I.M., Vethaak, A.D., 2012. Integrated Marine Environmental Monitoring of Chemicals and Their Effects. ICES Cooperative Research Report No. 315, 277 pp.

De Coen, W.M., Janssen, C.R., 2003. The missing biomarker link: relationship between effects on the cellular energy allocation biomarker of toxicant-stressed *Daphnia magna* and corresponding population characteristics. Environ. Toxicol. Chem. 22, 1632–1641.

De Kock, W.C., Kramer, K.J.M., 1994. Active biomonitoring (ABM) by translocation of bivalve mollusks. In: Kramer, K.J.M. (Ed.), Biomonitoring of Coastal Waters and Estuaries. CRC Press, Boca Raton (FL), pp. 51–84.

De Lafontaine, Y., Gagné, F., Blaise, C., et al., 2000. Biomarkers in zebra mussels (*Dreissena polymorpha*) for the assessment and monitoring of water quality of the St Lawrence River (Canada). Aquat. Toxicol. 50, 51–71.

De los Ríos, A., Pérez, L., Ortiz-Zarragoitia, M., et al., 2013. Assessing the effects of treated and untreated urban discharges to estuarine and coastal waters applying selected biomarkers on caged mussels. Mar. Pollut. Bull. 77, 251–265.

Delgado, L., Guerao, G., Ribera, C., 2011. Effects of different salinities on juvenile growth of *Gammarus aequicaudata* (Malacostraca: Amphipoda). Int. J. Zool. 6. http://dx.doi.org/10.1155/2011/248790. Article ID 248790.

Denoël, M., Bichot, M., Ficetola, G.F., et al., 2010. Cumulative effects of a road de-icing salt on amphibian behavior. Aquat. Toxicol. 99, 275–280.

Depledge, M.H., 1994. The rational basis for the use of biomarkers as ecotoxicological tools. In: Fossi, M.C., Leonzio, C. (Eds.), Nondestructive Biomarkers in Vertebrates. Lewis Publishers, Boca Raton, pp. 261–285.

Devaux, A., Fiat, L., Gillet, C., et al., 2011. Reproduction impairment following paternal genotoxin exposure in brown trout (*Salmo trutta*) and Arctic charr (*Salvelinus alpinus*). Aquat. Toxicol. 101, 405–411.

Devin, S., Burgeot, T., Giambérini, L., et al., 2014. The integrated biomarker response revisited: optimization to avoid misuse. Environ. Sci. Pollut. Res. 21, 2448–2454.

Di Poi, C., Darmaillacq, A.S., Dickel, L., et al., 2013. Effects of perinatal exposure to waterborne fluoxetine on memory processing in the cuttlefish *Sepia officinalis*. Aquat. Toxicol. 132–133, 84–91.

Dick, J.T.A., Faloon, S.E., Elwood, R.W., 1998. Active brood care in an amphipod: influences of embryonic development, temperature and oxygen. Anim. Behav. 56, 663–672.

Dimitriadis, V.K., Gougoula, C., Anestis, A., et al., 2012. Monitoring the biochemical and cellular responses of marine bivalves during thermal stress by using biomarkers. Mar. Environ. Res. 73, 70–77.

Duchemin, M.B., Fournier, M., Auffret, M., 2007. Seasonal variations of immune parameters in diploid and triploid pacific oysters, *Crassostrea gigas*. Aquaculture 264, 73–81.

Dupuy, C., Galland, C., Devaux, A., et al., 2014. Responses of the European flounder (*Platichthys flesus*) to a mixture of PAHs and PCBs in experimental conditions. Environ. Sci. Pollut. Res. 21, 13789–13803.

Durou, C., Mouneyrac, C., Amiard-Triquet, C., 2005. Tolerance to metals and assessment of energy reserves in the polychaete *Nereis diversicolor* in clean and contaminated estuaries. Environ. Toxicol. 20, 23–31.

Durou, C., Poirier, L., Amiard, J.C., et al., 2007a. Biomonitoring in a clean and a multi-contaminated estuary based on biomarkers and chemical analyses in the endobenthic worm *Nereis diversicolor*. Environ. Pollut. 148, 445–458.

Durou, C., Smith, B.D., Roméo, M., et al., 2007b. From biomarkers to population responses in *Nereis diversicolor*: assessment of stress in estuarine ecosystems. Ecotoxicol. Environ. Saf. 66, 402–411.

Durou, C., Mouneyrac, C., Amiard-Triquet, C., 2008. Environmental quality assessment in estuarine ecosystems: use of biometric measurements and fecundity of the ragworm *Nereis diversicolor* (Polychaeta, Nereididae). Water Res. 42, 2157–2165.

Edge, K.J., Dafforn, K.A., Simpson, S.L., et al., 2014. A biomarker of contaminant exposure is effective in large scale assessment of ten estuaries. Chemosphere 100, 16–26.

Eriksson Wiklund, A.K., Sundelin, B., 2004. Biomarker sensitivity to temperature and hypoxia—a seven year field study. Mar. Ecol. Prog. Ser. 274, 209–214.

Farcy, E., Burgeot, T., Haberkorn, H., et al., 2013. An integrated environmental approach to investigate biomarker fluctuations in the blue mussel *Mytilus edulis* L. in the Vilaine estuary, France. Environ. Sci. Pollut. Res. 20, 630–650.

Fielding, N.J., MacNeil, C., Dick, J.T.A., et al., 2003. Effects of the acanthocephalan parasite *Echinorhynchus truttae* on the feeding ecology of *Gammarus pulex* (Crustacea: Amphipoda). J. Zool. 261, 321–325.

Forbes, V.E., Plamqvist, A., Bach, L., 2006. The use and misuse of biomarkers in ecotoxicology. Environ. Toxicol. Chem. 25, 272–280.

Fossi Tankoua, O., Buffet, P.E., Amiard, J.C., et al., 2012a. Intersite variations of a battery of biomarkers at different levels of biological organisation in the estuarine endobenthic worm *Nereis diversicolor* (Polychaeta, Nereididae). Aquat. Toxicol. 114–115, 96–103.

Fossi Tankoua, O., Buffet, P.E., Amiard, J.C., et al., 2013. Integrated assessment of estuarine sediment quality based on a multi-biomarker approach in the bivalve *Scrobicularia plana*. Ecotoxicol. Environ. Saf. 88, 117–125.

Fossi Tankoua, O., Amiard-Triquet, C., Denis, F., et al., 2012b. Physiological status and intersex in the endobenthic bivalve *Scrobicularia plana* from thirteen estuaries in Northwest France. Environ. Pollut. 167, 70–77.

Fossi Tankoua, O., Buffet, P.E., Amiard, J.C., et al., 2011. Potential influence of confounding factors (size, salinity) on biomarker tools in the sentinel species *Scrobicularia plana* used in monitoring programmes of estuarine quality. Environ. Sci. Pollut. Res. 18, 1253–1263.

Fritsch, E.B., Connon, R.E., Werner, I., et al., 2013. Triclosan impairs swimming behavior and alters expression of excitation-contraction coupling proteins in fathead minnow (*Pimephales promelas*). Environ. Sci. Technol. 47, 2008–2017.

Frydman, J., 2001. Folding of newly translated proteins in vivo: the role of molecular chaperones. Ann. Rev. Biochem. 70, 603–647.

Gagnaire, B., Frouin, H., Moreau, K., et al., 2006. Effects of temperature and salinity on haemocyte activities of the Pacific oyster, *Crassostrea gigas* (Thunberg). Fish Shellfish Immun. 20, 536–547.

Gagné, F., 2014. Biochemical Ecotoxicology. Principles and Methods. Elsevier, Amsterdam.

Gagné, F., Blaise, C., Pellerin, J., et al., 2008. Relationships between intertidal clam population and health status of the soft-shell clam *Mya arenaria* in the St. Lawrence Estuary and Saguenay Fjord (Quebec, Canada). Environ. Int. 34, 30–43.

Gagné, F., Blaise, C., Pellerin, J., et al., 2009. Impacts of pollution in feral *Mya arenaria* populations: the effects of clam bed distance from the shore. Sci. Total Environ. 407, 5844–5854.

Gambardella, C., Mesarič, T., Milivojević, T., et al., 2014. Effects of selected metal oxide nanoparticles on *Artemia salina* larvae: evaluation of mortality and behavioural and biochemical responses (online) Environ. Monit. Assess. 186, 4249–4259.

Gamboa-Delgado, J., Molina-Poveda, C., Cahu, C., 2003. Digestive enzyme activity and food ingesta in juvenile shrimp *Litopenaeus vannamei* (Boone, 1931) as a function of body weight. Aquac. Res. 34, 1403–1411.

Giarratano, E., Gil, M.N., Malanga, G., 2011. Seasonal and pollution-induced variations in biomarkers of transplanted mussels within the Beagle Channel. Mar. Pollut. Bull. 62, 1337–1344.

Gillet, P., Mouloud, M., Durou, C., et al., 2008. Response of *Nereis diversicolor* population (Polychaeta, Nereididae) to the pollution impact—Authie and Seine estuaries (France). Estuar. Coast. Shelf Sci. 76, 201–210.

Gillis, P.L., Reynoldson, T.B., Dixon, D.G., 2004. Natural variation in a metallothionein-like protein in *Tubifex tubifex* in the absence of metal exposure. Ecotoxicol. Environ. Saf. 58, 22–28.

Gorokhova, E., Löf, M., Reutgard, M., et al., 2013. Exposure to contaminants exacerbates oxidative stress in amphipod *Monoporeia affinis* subjected to fluctuating hypoxia. Aquat. Toxicol. 127, 46–53.

Greco, L., Pellerin, J., Capri, E., et al., 2011. Physiological effects of temperature and herbicide water exposure on the soft-shell clam *Mya arenaria* (Mollusca: Bivalvia). Environ. Toxicol. Chem. 30, 132–141.

Gust, M., Buronfosse, T., André, C., et al., 2011. Is exposure temperature a confounding factor for the assessment of reproductive parameters of New Zealand mudsnails *Potamopyrgus antipodarum* (Gray)? Aquat. Toxicol. 101, 396–404.

Gust, M., Buronfosse, T., Geffard, O., et al., 2010. In situ biomonitoring of freshwater quality using the New Zealand mudsnail *Potamopyrgus antipodarum* (Gray) exposed to waste water treatment plant (WWTP) effluent discharges. Water Res. 44, 4517–4528.

Gust, M., Gagné, F., Berlioz-Barbier, A., et al., 2014. Caged mudsnail *Potamopyrgus antipodarum* (Gray) as an integrated field biomonitoring tool: exposure assessment and reprotoxic effects of water column contamination. Water Res. 54, 222–236.

Hagger, J.A., Lowe, D., Dissanayake, A., et al., 2010. The influence of seasonality on biomarker responses in *Mytilus edulis*. Ecotoxicology 19, 953–962.

Hallberg, M., Renman, G., Lundbom, T., 2007. Seasonal variations of ten metals in highway runoff and their partition between dissolved and particulate matter. Water Air Soil Pollut. 181, 183–191.

Halpin, P.M., Menge, B.A., Hofmann, G.E., 2004. Experimental demonstration of plasticity in the heat shock response of the intertidal mussel *Mytilus californianus*. Mar. Ecol. Prog. Ser. 276, 137–145.

Hartl, M.G.J., Coughlan, B.M., Sheehan, D., et al., 2004. Implications of seasonal priming and reproductive activity on the interpretation of Comet assay data derived from the clam, *Tapes semidecussatus* Reeves 1864, exposed to contaminated sediment. Mar. Environ. Res. 57, 295–310.

Hedgespeth, M.L., Anders Nilsson, P., Berglund, O., 2014. Ecological implications of altered fish foraging after exposure to an antidepressant pharmaceutical. Aquat. Toxicol. 151, 84–87.

van den Heuvel-Greve, M., Postma, J., Jol, J., et al., 2007. A chronic bioassay with the estuarine amphipod *Corophium volutator*: test method description and confounding factors. Chemosphere 66, 1301–1309.

Hontela, A., 2000. Endocrine biomarkers: hormonal indicators of sublethal toxicity in fishes. In: Lagadic, L., Caquet, T., Amiard, J.C., Ramade, F. (Eds.), Use of Biomarkers for Environmental Quality Assessment. Science Publishers, Enfield (NH), pp. 187–204.

Hylleberg, J., 1975. The effect of salinity and temperature on egestion in mud snails (Gastropoda: Hydrobiidae). Oecologia 21, 279–289.

Ivanković, D., Pavičić, J., Erk, M., et al., 2005. Evaluation of the *Mytilus galloprovincialis* Lam. digestive gland metallothionein as a biomarker in a long-term field study: seasonal and spatial variability. Mar. Pollut. Bull. 50, 1303–1313.

Jones, J.C., Reynolds, J.D., 1997. Effects of pollution on reproductive behaviour of fishes. Rev. Fish. Biol. Fish 7, 463–491.

Kadar, E., Tarran, G.A., Jha, A.N., et al., 2011. Stabilization of engineered zero-valent nanoiron with Na-acrylic copolymer enhances spermiotoxicity. Environ. Sci. Technol. 45, 3245–3251.

Kalman, J., Palais, F., Amiard, J.C., et al., 2009. Assessment of the health status of populations of the ragworm *Nereis diversicolor* using biomarkers at different levels of biological organisation. Mar. Ecol. Prog. Ser. 393, 55–67.

Kalman, J., Buffet, P.E., Amiard, J.C., et al., 2010. Assessment of the influence of confounding factors (weight, salinity) on the response of biomarkers in the estuarine polychaete *Nereis diversicolor*. Biomarkers 15, 461–469.

Kamel, N., Burgeot, T., Banni, M., et al., 2014. Effects of increasing temperatures on biomarker responses and accumulation of hazardous substances in rope mussels (*Mytilus galloprovincialis*) from Bizerte lagoon. Environ. Sci. Pollut. Res. 21, 6108–6123.

Kerambrun, E., Sanchez, W., Henry, F., et al., 2011. Are biochemical biomarker responses related to physiological performance of juvenile sea bass (*Dicentrarchus labrax*) and turbot (*Scophthalmus maximus*) caged in a polluted harbour? Comp. Biochem. Physiol. 154C, 187–195.

Kidd, K.A., Blanchfield, P.J., Mills, K.H., et al., 2007. Collapse of a fish population after exposure to a synthetic estrogen. Proc. Nat. Acad. Sci. 104, 8897–8901.

Kim, J.H., Yeom, D.H., An, K.G., 2014. A new approach of Integrated Health Responses (IHRs) modeling for ecological risk/health assessments of an urban stream. Chemosphere 108, 376–382.

Klobučar, G.I.V., Štambuk, A., Pavlica, M., et al., 2010. Genotoxicity monitoring of freshwater environments using caged carp (*Cyprinus carpio*). Ecotoxicology 19, 77–84.

Koehler, A., 2004. The gender-specific risk to liver toxicity and cancer of flounder (*Platichthys flesus* (L.)) at the German Wadden Sea coast. Aquat. Toxicol. 70, 257–276.

Kolarević, S., Knežević-Vukčević, J., Paunović, M., et al., 2013. Monitoring of DNA damage in haemocytes of freshwater mussel *Sinanodonta woodiana* sampled from the Velika Morava River in Serbia with the comet assay. Chemosphere 93, 243–251.

Kopecka, J., Pempkowiak, J., 2008. Temporal and spatial variations of selected biomarker activities in flounder (*Platichthys flesus*) collected in the Baltic proper. Ecotoxicol. Environ. Saf. 70, 379–391.

Körner, O., Kohno, S., Schönenberger, R., et al., 2008. Water temperature and concomitant waterborne ethinylestradiol exposure affects the vitellogenin expression in juvenile brown trout (*Salmo trutta*). Aquat. Toxicol. 90, 188–196.

Krell, B., Moreira-Santos, M., Ribeiro, R., 2011. An estuarine mudsnail in situ toxicity assay based on postexposure feeding. Environ. Toxicol. Chem. 30, 1935–1942.

Kumar, V., Nakada, N., Yamashita, N., et al., 2011. How seasonality affects the flow of estrogens and their conjugates in one of Japan's most populous catchments. Environ. Pollut. 159, 2906–2912.

Kuz'mina, V.V., 1996. Influence of age on digestive enzyme activity in some freshwater teleosts. Aquaculture 148, 25–37.

Lacaze, E., Geffard, O., Goyet, D., et al., 2011b. Linking genotoxic responses in *Gammarus fossarum* germ cells with reproduction impairment, using the Comet assay. Environ. Res. 111, 626–634.

Lacaze, E., Devaux, A., Bony, S., et al., 2013. Genotoxic impact of a municipal effluent dispersion plume in the freshwater mussel *Elliptio complanata*: an in situ study. J. Xenobiotics 3 (s1), e6. 14–16.

Lacaze, E., Devaux, A., Mons, R., et al., 2011a. DNA damage in caged *Gammarus fossarum* amphipods: a tool for freshwater genotoxicity assessment. Environ. Pollut. 159, 1682–1691.

Lacaze, E., Geffard, O., Bony, S., et al., 2010. Genotoxicity assessment in the amphipod *Gammarus fossarum* by use of the alkaline Comet assay. Mut. Res. 700, 32–38.

Landry, C.A., Steele, S.L., Manning, S., et al., 2007. Long-term hypoxia suppresses reproductive capacity in the estuarine fish, *Fundulus grandis*. Comp. Biochem. Physiol. A148, 317–323.

Lang, T., Wosniok, W., Baršienė, J., et al., 2006. Liver histopathology in Baltic flounder (*Platichthys flesus*) as indicator of biological effects of contaminants. Mar. Pollut. Bull. 53, 488–496.

Lange, U., Saborowski, R., Siebers, D., et al., 1998. Temperature as a key factor determining the regional variability of the xenobiotic-inducible ethoxyresorufin-O-deethylase activity in the liver of dab (*Limanda limanda*). Can. J. Fish. Aquat. Sci. 55, 328–338.

Laroche, J., Gauthier, O., Quiniou, L., et al., 2013. Variation patterns in individual fish responses to chemical stress among estuaries, seasons and genders: the case of the European flounder (*Platichthys flesus*) in the Bay of Biscay. Environ. Sci. Pollut. Res. 20, 738–748.

Leiniö, S., Lehtonen, K.K., 2005. Seasonal variability in biomarkers in the bivalves *Mytilus edulis* and *Macoma balthica* from the northern Baltic Sea. Comp. Biochem. Physiol. 140C, 408–421.

Lekube, X., Izagirre, U., Soto, M., et al., 2014. Lysosomal and tissue-level biomarkers in mussels cross-transplanted among four estuaries with different pollution levels. Sci. Total Environ. 472, 36–48.

Lesser, M.P., Bailey, M.A., Merselis, D.G., et al., 2010. Physiological response of the blue mussel *Mytilus edulis* to differences in food and temperature in the Gulf of Maine. Comp. Biochem. Physiol. A156, 541–551.

Letendre, J., Chouquet, B., Manduzio, H., Marin, M., Bultelle, F., Leboulenger, F., et al., 2009. Tidal height influences the levels of enzymatic antioxidant defences in *Mytilus edulis*. Mar. Environ. Res. 67, 69–74.

Lettini, S.E., Sukhdeo, M.V.K., 2010. The energetic cost of parasitism in isopods. Ecoscience 17, 1–8.

Leung, K.M.Y., Svavarsson, J., Crane, M., et al., 2002. Influence of static and fluctuating salinity on cadmium uptake and metallothionein expression by the dogwhelk *Nucella lapillus* (L.). J. Exp. Mar. Biol. Ecol. 274, 175–189.

Leung, K.M.Y., Taylor, A.C., Furness, R.W., 2000. Temperature-dependent physiological responses of the dogwhelk *Nucella lapillus* to cadmium exposure. J. Mar. Biol. Assoc. UK 80, 647–660.

Lewis, C., Galloway, T., 2008. Genotoxic damage in polychaetes: a study of species and cell-type sensitivities. Mutat. Res. Gen. Tox. En. 654, 69–75.

Lewis, C., Galloway, T., 2009. Reproductive consequences of paternal genotoxin exposure in marine invertebrates. Environ. Sci. Technol. 43, 928–933.

Liu, W., Huang, X., Lin, J., et al., 2012. Seawater acidification and elevated temperature affect gene expression patterns of the pearl oyster *Pinctada fucata*. PLoS One 7 (3), e33679.

Livingstone, D.R., Mitchelmore, C.L., Peters, L.D., et al., 2000. Development of hepatic CYP1A and blood vitellogenin in eel (*Anguilla anguilla*) for use as biomarkers in the Thames Estuary, UK. Mar. Environ. Res. 50, 367–371.

López-Galindo, C., Ruiz-Jarabo, I., Rubio, D., et al., 2014. Temperature enhanced effects of chlorine exposure on the health status of the sentinel organism *Mytilus galloprovincialis*. Environ. Sci. Pollut. Res. 21, 1680–1690.

Lozano, S.J., Gedeon, M.L., Landrum, P.F., 2003. The effects of temperature and organism size on the feeding rate and modeled chemical accumulation in *Diporeia* spp. for Lake Michigan sediments. J. Great Lakes Res. 29, 79–88.

Lushchak, V.I., 2011. Environmentally induced oxidative stress in aquatic animals. Aquat. Toxicol. 101, 13–30.

MacInnes, J.R., Calabrese, A., 1979. Combined effects of salinity, temperature, and copper on embryos and early larvae of the American oyster, *Crassostrea virginica*. Arch. Environ. Contam. Toxicol. 8, 553–562.

Maltby, L., Naylor, C., Calow, P., 1990. Effect of stress on a freshwater benthic detritivore: scope for growth in *Gammarus pulex*. Ecotoxicol. Environ. Saf. 19, 285–291.

Maltby, L., Clayton, S.A., Wood, R.M., et al., 2002. Evaluation of the *Gammarus pulex in situ* feeding assay as a biomonitor of water quality: robustness, responsiveness, and relevance. Environ. Toxicol. Chem. 21, 361–368.

Maranhão, P., Marques, J.C., 2003. The influence of temperature and salinity on the duration of embryonic development, fecundity and growth of the amphipod *Echinogammarus marinus* Leach (Gammaridae). Acta Oecol. 24, 5–13.

Marigómez, I., Zorita, I., Izagirre, U., et al., 2013a. Combined use of native and caged mussels to assess biological effects of pollution through the integrative biomarker approach. Aquat. Toxicol. 136–137, 32–48.

Marigómez, I., Garmendia, L., Soto, M., et al., 2013b. Marine ecosystem health status assessment through integrative biomarker indices: a comparative study after the Prestige oil spill ''Mussel Watch''. Ecotoxicology 22, 486–505.

Martineau, D., Lemberger, K., Dallaire, A., et al., 2002. Cancer in wildlife, a case study: Beluga from the St. Lawrence Estuary, Québec, Canada. Environ. Health Perspect. 110, 285–292.

Martínez-Álvarez, R.M., Hidalgo, M.C., Domezain, A., et al., 2002. Physiological changes of sturgeon *Acipenser naccarii* caused by increasing environmental salinity. J. Exp. Biol. 205, 3699–3706.

Matozzo, V., Binelli, A., Parolini, M., et al., 2012. Biomarker responses in the clam *Ruditapes philippinarum* and contamination levels in sediments from seaward and landward sites in the Lagoon of Venice. Ecol. Indic. 19, 191–205.

Matozzo, V., Boscolo, A., Marin, M.G., 2013. Seasonal and gender-related differences in morphometric features and cellular and biochemical parameters of *Carcinus aestuarii* from the Lagoon of Venice. Mar. Environ. Res. 89, 21–28.

McCahon, C.P., Brown, A.F., Pascoe, D., 1988. The effect of the acanthocephalan *Pomphorhynchus laevis* (Muller 1776) on the acute toxicity of cadmium to its intermediate host, the amphipod *Gammarus pulex* (L.). Arch. Environ. Contam. Toxicol. 17, 239–243.

Menezes, S., Soares, A.M.V.M., Guilhermino, L., et al., 2006. Biomarker responses of the estuarine brown shrimp *Crangon crangon* L. to non-toxic stressors: temperature, salinity and handling stress effects. J. Exp. Mar. Biol. Ecol. 335, 114–122.

Miller, D.H., Jensen, K.M., Villeneuve, D.L., et al., 2007. Linkage of biochemical responses to population-level effects: a case study with vitellogénine in the fathead minnow (*Pimephales promelas*). Environ. Toxicol. Chem. 26, 521–527.

Minguez, L., Meyer, A., Molloy, D.P., et al., 2009. Interactions between parasitism and biological responses in zebra mussels (*Dreissena polymorpha*): importance in ecotoxicological studies. Environ. Res. 109, 843–850.

Monari, M., Matozzo, V., Foschi, J., et al., 2007. Effects of high temperatures on functional responses of haemocytes in the clam *Chamelea gallina*. Fish Shellfish Immun. 22, 98–114.

Moreira, S.M., Moreira-Santos, M., Guilhermino, L., et al., 2006. An *in situ* postexposure feeding assay with *Carcinus maenas* for estuarine sediment-overlying water toxicity evaluations. Environ. Pollut. 139, 318–329.

Mouneyrac, C., Amiard-Triquet, C., 2013. Biomarkers of ecological relevance. In: Férard, J.F., Blaise, C. (Eds.), Encyclopedia of Aquatic Ecotoxicology. Springer, Dordrecht, pp. 221–236.

Mouneyrac, C., Linot, S., Amiard, J.C., et al., 2008. Biological indices, energy reserves, steroid hormones and sexual maturity in the infaunal bivalve *Scrobicularia plana* from three sites differing by their level of contamination. Gen. Comp. Endocrinol. 157, 133–141.

Mouneyrac, C., Mastain, O., Amiard, J.C., et al., 2003. Physico-chemical forms of storage and the tolerance of the estuarine worm *Nereis diversicolor* chronically exposed to trace metals in the environment. Mar. Biol. 143, 731–744.

Mouneyrac, C., Perrein-Ettajani, H., Amiard-Triquet, C., 2010. Influence of anthropogenic stress on fitness and behaviour of a key-species of estuarine ecosystems, the ragworm *Nereis diversicolor*. Environ. Pollut. 158, 121–128.

Mouneyrac, C., Buffet, P.E., Poirier, L., et al., 2014. Fate and effects of metal-based nanoparticles in two marine invertebrates, the bivalve mollusc *Scrobicularia plana* and the annelid polychaete *Hediste diversicolor*. Environ. Sci. Pollut. Res. 21, 7899–7912.

Mouneyrac, C., Amiard, J.C., Amiard-Triquet, C., 1998. Effects of natural factors: salinity and body weight on cadmium, copper, zinc and metallothionein-like protein levels in resident populations of oysters (*Crassostrea gigas*) from a polluted estuary. Mar. Ecol. Prog. Ser. 162, 125–135.

Myers, M.S., Anulacion, B.F., French, B.L., et al., 2008. Improved flatfish health following remediation of a PAH-contaminated site in Eagle Harbor, Washington. Aquat. Toxicol. 88, 277–288.

Nahrgang, J., Brooks, S.J., Evenset, A., et al., 2013. Seasonal variation in biomarkers in blue mussel (*Mytilus edulis*), Icelandic scallop (*Chlamys islandica*) and Atlantic cod (*Gadus morhua*)—implications for environmental monitoring in the Barents Sea. Aquat. Toxicol. 127, 21–35.

Naylor, C., Maltby, L., Calow, P., 1989. Scope for growth in *Gammarus pulex*, a freshwater benthic detritivore. Hydrobiologia 188/189, 715–723.

Neuparth, T., Costa, F.O., Costa, M.H., 2002. Effects of temperature and salinity on life history of the marine amphipod *Gammarus locusta*. Implications for ecotoxicological testing. Ecotoxicology 11, 61–73.

Nilsson, L.M., 1974. Energy budget of a laboratory population of *Gammarus pulex* (Amphipoda). Oikos 25, 35–42.

Oikari, A., 2006. Caging techniques for field exposures of fish to chemical contaminants. Aquat. Toxicol. 78, 370–381.

Padmini, E., Vijaya Geetha, B., Usha Rani, M., 2008. Liver oxidative stress of the grey mullet *Mugil cephalus* presents seasonal variations in Ennore estuary. Braz. J. Med. Biol. Res. 41, 951–955.

Paital, B., Chainy, G.B.N., 2012. Effects of salinity on O_2 consumption, ROS generation and oxidative stress status of gill mitochondria of the mud crab *Scylla serrata*. Comp. Biochem. Physiol. C155, 228–237.

Pascal, P.Y., Dupuy, C., Richard, P., et al., 2008. Influence of environment factors on bacterial ingestion rate of the deposit-feeder *Hydrobia ulvae* and comparison with meiofauna. J. Sea Res. 60, 151–156.

Pascoe, D., Kedwards, T.J., Blockwell, S.J., et al., 1995. *Gammarus pulex* (L.) feeding bioassay—effects of parasitism. Bull. Environ. Contam. Toxicol. 55, 629–632.

Peltzer, P.M., Junges, C.M., Attademo, A.M., et al., 2013. Cholinesterase activities and behavioral changes in *Hypsiboas pulchellus* (Anura: Hylidae) tadpoles exposed to glufosinate ammonium herbicide. Ecotoxicology 22, 1165–1173.

Peterson, C.L., Côté, J., 2004. Cellular machineries for chromosomal DNA repair. Gene Dev. 18, 602–616.

Pfeifer, S., Schiedek, D., Dippner, J.W., 2005. Effect of temperature and salinity on acetylcholinesterase activity, a common pollution biomarker, in *Mytilus* sp. from the south-western Baltic Sea. J. Exp. Mar. Biol. Ecol. 320, 93–103.

Pichaud, N., Briatte, S., Desrosiers, V., et al., 2009. Metabolic capacities and immunocompetence of sea scallops (*Placopecten magellanicus*, Gmelin) at different ages and life stages. J. Shellfish Res. 28, 865–876.

Pöckl, M., 1992. Effects of temperature, age and body size on moulting and growth in the freshwater amphipods *Gammarus fossarum* and *G. roeseli*. Freshwater Biol. 27, 211–225.

Pöckl, M., 1993. Reproductive potential and lifetime potential fecundity of the freshwater amphipods *Gammarus fossarum* and *G. roeseli* in Austrian streams and rivers. Freshwater Biol. 30, 73–91.

Pöckl, M., 2007. Strategies of a successful new invader in European fresh waters: fecundity and reproductive potential of the Ponto-Caspian amphipod *Dikerogammarus villosus* in the Austrian Danube, compared with the indigenous *Gammarus fossarum* and *G. roeseli*. Freshwater Biol. 52, 50–63.

Pöckl, M., Humpesch, U.H., 1990. Intra- and inter-specific variations in egg survival and brood development time for Austrian populations of *Gammarus fossarum* and *G. roeseli* (Crustacea: Amphipoda). Freshwater Biol. 23, 441–455.

Pöhlmann, K., Koenigstein, S., Alter, K., et al., 2011. Heat-shock response and antioxidant defense during air exposure in Patagonian shallow-water limpets from different climatic habitats. Cell Stress Chaperon. 16, 621–632.

Prato, E., Biandolino, F., Scardicchio, C., 2008. Implications for toxicity tests with amphipod *Gammarus aequicaudata*: effects of temperature and salinity on life cycle. Environ. Technol. 29, 1349–1356.

Rainbow, P.S., 1997. Trace metal accumulation in marine invertebrates: marine biology or marine chemistry? J. Mar. Biol. Assoc. UK 77, 195–210.

Ramos-Gómez, J., Martins, M., Raimundo, J., et al., 2011. Validation of *Arenicola marina* in field toxicity bioassays using benthic cages: biomarkers as tools for assessing sediment quality. Mar. Pollut. Bull. 62, 1538–1549.

Ricciardi, F., Binelli, A., Provini, A., 2006. Use of two biomarkers (CYP450 and acetylcholinesterase) in zebra mussel for the biomonitoring of Lake Maggiore (northern Italy). Ecotoxicol. Environ. Saf. 63, 406–412.

Richardson, N., Gordon, A.K., Muller, W.J., et al., 2011. A weight-of-evidence approach to determine estuarine fish health using indicators from multiple levels of biological organization. Aquat. Conserv. 21, 423–432.

Robillard, S., Beauchamp, G., Laulier, M., 2003. The role of abiotic factors and pesticide levels on enzymatic activity in the freshwater mussel *Anodonta cygnea* at three different exposure sites. Comp. Biochem. Physiol. C135, 49–59.

Russo, J., Madec, L., Brehélin, M., 2009. Haemocyte lysosomal fragility facing an environmental reality: a toxicological perspective with atrazine and *Lymnaea stagnalis* (Gastropoda, Pulmonata) as a test case. Ecotoxicol. Environ. Saf. 72, 1719–1726.

Sacchi, A., Mouneyrac, C., Bolognesi, C., et al., 2013. Biomonitoring study of an estuarine coastal ecosystem, the Sacca di Goro lagoon, using *Ruditapes philippinarum* (Mollusca: Bivalvia). Environ. Pollut. 177, 82–89.

Sanchez, W., Sremski, W., Piccini, B., et al., 2011. Adverse effects in wild fish living downstream from pharmaceutical manufacture discharges. Environ. Int. 37, 1342–1348.

Sandström, O., Larsson, A., Andersson, J., et al., 2005. Three decades of Swedish experience demonstrates the need for integrated long-term monitoring of fish in marine coastal areas. Water Qual. Res. J. Can. 40, 233–250.

Sänkiaho, L.K., 2009. Metals in Stormwater Pollution (Master's thesis). Helsinki University of Technology.

Schmidt, W., Power, E., Quinn, B., 2013. Seasonal variations of biomarker responses in the marine blue mussel (*Mytilus* spp.). Mar. Pollut. Bull. 74, 50–55.

Schmidt, W., O'Shea, T., Quinn, B., 2012. The effect of shore location on biomarker expression in wild *Mytilus* spp. and its comparison with long line cultivated mussels. Mar. Environ. Res. 80, 70–76.

Schmitt, C.J., Dethloff, G.M. (Eds.), 2000. Biomonitoring of Environmental Status and Trends (BEST) Program: Selected Methods for Monitoring Chemical Contaminants and Their Effects in Aquatic Ecosystems. Information and Technology Report USGS/BRD-2000–0005. p. 81. www.cerc.usgs.gov/pubs/BEST/methods.pdf (last accessed 02.09.14).

Seabra Pereira, C.D., Abessa, D.M., Choueri, R.B., et al., 2014. Ecological relevance of Sentinels' biomarker responses: a multi-level approach. Mar Environ. Res. 96, 118–126.

Selck, H., Aufderheide, J., Pounds, N., et al., 2006. Effects of food type, feeding frequency, and temperature on juvenile survival and growth of *Marisa cornuarietis* (Mollusca: Gastropoda). Invertebr. Biol. 125, 106–116.

Serafim, M.A., Company, R.M., Bebianno, M.J., et al., 2002. Effect of temperature and size on metallothionein synthesis in the gill of *Mytilus galloprovincialis* exposed to cadmium. Mar. Environ. Res. 54, 361–365.

Shang, E., Yu, R.M.K., Wu, R.S.S., 2006. Hypoxia affects sex differentiation and development, leading to a male-dominated population in zebrafish (*Danio rerio*). Environ. Sci. Technol. 40, 3118–3122.

Sheehan, D., Power, A., 1999. Effects of seasonality on xenobiotic and antioxidant defence mechanisms of bivalve molluscs. Comp. Biochem. Physiol. 123C, 193–199.

Short, S., Yang, G., Kille, P., et al., 2014. Vitellogenin is not an appropriate biomarker of feminisation in a Crustacean. Aquat. Toxicol. 153, 89–97.

Sieratowicz, A., Stange, D., Schulte-Oehlmann, U., et al., 2011. Reproductive toxicity of bisphenol A and cadmium in *Potamopyrgus antipodarum* and modulation of bisphenol A effects by different test temperature. Environ. Pollut. 159, 2766–2774.

Silva, C., Mattioli, M., Fabbri, E., et al., 2012. Benthic community structure and biomarker responses of the clam *Scrobicularia plana* in a shallow tidal creek affected by fish farm effluents (Rio San Pedro, SW Spain). Environ. Int. 47, 86–98.

Sleiderink, H.M., Beyer, J., Everaarts, J.M., et al., 1995a. Influence of temperature on cytochrome P450 1A in dab (*Limanda limanda*) from the Southern North Sea: results from field surveys and a laboratory study. Mar. Environ. Res. 39, 61–71.

Sleiderink, H.M., Beyer, J., Scholtens, E., et al., 1995b. Influence of temperature and polyaromatic contaminants on CYPlA levels in North Sea dab (*Limanda limanda*). Aquat. Toxicol. 32, 189–209.

Smit, M.G.D., Bechmann, R.K., Hendriks, A.J., et al., 2009. Relating biomarkers to whole-organism effects using species sensitivity distributions: a pilot study for marine species exposed to oil. Environ. Toxicol. Chem. 28, 1104–1109.

Sokolova, I.M., Lannig, G., 2008. Interactive effects of metal pollution and temperature on metabolism in aquatic ectotherms: implications of global climate change. Clim. Res. 37, 181–201.

Stentiford, G.D., Bignell, J.P., Lyons, B.P., et al., 2010. Effect of age on liver pathology and other diseases in flatfish: implications for assessment of marine ecological health status. Mar. Ecol. Prog. Ser. 411, 215–230.

Sutcliffe, D.W., 1992. Reproduction in *Gammarus* (Crustacea, Amphipoda): basic processes. Freshwater Forum 2, 102–128.

Thatje, S., Lovrich, G.A., Anger, K., 2004. Egg production, hatching rates, and abbreviated larval development of *Campylonotus vagans* Bate, 1888 (Crustacea: Decapoda: Caridea), in subantarctic waters. J. Exp. Mar. Biol. Ecol. 301, 15–27.

Thomas, P., Rahman, M.S., 2012. Extensive reproductive disruption, ovarian masculinization and aromatase suppression in Atlantic croaker in the northern Gulf of Mexico hypoxic zone. Proc. Roy. Soc. Lond. B279, 28–38.

Thomas, P., Rahman, M.S., Khan, I.A., Kummer, J.A., 2007. Widespread endocrine disruption and reproductive impairment in an estuarine fish population exposed to seasonal hypoxia. Proc. Roy. Soc. Lond. B274, 2693–2701.

Tlili, S., Minguez, L., Giamberini, L., et al., 2013. Assessment of the health status of *Donax trunculus* from the Gulf of Tunis using integrative biomarker indices. Ecol. Indic. 32, 285–293.

Townsend, M., Hewitt, J., Philips, N., et al., 2009. Interactions between Heavy Metals, Sediment and Cockle Feeding and Movement. Prepared by NIWA for Auckland Regional Council. Auckland Regional Council Technical Report TR 2010/023.

Triebskorn, R., Köhler, H.R., Honnen, W., et al., 1997. Induction of heat shock proteins, changes in liver ultra-structure, and alterations of fish behavior: are these biomarkers related and are they useful to reflect the state of pollution in the field? J. Aquat. Ecosyst. Stress Recovery 6, 57–73.

Tsangaris, C., Hatzianestis, I., Catsiki, V.A., et al., 2011. Active biomonitoring in Greek coastal waters: application of the integrated biomarker response index in relation to contaminant levels in caged mussels. Sci. Total Environ. 412–413, 359–365.

Tu, H.T., Silvestre, F., De Meulder, B., et al., 2012. Combined effects of deltamethrin, temperature and salinity on oxidative stress biomarkers and acetylcholinesterase activity in the black tiger shrimp (*Penaeus monodon*). Chemosphere 86, 83–91.

Turja, R., Höher, N., Snoeijs, P., et al., 2014. A multibiomarker approach to the assessment of pollution impacts in two Baltic Sea coastal areas in Sweden using caged mussels (*Mytilus trossulus*). Sci. Total Environ. 473–474, 398–409.

UNEP/MAP/MED POL, 2007. MED POL biological effects monitoring programme: achievements and future orientations. In: Proceedings of the Workshop, Alessandria, Italy, December 20–21, 2006. MAP Technical Report Series No. 166. UNEP/MAP, Athens.

Verlecar, X.N., Jena, K.B., Chainy, G.B.N., 2007. Biochemical markers of oxidative stress in *Perna viridis* exposed to mercury and temperature. Chem.Biol. Interact. 167, 219–226.

Vidal, M.L., Bassères, A., Narbonne, J.F., 2002. Influence of temperature, pH, oxygenation, water-type and substrate on biomarker responses in the freshwater clam *Corbicula fluminea* (Müller). Comp. Biochem. Physiol. C132, 93–104.

Vidal-Liñán, L., Bellas, J., Etxebarria, N., et al., 2014. Glutathione *S*-transferase, glutathione peroxidase and acetylcholinesterase activities in mussels transplanted to harbour areas. Sci. Total Environ. 470–471, 107–116.

Wang, C., Lu, G., Wang, P., et al., 2011. Assessment of environmental pollution of Taihu Lake by combining active biomonitoring and integrated biomarker response. Environ. Sci. Technol. 45, 3746–3752.

Wayne, N.L., 2001. Regulation of seasonal reproduction in mollusks. J. Biol. Rhythms 16, 391–402.

Weis, J.S., Bergey, L., Reichmuth, J., et al., 2011. Living in a contaminated estuary: behavioral changes and ecological consequences for five species. Bioscience 61, 375–385.

Weis, J.S., Smith, G.M., Zhou, T., et al., 2001. Effects of contaminants on behavior. Biochemical mechanisms and ecological consequences. Bioscience 51, 209–217.

Werner, I., 2004. The influence of salinity on the heat-shock protein response of *Potamocorbula amurensis* (Bivalvia). Mar. Environ. Res. 58, 803–807.

Wijnhoven, S., Van Riel, M.C., Van Der Velde, G., 2003. Exotic and indigenous freshwater gammarid species: physiological tolerance to water temperature in relation to ionic content of the water. Aquat. Ecol. 37, 151–158.

Wilson, J.T., Pascoe, P.L., Parry, J.M., et al., 1998. Evaluation of the comet assay as a method for the detection of DNA damage in the cells of a marine invertebrate, *Mytilus edulis* L. (Mollusca: Pelecypoda). Mutat. Res. 399, 87–95.

Winkler, G., Greve, W., 2002. Laboratory studies of the effect of temperature on growth, moulting and reproduction in the co-occurring mysids *Neomysis integer* and *Praunus flexuosus*. Mar. Ecol. Prog. Ser. 235, 177–188.

Wosniok, W., Broeg, K., Feist, S.W., 2005. On the applicability of the various available 'health indices' for the interpretation of data obtained from biological effects monitoring activities and associated research studies using pathology and disease endpoints. ICES CM 2005/F:02, Annex 8. In: Report of the Working Group on Pathology and Diseases of Marine Organisms (WGPDMO), March 8–12, 2005, La Tremblade, France, pp. 76–84.

Wu, R.S.S., 2009. Effects of hypoxia on fish reproduction and development. In: Richards, J.G., Farrell, A.P., Brauner, C.J. (Eds.), Fish Physiology. Academic Press, New York, pp. 79–141.

Wu, R.S.S., Zhou, B.S., Randall, D.J., et al., 2003. Aquatic hypoxia is an endocrine disruptor and impairs fish reproduction. Environ. Sci. Technol. 37, 1137–1141.

Xuereb, B., Chaumot, A., Mons, R., et al., 2009. Acetylcholinesterase activity in *Gammarus fossarum* (Crustacea Amphipoda). Intrinsic variability, reference levels, and a reliable tool for field surveys. Aquat. Toxicol. 93, 225–233.

Yang, Q.L., Yao, C.L., Wang, Z.Y., 2012. Acute temperature and cadmium stress response characterization of small heat shock protein 27 in large yellow croaker, *Larimichthys crocea*. Comp. Biochem. Physiol. C155, 190–197.

Yeom, D.H., Adams, S.M., 2007. Assessing effects of stress across levels of biological organization using an aquatic ecosystem health index. Ecotoxicol. Environ. Saf. 67, 286–295.

Zaccaron da Silva, A., Zanette, J., Ferreira, J.F., et al., 2005. Effects of salinity on biomarker responses in *Crassostrea rhizophorae* (Mollusca, Bivalvia) exposed to diesel oil. Ecotoxicol. Environ. Saf. 62, 376–382.

CHAPTER 8

Omics in Aquatic Ecotoxicology: The Ultimate Response to Biological Questions?

Patrice Gonzalez, Fabien Pierron

Abstract

In aquatic toxicology, it is now well-established that the chemical characterization of pollutants exposure is not sufficient and that multidisciplinary approaches coupling chemistry and biology have to be developed to allow linking of the presence of contaminants and their putative toxic impacts. Along with the improvement of molecular biology methods, the development of "omics" technologies is booming worldwide since the early 1990s when these techniques have begun to emerge. During the past 20 years, these techniques have been in turn described as a unique research opportunity to decrypt all the biological mechanisms. In this chapter, the major genomic, proteomic, metabolomic, and fluxomic approaches developed in aquatic ecotoxicology are described and illustrated by studies on fish and mollusks from the recent literature. The advantages and main limitations of these techniques will be discussed. Finally, some important points to be taken into account in future prospects will be discussed.

Keywords: Biomarkers; Biomonitoring; Cellular pathways; Genomics; Metabolomics; Molecular impacts; Proteomics.

Chapter Outline

Initially, ecotoxicology borrowed many concepts and methods from toxicology. The works carried out during the 1970s were originally mainly oriented toward understanding and predicting the behavior of chemical compounds in the environment and their ability to

Aquatic Ecotoxicology. http://dx.doi.org/10.1016/B978-0-12-800949-9.00008-5

interact with biological membranes. Subsequently, during the 1980s, many researchers have attempted to evaluate the toxic effects of contaminants on living organisms. Always using the precepts of toxicology, they have developed acute toxicity tests based on the dose–effect relationship. In the 1990s, many ecotoxicologists denounced the lack of representativeness and the poor ecological relevance of these approaches. Therefore, new tools called bioindicators and biomarkers were developed (Moore et al., 2004; Clements and Rohr, 2009). It is now well-established that the chemical characterization of pollutants exposure is not sufficient and that multidisciplinary approaches coupling chemistry and biology have to be developed to allow linking presence of contaminants and their putative toxic impacts (Van der Oost et al., 2003). In this aim, the use of early warning indicators, representative of biological effects of contaminants on organisms called biomarkers, has grown significantly over the past several decades. Indeed, the effects at high hierarchical levels are always preceded by early changes at the genome level (Van der Oost et al., 2003). Along with the improvement of molecular biology methods, the development of "omics" technologies is booming worldwide in all fields of biology since the early 1990s, when these techniques have begun to emerge. In recent decades, genomic and proteomic have been described as able to help biologists to answer any questions and thus decrypt all the mechanisms used by organisms to respond to changes in their environment. Molecular techniques have been largely applied to mammalian, notably human, but its use in aquatic ecotoxicology is very recent. In this context, the term "ecotoxicogenomics" has been proposed to describe studies analyzing the adaptive response to toxic exposure at the transcriptomic, proteomic, and metabolomic levels (Snape et al., 2004). In this chapter, we will briefly review the major omic approaches developed in aquatic ecotoxicology and give some significant results taken from the recent literature. The advantages and main limitations of each technique will be reviewed. Finally, some important points to be taken into account in future prospects will be discussed.

8.1 Genome and Its Applications

Studies at the genome level not only concerned the knowledge of the genome structure and gene sequences, but also the determination of the expression levels of genes and the epigenetic modifications that could influence genome organization and transcriptome expression.

8.1.1 Genome Sequencing

Genomic approaches firstly imply the knowledge of the structure and the sequence of the organisms' genome. Indeed, for example, sequencing of the target genes involved in cellular pathways used during adaptive response to environmental variations is essential for transcriptomic studies using quantitative polymerase chain reaction (qPCR) or complementary DNA (cDNA) microarrays analyses. Determination of genome sequence has largely increased during the past 5 years, mainly from improvement in sequencing technologies (next-generation

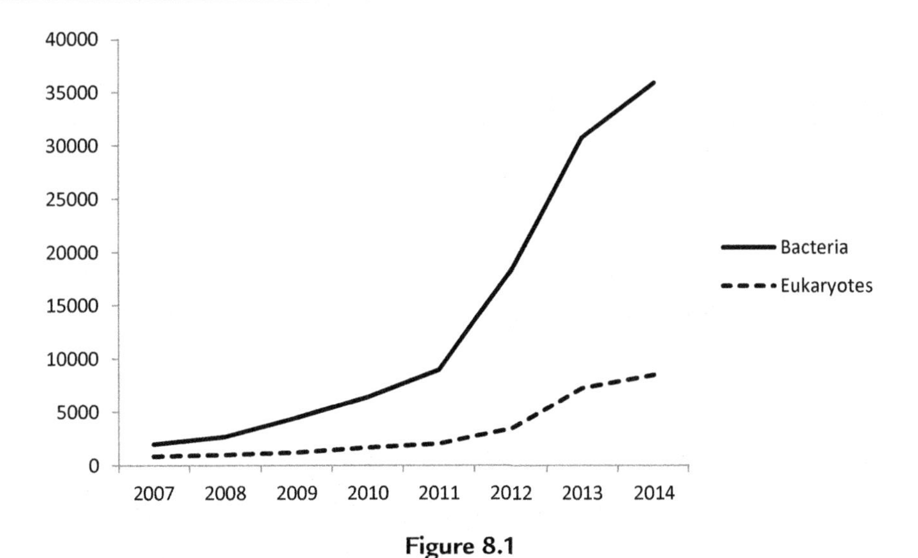

Figure 8.1
Cumulative number of organism sequences for bacteria and eukaryotes available on the GOLD Website from 2007 to present.

sequencing such as 454, Illumina, and PACBIO RS) and the decreasing cost of sequencing reactions (Figure 8.1).

To date, 49,632 organisms have been completely or partially sequenced and are available via the GOLD (Genome OnLine Database; http://www.genomesonline.org) Website. Among them, most represent transcriptomes of interesting organisms; the others are complete genome sequences. The major parts of organisms in this database are bacteria with around 79% of the available genomes, followed by eukaryotes (19%) and archae (2%). Within the eukaryotic sequences, a large part is represented by fungal genomes or transcriptomes belonging to the Ascomycota (2115 genomes; 26.9%) and the Basidiomycota (1291 genomes; 16.4%) phyla. On the contrary, chordate phylum which includes numerous species of interest or model species (e.g., human, rat, mouse) only reach 12% of the organisms sequenced to date (936 different species). This finding is more pronounced if only aquatic organisms are depicted. Indeed, only 112 fish genomes are available (1.7%) and mollusk genomes are scarce (35 genomes, 0.4%). Most of them concerned important economic species such as the Pacific oyster *Crassostrea gigas* (Zhang et al., 2012a) or model organisms such as the zebrafish *Danio rerio* (Howe et al., 2013) or the Japanese medaka *Oryzias latipes* (Kasahara et al., 2007).

A main limitation of genome sequencing was the difficulties to associate a gene name and function to certain sequences. In this context, a large part of these genomes has not been annotated and several gene products appeared as hypothetical proteins or unknown. This was mainly because of the short sequences determined by the next-generation sequencing methodologies—only a few hundred nucleotides compared with classical Sanger reactions. This has led to assembly false positives and has complicated the bioinformatics analysis of the generated data. Moreover, several interesting species, from an ecotoxicological point of view, are still missing. For example, this is the case for the freshwater bivalves *Corbicula fluminea* and *Dreissena polymorpha*, whereas these species are largely used as bioindicators to follow contamination or remediation of aquatic ecosystems (Achard-Jorris et al., 2006; Marie et al., 2006a; Palais et al., 2012; Paroloni et al., 2013; Arini et al., 2014a,b; Brandão et al., 2014). In this context, sequencing of aquatic model genomes or transcriptomes used in ecotoxicological studies represents a challenge for the future to extend our knowledge on the molecular responses of aquatic organisms facing variations (natural, anthropogenic) in their environment.

8.1.2 Transcriptomic

Determination of the transcription levels of target genes involved in essential cellular functions and pathways could be of great interest to evaluate the molecular impacts of environmental pollutants. In this context, in the past few decades, various techniques were developed to follow the variations of the expression capabilities of organisms (Pantanjali et al., 1991; Arnheim and Erlich, 1992; Schena et al., 1995; Velculescu et al., 1995; Rockett and Dix, 2000; Clark et al., 2002). This has led to an important increase of scientific publications. If we use the terms "gene expression" and "ecotoxicology" in classical databases algorithms, we can observed that this increase has been exponential since the beginning of the 21st century (Figure 8.2.). Indeed, although only 67 papers were published in 2000, more than 780 were produced in 2013.

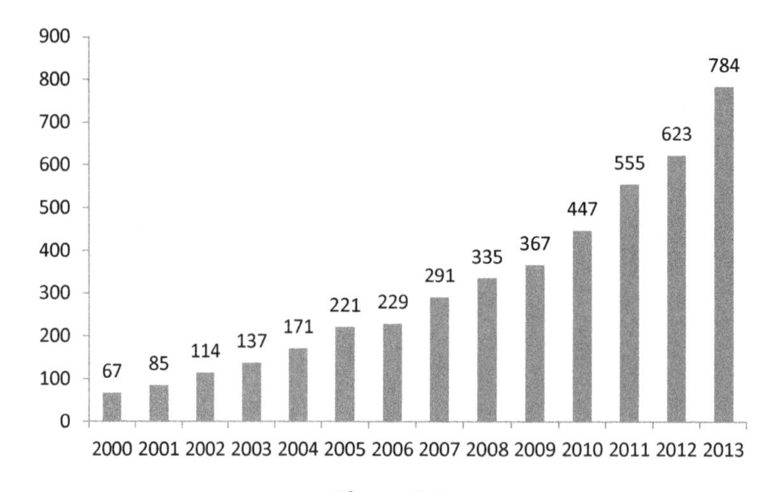

Figure 8.2
Number of publications dealing with gene expression in ecotoxicological studies since 2000.

8.1.2.1 Quantitative real-time polymerase chain reaction

Correctly performed, this method may be used for precise gene expression analysis in life science, medicine, and diagnostics and has become the standard method of choice for the quantification of messenger RNA (mRNA) levels. This technique appeared in the late 1990s with an improvement of the classical PCR. The addition of a fluorescent compound, mainly SYBR Green, allows determination of the cycle threshold (Ct). At this specific point, PCR enters in its exponential phase and quantification is possible (for a review, see Bustin, 2000, 2010). Two kinds of quantification could be applied: absolute and related quantification. Absolute quantification calculates the copy number of transcripts from a gene. In this aim, absolute quantification needs the use of standard range (mainly dilution of a plasmid DNA carrying a target gene) with known quantities. Ct from unknown samples is then compared with this standard range to determine their expression levels. This approach is time-consuming and only used with parsimony in aquatic ecotoxicology. For example, absolute quantification was used to compare the differential expression of metallothionein (MT) isoform genes, together with the biosynthesis of the total MT proteins, in the gills of triploid and diploid juvenile Pacific oyster *C. gigas* in response to Cd and Zn exposure (Marie et al., 2006b). Results showed similar response capacities to metal exposures in the two populations. Among the three MT isoform genes, *Cgmt2* appeared to be more expressed than *Cgmt1*, whereas *Cgmt3* appeared to be anecdotal (10^6 times lower than *Cgmt2*) and suggested to be expressed only during embryo-larval stages.

The second, and most popular, approach used to perform quantification is the use of relative quantification. It needs the use of a gene equally expressed whatever the conditions to normalize the quantification. Indeed, expression levels of target genes are normalized according to the reference gene expression used as an internal calibrator. Usually, these reference genes are mainly housekeeping genes (e.g., 18S, β actin, ribosomal proteins). Relative quantification could be calculated using three major methods. The most commonly used is the $2^{-\Delta Ct}$ method described by Livak and Schmittgen (2001) and Schmittgen and Livak (2008), in which ΔCt represents the difference between the Ct of a specific gene and the Ct of the reference gene. The two other methods are Normfinder and GeNorm (Vandesompele et al., 2002; Andersen et al., 2004; Heckmann et al., 2011) based on algorithms estimating PCR variance and calibrating gene expression levels. However, recently, Feng et al. (2013) compared these three different normalization methods and showed that all give congruent results. In the past decade, relative quantification has been largely applied in aquatic ecotoxicology and showed to be relevant to better understand molecular mechanisms involved in aquatic organisms' response to contaminants. Effects of dietary methylmercury (MeHg) on gene expression were examined in three organs (liver, skeletal muscle, and brain) of the zebrafish (*D. rerio*) (Gonzalez et al., 2005). Thirteen genes known to be involved in antioxidant defenses, metal chelation, active efflux of organic compounds, mitochondrial metabolism, DNA repair, and apoptosis were investigated by quantitative real-time reverse transcriptase PCR. No significant changes in the expression levels of these genes were observed in contaminated brain samples, although this organ accumulated the highest mercury concentration. This lack of genetic response could explain the high neurotoxicity of MeHg. Moreover, an impact on the mitochondrial metabolism and production of reactive oxygen species has been observed in liver and muscle, demonstrating that skeletal muscle was not only an important storage reservoir but was also affected by MeHg contamination.

Recently, in the Asiatic clam *Ruditapes philippinarum*, Liu et al. (2014) used qPCR to evaluate the relevance of target gene expression levels to constitute valuable biomarkers of benzo[a]pyrene exposure. Transcript levels of eight different genes were investigated in the gills and digestive glands. The results showed that mRNA expressions of *pgp*, *ahr*, *cyp4*, *gst-pi*, *gst-S2*, and *Mn-sod* were induced significantly by all the concentrations of benzo[a] pyrene tested. According to the correlation analysis, expressions of *ahr*, *gst-pi*, and *Mn-sod* in gills and *gst-pi* in digestive glands appeared to be useful molecular biomarkers of benzo[a] pyrene exposure.

One of the main limitations of qPCR is based on the lack of availability of gene sequences for certain species. In this case, this approach could be costly and time-consuming because cloning and sequencing of target genes are needed. In this context, analysis could be limited to a few genes and consequently not so informative.

Another point concerns the choice of the reference gene for relative quantification. For the past decade, this question has led to numerous conflicting studies, especially concerning classical reference genes such as *gapdh*, or beta actin. Indeed, although several authors do not consider beta actin a suitable reference gene because of its high variability, numerous other studies evidence that, on the contrary, this gene is among the putative reference genes, one of the most appropriated and stable for qPCR analysis in plants (Moura et al., 2012; Zhang et al., 2012b), insects (Lu et al., 2013), or fish (Li et al., 2010; Zheng and Sun, 2011). Among the entire tested reference genes, one proved to be mostly relatively constant between tissue and organisms, the elongation factor 1 α (*ef1a*) (Araya et al., 2008; Morga et al., 2010).

Recently, Dundas and Ling (2012) tried to respond to some of the questions surrounding reference genes and their reliability for quantitative experiments. They concluded that there is no universal reference gene that can be deemed suitable for all the experimental conditions and said that every experiment will require the scientific evaluation and selection of the best candidate gene for use as a reference gene to obtain reliable scientific results. At that time, the multiple reference gene approach is recommended because it eliminates any biased results, which represents the major flaw of a single reference gene. In this case, a minimum of three reference genes are used and their mean Ct values are employed to calibrate expression level of target genes.

Another limitation concerned the use of the internal standard in absolute qPCR. Indeed, Hou et al. (2010) demonstrated that the use of circular plasmid DNA carrying a target gene as standard can lead to overestimation of the expression level. They compared quantification of a ubiquitous cell nuclear gene (*pcna*) on the genome of five microalgae using circular and linear plasmids as standard. Sequencing of the algae *Thalassiosira pseudonana* showed that only one copy was present in the genome (Armbrust et al., 2004). Results showed that qPCR using the circular plasmid as standard yielded an estimate of 7.77 copies of *pcna* per genome, whereas that using the linear standard gave 1.02 copies per genome. In this context, they concluded that the use of circular plasmid DNA is probably unsuitable in absolute qPCR. The serious overestimation by the circular plasmid standard was related to the undetected lower efficiency of its amplification in the early stage of PCR when the supercoiled plasmid is the dominant template.

8.1.2.2 DNA microarrays

DNA microarrays appeared during the beginning of the 1990s. However, it has been extensively used during the past 5 years. Indeed, to date, the ArrayExpress site of the European Bioinformatics Institute (http://www.ebi.ac.uk/arrayexpress/) reported that 51,076 experiments have been conducted and more than 1 million assays (1,482,361). Among these data, almost more than a half were dedicated to humans. Concerning aquatic organisms, the most used is *D. rerio*. Mollusks are largely limited to assays on oyster species such as *Crassostrea virginica* and *C. gigas* or on mussels *Mytilus galloprovincialis* and *Mytilus edulis* (Rustici et al., 2013). This small number of aquatic organisms is mainly due to the technology itself. Indeed, microarrays are based on the knowledge of the gene sequences. Short sequences specific to the studied genes are spotted on a glass side. RNA from two different conditions (control and exposed to contaminant, for example) are harvested and differentially labeled with two fluorochromes: the green one, Cyanin 3, and the red one, Cyanin 5. cDNAs from the two conditions are hybridized on the glass side. Then a high-resolution scanner is used to analyze each gene spot at the excitation wave length of Cyan 3 and Cyan 5. Computer analysis of the result allows determining which genes are equally and/or differentially expressed in the two conditions. In this context, the major value of DNA microarrays is to investigate the expression level of a thousand genes in the same experiment. Consequently, this approach has been extensively used to determine the impacts of xenobiotics on model aquatic organisms or to analyze tissue specific expression levels of genes.

In the Pacific oyster *C. gigas*, the comparison of gene expression was performed between males and females during the reproductive cycle (Dheilly et al., 2012). Results show that 2482 genes were differentially expressed during the course of males and/or females gametogenesis, and 434 genes could be localized in either germ cells or somatic cells of the gonad; mainly genes involved in chromatin condensation, DNA replication and repair, mitosis and meiosis regulation, transcription, translation, and apoptosis. However, male-specific genes (bindin and a *dpy-30* homolog) and female-specific genes (*foxL2*, nanos homolog 3, a pancreatic lipase related protein, *cd63*, and vitellogenin) were also observed. In the blue mussel *M. galloprovincialis*, Negri et al. (2013) investigated the impacts of Cu concentrations and temperature in the gills. For this purpose, a cDNA microarray with 1673 sequences was used. In animals exposed to heat stress, gene expression profiles could be grouped in three different clusters: (1) downregulation of translation-related genes, (2) important upregulation of genes involved in protein folding, and (3) an increase of chitin metabolism-related genes. On the contrary, Cu led to the upregulation of genes involved in translation, as well as the downregulation of genes encoding for heat shock proteins and "microtubule-based movement" proteins.

In fish, the impacts of perfluorooctane sulfonate have been determined in the liver of the common carp *Cyprinus carpio* (Hagenaar et al., 2008). Custom microarrays were constructed from cDNA libraries obtained with suppression subtractive hybridization PCR experiments. Differentially expressed genes were implied in energy metabolism, reproduction, and stress response. From this study, authors concluded that increase in energy expenditure negatively affects processes vital to the survival of an organism, such as growth. In the same way, Yum et al. (2010) investigated the effects of Aroclor 1260 (a polychlorinated biphenyl mixture) in the liver tissue of the Japanese medaka, *Oryzias latipes*. Twenty-six differentially expressed genes were identified that were associated with the cytoskeleton, development, endocrine function/reproduction, immunity, metabolism, nucleic acid/protein binding, and signal transduction. Moreover, the expression levels of molecular biomarkers known to be involved in endocrine disruption (vitellogenin, choriogenin, and estrogen receptor alpha) were highly upregulated. Individuals of the fish *Lithognathus mormyrus* were exposed to Cd administered to the fish by feeding or injections (Auslander et al., 2008). During Cd contamination, differentially expressed genes were observed including elastase 4, carboxypeptidase B, trypsinogen, perforin, complement C31, cytochrome P450 2K5, ceruloplasmin, carboxyl ester lipase, and metallothioneins. Most genes were shown to be down regulated.

All these examples showed that cDNA microarrays could constitute a useful tool to better understand the molecular pathways involved in the cellular response to toxicants but also to evidence potential biomarkers of xenobiotics exposure. In this manner, by allowing the simultaneous study of more than a thousand genes, DNA microarray is analogous to a multibiomarker approach.

In this context, some researches try to investigate the potentialities of this approach for environmental biomonitoring or water and sediment quality assessment.

Falciani et al. (2008) built 13,000 cDNA microarrays for the European flounder *Platichthys flesus* and used it to discriminate individuals taken from six sampling sites presenting different pollution status. From this study and by comparison with laboratory experiment results obtained for selected toxicants, the authors showed 16 genes in which expression levels could be used to assign the site of origin of fish obtained from three of the sites in an independent sampling. They concluded that a microarray comprising these genes constitutes a potential tool for environmental impact assessment. Recently, the transcriptional response of European eels (*Anguilla anguilla*) chronically exposed to pollutants was investigated for different sites in Belgium with lowly, highly, and extremely polluted media (Pujolar et al., 2013). Results showed that among sites, genes involved in detoxification processes were upregulated (*cyp3a*, glutathione-s-transferase). Moreover, many genes involved in the mitochondrial respiratory chain and oxidative phosphorylation were downregulated in the highly polluted site, suggesting that pollutants may have a significant effect on energy metabolism in these fish.

The main limitation of the cDNA microarray technique is the knowledge of the species to be analyzed (Pina and Barata, 2011). Indeed, in spite of the increasing genomes and transcriptomes available in the database, ecotoxicological relevant species remained few or not described at the genomic level. Another problem is the difficulties to correctly associate the expression patterns and to correlate them to potential physiological disturbances. Hence, this approach could produce false positives and expression status sometimes needs to be controlled with qPCR analysis. Finally, this technique is also time-consuming and costly, notably because of the materials needed (e.g., scanner, fluorochrome, hybridization chamber); to date, some ecotoxicologists think that it will be progressively replaced by high throughput sequencing approaches.

8.1.2.3 High-throughput sequencing

As described in the first paragraphs of this chapter, high-throughput sequencing has largely increased during the past 5 years, mainly because of the improvement in sequencing technologies and decreasing cost of sequencing reactions. This approach could be applied to various species without a necessary knowledge of the genome sequence. Indeed, transcriptomic techniques such as RNA-Seq (also called whole transcriptome-shotgun sequencing) furnished a large set of data and gene sequences obtained by de novo assembling could be annotated by comparison with known sequences available in databases for other species (for review see Qian et al., 2014). The important value of this approach is that it is not limitative and can investigate the effects at a transcriptome-wide level without a priori and to putatively determine all the genes that are differentially expressed during exposure of aquatic organisms to a toxicant or to variations in their environment. Indeed, comparison of the transcriptome obtained with control organisms could be used to normalize transcriptome achieved with an exposed individual, allowing a high resolution of the molecular impacts and of the cellular pathways involved in the response to xenobiotics.

A serial analysis of gene expression was conducted on the skeletal muscle of *D. rerio* to evaluate the cellular impact of MeHg (Cambier et al., 2010). Among the 5280 different transcripts identified, 60 genes appeared upregulated and 15 downregulated by more than two times. A net impact of MeHg was noticed on 14 ribosomal protein genes, indicating a perturbation of protein synthesis. Several genes involved in mitochondrial metabolism, the electron transport chain, endoplasmic reticulum function, detoxification, and general stress responses were differentially regulated, suggesting an onset of oxidative stress and endoplasmic reticulum stress. Several other genes for which expression varied with MeHg contamination could be clustered in various compartments of the cell's life, such as lipid metabolism, calcium homeostasis, iron metabolism, muscle contraction, and cell-cycle regulation. Results allow the authors to propose a schematic representation of the cellular pathways involved in MeHg response. RNA-Seq by means of the 454 sequencing technology was used to examine the in situ impacts of chronic exposure to Cd and Cu on the yellow perch (*Perca flavescens*) transcriptome in fish sampled along a polymetallic gradient (Pierron et al., 2011). This work aimed to identify genes for which transcription levels were specifically affected by metals by carrying out correlation analyses between the transcription level of a given gene and the hepatic Cd and Cu concentrations in the fish. Chronic metal exposure was associated with a decrease in the transcription levels of numerous genes involved in protein biosynthesis, in the immune system, and in lipid and energy metabolism. The results suggested that this marked decrease could result from an impairment of bile acid metabolism by Cd and energy restriction but also from the recruitment of several genes involved in epigenetic modifications of histones and DNA that lead to gene silencing. This study successfully showed that RNA-Seq is a powerful approach to study the ecotoxicological response of fish to the polluted environments.

In mussel *M. edulis*, biomonitoring of contaminants rely on the determination of the bioaccumulation level of compounds. Recently, transcriptomic approach was achieved to link tissue concentrations to potential toxic effects (Poynton et al., 2014). Mussels were exposed to sublethal concentrations of Cd and Pb used alone or mixed. From this analysis, transcripts that expressed a level correlated with gill metal concentration were reported. Along with the metallothionein isoforms *mt10* and *mt20*, several genes belonged to the nucleoside phosphate biosynthesis processes, mRNA metabolic pathway, unfolded protein response and response to stress. Recently, Garcia et al. (2012) used salt marsh minnows *Fundulus grandis* to determine the impacts resulting from the blowout of the Deepwater Horizon platform. In this study, transcriptomes from individuals harvested in the impacted area and from a reference site were obtained by RNA-Seq approach and then compared. Results showed 1070 downregulated and 1251 upregulated genes. Among them, genes involved in the expected Aryl-hydrocarbon receptor-mediated response (*ahr1, ahr2*) were overexpressed, as were several cytochrome P450 genes. More surprisingly, genes suggesting a hypoxic response (*hif2a* and *E1A* binding protein p300) and an active immune response were also upregulated. On the contrary a large number of transcripts encoding for 40S and 60S ribosomal proteins and for choriogenins were strongly repressed. Authors concluded that this study suggested a limited oxygen exchange between air and water in oil-slick contaminated sites. In this way, the observed decreasing expression of genes implied in the protein synthesis would probably reflect a survival strategy during these hypoxic conditions.

However, the main limitation of this approach is, on one hand, that a large part of the Express Sequence Tag showed could not be attributed to a clear function or a gene name. On the other hand, analysis of the data generated by high-throughput sequencing is time-consuming and requires an extended knowledge in bioinformatics.

8.1.3 Epigenetics

Epigenetics is an area of recent research and is currently expanding. In recent years, many studies have been conducted to determine the epigenetic mechanisms and their effects on the phenotype of organisms, mainly in mammals' models. Various definitions have been given to this discipline and the most recent that reaches a major consensus is: "Epigenetics is the study of changes in gene function that are mitotically and/or meiotically heritable and that cannot be explained by changes in the DNA sequence." Thus, this definition focuses on the fact that epigenetic phenomena cannot be explained by mutations but through mechanisms acting on the DNA molecule. One of the major studied mechanisms is based on DNA methylation. Methylation of cytosine bases leads, in most cases described, to the condensation of DNA. This condensation triggers a decrease in the level of transcription of certain genes by avoiding the access to the cellular machinery on the DNA molecule (gene silencing; Liu et al., 2008; Suzuki and Bird, 2008). It was shown that methylation of the genome play an important role in various fundamental processes such as cell differentiation, embryonic development, or maintenance of chromosomal stability (Lister et al., 2011). Moreover, methylation of DNA is a long-term and stable mechanism. Indeed, epigenetic modifications acquired at a time of development can be maintained during the entire life-cycle of an organism and can even be transmitted to subsequent generations (Rakyan et al., 2003; Johanses et al., 2009; Skinner et al., 2010). A classic example, in mammals, is based on the work done by Anway et al. (2005) on female rats exposed during gestation to the fungicide vinclozolin. The rats of the first generation showed reduced spermatogenic ability that could be easily explained by the exposure conditions in the uterus. However, these effects were observed in the following three generations, which were never exposed to known contaminants. In parallel, the different generations all presented aberrations in their DNA methylation patterns. This discipline is still in its infancy and the mechanisms underlying epigenetic modification of DNA status are poorly known. However, evidence is accumulating to show that environmental factors, including exposure to organic and metallic pollutants may act strongly on the epigenetic status of cells (Jaenish and Bird, 2003; Jiang et al., 2008; Skinner et al., 2010; Pierron et al., 2011).

In this context, it appears obvious that the appearance of "errors" in the methylation patterns resulting from pollution could have important consequences for organisms (Das and Singal, 2004; Skinner et al., 2010). However, even if numerous studies showed that pollutants could affect the epigenetic status of mammalian cells in culture or in vivo, epigenetic studies in aquatic ecotoxicology are scarce and limited, to our knowledge, to experimental approaches (Vandegehuchte and Janssen, 2011).

DNA methylation mainly occurred in CpG-rich region located in the promoter or the first exon of the genes (Doherty and Couldrey, 2014). Quantification of the DNA methylation level has been performed by different techniques using methylation-sensitive enzymes (*HpaII*, *SmaI*), immunodetection (Suzuki et al., 2010), high-performance liquid chromatography detection (Fraga et al., 2005), and methylation-sensitive PCR (Pierron et al., 2014). Recently, a new methodology called reduced representation bisulfite sequencing has allowed for genome-wide DNA methylation analysis with reduced sequencing requirements. This methodology, developed by Meissner et al. (2005) and Gu et al. (2011), allows for preferential selection and sequencing of CpG-rich regions. Bisulfite treatment allows discriminating unmethylated cytosine and 5-methylated cytosine. Indeed, although unmethylated cytosine is transformed into uracil, 5-methylated cytosine remains unchanged. Subsequent sequencing permits to clearly discriminate methylated bases (Doherty and Couldrey, 2014).

In fish, Pierron et al. (2014) investigated the possible effects of low-dose Cd exposure on the DNA methylation profile in a critically endangered fish species, the European eel (*A. anguilla*). Eels were exposed to environmentally realistic concentrations of Cd and the global CpG methylation status of eel liver was determined by means of a homemade enzyme-linked immunosorbent assay. Authors used also a methylation-sensitive arbitrarily primed PCR method to identify genes that are differentially methylated between control and Cd-exposed eels. Results showed that Cd exposure is associated with DNA hypermethylation and with a decrease in total RNA synthesis. Among hypermethylated sequences identified, several fragments presented high homologies with genes encoding for proteins involved in intracellular trafficking, lipid biosynthesis, and phosphatidic acid signaling pathway. In addition, a few fragments presented high homologies with retrotransposon-like sequences. Other studies revealed that aberrant DNA methylation status was found to be an important factor in tumorigenesis, including in the development of hepatocellular adenoma tumors in wild fish from contaminated environments (Mirbahai et al., 2011).

A limitation of epigenetics study is difficult to evaluate because this approach is still in its infancy in aquatic ecotoxicology. However, despite increasing knowledge about DNA methylation, we still lack a complete understanding of its specific functions and association with the environment and gene expression in diverse organisms. We also lack studies of the putative transgenerational effects of environmental pollutants on aquatic organisms, which is still a matter of debate (Heard and Martienssen, 2014).

8.2 Proteomic

Proteomics described sciences dedicated to the knowledge of the proteome. It covered all the proteins present in the cells, an organite, a tissue, or an individual (Sanchez et al., 2011). Proteomics aims to identify proteins, determine their location within the cell, their posttranslational modification, but also their amount. It could also be applied to assess the changes in

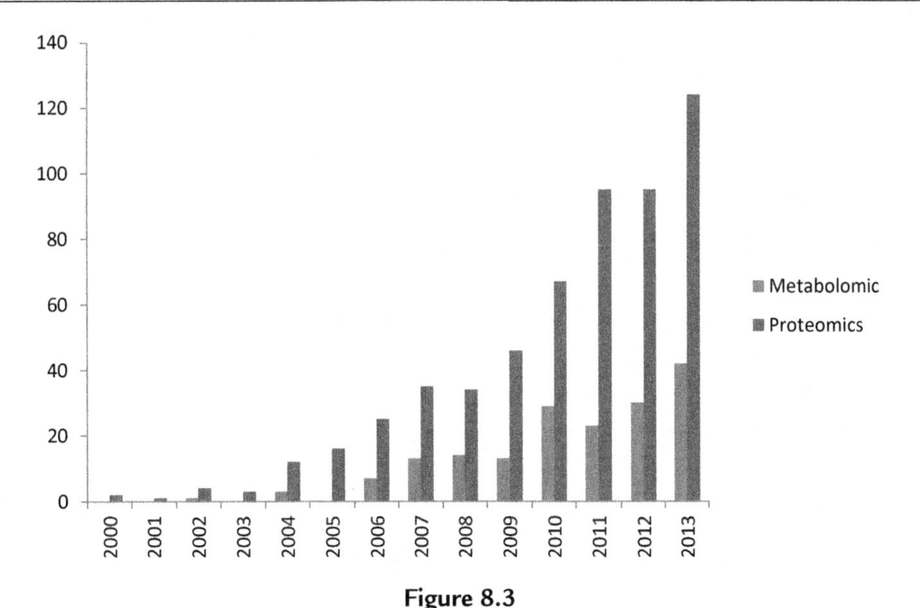

Figure 8.3
Evolution of the number of publications dealing with proteomics in ecotoxicological studies since 2000.

the expression level of the proteins according to the time, the age or the development status of the individuals and their physiological or pathological status, but also during variations in their environment. Proteomic applications in aquatic ecotoxicology are very recent. Indeed, before 2006, this kind of analysis remain confidential as shown by the number of publications retrieved with the terms "proteomics" and "ecotoxicology" (Figure 8.3). However, along with the increasing knowledge on organisms' genomes, proteomic studies have increased during the past 5 years.

Initially, proteomic studies used two-dimensional gel electrophoresis to reveal variations of protein expression under contaminant exposure (Bradley et al., 2002). However, many studies were not able to correctly identify the proteins that underwent significant changes. The use of liquid chromatography and mass spectrometry (MS) greatly improved protein identification (Monsinjon and Knigge, 2007). To date, more sophisticated techniques such as surface-enhanced laser desorption ionization time-of-flight-MS or LTQ-Orbitrap-MS (Walker et al., 2007) are used. These techniques allow determination of a part of the target protein and subsequent identification by comparison with proteins available in databases.

Recently a nano liquid chromatography tandem MS with a high-resolution Orbitrap hybrid mass spectrometer was used to determine the effects of three different xenobiotics: cadmium and the pesticides methoxyfenozide and pyriproxyfen on sperm production in the crustacean *Gammarus fossarum* (Trapp et al., 2014). A total of 57,558 tandem MS spectra were identified from the testis proteome, leading to 871 mRNA-translated products. Results showed that

70 proteins were significantly expressed during exposure and among them, 44 were common to the three toxicants. Most of them were involved in the cytoskeletal structure maintenance, the energy metabolism, calcium homeostasis, heat-shock response, and immunity. From this analysis, authors concluded that Cd mainly acts by interacting with transporter involved in the uptake and homeostasis of calcium, whereas pesticides lead to a disturbance of the iono-osmoregulation.

In mollusks, the effects of exposure to Cu and BaP were investigated in *Mytilus galloprovinciallis* using proteomic analysis (Maria et al., 2013). After 7 days of exposure to these two compounds used alone or mixed, proteomic analysis showed different protein expression patterns associated to each contamination condition. Proteins associated with adhesion and motility (catchin, twitchin, and twitchin-like protein), cytoskeleton and cell structure (α-tubulin and actin), stress response (heat shock cognate 71, heat shock protein 70, putative C1q domain-containing protein), transcription regulation (zinc-finger BED domain–containing and nuclear receptor subfamily 1G) and energy metabolism (ATP synthase F0 subunit 6 protein and mannose-6-phosphate isomerase) were assigned to all three conditions. Cu exposure alone altered proteins associated with oxidative stress (glutathione-S-transferase) and digestion, growth, and remodeling processes (chitin synthase). From this study, new potential biomarkers have been determined to follow field exposure by these compounds. In the same way, analysis of the combined effects of arsenic and salinity were studied in the Asiatic clam *R. philippinarum* (Wu et al., 2013). Results showed that salinity could greatly modulate the toxicological response of clams to arsenic. Indeed, exposure induced disturbance in energy metabolism and/or osmotic regulation under different salinities. Moreover, protein identification indicated oxidative stress, cellular injury, apoptosis and disturbance in energy metabolism.

All of these studies showed that proteomic approaches are of great interest as a biomarker discovery tool. It can also increase our knowledge for understanding the cellular mechanisms involved in stress response. However, the main limitation of this kind of approach is the lack of annotated proteomes. Indeed, only a few numbers of aquatic organisms are well-described. In this context, identification of nonmodel organisms' proteins could be fastidious. But, one could hypothesize that proteomic researches could be overcome in the future because of the extensive use of de novo sequencing.

8.3 Metabolomic and Fluxomic

Metabolomic and fluxomic are very recent omic applications (Winter and Kromer, 2013). Metabolomic analyze all the metabolites (sugar, amino acids, fatty acids) present in a cell, an organ, or organism. Fluxomic tries to determine metabolic fluxes. These two approaches are considered as endpoint biological techniques representing the end results of gene expression, protein concentration and kinetics, regulation and metabolite concentrations. In aquatic ecotoxicology, their applications remained largely limited, probably due to the analytical cost. Indeed, metabolomic and fluxomic studies need the use of mass spectroscopy, ultra-performance liquid chromatography, and/or nuclear magnetic resonance, but also the use of tracer compounds (Cubero-Leon et al., 2012).

For example, sex-specific differences were investigated in the mussel metabolome to further characterize the reproductive physiology of this species. A comparison of mantle tissue containing mature gonads from male and female revealed great differences in glycerophosphatidylcholine and lysophosphatidylcholine metabolites content. However, other differentially expressed metabolites could have not been identified.

The first limitation of metabolomic and fluxomic concerns the availability and the price of tracer compounds. Consequently. these approaches are more or less limited to simple systems with only a few carbon sources and mostly sugar-based processes (Cubero-Leon et al., 2012). The second limitation is the analytical process. Indeed, pattern analysis is fastidious and could lead to errors during interpretation of results.

8.4 Conclusion

Omic approaches have become more and more popular in aquatic ecotoxicology during the past 20 years. They have brought reliable tools to the determination of molecular pathways used by organisms in response to xenobiotic exposure. Some of their applications are also useful to identify taxa or individuals variations. Indeed, DNA barcoding, metabarcoding, or analysis of short nucleotide polymorphisms are of great interest and could lead to pertinent indicators to evaluate the health status of aquatic organisms or be useful for the assessment of the good environmental status needed by the European Water Framework Directive and the Marine Strategy Framework Directive (Bourlat et al., 2013). The title of this chapter questions if omics could constitute the ultimate answer to biological questions. One could say yes. Indeed, transcriptomic and proteomic approaches allow a better understanding of the molecular mechanisms involved in aquatic organisms' response to environmental pollutants. However, future prospects should have to take into account several recommendations to avoid misinterpretation of the results.

The first essential point concerns the experimental design. Indeed, as described by Pina and Barata (2011), several ecotoxicogenomic studies have neglected differences between chronic versus acute effects as well as low versus high exposure levels. Future prospects should use environmentally realistic concentrations of pollutants. Indeed, acute contamination is not valuable and only represents a rare contamination phenomena (Figure 8.4). The contamination pressure used in experimental condition will greatly modulate the results. Although moderate contamination will allow discriminating adaptive response, high doses will only represent acute toxicity mechanisms mainly leading to apoptosis and cellular death.

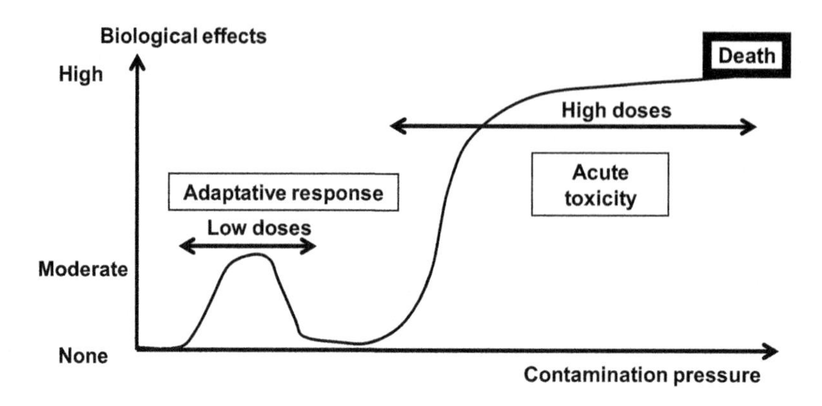

Figure 8.4
Schematic representation of the putative biological effects shown with omic approaches according to the contamination pressure used.

Along with chemical concentration, time could modify the response of aquatic organisms to toxicants. Indeed, classically response to pollutants could follow three successive phases (Pina and Barata, 2011): (1) the early response from the interaction between stressors and its cell target; (2), the stress response corresponding to cellular activation of genes to fight against toxicant damages; and (3) the adaptive response in which only a few numbers of genes are differentially expressed to restore the normal function of the cell (i.e., homeostasis) in the presence of the stressor. In this context, ecotoxicologists should keep in mind that transcriptomic approaches give only an instant picture of the expression. Consequently, future experiments should investigate the transcriptomic effects at short, moderate, and long term.

Deep transcriptome sequencing using next-generation sequencing is emerging as a powerful approach to rapidly increase sequence information in nonmodel species. Such information is essential to allow the discrimination of gene products and functions and consequently improve transcriptomic analysis. It is also important for species located at the bottom of the trophic web-like diatoms. Only *Thalassiosira pseudonana* and *Phaeodactylum tricornutum*—two marine diatom species—are yet to be entirely sequenced, but genomic information about freshwater species is still very scarce (Kim Tiam et al., 2012).

qPCR has been useful in determining response to xenobiotics. However, reference gene use in these analyses should be selected with care. There have been a large number of publications about the reference gene for use in qPCR for aquatic organisms (for a review, see Kozera and Rapacz, 2013). At this time, the multiple reference gene approach is recommended because it eliminates any biased results, which represented a major flaw of single-reference genes. According to Nikinmaa and Rytkönen (2011) in their review on functional genomics, the best normalization for qPCR experiments will be to add a known amount of an external mRNA to the sample. Unfortunately, this approach, first proposed by Ellefsen et al. (2008), has not been routinely used, probably because of the difficulties in guaranteeing the quantity of the external mRNA.

In spite of the advent of omics, numerous questions remain in aquatic ecotoxicology. How should the observed effects be linked at the molecular level to effects at the cellular level, tissue, individual, or population? What are the possible interactions between contaminants, but also between contaminant and natural factors, on the final toxicity of these compounds? What are the effects of a cocktail of contaminants on populations or ecosystems? What are the transgenerational effects of these compounds? To answer these questions, omics are clearly powerful tools but they should not be used alone. Indeed, mRNA expression does not necessarily reflect the protein expression (Nikinmaa and Rytkönen, 2011). Xenobiotics could affect mRNA or protein stability. Moreover, not all mRNAs produced during transcription are translated into proteins. Hence, the time course of mRNAs and proteins are very different. Although mRNA peaks are rapidly produced (minute to hours), hours to days are required to obtain the corresponding protein peaks (Buckley et al., 2006). In this context, numerous studies have pointed out discrepancies between expression gene levels and protein amounts (Marie et al., 2006b; Lucia et al., 2010). In rare cases, such as in the amphibian *Xenopus laevis*, an identical amount of mRNAs and corresponding protein was found for metallothioneins during Cd exposure (Mouchet et al., 2006). This indicated that MTs in *X. laevis* are only regulated at the transcriptional and not at the translational level. This suggests that *X. laevis* larvae possess a high metabolic activity and that probably all transcribed genes are translated into proteins; this could be in relation with the high levels of metabolic activity occurring in the premetamorphic stages of these organisms.

Only a few number of articles used several omic approaches in the same study. Hagenaars et al. (2013), with transcriptomic, proteomic, and biochemical parameters, demonstrated that mitochondrial dysfunction plays an important role in perfluorooctanoic acid toxicity in zebrafish. Indeed, an increase in the mitochondrial membrane permeability was linked to an impairment of the ATP production, leading to an oxidative stress and apoptotic mechanisms. Mitochondria were shown to be the major target of MeHg in *D. rerio*. Indeed, transcriptomic, bioenergetic, and histological analysis revealed a clear disturbance of mitochondria, also leading to ATP production decrease, oxidative stress, and apoptotic pathway (Cambier et al., 2009, 2010; 2012). As reported by Nikinmaa and Rytkönen (2011), we believe that an important question for the future is how contaminants affect reactive oxygen species–dependent signaling and ROS production. We could hypothesize that mitochondria will be a key target in cells.

To conclude, omic techniques have already greatly improved our knowledge of the impacts of contaminants on aquatic organisms. In spite of their efficiency, these approaches should be used with care. The next important step is to be able to link the expression of mRNA and protein profiles with their impacts at the individual and population levels to use judiciously these tools in an ecotoxological context (Fent and Sumpter, 2011).

References

Achard-Jorris, M., Gonzalez, P., Marie, V., et al., 2006. cDNA cloning and gene expression of ribosomal S9 protein gene in the mollusc *Corbicula fluminea*: a new potential biomarker of metal contamination up-regulated by cadmium and repressed by zinc. Environ. Toxicol. Chem. 25, 527–533.

Andersen, C.L., Jensen, J.L., Orntoft, T.F., 2004. Normalization of real-time quantitative reverse transcription-PCR data: a model- based variance estimation approach to identify genes suited for normalization, applied to bladder and colon cancer data sets. Cancer Res. 64, 5245–5250.

Anway, M.D., Cupp, A.S., Uzumcu, M., et al., 2005. Epigenetic transgenerational actions of endocrine disruptors and male fertility. Science 308, 1466–1469.

Araya, M.T., Siah, A., Mateo, D., et al., 2008. Selection and evaluation of housekeeping genes for haemocytes of soft-shell clams (*Mya arenaria*) challenged with *Vibrio splendidus*. J. Invertebr. Pathol. 99, 326–331.

Arini, A., Daffe, C., Gonzalez, P., et al., 2014a. What are the outcomes of an industrial remediation on a metal-impacted hydrosystem? A 2-year field biomonitoring of the filter-feeding bivalve *Corbicula fluminea*. Chemosphere 108, 214–224.

Arini, A., Daffe, G., Gonzalez, P., et al., 2014b. Detoxification and recovery capacities of *Corbicula fluminea* after an industrial metal contamination (Cd and Zn): a one-year depuration experiment. Environ. Pollut. 192, 74–82.

Armbrust, E.V., Berges, J.A., Bowler, C., et al., 2004. The genome of the diatom *Thalassiosira pseudonana*: ecology, evolution, and metabolism. Science 306, 79–86.

Arnhein, N., Erlich, H., 1992. Polymerase chain reaction strategy. Ann. Rev. Biochem. 61, 131–156.

Auslander, M., Yudkovski, Y., Chalifa-Caspi, V., et al., 2008. Pollution-affected fish hepatic transcriptome and its expression patterns on exposure to cadmium. Mar. Biotechnol. (NY) 10, 250–261.

Bourlat, S.J., Borja, A., Gilbert, J., et al., 2013. Genomics in marine monitoring: new opportunities for assessing marine health status. Mar. Pol. Bul. 574, 19–31.

Bradley, B.P., Schrader, E.A., Kimmel, D.G., et al., 2002. Protein expression signatures: an application of proteomics. Mar. Environ. Res. 54, 373–377.

Brandão, F.P., Pereira, J.L., Gonçalves, F., et al., 2014. The impact of paracetamol on selected biomarkers of the mollusc species *Corbicula fluminea*. Environ. Toxicol. 29, 74–83.

Buckley, B.A., Gracey, A.Y., Somero, G.N., 2006. The cellular response to heat stress in the goby *Gillichthys mirabilis*: a cDNA microarray and protein-level analysis. J. Exp. Biol. 209, 2660–2677.

Bustin, S.A., 2000. Absolute quantification of mRNA using real-time reverse transcription polymerase chain reaction assays. J. Mol. Endocrinol. 25, 169–193.

Bustin, S.A., 2010. Why the need for qPCR publication guidelines? The case for MIQE. Methods 50, 217–226.

Cambier, S., Bénard, G., Mesmer-Dudons, N., et al., 2009. At environmental doses, dietary methylmercury inhibits mitochondrial energy metabolism in skeletal muscles of the zebra fish (*Danio rerio*). Int. J. Biochem. Cell. Biol. 41, 791–799.

Cambier, S., Gonzalez, P., Durrieu, G., et al., 2010. Serial analysis of gene expression in the skeletal muscles of zebrafish fed with a methylmercury-contaminated diet. Environ. Sci. Technol. 44, 469–475.

Cambier, S., Gonzalez, P., Mesmer-Dudons, N., et al., 2012. Effects of dietary methylmercury on the zebrafish brain: histological, mitochondrial and gene transcription analyses. Biometals 25, 165–180.

Clark, T., Lee, S., Ridgway Scott, L., et al., 2002. Computational analysis of gene identification with SAGE. J. Comput. Biol. 9, 513–526.

Clements, W.H., Rohr, J.R., 2009. Community responses to contaminants: using basic ecological principles to predict ecotoxicological effects. Environ. Toxicol. Chem. 28, 1789–1800.

Cubero-Leon, E., Minier, C., Rotchell, J.M., et al., 2012. Metabolomic analysis of sex specific metabolites in gonads of the mussel, *Mytilus edulis*. Comp. Biochem. Physiol. D7, 212–219.

Das, P.M., Singal, R., 2004. DNA methylation and cancer. J. Clin. Oncol. 22, 4632–4642.

Dheilly, N.M., Lelong, C., Huvet, A., et al., 2012. Gametogenesis in the Pacific oyster *Crassostrea gigas*: a microarrays-based analysis identifies sex and stage specific genes. PLoS One 7 (5), e36353.

Doherty, R., Couldrey, C., 2014. Exploring genome wide bisulfite sequencing for DNA methylation analysis in livestock: a technical assessment. Front Genet. 5, 126. http://dx.doi.org/10.3389/fgene.

Dundas, J., Ling, M., 2012. Reference genes for measuring mRNA expression. Theory Biosci. 131, 215–223.

Ellefsen, S., Stenslokken, K.O., Sanvik, G.K., et al., 2008. Improved normalization of real-time reverse transcriptase polymerase chain reaction data using an external RNA control. Anal. Biochem. 376, 83–93.

Falciani, F., Diab, A.M., Sabine, V., et al., 2008. Hepatic transcriptomic profiles of European flounder (*Platichthys flesus*) from field sites and computational approaches to predict site from stress gene responses following exposure to model toxicants. Aquat. Toxicol. 90, 92–101.

Feng, L., Yu, Q., Li, X., et al., 2013. Identification of reference genes for qRT-PCR analysis in Yesso scallop *Patinopecten yessoensis*. PLoS One 8 (9), e75609.

Fent, K., Sumpter, J.P., 2011. Progress and promises in toxicogenomics in aquatic toxicology: is technical innovation driving scientific innovation? Aquat. Toxicol. 105 (3–4 Suppl.), 25–39.

Fraga, M.F., Ballestar, E., Paz, M.F., et al., 2005. Epigenetic differences arise during the lifetime of monozygotic twins. Proc. Natl. Acad. Sci. U.S.A. 102, 10604–10609.

Garcia, T.I., Shen, Y., Crawford, D., et al., 2012. RNA-seq reveals complex genetic response to deepwater horizon oil release in *Fundulus grandis*. BMC Genomics 13, 474.

Gonzalez, P., Dominique, Y., Massabuau, J.C., et al., 2005. Comparative effects of dietary methylmercury on gene expression in liver, skeletal muscle, and brain of the zebrafish (*Danio rerio*). Environ. Sci. Technol. 39, 3972–3980.

Gu, H., Smith, Z.D., Bock, C., et al., 2011. Preparation of reduced representation bisulfate sequencing libraries for genome-scale DNA methylationprofiling. Nat.Protoc. 6, 468–481.

Hagenaars, A., Knapen, D., Meyer, I.J., et al., 2008. Toxicity evaluation of perfluorooctane sulfonate (PFOS) in the liver of common carp (*Cyprinus carpio*). Aquat. Toxicol. 88, 155–163.

Hagenaars, A., Vergauwen, L., Benoot, D., et al., 2013. Mechanistic toxicity of perfluorooctanois acid in zebrafish suggests mitochondrial dysfunction to play a key role in PFOA toxicity. Chemosphere 91, 844–856.

Heard, E., Martienssen, R.A., 2014. Transgeneration at epigenetic inheritance: myths and mechanisms. Cell 157, 95–109.

Heckmann, L.H., Sørensen, P.B., Krogh, P.H., et al., 2011. NORMA-Gene: a simple and robust method for qPCR normalization based on target gene data. BMC Bioinform. 12, 250.

Hou, Y., Zhang, H., Miranda, L., Lin, S., 2010. Serious overestimation in quantitative PCR by circular (supercoiled) plasmid standard: microalgal pcna as the model gene. PLoS One 5 (3), e9545.

Howe, K., Clark, M.D., Torroja, C.F., et al., 2013. The zebrafish reference genome sequence and its relationship to the human genome. Nature 496, 498–503.

Jaenisch, R., Bird, A., 2003. Epigenetic regulation of gene expression: how the genome integrates intrinsic and environmental signals. Nat. Genet. 33, 245–254.

Jiang, G., Xu, L., Song, S., et al., 2008. Effects of long-term low-dose cadmium exposure on genomic DNA methylation in human embryo lung fibroblast cells. Toxicology 244, 49–55.

Johannes, F., Porcher, E., Teixeira, F.K., et al., 2009. Assessing the impact of transgenerational epigenetic variation on complex traits. PLoS Genet. 5 (6), 1–11.

Kasahara, M., Naruse, K., Sasaki, S., et al., 2007. The medaka draft genome and insights into vertebrate genome evolution. Nature 447, 714–719.

Kim Tiam, S., Feurtet-Mazel, A., Delmas, F., et al., 2012. Development of q-PCR approaches to assess water quality: effects of cadmium on gene expression of the diatom *Eolimna minima*. Water Res. 46, 934–942.

Kozera, B., Rapacz, M., 2013. Reference genes in real-time PCR. J. Appl. Genet. 54, 391–406.

Li, Z., Yang, L., Wang, J., et al., 2010. beta-Actin is a useful internal control for tissue-specific gene expression studies using quantitative real-time PCR in the half-smooth tongue sole *Cynoglossus semilaevis* challenged with LPS or *Vibrio anguillarum*. Fish Shellfish Immunol. 29, 89–93.

Lister, R., Pelizzola, M., Kida, Y.S., et al., 2011. Hotspots of aberrant epigenomic reprogramming in human induced pluripotent stem cells. Nature 471, 68–73.

Liu, L., Li, Y., Tollefsbol, T.O., 2008. Gene-environment interactions and epigenetic basis of human diseases. Curr. Issue. Mol. Biol. 10, 25–36.

Liu, D., Pan, L., Cai, Y., et al., 2014. Response of detoxification gene mRNA expression and selection of molecular biomarkers in the clam *Ruditapes philippinarum* exposed to benzo[a]pyrene. Environ. Pollut. 189, 1–8.
Livak, K.J., Schmittgen, T.D., 2001. Analysis of relative gene expression data using real-time quantitative PCR and the $2^{-\Delta Ct}$ method. Methods 25, 402–408.
Lu, Y., Yuan, M., Gao, X., et al., 2013. Identification and validation of reference genes for gene expression analysis using quantitative PCR in *Spodoptera litura* (Lepidoptera: Noctuidae). PLoS One 8 (7), e68059.
Lucia, M., Andre, J.M., Gonzalez, P., et al., 2010. Effects of dietary cadmium contamination on bird anas platyrhynchos – comparison with species *Cairina moschata*. Ecotoxicol. Environ. Saf. 73, 2010–2016.
Maria, V.L., Gomes, T., Barreira, L., et al., 2013. Impact of benzo(a)pyrene, Cu and their mixture on the proteomic response of *Mytilus galloprovincialis*. Aquat. Toxicol. 144–145, 284–295.
Marie, V., Gonzalez, P., Baudrimont, M., et al., 2006a. Metallothionein response to cadmium and zinc exposures compared in two freshwater bivalves, *Dreissena polymorpha* and *Corbicula fluminea*. Biometals 19, 299–307.
Marie, V., Gonzalez, P., Baudrimont, M., et al., 2006b. Metallothionein gene expression and protein levels in triploid and diploid oysters *Crassostrea gigas* after exposure to cadmium and zinc. Environ. Toxicol. Chem. 25, 412–418.
Meissner, A., Gnirke, A., Bell, G.W., et al., 2005. Reduced representation bisulfate sequencing for comparative high- resolution DNA methylation analysis. Nucl. Acids Res. 33, 5868–5877.
Mirbahai, L., Yin, G., Bignell, J.P., et al., 2011. DNA methylation in liver tumorigenesis in fish from the environment. Epigenetics 6, 1319–1333.
Monsinjon, T., Knigge, T., 2007. Proteomic applications in ecotoxicology. Proteomics 7, 2997–3009.
Moore, M.N., Depledge, M.H., Readman, J.W., Paul Leonard, D.R., 2004. An integrated biomarker-based strategy for ecotoxicological evaluation of risk in environmental management. Mutat. Res. 552, 247–268.
Morga, B., Arzul, I., Faury, N., et al., 2010. Identification of genes from flat oyster *Ostrea edulis* as suitable housekeeping genes for quantitative real time PCR. Fish Shellfish Immunol. 29, 937–945.
Mouchet, F., Baudrimont, M., Gonzalez, P., et al., 2006. Genotoxic and stress inductive potential of cadmium in *Xenopus laevis* larvae. Aquat. Toxicol. 78, 157–166.
Moura, J.C., Araújo, P., Brito Mdos, S., et al., 2012. Validation of reference genes from Eucalyptus spp. under different stress conditions. BMC Res. Notes 14, 634. http://dx.doi.org/10.1186/1756-0500-5-634.
Negri, A., Oliveri, C., Sforzini, S., et al., 2013. Transcriptional response of the mussel *Mytilus galloprovincialis* (Lam.) following exposure to heat stress and copper. PLoS One 8 (6), e66802.
Nikinmaa, M., Rytkönen, K.T., 2011. Functional genomics in aquatic toxicology-do not forget the function. Aquat. Toxicol. 105 (3–4 Suppl.), 16–24.
Palais, F., Dedourge-Geffard, O., Beaudon, A., et al., 2012. One-year monitoring of core biomarker and digestive enzyme responses in transplanted zebra mussels (*Dreissena polymorpha*). Ecotoxicology 21, 888–905.
Pantanjali, S.R., Parimoo, S., Weissman, S.M., 1991. Construction of a uniform-abundance (normalized) cDNA library. Proc. Natl. Acad. Sci. U.S.A. 88, 1943–1947.
Parolini, M., Pedriali, A., Binelli, A., 2013. Chemical and biomarker responses for site-specific quality assessment of the Lake Maggiore (Northern Italy). Environ. Sci. Pollut. Res. Int. 20, 5545–5557.
Pierron, F., Normandeau, E., Defo, M., et al., 2011. Effects of chronic metal exposure on wild fish populations revealed by high-throughput cDNA sequencing. Ecotoxicology 20, 1388–1399.
Pierron, F., Baillon, L., Sow, M., et al., 2014. Ecotoxicology and epigenetics: effect of low-dose exposure on DNA methylation in the endangered European eel. Environ. Sci. Technol. 48, 797–803.
Pina, B., Barata, C., 2011. A genomic and ecotoxicological perspective of DNA array studies in aquatic environmental risk assessment. Aquat. Toxicol. 105S, 40–49.
Poynton, H.C., Robinson, W.E., Blalock, B.J., et al., 2014. Correlation of transcriptomic responses and metal bioaccumulation in *Mytilus edulis* L. reveals early indicators of stress. Aquat. Toxicol. 155, 129–141.
Pujolar, J.M., Milan, M., Marino, I.A., et al., 2013. Detecting genome-wide gene transcription profiles associated with high pollution burden in the critically endangered European eel. Aquat. Toxicol. 132-133, 157–164.
Qian, X., Ba, Y., Zhuang, Q., et al., 2014. RNA-seq technology and its application in fish transcriptomics. OMICS 18, 98–110.

Rakyan, V.K., Chong, S., Champ, M.E., et al., 2003. Transgenerational inheritance of epigenetic states at the murine AxinFu allele occurs after maternal and paternal transmission. Proc. Natl. Acad. Sci. U.S.A. 100, 2538–2543.

Rockett, J.C., Dix, D.J., 2000. DNA arrays: technology, options and toxicological applications. Xenobiotica 30, 155–177.

Rustici, G., Kolesnikov, N., Brandizi, M., et al., 2013. ArrayExpress update–trends in database growth and links to data analysis tools. Nucleic Acids Res. 41, D987–D990.

Sanchez, B.C., Ralston-Hooper, K., Sepúlveda, M.S., 2011. Review of recent proteomic applications in aquatic toxicology. Environ. Toxicol. Chem. 30, 274–282.

Schmittgen, T.D., Livak, K.J., 2008. Analyzing real-time PCR data by the comparative C(T) method. Nat. Protoc. 3, 1101–1108.

Schena, M., Shalon, D., Davis, R.W., et al., 1995. Quantitative monitoring of gene expression patterns with a complementary DNA microarray. Science 270, 467–470.

Skinner, M.K., Manikkam, M., Guerrero-Bosagna, C., 2010. Epigenetic transgenerational actions of environmental factors in disease etiology. Trends Endocrinol. Metab. 21, 214–222.

Snape, J.R., Maund, S.J., Pickford, D.B., et al., 2004. Ecotoxicogenomics: challenge of integrating genomics into aquatic and terrestrial ecotoxicology. Aquat. Toxicol. 67, 143–154.

Suzuki, M.M., Bird, A., 2008. DNA methylation landscapes: provocative insights from epigenomics. Nat. Rev. Genet. 9, 465–476.

Suzuki, G., Shiomi, M., Morihana, S., et al., 2010. DNA methylation and histone modification in onion chromosomes. Genes Genet. Syst. 85, 377–382.

Trapp, J., Armengaud, J., Pible, O., et al., 2014. Proteomic investigation of male Gammarus fossarum, a freshwater crustacean, in response to endocrine disruptors. J. Proteome Res. 14, 292–303.

Van der Oost, R., Beyer, J., Vermeulen, N.P.E., 2003. Fish bioaccumulation and biomarkers in environmental risk assessment: a review. Environ. Toxicol. Pharmacol. 13, 57–149.

Vandegehuchte, M., Janssen, C., 2011. Epigenetics and its implications for ecotoxicology. Ecotoxicology 20, 607–624.

Vandesompele, J., De Preter, K., Pattyn, F., et al., 2002. Accurate normalization of real-time quantitative RT-PCR data by geometric averaging of multiple internal control genes. Genome Biol. 3 (7) Research 0034.1–0034.11.

Velculescu, V.E., Zhang, L., Vogelstein, B., et al., 1995. Serial analysis of gene expression. Science 270, 484–487.

Walker, C.C., Salinas, K.A., Harris, P.S., et al., 2007. A proteomic (SELDI-TOF-MS) approach to estrogen agonist screenin. Toxicol. Sci. 95, 74–81.

Winter, G., Krömer, J.O., 2013. Fluxomics-connecting omics analysis and phenotype. Env. Microbiol. 15, 1901–1916.

Wu, H., Liu, X., Zhang, X., et al., 2013. Proteomic and metabolomic responses of clam *Ruditapes philippinarum* to arsenic exposure under different salinities. Aquat. Toxicol. 136-137, 91–100.

Yum, S., Woo, S., Kagami, Y., et al., 2010. Changes in gene expression profile of medaka with acute toxicity of Arochlor 1260, a polychlorinated biphenyl mixture. Comp. Biochem. Physiol. C151, 51–56.

Zheng, W.J., Sun, L., 2011. Evaluation of housekeeping genes as references for quantitative real time RT-PCR analysis of gene expression in Japanese flounder (*Paralichthys olivaceus*). Fish Shellfish Immunol. 30, 638–645.

Zhang, G., Fang, X., Guo, X., et al., 2012a. The oyster genome reveals stress adaptation and complexity of shell formation. Nature 490, 49–54.

Zhang, G., Zhao, M., Song, C., et al., 2012b. Characterization of reference genes for quantitative real-time PCR analysis in various tissues of *Anoectochilus roxburghii*. Mol. Biol. Rep. 39, 5905–5912.

CHAPTER 9

Reference Species

Brigitte Berthet

Abstract

Species used in ecotoxicological studies are called reference species. According to the objectives under consideration, different species can be chosen. There are, however, several parameters that may lead to misinterpretation of results, and numerous reservations occur when extrapolating beyond the studied species to the whole ecosystem. Thus we will focus on (1) the genetic diversity of organisms used in bioassays and their extrapolation to wild populations; (2) criteria to be taken into account for biomonitoring, and to approach the structure and functioning of the whole ecosystem; (3) the variability of responses through taxonomic levels, and the difficulties in clearly defining on what (species, strain, ecotype...) the studies will be carried out; and (4) the variability of responses as a result of confounding factors.

Keywords: Bioassays; Biomonitoring; Cryptic species; Field organisms; Functional role; Genetic diversity; In-house organisms; Representativeness.

Chapter Outline

The improvement of analytical techniques has allowed the identification and the quantification of an increasing number of xenobiotics, but chemical data do not inform us about the toxicological and ecotoxicological risks of these contaminants. Bioassays carried out in the laboratory or in situ are used for prospective assessments to determine predicted no effect concentrations, according to legislation controlling the input of new chemicals into

Aquatic Ecotoxicology. http://dx.doi.org/10.1016/B978-0-12-800949-9.00009-7

the environment (Chapter 6). To optimize the reproducibility of standardized methods, official bodies recommend the use of laboratory strains of organisms. It has often been shown, however, that maintaining populations in the laboratory leads to a genetic impoverishment, and this has to be considered when extrapolating from laboratory strains to field populations. Thus, the use of wild organisms seems a good alternative to take into account genetic variability and to improve the ecological value of risk assessments. Nevertheless, the choice of the field sites where the organisms will be collected is important because these organisms may have developed tolerance resulting from chronic exposure, thus leading to an undervaluation of any toxicological risks. Moreover, biotic or abiotic parameters may act as confounding factors (Chapter 7), so the biology of species has to be well known to differentiate the signal caused by contaminant exposure from the background. In addition, the choice of organisms (e.g., sex, age, size class) must be as uniform as possible. A difficulty will be to extrapolate the results of bioassays to the field.

Biomonitoring using sentinel species precociously reveals the presence of contaminants in the environment, their bioavailability to organisms (Chapter 3), and the changes that they may provoke in biological parameters at various levels of biological organization (e.g., molecular, cellular, physiological, individual), leading to dysfunctions, changes in health status, or modification of behavior (Chapter 7). The sentinel species must be chosen for its representativeness of their environment. As pointed out by Chapman (2002), this ecological approach has to be carried out with a focus on keystone species that have a functional role in the ecosystem, their disturbance causing important perturbations of the whole ecosystem (e.g., structure, functioning, organization, temporal successions). As an example, such a keystone species might be an endobenthic species carrying out bioturbation, thereby playing a key role in the transfer of pollutants (Chapter 10).

Among the criteria for the selection of a reference species, the ease of its identification is important. This, however, may be difficult for several species as a consequence of hybridization or the presence of different ecotypes or morphotypes in the population. Moreover, the validation of sentinel species is based on the possibility of extrapolating the results obtained to all the species of the ecosystem, but sensitivity to xenobiotics can vary by several orders of magnitude between different taxa, making extrapolation difficult.

In the present chapter, we will first review the biological models used for bioassays, focusing on questions about the genetic diversity of test organisms and the implications for the interpretation of bioassays; then, we will discuss the extrapolation of results obtained in laboratory studies to the field. A second part of this chapter will be dedicated to the species used for the determination of pollutant uptake and accumulation and their effects in situ. Finally, we will examine how it is possible to extrapolate from the results obtained with a single sentinel species to other taxonomic groups and eventually to the ecosystem.

9.1 Biological Models Used for Bioassays

Bioassays can be implemented to: (1) analyze the toxicological risk associated with chemicals new to the marketplace and (2) estimate the quality of effluents or sediments from natural areas already contaminated. They are generally performed in compliance with international guidelines, including the US Environmental Protection Agency (USEPA), ASTM, Organization for Economic Co-operation and Development, OSPAR, and International Organization for Standardization. These tests have been developed through a process of harmonization to minimize variations among testing procedures, using groups of selected organisms available from inhouse or commercial sources, exposed to chemicals under well-defined conditions.

9.1.1 Test Organisms

Standardized tests with microorganisms can be performed with several species of unicellular marine or freshwater green algae (*Pseudokirchneriella subcapitata*, formerly known as *Selenastrum capricornutum*, *Scenedesmus subspicatus*, *Chlorella vulgaris*), diatoms (*Navicula pelliculosa*, *Skeletonema costatum*), or cyanobacteria (*Anabaena flos-aquae*, *Synechococcus leopoldensis*). The endpoint is generally the inhibition of their growth (USEPA, 1996; OECD, 2002; OEHHA, 2004). Specific strains available from culture collections are recommended (Table 9.1).

Cultures of these microorganisms are standardized and can be easily maintained in the laboratory all year round, although this is not the case for macroalgae that are underrepresented among aquatic toxicity test organisms (Han et al., 2011). Nevertheless, 48-h spore germination success and length of gametophyte germ tubes of the giant kelp *Macrocystis*

Table 9.1: Recommended Strains of Algae OECD (2002)

Algal Species	Habitat	Recommended Strains
	Green Algae	
Pseudokirchneriella subcapitata,	Freshwater	ATCC 22662, CCAP278/4, 61.81 SAG
Scenedesmus subspicatus	Freshwater	86.81 SAG
Chlorella vulgaris	Freshwater	
	Diatoms	
Navicula pelliculosa	Freshwater	UTEX 664
Skeletonema costatum	Marine	
	Cyanobacteria	
Anabaena flos-aquae	Freshwater	UTEX 1444, ATCC 29413, CCAP 1403/13A
Synechococcus leopoliensis	Marine	UTEX 625, CCAP 1405/1

pyrifera have been developed as marine standardized toxicity tests (OEHHA, 2004), and, in Australia and New Zealand, assays related to the fertilization of *Hormosira banksii*, and the germination and growth of *Ecklonia radiata* have also been standardized (ANZECC and ARMCANZ, 2000).

As mentioned in Chapter 6, analyses of ecotoxicity data submitted within the framework of Registration, Evaluation, and Authorization of Chemicals in Europe regulation have been mainly based on single-species bioassays. Within invertebrates, priority has been given to *Daphnia magna* and only secondarily to other daphnids for both short- and long-term aquatic invertebrate studies. Nevertheless, invertebrates used in toxicity tests were first chosen on criteria of practicability, such as easy collection and/or ease of breeding in a laboratory. The example of the freshwater crustacean *Daphnia pulex* is significant. Widely used to test the toxicity of effluents in the 1970s, it has been replaced by other species of daphnid, such as *D. magna* and *Ceriodaphnia dubia* (Figure 9.1), which are easier to breed in the laboratory, although Chapman et al. (1994) showed that *D. pulex* was more sensitive to pollutants than other daphnids.

For both short- and long-term fish studies (Figure 9.1), *Pimephales promelas*, *Oncorhynchus mykiss*, and *Danio rerio* have been the most frequently used species (more than 10% of studies) (Tarazona et al., 2014). For experimental sediment toxicity assays (Chapter 10), *Corophium volutator* was used in 35.1% of the studies, followed by *Chironomus* sp. (15.1%) and *Lumbriculus variegatus* (7.4%) (Cesnaitis et al., 2014). All of these species are available from inhouse or commercial sources.

9.1.2 Genetic Diversity of Test Organisms

If the choice of laboratory strain organisms optimizes the reproducibility of results, and then reduces the number of required animals, numerous reservations can be are expressed as regards extrapolation beyond the studied strains to the species level. Siaut et al. (2011) have shown that different widely used laboratory strains of *Chlamydomonas reinhardtii* have up to five-fold variation in their capacity to accumulate oil. They speculate that the low selective pressure on triacylglycerols (energy-rich reserve compounds that might have important roles for cell survival in natural environments) that is applicable to strains cultured under standard conditions might have resulted in the accumulation of mutations lowering oil content in the haploid cells of *C. reinhardtii*. In a review, Lakeman et al. (2009) have examined the processes that cause laboratory cultures to evolve and diverge over time. They describe changes in the properties of phytoplankton cultures in various taxa, including changes in life-cycle phase, loss of sexual phase expression, decline in heterotrophic ability, phototrophic efficiency and bioluminescence, slower growth rate, modification of circadian rhythm, changes in ploidy, or loss of heterozygosity. They suggest, a strain really represents, over time, just one single trajectory manifested from the vast evolutionary potential possessed by the original

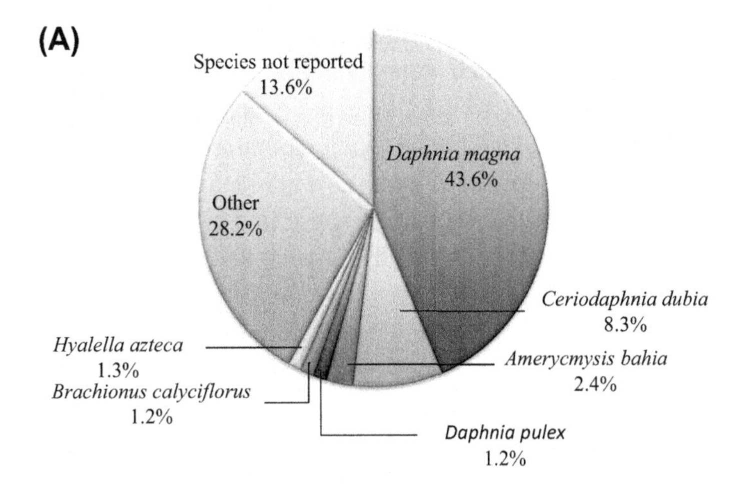

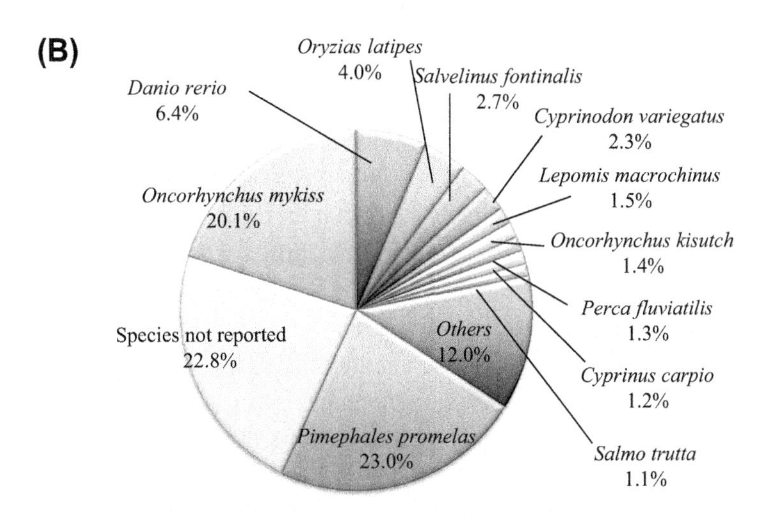

Figure 9.1
Distribution of species reported by Tarazona et al. (2014) for long-term aquatic invertebrate (A) and fish (B) studies (with permission).

isolate, it is necessary to take into account the time that an isolate has been in culture, as is already done for tissue or bacterial cultures.

Weston et al. (2013) examined the effects of a pyrethroid insecticide on a nontarget aquatic amphipod, *Hyalella azteca*. They demonstrated that the 10 populations tested, three strains from laboratory cultures and seven from natural sites in California, differed more than 550-fold in sensitivity to the pyrethroid. Weston et al. (2013) concluded that even a single species can have dramatically varying pesticide sensitivity if resistance has been acquired by some populations, with important implications when testing the toxicity of sediment.

Comparing selenium toxicity on laboratory-reared and field-collected *H. azteca*, Pieterek and Pietrock (2012) has shown that the former were about one order of magnitude more sensitive than the latter. Major et al. (2013) determined the genetic relationships of 38 populations of *H. azteca* from US and Canadian toxicology research laboratories and from North American field sites. Apart from a single exception, all samples were identified as conspecific, and the commonly occurring laboratory strains were no longer found in the natural environment (except in northern Florida). Warming et al. (2009) studied the acute and chronic physiological effects of a pesticide (azoxystrobin) on three clones of *D. magna* originating from different Danish lakes. Their results have shown significant clonal variation in the sensitivity of *D. magna*. The less sensitive clone (in terms of acute toxicity) had been kept in laboratory culture for much longer (28 years) than the two others (3 years and 1 year). Thus, the duration of culture may influence the susceptibility to chemicals, the older the culture, the greater the risk of deviation from responses of field animals in situ.

Genetic selection resulting from laboratory housing has also been studied in vertebrates, particularly in fish. Athrey et al. (2007) have compared genetic variation in killifish *Heterandria formosa* (1) selected in the laboratory over eight generations for an increased resistance to cadmium; (2) from a laboratory control population; and (3) from the original source population. Comparing microsatellite markers, they observed much lower heterozygosity in laboratory populations, both selected and control, than in the original population. A consequence of this loss of genetic variability may be an inability of the laboratory population to cope with other stressors in the future, whereas wild ones could succeed. Because the loss of genetic variation in laboratory populations appeared to be due to lower population sizes than in the wild population, Athrey et al. (2007) concluded that there is a necessity to maintain large enough laboratory populations to reflect the behavior of natural populations in environmental toxicology studies. A similar study was carried out on the zebrafish *Danio rerio* by Coe et al. (2009), with "wild-type" strains (being the "normal" zebrafish without genetic alteration) obtained from research institutions (Max Planck Institute and University of Harvard) as well as a local aquarium store and wild fish from a Bangladesh canal. They evaluated genetic variation with several indicators including allelic richness (Figure 9.2). For all the indicators of genetic variability, the laboratory strains had lower levels than wild fish. Moreover, the same strain from different institutes (i.e., strain AB) differed in their levels of genetic variation, and the variability declined over time (strain WIK obtained from Max Planck Institute in 2006 and 2007).

Thus, as a result of practical constraints limiting effective population sizes (i.e., number of breeding individuals), laboratory animal strains are generally more inbred, less genetically diverse, and more susceptible to genetic drift than their wild counterparts. So the use of these laboratory strains may affect the outcome of many toxicological tests, as demonstrated by Brown et al. (2011) in a study intended to test if inbred zebrafish differ in their susceptibility to endocrine disruptor effects (clotrimazole at high-level exposure of 43.7 $\mu g.L^{-1}$) compared

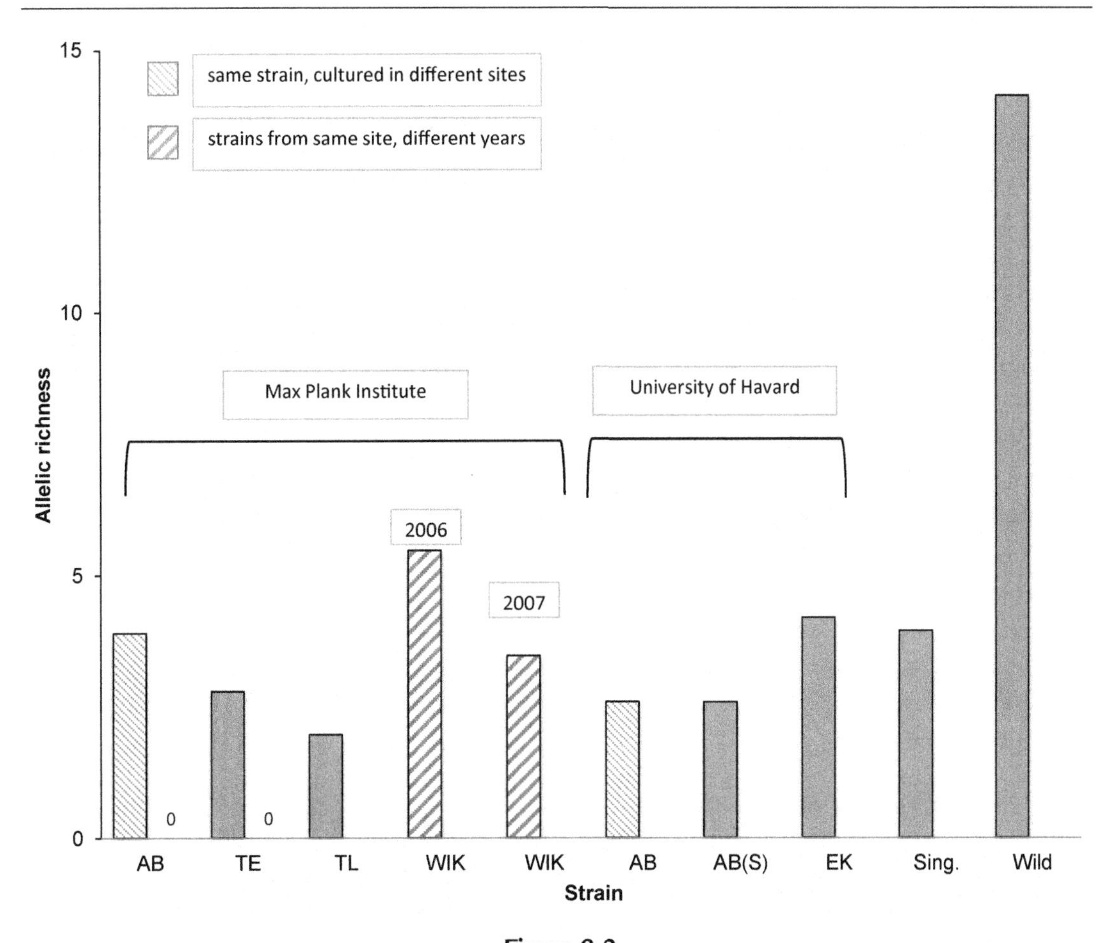

Figure 9.2

Comparison of allelic richness of the zebrafish *Danio rerio* (based on a minimum sample size of 20 diploid individuals) in a few laboratory strains from the Max Planck Institute (AB, TE, TL, WIK) and the University of Harvard (AB, AB(S), EK), a strain from a commercial dealer (Sing.) and wild fish from a Bangladesh canal (wild). *Modified after Coe et al. (2009).*

with wild populations. Their results indicated that the effects of clotrimazole on testis development differed between inbred and outbred zebrafish (representing wild populations). They concluded that there is a need to report pedigree/genetic information for laboratory animals in chemical testing. To go farther in the understanding of inbreeding on phenotypic endpoints (growth and sex ratio), Brown et al. (2012) compared a WIK zebrafish strain (standard reference) and a WIK/wild hybrid zebrafish strain with three levels of inbreeding (Inbreeding coefficients $F_{IT}=n$, $n+0.25$, $n+0.375$) under standardized control conditions following the Fish Sexual Development Test (OECD, 2011). Compared with WIKs, WIK/wild hybrids were significantly larger in size, with more advanced germ cell development after 63 days postfertilization (end of the test). Moreover, in the context of the Fish Sexual

Development Test, the WIK strain failed to comply with acceptable limits for sex ratio (30% to 70% females).

Because the reliability of results in ecotoxicity testing can be altered depending on the genetic characteristics of the organisms used in tests, the natural variability of organisms can be theoretically surpassed through the use of genetically similar individuals, held under strict control and artificial abiotic conditions—i.e., clones of *Artemia* species (Nunes et al., 2006), rotifers (Snell and Carmona, 1995), or daphnid species (Haap and Köhler, 2009). However, Sukumaran and Grant (2013a, b) have shown that sexual reproduction mitigates significantly the long-term consequences of genetic damage on *Artemia* (the sexual *Artemia franciscana* and the asexual *Artemia parthenogenetica*) submitted to a reference mutagen (ethyl methane sulfonate) over three generations. They concluded that bioassays should include information from multigenerational studies, and that single-generation studies may under or overestimate toxicological risks. Snell and Carmona (1995) have compared the effects of four toxicants on the sexual and asexual reproduction of the rotifer *Brachionus calyciflorus*. They observed that the former was more sensitive than the latter, because at the lowest observed effect concentration, the inhibition was 2–68 times higher for sexual than for asexual reproduction. Thus, even if asexual reproduction is generally more productive and provides organisms with less variability in responses to toxicants, different choices must be made according to the endpoints of the study.

9.1.3 From the Laboratory to the Field

Extrapolating from findings from laboratory studies to wild populations may be difficult, not to say wrong, partly as a consequence of reduced or unrepresentative levels of genetic variation in laboratory strains of species as shown previously. In addition, the different conditions prevailing in the field and in the laboratory can also explain differences in biological responses. For instance, testing the activity of three antifouling paints in bioassays and field tests using marine bacteria, microalgae, and barnacles, Bressy et al. (2010) have shown that results from laboratory assays do not fully concur with the antifouling activity in the field trial, because of the absence of film conditioning process. As mentioned previously, Major et al. (2013) found that the diversity of the *H. azteca* species detected in the field is not represented in American laboratories, and they question the extrapolation of results from laboratories to the field. They also recommend laboratories to determine if the strains they use are appropriate for their region. Menchaca et al. (2010) have compared the responses of amphipods *Corophium multisetosum* cultured in the laboratory with those of field amphipods in sediment toxicity tests. They considered that the much lower sensitivity of the cultured amphipods to cadmium may be attributed to several parameters. The first one was that cultured animals were fed ad libitum for 1 year before the bioassay, preventing them from feeding on tested sediment, whereas field amphipods were more hungry and fed on contaminated sediment. The second one was that cultured amphipods could have been genetically selected, favoring the survival of the most resistant individuals. The last hypothesis was that field individuals may have been submitted to multiple stressors increasing their overall sensitivity. In conclusion, these authors advise against the use of *Corophium* laboratory strains to evaluate the quality of marine sediments. Their last hypothesis, also called tolerance, has been developed in a recent book entitled *Tolerance to Environmental Contaminants* (Amiard-Triquet et al., 2011). When differences in tolerance ratios between populations are relatively small, application of uncertainty or safety factors when extrapolating results from the laboratory to the field are sufficient; however, when large ranges are observed, it may not be adequate for all contaminants and further investigations are needed to assess the extent to which tolerance can modify the responses of organisms (Johnston, in Amiard-Triquet et al., 2011). Particular attention should be given to the populations under extreme field conditions or species with specific traits that are usually expected to possess a wide tolerance. Hummel et al. (in Amiard-Triquet et al., 2011) have revisited this assumption and, on the contrary, concluded that "organisms living under conditions close to their environmental tolerance limits appeared to be more vulnerable to additional chemical stress than their conspecifics living under more optimal conditions."

Not only do the origins of organisms used in laboratory tests have to be taken into account to minimize the under- or overestimation of the toxicological risk for natural populations, but it is also advisable to use conditions in the laboratory that are comparable with those prevailing in the field, as recommended by Chapman (2002) to put more "eco" in ecotoxicology. Moreover, Janssen and Heijerick (2003) point out that the limited number of species tested does not reflect natural phytoplankton communities. So they suggest that bioassays for assessing the environmental impact of contaminants in a specified ecoregion should be based on a battery of algal tests with species adapted to and tested under the specific natural conditions of the region. Relatively little is known about the sensitivity of tropical species, thus standard models that are mainly organisms from temperate regions (Leboulanger et al., 2011) are not useful to assess chemical toxicity in tropical marine waters. The same conclusions were drawn by Van Dam et al. (2008), who reviewed the literature dealing with marine plant and animal testing. They found a paucity of regionally relevant marine toxicity testing methods for Australian tropical marine species and recommended the development of additional toxicity tests for site-specific assessments. The opposite appears to be the case for bioassays of antifouling activity. Because the main maritime routes pass through tropical zones, paint manufacturers are in search of new compounds with activity toward tropical fouling organisms, whereas cold or temperate fouling species are neglected (Maréchal and Hellio, 2011). However, with the development of Arctic maritime transport, it seems necessary to develop new antifouling bioassays that specifically target cold and/or temperate species.

Standardized protocols for measuring the accumulation of chemicals from sediments or water in the laboratory are also used to establish bioaccumulation factors (BAF) for contaminated sites, but BAFs from field samples can differ by up to several orders of magnitude from laboratory bioconcentration factor measurements for some chemicals (Burkhard et al., 2011a). To cope with the need for a better understanding of this variability, a Lab-Field Bioaccumulation Workshop was held in 2009, and the key findings and recommendations are exposed in five articles summarized by Burkhard et al. (2011a). One of these studies (Burkhard et al., 2011b) has compared laboratory- and field-measured biota-sediment accumulation factors using oligochaetes from the genus *Lumbriculus* and bivalves from different genera For oligochaetes, BAFs from the laboratory tests and field BAFs did not differ considerably for the majority of the 10 contaminated sediment sites (within a factor of 2). Discrepancies were attributed to field sampling (sediment not representative of the habitat of field oligochaetes) and sample handling procedures (loss of the more soluble lower molecular weight polyaromatic hydrocarbons [PAHs]). Therefore, Burkhard et al. (2011b) concluded that *Lumbriculus variegatus* is a reliable indicator of contaminant accumulation for oligochaetes in the field. As regards the bivalves, the difference between BAFs from laboratory tests and field BAFs was much more important (up to six times), but laboratory BAFs can also provide estimates within a factor of two of field BAFs if precautions are taken: (1) the use of steady-state adjustment factors seems to be necessary; (2) extrapolation of laboratory BAFs between various species should be done with great caution (see Section 9.3); (3) laboratory exposure chambers have to be improved to better mimic contaminant exposures in the field, because in the latter, bivalves can be exposed to both sediment and overlying water; and (4) particular attention has to be given to chlorinated polychlorinated biphenyls and certain pesticides. Casado-Martinez et al. (2013) have compared metal bioaccumulation from estuarine sediments in *Arenicola marina* over 30 days under controlled laboratory conditions and in field-collected worms. Because among deposit-feeding polychaetes, symptoms of stress in feeding are frequently associated with laboratory handling and acclimation, these authors concluded that the most usual exposure duration of 28 days indicated in standardized laboratory protocols is too short, and that a longer time is necessary to ensure steady-state concentrations.

According to the USEPA (2002), the use of test organisms taken from potentially contaminated "receiving waters" in the field are impractical for several reasons: (1) tolerance acquired by organisms chronically exposed to contaminants in their environment; (2) heterogeneity of age and quality of field-collected organisms; (3) collection permits may be required and sometimes difficult to obtain; (4) the required quality assurance/quality control records, such as the single laboratory precision data, would not be available; (5) the identification of test organisms to the species level may be labor-intensive and stressful to the organisms; and (6) field-collected organisms must be acclimated in the laboratory for a minimum of 1 week before use, to assure that they are in good health, and fish captured by electroshocking must not be used in toxicity testing. Despite these limits, the use of test organisms collected from receiving waters has strong appeal, and would seem to be the logical approach because of their representativeness toward their environment and more generally, toward the real world. To optimize the extrapolation of laboratory results to the field, in situ bioassays and mesocosms can be used with cultured organisms or with wild ones. This issue is developed and discussed in Chapter 6.

9.2 Sentinel Species for the Determination of Contaminant Uptake and Effects

Many criteria should be met to qualify a species as suitable for use as a sentinel species in contaminant monitoring, including features widely developed in previous publications (Berthet and literature cited in Amiard-Triquet et al., 2013) such as:

- Ease of identification (see sub Section 9.3.1);
- Sedentary nature for the representativeness of pollution at the site of collection;
- Sufficient population size and easy sampling to ensure practicability;
- Longevity of several years and sufficient resistance to pollutants (integrative indicators);
- Wide and known distribution range, or distribution involving ecological analogs, availability for capture all year round, and a well-known biology and ecology;
- Dose–effect and cause-and-effect relationships consistently established, that mean that sentinel species have to show sufficient sensitivity so that dysfunctions may be measurable; and
- Representativeness toward their environment.

The most commonly used species for the determination of contaminant uptake and effects (biomarkers) are shown in Chapters 3 and 7, respectively. The uses of specific taxonomic or ecological groups in ecotoxicology are reviewed in Chapters 10 (endobenthic invertebrates), 11 (gammarids in freshwater), 12 (copepods in estuarine and marine waters), and 13 (fish as reference species in different water masses).

9.2.1 Sentinels in Current Use in Monitoring Programs

Considering either bioaccumulation (see Chapter 3, sub Section 3.2.3) or biomarkers (Table 7.1), bivalves and fish are the groups that provide the majority of biological models, both for freshwater and marine environments. A guidance document launched by the USEPA (2000) underlined that the greater sedentary nature of bivalves was a guarantee of their representativeness of the degree of contamination of their medium of origin. However, fish are sometimes preferred or used concurrently for different reasons such as the fact that they are also important food products that should remain free of contamination above food security standards. For biomarkers, fish as vertebrates share a number of common features with mammals in which the mechanisms of action are far better known, thus improving the interpretation of results. Migrations complicate any interpretation as evoked for histopathological biomarkers in flounders (Stentiford et al., 2003) and sole *Solea senegalensis* (Olivia et al., 2013). Kirby et al. (2000) have recommended the incorporation of species with limited migrational tendencies. This is the case for eelpout *Zoarces viviparus* recommended for the assessment of reproductive success by the International Council for the Exploration of the Sea (Table 7.1), and also used in the application of biomarkers of apoptosis and DNA damage (Lyons et al., 2004). Another strategy consists in active biomonitoring with caged fish, used to impose a sedentary nature as exemplified with juvenile coho salmon (*Oncorhynchus kisutch*) exposed to PAH-contaminated sediments in a freshwater lake (Barbee et al., 2008). Despite that fish caging is widely used, the pros and cons of this practice need to be carefully evaluated (Chapter 13). Migratory habits are also important when using seabird eggs (Miller et al., 2014), and updated guidelines have been launched under the auspices of the OSPAR commission (2012) for the sampling and analysis of contaminants in seabird eggs in addition to fish and shellfish.

In the specific case of the impact of ionizing radiation on nonhuman species, a framework has been designed by the International Commission on Radiological Protection (ICPR, 2008). It is recommended to develop a small set of reference fauna and flora to serve as a basis for a more fundamental understanding and interpretation of the relationships between exposure and dose and between dose and certain categories of effect for a few, clearly defined types of animals and plants. These reference organisms must be typical of their environment, wild rather than domesticated and ubiquitous. For the aquatic environment, a bird (a duck of the family *Anatidae*), an amphibian (a frog of the family *Ranidae*), a freshwater salmonid (trout rather salmon to avoid the problems of migratory effects), a marine flatfish (such as plaice, founder, dab, and halibut), and a marine crustacean (a crab of the family *Cancridae*) have been chosen. Additional biological models can be developed to assess and manage risks to nonhuman species in more area- and situation-specific approaches.

9.2.2 The Concept of an Ecosystem Approach

Species used as sentinels have too often been selected for their ease of collection or manipulation, whereas their roles in the ecosystem have been neglected (Chapman, 2002). Initiated in 2001, "The ECOMAN project" (Galloway et al., 2006) recommends a multispecific approach in which estuarine and coastal species are selected taking into account the variety of their habitats and food strategies. However, as pointed out by Chapman (2002), this ecological approach, although essential, is not sufficient, and sentinel species must also be chosen from keystone species. These species interact with numerous components of the ecosystem, and if no ecological equivalents are present, their disturbance can cause modifications of the structure and the functioning of the ecosystem. Berthet (and literature cited, in Amiard-Triquet et al., 2013) have reviewed some of these interactions:

- Roles in food webs because these species have a particular importance through their activities of predation, or, on the contrary, because of their biomass, as prey themselves; and
- Roles in biogeochemical cycles, as for example that of the primary producers playing a major role in the transfer of xenobiotics from the physical compartment to consumers, or that of filter-feeding species which are responsible for an important part of transfer of chemicals between the water column and the sediment.

As sediments are the main reservoirs for most organic and inorganic chemicals (Bernes and Naylor, 1998), particular attention should be paid to keystone endobenthic species. This can be illustrated with species of interest for intertidal mudflats in Europe (Figure 9.3). The tellinid clam *Scrobicularia plana* and the ragworm *Nereis diversicolor* represent a high biomass in these ecosystems with population densities reaching several thousand individuals per square meter (Orvain, 2005; Gillet et al., 2008). They are major links in food webs, being important prey for crabs, flatfish, and wading birds. Because they are in close contact with the sediment, they can act as a vector of contaminant transfer from sediment to higher trophic levels (Coelho et al., 2008; Cardoso et al., 2009). According to the studies of Davey (1993) in the Tamar estuary (England), they are among the main macrofaunal species responsible for most of the bioturbation. Their influence on oxygen diffusion via their burrows (Figure 9.3) has effects on the fluxes of nutrients and contaminants between sediment and overlying water (e.g., Banta and Andersen, 2003; Orvain, 2005), and on the distribution of other species (Bouchet et al., 2009); therefore, they have clearly a key role in the structure and functioning of estuarine and coastal ecosystems.

The larvae of the midge *Chironomus plumosus* are very abundant in lake sediments, and as for the species mentioned previously, they have a high bioturbation/bioirrigation potential. Schaller (2014) has used these insect larvae in a laboratory experiment to assess the effects of bioturbation/bioirrigation on the remobilization of several metal/metalloid/rare earth/radionuclide elements from sediments to water. An impact was observed for 18 elements, ranging from strong influence (dissolved organic carbon, N, Mg, Ca, Sr, Mo, and U) to influence only at start when the larvae dig into the sediments (Al, As, Fe, Cd, Co, Cu, Mn, Ni, Zn, Ce, and

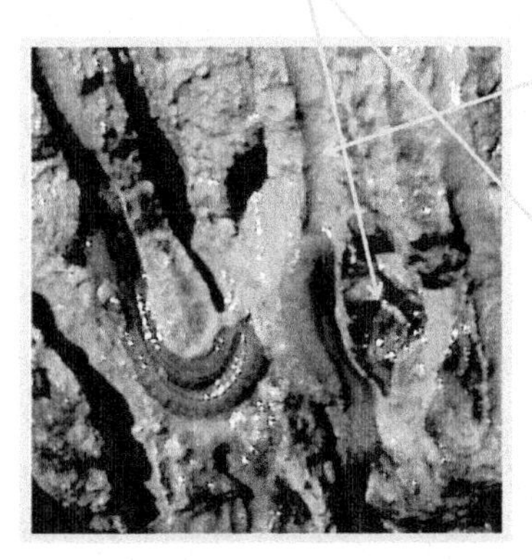

Figure 9.3
Reworking of sediment by ecosystem engineers is important for the normal course of biogeochemical cycles (nutrients and contaminants).

Cs), depending on the chemical characteristics of the element. Moreover, accumulation of most elements was found in *C. plumosus*. Using *L. variegatus*, a freshwater oligochaete, Pang et al. (2012) have evaluated the influence of bioturbation on the bioavailability of PAHs in field-contaminated sediments, using different densities of worms to test different levels of bioturbation. According to the densities of the worms, they have observed no significant differences in bioavailability of PAHs, whereas concentrations in sediment and *L. variegatus* decreased with increasing density. Moreover, when the amphipod *H. azteca* was added simultaneously, its mortality increased at the highest density of worms, suggesting that bioturbation can modify the toxicity of contaminants to other organisms. Studying the influence of another oligochaete *Tubifex tubifex* on sediments contaminated with fungicides (respectively epoxiconazole and metalaxyl), Liu et al. (2013) and Di et al. (2013) observed positive effects of these oligochaetes on the diffusion of fungicides from sediments to overlying water, and in the case of metalaxyl, decreased concentrations in sediments indicating its degradation by the worms. *T. tubifex* and *Chironomus riparius* were shown to increase the diffusive oxygen uptake (DOU) of sediments (13–14 %) after 72 h. Their bioturbation activity was reduced in uranium-contaminated sediment, but remained sufficient to induce uranium release toward water and an increase of diffusive oxygen uptake (53%) in case of case *T. tubifex* (Lagauzère et al., 2009). Uranium favors the aerobic reactions in sediments, and Lagauzère et al. (2013) suggest that the phenomenon is amplified by bioturbation that induces biogeochemical modifications in the contaminated sediment. Moreover, these authors have shown a clear impact of uranium and/or bioturbation on microbial-driven diagenetic reactions such as denitrification, sulfate-reduction, and iron dissolutive reduction.

9.3 Variability of Ecotoxicological Responses through Taxonomic Levels

9.3.1 Problems in Species Identification

Among the criteria for the selection of reference species, ease of identification is put forward, but several species may present plasticity according to their environment, whereas different species, for which identification is difficult, may coexist in sympathy, with the possibility of hybridization, leading to complexes of cryptic species. Ecotypes of a single species, and the different species in a species complex, may have different responses to pollutants.

Mytilus edulis is a filter-feeder widely used in monitoring programs (i.e., Mussel Watch), because it presents most of the criteria of sentinel species listed above. Nevertheless, three genetic groups of *M. edulis* may coexist along the European coast, namely *M. edulis edulis* along the Atlantic coast, *M. edulis trossulus* in the Baltic Sea, and *M. edulis galloprovincialis* in the Mediterranean Sea and along the Spanish coast (Hummel et al., 2001). Hybridization between *M. edulis edulis* and *M. edulis trossulus*, as well as between *M. edulis edulis* and *M. edulis galloprovincialis*, has been established respectively in Mecklenburg Bay, Germany, and around Ré Island, France (Przytarska et al., 2010). Thus these authors use the term *M. edulis* complex in a study using mussels to compare Fe, Mn, Pb, Zn, and Cu bioavailability at 17 sampling sites in coastal waters around the European continent. Except for Zn and Cu that presented low geographic variation because of their partial regulation by mussels, the bioavailabilities of metals were correlated to riverine discharge and industry proximity. Brooks and Farmen (2013) have studied the distribution of *Mytilus* "species" along the Norwegian coast, and have identified the presence of all the three *Mytilus* "species" as well as both hybrids of *M. edulis/M. galloprovincialis* and *M. trossulus/M. galloprovincialis*, with an impossibility to distinguish between them by mere visual inspection. Although few studies have examined comparatively the responses of the three "species" to contaminants, differences in bioaccumulation, genotoxic responses, or histological parameters have been documented (Brooks and Farmen, 2013 and literature quoted therein). Thus these authors highly recommend the correct identification of the mussel "species" before their use in biomonitoring programs or laboratory experiments.

The suitability of the green crab *Carcinus maenas* in ecotoxicological studies has been recently reviewed (Rodrigues and Pardal, 2014). The green crab may have two color morphotypes, green and red, that have several ecological differences and present different responses to the same pollutant (Cd, or pyrene). Thus it is important to take into account this parameter in experimental design and when comparing results, but according to Rodrigues and Pardal (2014), information about the *Carcinus* sp. morphotype is missing in most publications.

Hyalella azteca has generally been considered as a single species, but Weston et al. (2013) have indicated that they have to be treated as a complex of cryptic species. These species often coexist in sympatry and their identification is rather difficult, except by using molecular markers (Wellborn and Cothran, 2004). These findings can explain the differences observed in the resistance to pyrethroid (see sub Section 9.1.2). Leung (2014) tested the responses of different members of the *H. azteca* species complex to copper and nickel exposure, and similarly found significant differences of mortality and juvenile production between clades, and thus concluding that genetically characterized cultures of *H. azteca* should be used in toxicity tests to prevent misinterpretation of the results. Another amphipod, *Gammarus fossarum*, has been submitted to ammonium, a common anthropogenic discharge by Feckler et al. (2014). They differentiated two cryptic lineages, and observed a factor greater than two between the ammonium tolerances of these lineages at environmentally relevant test concentrations (for other examples of cryptism in gammarids, see Chapter 11). As for *H. azteca*, they concluded that environmental monitoring data, which are usually based on the morphological characteristics of species, should be interpreted carefully.

Rocha-Olivares et al. (2004) have conducted toxicity bioassays with both polynuclear aromatic hydrocarbon and metal contamination on *Cletocamptus fourchensis* and *C. stimpsoni*, segregated cryptic species from the cosmopolitan marine and brackish-water copepod *Cletocamptus deitersi* commonly used as a laboratory toxicity test species. Their study suggested differential tolerances to a mixture of metals, but similar ones to phenanthrene, and again the conclusion was that members of a cryptic species complex should not be used in laboratory toxicity tests unless populations are genetically characterized.

9.3.2 Intraspecific Variability

It has been recognized for a long time that contaminant bioaccumulation depends upon intrinsic factors such as the complex of size/weight/age, reproductive status, and stage of development (NAS, 1980), and it is commonly accepted that the reproductive and young phases of organisms are the most sensitive stages, and thus attract the attention of ecotoxicologists. Although in many cases this assumption is verified, in a recent review, Berthet (in Amiard-Triquet et al., 2013) quoted several examples of bivalves, crustaceans, or fish that do not follow this rule. Confounding factors also affect the biological responses to toxicants as developed in Chapter 7. Tolerance acquired in populations previously exposed to pollutants is also a source of intraspecific variability (Amiard-Triquet et al., 2011). Moreover, as we have shown previously, even when a species is well determined, it is necessary to pay attention to morphotypes or ecotypes, and to the origins of the individuals.

9.3.3 Variability Between Species and Higher Taxonomic Groups

Because findings on reference species tested in laboratories are often extrapolated with the attempt to protect endangered and threatened species, Jorgenson et al. (2014) have tested the model endocrine active compound 17 β-estradiol (E2) on two cyprinids (the endangered Rio Grande silvery minnow and the fathead minnow, commonly used in bioassays), both species being found in the Middle Rio Grande. They have also assessed exposure effects in bluegill sunfish, a centrarchid sometimes used in field studies, and phylogenetically distant from cyprinids. Results indicate sensitivity differences between the minnow species according to the biomarker tested, but these two species responded to estrogenic exposure more similarly than the phylogenetically more distant bluegill sunfish. They concluded that it is possible to use surrogate species for initial implementation of water quality if phylogenetic relationships are taken into consideration.

Nevertheless, for many chemical contaminants, it is generally unclear whether one of these four taxa is more sensitive than others. This is because differences in sensitivity are as high within samples of each of them as between them. On the other hand, clear differences may be present between zoological groups when a given chemical such as the organophosphate insecticide chlorpyrifos is considered (Figure 9.4). In this case, insects, as well as other arthropods, appear to be two orders of magnitude more sensitive than fish. By comparison, the distribution of median lethal concentrations for arthropod species exposed to dissolved cadmium covers a range of concentrations from <0.01 to $>10\,mg\,L^{-1}$ (http://www.epa.gov/caddis/).

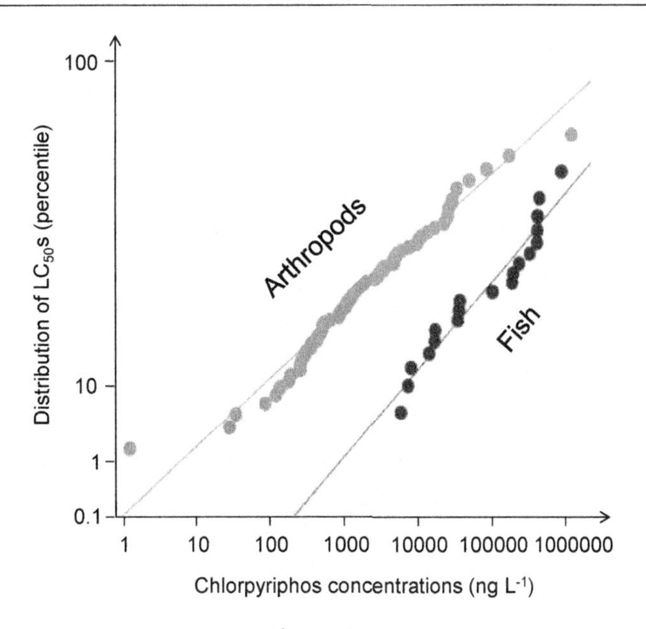

Figure 9.4
A species sensitivity distribution (SSD) plot showing the distribution of median lethal concentrations for arthropod versus fish species exposed to chlorpyrifos. *Modified after Giddings and Hendley* http://www.epa.gov/oppefed1/ecorisk/ecofram/sra.htm.

Even in the restricted group of benthic diatoms, Larras et al. (2012) have shown contrasted responses following exposure to several herbicides, according to their trophic mode and their ecological guild: for photosystem-II inhibitor herbicides, N-heterotroph and "motile" guild diatom species were more tolerant, whereas N-autotroph and "low profile" guild species were more sensitive. Herlory et al. (2013) have conducted a 7-month survey of diatom assemblages to biomonitor the impact of treated uranium mining effluent discharge. Neither the biomass nor the photosynthetic activity of periphyton exhibited changes in response to the stress induced by mining effluents, whereas the composition of diatom communities was clearly impacted, with *Neidium alpinum* and several species of *Gomphonema* appearing as the most tolerant to this contamination. This differential sensitivity is the basis of several diatom indices (Kelly and Whitton, 1995; Kelly, 1998; Kelly et al., 2009) currently used in freshwaters, but still little used in transitional waters, and only the trophic diatom index, seems adaptable to estuarine ecosystems after some adjustments in their species indicator values (Rovira et al., 2012).

As for diatoms, highly variable responses to pollutants are observed in microbial communities (algae, bacteria, protozoa, fungi, viruses), resulting in the selection of the most tolerant strains. Consequently, Blanck et al. (1988) have proposed the concept of pollution-induced community tolerance. The objectives and the application of the pollution-induced community tolerance concept have been developed by Tlili and Montuelle (in Amiard-Triquet et al., 2011).

Differential sensitivity between groups of invertebrates has been widely exemplified in several recent reviews dedicated to the interspecific variability of tolerance with reference to tributyltin, metals, or hydrocarbons (Berthet and literature cited, in Amiard-Triquet et al., 2011), or to interspecific variability in biological responses that can guide the choice of the most relevant species for the determination of biomarkers (Berthet and literature cited, in Amiard-Triquet et al., 2013). In the case of an oil spill, copepods are more sensitive than nematodes, thus leading Carman et al. (2000) to propose the "nematode/copepod ratio" as an index of petroleum exposure. Similarly, polychaetes appear to be resistant to petroleum hydrocarbons, whereas amphipods are greatly affected, until approaching local extinction in case of an oil spill (Dauvin and Ruellet, 2007). Thus these authors have proposed the use of a "polychaete/amphipod ratio" (Benthic Opportunistic Polychaetes Amphipods Index) in coastal waters, in a parallel of the nematode/copepod ratio. This index has been successfully tested by Andrade and Renaud (2011) along the Norwegian continental shelf as an indicator of impact of offshore oil and gas production. Because of the spatial heterogeneity and complexity of estuaries, the Benthic Opportunistic Annelida Amphipods index has been adapted from the Benthic Opportunistic Polychaetes Amphipods index for transitional waters, adding worms from the taxon of clitellates (oligochaetes and leeches), which are common in muddy sediments and estuarine waters (Dauvin and Ruellet, 2009).

Species-sensitivity distributions are widely used to establish environmental quality guidelines, in particular for the calculation of predicted no effect concentrations (Dowse et al., 2013), but may be used only when many no observed effect concentrations, determined in different taxa (e.g., algae, invertebrates, fish, amphibians), are available (Chapter 2). Concentration-effect curves are frequently established for different taxa including algae, invertebrates, fish and amphibians as available (Environment Canada Website: http://www.ec.gc.ca/) for federal water quality guidelines established under the Canadian Environmental Protection Act (1999).

9.4 Conclusions

Whatever the tools used (bioassays, bioaccumulation determination, biomarkers), the selection of a single species as unique sentinel organism is unrealistic, and a multispecies approach is recommended. The choice has to turn toward a set of species that can represent at best the various functional aspects of the ecosystem (e.g., role in the food web, in the biogeochemical cycle of contaminants, and nutrients). Bioassays are carried out according to methodologies approved by national and international authorities (Chapter 2). Safety factors are used in ecological risk assessment to extrapolate from the toxic responses of laboratory test species to all species representing that group in the environment (Figure 9.5).

Organisms within a species are expected to have similar ecological needs and physiological functioning, but in fact intraspecific differences are important and it is necessary to move from a too rigid concept of species. Care should be taken when extrapolating results of bioassays obtained from few strains, ecotypes, or morphotypes to an entire species, a complex

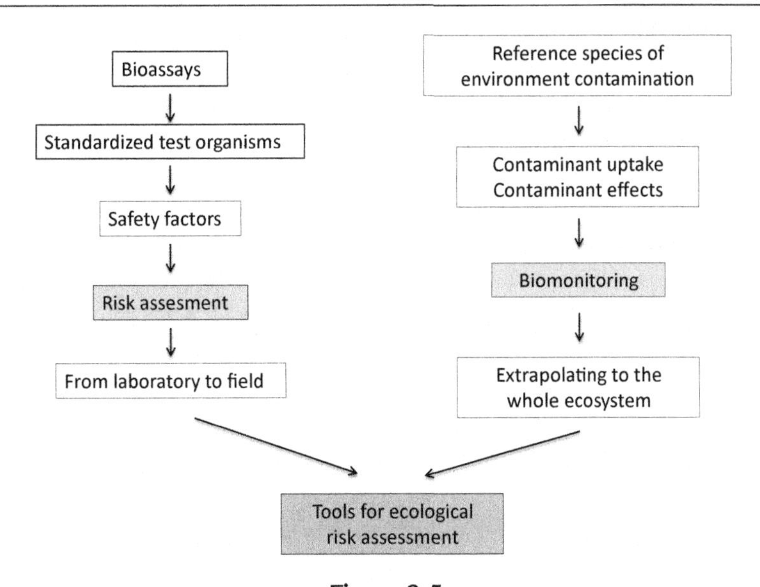

Figure 9.5

Different uses of reference species according to the objectives under consideration.

of species or to field populations living across contaminated gradients. To prevent misinterpretation of results, the use of genetically characterized strains in toxicity tests is preferable, and, more generally, identifying populations with molecular markers before their use in biomonitoring would be a clear improvement.

When the objectives are the evaluation of the potential risk associated with the input of a new product in the environment, selected organisms available from inhouse or commercial sources, exposed to chemicals under well-defined conditions should be used to optimized the reproducibility of results, and reduce the number of required individuals. On the other hand, when the objectives are contaminant uptake in and effects on organisms, and thus survival of species and ecosystem functioning (Figure 9.5), reference species should be chosen from wild populations for their representativeness of their environment and their roles in biogeochemical cycles.

Acknowledgments

Thanks are due to Pr. Rainbow, Keeper of Zoology, Natural History Museum, London, for his kind revision.

References

Amiard-Triquet, C., Amiard, J.C., Rainbow, P.S., 2013. Ecological Biomarkers: Indicators of Ecotoxicological Effects. CRC Press, Boca Raton.

Amiard-Triquet, C., Rainbow, P.S., Romeo, M., 2011. Tolerance to Environmental Contaminants. CRC Press, Boca Raton.

Andrade, H., Renaud, P.E., 2011. Polychaete/amphipod ration as an indicator of environmental impact related to offshore oil and gas production along the Norwegian continental shelf. Mar. Pollut. Bull. 62, 2836–2844.

ANZECC and ARMCANZ (Australian and New Zealand Environment Conservation Council and Agriculture and Resource Management Council of Australia and New Zealand), 2000. Australian and New Zealand guidelines for fresh and marine water quality. Aquatic Ecosystems – Rationale and Background Information. Paper No. 4, vol. 2, Chapter 8, 678 p.

Athrey, N.R.G., Leberg, P.L., Klerks, P.L., 2007. Laboratory culturing and selection for increased resistance to cadmium reduce genetic variation in the least killifish, *Heterandria formosa*. Environ. Toxicol. Chem. 26, 1916–1921.

Banta, G.T., Andersen, O., 2003. Bioturbation and the fate of sediment pollutants. Vie Milieu 53, 233–248.

Barbee, G.C., Barich, J., Duncan, B., et al., 2008. *In situ* biomonitoring of PAH-contaminated sediments using juvenile coho salmon (*Oncorhynchus kisutch*). Ecotox. Environ. Saf. 71, 454–464.

Bernes, C., Naylor, M., 1998. Persistent Organic Pollutants. A Swedish View of an International Problem. Swedish Environmental Protection Agency, Stockholm.

Blanck, H., Wängberg, S.A., Molander, S., 1988. Pollution-induced community tolerance – a new tool. In: Cairns, J.J., Pratt, J.R. (Eds.), Function Testing of Aquatic Biota for Estimating Hazards of Chemicals. American Society for Testing and Materials, Philadelphia, pp. 219–230.

Bouchet, V.M.P., Sauriau, P.G., Debenay, J.P., et al., 2009. Influence of the mode of macrofauna-mediated bioturbation on the vertical distribution of living benthic foraminifera: first insight from axial tomodensitometry. J. Exp. Mar. Biol. Ecol. 371, 20–33.

Bressy, C., Hellio, C., Maréchal, J.P., et al., 2010. Bioassays and field immersion tests: a comparison of the antifouling activity of copper-free poly(methacrylic)-based coatings containing tertiary amines and ammonium salt groups. Biofouling 26, 769–777.

Brooks, S.J., Farmen, E., 2013. The distribution of the mussel *Mytilus* species along the norwegian coast. J. Shellfish Res. 32, 265–270.

Brown, A.R., Bickley, L.K., Le Page, G., et al., 2011. Are toxicological responses in laboratory (inbred) zebrafish representative of those in outbred (wild) populations? – A case study with an endocrine disrupting chemical. Environ. Sci. Technol. 45, 4166–4172.

Brown, A.R., Bickley, L.K., Ryan, T.A., 2012. Differences in sexual development in inbred and outbred zebrafish (*Danio rerio*) and implications for chemical testing. Aquat. Toxicol. 112–113, 27–38.

Burkhard, L.P., Cowan-Ellsberry, C., Embry, M.R., et al., 2011a. Bioaccumulation data from laboratory and field studies: are they comparable? Integr. Environ. Assess. Manag. 8, 13–16.

Burkhard, L.P., Arnot, J.A., Embry, M.R., et al., 2011b. Comparing laboratory- and field-measured biota–sediment accumulation factors. Integr. Environ. Assess. Manag. 8, 32–41.

Cardoso, P.G., Lillebø, A.I., Pereira, E., et al., 2009. Different mercury bioaccumulation kinetics by two macrobenthic species: the bivalve *Scrobicularia plana* and the polychaete *Hediste diversicolor*. Mar. Environ. Res. 68, 12–18.

Carman, K.R., Fleeger, J.W., Pomarico, S.M., 2000. Does historical exposure to hydrocarbon contamination alter the response of benthic communities to diesel contamination. Mar. Environ. Res. 49, 255–278.

Casado-Martinez, M.C., Smith, B.D., Rainbow, P.S., 2013. Assessing metal bioaccumulation from estuarine sediments: comparative experimental results for the polychaete *Arenicola marina*. J. Soils Sediments 13, 429–440.

Cesnaitis, R., Sobanska, M.A., Versonnen, B., et al., 2014. Analysis of the ecotoxicity data submitted within the framework of the REACH Regulation. Part 3. Experimental sediment toxicity assays. Sci. Total Environ. 475, 116–122.

Chapman, P.M., 2002. Integrating toxicology and ecology: putting the "eco" into ecotoxicology. Mar. Pollut. Bull. 44, 7–15.

Chapman, P.M., Paine, M.D., Moran, T., et al., 1994. Refinery water (intake and effluent) quality: update of 1970s with 1990s toxicity testing. Environ. Toxicol. Chem. 13, 897–909.

Coe, T.S., Hamilton, P.B., Griffiths, A.M., et al., 2009. Genetic variation in strains of zebrafish (*Danio rerio*) and the implications for ecotoxicology studies. Ecotoxicology 18, 144–150.

Coelho, J.P., Nunes, M., Dolbeth, M., et al., 2008. The role of two sediment-dwelling invertebrates on the mercury transfer from sediments to the estuarine trophic web. Estuar. Coast. Shelf Sci. 78, 505–512.

Dauvin, J.-C., Ruellet, T., 2007. Polychaete/amphipod ratio revisited. Mar. Pollut. Bull. 55, 215–224.

Dauvin, J.-C., Ruellet, T., 2009. The estuarine quality paradox: Is it possible to define an ecological quality status for specific modified and naturally stressed estuarine ecosystems? Mar. Pollut. Bull. 59, 38–47.

Davey, J.T., 1993. Macrofaunal community bioturbation along an estuarine gradient. Neth. J. Aquat. Ecol. 27, 147–153.

Di, S., Liu, T., Diao, J., et al., 2013. Enantioselective bioaccumulation and degradation of sediment-associated metalaxyl enantiomers in *Tubifex tubifex*. Agric. Food Chem. 61, 4997–5002.

Dowse, R., Tang, D., Palmer, C.G., et al., 2013. Risk assessment using the species sensitivity distribution method: data quality versus data quantity. Environ. Toxicol. Chem. 32, 1360–1369.

Feckler, A., Zubrod, J.P., Thielsch, A., et al., 2014. Cryptic species diversity: an overlooked factor in environmental management? J. Appl. Ecol. 51, 958–967.

Galloway, T.S., Brown, R.J., Browne, M.A., et al., 2006. The ECOMAN project: a novel approach todefining sustainable ecosystem function. Mar. Pollut. Bull. 53, 186–194.

Gillet, P., Mouloud, M., Durou, C., et al., 2008. Response of *Nereis diversicolor* population (Polychaeta, Nereididae) to the pollution impact – Authie and Seine estuaries (France). Estuar. Coast. Shelf Sci. 76, 201–210.

Han, T., Kong, J.A., Kang, H.G., et al., 2011. Sensitivity of spore germination and germ tube elongation of *Saccharina japonica* to metal exposure. Ecotoxicology 20, 2056–2068.

Haap, T., Köhler, H.-R., 2009. Cadmium tolerance in seven *Daphnia magna* clones is associated with reduced hsp70 baseline and induction. Aquat. Toxicol. 94, 131–137.

Herlory, O., Bonzom, J.-M., Gilbin, R., et al., 2013. Use of diatom assemblages as biomonitor of the impact of treated uranium mining effluent discharge on a stream: case study of the Ritord watershed (Center-West France). Ecotoxicology 22, 1186–1199.

Hummel, H., Colucci, F., Bogaards, R.H., et al., 2001. Genetic traits in the bivalve *Mytilus* from Europe, with an emphasis on Artic populations. Polar Biol. 24, 44–52.

ICPR, 2008. Environmental Protection: The Concept and Use of Reference Animals and Plants. ICPR Publication 108. Ann. ICPR 38.

Janssen, C.R., Heijerick, D.G., 2003. Algal toxicity tests for environmental risk assessment of metals. Rev. Environ. Contam. Toxicol. 178, 23–52.

Jorgenson, Z.G., Buhl, K., Bartell, S.E., et al., 2014. Do laboratory species protect endangered species? Interspecies variation in responses to 17β-estradiol, a model endocrine active compound. Arch. Environ. Contam. Toxicol. http://dx.doi.org/10.1007/s00244-014-0076-9.

Kelly, M.G., 1998. Use of the trophic diatom index to monitor eutrophication in rivers. Wat. Res. 32, 236–242.

Kelly, M.G., Whitton, B.A., 1995. The trophic diatom index :a new index for monitoring eutrophication in rivers. J. Appl. Phycol. 7, 433–444.

Kelly, M., Bennett, C., Coste, M., et al., 2009. A comparison of national approaches to setting ecological status boundaries in phytobenthos assessment for the European Water Framework Directive: results of an intercalibration exercise. Hydrobiologia 621, 169–182.

Kirby, M.F., Lyons, B.P., Waldock, M.J., et al., 2000. Biomarkers of polycyclic aromatic hydrocarbon (PAH) exposure in fish and their application in marine monitoring. Sci. Ser. Tech. Rep. CEFAS, Lowestoft. 110, 30.

Lakeman, M.B., Von Dassow, P., Cattolico, R.A., 2009. The strain concept in phytoplankton ecology. Harmful Algae 8, 746–758.

Lagauzère, S., Pischedda, L., Cuny, P., et al., 2009. Influence of *Chironomus riparius* (Diptera, Chironomidae) and *Tubifex tubifex* (Annelida, Oligochaeta) on oxygen uptake by sediments. Consequences of uranium contamination. Environ. Pollut. 157, 1234–1242.

Lagauzère, S., Motelica-Heino, M., Viollier, E., et al., 2013. Remobilisation of uranium from contaminated freshwater sediments by bioturbation. Biogeosciences 11, 3381–3396.

Larras, F., Bouchez, A., Rimet, F., et al., 2012. Using bioassays and species sensitivity distributions to assess herbicide toxicity towards benthic diatoms. PLoS One 7 (8), e44458. http://dx.doi.org/10.1371/journal.pone.0044458.

Leboulanger, C., Schwartz, C., Somville, P., et al., 2011. Sensitivity of two mesocyclops (Crustacea, Copepoda, Cyclopidae), from tropical and temperate origins, to the herbicides, diuron and paraquat, and the insecticides temephos and fenitrothion. Bull. Environ. Contam. Toxicol. 87, 487–493.

Leung, J., 2014. Implications of Copper and Nickel Exposure to Different Members of the *Hyalella Azteca* Species Complex. Master of Science in biology thesis. Univ. of Waterloo, Ontario, Canada. 92 pp.
Liu, C., Wang, B., Xu, P., et al., 2013. Enantioselective determination of triazole fungicide epoxiconazole bioaccumulation in Tubifex based on HPLC-MS/MS. J. Agric. Food Chem. 15, 360–367.
Lyons, B.P., Bignell, J., Stentiford, G.D., et al., 2004. The viviparous blenny (*Zoarces viviparus*) as a bioindicator of contaminant exposure: application of biomarkers of apoptosis and DNA damage. Mar. Environ. Res. 58, 757–761.
Major, K., Soucek, D.J., Giordano, R., et al., 2013. The common ecotoxicology laboratory strain *Hyalella azteca* is genetically distinct from most wild strains sampled in Eastern North America. Environ. Toxicol. Chem. 32, 2637–2647.
Maréchal, J.-P., Hellio, C., 2011. Antifouling activity against barnacle cypris larvae: do target species matter (*Amphibalanus amphitrite* versus *Semibalanus balanoides*)? Int. Biodeter. Biodegrad. 65, 92–101.
Menchaca, I., Belzunce, M.J., Franco, J., et al., 2010. Sensitivity comparison of laboratory-cultured and field-collected Amphipod *Corophium multisetosum* in toxicity tests. Bull. Environ. Contam. Toxicol. 84, 390–394.
Miller, A., Nyberg, E., Danielsson, S., et al., 2014. Comparing temporal trends of organochlorines in guillemot eggs and Baltic herring: advantages and disadvantages for selecting sentinel species for environmental monitoring. Mar. Environ. Res. 100, 38–47.
NAS, 1980. The International Mussel Watch: Report of a Workshop Sponsored by the Environment Studies Board. Natural Resources Commission of National Academy of Sciences, Washington, DC.
Nunes, B.S., Carvalho, F.D., Guilhermino, L.M., et al., 2006. Use of the genus *Artemia* in ecotoxicity testing. Environ. Pollut. 144, 453–462.
OECD (Organization for Economic Co-operation and Development), 2002. OECD guidelines for the testing of chemicals. In: Proposal for Updating Guideline 201. Freshwater Alga and Cyanobacteria, Growth Inhibition Test. Draft Revised Guideline 201, July 2002.
OECD (Organization for Economic Co-operation and Development), 2011. OECD guideline for testing of chemicals. In: Guideline 234: Fish, Sexual Development Test. Paris, France, Adopted 28 July 2011.
OEHHA (Office of Environmental Health Hazard Assessment), 2004. Overview of Freshwater and Marine Toxicity Tests: A Technical Tool for Ecological Risk Assessment. http://oehha.ca.gov/ecotox/pdf/marinetox3.pdf (last accessed 09.01.14).
Olivia, M., Vicente-Martorell, J.J., Galindo-Riaño, M.D., et al., 2013. Histopathological alterations in Senegal sole, *Solea senegalensis*, from a polluted Huelva estuary (SW, Spain). Fish Physiol. Biochem. 39, 523–545.
OSPAR Commission, 2012. JAMP Guidelines for Monitoring Contaminants in Biota. http://www.ospar.org/v_measures/browse.asp?menu=01290301120125_000002_000000.
Orvain, F., 2005. A model of sediment transport under the influence of surface bioturbation: generalisation to the facultative suspension-feeder *Scrobicularia plana*. Mar. Ecol. Prog. Ser. 286, 43–56.
Pang, J., Sun, B., Li, H., et al., 2012. Influence of bioturbation on bioavailability and toxicity of PAHs in sediment from an electronic wasre recycling site in South China. Ecotoxicol. Environ. Saf. 84, 227–233.
Pieterek, T., Pietrock, M., 2012. Comparative selenium toxicity to laboratory-reared and field-collected *Hyalella azteca* (Amphipoda, Hyalellidae). Water Air Soil Pollut. 223, 4245–4252.
Przytarska, J.E., Sokolowski, M.W., Hummel, H., et al., 2010. Comparison of trace metal bioavailabilities in European coastal waters using mussels from *Mytilus edulis* complex as biomonotors. Environ. Monit. Assess. 166, 461–476.
Rocha-Olivares, A., Fleeger, J.W., Foltz, D.W., 2004. Differential tolerance among cryptic species: a potential cause of pollutant-related reductions in genetic diversity. Environ. Toxicol. Chem. 23, 2132–2137.
Rodrigues, E.T., Pardal, M.Â., 2014. The crab *Carcinus maenas* as a suitable experimental model in ecotoxicology. Environ. Int. 70, 158–182.
Rovira, L., Trobajo, R., Ibáñez, C., 2012. The use of diatom assemblages as ecological indicators in highly stratified estuaries and evaluation of existing diatom indices. Mar. Pollut. Bull. 64, 500–511.
Schaller, J., 2014. Bioturbation/bioirrigation by *Chironomus plumosus* as main factor controlling elemental remobilization from aquatic sediments. Chemosphere 107, 336–343.

Siaut, M., Cuiné, S., Cagnon, C., et al., 2011. Oil accumulation in the model green alga *Chlamydomonas reinhardtii*: characterization, variability between common laboratory strains and relationship with starch reserves. BMC Biotechnol. 11, 7–27.

Snell, T.W., Carmona, M.J., 1995. Comparative toxicant sensitivity of sexual and asexual reproduction in the rotifer *Brachionus calyciflorus*. Environ. Toxicol. Chem. 14, 415–420.

Stentiford, G.D., Longshaw, M., Lyons, B.P., et al., 2003. Histopathological biomarkers in estuarine fish species for the assessment of biological effects of contaminants. Mar. Environ. Res. 55, 137–159.

Sukumaran, S., Grant, A., 2013a. Multigenerational demographic responses of sexual and asexual *Artemia* to chronic genotoxicity by a reference mutagen. Aquat. Toxicol. 144–145, 66–74.

Sukumaran, S., Grant, A., 2013b. Differential responses of sexual and asexual *Artemia* to genotoxicity by a reference mutagen: Is the comet assay a reliable predictor of population level responses? Ecotox. Environ. Toxicol. 91, 110–116.

Tarazona, J.V., Sobanska, M.A., Cesnaitis, R., et al., 2014. Analysis of the ecotoxicity data submitted within the framework of the REACH Regulation. Part 2. Experimental aquatic toxicity assays. Sci. Tot. Environ. 472, 137–145.

USEPA, 1996. Ecological Effects Test Guidelines. OPPTS 850.5400. Algal Toxicity, Tiers I and II. 11 pp.

USEPA, 2000. Guidance for Assessing Chemical Contaminant Data for Use in Fish Advisories. Fish Sampling and Analysis. EPA 823-B-00–007. vol. 1. http://water.epa.gov/scitech/swguidance/fishshellfish/techguidance/risk/upload/2009_04_23_fish_advice_volume1_v1cover.pdf.

USEPA, 2002. Methods for Measuring the Acute Toxicity of Effluents and Receiving Waters to Freshwater and Marine Organisms. http://water.epa.gov/scitech/methods/cwa/wet/upload/2007_07_10_methods_wet_disk2_atx1-6.pdf.

Van Dam, R.A., Harford, A.J., Houston, M.A., et al., 2008. Tropical marine toxicity testing in Australia: a review and recommendations. Australas. J. Ecotoxicol. 14, 55–88.

Warming, T.P., Mulderij, G., Christoffersen, K.S., 2009. Clonal variation in physiological responses of *Daphnia magna* to the strobilurin fungicide azoxystrobin. Env. Toxicol. Chem. 28, 374–380.

Wellborn, G.A., Cothran, R.D., 2004. Phenotypic similarity and differentiation among sympatric cryptic species in a freshwater amphipod complex. Freshwater Biol. 4, 1–13.

Weston, D.P., Poynton, H.C., Wellborn, G.A., et al., 2013. Multiple origins of pyrethroid insecticide resistance across the species complex of a nontarget aquatic crustacean, *Hyalella azteca*. Proc. Natl. Acad. Sci. USA 110, 16532–16537.

CHAPTER 10

Endobenthic Invertebrates as Reference Species

Claude Amiard-Triquet, Brigitte Berthet

Abstract

Endobenthic species living in close contact with sediment are particularly at risk in contaminated environments. Being relatively sedentary, they cannot avoid deteriorating sediment quality conditions. They are representative of different zoological groups that contribute greatly to ecosystem structure and functioning. Bioaccumulation studies give access to the lability of sediment-bound contaminants, whereas associated risks for biota can be revealed by using biomarkers. Endobenthic invertebrates feed at a low level of the trophic web, but as preys, they have a crucial role in the trophic transfer of contaminants and toxicity. Diverse species exhibit different tolerances to stress, thus providing useful indicators of the health status of communities. For standardized bioassays, numerous species are commercially available, but the use of local species may be more relevant for a precise identification of chemical risk in a local area.

Keywords: Bioaccumulation; Bioassays; Bioavailability; Bioindicators; Biomarkers; Trophic transfer.

Chapter Outline

Aquatic sediments represent the major sink for the storage of chemical contaminants entering the environment with only few exceptions (e.g., perfluorooctanoic acid, water-soluble pesticides such as alachlor, atrazine, diuron). Consequently, organisms living in close contact with sediments in contaminated environments are particularly at risk, even if many factors are able

Aquatic Ecotoxicology. http://dx.doi.org/10.1016/B978-0-12-800949-9.00010-3

to affect the bioavailability of contaminants (Chapter 5). Approximately 10% of sediments underlying US surface water are sufficiently contaminated with toxic pollutants to pose potential risks to wildlife (ITRC, 2011). This is of particular concern because many endobenthic species are important in the structure and functioning of aquatic ecosystems: (1) exhibiting high densities and biomasses; (2) used as food by many other species, among which some are important for commerce (e.g., fish) or ecotourism and cultural heritage (e.g., wading birds); and (3) having a role as ecosystem engineers (e.g., influence of bioturbation on environmental cycles of nutrients and contaminants).

Endobenthic organisms are relatively sedentary, so they cannot avoid deteriorating sediment quality conditions. They include diverse species that exhibit different tolerances to stress. They are representative of different zoological groups including annelids, bivalves, and crustaceans that contribute greatly to aquatic ecosystems. Among annelids, oligochaetes are mainly present in freshwater, whereas polychaetes are mainly marine organisms. Insects, among which *Chironomus* larvae are widely used in ecotoxicological studies, are mainly present in freshwater, whereas echinoderms are strictly marine. On the other hand, nematodes, bivalves, and crustaceans (particularly amphipods, see Chapter 11) are ubiquitously distributed.

In this chapter, the use of endobenthic species will be reviewed considering the different tools used in ecotoxicological studies: (1) chemical measurements of contaminants in environmental matrices; (2) experimental sediment toxicity assays; (3) biological responses at infraindividual and individual levels of biological organization used as biomarkers of presence/effects of contaminants; and (4) in situ biological assessment to measure effects at the supraindividual level.

10.1 Endobenthic Species as Bioaccumulator Species

10.1.1 Bioavailability of Sediment-Bound Contaminants

Many reports have established that the total concentration of a given contaminant in aquatic sediments is not predictive of the bioaccumulation and effects that may be observed in biota living in this sediment (Chapter 5). A striking example is provided by Berthet et al. (2003), who studied metal bioaccumulation in the ragworm *Nereis* (*Hediste*) *diversicolor* living in sediments contrasting by their level of total metal concentrations. Metal concentrations in sediments at the contaminated site were orders of magnitude higher than in the comparatively clean site, whereas metal concentrations in biota were similar (Table 10.1). This discrepancy was due to the lack of metal bioaccessibility (usually called bioavailability; see Chapter 5) because no metal release was observed when these sediments were exposed at pH as low as 2 for Cd and Cu and 4 for Zn.

Endobenthic organisms can be exposed to sediment according to different modes. Deposit-feeders obtain their food mainly from deposited sediment, even if some bivalves are also able

Table 10.1: Metal Concentrations (μg g^{-1} d.w.) in Sediments and Ragworms *Nereis (Hediste) diversicolor* from a Relatively Clean (Bay of Somme) or a Contaminated Site (Boulogne Harbor)

	Sediment		Ragworms		Biota-Sediment Accumulation Factor	
Metal	Bay of Somme	Boulogne Harbor	Bay of Somme	Boulogne Harbor	Bay of Somme	Boulogne Harbor
Cd	0.05	0.38	0.049	0.055	0.98	0.14
Cu	0.64	153	11	15	17.19	0.09
Zn	7.3	1092	105	88	14.38	0.08

After Berthet et al. (2003).

to collect particles from the water column at high tide (e.g., *Macoma balthica*, *Scrobicularia plana*). Filter-feeders obtain their food mainly from the water column but in doing so, they also ingest sedimentary particles that are resuspended through a range of natural and anthropogenic processes (Roberts, 2012).

Direct determination of bioavailability (*stricto sensu*; see Chapter 5) through the assessment of bioaccumulation is preferred to chemical-based approaches by several authors. Using the traditional 28-d bioaccumulation test with the aquatic oligochaete *Lumbriculus variegatus*, Pickard and Clarke (2008) have quantified the benthic bioavailability of 11 polychlorinated dibenzo-*p*-dioxin/F congeners in surficial Lake Ontario sediments through the calculation of biota to sediment accumulation factors. The same procedure was applied for perfluorochemicals (Higgins et al., 2007) and polybrominated diphenyl ethers (Lotufo and Pickard, 2010). For metals, Tan et al. (2013) have proposed to assess bioavailability through the assimilation into the systemic circulation of endobenthic sipunculan worms. In fact, these strategies are at the basis of biomonitoring based on the use of bioaccumulator species.

10.1.2 Biomonitoring

The use of biological matrices in environmental monitoring has been widely developed in the water column (The International Mussel Watch initiated by Goldberg, 1975). In complementarity with biomonitors of metal availabilities in the water column (macroalgae, filter-feeding bivalves), sediment-dwelling deposit feeders are able to reveal the bioavailable concentrations of metals in sediments (Coelho et al., 2010; Rainbow et al., 2011). To ensure practicability, the sentinel species must be present in abundance, of reasonable size, easy to identify and collect, and tolerant of exposure to environmental variations in physicochemical parameters (Chapter 9). In the case of metals, biomonitors should not regulate the total concentration of an element in their body tissues exposed to different metal bioavailabilities, a defense mechanism that has been shown, for instance, in *N. diversicolor* exposed to zinc (Berthet et al., 2003; Rainbow et al., 2011).

Polychaetes are frequently considered for sediment biomonitoring purposes, particularly *Nereis* spp. The absence of bioaccumulation of Cd, Cu, and Zn over background was observed at some sites with very high metal concentrations in sediments (Saiz-Salinas and Francès-Zubillaga, 1997; Berthet et al., 2003; Rainbow et al., 2011), depending either on a lack of bioavailability or on body level regulation. In agreement with the known potential of organometallics to bioaccumulate preferentially, sediment methylmercury was more bioavailable than inorganic Hg in *N. diversicolor* and *N. virens* (Muhaya et al., 1997; Amirbahman et al., 2013). Despite it living in less muddy sediments, with lower metal concentrations, *Arenicola marina* has also been used as biomonitor species concurrently with *N. diversicolor* (e.g., Bird et al., 2011). The polychaete *Chaetopterus variopedatus* has been validated as a good biomonitor organism for many trace and major elements in coastal regions in Brazil (Eça et al., 2013).

Both *Nereis* species as well as *Neanthes arenaceodentata* exposed to native organic contaminants (polyaromatic hydrocarbons [PAHs], polychlorinated biphenyl [PCBs]) bound to natural sediments collected from contaminated sites seem also able to reveal the bioavailable concentrations in sediments (Vinturella et al., 2004; Cornelissen et al., 2006; Janssen et al., 2010, 2011). As already mentioned for metals, the tissues concentrations are not only influenced by the bioavailability of contaminants but also by the biological responses enabling the organisms to cope with the presence of contaminants in their environment. Marine polychaetes exhibit a high biotransformation capacity for PAHs favoring their elimination (Driscoll and Mc Elroy, 1996; Christensen et al., 2002; Giessing et al., 2003; Jørgensen et al., 2005, 2008), thus leading to relatively low concentrations in worms compared with sediment (Durou et al., 2007). On the other hand, biotransformation of PCBs is much less efficient and concentration in worms is thus more higher than concentrations in sediments, particularly for the most stable congeners (CB-138, CB-153) compared with CB-50, -66, -77, and -101 that are much more reactive in biotransformation processes (Durou et al., 2007 and literature cited therein).

In freshwater, despite the practical problems linked to their smaller size, oligochaetes (*L. variegatus*) and chironomid larvae have been used for the monitoring of sediment-bound contaminants (e.g., Lasier et al., 2011; De Jonge et al., 2012). Because biotransformation of the PAH pyrene by *L. variegatus* is slow, the use of this species in standard bioaccumulation tests (Carrasco Navarro et al., 2011) and in biomonitoring programs may be recommended.

Many bivalves are bioaccumulators but several species, despite living within the sediment, are filter-feeders (freshwater bivalves: *Unio* spp., *Anodonta* spp., *Corbicula* spp.; estuarine and coastal bivalves: *Cerastoderma edule*, *Mercenaria mercenaria*, *Mya arenaria*, *Ruditapes* spp.) and consequently have less interest for the assessment of sediment-bound contaminants (Coelho et al., 2010). Deposit-feeding bivalves such as *S. plana* or *Macoma* spp. have been widely used for monitoring metal contamination in estuarine and coastal environments (Bryan et al., 1980; Amiard et al., 1987; Ruiz and Saiz-Salinas, 2000; Ridgway et al., 2003;

Cheggour et al., 2005; Hendozko et al., 2009; Bird et al., 2011; Rainbow et al., 2011; Amirbahman et al., 2013; Coelho et al., 2014), as have organochlorine contaminants (Grilo et al., 2013). Other deposit-feeding bivalves were selected for use in various local contexts (*Brotia costula* and *Melanoides tuberculata* in a Malaysian river, Lau et al., 1998; *Asaphis deflorata* in Hong Kong coastal sediments, Moganti et al., 2008; and *Amiantis umbonella* in Kuwait Bay, Tarique et al., 2013).

In the case of aquatic arthropods, the effect of molting on accumulated concentrations may be crucial (e.g., Bergey and Weis, 2007), thus it is needed to select intermolt individuals, a procedure that may be labor-intensive, particularly when species exhibiting small size are used (e.g., talidrid amphipods used as metal monitors, Fialkowski et al., 2009, and literature quoted therein).

In addition to legacy contaminants, benthic infaunal organisms could also encounter relatively high concentrations of emerging contaminants such as nanoparticles (Chapter 17). Thus several species commonly used for biomonitoring have already been tested for their ability to bioaccumulate metal-based nanoparticles such as *N. diversicolor* and *S. plana* (e.g., Cong et al., 2011; Mouneyrac et al., 2014) or the estuarine amphipod *Leptocheirus plumulosus* (Hanna et al., 2013).

10.1.3 Confounding Factors, Correction Factors

"Design and interpretation of biomonitoring studies require consideration of sample size, gut content, animal size, temporal variability and systematics" (Luoma and Rainbow, 2008). Because of their feeding regimen, when endobenthic species are collected in the field, their gut content includes an important fraction of sediment in which the contaminant concentrations are often higher than in the tissues. Thus, gut purging is indispensable for a correct assessment of contaminants really incorporated in organisms' tissues (Ingersoll et al., 1998). A relevant handling of confounding factors is crucial for a sound assessment of local bioavailabilities because they can be revealed by using bioaccumulators. The sources of variations and the remedies have been investigated in deep in Mussel Watch programs (NAS, 1980; Andral et al., 2004). The main sources of fluctuations include both extrinsic (temperature, salinity, seasonal fluctuations of exposure, etc.) and intrinsic (size, weight, age, reproductive status, etc.) factors.

From a review of the available literature on *S. plana*, Coelho et al. (2010) report that each researcher often uses just one single size class or does not even make reference to the size class studied, thus hampering intercomparison of different studies. They suggests that "having some knowledge of the lifespan and the annual bioaccumulation patterns of each species, results can then be extrapolated from one specific size class to another, and meaningful comparison between different studies becomes possible." Active biomonitoring is part of the response because it is possible to cage specimens of the same origin, same size, and possibly the same reproductive status (De Kock and Kramer, 1994). However, comparing sites with different trophic characteristics requires the use of modeling to adjust raw data on contamination for a reference individual, by making a clear distinction between physiological (growth) and contamination factors (Andral et al., 2004). This strategy has also been adapted for endobenthic polychaetes (Poirier et al., 2006).

Cautions concerning the choice of the sampling period in agreement with the known effect of reproductive status that has been well-established in Mussel Watch programs may be applied provided that the reproductive cycle has been fully described as it is the case for *S. plana* (Rodríguez-Rúa et al., 2003; Mouneyrac et al., 2008) and is not too dissimilar along geographic clines as shown for the ragworm *N. diversicolor* (Gillet et al., 2008).

For species commonly used in Mussel Watch programs, background assessment concentrations are available for naturally occurring substances, whereas for man-made substances the background must be close to zero (OSPAR, 2012). Despite a number of endobenthic species having already been widely used, the dataset is much more limited and, in many cases, it is impossible to avoid the use of a reference. This may be either organisms commercially available or organisms collected from the wild at a reference site, as clean as possible, because "pristine" sites no longer exist, and showing a maximum similarity with the site under investigation.

10.1.4 Trophic Transfer

In a recent book, Coelho et al. (2010) advocate the importance of using sediment-dwelling biomonitors, highlighting their role in transferring contaminants from sediment to higher trophic levels. Kalantzi et al. (2014) have highlighted that Hg is biomagnified and P is bioaccumulated in wild fish through zoobenthos. Boyle et al. (2010) have shown that Pb naturally incorporated in the ragworm *N. diversicolor* collected from an estuary with legacy Pb is bioavailable to the zebrafish *Danio rerio*. Relationships between subcellular metal distribution (metal-rich granules, cellular debris, organelles, metallothionein-like proteins, and other (heat-sensitive) proteins) of prey and metal trophic transfer to a predator gave rise to an important body of literature. Many endobenthic species have been studied from this point of view such as the freshwater oligochaetes *Limnodrilus hoffmeisteri* (Wallace and Lopez, 1997; Wallace et al., 1998) and *L. variegatus* (Ng and Wood, 2008), the larval chironomids *Chironomus riparius* (Béchard et al., 2008), the marine polychaete *N. diversicolor* (Rainbow et al., 2006a; Rainbow and Smith, 2013), and the bivalves *M. balthica* and *Potamocorbula amurensis* (Wallace and Luoma, 2003).

In addition to metal trophic transfer, a trophic transfer of toxicity has been frequently observed in predators fed the polychaete *N. diversicolor* collected from metal-impacted estuaries. This includes the breakdown of the regulation of body copper concentrations by the decapod crustacean *Palaemonetes varians* (Rainbow and Smith, 2013) and in some cases significant mortality (Rainbow et al., 2006b); the decreased reproductive output of zebrafish *D. rerio* attributed to arsenic uptake (Boyle et al., 2008) and significant mortality of the marine fish *Terapon jurbua terepon* (Dang et al., 2012). These authors have also shown that more fish died after exposure to metals naturally incorporated in the clam *S. plana* than in those fed *N. diversicolor* originating from the same metal-impacted estuaries. This difference was attributed to interspecies differences in metal doses. The oligochaetes *L. hoffmeisteri* and *L. variegatus* not only transferred cadmium to their predators the grass shrimp *Palaemonetes pugio* and the rainbow trout *Oncorhynchus mykiss*, respectively, but also Cd toxicity, reducing prey capture in grass shrimp (Wallace et al., 2000) and reducing growth by 50% in the trout exposed to the highest Cd dose (Ng and Wood, 2008).

Among organic contaminants, sediment-associated fluoranthene can be transferred through the diet as shown for grass shrimp *P. pugio* fed tubificid oligochaetes *Monopylephorus rubroniveus* (Filipowicz et al., 2007). Using the invertebrates *L. variegatus* or *C. riparius* as prey, Carrasco Navarro et al. (2012, 2013) have shown a trophic transfer of pyrene metabolites to predators *Gammarus setosus* and *Salmo trutta*.

Using artificial sediment spiked with ^{14}C-labeled hexachlorobenzene, Egeler et al. (2001) have shown that hexachlorobenzene was biomagnified in a laboratory food chain including the oligochaete *Tubifex tubifex* and a fish, the three-spined stickleback *Gasterosteus aculeatus*. In agreement with the well-established potential of PCBs to biomagnify (Abarnou, 2009), their dietary transfer from naturally contaminated sediments has been shown in laboratory food chains consisting of sediments, polychaetes (*N. virens*) and the lobster *Homarus americanus* (Pruell et al., 2000), the predatory fish (*Leiostomus xanthurus*) (Rubinstein et al., 1984), and the Atlantic cod *Gadus morhua* (Ruus et al., 2012).

10.2 Biological Testing with Endobenthic Species

10.2.1 Predictive Risk Assessment of New and Existing Chemicals

Persistency, toxicity, and bioaccumulation are major criteria for the risk assessment of chemicals. Endobenthic species may be used for both bioaccumulation and toxicity tests. Sediment bioaccumulation tests have been standardized for marine (bivalves *Macoma* spp., *Yoldia limatula*, polychaetes *Nereis* spp., amphipods *Diporeia* spp.) and freshwater (oligochaete *L. variegatus*, *Stylodrilus heringianus*, *L. hoffmeisteri*, *Tubifex tubifex*, and *Pristina leidyi*, amphipod *Hyalella azteca*, chironomid insects) species. Bioaccumulation tests with sediments are usually conducted for 28 days with sediments either field-collected or spiked because the steady-state is generally reached within this duration (USEPA, 2000; ASTM, 2002). In fact, many other species were harnessed with the aim of linking bioaccessibility and bioaccumulation.

An analysis of the ecotoxicity data submitted within the framework of the European regulation Registration, Evaluation, Authorization and Restriction of Chemicals has been recently published (Cesnaitis et al., 2014). In the marine context, the amphipod crustacean *Corophium volutator* is the most commonly used species (Figure 10.1). Freshwater endobenthic species are representative of different taxa including nematodes (*Caenorhabditis elegans*), oligochaetes (*L. variegatus, T. tubifex*), the amphipod crustacean *H. azteca*, and insect larvae belonging to different *Chironomus* species (Figure 10.1).

All these species are used in standard toxicity tests listed in Table 10.2. The exposure pathway may be sediments and in some cases, the test substance can be introduced via overlying water (intended to simulate a chemical spray event and to cover the initial peak of concentrations in porewater). According to the Technical Guidance Document on risk assessment in support of European legislation (TGD, 2003), only whole-sediment tests using benthic organisms are suitable for a realistic risk assessment of the sediment

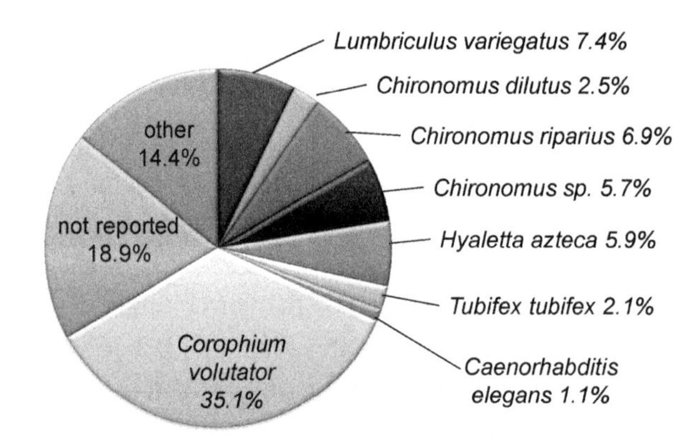

Figure 10.1
Endobenthic species used in single-species bioassays within the framework of the Registration, Evaluation, Authorization and Restriction of Chemicals regulation in Europe. *After Cesnaitis et al. (2014), with permission.*

Table 10.2: Endobenthic Species Used in Standard Toxicity Tests

Test Organism	Standard Toxicity Test
Corophium spp. (including *volutator*)	OSPAR Commission (2005)
Chironomus spp.	OECD 218, OECD 219
	USEPA OPPTS 850.1735, USEPA OPPTS 850.1790
	ASTM E1706-04, ASTM E1706-95b
Lumbriculus variegatus	OECD 225
Hyalella azteca	USEPA OPPTS 850. 1735,
	ASTM E1706-04, ASTM E1706-95b
Tubifex tubifex	ASTM E1706-04, ASTM E1706-95b
Caenorhabditis elegans	ISO 10872

ISO, International Organization for Standardization; OECD, Organization for Economic Co-operation and Development; OPPTS, Office of Chemical Safety and Pollution Prevention; USEPA, US Environmental Protection Agency.

compartment. However, contaminants present in porewater or interstitial water that occupy the spaces between sediment particles are also of interest because they represent the water-soluble, bioavailable fraction, which is a major route of exposure to benthic organisms (Hartl, 2010). Thus, procedures for the collection of interstitial water have been launched by the USEPA (2001).

Some recent articles have integrated procedures allowing an improvement of toxicity assessment (Chapter 6) as underlined, for instance, by Du et al. (2013): (1) characterization of exposure considering the bioavailability of the substance tested; (2) chronic exposures to low levels of contaminants that are more environmentally relevant; and (3) whole life-cycle toxicity testing examining not only survival, but also growth, emergence, and reproduction.

Apart from accidents and pulse exposures (intermittent discharge of chemical in the environment) occur as a consequence of surface runoff in agricultural or urban areas. Because exposure is not constant, standardized testing is not appropriate in this case (Chèvre and Valloton, 2013). Lizotte et al. (2012) have proposed a strategy associating a spatiotemporal study of the fate of sediment-bound pesticides based on an experiment in the field and an assessment of the effects of these contaminated sediments by using a traditional 10-day sediment toxicity to *H. azteca*.

10.2.2 Retrospective Risk Assessment Based on Sediment Bioassays

The various types of aquatic ecotoxicological tests that may be used in environmental assessment have been described in Chapter 6. Toxicity bioassays using whole sediment collected from the environment have been developed for many relevant taxa; carrying this type of bioassays needs a careful approach because sediment handling modifies the natural characteristics of sediment, then being able to influence the bioavailability of sediment-bound contaminants and their toxicity (Hartl, 2010). Aging of sediment in the laboratory before use in the tests may also influence biological effects as shown in behavioral tests with the clam *S. plana*. Burrowing speed decreased significantly when clams were exposed to sediment kept in the laboratory for 4, 7, and 23 days (Boldina-Cosqueric et al., 2010). The presence of elevated concentrations of ammonia (resulting from aging and bacterial decomposition of organic matter) can interfere with the effects of the tested contaminants (Galluba et al., 2012).

Despite these technical problems, sediment bioassays are sensitive tools able to reveal the presence of sediments at risk in natural areas. Some examples among many include toxicity of sediment-associated pesticides in agriculture-dominated water bodies of California's central valley (Weston et al., 2004), an urban stream in Guangzhou, China (Li et al., 2013), and in a multipolluted sediment toxicity assessment of Hessian (Germany) surface waters (Galluba et al., 2012) or coastal harbors (Gomes et al., 2014).

10.3 Biomarkers in Endobenthic Invertebrates

The different kinds of biomarkers that may be determined in aquatic biota have been reviewed in Chapters 7 and 8. For endobenthic species, relationships with their substrate are particularly important: avoidance toward contaminated sediments or impairment of burrowing may have consequences for the survival of individuals. Behavioral disturbances have been frequently observed in sediment contaminated in the field or at experimental concentrations that can be encountered in impacted environments (Amiard-Triquet and Amiard, in Amiard-Triquet et al., 2013a). Possibilities of automated behavioral monitoring is an additional argument favoring the use of behavioral biomarkers (Bae and Park, 2014) such as the Multispecies Freshwater Biomonitor (LimCo International GmbH, http://www.limco-int.com) used for instance to study the effects of the water accommodated fraction of weathered crude oil on the behavior of the amphipod crustacean *C. volutator* (Kienle and Gerhardt, 2008) or of the pesticide imidacloprid on the behavior of the aquatic oligochaete *L. variegatus* (Sardo and Soares, 2010).

Numerous biomarkers may be determined in bivalve species, most often using mussels (Davies and Vethaak, 2012). However, biomarkers at different levels of biological organization (subcellular, tissue, or individual responses) may be suitable in endobenthic species. Because Byrne and O'Halloran (2001) and Bebianno et al. (2004) have reviewed the responsiveness of many different biomarkers in bivalve molluscs (the deposit-feeding clam *S. plana* and the filter-feeding *Tapes semidecussatus* and *Ruditapes decussatus*), these clams or nearly allied species gave rise to many studies examining their effectiveness for estuarine and coastal sediment toxicity testing (Table 10.3).

The estuarine clam *Mya arenaria* has been widely used as a matrix for the determination of biomarkers, particularly in the St. Lawrence estuary and Saguenay Fjord, as reviewed by Gagné et al. (2009). New developments are still in progress, including the influence of confounding factors on core biomarkers (Greco et al., 2011) and DNA integrity assessment in clams collected at sites under anthropogenic pollution in the Saguenay (Debenest et al., 2013).

In freshwater, different species of unionids are present worldwide and have been used by many researchers, both in experiments and field studies, for the determination of a large range of biomarkers. However, because their behavior toward the sediment is ambiguous (small depth, depending on the age of specimens and hydrodynamism of the medium), different researchers have not given the same credit to these species concerning their representativeness toward the sediment compartment. Cossu et al. (2000), Guidi et al. (2010), and Marasinghe Wadige et al. (2014) consider that *Unio* spp. and *Hyridella australis* have a strong significance for assessing effects of sediment-bound contaminants. On the other hand, the term "sediment" does not appear in many articles reporting biomarker responses in *Unio* spp., *Anodonta cygnea, Ellipsio complanata, Batissa violacea*, or *Lasmigona costata* (Štambuk

Table 10.3: Responsiveness of Different Biomarkers in Deposit-Feeding Endobenthic Bivalves

Biological Model	Mode of Exposure	Contaminant	Biological Response	References
Ruditapes decussatus	Spiked water	Cadmium Zinc	MT	Serafim and Bebianno (2007a,b)
R. decussatus	Spiked sediment	Phenanthrene Benzo[b] fluoranthene	Oxidative stress DNA damage	Martins et al. (2013)
R. decussatus	Field study		Genotoxicity	Almeida et al. (2013)
R. decussatus	Field study	PAHs	Oxidative stress	Barreira et al. (2007)
R. decussatus	Field study	Impacted estuaries vs cleaner sites	Histopathological alterations	Costa et al. (2013)
Ruditapes philippinarum	Spiked water	Triclosan Fluoxetine Endosulfan	VTG, SOD, AChE Immune parameters, AChE EROD, oxidative stress	Matozzo et al. (2012) Munari et al. (2014) Tao et al. (2013)
R. philippinarum	Whole sediment bioassay	Oil from oil spills	EROD	Moralles-Caselles et al. (2008)
R. philippinarum	Whole sediment bioassay	Nonylphenol, metals	Antioxidant enzymes	Won et al. (2012b)
Tapes semidecussatus	Whole sediment bioassay		Genotoxicity	Hartl et al. (2004)
Scrobicularia plana	Spiked water	Lindane	Histopathological alterations	Gonzalez De Canales et al. (2009)
S. plana	Field study	Metals Sewage discharge, industrial effluent Mercury gradient Fish farm effluent Different patterns of estuarine pollution Multipolluted estuary	MT, glyoxalase II inhibition Oxidative stress (defense/damage) Oxidative stress, immunosuppression GPx, LPO, DNA damage Intersex (histology and transcriptomic analysis) GST, behavior, condition, GSI	Romero-Ruiz et al. (2008) Bergayou et al. (2009) Ahmad et al. (2011a,b) Silva et al. (2012) Fossi Tankoua et al. (2012b) Ciocan et al. (2013) Fossi Tankoua et al. (2013)
Macoma balthica	Spiking	HBCDD (BFR)	MN, nuclear abnormalities	Smolarz and Berger (2009)
M. balthica	Spiked water	Ag, Cd, Hg	MT	Mouneyrac et al. (2000)
M. balthica	Whole sediment bioassay	Harbor sediment	Histopathology	Tay et al. (2003)

Continued

Table 10.3: Responsiveness of Different Biomarkers in Deposit-Feeding Endobenthic Bivalves—cont'd

Biological Model	Mode of Exposure	Contaminant	Biological Response	References
M. balthica	Field study	Metals in Arctic sites PCBs, DDTs, metals Different patterns of pollution in the Baltic Sea	Antioxidant responses MT, GST, CAT MN	Regoli et al. (1998) Lehtonen et al. (2006) Baršienė et al. (2008)
Macoma nasuta	Whole sediment bioassay	Multipolluted San Francisco Bay	LMD, HSP 70, histopathology	Werner et al. (2004)
Tellina deltoidalis	Spiked sediment	Cadmium, lead	Oxidative stress	Taylor and Maher (2013, 2014)
Donax trunculus	Field study	Multiple pollution sources	Oxidative stress, GST, AChE, reproductive endpoints	Tlili et al. (2010, 2011), Sifi et al. (2013) and Bensouda and Soltani-Mazouni (2014)
Meretrix meretrix	Field study	Various anthropogenic inputs	GSHt, AChE, GPx, GST, CAT, SOD, TBARS	Meng et al. (2013)
Anadara trapezia	Spiked sediment	Lead	Oxidative stress	Taylor and Maher (2012)
Cerastoderma edule	Field study	Sewage discharges, industrial effluents	LPO, oxidative stress	Bergayou et al. (2009)

AChE, acetylcholinesterase activity; BFR, brominated flame retardant; CAT, catalase activity; EROD, ethoxyresorufin O-deethylase; GPx, glutathione peroxidase; GSHt, total glutathione; GSI, gonado-somatic index; GST, glutathione *S*-transferase activity; HBCDD, hexabromocyclododecane; HSP, heat-shock proteins; LMD, lysosomal membrane damage; LPO, lipid peroxidation; MN, micronucleus assay; MT, metallothionein; PAH, polyaromatic hydrocarbon; PCB, polychlorinated biphenyl; SOD, superoxide dismutase; TBARS, thiobarbituric acid reactive substances; VTG, vitellogenin.

et al., 2009; Köpröcö et al., 2010; Falfushynska et al., 2010, 2013; Farcy et al., 2011; Irinco-Salinas, 2012; Borkovíc-Mitíc et al., 2013; Vuković-Gačić et al., 2014; Irinco-Salinas and Pocsidio, 2014; Gillis et al., 2014).

The major categories of biomarkers corresponding to different classes of contaminants that may be determined in polychaetes have been recently reviewed (Amiard-Triquet et al., 2013b). Nereididae were among the most often used biological models. *Perinereis gualpensis* has been used in caging experiments in polluted and nonpolluted estuarine sediments (Díaz-Jaramillo et al., 2013) and *N. diversicolor* in a large field study with monthly sample collection from a reference and a contaminated site (Fossi Tankoua et al., 2012a). Using a battery of biomarkers, Catalano et al. (2012) concluded that *H. (Nereis) diversicolor* is an appropriate bioindicator of PAH contamination. In the polychaete *Perinereis nuntia* exposed either to copper or field sediment, several potential biomarker genes have been recognized as early warning signals for oxidative stress (Won et al., 2012a). Biomarkers sufficiently sensitive to identify exposure to PAHs and/or pharmaceuticals and their toxicity were shown in caged *A. marina* (Ramos-Gomez et al., 2011) and *H. diversicolor* (Maranho et al., 2014). In freshwater, in the absence of polychaetes, oligochaetes have been tested. From an experimental exposure to the herbicide isoproturon, Paris-Palacios et al. (2010) have suggested that autotomy may be a useful biomarker in *T. tubifex*. Many biomarkers at different levels of biological organization were shown as responsive in *L. variegatus* exposed in the laboratory to B[a]P, Cd, pentachlorophenol (transcriptional and metabolic changes), the carbamate pesticide carbaryl (cholinesterase, carboxylesterase), Pb (growth, behavior), and nanomaterials (catalase) (Kristoff et al., 2010; Sardo and Soares, 2011; Agbo et al., 2013; Wang et al., 2014). An additional step toward field application has been taken by exposing *L. variegatus* to atrazine not only in semistatic tests but also to urban river sediments in whole sediment bioassays, thus inducing responses of glutathione *S*-transferase activity (phase II detoxifying enzyme) and catalase activity (antioxidant enzyme) (Contardo-Jara and Wiegand, 2008). Field and laboratory studies have shown the responsiveness of different core biomarkers (cholinesterase, ethoxyresorufin O-deethylase, glutathione content glutathione, glutathione *S*-transferase) in *Limnodrilus* spp. (Ozdemir et al., 2011; Oztetik et al., 2013).

Among arthropods, the amphipod crustacean *H. azteca* is commonly used in applying the methodology of biomarkers (e.g., Vandenbergh and Janssen, 2001; Oviedo-Gómez et al., 2010; Bundschuh et al., 2011; Palmquist et al., 2011; Giusto et al., 2014), but its status as an endobenthic or an epibenthic species also seems unclear. Chironomid larvae have also been tested as biological models for the determination of biomarkers. Biomarkers of oxidative and genotoxic stress and larval growth were used to assess the effects of nanomaterials (Oberholster et al., 2011). Biotransformation enzymes and cholinesterases have been investigated for their responses to pesticides (e.g., Sturm and Hansen, 1999; Fisher et al., 2003), then endpoints linked to development and reproduction were considered as also gene expression and genotoxicity (e.g., Crane et al., 2002; Lee et al., 2008; Park and Choi, 2009). However, in situ assays remain limited (e.g., Maycock et al., 2003; Soares et al., 2005).

The use of integrated biomarker approaches and multimarker pollution indices has been described with some details in Chapter 7. This methodology has been developed for several species already mentioned in this section such as *N. diversicolor* (Bouraoui et al., 2010), *Ruditapes decussatus* (Cravo et al., 2012), *S. plana* (Fossi Tankoua et al., 2013) and *Macoma balthica* (Barda et al., 2014) as also other bivalve species such as *Meretrix meretrix* (Meng et al., 2013) and *Donax trunculus* (Tlili et al., 2013). Also in Chapter 7, the problem of confounding factors has been reviewed in full. For endobenthic species as for others, it is crucial to be able to distinguish between changes of biomarker responses because of natural versus contamination factors.

10.4 Endobenthic Species as Benthic Indicators

Biological responses at supraindividual levels are reviewed in Chapter 14. Under the auspices of the International Council for the Exploration of the Sea, the Benthos Ecology Working Group launches a report every year dealing with works about benthos-related quality assessment or toward linking biodiversity and ecosystem functioning (ICES, 2013). The advantages of using macroinvertebrates to assess ecological quality include their relative sedentarity; their relatively long lifespans, the existence of diverse species that exhibit different tolerances to natural and chemical stress, and their important role in biogeochemical cycles. Ecological characteristics of benthic species allow the distinction between sensitive versus tolerant species, opportunistic species (that can quickly exploit new resources or ecological niches as they become available), species linked to a particular community, and sentinel or indicative species, the presence or relative abundance of which warns about possible imbalances in the surrounding environment and/or alterations of the community functions.

The ordination of soft-bottom macrofauna species into ecological groups and the determination of the relative proportion of abundance of each group in a community allow the calculation of a biotic index. The biotic index was initially designed to assess the impact of increasing levels of organic matter, but it seems relevant for chemical impacts since a relationship may be established with the effects range-low (representing concentrations above which adverse biological effects are expected to occur) of organic and inorganic contaminants (Borja et al., 2000).

However, using macroinvertebrates to assess ecological quality may not be specific enough in terms of the different kinds of stress. They cannot be applied consistently across all the biogeographic regions. In an article entitled provocatively "On the myths of indicator species," the authors (Zettler et al., 2013) concluded that when applying static indicator-based quality indices, it was necessary (1) to considerer species tolerances and preferences that may change along environmental gradients and between different biogeographic regions (2) to adjust indicator species lists along major environmental gradients as environment modifies species autecology. Another disadvantage is that using benthic indicators can be labor-intensive and a practical problem arises in relation to the number of taxonomists that has decreased dramatically. Thus it has been proposed that the taxonomic identification may be done at levels higher than the species (e.g., at the genus or family levels), according to the taxonomic sufficiency concept (de-la-Ossa-Carretero et al., 2012; Brind'Amour et al., 2014). More specific indices may be used such as the polychaete/amphipod and the nematode/copepod ratios in the case of oil spills (Chapter 9).

Advantages of meioorganisms versus macroorganisms include (1) their abundance, allowing investigations based on small samples, relatively simple equipment and minimized impacts on study sites; (2) their higher diversity increasing the range of sensitivity to stress, particularly chemical stress; and (3) the collection of abundant data, improving the statistical treatment (Amiard-Triquet and Rainbow, 2009). The different generation times of meio- and macrofauna can be chosen depending on the objectives (chronic assessment or short-term responses). However, the taxonomic impediment previously mentioned for macrofauna is even worse for meiofauna.

10.5 Conclusions

The interest of endobenthic species in ecotoxicology arises first from their importance in the structure and functioning of aquatic ecosystems and their close contact with deposited sediments that are the final sink for most of the chemical contaminants entering the environment. This is also true for emerging contaminants or legacy contaminants of emerging concern such as nanomaterials (Chapter 17), endocrine disrupting chemicals (Chapter 15), and pharmaceuticals (Chapter 16). Most of them, as deposit- or filter-feeders, feed at a low level of the trophic web but as preys for many different species (fish, birds), they represent a major entry way of contaminants in food chains. They are representative of the main taxa living in aquatic media.

The choice of species interesting for ecotoxicological studies includes the "classical" species, as cosmopolitan as possible, such as those recommended for both bioaccumulation and toxicity standard tests, but other species may have an interest for the more specific assessment of a local area. For sediment-dwelling species, sampling is quite labor-intensive (Byrne and O'Halloran, 2001; Coelho et al., 2010), but many of those that are recommended as standardized test species are commercially available. Otherwise, the condition of the individuals is highly variable and information on pretest life history is unavailable, a situation that complicates the possibilities of mastering confounding factors.

Among the criteria for the selection of biomonitors, Berthet (Chapter 9) mentions that they should be easy to identify. A special problem is that of cryptic species, which have been recognized in popular test organisms, such as *H. azteca* (Weston et al. 2013) and *Gammarus fossarum* (Feckler et al., 2014). These species complex are often treated as a single species but major differences in ecotoxicological responses can occur with far-reaching implications for monitoring and environmental policy decisions (Chapter 9).

As conceptualized in this chapter, Bettinetti et al. (2012) have applied the different uses of benthic organisms as a tool for assessing the quality of sediment in lakes, considering responses from a single species to the community. According to these authors, all ecological quality elements, from chemical analyses to more complicated biological analyses (from single laboratory tests to in situ community studies) may together provide, according to a weight-of-evidence approach, the information needed to allow decision-making. In addition, they conclude that developing mechanistic models to predict the response of natural communities seems to be particularly powerful for community ecology.

References

Abarnou, A., 2009. Organic contaminants in coastal and estuarine foodwebs. In: Amiard-Triquet, C., Rainbow, P.S. (Eds.), Environmental Assessment of Estuarine Ecosystems. A Case Study. CRC Press, Boca Raton, pp. 107–134.

Agbo, S.O., Lemmetyinen, J., Keinänen, M., et al., 2013. Response of *Lumbriculus variegatus* transcriptome and metabolites to model chemical contaminants. Comp. Biochem. Physiol. C157, 183–191.

Ahmad, I., Coelho, J.P., Mohmood, I., et al., 2011b. Immunosuppression in the infaunal bivalve *Scrobicularia plana* environmentally exposed to mercury and association with its accumulation. Chemosphere 82, 1541–1546.

Ahmad, I., Mohmood, I., Mieiro, C.L., et al., 2011a. Lipid peroxidation vs. antioxidant modulation in the bivalve *Scrobicularia plana* in response to environmental mercury–organ specificities and age effect. Aquat. Toxicol. 103, 150–158.

Almeida, C., Pereira, C.G., Gomes, T., et al., 2013. Genotoxicity in two bivalve species from a coastal lagoon in the south of Portugal. Mar. Environ. Res. 89, 29–38.

Amiard, J.C., Amiard-Triquet, C., Berthet, B., et al., 1987. Comparative study of the patterns of bioaccumulation of essential (Cu, Zn) and non-essential (Cd, Pb) trace metals in various estuarine and coastal organisms. J. Exp. Mar. Biol. Ecol. 106, 73–89.

Amiard-Triquet, C., Amiard, J.C., Rainbow, P.S., 2013a. Ecological biomarkers: indicators of ecotoxicological effects. CRC Press, Boca Raton.

Amiard-Triquet, C., Mouneyrac, C., Berthet, B., 2013b. Polychaetes in ecotoxicology. In: Férard, J.F., Blaise, C. (Eds.), Encyclopedia of Aquatic Ecotoxicology. Springer Reference, Dordrecht, pp. 893–908.

Amiard-Triquet, C., Rainbow, P.S., 2009. In: Environmental Assessment of Estuarine Ecosystems. A Case Study. CRC Press, Boca Raton.

Amirbahman, A., Massey, D.I., Lotufo, G., et al., 2013. Assessment of mercury bioavailability to benthic macroinvertebrates using diffusive gradients in thin films (DGT). Environ. Sci. Process. Impacts 15, 2104–2114.

Andral, B., Stanisière, J.Y., Sauzade, D., et al., 2004. Monitoring chemical contamination levels in the Mediterranean based on the use of mussel caging. Mar. Pollut. Bull. 49, 704–712.

ASTM (American Society for Testing and Materials), 2002. Standard Guide for Determination of the Bioaccumulation of Sediment-associated Contaminants by Benthic Invertebrates. E1688–00a. ASTM 2002 Annual Book of Standards, vol. 11.05, West Conshohocken, Pennsylvania.

Bae, M.J., Park, Y.S., 2014. Biological early warning system based on the responses of aquatic organisms to disturbances: a review. Sci. Total Environ. 466–467, 635–649.

Barda, I., Purina, I., Rimsa, E., et al., 2014. Seasonal dynamics of biomarkers in infaunal clam *Macoma balthica* from the Gulf of Riga (Baltic Sea). J. Marine Syst. 129, 150–156.

Barreira, L.A., Mudge, S.M., Bebianno, M.J., 2007. Oxidative stress in the clam *Ruditapes decussatus* in relation to polycyclic aromatic hydrocarbons body burden. Environ. Toxicol. 22, 203–221.

Baršienė, J., Andreikėnaitė, L., Garnaga, G., et al., 2008. Genotoxic and cytotoxic effects in bivalve mollusks *Macoma balthica* and *Mytilus edulis* from the Baltic Sea. Ekologija 54, 44–50.

Bebianno, M.J., Hoarau, P., Geret, F., et al., 2004. Biomarkers in *Ruditapes* sp. Biomarkers 9, 305–330.

Béchard, K.M., Gillis, P.L., Wood, C.M., 2008. Trophic transfer of Cd from larval chironomids (*Chironomus riparius*) exposed via sediment or waterborne routes, to zebrafish (*Danio rerio*): tissue-specific and subcellular comparisons. Aquat. Toxicol. 90, 310–321.

Bensouda, L., Soltani-Mazouni, N., 2014. Measure of oxidative stress and neurotoxicity biomarkers in *Donax trunculus* from the Gulf of Annaba (Algeria): case of the year 2012. Annu. Res. Rev. Biol. 4, 1902–1914.

Bergayou, H., Mouneyrac, C., Pellerin, J., et al., 2009. Oxidative stress responses in bivalves (*Scrobicularia plana, Cerastoderma edule*) from the Oued Souss estuary (Morocco). Ecotoxicol. Environ. Saf. 72, 765–769.

Bergey, L.L., Weis, J.S., 2007. Molting as a mechanism of depuration of metals in the fiddler crab, *Uca pugnax*. Mar. Environ. Res. 64, 556–562.

Berthet, B., Mouneyrac, C., Amiard, J.C., et al., 2003. Accumulation and soluble binding of cadmium, copper, and zinc in the polychaete *Hediste diversicolor* from coastal sites with different trace metal bioavailabilities. Arch. Environ. Contam. Toxicol. 45, 468–478.

Bettinetti, R., Ponti, B., Marziali, L., et al., 2012. Biomonitoring of lake sediments using benthic macroinvertebrates. Trend. Anal. Chem. 36, 92–102.

Bird, D.J., Duquesne, S., Hoeksema, S.D., et al., 2011. Complexity of spatial and temporal trends in metal concentrations in macroinvertebrate biomonitor species in the Severn Estuary and Bristol Channel. J. Mar. Biol. Ass. UK 91, 139–153.

Boldina-Cosqueric, I., Amiard, J.C., Amiard-Triquet, C., et al., 2010. Biochemical, physiological and behavioural markers in the endobenthic bivalve *Scrobicularia plana* as tools for the assessment of estuarine sediment quality. Ecotoxicol. Environ. Saf. 73, 1733–1741.

Borja, A., Franco, J., Pérez, V., 2000. A marine biotic index to establish the ecological quality of soft-bottom benthos within european estuarine and coastal environments. Mar. Pollut. Bull. 40, 1100–1114.

Borkovíc-Mitíc, S., Pavlovíc, S., Perendija, B., et al., 2013. Influence of some metal concentrations on the activity of antioxidant enzymes and concentrations of vitamin E and SH-groups in the digestive gland and gills of the freshwater bivalve *Unio tumidus* from the Serbian part of Sava River. Ecol. Indic. 32, 212–221.

Bouraoui, Z., Banni, M., Chouba, L., et al., 2010. Monitoring pollution in Tunisian coasts using a scale of classification based on biochemical markers in worms *Nereis (Hediste) diversicolor*. Environ. Monit. Assess. 164, 691–700.

Boyle, D., Amlund, H., Lundebye, A.K., et al., 2010. Bioavailability of a natural lead-contaminated invertebrate diet to zebrafish. Environ. Toxicol. Chem. 29, 708–714.

Boyle, D., Brix, K.V., Amlund, H., et al., 2008. Natural arsenic contaminated diets perturb reproduction in fish. Environ. Sci. Technol. 42, 5354–5360.

Brind'Amour, A., Laffargue, P., Morin, J., et al., 2014. Morphospecies and taxonomic sufficiency of benthic megafauna in scientific bottom trawl surveys. Cont. Shelf Res. 72, 1–9.

Bryan, G.W., Langston, W.J., Hummerstone, L.G., 1980. The use of biological indicators of heavy metal contamination in estuaries. Mar. Biol. Ass. UK, Occ Publ n° 1, 73 pp.

Bundschuh, M., Zubrod, J.P., Seitz, F., et al., 2011. Mercury-contaminated sediments affect amphipod feeding. Arch. Environ. Contam. Toxicol. 60, 437–443.

Byrne, P.A., O'Halloran, J., 2001. The role of bivalve molluscs as tools in estuarine sediment toxicity testing: a review. Hydrobiologia 465, 209–217.

Carrasco Navarro, V., Leppänen, M.T., Honkanen, J.O., et al., 2012. Trophic transfer of pyrene metabolites and nonextractable fraction from Oligochaete (*Lumbriculus variegatus*) to juvenile brown trout (*Salmo trutta*). Chemosphere 88, 55–61.

Carrasco Navarro, V., Brozinski, J.M., Leppänen, M.T., et al., 2011. Inhibition of pyrene biotransformation by piperonyl butoxide and identification of two pyrene derivatives in *Lumbriculus variegatus* (Oligochaeta). Environ. Toxicol. Chem. 30, 1069–1078.

Carrasco Navarro, V., Leppänen, M.T., Kukkonen, J.V.K., et al., 2013. Trophic transfer of pyrene metabolites between aquatic invertebrates. Environ. Pollut. 173, 61–67.

Catalano, B., Moltedo, G., Martuccio, G., et al., 2012. Can *Hediste diversicolor* (Nereidae, Polychaete) be considered a good candidate in evaluating PAH contamination? A multimarker approach. Chemosphere 86, 875–882.

Cesnaitis, R., Sobanska, M.A., Versonnen, B., et al., 2014. Analysis of the ecotoxicity data submitted within the framework of the REACH regulation. Part 3. Experimental sediment toxicity assays. Sci. Total Environ. 475, 116–122.

Cheggour, M., Chafik, A., Fisher, N.S., et al., 2005. Metal concentrations in sediments and clams in four Moroccan estuaries. Mar. Environ. Res. 59, 119–137.

Chèvre, N., Valloton, N., 2013. Pulse exposure in ecotoxicology. In: Férard, J.F., Blaise, C. (Eds.), Encyclopedia of Aquatic Ecotoxicology. Springer Reference, Dordrecht, pp. 917–925.

Christensen, M., Andersen, O., Banta, G.T., 2002. Metabolism of pyrene by the polychaetes *Nereis diversicolor* and *Arenicola marina*. Aquat. Toxicol. 58, 15–25.

Ciocan, C.M., Cubero-Leon, E., Peck, M.R., et al., 2013. Intersex in *Scrobicularia plana*: transcriptomic analysis reveals novel genes involved in endocrine disruption. Environ. Sci. Technol. 46, 12936–12942.

Coelho, J.P., Lillebø, A.I., Pacheco, M., et al., 2010. Biota analysis as a source of information on the state of aquatic environments. In: Namieśnik, J., Szefer, P. (Eds.), Analytical Measurements in Aquatic Environments. CRC Press, Boca Raton, pp. 103–120.

Coelho, J.P., Duarte, A.C., Pardal, M.A., et al., 2014. *Scrobicularia plana* (Mollusca, Bivalvia) as a biomonitor for mercury contamination in Portuguese estuaries. Ecol. Indic. 46, 447–453.

Cong, Y., Banta, G.T., Selck, H., et al., 2011. Toxic effects and bioaccumulation of nano-, micronand ionic-Ag in the polychaete *Nereis diversicolor*. Aquat. Toxicol. 105, 403–411.

Contardo-Jara, V., Wiegand, C., 2008. Biotransformation and antioxidant enzymes of *Lumbriculus variegates* as biomarkers of contaminated sediment exposure. Chemosphere 70, 1879–1888.

Cornelissen, G., Breedveld, G., Kristoffer, N., et al., 2006. Bioaccumulation of native polycyclic aromatic hydrocarbons from sediment by a polychaete and a gastropod: freely dissolved concentrations and activated carbon amendment. Environ. Toxicol. Chem. 25, 2349–2355.

Cossu, C., Doyotte, A., Babut, M., et al., 2000. Antioxidant biomarkers in freshwater bivalves, *Unio tumidus*, in response to different contamination profiles of aquatic sediments. Ecotoxicol. Environ. Saf. 45, 106–121.

Costa, P.M., Carreira, S., Costa, M.H., et al., 2013. Development of histopathological indices in a commercial marine bivalve (*Ruditapes decussatus*) to determine environmental quality. Aquat. Toxicol. 126, 442–454.

Crane, M., Sildanchandra, W., Kheir, R., et al., 2002. Relationship between biomarker activity and developmental endpoints in *Chironomus riparius* Meigen exposed to an organophosphate insecticide. Ecotoxicol. Environ. Saf. 53, 361–369.

Cravo, A., Pereira, C., Gomes, T., et al., 2012. A multibiomarker approach in the clam *Ruditapes decussatus* to assess the impact of pollution in the Ria Formosa lagoon, South Coast of Portugal. Mar. Environ. Res. 75, 23–34.

Dang, F., Rainbow, P.S., Wang, W.X., 2012. Dietary toxicity of field-contaminated invertebrates to marine fish: effects of metal doses and subcellular metal distribution. Aquat. Toxicol. 120–121, 1–10.

Davies, I.M., Vethaak, D., 2012. Integrated marine environmental monitoring of chemicals and their effects. In: ICES Cooperative Research Report, No. 315, p. 289.

De Jonge, M., Belpaire, C., Geeraerts, C., et al., 2012. Ecological impact assessment of sediment remediation in a metal-contaminated lowland river using translocated zebra mussels and resident macroinvertebrates. Environ. Pollut. 171, 99–108.

De Kock, W.C., Kramer, K.J.M., 1994. Active biomonitoring (ABM) by translocation of bivalve mollusks. In: Kramer, K.J.M. (Ed.), Biomonitoring of Coastal Waters and Estuaries. CRC Press, Boca Raton (FL), pp. 51–84.

Debenest, T., Gagné, F., Burgeot, T., et al., 2013. DNA integrity assessment in hemocytes of soft-shell clams (Mya arenaria) in the Saguenay Fjord (Québec, Canada). Environ. Sci. Pollut. Res. 20, 621–629.

de-la-Ossa-Carretero, J.A., Simboura, N., Del-Pilar-Ruso, Y., et al., 2012. A methodology for applying taxonomic sufficiency and benthic biotic indices in two Mediterranean areas. Ecol. Indic. 23, 232–241.

Díaz-Jaramillo, M., Martins da Rocha, A., Chiang, G., et al., 2013. Biochemical andbehavioralresponsesintheestu-arinepolychaete *Perinereis gualpensis* (Nereididae) after *in situ* exposure to polluted sediments. Ecotoxicol. Environ. Saf. 89, 182–188.

Driscoll, S.K., McElroy, A.E., 1996. Bioaccumulation and metabolism of benzo[a]pyrene in three species of polychaete worms. Environ. Toxicol. Chem. 15, 1401–1410.

Du, J., Pang, J., You, J., 2013. Bioavailability-based chronic toxicity measurements of permethrin to *Chironomus dilutus*. Environ. Toxicol. Chem. 32, 1403–1411.

Durou, C., Poirier, L., Amiard, J.C., et al., 2007. Biomonitoring in a clean and a multi-contaminated estuary based on biomarkers and chemical analyses in the endobenthic worm *Nereis diversicolor*. Environ. Pollut. 148, 445–458.

Eça, G.F., Pedreira, R.M., Hatje, V., 2013. Trace and major elements distribution and transfer within a benthic system: polychaete *Chaetopterus variopedatus*, commensal crab *Polyonyx gibbesi*, worm tube, and sediments. Mar. Pollut. Bull. 74, 32–41.

Egeler, P., Meller, M., Roembke, J., et al., 2001. *Tubifex tubifex* as a link in food chain transfer of hexachlorobenzene from contaminated sediment to fish. Hydrobiologia 463, 171–184.

Falfushynska, H.I., Gnatyshyna, L.L., Farkas, A., et al., 2010. Vulnerability of biomarkers in the indigenous mollusk *Anodonta cygnea* to spontaneous pollution in a transition country. Chemosphere 81, 1342–1351.

Falfushynska, H.I., Gnatyshyna, L.L.L., Stoliar, O.B., 2013. *In situ* exposure history modulates the molecular responses to carbamate fungicide Tattoo in bivalve mollusk. Ecotoxicology 22, 433–445.

Farcy, E., Gagné, F., Martel, L., et al., 2011. Short-term physiological effects of a xenobiotic mixture on the freshwater mussel *Elliptio complanata* exposed to municipal effluents. Environ. Res. 111, 1096–1106.

Feckler, A., Zubrod, J.P., Thielsch, A., et al., 2014. Cryptic species diversity: an overlooked factor in environmental management?. J. Appl. Ecol. 51, 958–967.

Fialkowski, W., Calosi, P., Dahlke, S., et al., 2009. The sandhopper *Talitrus saltator* (Crustacea: Amphipoda) as a biomonitor of trace metal bioavailabilities in European coastal waters. Mar. Pollut. Bull. 58, 39–44.

Filipowicz, A.B., Weinstein, J.E., Sanger, D.M., 2007. Dietary transfer of fluoranthene from an estuarine oligochaete (*Monopylephorus rubroniveus*) to grass shrimp (*Palaemonetes pugio*): Influence of piperonyl butoxide. Mar. Environ. Res. 63, 132–145.

Fisher, T., Crane, M., Callaghan, A., 2003. Induction of cytochrome P-450 activity in individual *Chironomus riparius* Meigen larvae exposed to xenobiotics. Ecotoxicol. Environ. Saf. 54, 1–6.

Fossi Tankoua, O., Buffet, P.E., Amiard, J.C., et al., 2012a. Intersite variations of a battery of biomarkers at different levels of biological organisation in the estuarine endobenthic worm *Nereis diversicolor* (Polychaeta, Nereididae). Aquat. Toxicol. 114–115, 96–103.

Fossi Tankoua, O., Buffet, P.E., Amiard, J.C., et al., 2013. Integrated assessment of estuarine sediment quality based on a multi-biomarker approach in the bivalve *Scrobicularia plana*. Ecotoxicol. Environ. Saf. 88, 117–125.

Fossi Tankoua, O., Amiard-Triquet, C., Denis, F., et al., 2012b. Physiological status and intersex in the endobenthic bivalve *Scrobicularia plana* from thirteen estuaries in Northwest France. Environ. Pollut. 167, 70–77.

Gagné, F., Blaise, C., Pellerin, J., et al., 2009. Études de biomarqueurs chez la mye commune (*Mya arenaria*) du fjord du Saguenay: bilan de recherches (1997 à 2006). Rev. Sci. Eau/J. Water Sci. 22, 253–269.

Galluba, S., Oetken, M., Oehlmann, J., 2012. Comprehensive sediment toxicity assessment of Hessian surface waters using *Lumbriculus variegatus* and *Chironomus riparius*. Environ. Sci. Health A Tox. Hazard Subst. Environ. Eng. 47, 507–521.

Giessing, A.M.B., Mayer, L.M., Forbes, T.L., 2003. 1-hydroxypyrene glucuronide as the major aqueous pyrene metabolite in tissue and gut fluid from the marine deposit-feeding polychaete *Nereis diversicolor*. Environ. Toxicol. Chem. 22, 1107–1114.

Gillet, P., Mouloud, M., Durou, C., et al., 2008. Response of *Nereis diversicolor* population (Polychaeta, Nereididae) to the pollution impact – Authie and Seine estuaries (France). Estuar. Coast. Shelf Sci. 76, 201–210.

Gillis, P.L., Higgins, S.K., Jorge, M.B., 2014. Evidence of oxidative stress in wild freshwater mussels (*Lasmigona costata*) exposed to urban-derived contaminants. Ecotoxicol. Environ. Saf. 102, 62–69.

Giusto, A., Salibián, A., Ferrari, L., 2014. Biomonitoring toxicity of natural sediments using juvenile *Hyalella curvispina* (Amphipoda) as test species: evaluation of early effect endpoints. Ecotoxicology 23, 293–303.

Goldberg, E.D., 1975. The mussel watch. A first step in global marine monitoring. Mar. Pollut. Bull. 6, 111–113.

Gomes, I.D.L., Lemos, M.F.L., Soares, A.M.V.M., et al., 2014. Effects of Barcelona harbor sediments in biological responses of the polychaete *Capitella teleta*. Sci. Total Environ. 485–486, 545–553.

González De Canales, M.L., Oliva, M., Garrido, C., 2009. Toxicity of lindane (γ-hexachloroxiclohexane) in *Sparus aurata, Crassostrea angulata* and *Scrobicularia plana*. J. Environ. Sci. Health Part B: Pesticides, Food Contaminants, Agric. Wastes 44, 95–105.

Greco, L., Pellerin, J., Capri, E., et al., 2011. Physiological effects of temperature and a herbicide mixture on the soft-shell clam *Mya arenaria* (Mollusca, Bivalvia). Environ. Toxicol. Chem. 30, 132–141.

Grilo, T.F., Cardoso, P.G., Pato, P., et al., 2013. Organochlorine accumulation on a highly consumed bivalve (*Scrobicularia plana*) and its main implications for human health. Sci. Total Environ. 461–462, 188–197.

Guidi, P., Frenzilli, G., Benedetti, M., et al., 2010. Antioxidant, genotoxic and lysosomal biomarkers in the freshwater bivalve (*Unio pictorum*) transplanted in a metal polluted river basin. Aquat. Toxicol. 100, 75–83.

Hanna, S.K., Miller, R.J., Zhou, D.X., et al., 2013. Accumulation and toxicity of metal oxide nanoparticles in a soft-sediment estuarine amphipod. Aquat. Toxicol. 142–143, 441–446.

Hartl, M.G.J., 2010. Biomarkers in integrated ecotoxicological sediment assessment. Sedimentology of Aqueous Systems. In: Poleto, C., Charlesworth, S. (Eds.), Sedimentology of Aqueous Systems, first ed. Blackwell Publishing, Oxford, pp. 147–170.

Hartl, M.G.J., Coughlan, B.M., Sheehan, D., et al., 2004. Implications of seasonal priming and reproductive activity on the interpretation of Comet assay data derived from the clam, *Tapes semidecussatus* Reeves 1864, exposed to contaminated sediment. Mar. Environ. Res. 57, 295–310.

Hendozko, E., Szefer, P., Warzocha, J., 2009. Heavy metals in *Macoma balthica* and extractable metals in sediments from the southern Baltic Sea. Ecotoxicol. Environ. Saf. 73, 152–163.

Higgins, C.P., McLeod, P.B., MacManus-Spencer, L.A., et al., 2007. Bioaccumulation of perfluorochemicals in sediments by the aquatic oligochaete *Lumbriculus variegatus*. Environ. Sci. Technol. 41, 4600–4606.

ICES, 2013. Report of the Benthos Ecology Working Group (BEWG), 22–25 April 2013. A Coruña, Spain. ICES CM 2013/SSGEF:09. 39 p.

Ingersoll, C.G., Brunson, E.L., Dwyer, F.J., 1998. Methods for assessing bioaccumulation of sediment-associated contaminants with freshwater invertebrates. In: National Sediment Bioaccumulation Conference. USEPA, Columbia. http://water.epa.gov/polwaste/sediments/cs/upload/ingersol.pdf (last accessed 18.06.14).

Irinco-Salinas, R., 2012. Histopathology in the digestive gland of *Batissa violaceae* Lamark as a biomarker of pollution in the Catubig River, Northern Samar,Philippines. International Conference on Environment, Chemistry and Biology IPCBEE, vol. 49, pp. 20–24.

Irinco-Salinas, R., Pocsidio, G.N., 2014. Lipid peroxidation as a biomarker of field exposure in the gills and digestive gland of the freshwater bivalve *Batissa violaceae* Lamarck. J. Med. Bioeng. 3, 207–211.

ITRC (Interstate Technology & Regulatory Council), 2011. Incorporating Bioavailability Considerations into the Evaluation of Contaminated Sediment Sites. CS-1. Interstate Technology & Regulatory Council, Contaminated Sediments Team, Washington, D.C. http://www.itrcweb.org/contseds-bioavailability/cs_1.pdf.

Janssen, E.M., Croteau, M.N., Luoma, S.N., et al., 2010. Polychlorinated biphenyl bioaccumulation from sediment for the marine polychaete *Neanthes arenaceodentata* and response to sorbent amendment. Environ. Sci. Technol. 44, 2857–2863.

Janssen, E.M., Oen, A.M., Luoma, S.N., et al., 2011. Assessment of field-related influences on polychlorinated biphenyl exposures and sorbent amendment using polychaete bioassays and passive sampler measurements. Environ. Toxicol. Chem. 30, 173–180.

Jørgensen, A., Giessing, A.M.B., Rasmussen, L.J., et al., 2005. Biotransformation of the polycyclic aromatic hydrocarbon pyrene in the marine polychaete *Nereis virens*. Environ. Toxicol. Chem. 24, 2796–2805.

Jørgensen, A., Giessing, A.M.B., Rasmussen, L.J., et al., 2008. Biotransformation of polycyclic aromatic hydrocarbons in marine polycheates. Mar. Environ. Res. 65, 171–186.

Kalantzi, I., Papageorgiou, N., Sevastou, K., et al., 2014. Metals in benthic macrofauna and biogeochemical factors affecting their trophic transfer to wild fish around fish farm cages. Sci. Total Environ. 470–471, 742–753.

Kienle, C., Gerhardt, A., 2008. Behavior of *Corophium volutator* (crustacea, amphipoda) exposed to the water-accommodated fraction of oil in water and sediment. Environ. Toxicol. Chem. 27, 599–604.

Köprücö, K., Yonar, S.M., Şeker, E., 2010. Effects of cypermethrin on antioxidant status, oxidative stress biomarkers, behavior, and mortality in the freshwater mussel *Unio elongatulus eucirrus*. Fish. Sci. 76, 1007–1013.

Kristoff, G., Guerrero, N.R., Cochón, A.C., 2010. Inhibition of cholinesterases and carboxylesterases of two invertebrate species, *Biomphalaria glabrata* and *Lumbriculus variegatus*, by the carbamate pesticide carbaryl. Aquat. Toxicol. 96, 115–123.

Lasier, P.J., Washington, J.W., Hassan, S.M., et al., 2011. Perfluorinated chemicals in surface waters and sediments from northwest Georgia, USA, and their bioaccumulation in *Lumbriculus variegatus*. Environ. Toxicol. Chem. 30, 2194–2201.

Lau, S., Mohamed, M., Tan Chi Yen, A., et al., 1998. Accumulation of heavy metals in freshwater mollusks. Sci. Total Environ. 214, 113–121.

Lee, S.W., Park, K., Hong, J., et al., 2008. Ecotoxicological evaluation of octachlorostyrene in fourth instar larvae of *Chironomus riparius* (Diptera, Chironomidae). Environ. Toxicol. Chem. 27, 1118–1127.

Lehtonen, K.K., Leiniö, S., Schneider, R., et al., 2006. Biomarkers of pollution effects in the bivalves *Mytilus edulis* and *Macoma balthica* collected from the southern coast of Finland (Baltic Sea). Mar. Ecol. Prog. Ser. 322, 155–168.

Li, H., Sun, B., Lydy, M.J., et al., 2013. Sediment-associated pesticides in an urban stream in guangzhou, china: implication of a shift in pesticide use patterns. Environ. Toxicol. Chem. 32, 1040–1047.

Lizotte, R.E., Shields, F.D., Testa III, S., 2012. Effects of a simulated agricultural runoff event on sediment toxicity in a managed backwater wetland. Water Air Soil Pollut. 223, 5375–5389.

Lotufo, G.R., Pickard, S.W., 2010. Benthic Bioaccumulation and bioavailability of polybrominated diphenyl ethers from surficial Lake Ontario sediments near Rochester, New York, USA. Bull. Environ. Contam. Toxicol. 85, 348–351.

Luoma, S., Rainbow, P.S., 2008. Metal Contamination in Aquatic Environments: Science and Lateral Management. Cambridge University Press, Cambridge.

Maranho, L.A., Baena-Nogueras, R.M., Lara-Martín, P.A., 2014. Bioavailability, oxidative stress, neurotoxicity and genotoxicity of pharmaceuticals bound to marine sediments. The use of the polychaete *Hediste diversicolor* as bioindicator species. Environ. Res. 134, 353–365.

Marasinghe Wadige, C.P.M., Taylor, A.M., Maher, W.A., et al., 2014. Effects of lead-spiked sediments on freshwater bivalve, *Hyridella australis*: linking organism metal exposure-dose-response. Aquat. Toxicol. 149, 83–93.

Martins, M., Costa, P.M., Ferreira, A.M., et al., 2013. Comparative DNA damage and oxidative effects of carcinogenic and non-carcinogenic sediment-bound PAHs in the gills of a bivalve. Aquat. Toxicol. 142–143, 85–95.

Matozzo, V., Formenti, A., Donadello, G., et al., 2012. A multi-biomarker approach to assess effects of Triclosan in the clam *Ruditapes philippinarum*. Mar. Environ. Res. 74, 40–46.

Maycock, D.S., Prenner, M.M., Kheir, R., et al., 2003. Incorporation of *in situ* and biomarker assays in higher-tier assessment of the aquatic toxicity of insecticides. Water Res. 37, 4180–4190.

Meng, F.P., Wang, Z.F., Cheng, F.L., et al., 2013. The assessment of environmental pollution along the coast of Beibu Gulf, Northern South China Sea: an integrated biomarker approach in the clam *Meretrix meretrix*. Mar. Environ. Res. 85, 64–75.

Moganti, S., Richardson, B.J., McClellan, K., et al., 2008. Use of the clam *Asaphis deflorata* as a potential indicator of organochlorine bioaccumulation in Hong Kong coastal sediments. Mar. Pollut. Bull. 57, 672–680.

Morales-Caselles, C., Martín-Díaz, M.L., Riba, I., et al., 2008. The role of biomarkers to assess oil-contaminated sediment quality using toxicity tests with clams and crabs. Environ. Toxicol. Chem. 27, 1309–1316.

Mouneyrac, C., Buffet, P.E., Poirier, L., et al., 2014. Fate and effects of metal-based nanoparticles in two marine invertebrates, the bivalve mollusc *Scrobicularia plana* and the annelid polychaete *Hediste diversicolor*. Environ. Sci. Pollut. Res. 21, 7899–7912.

Mouneyrac, C., Geffard, A., Amiard, J.C., et al., 2000. Metallothionein-like proteins in *Macoma balthica*: effects of metal exposure and natural factors. Can. J. Fish. Aquat. Sci. 57, 34–42.

Mouneyrac, C., Linot, S., Amiard, J.C., et al., 2008. Biological indices, energy reserves, steroid hormones and sexual maturity in the infaunal bivalve *Scrobicularia plana* from three sites differing by their level of contamination. Gen. Comp. Endocrinol. 157, 133–141.

Muhaya, B.B.M., Leermakers, M., Baeyens, W., 1997. Total mercury and methylmercury in sediments and in the polychaete *Nereis diversicolor* at Groot Buitenschoor (Scheldt estuary, Belgium). Water Air Soil Pollut. 94, 109–123.

Munari, M., Marin, M.G., Matozzo, V., 2014. Effects of the antidepressant fluoxetine on the immune parameters and acetylcholinesterase activity of the clam *Venerupis philippinarum*. Mar. Environ. Res. 94, 32–37.

NAS, 1980. The International Mussel Watch. Report of a workshop sponsored by the environment studies board. Vol. Natural Resources Commission National Research Council National Academy of Sciences, Washington, DC.

Ng, T.Y.T., Wood, C.M., 2008. Trophic transfer and dietary toxicity of Cd from the oligochaete to the rainbow trout. Aquat. Toxicol. 87, 47–59.

Oberholster, P.J., Musee, N., Botha, A.M., et al., 2011. Assessment of the effect of nanomaterials on sediment-dwelling invertebrate *Chironomus tentans* larvae. Ecotoxicol. Environ. Saf. 74, 416–423.

OSPAR Commission, 2005. Protocols on Methods for the Testing of Chemicals Used in the Offshore Oil Industry. (reference number: 2005–11 (a revised version of agreement 1995–07)) http://www.ospar.org/content/content.asp?menu=00850305120000_000000_000000.

OSPAR Commission, 2012. CEMP 2011 Assessment Report. Monitoring and Assessment Series. 35 p.

Oviedo-Gómez, D.G., Galar-Martínez, M., García-Medina, S., et al., 2010. Diclofenac-enriched artificial sediment induces oxidative stress in *Hyalella azteca*. Environ. Toxicol. Phar. 29, 39–43.

Ozdemir, A., Duran, M., Sem, A., 2011. Potential use of the oligochaete *Limnodrilus profundicola* V., as a bioindicator of contaminant exposure. Environ. Toxicol. 26, 37–43.

Oztetik, E., Cicek, A., Arslan, N., 2013. Early antioxidative defence responses in the aquatic worms (*Limnodrilus* sp.) in Porsuk Creek in Eskisehir (Turkey). Toxicol. Ind. Health 29, 541–554.

Palmquist, K., Fairbrother, A., Salatas, J., et al., 2011. Environmental fate of pyrethroids in urban and suburban stream sediments and the appropriateness of *Hyalella azteca* model in determining ecological risk. Integr. Environ. Assess. Manag. 7, 325–335.

Paris-Palacios, S., Mosleh, Y.Y., Almohamad, M., et al., 2010. Toxic effects and bioaccumulation of the herbicide isoproturon in *Tubifex tubifex* (Oligocheate, Tubificidae): a study of significance of autotomy and its utility as a biomarker. Aquat. Toxicol. 98, 8–14.

Park, S.Y., Choi, J., 2009. Genotoxic effects of nonylphenol and bisphenol a exposure in aquatic biomonitoring species: freshwater crustacean, *Daphnia magna*, and aquatic midge, *Chironomus riparius*. Bull. Environ. Contam. Toxicol. 83, 463–468.

Pickard, S.W., Clarke, J.U., 2008. Benthic bioaccumulation and bioavailability of polychlorinated dibenzo-p-dioxins/dibenzofurans from surficial Lake Ontario sediments. J. Great Lakes Res. 34, 418–433.

Poirier, L., Berthet, B., Amiard, J.C., et al., 2006. A suitable model for the biomonitoring of trace metals bioavailabilities in estuarine sediment: the annelid polychaete *Nereis diversicolor*. J. Mar. Biol. Ass. UK 86, 71–82.

Pruell, R.J., Taplin, B.K., McGovern, D.G., et al., 2000. Organic contaminant distributions in sediments, polychaetes (*Nereis virens*) and American lobster (*Homarus americanus*) from a laboratory food chain experiment. Mar. Environ. Res. 49, 19–36.

Rainbow, P.S., Poirier, L., Smith, B.D., et al., 2006a. Trophic transfer of trace metals: subcellular compartmentalization of Ag, Cd and Zn in the polychaete worm *Nereis diversicolor* and assimilation by the decapod crustacean *Palaemonetes varians*. Mar. Ecol. Progr. Ser. 308, 91–100.

Rainbow, P.S., Kriefman, S., Smith, B.D., et al., 2011. Have the bioavailabilities of trace metals to a suite of biomonitors changed over three decades in SW England estuaries historically affected by mining? Sci. Total Environ. 409, 1589–1602.

Rainbow, P.S., Smith, B.D., 2013. Accumulation and detoxification of copper and zinc by the decapod crustacean *Palaemonetes varians* from diets of field-contaminated polychaetes *Nereis diversicolor*. J. Exp. Mar. Biol. Ecol. 449, 312–320.

Rainbow, P.S., Poirier, L., Smith, B.D., et al., 2006b. Trophic transfer of trace metals from the polychaete worm *Nereis diversicolor* to the polychaete *N. virens* and the decapod crustacean *Palaemonetes varians*. Mar. Ecol. Prog. Ser. 321, 167–181.

Ramos-Gómez, J., Martins, M., Raimundo, J., et al., 2011. Validation of *Arenicola marina* in field toxicity bioassays using benthic cages: biomarkers as tools for assessing sediment quality. Mar. Pollut. Bull. 62, 1538–1549.

Regoli, F., Hummel, H., Amiard-Triquet, C., et al., 1998. Trace metals and variations of antioxidant enzymes in Arctic bivalve populations. Arch. Environ. Contam. Toxicol. 35, 594–601.

Ridgway, J., Breward, N., Langston, W.J., et al., 2003. Distinguishing between natural and anthropogenic sources of metals entering the Irish Sea. Appl. Geochem. 18, 283–309.

Roberts, D.A., 2012. Causes and ecological effects of resuspended contaminated sediments (RCS) in marine environments. Environ. Int. 40, 230–243.

Rodríguez-Rúa, A., Prado, M.A., Romero, Z., et al., 2003. The gametogenic cycle of *Scrobicularia plana* (da Costa, 1778) (Mollusc: Bivalve) in Guadalquivir estuary (Cadiz, SW Spain). Aquaculture 217, 157–166.

Romero-Ruiz, A., Alhama, J., Blasco, J., et al., 2008. New metallothionein assay in *Scrobicularia plana*: heating effect and correlation with other biomarkers. Environ. Pollut. 156, 1340–1347.

Rubinstein, N.I., Gilliam, W.T., Gregory, N.R., 1984. Dietary accumulation of PCBs from a contaminated sediment source by a demersal fish (*Leiostomus xanthurus*). Aquat. Toxicol. 5, 331–342.

Ruiz, J.M., Saiz-Salinas, J.I., 2000. Extreme variation in the concentration of trace metals in sediments and bivalves from the Bilbao estuary (Spain) caused by the 1989–90 drought. Mar. Environ. Res. 49, 307–317.

Ruus, A., Daae, I.A., Hylland, K., 2012. Accumulation of polychlorinated biphenyls from contaminated sediment by Atlantic cod (*Gadus morhua*): direct accumulation from resuspended sediment and dietary accumulation via the polychaete *Nereis virens*. Environ. Toxicol. Chem. 31, 2472–2481.

Saiz-Salinas, J.I., Francès-Zubillaga, G., 1997. *Nereis diversicolor*: an unreliable biomonitor of metal contamination in the ''Ria de Bilbao'' (Spain). Mar. Ecol. Prog. Ser. 18, 113–125.

Sardo, A.M., Soares, A.M.V.M., 2011. Short- and long-term exposure of *Lumbriculus variegatus* (Oligochaete) to metal lead: ecotoxicological and behavioral effects. Hum. Ecol. Risk Assess. 17, 1108–1123.

Sardo, A.M., Soares, A.M.V.M., 2010. Assessment of the effects of the pesticide imidacloprid on the behaviour of the aquatic oligochaete *Lumbriculus variegatus*. Arch. Environ. Contam. Toxicol. 58, 648–656.

Serafim, A., Bebianno, M.J., 2007a. Kinetic model of cadmium accumulation and elimination and metallothionein response in Ruditapes decussatus. Environ. Toxicol. Chem. 26 (5), 960–969.

Serafim, A., Bebianno, M.J., 2007b. MT involvement in metal metabolism in the tissues of *Ruditapes decussatus*: II. Zn accumulation and elimination. Arch. Environ. Contam. Toxicol. 52, 189–199.

Sifi, K., Amira, A., Soltani, N., 2013. Oxidative stress and biochemical composition in *Donax trunculus* (Mollusca, Bivalvia) from the gulf of Annaba (Algeria). Adv. Environ. Biol. 7, 594–604.

Silva, C., Mattioli, M., Fabbri, E., et al., 2012. Benthic community structure and biomarker responses of the clam *Scrobicularia plana* in a shallow tidal creek affected by fish farm effluents (Rio San Pedro, SW Spain). Environ. Int. 47, 86–98.

Smolarz, K., Berger, A., 2009. Long-term toxicity of hexabromocyclododecane (HBCDD) to the benthic clam *Macoma balthica* (L.) from the Baltic Sea. Aquat. Toxicol. 95, 239–247.

Soares, S., Cativa, I., Moreira-Santos, M., et al., 2005. A short-term sublethal *in situ* sediment assay with *Chironomus riparius* based on postexposure feeding. Arch. Environ. Contam. Toxicol. 49, 163–172.

Štambuk, A., Pavlica, M., Vignjević, G., et al., 2009. Assessment of genotoxicity in polluted freshwaters using caged painter's mussel, *Unio pictorum*. Ecotoxicology 18, 430–439.

Sturm, A., Hansen, P.D., 1999. Altered cholinesterase and monooxygenase levels in *Daphnia magna* and *Chironomus riparius* exposed to environmental pollutants. Ecotoxicol. Environ. Saf. 42, 9–15.

Tan, Q.G., Ke, C., Wang, W.X., 2013. Rapid assessments of metal bioavailability in marine sediments using coelomic fluid of sipunculan worms. Environ. Sci. Technol. 47, 7499–7505.

Tao, Y., Pan, L., Zhang, H., et al., 2013. Assessment of the toxicity of organochlorine pesticide endosulfan in clams *Ruditapes philippinarum*. Ecotoxicol. Environ. Saf. 93, 22–30.

Tarique, Q., Burger, J., Reinfelder, J.R., 2013. Relative importance of burrow sediment and porewater to the accumulation of trace metals in the clam *Amiantis umbonella*. Arch. Environ. Contam. Toxicol. 65, 89–97.

Tay, K.L., Teh, S.J., Doe, K., et al., 2003. Histopathologic and histochemical biomarker responses of Baltic clam, *Macoma balthica*, to contaminated Sydney harbour sediment, Nova Scotia, Canada. Environ. Health Perspect. 111, 273–280.

Taylor, A.M., Maher, W.A., 2012. Exposure–dose–response of *Anadara trapezia* to metal contaminated estuarine sediments. 2. Lead spiked sediments. Aquat. Toxicol. 116–117, 79–89.

Taylor, A.M., Maher, W.A., 2013. Exposure–dose–response of *Tellina deltoidalis* to metal contaminated estuarine sediments. 2. Cadmium spiked sediments. Comp. Biochem. Physiol. C158, 44–55.

Taylor, A.M., Maher, W.A., 2014. Exposure–dose–response of *Tellina deltoidalis* to metal contaminated estuarine sediments. 2. Lead spiked sediments. Comp. Biochem. Physiol. C159, 52–61.

TGD, 2003. Technical Guidance Document on Risk Assessment in Support of Commission Directive 93/67/EEC on Risk Assessment for New Notified Substances, Commission Regulation (EC) N° 1488/94 on Risk Assessment for Existing Substances and Directive 98/8/EC of the European Parliament and of the Council Concerning the Placing of Biocidal Products on the Market. European Commission. Joint Research Centre. EUR 20418 EN/2.

Tlili, S., Métais, I., Boussetta, H., et al., 2010. Linking changes at sub-individual and population levels in *Donax trunculus*: assessment of marine stress. Chemosphere 81, 692–700.

Tlili, S., Métais, I., Ayache, N., et al., 2011. Is the reproduction of *Donax trunculus* affected by their sites of origin contrasted by their level of contamination? Chemosphere 84, 1362–1370.

Tlili, S., Minguez, L., Giamberini, L., et al., 2013. Assessment of the health status of *Donax trunculus* from the Gulf of Tunis using integrative biomarker indices. Ecol. Indic. 32, 285–293.

USEPA, 2000. Bioaccumulation Testing and Interpretation for the Purpose of Sediment Quality Assessment – Status and Needs. EPA-823-R-00–001.

USEPA, 2001. Methods for the Collection, Storage and Manipulation of Sediments for Chemical and Toxicological Analyses: Technical Manual. EPA 823-B-01–002. 208p. http://water.epa.gov/polwaste/sediments/cs/upload/collectionmanual.pdf (last accessed 21.05.14).

Vandenbergh, G.F., Janssen, C., 2001. Vitellogenesis as a biomarker of endocrine disruption in the freshwater amphipod *Hyalella azteca*. In: Perspectives in Comparative Endocrinology: Unity and Diversity. Monduzzi Editore, Sorrento (Italia), pp. 261–267.

Vinturella, A.E., Burgess, R.M., Coull, B.A., et al., 2004. Importance of black carbon in distribution and bioaccumulation models of polycyclic aromatic hydrocarbons in contaminated marine sediments. Environ. Toxicol. Chem. 23, 2578–2586.

Vuković-Gačić, B., Kolarević, S., Sunjog, K., et al., 2014. Comparative study of the genotoxic response of freshwater mussels *Unio tumidus* and *Unio pictorum* to environmental stress. Hydrobiologia 735, 221–231.

Wallace, W.G., Lopez, G.R., Levinton, J.S., 1998. Cadmium resistance in an oligochaete and its effect on cadmium trophic transfer to an omnivorous shrimp. Mar. Ecol. Prog. Ser. 172, 225–237.

Wallace, W.G., Lopez, G.R., 1997. The trophic transfer of biologically sequestered cadmium from an oligochaete to grass shrimp. Mar. Ecol. Prog. Ser. 147, 149–157.

Wallace, W.G., Luoma, S.N., 2003. Subcellular compartmentalization of Cd and Zn in two bivalves. II. Significance of trophically available metal (TAM). Mar. Ecol. Prog. Ser. 257, 125–137.

Wallace, W.G., Hoexum Brouwer, T.M., Brouwer, M., et al., 2000. Alterations in prey capture and induction of metallothioneins in grass shrimp fed cadmium-contaminated prey. Environ. Toxicol. Chem. 19, 962–971.

Wang, J., Wages, M., Yu, S., et al., 2014. Bioaccumulation of fullerene (C60) and corresponding catalase elevation in *Lumbriculus variegatus*. Environ. Toxicol. Chem. 33, 1135–1141.

Werner, I., The, S.J., Datta, S., et al., 2004. Biomarker responses in *Macoma nasuta* (Bivalvia) exposed to sediments from northern San Francisco Bay. Mar. Environ. Res. 58, 299–304.

Weston, D.P., Poynton, H.C., Wellborn, G.A., et al., 2013. Multiple origins of pyrethroid insecticide resistance across the species complex of a nontarget aquatic crustacean, *Hyalella azteca*. Proc. Nat. Acad. Sci. 110, 16532–16537.

Weston, D.P., You, J., Lydy, M.J., 2004. Distribution and toxicity of sediment-associated pesticides in agriculture-dominated water bodies of California's Central Valley. Environ. Sci. Technol. 38, 2752–2759.

Won, E.J., Rhee, J.S., Kim, R.O., et al., 2012a. Susceptibility to oxidative stress and modulated expression of antioxidant genes in the copper-exposed polychaete *Perinereis nuntia*. Comp. Biochem. Physiol. C155, 344–351.

Won, E.J., Hong, S., Ra, K., et al., 2012b. Evaluation of the potential impact of polluted sediments using Manila clam *Ruditapes philippinarum*: bioaccumulation and biomarker responses. Environ. Sci. Pollut. Res. 19, 2570–2580.

Zettler, M.L., Proffitt, C.E., Darr, A., et al., 2013. On the myths of indicator species: issues and further consideration in the use of static concepts for ecological applications. PLoS One 8 (10), e78219.

CHAPTER 11

Gammarids as Reference Species for Freshwater Monitoring

Arnaud Chaumot, Olivier Geffard, Jean Armengaud, Lorraine Maltby

Abstract

This chapter illustrates how *Gammarus* can be used for assessing the impact of chemical stressors in freshwaters. Organism-level in situ tests can address some of the limitations of community-based biomonitoring methods, whether we understand the implications of changes in the biochemistry or physiology of an organism for the population and community to which it belongs. To this end, we must link effects at different levels of biological organization. We demonstrate how modeling is used to address some of the limitations of the in situ tests, namely: (1) the definition of reference values to address the influence of confounding factors and enable implementation at large spatial and temporal scales and (2) the use of population modeling to improve the ecological relevance of transplantation methodologies for water quality assessment. Finally, we report recent developments in the study of phylogenetic diversity within *Gammarus* and its implications for their use in assessing water quality.

Keywords: Gammarids; In situ assay; Life history traits; Modeling, interpopulation variability; Molecular markers; Physiological markers; Reference and threshold values; Toxicity assessment; Upscaling between biological levels.

Chapter Outline

Aquatic Ecotoxicology. http://dx.doi.org/10.1016/B978-0-12-800949-9.00011-5

Introduction

Gammarus species (i.e., gammarids) are a diverse group of amphipod crustaceans in the family Gammaridae. More than 200 species of *Gammarus* occur across the inland and coastal waters of the Holarctic (Väinölä et al., 2008), and they represent the dominant macroinvertebrate species, in terms of biomass, in many running freshwater ecosystems (MacNeil et al., 1997). Gammarids represent important keystone species in aquatic ecosystems because they play a central role in the detritus cycle (e.g., litter breakdown processes) and constitute an important element in food webs by providing prey for secondary consumers. Furthermore, they are known to be sensitive to a wide range of chemical stressors. Together with crustacean, cladocerans (e.g., *Daphnia*), amphipods are ranked among the most sensitive aquatic invertebrate species, both for metal and organic toxic compounds (von der Ohe and Liess, 2004) and oil spills (Dauvin and Ruellet, 2007).

Gammarids are suitable organisms for use in laboratory and field ecotoxicological studies (review in Kunz et al., 2010) and they have been used to investigate chemical stressors for more than 88 years. However, their use in ecotoxicological studies did not really take off until the 1970s. A review of the literature published between 1970 and 2014 indicated that at least 20 different species of *Gammarus* have been used in ecotoxicological studies (Figure 11.1), but by far the dominant study species has been *Gammarus pulex* (51% of studies) followed by *Gammarus fossarum* (19% of studies). Throughout the world, amphipods are used as bioindicator species for marine, estuarine, and freshwater ecotoxicological assessment. Laboratory toxicity tests with amphipods have been employed to assess the toxicity of natural and spiked sediments (Neuparth et al., 2005; Costa et al., 2005; Gaskell et al., 2007) and waters (Maltby et al., 1990a; Lawrence and Poulter, 2001; Wilding and Maltby, 2006; Beketov and Liess, 2008; Xuereb et al., 2009a). Passive biomonitoring has been used for assessing the impact of effluents (Gross et al., 2001; Schirling et al., 2005). Gammarids have also been deployed in cages in the field (i.e., in situ bioassays) and used to assess effluents, surface waters, and sediments (Crane et al., 1996; Maltby et al., 2002).

Drawing extensively on our own experience, this chapter illustrates how *Gammarus* can be used for assessing the impact of chemical stressors on freshwaters. We begin with an overview of biological responses useful for assessing toxicity before describing how these responses may be incorporated into in situ bioassays. Organism-level in situ bioassays can address some of the limitations of community-based biomonitoring methods by providing a sensitive and ecologically relevant early-warning measure of potential impact, coupled with improved diagnosis of the causes of environmental degradation (Maltby, 1999; Burton et al., 2002). However, one of the challenges of using organism- or suborganism-level stress responses is understanding the implications of changes in the biochemistry or physiology of an organism for the population and community to which it belongs. To evaluate the ecological relevance of organism-based bioassays, we must link effects at different levels of

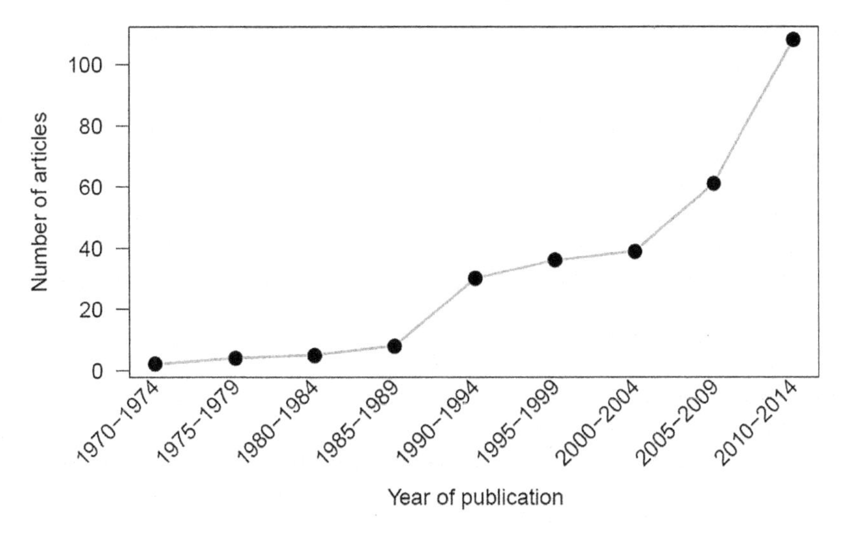

Figure 11.1
Temporal change in the number of journal articles reporting the use of *Gammarus* in ecotoxicological studies (1970–2014). The literature search was conducted using Web of Science; a total of 293 articles were identified.

biological organization in a mechanistic way, and ecological modeling can constitute a supporting framework for this (Maltby, 1999; Baird et al., 2007a). The use of models as both interpretive and predictive tools is advanced for *Gammarus*. We demonstrate how modeling has been employed to address some of the limitations of the in situ bioassays, namely: (1) the definition of reference values to address the influence of confounding factors, allowing implementation at large spatial and temporal scales and (2) the use of population modeling to improve the ecological relevance of transplantation methodologies for water quality assessment. Finally, we report recent developments in the study of ecological and phylogenetic diversity within *Gammarus* and discuss implications for their use in assessing water quality.

11.1 A Large Suite of Biological Responses Is Available for Toxicity Assessment in Gammarus Species

11.1.1 Life History Traits

In Europe, two closely related species, *G. fossarum* and *Gammarus pulex*, are intensively used in ecotoxicology, and their biology is relatively well-known. They are both gonochoristic species with a reproductive period that extends throughout the year. The female reproductive cycle is driven by the molting cycle (Charniaux-Cotton, 1965). In sexually active females, gonad maturation (i.e., oocyte growth) and the development of embryos in the marsupium are perfectly synchronized. Newly hatched juveniles leave the marsupium shortly before the female molts, after which the females lays a new batch of mature oocytes that are immediately fertilized by a male. At the same time, a new batch of oocytes enters vitellogenesis. The male reproductive cycle or spermatogenesis is not connected directly to the molting cycle and is much shorter than that of the female; 7 days being sufficient for the maximum stock sperm to be restored after mating (Lacaze et al., 2011a).

Several studies have demonstrated the impact of chemicals on gammarid reproduction, including effects on fertility (Maltby and Naylor, 1990; Cold and Forbes, 2004; Bloor et al., 2005), embryo development (Sundelin and Eriksson, 1998), copulatory behavior (Lawrence and Poulter, 2001), and gonadal anomalies (Gross et al., 2001; Schirling et al., 2005). Most reproductive toxicity tests cannot discriminate or assess whether observed impairments result from either a decrease in the number of oocytes produced, or an impact related to embryonic impairment, or a delay in organism development. Geffard et al. (2010) developed a bioassay that assessed the reprotoxic effects of pollutants and provided information on the potential toxic mode of action involved. No endocrine disruptor biomarkers are currently available for amphipods (Trapp et al., 2014a) and therefore the bioassay developed by Geffard et al. (2010) measures the impact of potential endocrine disrupters on physiological processes under endocrine regulation (e.g., molting, embryonic development, vitellogenesis, fertility) (Verslycke et al., 2007). Molting, reproductive cycles, maturation of oocytes, and embryonic development in the marsupium of amphipods are highly synchronized and predictable under control conditions (Geffard et al., 2010). For example in intermolting stage females, 90% of the embryos were in stage 3 (characterized by the presence of cephalothorax) and the mean surface of the developing ovarian follicles was 106,000 μm^2. The bioassay proposed by Geffard et al. (2010) uses disruption of the synchronization of these endpoints to highlight specific mode of action and assess the impact of endocrine disruptors.

Other life history traits used to assess the effects of chemical stressors include survival (Taylor et al., 1991) and growth (Maltby, 1994; Blockwell et al., 1996; Bloor et al., 2005). Survival, growth and reproduction drive population dynamics (Maltby et al., 2001) and population models can be used to link effects on life history traits with population-level responses (Maltby, 1999) as well as affording the opportunity to integrate probabilistic approaches into ecological risk assessment (Raimondo and McKenney, 2005).

11.1.2 Physiological Measurements

Several physiological responses have been used to investigate the effect of chemical stressors on gammarids (i.e., osmoregulation, respiration, feeding and scope for growth). The osmotic pressure of the hemolymph freshwater gammarids is maintained at a level much greater than that found in the surrounding medium and impairment of osmoregulatory function could prove fatal. Water regulation and the exchange of dissolved ions and gases between the external water and the blood of *Gammarus* occurs largely across the thin epithelium of the gills (Sutcliffe, 1971). Consequently, toxicant-induced gill damage may result in impairment of osmoregulatory and respiratory functions. Several metals are known to affect osmoregulation in gammarids including zinc (Spicer et al., 1998), copper (Brooks and Mills, 2003), cadmium (Felten et al., 2008a; Issartel et al., 2010), silver (Funck et al., 2013), and low pH (Felten et al., 2008b). Studies have also investigated the response of *Gammarus* respiration rate to metal exposure (e.g., Naylor et al., 1989; Maltby and Naylor, 1990; Kedwards et al., 1996; Aronsson and Ekelund, 2005), pesticide exposure (Maltby et al., 1990a; Lukancic et al., 2010a,b), low pH (Naylor et al., 1989; Felten et al., 2008b), ammonia exposure, and hypoxia (Maltby et al., 1990b).

Respiration rate is an integration of the metabolic energy demands of an organism and is generally quantified by measuring oxygen uptake (Sutcliffe, 1984). Energy acquisition and allocation determine an organism's fitness and provide a link between individual and population level responses (Maltby et al., 2001). Naylor et al. (1989), building on the work of Nilsson (1974), developed a method for determining the effect of chemicals on the energy budget of *G. pulex*. The method involved to determine the amount of energy absorbed (i.e., food intake—fecal production) over 6 days, after which energy metabolized (i.e., respiration) was estimated. Short-term measurements of energy intake and energy expenditure were used to calculate "scope for growth" (a measure of the amount of energy available for production), which correlates well with longer-term measures of growth (Maltby, 1994).

Gammarus scope for growth is a sensitive indicator of a wide range of stressors including metals (Maltby and Naylor, 1990; Stuhlbacher and Maltby, 1992; Maltby, 1994), organic compounds (Maltby et al., 1990a; Maltby, 1992, 1994), dissolved gases (Naylor et al., 1989; Maltby, 1994), and low pH (Maltby et al., 1990a). However, in all these cases, changes in scope for growth were primarily driven by changes in energy intake, and in all cases but one, feeding rate was the most sensitive energy budget component. The exception was chromium, where respiration and feeding rates were equally sensitive (Maltby, 1994). Given these observations, in 1991, Crane and Maltby proposed that *Gammarus* feeding rate alone could be used as a sensitive and general stress; since then, it has been used widely in laboratory and field studies (Kunz et al., 2010).

Gammarus feeding rate is influenced by a number of intrinsic and extrinsic factors including body size (Nilsson, 1974; Coulaud et al., 2011), parasitism (Pascoe et al., 1995), population (Crane and Maltby, 1991; Maltby and Crane, 1994), temperature (Nilsson, 1974; Maltby et al., 2002; Coulaud et al., 2011), food type (Nilsson, 1974), and food quality (Graça et al., 1993, 1994). These factors must therefore be either controlled or accounted for when designing feeding rate studies, for example, using nonparasitized male animals within a limited size range, obtained from a single reference population, fed leaf material conditioned with a known fungal species under controlled temperature conditions using a standardized test medium (Naylor et al., 1989). Although some early studies used *Artemia salina* eggs as a food source (Taylor et al., 1993), most studies of *Gammarus* feeding rate use leaf material as a food source. *Gammarus* feeding rate has primarily been used in laboratory studies to assess the toxicity of waterborne contaminants, although it can also be used to assess contaminated sediment toxicity (Forrow and Maltby, 2000). In addition to the studies mentioned previously, laboratory-based feeding assays have also been used to assess the toxicity of heavy metals (Blockwell et al., 1998; Wilding and Maltby, 2006; Dedourge-Geffard et al., 2009), insecticides (Malbouisson et al., 1995; Blockwell et al., 1998; Maltby and Hills, 2008; Agatz et al., 2014), and fungicides (Zubrod et al., 2014).

Measuring *Gammarus* feeding rate in the field (Maltby et al., 1990a) provides a short-term robust, responsive, and ecologically relevant assay of water and sediment quality that is indicative of longer-term changes in community structure and functioning (Maltby et al., 2002). Field deployments, and the challenges associated with them, are discussed in Section 11.2.

11.1.3 Molecular Markers

In a recent review, Trapp et al. (2014a) summarized molecular markers developed in *Gammarus* genus. Methodologies are available for measuring a total of 21 biomarkers, such as the metallothionein-like proteins, indicative of metal exposure (Stuhlbacher and Maltby, 1992; Correia et al., 2001; Geffard et al., 2007; Gismondi et al., 2012), acetylcholinesterase activity (McLoughlin et al., 2000; Xuereb et al., 2007, 2009a) related to pesticide exposure, and neurotoxicity, the phase II xenobiotic transformation enzyme (glutathione *S*-transferase, glutathione peroxidase, catalase, and superoxide dismutase) (Maltby and Hills, 2008; Bedulina et al., 2010a; Turja et al., 2014), which signal detoxification activity. Effective biomarkers determined in *Gammarus* genus include catalase (Sroda and Cossu-Leguille, 2011), total glutathione peroxidase (Turja et al., 2014), peroxidase (Bedulina et al., 2010b), and superoxide dismutase (Turja et al., 2014) activities, all indicative of antioxidative defense induction; the increase of heat shock protein from different families (Schirling et al., 2006; Scheil et al., 2008; Bedulina et al., 2010a), related to the general stress response; changes in digestive enzymes including amylase, cellulase, endoglucanase, esterase, trypsin, β-galactosidase, and β-glucosidase (Dedourge-Geffard et al., 2009; Charron et al., 2013); and, finally, activity of the sodium pump Na^+/K^+ adenosine triphosphatase (Felten et al., 2008a,b; Issartel et al., 2010) linked to alteration of iono-osmoregulation. Three-quarters of the studies on biomarkers have been conducted under laboratory conditions in the context of environmental risk assessment of chemical contamination and very few in field surveys (Crane et al., 1995; Maltby and Hills, 2008; Sanchez and Porcher, 2009).

The use of biomarkers in environmental risk assessment has been criticized, and particular concerns have been raised about their ecological relevance and the impact of confounding factors (Forbes et al., 2006). Xuereb et al. (2009a) and Charron et al. (2013, 2014) showed that physiological parameters such as gender, reproductive status, and energetic status influence some biomarkers and consequently, their impact should be taken into account when establishing relationships between biomarker changes and contamination levels. There are also concerns about the specificity and sensitivity of the so-called "specific biomarkers," which should be considered when using biomarkers to assess the impacts of chemical exposure (McLoughlin et al., 2000).

The lack of genomic and/or proteomic data for the majority of aquatic invertebrates means that direct methods for the monitoring and the quantification of molecular biomarkers are scarce. Indirect assays for monitoring protein activity (i.e., enzymatic activities by spectrophotometry) are therefore widely used in invertebrates. Biomarkers currently used in invertebrate studies have been mainly developed and validated in vertebrates for which molecular structure and characterization are well-known (Trapp et al., 2014a). In general, enzymatic biomarkers developed for vertebrates have been transferred to invertebrates without systematic molecular characterization. For proteins that are highly conserved during evolution, this transfer leads to robust and accurate biomarkers, such as the measurement of cholinesterase and NaKTPase activities (Xuereb et al., 2007; Lemos et al., 2010). However, for proteins which sequence and/or physiological function have significantly changed during evolution, this transfer approach leads to inconsistent results in invertebrates. For example the use and relevance of anti-vitellogenin (VTG) antibodies as specific biomarkers of endocrine disruption in vertebrates is widely recognized. However, anti-VTG antibodies available for vertebrates (fish) and invertebrates (including crustaceans) do not permit identification of yolk proteins in *G. fossarum* (Boulangé-Lecomte et al., 2013). Taking their lead from work with vertebrates, most researchers have focused on the development of VTG measurement as an endocrine disruption biomarker in crustaceans. A specific method, based on liquid chromatography coupled with tandem mass spectrometry, has recently been developed and validated for identifying and quantifying VTG in invertebrates and in particular in freshwater amphipods (Simon et al., 2010; Jubeaux et al., 2012a,b). This methodology has been used to investigate whether VTG measurement could be a specific endocrine disruption biomarker in males and used as an indicator of feminization (Jubeaux et al., 2012c; Short et al., 2014). Results showed that its potential use as an endocrine disruption biomarker in males is compromised by large interindividual variability, a very low or no induction under contamination exposure and a large confounding effect of unidentified environmental factors. According to findings of Short et al. (2014), these recent studies highlighted the need for specific identification of new key molecular markers in the reproductive biology of gammarids for use as relevant biomarkers in ecotoxicology.

11.1.4 Next-Generation Omics in Gammarus

Proteomics emerged in the 1980s with the development of high-resolution gel electrophoresis, two-dimensional polyacrylamide gel electrophoresis, which allowed the separation and visualization of proteins from complex mixtures (Gevaert and Vandekerckhove, 2000). However, the protein expression signature was only descriptive and did not explain the molecular mechanisms linked to the disturbance because proteins were not identified. The development of mass spectrometry technology in protein chemistry, combined with large-scale nucleotide sequencing (expressed sequence tags and genomic DNA), enabled the development of whole-genome proteomics (Armengaud, 2013), based on "peptide mass fingerprinting." Protein spots can be selected and excised from the two-dimensional polyacrylamide gel electrophoresis gel, digested by trypsin and the resulting peptides analyzed using matrix-assisted laser desorption/ionization-time of flight mass spectrometry. A list of experimental peptide masses with their corresponding intensities has been produced and compared with theoretical peptide masses, which are generated from in silico digestion of protein sequences present in the database and annotated from DNA sequencing information (Gevaert and Vandekerckhove, 2000). This is still a developing field but proteomics holds promises for improving ecotoxicological assessment at the subindividual level.

A major current limit in the use of proteomic approaches for nonmodel species is the lack of appropriate protein sequence database. Unfortunately, for most species used in ecotoxicology, protein sequence information is lacking. Homology-driven proteomics using protein cross-species matching is an alternative to the lack of a properly annotated genome, but this approach only detects the most conserved proteins, which are generally related to protein synthesis and folding, general metabolism, and structural proteins (Leroy et al., 2010). Many studies conducted over the past decade have had problems matching peptide masses because of low scores or false-positive matches (Armengaud et al., 2014). The most frequently identified proteins in invertebrates are those involved in adenosine triphosphate supply (adenosine triphosphate synthase, glyceraldehyde-3-phosphate dehydrogenase, enolase, and arginine kinase) and in cytoskeleton structure maintenance (actin, myosin, and tubulin) (Trapp et al., 2014b). Moreover, the protein function is predicted on the basis of sequence conservation by combining available known sequences and functional information from various taxonomic groups. This information is mainly derived from vertebrates, for which annotated genomes are more numerous. Confidence in these predictions for invertebrates is relatively limited because high diversification rates and longer evolutionary times mean that homologues in distant organisms may have acquired different functions (Studer et al., 2009; Nikinmaa and Rytkönen, 2011). In addition, lineage-specific genes have been shown to be among the most responsive to environmental challenges, as demonstrated by transcriptomic studies on *Daphnia pulex* (Colbourne et al., 2011). Consequently, cross-species matching for protein identification does not allow relevant and specific modes of action to be deciphered in ecophysiological and ecotoxicological studies.

An alternative to homology-driven proteomics should be developed. Although sequencing the genome of organisms is the most appropriate approach for creating a comprehensive protein sequence database, sequencing only mature RNAs is an attractive alternative for rapid identification of protein sequences for nonmodel organisms. RNA-Seq, a next-generation sequencing technology, is a cost-effective technique for rapidly identifying protein-encoding genes among mature RNAs through a comprehensive coverage of the transcriptome by deep sequencing. The combination of genomics and proteomics, the so-called proteogenomics approach, is a strategy for discovering proteins in nonmodel organisms employed in environmental science (review by Armengaud et al., 2014). This approach has been applied to *G. fossarum*, leading to the identification of specific proteins involved in reproduction (Trapp et al., 2014b). A total of 1873 proteins from *G. fossarum* have been certified by tandem mass spectrometry; 75 of which are classified as female-specific and 28 are classified as male-specific. However, although the proteogenomic approach greatly improved the identification of proteins in nonmodel species, functional annotation by sequence similarities was limited because of the absence of other datasets for this lineage. In this study, functional annotation or validation was performed by comparative analysis of the abundance of proteins in male and female reproductive tissue-specific candidates involved in the reproductive process identified and characterized by morphological and physiological parameters (Trapp et al., 2014b,c).

11.2 In situ Biotests Are Operational in Gammarus *Species*

11.2.1 In situ Assay

In situ effect measures, which include the use of caged organisms, have several advantages over passive biomonitoring approaches, including better control of stressor exposure to a defined population of organisms under natural conditions (Maltby and Burton, 2006; Baird et al., 2007b). In situ approaches improve the link between toxic effects and field contamination levels (Crane et al., 2007) and integrate the effects of environmental parameters that influence bioavailability (e.g., temperature, pH, hardness). By using transplanted organisms, intrinsic factors that are known to affect sensitivity to toxicants (i.e., age, gender, nutritional status, reproductive status, population, exposure history) can be standardized and therefore their impact on the interpretation of effects minimized (Liber et al., 2007). In situ assays may be based on life-history traits (survival, growth, and reproduction), physiological measurements (scope for growth, feeding) or molecular markers. Baird et al. (2007a) review the selection of test organisms and measurement endpoints for in situ assays and provide a framework for linking effects at different levels of biological organization, using *G. pulex* as an example. In situ assays using gammarids have been used to assess a range of contaminant sources, such as wastewater treatment plants (Lacaze et al., 2011b) and industrial inputs (Dedourge-Geffard et al., 2009) or for large-scale surveys (Coulaud et al., 2011; Jubeaux et al., 2012c). The most widely used endpoint is feeding rate, which has been employed to investigate the toxicity of farm waste (Veerasingham and Crane, 1992), pesticide spray drift (Maltby and Hills, 2008), municipal wastewaters (Maltby et al., 2002; Bundschuh et al., 2011), coal mine effluents (Maltby et al., 2002), industrial effluents (Maltby et al., 2002), and landfill leachates (Bloor and Banks, 2006), among others.

11.2.2 Dealing with Environmental Variability: Definition of Reference Values

Numerous studies have demonstrated that abiotic factors can affect the basal levels of biomarkers or individual responses. In the great majority of in situ assays, abiotic confounding factors are not explicitly taken into account and applications are generally restricted to comparisons between reference and assumed impacted sites with similar physicochemical parameters (e.g., upstream and downstream of identified point sources). Such designs, however, do not necessarily prevent confounding effects occurring, even at small spatial scales (Coulaud et al., 2011). Hanson et al. (2010) demonstrated how defining the natural variability of toxicity markers can improve their reliability and reduce the risk of interpreting natural variation as impact. The use of independent reference benchmarks (based on the range of natural variability of markers) also offers the possibility to assess water quality in isolated stations as part of large scale surveys, notably in non–point-source pollution contexts, for instance for regulatory purpose of site prioritization.

11.2.2.1 Empirically Defining the Range of Biomarker Natural Variability

One approach to limiting false positives induced by confounding environmental factors is to define a range of reference values that incorporate annual and spatial variability observed in reference stations (Hagger et al., 2008). This has been developed for instance for marine environmental monitoring, as part of the framework of the OSPAR Convention. Background levels of response of biomarkers and associated threshold values have been defined by extending the development of assessment criteria for contaminant concentrations, namely background concentrations, and background assessment concentrations from the OSPAR Hazardous Substances Strategy (Davies and Vethaak, 2012). This data-driven approach based on the compilation of data acquired in stations defined a priori as references was also applied and combined with information from laboratory exposures to define the natural range of marker variability and determine reference and threshold values for in situ biotest measurements with *G. fossarum*. The spatiotemporal variability of basal whole-body acetylcholinesterase activity (AChE) (Xuereb et al., 2009a) or DNA damage in *Gammarus* sperm (Comet assay) (Lacaze et al., 2011a) of *Gammarus* deployed in reference stations at different seasons and throughout watersheds, was very limited relative to level of responses of these biomarkers to contaminants. This pattern agrees with the absence of effect of tested abiotic factors (i.e., temperature, water hardness) in laboratory experiments and an empirical definition of a "normal" range of variability for these markers has therefore been proposed based on the distribution of multiple measurements in reference stations, corresponding to a range of measured values covering seasonal and spatial variability (Figure 11.2). For these experiments, only the deviation in Ardiere (2007) should be interpreted as pollution impact, contrary to outcomes in Amous (2008), always within the range of natural variability. In the case of Bourbre (2007), degraded levels of AChE only revealed by the comparison to reference values are unrelated with the source of pollution.

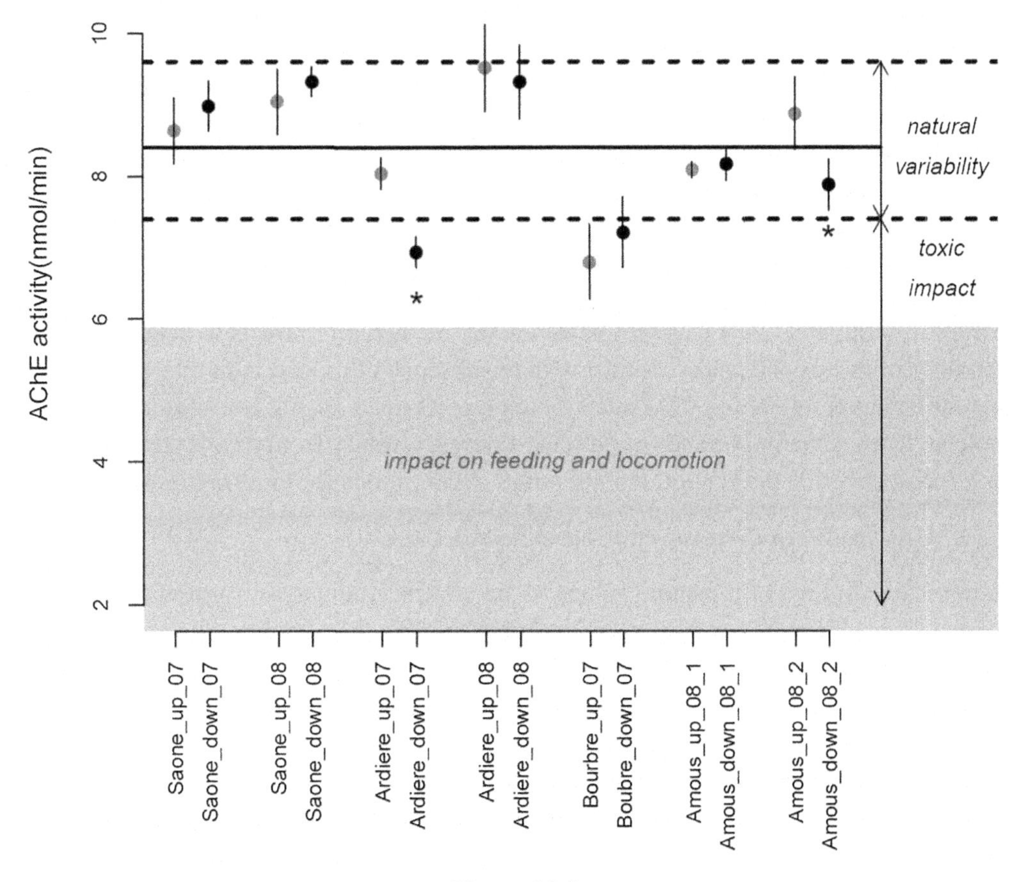

Figure 11.2

Use of reference values for the interpretation of AChE biomarker measurements in *Gammarus*. Here are the results of seven campaigns of in situ caging (2 weeks) in paired stations upstream (gray) downstream (black) of pollution sources. Stars indicate significant differences between upstream and downstream records. The range of natural variability of the AChE measurement was established in Xuereb et al. (2009a). Levels of inhibition impacting fitness traits are assessed from Xuereb et al. (2009b).

In contrast to AChE and DNA damage, digestive enzymes in *G. fossarum* (amylase, cellulase, and trypsin) are weakly influenced by water hardness and there is a strong negative effect of low temperatures on enzymatic biomarkers under laboratory conditions (Charron et al., 2013). Although the influence of water hardness could be taken into account in the empirical definition of variability range derived from in situ measurements at five noncontaminated stations, the effect of low temperature prevents to use the defined reference and threshold values for deployments occurring during cold conditions.

Another important issue is the number of reference measurements needed to validate the generality of established basal levels of biological responses. Studies on the possible induction of VTG in *Gammarus* illustrate this problem. Initially, a VTG assay was developed based on the expression of the VTG gene using calibrated real-time reverse transcription polymerase chain reaction (Xuereb et al., 2011). Deploying caged animals upstream and downstream of wastewater treatment plants allowed reference levels to be defined based on measurements in four upstream stations. A second series of experiments quantified VTG protein in males deployed in reference and contaminated stations at different seasons (Jubeaux et al., 2012c). An unexpected high seasonal variability of VTG levels was observed in caged male organisms under reference conditions (seven rivers, four seasons), which prohibited the definition of a relevant reference level for this marker, thereby severely limiting the use of the VTG biomarker in *Gammarus* to detect significant endocrine disruptor effects in field monitoring. The findings of the second study contrast with the absence of induction in reference stations during our first experiment for VTG quantification performed only at one season. This example underlines the importance of defining reference values from a consistent set of reference data covering the spatiotemporal range of conditions for bioassay implementation.

11.2.2.2 Modeling When Environmental Influences Are Large

The empirical definition of threshold values based on observed natural variability of biological markers is problematic for most individual responses such as feeding rate or other life history traits (e.g., growth, intermolting interval) because of their high natural variability (e.g., induced by temperature). This variability results in reduced statistical power and thus to difficulties in discriminating inhibitions related to contaminants (see Figure 11.3). As an alternative, modeling the influence of abiotic confounding factors provides a way of correcting for environmental variables and producing comparable measurements even in variable environmental conditions. This method has been employed to model the influence of temperature on the in situ feeding rate of *G. pulex* and *G. fossarum* (Maltby et al., 2002; Coulaud et al. 2011). Both studies use a single-source population to minimize intrinsic confounding factors and, whereas Coulaud et al. (2011) used laboratory data to model the influence of temperature, Maltby et al. (2002) used field data. Both approaches greatly improved the reliability of in situ toxicity assessment and by eliminating temperature as a confounding factor, enabled the establishment of reference levels for multiple in situ deployments throughout the year and among geographically dispersed watersheds. Maltby et al. (2002) modeled the influence of temperature and other environmental factors on gammarid feeding rate, accounting for almost 80% of natural variability in their statistical model. Coulaud et al. (2011) observed similar results with the two approaches for calibrating their temperature model (either laboratory assays or in situ reference deployments). These studies emphasize the tractability of *Gammarus* for both laboratory and field testing and demonstrate the combined value of in situ measurements and laboratory studies for establishing reference and threshold values.

The influence of temperature on reproductive molting dynamics in *G. fossarum* females (Geffard et al., 2010) has recently been investigated using a modeling approach that builds on work by Pöckl and Timischl (1990). This modeling approach predicts female molting stages after exposure to variable temperatures in reference rivers, thereby enabling an in situ application of

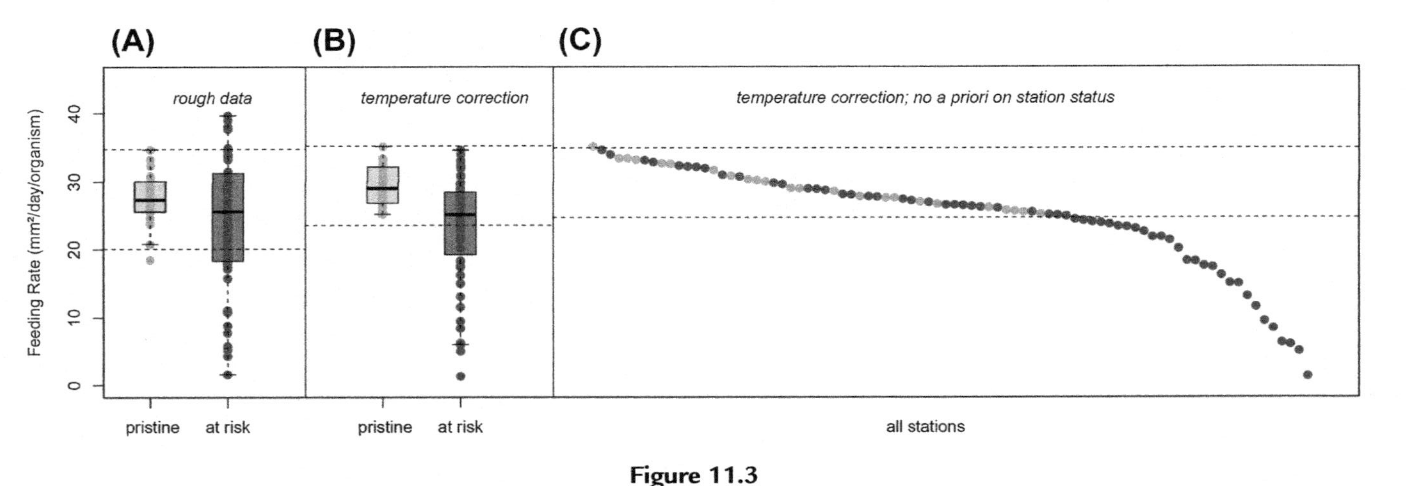

Figure 11.3

Definition of reference values for feeding assay with *Gammarus*. Here are data of 80 in situ feeding assays from Coulaud et al. (2011), in either reference "pristine" stations or "at risk" stations potentially impacted by various contaminations (expert a priori). Feeding rate is assessed via alder leaf consumption evaluated by reduction of foliar disc surfaces (1 week). (A) Only a weak discrimination of the two types of stations is possible when direct measurements of feeding rate are considered. (B) The confounding effect of temperature on feeding activity is removed by the methodology exposed in Coulaud et al. (2011); here values are presented for 14 °C for all stations. (C) Statistical delimitation of a homogenous set of values defined as reference distribution following the methodology presented in Besse et al. (2013) applied on pooled data from all stations.

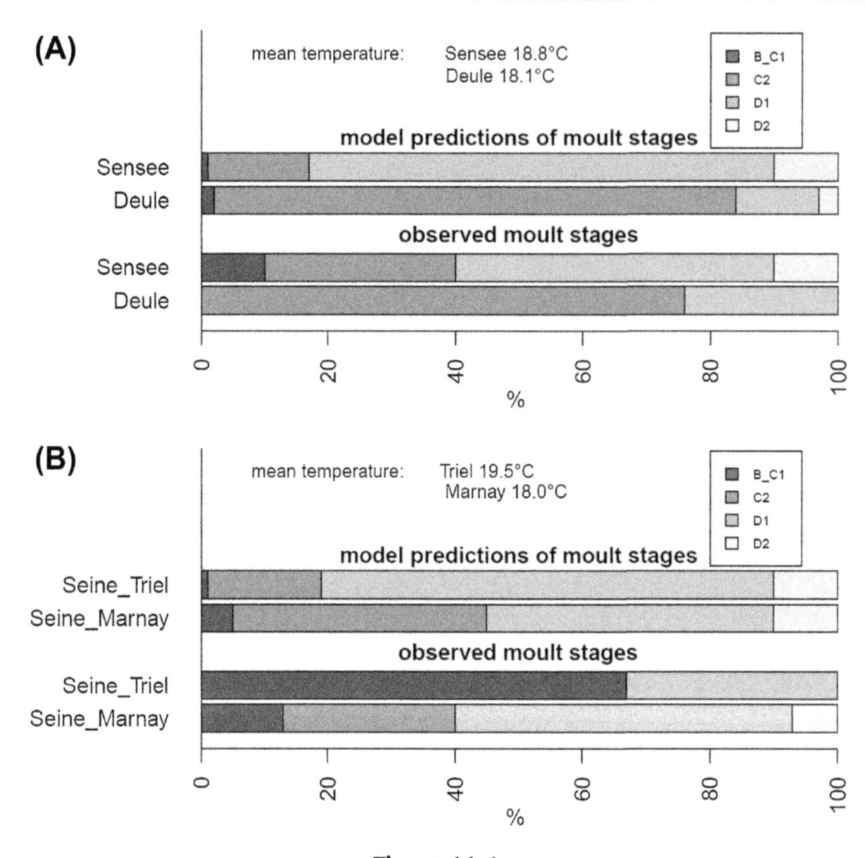

Figure 11.4
Use of a temperature model as reference of molting dynamics in in situ caged *G. fossarum* females. (A, B) The comparison of distribution of molt stages after 15 days of exposure in two close stations (definition of molt stages in Geffard et al., 2010). (A) Differences in molting advance between Sensee and Deule rivers are explained by temperature heterogeneity between stations. (B) The delay recorded in the Triel station in comparison to the Marnay station on the Seine river is clearly an impact of degraded water quality in this station.

the reproductive bioassay developed in *G. fossarum* (Geffard et al., 2010). The case studies illustrated in Figure 11.4 demonstrate how the use of model predictions allowed the assessment of whether a molting delay recorded between two stations should to be interpreted as an effect of a decrease in water quality or only as an effect of temperature heterogeneity between stations.

11.2.2.3 How to Proceed with No a Priori about the "Reference Status" of Stations Selected as Control

One major issue is the selection of the stations, on which reference levels of responses are calibrated. But this selection is often problematic because contamination status of a station can change in time, and because it prevents from benefiting of a large dataset, whereas one could take advantage of a wealth of ignored data in which a biomarker does not respond to exposure in suspected impacted stations at one date because of the typology or the temporal variability of toxic pressure. This situation occurred in the study by Besse et al. (2013), which investigated the suitability of an active biomonitoring approach using *G. fossarum* to assess trends of bioavailable contamination in continental waters with regard to the priority substances of the European Water Framework Directive. After 1 week of exposure on 27 stations, concentrations of 11 metals/metalloids and 38 hydrophobic organic substances were measured in caged male gammarids. To propose an operational tool for chemical monitoring, the authors defined threshold values of contamination, following the philosophy background concentrations and background assessment concentrations of the OSPAR Hazardous Substances Strategy for marine monitoring. But opposite to the OSPAR strategy, thresholds were not computed from a set of a priori pristine stations. In fact, this was not possible because for metals, for instance, numerous discrepancies were observed with regard to the a priori expert classification of the regional water agency, with a high level of bioaccumulation in expected pristine stations. Such discrepancies may stem from the presence of an anthropogenic source of contamination not previously identified, but certainly also from local geochemical backgrounds. That is why we proposed in this study to define reference and threshold values using a statistical approach, which delineates a homogenous set of lowest measurements, namely a reference distribution, among the pooled data from all stations regardless of expert classification. The robustness of baseline levels was further validated during in situ deployments at the French national scale in 130 stations selected to represent different physicochemical characteristics and various anthropic pressures. This blind approach can be extended to any biological markers, because one gathers a sufficient dataset of measurements implemented strictly following a fixed protocol to reduce source of variability (e.g., source population, duration of exposure, selection of test organisms). Figure 11.3 illustrates how this approach could be applied to a feeding assay (after temperature correction) based on data presented in Coulaud et al. (2011) and underlines that no questionable a priori data are necessary to define robust reference and threshold values for this marker, thus offering a basis for further implementation of quality grids to rank sites, or qualify temporal trend of restoration or degradation of specific stations.

11.3 Linking Biological Scales and Ecological Relevance of In situ–Based Effect Assessment with Gammarus

The application of biomarkers in ecotoxicology and ecological risk assessment has been questioned, especially the extent to which biomarkers are able to provide specific and ecologically relevant indicators of effects of toxicants (Forbes et al., 2006). Gammarids are good candidates to illustrate how one could address questions concerning the ecological relevance of molecular biomarkers and suborganism responses to stressors. Indeed, by concentrating on key organisms and processes, the design of single-species tests can provide real insight into the effects of stress and disturbance on ecosystems. For example, it has been shown that reduction of feeding activities of transplanted *G. pulex* are correlated with reductions in decomposition processes in streams and changes in macroinvertebrate community structure (Maltby et al., 2002). We describe here (1) how the suite of biomarkers and bioassays developed for gammarids can be used in a comprehensive mechanistic framework that integrates them into a measure of individual fitness and (2) how advances in ecological modeling in these species support the mechanistic understanding of contaminant impacts across biological scales, potentially improving the predictive ability of specific infraindividual markers of toxicant actions.

11.3.1 From Suborganism to Individual Levels of Biological Organization

Molecular biomarkers can be helpful for gaining insight into the mechanisms causing effects of chemicals on whole-organism performance. But in many cases, responses of biomarkers to contaminants are not assessed in terms of the severity of deviation from baseline levels and therefore potential impacts on individual fitness are unclear. Some advances have been operated in marine monitoring, in which several assessment criteria (i.e., representing levels of response below which unacceptable responses at higher [e.g., organism or population] levels would not be expected), analogous to environmental assessment concentrations applied in chemical monitoring of hazardous substances, are proposed for some biological effects measurements (Davies and Vethaak, 2012). Some studies in *Gammarus* have also addressed this limitation of the ecological relevance of biomarkers by drawing links between biomarker responses and impacts on life history traits. For instance, quantitative relationships between AChE inhibition and changes in feeding and locomotor behaviors have been investigated in *G. fossarum* for two model insecticides (Xuereb et al., 2009b). The feeding activity and locomotion impairment in adult males were directly correlated to levels of AChE inhibition for both chemicals even though they have different modes of action. Inhibition of AChE in gammarids caged for 2 weeks rarely exceeds 30% (e.g., Figure 11.2.), which appears to be a threshold below which no effects on feeding and locomotion occur. Consequently, inhibition greater than 30% is considered to be evidence of toxic impact with potential fitness implications. Another example is provided by the Comet assay using *G. fossarum* spermatozoa and the consequences of germ cell DNA damage resulting from parental exposure to genotoxic contaminants on reproduction quality (Lacaze et al., 2011c). A significant correlation between the level of damage to the sperm DNA of exposed parents in laboratory exposure to chemicals, and the abnormality rate in embryos that had developed from these gametes was established. From this, it appeared that the levels of genotoxicity recorded during field deployment did not reach the minimal threshold defined as inducing reproductive defects. These results were corroborated by assessing the reproduction of caged animals during this monitoring program.

Physiological energetics is another approach for examining the links between suborganism responses and life history traits (Maltby, 1999). Implicit in the use of *Gammarus* scope for growth as a stress indicator is the assumption that perturbations in energy budgets translate into changes in either growth and/or reproduction, which are related, in a rigorous and mechanistic way, to effects at higher levels of biological organization. The acquisition and allocation of energy determines developmental rates, growth rates, fecundity and survival. All are important components of fitness and determinants of population structure and dynamics. *Gammarus* energy budgets have been successfully used to predict the concentrations of pollutants at which growth and reproduction will be impaired as well as the magnitude of this impairment. For example, knowledge of the effect of unionized ammonia on the energy budget of juvenile *G. pulex* was used to predict the effect of this stressor on growth rates (Maltby, 1994). Furthermore, stress-induced reductions in scope for growth have been shown to be correlated with reductions in reproductive output, both in the form of reduced offspring size and brood viability (Maltby and Naylor, 1990). Energy budget models have been used to translate changes in *Gammarus* feeding rate, the most sensitive driver of changes in scope for growth, to effects on vital rates (i.e., fecundity, survival), which have then been combined with matrix population models to predict population growth rate (Baird et al., 2007a).

11.3.2 Population Modeling to Extrapolate Impacts on Population Dynamics

Population models are mechanistic models that relate individual-level responses (vital rates in demographic terminology or life history traits in eco-evolutionary terms) to changes in population density and structure (Maltby et al., 2001). Population modeling can add value to ecological risk assessment by reducing uncertainty when extrapolating from ecotoxicological observations (suborganismal or individual responses) to relevant ecological effects (Forbes et al., 2008). Such population models are currently increasingly used in predictive approaches (Galic et al., 2010; Kramer et al., 2011), but their use in the diagnostic framework remains limited. Yet, different types of population models have been developed in European freshwater *Gammarus*, namely *G. pulex* and *G. fossarum*, offering the possibility to apply population extrapolation on field-based data. Indeed, stage- or size-structured population dynamic models, following either matrix or differential equations formalism (Baird et al., 2007a; Kupisch et al., 2012; Coulaud et al., 2014) as well as individual-based models (Galic et al., 2014) have been proposed in these species. All these models allow the computation of population endpoints (e.g., asymptotic population growth rate, population densities, recovery times), based on life history traits and therefore evaluate the ecological relevance of individual ecotoxicity endpoints.

Population models developed for *Gammarus* take into account species phenology and seasonal variability in population dynamics, driven by temperature or variability of food availability (Kupisch et al., 2012; Coulaud et al., 2014). In the majority of studies, population models are based on laboratory data with species that are not representative of ecosystems. Consequently, by incorporating greater levels of ecological relevance and complexity, the development of *Gammarus* population models offers the rare opportunity to integrate field-based input data to anticipate what would happen in the field in native populations. This clearly responds to the undisputed call for improved ecological realism of population models in ecological risk assessment (Forbes et al., 2008). Coulaud et al. (2014) used a periodic matrix model to investigate the drivers of temporal variation in the demographics of a *G. fossarum* population during a year. Population dynamics were highly sensitive to reductions in adult survival in winter and to fertility inhibitions in spring and summer. This pattern of demographic impacts is governed by the life history of the studied *Gammarus* population and modeling provides a mechanistic understanding of the complexity of demographic impacts occurring in wild populations. Therefore, as exemplified with a case study watershed impacted by mining activities, this approach can therefore be used to anticipate population impacts of life history trait alterations caused by pulsed or continuous exposure to toxic contaminants (Coulaud et al., 2014). A similar approach, combining a dynamic mass budget model with a stage-structured matrix population model, was used by Baird et al. (2007a) to predict changes in population growth rate resulting from stressor-induced reductions in *Gammarus* feeding rate measured in situ. The models predicted that a 50% impairment of feeding rate was sufficient to produce net population growth rate values <1 (i.e., population in decline). Consequently a 50% impairment in feeding rate as measured in a *Gammarus* in situ bioassay should be associated with a reduced density of *Gammarus* populations in the field. Model predictions were tested at nine sites receiving contaminated discharges (Maltby et al., 2002) and there was strong support for the 50% in situ feeding response threshold being predictive of population impairment (Baird et al., 2007a).

The combination of in situ–based effects measures and modeling strengthens the ecological relevance of environmental monitoring (Baird et al., 2007a). The extrapolation of in situ toxicity data to the population level could help to fill the gap between the detection of chemical compounds in the field and alteration of ecological communities, namely regulatory protection goals. In their view point, discussing ecological relevance of in situ–based effects measures, Baird et al. (2007a) provide a framework that makes explicit the linkages between effects at lower levels of biological organization and higher-order ecological effects at the population and ecosystem levels, which they illustrate with a *Gammarus* case study. The *Gammarus* case study presented by Baird et al. (2007a) paves the way for quantitative links between suborganismal responses and population impacts such as those advocated by the Adverse Outcome Pathway approach (Kramer et al., 2011). One cannot envisage that all biomarkers can be integrated in such an extrapolation scheme, because they do not share all the same predictive ability. However, in *Gammarus*, examples presented in the previous section (AChE, Comet assay in spermatozoa) are good candidates, because links to fitness have been established. Measurements of the state of energy reserves and metabolism (e.g., Maltby and Naylor, 1990; Charron et al., 2013) are also of great value for assessing changes in individual performance. In that sense, feeding inhibition is of great interest for such multiscale assessment. Feeding rate can be related to alterations in life history traits (see the previous section) and feeding inhibition has been correlated with the impact of contaminants with diverse modes of actions as indicated by the modulation of specific molecular biomarkers (Xuereb et al., 2009b; Coulaud et al., 2011). Therefore, mechanistic modeling can be proposed to fill the gap between feeding inhibition and impaired population dynamics (Baird et al., 2007a). The position of feeding responses in the biological hierarchy between molecular biomarkers and fitness traits offers the opportunity to describe in *Gammarus* adverse outcome pathways in multiscale assessment schemes, which could reinforce weight of evidence approaches for the diagnosis of contaminant impacts in freshwater ecosystems (Maltby, 1999; Damásio et al., 2011). Similarly, development of high-throughput omics biomarkers in *Gammarus* species could contribute to this multiscale scheme for the diagnostic of water quality, with the condition that these developments are carefully conducted seeking to maintain a cross-talk with eco-physiology.

11.4 Water Quality Assessment and Interpopulation Variability in Gammarids

Laboratory and in situ bioassays are currently performed with species obtained from laboratory cultures or control native populations. The choice of the test species is a very important issue in the risk assessment of environmental contaminants because toxicity data obtained from one population of this species should, in principle, protect all populations of the same species, but also closely related species. Gauging the uncertainty induced by the choice of a unique experimental population for toxicity assessment is thus a major issue for the relevance toward the predicted potential impacts within the diversity of natural communities. These questions about the generalization of outcomes obtained with one source population are currently tackled in gammarids.

Divergence in life history traits and toxic sensitivity between gammarid populations are potentially explained by two determinants: a phylogenetic component resulting from the conservation and modification of these traits during gammarid evolutionary history at all timescales, and an ecological component because of the possible correlation of life styles in highly diversified habitats of gammarids with vulnerability to contaminants. First, several studies, particularly in crustaceans, have underlined that one of the main sources of variability in toxic sensitivity is the genetic diversity: between families of related individuals (Chaumot et al., 2009; Pease et al., 2010), between clones (Baird et al., 1991; Soares et al., 1992; Coors et al., 2009; Warming et al., 2009), between populations (Maltby and Crane, 1994; Barata et al., 2002; Lopes et al., 2004), or between sibling species (Alonso et al., 2010). As with most amphipods, *Gammarus* populations are better described as belonging to "species complex," with elusive boundaries between intra- and interspecific variations (e.g., Lefébure et al., 2006; Hou and Li, 2010; Westram et al., 2011). For instance, a complex of at least three cryptic species (*G. fossarum* types A, B, C; Müller, 2000; Westram et al., 2011) with overlapping distribution ranges belonging to nominal

G. fossarum has been uncovered by genetic methods. The morphological differentiation between the forms is negligible as compared to the genetic differentiation, leading to the qualification of "cryptism" for this delineation of homogeneously genetic subgroups of populations. Thus, genetic data (barcoding) revealed high cryptic genetic diversity in the *G. fossarum/G. pulex* groups, even at a small spatial scale. More precisely, thorough studies demonstrated additional genetic differentiation within the three lineages of the *G. fossarum* complex comparable to an interspecific level (Weiss et al., 2014). This molecular evidence for genetically divergent lineages correlates in a gradual manner with mate discrimination in natural species (Lagrue et al., 2014), leading to consider that *G. fossarum* or *G. pulex* may be separated themselves into (sub)species.

As a consequence of this intricate genetic structure at intra- and interspecific levels in *Gammarus*, some authors have begun to reevaluate variability of sensitivity to contaminants among species and populations. Alonso et al. (2010) and Boets et al. (2012) investigated short-term toxicity of cadmium to species of the genus *Gammarus* or of the Gammaridae family. However, whereas Alonso et al. (2010) found that *G. fossarum* was significantly greater sensitive to cadmium compared with *G. pulex*, Boets et al. (2012) found no significant difference between the two species. Alonso et al. (2010) also concluded that *G. pulex* was less tolerant to the pharmaceutical ivermectin than *G. fossarum*. Feckler et al. (2012) also suggest that there are differences in the sensitivity of the two cryptic *G. fossarum* lineages A and B to exposure to the fungicide tebuconazole and the insecticide thiacloprid. However, it is important to underline that all these studies used a single reference population for each species; therefore, other reasons, such as different habitats, in geochemical context of sampling sites, and physiological characteristics of organisms, could not be excluded as explaining factors for the reported intergroup variability.

The characterization and quantification of the between-population and between–species variability of toxicity markers at subindividual and individual levels would help address the uncertainties of ecotoxicological assessment. Figure 11.5 presents the result of a study that assess differences in the basal levels and inhibition rate of AChE by a carbamate insecticide in 12 gammarid populations belonging to either *G. pulex*, *G. fossarum* species complexes, or *G. roeseli* species. These populations were selected from a large geographical area in southeastern France (from Burgundy, the Alps, or the Cevennes), covering different types of noncontaminated rivers (from mountain brooks, large plain flood rivers, or Mediterranean streams) and different geological contexts (crystalline, limestone). From this study, it appeared that basal levels of the AChE biomarkers were variable, but that this variability was explained fully by heterogeneity in body size between populations; a biotic influence previously quantified in *G. fossarum* (Xuereb et al., 2009a). Strikingly, a very limited range of AChE response was observed within this diverse set of populations/species, with a very narrow range of variability of inhibition recorded after exposure to methomyl insecticide.

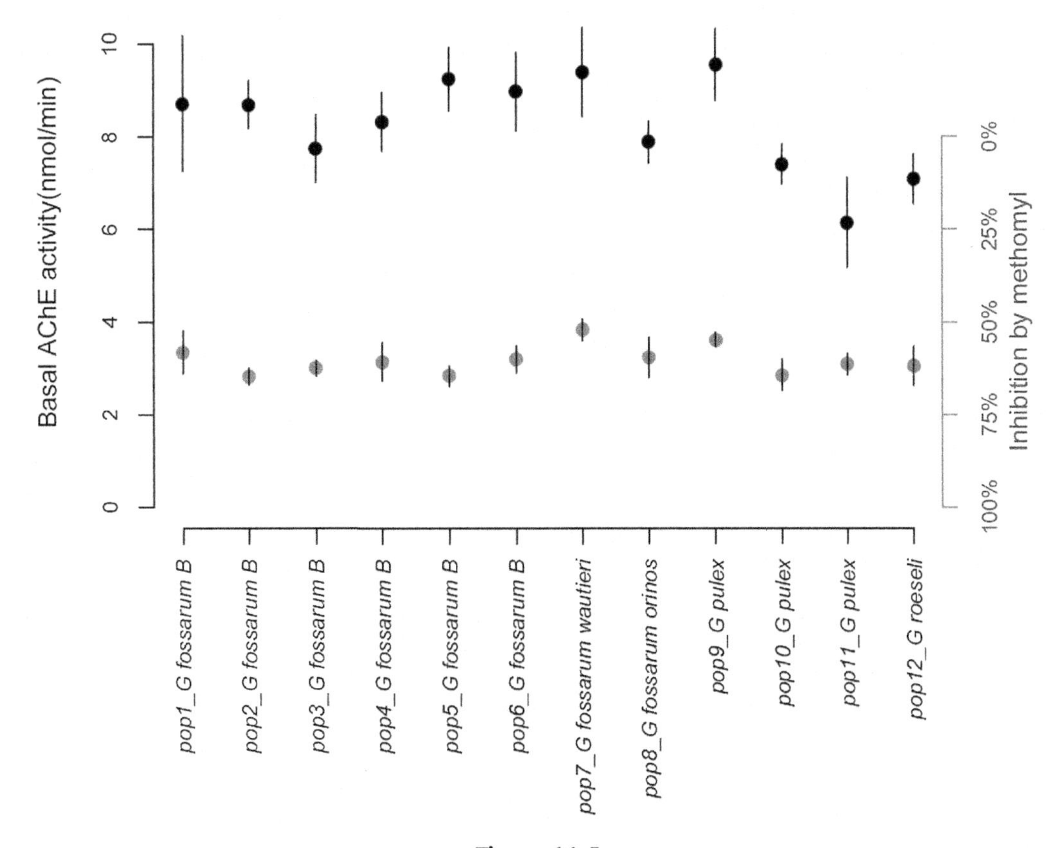

Figure 11.5

Comparison of basal levels of AChE activity and inhibition rate by the insecticide methomyl (1 week; 80 $\mu g L^{-1}$) in 12 *Gammarus* populations belonging to either *G. fossarum* or *G. pulex* species complexes and *G. roeseli* species. These populations were selected on a large geographical scale in southeastern France, covering diversified types of "pristine" stations (mountain brooks, large plain flood rivers, Mediterranean streams) and different geological contexts (crystalline, limestone). Black symbols: controls; gray symbols: exposed specimens.

11.5 Conclusion

This chapter illustrates how *Gammarus* species constitute relevant and well-studied biological models for multiscale assessment of contaminations in freshwater lotic ecosystems. Multiple biomarkers and bioassays using gammarids are available for field testing of contaminant impacts. In addition, complementary modeling developments quantifying the natural variability of these toxicity markers in relation to abiotic factors enhance the reliability of in situ methodology and permit its implementation at different spatial and temporal scales in monitoring programs. Next-generation "omics" are potential complements to this collection of ecotoxicological tools and offer the possibility to develop sensitive responses to detect specific contaminant effects in native invertebrate communities (e.g., endocrine disruption). One of the most promising developments with gammarids is the integration of toxicological endpoints within upscaling frameworks linking to ecological levels. Indeed, by coupling in situ assays and population modeling, an ecologically relevant assessment based on criteria yielding specific typology of field toxicity can be identified and adverse outcome pathways from molecular to population effects described for *Gammarus*. Finally, gammarids are characterized by a large phylogenetic and ecological diversity. Therefore, the establishment of gammarids as sentinel species for freshwater monitoring also provides an opportunity to gain insight into the source of uncertainty imposed by biodiversity for ecotoxicological evaluations.

Acknowledgments

Authors thank the Agence Nationale de la Recherche for financial support through two projects, Gamma (ANR 2011 CESA 021 01) and Proteogam (ANR-14-CE21-0006-01).

References

Agatz, A., Ashauer, R., Brown, C.D., 2014. Imidacloprid perturbs feeding of *Gammarus pulex* at environmentally relevant concentrations. Environ. Toxicol. Chem. 33, 648–653.

Alonso, Á., De Lange, H., Peeters, E., 2010. Contrasting sensitivities to toxicants of the freshwater amphipods *Gammarus pulex* and *G. fossarum*. Ecotoxicology 19, 133–140.

Armengaud, J., 2013. Microbiology and proteomics, getting the best of both worlds!. Environ. Microbiol. 15, 12–23.

Armengaud, J., Trapp, J., Pible, O., et al., 2014. Non-model organisms, a species endangered by proteogenomics. J. Proteomics 105, 5–18.

Aronsson, K.A., Ekelund, N.G.A., 2005. Respiration measurements can assess the fitness of *Gammarus pulex* (L.) after exposure to different contaminants; experiments with wood ash, cadmium and aluminum. Archiv für Hydrobiologie 164, 479–491.

Baird, D.J., Barber, I., Bradley, M., et al., 1991. A comparative study of genotype sensitivity to acute toxic stress using clones of *Daphnia magna* straus. Ecotoxicol. Environ. Saf. 21, 257–265.

Baird, D.J., Brown, S.S., Lagadic, L., et al., 2007a. *In situ* based effects measures: determining the ecological relevance of measured responses. Integr. Environ. Manag. Prot. 3, 259–267.

Baird, D.J., Burton, G.A., Culp, J.M., et al., 2007b. Summary and recommendations from a SETAC Pellston workshop on in situ measures of ecological effects. Integr. Environ. Manag. Prot. 3, 275–278.

Barata, C., Baird, D.J., Soares, A., 2002. Determining genetic variability in the distribution of sensitivities to toxic stress among and within field populations of *Daphnia magna*. Environ. Sci. Technol. 36, 3045–3049.

Bedulina, D.S., Timofeyev, M.A., Zimmer, M., et al., 2010a. Different natural organic matter isolates cause similar stress response patterns in the freshwater amphipod, *Gammarus pulex*. Environ. Sci. Pollut. Res. 17, 261–269.

Bedulina, D.S., Zimmer, M., Timofeyev, M.A., 2010b. Sub-littoral and supra-littoral amphipods respond differently to acute thermal stress. Comp. Biochem. Physiol. 155B, 413–418.

Beketov, M.A., Liess, M., 2008. Potential of 11 pesticides to initiate downstream drift of stream macroinvertebrates. Arch. Environ. Contam. Toxicol. 55, 247–253.

Besse, J.P., Coquery, M., Lopes, C., et al., 2013. Caged *Gammarus fossarum* (Crustacea) as a robust tool for the characterization of.bioavailable contamination levels in continental waters. Toward the determination of threshold values. Water Res. 47, 650–660.

Blockwell, S.J., Pascoe, D., Taylor, E.J., 1996. Effects of lindane on the growth of the freshwater amphipod *Gammarus pulex* (L). Chemosphere 32, 1795–1803.

Blockwell, S.J., Taylor, E.J., Jones, I., et al., 1998. The influence of fresh water pollutants and interaction with *Asellus aquaticus* (L.) on the feeding activity of *Gammarus pulex* (L.). Arch. Environ. Contam. Toxicol. 34, 41–47.

Bloor, M.C., Banks, C.J., 2006. An evaluation of mixed species *in-situ* and *ex-situ* feeding assays: the altered response of *Asellus aquaticus* and *Gammarus pulex*. Environ. Intern. 32, 22–27.

Bloor, M.C., Banks, C.J., Krivtsov, V., 2005. Acute and sublethal toxicity tests to monitor the impact of leachate on an aquatic environment. Environ. Intern. 31, 269–273.

Boets, P., Lock, K., Goethals, P.L.M., et al., 2012. A comparison of the short-term toxicity of cadmium to indigenous and alien gammarid species. Ecotoxicology 21, 1135–1144.

Boulangé-Lecomte, C., Chaumot, A., Fonbonne, C., et al., 2013. La vitellogénine comme biomarqueur d'exposition et d'effet aux perturbateurs endocriniens chez Gammarus fossarum et Eurytemora affinis : développement et application in situ. Rapport final, projet PNRPE. p. 33.

Brooks, S.J., Mills, C.L., 2003. The effect of copper on osmoregulation in the freshwater arnphipod *Gammarus pulex*. Comp. Biochem. Physiol. 135A, 527–537.

Bundschuh, M., Zubrod, J.P., Seitz, F., et al., 2011. Ecotoxicological evaluation of three tertiary wastewater treatment techniques via meta-analysis and feeding bioassays using *Gammarus fossarum*. J. Hazard. Mat 192, 772–778.

Burton, G.A., Chapman, P.M., Smith, E.P., 2002. Weight-of-evidence approaches for assessing ecosystem impairment. Hum. Ecol. Risk Ass. 8, 1657–1673.

Charniaux-Cotton, H., 1965. In: Dehaan, R.L., Ursprung, H. (Eds.), Hormonal Control of Sex Differentiation in Invertebrates. Organogenesis. Holt, Rinehart, and Winston, New York, pp. 701–740.

Charron, L., Geffard, O., Chaumot, A., et al., 2013. Effect of water quality and confounding factors on digestive enzyme activities in *Gammarus fossarum*. Environ. Sci. Pollut. Res. 20, 9044–9056.

Charron, L., Geffard, O., Chaumot, A., et al., 2014. Influence of molting and starvation on digestive enzyme activities and energy storage in *Gammarus fossarum*. PLoS One 9 (4), e96393.

Chaumot, A., Gos, P., Garric, J., et al., 2009. Additive vs non-additive genetic components in lethal cadmium tolerance of *Gammarus* (Crustacea): novel light on the assessment of the potential for adaptation to contamination. Aquat. Toxicol. 94, 294–299.

Colbourne, J.K., Pfrender, M.E., Gilbert, D., et al., 2011. The ecoresponsive genome of *Daphnia pulex*. Science 331, 555–561.

Cold, A., Forbes, V.E., 2004. Consequences of a short pulse of pesticide exposure for survival and reproduction of *Gammarus pulex*. Aquat. Toxicol. 67, 287–299.

Coors, A., Vanoverbeke, J., De Bie, T., et al., 2009. Land use, genetic diversity and toxicant tolerance in natural populations of *Daphnia magna*. Aquat. Toxicol. 95, 71–79.

Correia, A.D., Costa, F.O., Neuparth, T., et al., 2001. Sub-lethal effects of copper-spiked sediments on the marine amphipod *Gammarus locusta*: evidence of hormesis? Ecotoxicol. Environ. Saf. 2, 32–38.

Costa, F.O., Neuparth, T., Correia, A.D., et al., 2005. Multi-level assessment of chronic toxicity of estuarine sediments with the amphipod *Gammarus locusta*: II. Organism and population-level endpoints. Mar. Environ. Res. 60, 93–110.

Coulaud, R., Geffard, O., Coquillat, A., et al., 2014. Ecological modeling for the extrapolation of ecotoxicological effects measured during *in situ* assays in *Gammarus*. Environ. Sci. Technol. 48, 6428–6436.

Coulaud, R., Geffard, O., Xuereb, B., et al., 2011. *In situ* feeding assay with *Gammarus fossarum* (Crustacea): modelling the influence of confounding factors to improve water quality biomonitoring. Water Res. 45, 6417–6429.

Crane, M., Burton, G.A., Culp, J.M., et al., 2007. Review of aquatic *in situ* approaches for stressor and effect diagnosis. Integr. Environ. Ass. Manag. 3, 234–245.

Crane, M., Delaney, P., Watson, S., et al., 1995. The effect of malathion-60 on *Gammarus pulex* (L) below watercress beds. Environ. Toxicol. Chem. 14, 1181–1188.

Crane, M., Johnson, I., Maltby, L., 1996. *In situ* assays for monitoring the toxic impacts of waste in rivers. In: Trapp, J.F., Hunt, S.M., Wharfe, J.R. (Eds.), Toxic Impacts of Wastes on the Aquatic Environment. The Royal Society of Chemistry, Cambridge, pp. 116–124.

Crane, M., Maltby, L., 1991. The lethal and sublethal responses of *Gammarus pulex* to stress: sensitivity and sources of variation in an in situ assay. Environ. Toxicol. Chem. 10, 1331–1339.

Damásio, J., Fernández-Sanjuan, M., Sánchez-Avila, J., et al., 2011. Multi-biochemical responses of benthic macroinvertebrate species as a complementary tool to diagnose the cause of community impairment in polluted rivers. Water Res. 45, 3599–3613.

Dauvin, J.C., Ruellet, T., 2007. Polychaete/amphipod ratio revisited. Mar. Poll. Bull. 55, 215–224.

Davies, I.M., Vethaak, A.D., 2012. Integrated Marine Environmental Monitoring of Chemicals and Their Effects. Report No. 315. 277 p. ICES Cooperative Research.

Dedourge-Geffard, O., Palais, F., Biagianti-Risbourg, S., et al., 2009. Effects of metals on feeding rate and digestive enzymes in *Gammarus fossarum*: an *in situ* experiment. Chemosphere 77, 1569–1576.

Feckler, A., Thielsch, A., Schwenk, K., et al., 2012. Differences in the sensitivity among cryptic lineages of the *Gammarus fossarum* complex. Sci. Total Environ. 439, 158–164.

Felten, V., Charmantier, G., Charmantier-Daures, M., et al., 2008b. Physiological and behavioural responses of *Gammarus pulex* exposed to acid stress. Comp. Biochem. Physiol. 147C, 189–197.

Felten, V., Charmantier, G., Mons, R., et al., 2008a. Physiological and behavioural responses of *Gammarus pulex* (Crustacea: Amphipoda) exposed to cadmium. Aquat. Toxicol. 86, 413–425.

Forbes, V.E., Calow, P., Sibly, R.M., 2008. The extrapolation problem and how population modeling can help. Environ. Toxicol. Chem. 27, 1987–1994.

Forbes, V.E., Palmqvist, A., Bach, L., 2006. The use and misuse of biomarkers in ecotoxicology. Environ. Toxicol. Chem. 25, 272–280.

Forrow, D.M., Maltby, L., 2000. Toward a mechanistic understanding of contaminant-induced changes in detritus processing in streams: direct and indirect effects on detritivore feeding. Environ. Toxicol. Chem. 19, 2100–2106.

Funck, J.A., Danger, M., Gismondi, E., et al., 2013. Behavioural and physiological responses of *Gammarus fossarum* (Crustacea Amphipoda) exposed to silver. Aquat. Toxicol. 142, 73–84.

Galic, N., Ashauer, R., Baveco, H., Nyman, A.-M., Barsi, A., Thorbek, P., Bruns, E. and Van den Brink, P.J., 2014. Modeling the contribution of toxicokinetic and toxicodynamic processes to the recovery of Gammarus pulex populations after exposure to pesticides. Environmental Toxicology And Chemistry. 33, 1476–1488.

Galic, N., Hommen, U., Baveco, J.M., et al., 2010. Potential application of population models in the European ecological risk assessment of chemicals II: review of models and their potential to address environmental protection aims. Integr. Environ. Ass. Manag. 6, 338–360.

Gaskell, P.N., Brooks, A.C., Maltby, L., 2007. Variation in the bioaccumulation of a sediment-sorbed hydrophobic compound by benthic macroinvertebrates: patterns and mechanisms. Environ. Sci. Technol. 41, 1783–1789.

Geffard, A., Quéau, H., Dedourge, O., et al., 2007. Influence of biotic and abiotic factors on metallothionein level in *Gammarus pulex*. Comp. Biochem. Physiol. 145C, 632–640.

Geffard, O., Xuereb, B., Chaumot, A., et al., 2010. Ovarian cycle and embryonic development in *Gammarus fossarum*: application for reproductive toxicity assessment. Environ. Toxicol. Chem. 29, 2249–2259.

Gevaert, K., Vandekerckhove, J., 2000. Protein identification methods in proteomics. Electrophoresis 21, 1145–1154.

Gismondi, E., Rigaud, T., Beisel, J.N., et al., 2012. Microsporidia parasites disrupt the responses to cadmium exposure in a gammarid. Environ. Pollut. 160, 17–23.

Graça, M.A.S., Maltby, L., Calow, P., 1994. Comparative ecology of *Gammarus pulex* (L.) and *Asellus aquaticus* (L.). II: fungal preferences. Hydrobiologia 281, 163–170.

Graça, M.A.S., Maltby, L., Calow, P., 1993. Importance of fungi in the diet of *Gammarus pulex* and *Asellus aquaticus*. II. Effects on growth, reproduction and physiology. Oecologia 96, 304–309.

Gross, M.Y., Maycock, D.S., Thorndyke, M.C., et al., 2001. Abnormalities in sexual development of the amphipod *Gammarus pulex* (L.) found below sewage treatment works. Environ. Toxicol. Chem. 20, 1792–1797.

Hagger, J.A., Jones, M.B., Lowe, D., et al., 2008. Application of biomarkers for improving risk assessments of chemicals under the Water Framework Directive: a case study. Mar. Pollut. Bull. 56, 1111–1118.

Hanson, N., Förlin, L., Larsson, Å., 2010. Spatial and annual variation to define the normal range of biological endpoints: an example with biomarkers in perch. Environ. Toxicol. Chem. 29, 2616–2624.

Hou, Z., Li, S.H., 2010. Intraspecific or interspecific variation: delimitation of species boundaries within the genus *Gammarus* (Crustacea, Amphipoda, Gammaridae), with description of four new species. Zool. J. Linnean Soc. 160, 215–253.

Issartel, J., Boulo, V., Walton, S., et al., 2010. Cellular and molecular osmoregulatory responses to cadmium exposure in *Gammarus fossarum* (Crustacea, Amphipoda). Chemosphere 81, 701–710.

Jubeaux, G., Audouard-Combe, F., Simon, R., et al., 2012b. Vitellogenin-like protein among invertebrate species diversity: Potentiel of proteomic mass spectrometry (LC-MS/MS) for biomarker development. Environ. Sci. Technol. 46, 6315–6323.

Jubeaux, G., Simon, R., Salvador, A., et al., 2012a. Vitellogenin-like proteins in the freshwater amphipod *Gammarus fossarum* (Koch, 1835): functional characterization throughout reproductive process, potential for use as an indicator of oocyte quality and endocrine disruption biomarker in males. Aquat. Toxicol. 112–113, 72–82.

Jubeaux, G., Simon, R., Salvador, A., et al., 2012c. Vitellogenin-like protein measurement in caged *Gammarus fossarum* males as a biomarker of endocrine disruptor exposure: inconclusive experience. Aquat. Toxicol. 122–123, 9–18.

Kedwards, T.J., Blockwell, S.J., Taylor, E.J., et al., 1996. Design of an electronically operated flow-through respirometer and its use to investigate the effects of copper on the respiration rate of the amphipod *Gammarus pulex* (L). Bull. Environ. Contam. Toxicol. 57, 610–616.

Kramer, V.J., Etterson, M.A., Hecker, M., et al., 2011. Adverse outcome pathways and ecological risk assessment: bridging to population-level effects. Environ. Toxicol. Chem. 30, 64–76.

Kunz, P.Y., Kienle, C., Gerhardt, A., 2010. *Gammarus* spp. in aquatic ecotoxicology and water quality assessment: toward integrated multilevel tests. Rev. Environ. Contam. Toxicol. 205, 1–76.

Kupisch, M., Moenickes, S., Schlief, J., et al., 2012. Temperature-dependent consumer-resource dynamics: a coupled structured model for *Gammarus pulex* (L.) and leaf litter. Ecol. Model. 247, 157–167.

Lacaze, E., Devaux, A., Jubeaux, G., et al., 2011a. DNA damage in *Gammarus fossarum* sperm as a biomarker of genotoxic pressure: intrinsic variability and reference level. Sci. Total Environ. 409, 3230–3236.

Lacaze, E., Devaux, A., Mons, R., et al., 2011b. DNA damage in caged *Gammarus fossarum* amphipods: a tool for freshwater genotoxicity assessment. Environ. Pollut. 159, 1682–1691.

Lacaze, E., Geffard, O., Goyet, D., 2011c. Linking genotoxic responses in *Gammarus fossarum* germ cells with reproduction impairment, using the Comet assay. Environ. Res. 111, 626–634.

Lagrue, C., Wattier, R., Galipaud, M., et al., 2014. Confrontation of cryptic diversity and mate discrimination within *Gammarus pulex* and *Gammarus fossarum* species complexes. Freshwater Biol. 59, 2555–2570.

Lawrence, A.J., Poulter, C., 2001. Impact of copper, pentachlorophenol and benzo[a]pyrene on the swimming efficiency and embryogenesis of the amphipod *Chaetogammarus marinus*. Mar. Ecol. Progr. Ser. 223, 213–223.

Lefébure, T., Douady, C.J., Gouy, M., et al., 2006. Phylogeography of a subterranean amphipod reveals cryptic diversity and dynamic evolution in extreme environments. Mol. Ecol. 15, 1797–1806.

Lemos, M.F.L., van Gestel, C.A.M., Soares, A.M.V.M., 2010. Reproductive toxicity of the endocrine disrupters vinclozolin and bisphenol A in the terrestrial isopod *Porcellio scaber* (Latreille, 1804). Chemosphere 78, 907–913.

Leroy, D., Haubruge, E., De Pauw, E., et al., 2010. Development of ecotoxicoproteomics on the freshwater amphipod *Gammarus pulex*: Identification of PCB biomarkers in glycolysis and glutamate pathways. Ecotoxicol. Environ. Saf. 73, 343–352.

Liber, K., Goodfellow, W., den Besten, P., et al., 2007. *In situ*-based effects measures: considerations for improving methods and approaches. Integr. Environ. Ass. Manag. 3, 246–258.

Lopes, I., Baird, D.J., Ribeiro, R., 2004. Genetic determination of tolerance to lethal and sublethal copper concentrations in field populations of *Daphnia longispina*. Arch. Environ. Contam. Toxicol. 46, 43–51.

Lukancic, S., Zibrat, U., Mezek, T., et al., 2010a. A new method for early assessment of effects of exposing two non-target crustacean species, *Asellus aquaticus* and *Gammarus fossarum*, to pesticides, a laboratory study. Toxicol. Ind. Health 26, 217–228.

Lukancic, S., Zibrat, U., Mezek, T., et al., 2010b. Effects of exposing two non-target crustacean species, *Asellus aquaticus* L., and *Gammarus fossarum* Koch., to atrazine and imidacloprid. Bull. Environ. Contam. Toxicol. 84, 85–90.

Macneil, C., Dic, J.T.A., Elwoo, R.W., 1997. The trophic ecology of freshwater *Gammarus* spp. (Crustacea: Amphipoda): problems and perspectives concerning the functional feeding group concept. Biol. Rev. Cambridge Phil. Soc. 72, 349–364.

Malbouisson, J.F.C., Young, T.W.K., Bark, A.W., 1995. Use of feeding rate and re-pairing of precopulatory *Gammarus pulex* to assess toxicity of gamma-hexachlorocyclohexane (lindane). Chemosphere 30, 1573–1583.

Maltby, L., 1992. The use of the physiological energetics of *Gammarus pulex* to assess toxicity. A study using artificial streams. Environ. Toxicol. Chem. 11, 79–85.

Maltby, L., 1994. Stress, shredders and streams: using *Gammarus* energetics to assess water quality. In: Sutcliffe, D.W. (Ed.), Water Quality and Stress Indicators in Marine and Freshwater Ecosystems. Linking Levels of Organisation (Individuals, Population, Communities). Freshwater Biological Association, Ambleside, UK, pp. 98–110.

Maltby, L., 1999. Studying stress: the importance of organism-level responses. Ecol. Appl. 9, 431–440.

Maltby, L., Burton Jr., G.A., 2006. Field-based effects measures. Environ. Toxicol. Chem. 25, 2261–2262.

Maltby, L., Clayton, S.A., Wood, R.M., et al., 2002. Evaluation of the *Gammarus pulex* in situ feeding assay as a biomonitor of water quality: robustness, responsiveness, and relevance. Environ. Toxicol. Chem. 21, 361–368.

Maltby, L., Crane, M., 1994. Responses of *Gammarus pulex* (Amphipoda, Crustacea) to metalliferous effluents: Identification of toxic components and the importance of interpopulation variation. Environ. Pollut. 84, 45–52.

Maltby, L., Hills, L., 2008. Spray drift of pesticides and stream macroinvertebrates: experimental evidence of impacts and effectiveness of mitigation measures. Environ. Pollut. 156, 1112–1120.

Maltby, L., Kedwards, T.J., Forbes, V.E., et al., 2001. Linking individual-level responses and population-level consequences. In: Baird, D.I., Burton Jr., G.A. (Eds.), Ecological Variability: Separating Natural from Anthropogenic Causes of Ecosystem Impairment. Society of Environmental Toxicology and Chemistry (SETAC), Pensacola, FL, USA, pp. 27–82.

Maltby, L., Naylor, C., 1990. Preliminary observations on the ecological relevance of the *Gammarus* 'scope for growth' assay: effect of zinc on reproduction. Funct. Ecol. 4, 393–397.

Maltby, L., Naylor, C., Calow, P., 1990a. Effect of stress on a freshwater benthic detritivore: scope for growth in *Gammarus pulex*. Ecotoxicol. Environ. Saf. 19, 285–291.

Maltby, L., Naylor, C., Calow, P., 1990b. Field deployment of a scope for growth assay involving *Gammarus pulex*, a freshwater benthic invertebrate. Ecotoxicol. Environ. Saf. 19, 292–300.

McLoughlin, N., Yin, D.Q., Maltby, L., et al., 2000. Evaluation of sensitivity and specificity of two crustacean biochemical biomarkers. Environ. Toxicol. Chem. 19, 2085–2092.

Müller, J., 2000. Mitochondrial DNA variation and the evolutionary history of cryptic *Gammarus fossarum* types. Mol. Phylogen. Evol. 15, 260–268.

Naylor, C., Maltby, L., Calow, P., 1989. Scope for growth in *Gammarus pulex*, a freshwater benthic detritivore. Hydrobiologia 188, 517–523.

Neuparth, T., Correia, A.D., Costa, F.O., et al., 2005. Multi-level assessment of chronic toxicity of estuarine sediments with the amphipod *Gammarus locusta*: I. Biochemical endpoints. Mar. Environ. Res. 60, 69–91.

Nikinmaa, M., Rytkönen, K.T., 2011. Functional genomics in aquatic toxicology – do not forget the function. Aquat. Toxicol. 105 (3–4, Suppl.), 16–24.

Nilsson, L.M., 1974. Energy budget of a laboratory population of *Gammarus pulex* (Amphipoda). Oikos 25, 35–42.

Pascoe, D., Kedwards, T.J., Blockwell, S.J., et al., 1995. *Gammarus pulex* (L) feeding bioassay: effects of parasitism. Bull. Environ. Contam. Toxicol. 55, 629–632.

Pease, C.J., Johnston, E.L., Poore, A.G.B., 2010. Genetic variability in tolerance to copper contamination in a herbivorous marine invertebrate. Aquat. Toxicol. 99, 10–16.

Pöckl, M., Timixchil, W., 1990. Comparative study of mathematical models for the relationship between water temperature and brood development time of *Gammarus fossarum* and *G. roeseli* (Crustacea: Amphipoda). Freshwater Biol. 23, 433–440.

Raimondo, S., McKenney, C.L., 2005. Projecting population-level responses of mysids exposed to an endocrine disrupting chemical. Integr. Comp. Biol. 45, 151–157.

Sanchez, W., Porcher, J.M., 2009. Fish biomarkers for environmental monitoring within the Water Framework Directive of the European Union. TrAC 28, 150–158.

Scheil, V., Triebskorn, R., Kohler, H.R., 2008. Cellular and stress protein responses to the UV filter 3-benzylidene camphor in the amphipod crustacean *Gammarus fossarum* (Koch 1835). Arch. Environ. Contam. Toxicol. 54, 684–689.

Schirling, M., Jungmann, D., Ladewig, V., et al., 2005. Endocrine effects in *Gammarus fossarum* (Amphipoda): Influence of wastewater effluents, temporal variability, and spatial aspects on natural populations. Arch. Environ. Contam. Toxicol. 49, 53–61.

Schirling, M., Jungmann, D., Ladewig, V., et al., 2006. Bisphenol A in artificial indoor streams: II. Stress response and gonad histology in *Gammarus fossarum* (Amphipoda). Ecotoxicology 15, 143–156.

Short, S., Yang, G., Kille, P., et al., 2014. Vitellogenin is not an appropriate biomarker of feminisation in a crustacean. Aquat. Toxicol. 153, 89–97.

Simon, R., Jubeaux, G., Chaumot, A., et al., 2010. Mass spectrometry assay as an alternative to the enzyme-linked immunosorbent assay test for biomarker quantitation in ecotoxicology: application to vitellogenin in Crustacea (*Gammarus fossarum*). J. Chromatogr. 1217A, 5109–5115.

Soares, A.M.V.M., Baird, D.J., Calow, P., 1992. Interclonal variation in the performance of *Daphnia magna* straus in chronic bioassays. Environ. Toxicol. Chem. 11, 1477–1483.

Spicer, J.I., Morritt, D., Maltby, L., 1998. Effect of water-borne zinc on osmoregulation in the freshwater amphipod *Gammarus pulex* (L) from populations that differ in their sensitivity to metal stress. Funct. Ecol. 12, 242–247.

Sroda, S., Cossu-Leguille, C., 2011. Effects of sublethal copper exposure on two gammarid species: which is the best competitor? Ecotoxicology 20, 264–273.

Studer, R.A., Robinson-Rechavi, M., 2009. How confident can we be that orthologs are similar, but paralogs differ? Trends Genet. 25, 210–216.

Stuhlbacher, A., Maltby, L., 1992. Cadmium resistance in *Gammarus pulex* (L). Arch. Environ. Contam. Toxicol. 22, 319–324.

Sundelin, B., Eriksson, A.K., 1998. Malformations in embryos of the deposit-feeding amphipod *Monoporeia affinis* in the Baltic Sea. Mar. Ecol. Progr. Ser. 171, 165–180.

Sutcliffe, D.W., 1971. Regulation of water and some ions in *Gammarus* (Amphipoda) II *Gammarus pulex* (L.). J. Exp. Biol. 55, 345–355.

Sutcliffe, D.W., 1984. Quantitative aspects of oxygen uptake by *Gammarus* (Crustacea, Amphipoda): a critical review. Freshwater Biol. 14, 443–489.

Taylor, E.J., Jones, D.P.W., Maund, S.J., et al., 1993. A new method for measuring the feeding activity of *Gammarus pulex* (L). Chemosphere 26, 1375–1381.

Taylor, E.J., Maund, S.J., Pascoe, D., 1991. Toxicity of four common pollutants to the freshwater macroinvertebrates *Chironomus riparius* Meigen (Insecta:Diptera) and *Gammarus pulex* (L.) (Crustacea:Amphipoda). Arch. Environ. Contam. Toxicol. 21, 371–376.

Trapp, J., Armengaud, J., Pible, O., et al., 2015. Proteomic investigation of male *Gammarus fossarum*, a freshwater crustacean, in response to endocrine disruptors. J. Proteome Res. 14 (1), 292–303.

Trapp, J., Armengaud, J., Salvador, A., et al., 2014a. Next-generation proteomics: toward customized biomarkers for environmental biomonitoring. Environ. Sci. Technol. 48, 13560–13572.

Trapp, J., Geffard, O., Imbert, G., et al., 2014b. Proteogenomics of *Gammarus fossarum* to ocument the reproductive system from amphipods. Mol. Cell Proteomics 13, 3612–3625.

Turja, R., Guimaraes, L., Nevala, A., et al., 2014. Cumulative effects of exposure to cyanobacteria bloom extracts and benzo[a]pyrene on antioxidant defence biomarkers in *Gammarus oceanicus* (Crustacea: Amphipoda). Toxicon 2014 (78), 68–77.

Väinölä, R., Witt, J.D.S., Gabrowski, M., et al., 2008. Global diversity of amphipods (Amphipoda; Crustacea) in freshwater. Hydrobiologia 595, 241–255.

Veerasingham, M., Crane, M., 1992. Impact of farm waste on freshwater invertebrate abundance and the feeding rate of *Gammarus pulex* L. Chemosphere 25, 869–874.

Verslycke, T., Ghekiere, A., Raimondo, S., et al., 2007. Mysid crustaceans as standard models for the screening and testing of endocrine-disrupting chemicals. Ecotoxicology 16, 205–219.
Von der Ohe, P.C., Liess, M., 2004. Relative sensitivity distribution of aquatic invertebrates to organic and metal compounds. Environ. Toxicol. Chem. 23, 150–156.
Warming, T.P., Mulderij, G., Christoffersen, K.S., 2009. Clonal variation in physiological responses of *Daphnia magna* to the strobilurin fungicide azoxystrobin. Environ. Toxicol. Chem. 28, 374–380.
Weiss, M., Macher, J.N., Seefeldt, M.A., et al., 2014. Molecular evidence for further overlooked species within the *Gammarus fossarum* complex (Crustacea: Amphipoda). Hydrobiologia 721, 165–184.
Westram, A.M., Jokela, J., Baumgartner, C., et al., 2011. Spatial distribution of cryptic species diversity in European freshwater amphipods (*Gammarus fossarum*) as revealed by pyrosequencing. Plos One 6 (8), e23879.
Wilding, J., Maltby, L., 2006. Relative toxicological importance of aqueous and dietary metal exposure to a freshwater crustacean: implications for risk assessment. Environ. Toxicol. Chem. 25, 1795–1801.
Xuereb, B., Bezin, L., Chaumot, A., et al., 2011. Vitellogenin-like gene expression in freshwater amphipod *Gammarus fossarum* (Koch, 1835): functional characterization in females and potential for use of use as an endocrine disruption biomarker in males. Ecotoxicology 20, 1286–1299.
Xuereb, B., Chaumot, A., Mons, R., et al., 2009a. Acetylcholinesterase activity in *Gammarus fossarum* (Crustacea Amphipoda): intrinsic variability, reference levels, and a reliable tool for field surveys. Aquat. Toxicol. 93, 225–233.
Xuereb, B., Lefèvre, E., Garric, J., et al., 2009b. Acetylcholinesterase activity in *Gammarus fossarum* (Crustacea Amphipoda): linking AChE inhibition and behavioural alteration. Aquat. Toxicol. 94, 114–122.
Xuereb, B., Noury, P., Felten, V., et al., 2007. Cholinesterase activity in *Gammarus pulex* (Crustacea Amphipoda): characterization and effects of chlorpyrifos. Toxicology 236, 178–189.
Zubrod, J.P., Baudy, P., Schulz, R., et al., 2014. Effects of current-use fungicides and their mixtures on the feeding and survival of the key shredder *Gammarus fossarum*. Aquat. Toxicol. 150, 133–143.

CHAPTER 12

Copepods as References Species in Estuarine and Marine Waters

Kevin W.H. Kwok, Sami Souissi, Gael Dur, Eun-Ji Won, Jae-Seong Lee

Abstract

Copepods are a class of small crustaceans that are widespread and naturally abundant in all aquatic ecosystems as a key link between primary producers and higher predators. They have 13 distinct life stages (i.e., egg, six naupliar stages, five copepodite stages, adult), making development easy to trace, and many species are sexually dimorphic. These life-history traits and its importance in the aquatic food web make copepods a suitable candidate as reference organisms for ecotoxicology. Copepods are used particularly for full life-cycle toxicity evaluation. This potential is recognized by researchers, government agencies, and international bodies. This chapter provides a summary of the current state of science in copepod toxicity testing, published protocols of copepod tests, and identifies key research area for further development of copepod testing.

Keywords: Genetics; Life-cycle toxicity; Population effect; Toxicity testing.

Chapter Outline

Aquatic Ecotoxicology. http://dx.doi.org/10.1016/B978-0-12-800949-9.00012-7

Copepod toxicity tests play pivotal roles in aquatic ecotoxicology. They are particularly useful in evaluating full life-cycle effect and population-level toxicity of chemicals. The importance of using copepods in freshwater ecotoxicology was reviewed in a recent article (Kulkarni et al., 2013a); therefore, this chapter will focus on discussing why copepods should become reference species in estuarine and marine ecotoxicology, taking into account their basic ecology and biology. The current state of science on copepod toxicity tests, and efforts from regulatory bodies and international organizations on test standardization, will be reviewed. In addition, a brief summary on available tools for understanding chemical ecotoxicity using copepods will be reviewed. Finally, new development areas and key areas of future research will be highlighted.

12.1 Ecological Importance of Copepods in Estuarine and Marine Environments

Copepods (class: Maxillopoda; subclass: Copepoda) are a group of small crustaceans that can be found in almost all aquatic environments. Copepoda is made up of a total 10 Orders, and Calanoida, Cyclopoida, and Harpacticoida are its dominant Orders. Currently there are about 12,000 identified copepod species (Bron et al., 2011) with 2500 known marine species. New species are being identified regularly (Blanco-Bercial et al., 2014).

Copepods can be found in both the pelagic and benthic compartments of marine and estuarine environments. In pelagic environments, they are the most dominant form of zooplankton in terms of number. Ecologically, they are the principal link between primary production of phytoplankton and higher predators such as shrimps, fish juveniles, and even whales. They support population fisheries of important species such as cod, herring, and salmon (Beaugrand et al., 2003). In oligotrophic waters, copepods also play an important role in grazing heterotrophic bacteria and picoplankton and transferring energy to higher trophic levels (Roff et al., 1995). Because of their sheer number, copepod carcasses and fecal pellets also play important roles in the food web and nutrient cycles (Frangoulis et al., 2011). Copepods play major role in global carbon budget through daily vertical migration and carbon transfer into the deep sea (Frangoulis et al., 2005). Some pathogenic bacteria, such as *Vibrio cholera* and *Salmonella* spp., were also found to attach to surfaces of copepods (Huq et al., 1983; Venkateswaran et al., 1989). Therefore copepod activity and distribution could potentially influence how bacteria are distributed in the marine and estuarine environments. In benthic environments, copepods are also a key group of the meiofauna living in the sediment. In terms of providing biomass for the food chain in the benthic environment, meiofauna are about five times more important than macrofauna (Gerlach, 1971).

Given the high natural abundance, universal presence, and importance in the food web, understanding how chemical substances can impact this important class of animals is central to estuarine and marine ecotoxicology.

12.2 Life History of Copepods and Suitability as Reference Species

Copepods may be free-living (pelagic or benthic) or parasitic. They are a very diverse group of crustaceans, exhibiting a wide range of forms and sizes. Nevertheless, most copepods have a bullet shaped body about 0.5–15 mm long, with the largest parasitic copepods reaching as much as 30 cm (Bron et al., 2011).

Development from egg to adult stage involves six naupliar and five copepodite stages. Copepods molt in between each stage with a pronounced metamorphosis between the last naupliar and first copepodite stage. Development of copepods is rapid, typically requiring a few days to a few weeks to become sexually mature. For a given species, development time depends primarily on environmental temperature but is also influenced by other factors such as salinity (Devreker et al., 2004), food (Souissi et al., 1997), and population density (Zhang and Uhlig, 1993). For some copepod species, such as *Calanus* spp., development may be interrupted by a natural seasonal diapause occurring usually at the C5 stage. Moreover, copepods can produce resting eggs that can be accumulated in the sediment leading to the formation of an "egg bank" (Glippa et al., 2011).

Most copepods reproduce sexually and many species are sexually dimorphic. Females of many species are larger than males (Gilbert and Williamson, 1983). Some calanoids and cyclopoids show variation of color between the two sexes (Wilson, 1932). Other differences include different number of urosomal segments in the female and modification of certain male appendages for grasping of female during mating (e.g., Woodhead and Riley, 1959; Conover, 1965). These differences are normally not apparent until late copepodite or adult stages. For certain species (such as *Tigriopus* spp.), mate-guarding behavior can be observed in which males will grasp onto immature females for an extended period until the female becomes mature for mating (Fraser, 1936). The studies of mating behavior in copepods have been improved by using video-recording techniques, allowing the observation of the mate selection process (Dur et al., 2012). Females of cyclopoids and harpacticoids can produce multiple batches of eggs after a single mating event, whereas female calanoids develop multiple egg broods through repeated mating (Gilbert and Williamson, 1983). It was suggested that sex determination in some cyclopoids and calanoids was genetic, but no conclusive evidence is available to date (Conover, 1965). In addition, environmental factors can also influence sex ratio to different degrees depending on copepod species (Gusmao and McKinnon, 2009) and leading to the appearance of intersexual individuals (Souissi et al., 2010). The typical copepod population is female-biased. Although the exact reason behind this phenomenon is not known, many researchers have suggested that longer life expectancy of female copepods may be the explanation (Gilbert and Williamson, 1983).

All these natural features of copepods made them good candidates to become reference species in marine ecotoxicology. Important characteristics of a reference species for ecotoxicology should include availability, relevance, ease of handling, culturability, sensitivity, and support of background biological data information (Raisuddin et al., 2007). Copepods fulfill all these requirements. As discussed in the previous section, copepods play essential roles in marine ecosystems. They are also abundant naturally as the dominant member of the zooplankton community or benthic environment, making them easy to sample. Identification of field copepods may be difficult for nontaxonomists because of their small size and similar morphology, so consultation with a specially trained taxonomist is recommended (Bron et al., 2011). Nevertheless, many copepod species have been cultured in single-species colonies as live feed for aquaculture (for more information, see Cutts, 2003). Laboratory culture techniques are also common for many copepod species already used in toxicity tests (e.g., *Acartia tonsa*; Gorbi et al., 2012). These sources can provide copepod species with known identification for toxicity tests. Because of their ecological importance, background biological information of many copepod species are available, such as natural life-cycle information, diet, morphology, behavior, and tolerance to environmental variables (e.g., *Tigriopus* spp., see Raisuddan et al., 2007; *Eurytemora affinis*, see Lee and Petersen, 2002).

Intrinsic life-cycle features of copepods also make them desirable as reference species in ecotoxicology. Their distinct life developmental stages provide well-defined endpoints in ecotoxicity tests and allow easy tracking of development. Sexual dimorphism of copepods makes it possible to understand gender-specific chemical toxicity and designation of mating pairs in reproduction tests, which is of particular importance for endocrine disrupting chemicals (EDCs). Rapid development of copepods also makes them good candidates for performing life-cycle or even multigeneration toxicity tests.

A recent meta-analysis study showed that crustaceans are often the most sensitive saltwater taxa to different chemicals, including metals (e.g., copper, cadmium) and biocides (e.g., tributyltin) (Wang et al., 2014). The high sensitivity of crustacean species holds true even when the comparison is made consisting of exclusively temperate species or tropical species only (Wang et al., 2014). Copepod is a main group of crustaceans species used in this analysis, suggesting that they are highly sensitive species. Experimentally, the copepod *Amphiascus tenuiremi* was also the most sensitive species compared with clam, oyster, polychaete, and amphipod when challenged with field-collected sediments from stations in California (Greenstein et al., 2008).

12.3 Copepods in Ecotoxicology

12.3.1 A Brief History

Scientists used marine copepod in ecotoxicology for decades. A search of keywords (copepod AND tox* AND marine) on the ISI Web of Science generated 639 publications in the past 40 years. Interest in ecotoxicity with marine copepods increased rapidly in the 1990s. The majority of the studies were single-species toxicity tests, but a small number used a mixture of field-collected copepods to study community-level toxicity (e.g., Chandler et al., 1997; Gustafsson et al., 2010). Harpacticoids appeared to be the earliest group to be used (*Nitocra spinipes* as in Barnes and Stanbury, 1948). Although early literature were mostly limited to acute lethal toxicity of chemicals on adult copepods, the focus of toxicity studies soon shifted to complete life-cycle or even multigeneration effects of toxic chemicals (e.g., D'Agostino and Finney, 1974). The most extreme example is a 12-generation toxicity study of copper and tributyltin using *Tigriopus californicus* (Sun et al., 2014). Investigation of sublethal effects such as genetic and biochemical toxicity have also become more common in the past decade (see review by Lauritano et al., 2012).

Because of the small size of copepods, most toxicity studies were conducted with static or static-renewal systems. The only exception appeared to be a flow-through test system developed for *N. spinipes* by Bengtsson and Bergstrom (1987).

Scientists have recommended using copepods for toxicity tests for almost 4 decades (Bengtsson, 1978). Many copepod species have been recommended (e.g., *Tigriopus japonicus*, Kwok et al., 2008; *A. tonsa*, Gorbi et al., 2012). Outside academia, regulatory organizations are also looking into standardization of copepod toxicity test procedures (e.g., OECD, 2013). Test standardization by regulatory agencies will be discussed in a later section (section 12.6).

12.3.2 Copepod Species Commonly Used in Ecotoxicology

So far more than 30 copepod species have been used in marine ecotoxicology. Commonly used species include *A. tonsa, A. tenuiremis, Tisbe battagliai, N. spinipes, E. affinis,* and three species from the genus *Tigriopus* (Table 12.1).

12.3.2.1 Acartia tonsa

This estuarine planktonic calanoid is present in temperate and subtropical coastal waters and often a dominant member of the zooplankton community. It has a mean generation time of

Table 12.1: Ecotoxicological Tests Performed with Common Copepod Species

	Type of Test Performed				
	Acute	**Chronic**	**Full Life-Cycle**	**Multigeneration**	**Suborganismal Responses**
			Calanoid		
Acartia tonsa	✓	✓	✓		
Eurytemora affinis	✓	✓	✓		
			Harpacticoid		
Amphiascus tenuiremis	✓		✓		
Tisbe battagliai	✓		✓		
Nitocra spinipes	✓		✓		
Tigriopus spp.	✓		✓	✓	✓

25 days at 17 °C (Zillioux and Wilson, 1966). Since 1977, this species has been used in toxicity tests. In 1978, the US Environmental Protection Agency published an acute toxicity testing protocol for this species (Sosnowski and Gentile, 1978). Large-scale culture method is available because of interest in using this species as live feed in aquaculture (Hammervold et al., 2015; Stottrup et al., 1986). Adult females of *A. tonsa* reproduce by shedding eggs singly in water.

OECD (2007) has validated a full life-cycle bioassay with *A. tonsa* by using a ring test and concluded that consultancy laboratories are able to perform this type of test. This species showed consistent development in different laboratories and can generate data with low variability. It is also one of the species recommended by the International Organization for Standardization (ISO) for acute evaluation of the marine contaminants (ISO/TC 147/SC 5). Moreover, this species is highly sensitive to chemicals: Sverdrup et al. (2002) found that *A. tonsa* is >2 times more sensitive to 25 of 30 tested chemicals than *Daphnia magna*.

12.3.2.2 Eurytemora affinis

E. affinis is another calanoid frequently used in laboratory experiments and is often the most abundant calanoid in both contaminated and uncontaminated estuaries in Europe and North America (Lee et al., 2007; Devreker et al., 2008). Some clades of *E. affinis* can also be found in freshwater environment from North America to Japan (Lee, 1999). Contrary to *A. tonsa*, adult females of *E. affinis* carry an egg pouch with them until hatching of nauplii. The total duration of the cycle ranges from 18 to 20 days, depending on abiotic environmental parameters (i.e., temperature and salinity; Devreker et al., 2009). As with other copepods species in this section, various toxicity assays have been developed for this species (e.g., Lesueur et al., 2013). Several investigations were conducted on *E. affinis,* including the seasonal variation in contaminant concentration in the field (Cailleaud et al., 2007b,c), the uptake and elimination of contaminants (Cailleaud et al., 2009), the effect of several contaminants on its survival (Forget-Leray et al., 2005), reproduction (Wright et al., 1996; Forget-Leray et al., 2005), development (Forget-Leray et al., 2005), and behavior (Cailleaud et al., 2011; Michalec et al., 2013). Moreover, the phylogeny of the species is well-documented with the description of different clades (Lee, 1999), and even three lineages in Europe as a zoom on the European clade (Winkler et al., 2011). Intensive studies with morphological and molecular tools allowed identifying even very subtle invasions (Sukhikh et al., 2013). All these studies led to the conclusion that *E. affinis* has a great potential to be used as a standard bioindicator of water quality.

12.3.2.3 Amphiascus tenuiremis

This marine benthic harpacticoid has a generation time of 21 days at 20 °C (Woods and Coull, 1992). Laboratory culture of *A. tenuiremis* appeared to start in 1986 (Chandler, 1986). Although *A. tenuiremis* lives in the sediment in its natural environment, it is most often used in water-based toxicity tests (e.g., Chandler et al., 2004). It is occasionally used for sediment tests (e.g., Kovatch et al., 1999). Full life-cycle bioassay has been developed for this species using 96-well plates (ASTM, 2004; OECD, 2013).

12.3.2.4 Nitocra spinipes

This benthic harpacticoid was likely the first copepod species used in toxicity test (Barnes and Stanbury, 1948) and is one of the first species to be used in reproduction toxicity tests (Renberg et al., 1980). It has a generation time of about 3 weeks (Perez-Landa and Simpson, 2011). Bengtsson (1978) proposed an acute toxicity test method for adults of this species that was used to test 78 compounds (Linden et al., 1979). The ISO is currently reviewing a larval development test guideline for this species (ISO/CD 18220). Bengtsson and Bergstrom (1987) had also developed a flow-through test system that is unique for copepod toxicity tests.

12.3.2.5 Tisbe battagliai

This is another harpacticoid living in marine sediment. Contrary to *A. tenuiremis*, this species is typically used in sediment toxicity test. Testing with this species appeared to begin in 1987 (Hutchinson and Williams, 1989; Williams, 1992). Even in these early studies, cross-generation toxicity was already investigated. Adult females were held individually and survivorship, reproductive output and offspring survivorship were used as measurements of toxicity (Hutchinson and Williams, 1989; Williams, 1992). Other *Tisbe* species are also used in toxicity studies although not as frequently as *T. battagliai.*

12.3.2.6 Tigriopus *Species*

An extensive review on this genus is available (Raisuddin et al., 2007). Three species of this genus have been used in ecotoxicity studies (*Tigriopus brevicornis, T. californicus* and *T. japonicus*). *Tigriopus* lives in high density in supratidal rock pools and are commonly associated with macroalgae like *Enteromorpha* spp., making them easy to collect. They are tolerant to an extremely wide range of salinity, temperature, and dissolved oxygen levels (Raisudden et al., 2007). Developmental speeds of *Tigriopus* copepods depend on environmental factors. Under normal marine conditions, *T. japonicus* becomes reproductively active after 21 days (Ito, 1970). *T. japonicus* is the only common copepod for toxicity test that has genomic information published (Lee et al., 2010).

12.3.3 Copepod Tests for Polar Region

Understanding of how chemicals impact polar environment, especially for lower trophic level organisms, is very limited. Polar animals possess unique adaptations to this extreme environment and therefore could have different sensitivity toward chemicals. For example, polar copepods *Calanus glacialis* are typically larger than their temperate counterparts and with much longer time span (Atkinson, 1991; Kosobokova, 1999).

Meta-analyses have shown that sensitivity differed between polar versus temperate and tropical species so toxicity data from one geographic region cannot be extrapolated to other regions (Chapman et al., 2006). However, there are no established test crustaceans available from the polar region, although some toxicological studies on polar amphipods had been reported (e.g., *Gammarus wilkitizkii* in Hatlen et al., 2009). Copepods have great potential to fulfill this important role. Effects of crude oil, PAHs and surfactants on Arctic copepod *C. gracialis* have been extensively studied (e.g., Hansen et al., 2011; Gardiner et al., 2013), as has mercury toxicity (Overjordet et al., 2014). In another Arctic copepod *Calanus hyperboreus*, Norregaard et al. (2014) examined the effects of pyrene on reproductive effort, egg hatching success, fecal pellet production, and survivorship. In another recent study, the effect of metal exposure was assessed in three common Antarctic copepods: the calanoids *Paralabidocera antarctica* and *Stephos longipes* and the cyclopoid *Oncaea curvata* (Marcus et al., 2015).

12.3.4 Common Endpoints Used in Copepod Toxicity Tests

Raisuddin et al. (2007) provided a summary of marine copepod toxicity studies. The most common endpoint used in copepod toxicity test is acute mortality. Other endpoints used in copepod toxicity tests include developmental speed and success, fecal production, and reproductive output. For full life-cycle and multigeneration tests, offspring viability, sex ratio, and development are also commonly recorded (e.g., Kwok et al., 2009). Because copepods are extensively used in full life-cycle toxicity tests, endpoints based on population model estimates such as population growth rate (r) and generation time are also common.

12.4 Genetic and Biochemical Techniques Adopted and Developed for Copepods for Environmental Studies

Genetic and biochemical techniques have been widely applied for developing biomarkers to evaluate the effects of chemicals for about 15 years (Table 12.2). Despite copepods having been used in environmental monitoring and toxicological studies since the 1940s, mechanistic understanding of toxicity had been largely missing. Development of biochemical techniques and genomic approaches to understand the mechanisms of toxicity in copepods will be an important area in the future.

Early studies mostly focused on adaptability of established biomarkers (Chapter 7) to copepods. Barka et al. (2001) investigated acute toxicity of several metals (Cu, Zn, Ni, Cd, Ag, Hg) and concentrations of metallothionein-like proteins (MTLPs) using *T. brevicornis* (Barka et al., 2001). Increased concentration of MTLPs was recorded as early as after 24 h of exposure to metals and different concentrations of MTLPs in copepods generally reflected the toxicity of metals (Barka et al., 2001), indicating applicability of using this response

Table 12.2: Summary of Ecotoxicological Studies on Copepods Using Genetic and Biochemical Techniques

Species	Exposure	Target Mechanisms or Research Objectives	Responses (g, gene expression; e, enzyme activities; b, gene product or other biochemical analysis)	References
Tigriopus japonicus (*H*)	Salinity, Cu, Mn, H_2O_2	Cellular protective mechanisms	*GR* (g)	Seo et al. (2006a)
	Temperature	Chaperoning role in thermal stress	*Hsp20* (g)	Seo et al. (2006b)
	EDCs	Evaluate *hsp20* as a biomarker for EDCs	*Hsp20* (g)	Seo et al. (2006c)
	OP, polychlorinated biphenyls	Different mode of toxicity	*GSTs*	Lee et al. (2006)
	H_2O_2	Antioxidant defense system	*GSTs* (g)	Lee et al. (2007)
	H_2O_2, Ag, As, Cd, Cu	Antioxidant defense system and apoptosis	*GSTs* (g), TUNEL (b)	Lee and Raisuddin (2008)
	Ag, As, Cd, Cu	Endocrine system	*Vitellogenin* (g)	Lee et al. (2008a)
	Salinity, temperature	Stress-response regulation system	*Corticotropin-releasing hormone binding protein (g)*	Lee et al. (2008b)
	Cd	Cd effect of detoxification process and antioxidant system	SOD, GPx, GST, AChE, LPO, GSH/GSSG, MT (e), GSH (b)	Wang and Wang (2009)
	Cu, Zn, Ag, EDCs, temperature	Function of *hsp* gene in chaperoning and detoxification	*Hsps* (g)	Rhee et al. (2009)
	Cu	Detoxification and antioxidant system	6K oligochip (g)	Ki et al. (2009)
	EDCs	Function of P53 gene in EDC exposure	*P53* (g)	Hwang et al. (2010a)
	Ni	Ni effect of detoxification process and antioxidant system	SOD, GPx, GST, AChE, GSH/GSSG, LPO, MT (e), GSH (b)	Wang and Wang (2010)
	Cu, Zn, Ag, EDCs	Antioxidant system	*Cu/Zn SOD, MnSOD* (g)	Kim et al. (2011)
	Ag, As, Cd, Cu, Zn, UV radiation, WAF	Endocrine metabolic energy regulation	*Hyperglycemic hormone* (g)	Kim et al. (2013)
	Ag, As, Cd, Cu, B[*a*]P, polychlorinated biphenyls, tributyltin	Confirm ELISA for an indicator system	GST-S (ELISA) (b)	Rhee et al. (2013a)
	Cu	Apoptotic cell death	GSH, GPx, GR, GST, SOD (e), ROS, apoptosis (TUNEL/BrdU) (b) *GST, GR, GPx, SODs, CAT, Hsps, caspase, Bcl2, p53*(g)	Rhee et al. (2013b)
	Gamma radiation	DNA repair system, antioxidant system	ROS, GST, GR, GPx (e), *p53, Ku70, Ku80, DNA-PK, PCNA, GST, GPx, GR, SODs, CAT, Hsps* (g)	Han et al. (2014a)
	WAF	Detoxification mechanisms	Whole *CYP* genes (g), GSH, GST, GR, GPx, CAT (e)	Han et al. (2014b)
	Cu, Cd, Zn	Multidrug-resistance transport activity	*P-gp* (g)	Jeong et al. (2014a)

Continued

Table 12.2: Summary of Ecotoxicological Studies on Copepods Using Genetic and Biochemical Techniques —cont'd

Species	Exposure	Target Mechanisms or Research Objectives	Responses (g, gene expression; e, enzyme activities; b, gene product or other biochemical analysis)	References
	LPS, *Vibrio* spp.	Immune system	*Rel/NF-B, IB, lipopolysaccharide-induced TNF-a (LITAF)*(g)	Kim et al. (2014a)
	Ag, As, Cd, Cu, Zn	Oxidative-related chaperoning system	*Hsp* (g), ROS, GSH (b), GST, GR, GPx, SOD (e)	Kim et al. (2014b)
	Triphenyltin	Evaluating the toxic effect and mechanisms of triphenyltin chloride	*Hsp70, Hsp90, Hsp90b, GST-O, GST-S, RXR1-6* (g), ROS (b)	Yi et al. (2014)
	UV radiation	Cell signaling pathway and antioxidant system	*P53, hsps, p38 MAPK* (g), ROS, GSH (b), GST, GPx, SOD (e)	Kim et al. (2015)
	Biocides	Nervous signal transmission system	*AChE* (g), AChE inhibition (e)	Lee et al. (2015)
Tigriopus brevicornis (H)	As, Cu, Cd + pesticides (carbofuran, dichlorvos, or malathion)	Neurotoxic effects	AChE inhibition (e)	Forget et al. (1999)
	Ag, Cd, Cu, Hg, Ni, Zn	Detoxificatory mechanisms	MTLP (b)	Barka et al. (2001)
	Pesticides in water	Evaluate neurotoxicity in field	AChE inhibition (e)	Forget et al. (2003)
Tigriopus californicus (H)	Salinity	Osmolyte regulation system	*δ1-pyrroline-5-carboxylase synthase* (g)	Willet and Burton (2002)
	Salinity	Glutamate synthase system	Glutamate dehydrogenase activity (e)	Willet and Burton (2003)
Tigriopus fulvus (H)	Malathione, Pestanal®	Evaluate pesticide effects on enzyme and survival	AChE inhibition, GST (e)	Agrone et al. (2011)
Calanus finmarchicus (Ca)	Temperature	Thermal stress and physiological state	*Hsp70* (g)	Voznesensky et al. (2004)
	Mono ethanol amine, WSF, Cu, temperature	EST screening to each stress condition	SSH-PCR (189 ESTs) (g)	Hansen et al. (2007)
	naphthalene	Mode of naphthalene toxicity	*CYP330A1, CYP1A2, GST, SOD, CAT, Hsps* (g)	Hansen et al. (2008)
	Dispersed oil (WSF)	Toxic mode according to oil droplet	*GST, CYP330A1* (g)	Hansen et al. (2009)
	Diethanolamine	Mode of toxicity	*GCS, ferritin, SOD, Tp-beta, GSH-S, UBQ-SP7, GST, CYP330A1* (g)	Hansen et al. (2010)
	WAF	Investigate the potential toxicity	*GST* (g), lipid contents (b)	Hansen et al. (2011)
	Food shortage	Select target genes that seem to be responsive to environmental factors	EST database (g)	Lenz et al. (2012)
Calanus glacialis (Ca)	WAF	Investigate the potential toxicity	*GST* (g), lipid contents (b)	Hansen et al. (2011)

Calanus helgolandicus (Ca)	Diet effect	Select putative reference gene for further toxicogenomics	*Actin, Elongation factor a, GAPDH, 18S, S7, S20, ATPs, UBI, IST* (g)	Lauritano et al. (2011a)
	Toxic diet	aldehyde detoxification, apoptosis, cytoskeleton structure, and stress response	*Hsp70, Hsp40, CYP4, GST, GST-s, CAT, SOD, ALDHs, CARP, CAS, IAP, Tublins* (g)	Lauritano et al. (2011b)
Eurytemora affinis (Ca)	Salinity, temperature	Establish effective enzymatic biomarkers for in situ early-warning signal	AChE, GST (e)	Cailleaud et al. (2007a)
	Salinity, temperature	To test synergy effect of thermal and salinity and apply for field condition	Glucose-regulated protein 78, hsp90A (g)	Xuereb et al. (2012)
Hemidiaptomus toubaui (Ca)	Cd, heat stress	Cellular oxidative stress	Hsp70, metallothionein, SOD (b)	Liberge and Barthélémy (2007)
Eudiaptomus gracilis (Ca)	UV radiation	Oxidative stress related cell death and neurotransmission system	GST, caspase-3, ChE (e)	Souza et al. (2012)
Paracyclopina nana (Cy)	As, Cd, Cu	Endocrine disruption	*Vitellogenin* (g)	Hwang et al. (2010b)
	Gamma radiation	Oxidative stress and DNA repair	ROS, GSH (b), SOG, CAT, GT, GST (e), *SOD, GR, GST, Hsp p53, DNA-PK, Ku70, Ku80* (g)	Won and Lee (2014)
	UV radiation	Antioxidant system	GST, SOD (e), ROS, LPO, fatty acid composition (b)	Won et al. (2014)
Amphiascus tenuiremis (H)	Fipronil	Endocrine system	Vitellin (e)	Volz and Chandler (2004).
Arcatia tonsa (Ca)	UV radiation	Protein synthesis system	Hsp (b)	Tartarotti and Torres (2009)
	Toxic diet	Detoxifying mechanisms	GST, EROD (e)	Kozlowsky-Suzuki et al. (2009)
Temora longicornis (Ca)	Toxic diet	Detoxifying mechanisms	GST, EROD (e)	Kozlowsky-Suzuki et al. (2009)
Lepeophtheirus salmonis	Emamectin benzoate	Multidrug transport system	*Pgp* (g)	Tribble et al. (2007)
Pseudodiaptomus annandalei (Ca)	Ni	molecular mechanisms of nickel toxicity	SHH-PCR (474 ESTs), *Cathepsin D, ferritin, SODs, Myohemerythrin-1, ribosomal protein, Separase, Vitellogenin 2* (g)	Jiang et al. (2013)

Ca, calanoid; Cy, cyclopoid; ELISA, enzyme-linked immunosorbent assay; H, harpacticoid; SHH, suppression subtractive hybridization; UV, ultraviolet; WSF, water-soluble fraction.

as a biomarker of metal exposure and toxicity. Acetylcholinesterase (AChE) activities in *T. brevicornis* has also been successfully used to evaluate combined effects of insecticides in the field (Forget et al., 1999).

Other techniques and approaches applied to understand how copepods respond to environmental stressors such as temperature and salinity may be applicable for copepod toxicity studies. The two most common measurements in these stressor studies are oxygen consumption (e.g., McAllen and Taylor, 2001) and osmolality changes (Hansen et al., 2012). In a number of studies, combined effect of environmental stress and chemicals were studied. For instance, in the copepod *E. affinis*, enzymatic biomarkers (AChE and glutathione *S*-transferase, GST) were applied to measure in situ contaminated estuarine environment in combination with wide ranges of salinity and temperature (Cailleaud et al., 2007a). The authors demonstrated that understanding of combined effect of chemicals with these environmental factors would be important to interpret the fluctuations of enzymatic activities in *E. affinis* and most likely other copepods.

In the copepod *Eudiaptomus gracialis*, diverse enzyme activities (GST, AChE, and caspase-3) were useful to examine how organisms respond to ultraviolet radiation (Souza et al., 2012). Also some studies demonstrated that biochemical analyses can support biological alterations of copepod observed at individual levels such as reproduction impairment, retarded growth and reduced survival rate under laboratory-based stress condition (Han et al., 2014a,b; Won et al., 2014; Won and Lee, 2014).

In addition to the identification of potential biomarker genes in response to chemical exposure, application of gene expression techniques allow the evaluation of transcriptional changes of important genes in response to chemicals in target species, aiming at a better understanding of the mechanisms of environmental stress. Omics approaches will allow the identification of new genes which will give a more comprehensive overview of how copepods respond to specific stressors in laboratory and/or field conditions (Lauritano et al., 2012). Gene expression patterns of marine copepods in response to chemicals and several environmental stresses have been reviewed earlier by Lauritano et al. (2012). Gene expression endpoints were applied so far in studies covering only eight copepod species (e.g., *T. japonicus*, *T. californicus*, *Calanus finmarchicus*, *C. glacialis*, *Calanus helgolandicus*, *Paracyclopina nana*, *Lepeopththeirus salmonis*, and *Pseudodiaptomus annandalei*). The majority of the studies were using *T. japonicus* or *C. finmarchicus*.

The planktonic copepod *C. finmarchicus* has received a great deal of attention because they play an important role in the Northern Atlantic food web as the dominant zooplankton in spring and summer (Meise and O'Reilly, 1996). For *C. finmarchicus*, most of the studies focused on either oxidative stress generated by temperature (Voznesensky et al., 2004) and chemicals related to oil and gas production such as naphthalene (Hansen et al., 2008) or different oil fractions (Hansen et al., 2009, 2011). For example, the expression of *GST* and some cytochrome P450 (*CYP*) genes were studied to understand the toxicity of different chemicals to copepods under different physiological states. Increased *GST* and decreased

CYP330A1 expression levels were found to relate respectively to lipid peroxidation and ecdysteroidogenesis in *C. finmarchicus* after naphthalene exposure (Hansen et al., 2008). In *C. finmarchicus*, the Expressed Sequence Tags (ESTs) (11,859 ESTs as of December 31, 2014, in the National Center for Biotechnology Information database) are also recently available. This milestone of genomics knowledge of the species allow future studies to identify putative target genes that are responsive to environmental stressors to provide more mechanistic understanding of chemical and environmental stress (Lenz et al., 2012).

Tigriopus copepods, in particular *T. japonicus*, were also heavily used in studies of molecular response to toxicity. Target molecules include detoxification enzymes and proteins (e.g., GST, heat shock proteins (HSPs), superoxide dismutase (SOD), and catalase) in copepods at the molecular level. The data indicate high inter- and intraspecific variability in copepod response, depending on the type of stressor tested, the concentration and exposure time, and the enzyme isoform studied. Regarding oxidative stress, responses of GST, GR, SOD were evaluated in copepods in response to environmental stressors (Kozlowsky-Suzuki et al., 2009; Wang and Wang, 2009; Agrone et al., 2011; Han et al., 2014a;b; Won et al., 2014; Won and Lee, 2014; Kim et al., 2015). For example, cadmium- and nickel-induced oxidative stress was measured by enzymatic activities of SOD, glutathione peroxidase (GPx), GST, and AChE ratios that were reduced to oxidized glutathione (GSH/GSSG) along with MTLP expression and thiobarbituric reactive species assay to determine lipid peroxidation in the copepod *T. japonicus* (Wang and Wang, 2009, 2010). In cadmium- and nickel-exposed *T. japonicus*, all the biomarkers responded sensitively.

An enzyme-linked immunosorbent assay kit for measuring the antioxidant GST-S protein expression was recently developed for *T. japonicus* (Rhee et al., 2013a). The Tj-GST-S protein was upregulated in a dose-dependent manner in response to four metals and at high concentration of benzo[*a*]pyrene, polychlorinated biphenyls, and tributyltin, indicating oxidative stress can be caused by these chemicals (Rhee et al., 2013a). Biomarkers such as proliferating cell nuclear antigen and TUNEL assay for cell proliferation incorporating BrdU and apoptosis, respectively, were also used to understand mechanistic toxicity of copper (Rhee et al., 2013b). They found a significant induction of enzymatic activities of antioxidant proteins with increased intracellular reactive oxygen species (ROS) as well as an increase of TUNEL-positive cells but a decrease of BrdU-positive cells.

T. japonicus is a candidate reference species for understanding molecular aspect of toxicity because of the amount of genomic information available (Lee et al., 2010). A total of 686 ESTs including vitellin, several kinases and potential detoxification-related genes showed a possibility to develop into biomarkers for environmental research (Lee et al., 2005). Using *T. japonicus*, changes of gene expression such as *GST*, *GR*, and *hsp20* genes were examined in response to chemicals (e.g., heavy metals, EDCs) alone and in combination with environmental stressors (e.g., salinity and temperature) (Lee et al., 2006; Seo et al., 2006a,b,c). Also, mRNA expression patterns of *T. japonicus* vitellogenin (*Vtg*) genes were examined to develop a biomarker index for monitoring EDCs including nonylphenol, octylphenol, and bisphenol A (Lee et al., 2008a). Finally, a Tj-6K oligochip was successfully developed for this species to understand potential toxic pathways of toxic metals (Ki et al., 2009) and more precisely the specific mechanisms of toxicity in copepods along with obtaining early warning signals in response to toxic chemical exposure.

With the advance and popularization of next-generation sequencing (NGS) techniques and facilities, it can be predicted that more copepod genomes will become available. The NGS has been developed as an emerging and useful tool to obtain massive whole genome (Ekblom and Galido, 2011) and RNA-Seq information in copepods (Lee et al., 2011; Rhee et al., 2012). Using whole genome and RNA-Seq information, the finding of novel molecular biomarkers for ecotoxicological testing can be easily combined with mechanistic studies of toxicity (e.g., detoxification and defense mechanisms) covering the full life-cycle of target species. Recently, in *T. japonicus*, the genome sequence (185 Mb) and RNA-Seq information (54,761 contigs; average length 1515 bp; total length of contigs 82,984,914 bp) became available, whereas the copepod *T. californicus* whole genome (scaffold no. 2365; total length of scaffolds 184,634,130 bp; gene number 14,536; Figure 12.1) and RNA-Seq was also available at the National Center for Biotechnology Information. For example, in *T. japonicus*, a multidrug resistance system and two phases of metabolic system (phase I and phase II) were published after obtaining all the gene families from whole genome and RNA-Seq databases to better understand the defense mechanisms in copepods (Han et al., 2014a; Jeong et al., 2014b). Particularly, entire ATP-binding cassette superfamily genes that act as active efflux pumps of xenobiotic substrates were identified from *T. japonicus* (Jeong et al., 2014b). They demonstrated the reason why nauplii *T. japonicus* (N5 and N6 stage) was sensitive on transcription of ATP-binding cassette transporter genes in response to diverse environmental conditions including xenobiotic exposure, affecting molting and growth.

In *T. japonicus*, 52 different *CYP* genes were identified and their transcriptions in response to water-accommodated fractions (WAFs) of crude oil were evaluated (Han et al., 2014b). Entire *T. japonicus CYP* genes study demonstrated that three *CYP* genes belonging to specific *CYP* clan (clan 3) were significantly induced by WAF, playing a crucial role in detoxification mechanisms of low molecular weight PAHs, whereas clan 2 and mitochondrial *CYP* genes also played a major role in naphthalene metabolism in *C. finmarchicus* (Hansen et al., 2008). Thus, gene batteries of key defense-related mechanisms obtained from whole genome and RNA-Seq information can provide more accurate and informative knowledge to fill out our knowledge gaps in aquatic toxicology. However, the knowledge on copepod whole genomics is still largely lacking. Genome data using the transcriptome sequencing by 454 GS FLX of another copepod *Calanus sinicus* was recently released (Ning et al., 2013), but it is not a species typically used in toxicity tests. Only recently whole genome assembly data from *T. californicus* was released by the National Center for Biotechnology Information bioproject (https://i5k.nal.usda.gov/Tigriopus_californicus); this information may be adaptable to *T. japonicus* and *T. brevicornis*.

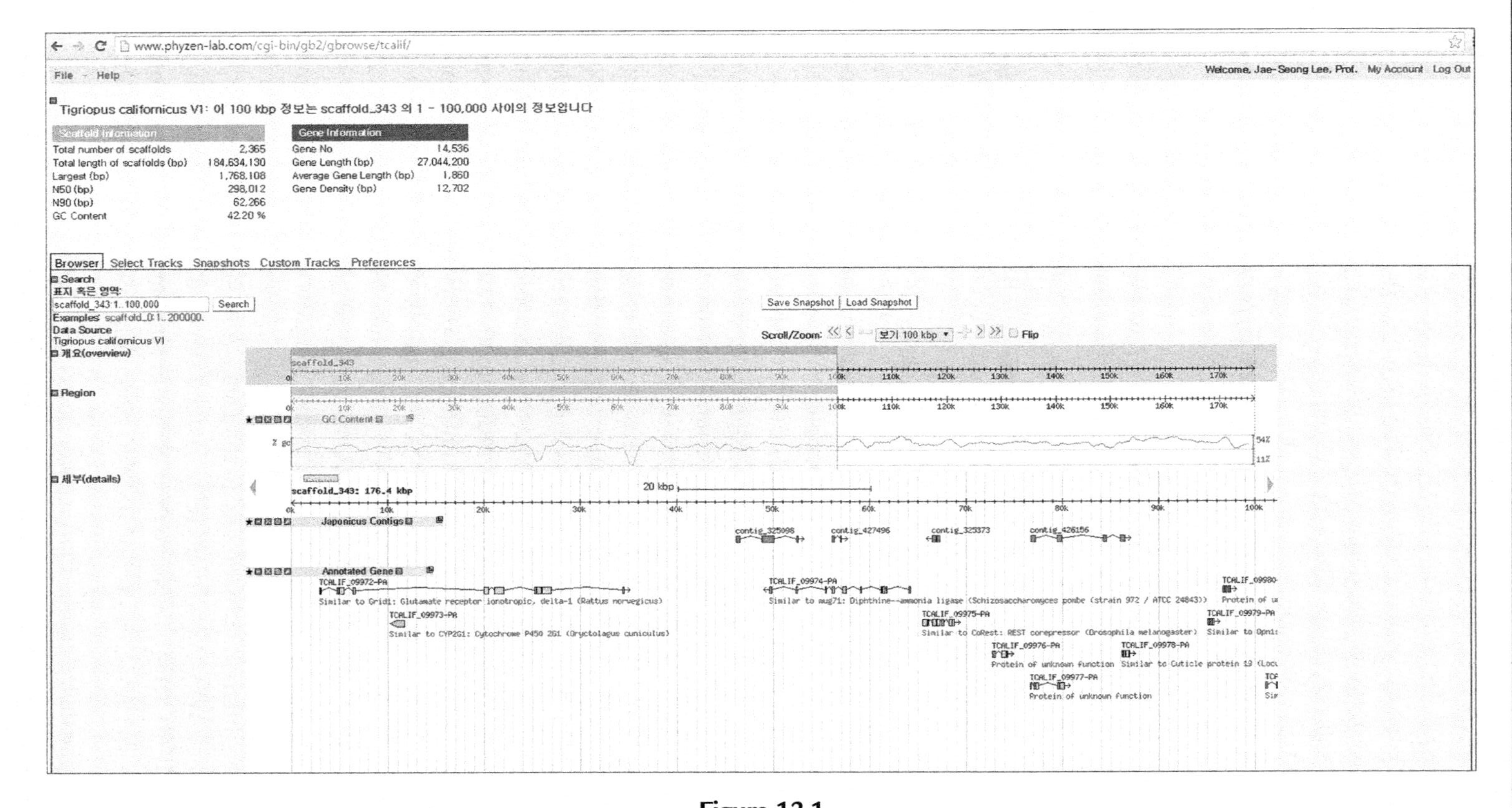

Figure 12.1
Genome structure of the copepod *Tigriopus californicus.*

Reverse genetics approaches such as knockdown of target genes via RNA interference was successfully applied to *T. californicus* (Barreto et al., in press). Such techniques are necessary to elucidate underlying molecular mechanisms of toxicity. These approaches will also enable functional study in copepods elucidating physiological function of each gene or large gene family in the near future to deepen background biological knowledge of copepods.

12.5 Use of Modeling as Tools to Integrate Effects from Individual to Populations

Even some are standardized (see the next section), it is most unusual that full life-cycle toxicity tests have been developed and performed for all common copepod test species.

Full life-cycle tests allow us to extrapolate individual-based effects to population level effects with help of modeling tools. This step is not only critical for us to understand higher level ecological impact of toxic chemicals, but it also allows integration of multiple life-cycle tests endpoints into one to two indices. This facilitates more effective comparison and communication of potential hazard with regulators, managers, and the general public.

Ecological models are increasingly used to highlight ecological risk, extrapolating individual-level laboratory-based scenarios to realistic population-level conditions in the field (Preuss et al., 2010; Forbes et al., 2011; Kulkami et al., 2014; Gasbi et al., 2014). Models also represent relevant tools to optimize experimental design (Preuss et al., 2011). There are different modeling methods to study emerging phenomenon in ecotoxicology (Mooij et al., 2003). Among these methods, those that describe individual organisms (i.e., Individual-Based Model, IBM) or agents (Agent-Based Model) are viewed as the most appropriate to mechanistically model individual-level processes and extrapolate them to the population level (Grimm and Railsback, 2005; DeAngelis and Grimm, 2014). IBMs allow considering aspects that are usually ignored in classical analytical models: individual variability, local interaction, complete life-cycle and animal adaptation according to its state and external environment. These additional considerations lead to better abilities to project effect at the population level as demonstrated for birds (Topping et al., 2005), soil arthropods (Meli et al., 2014), fish (Hanson and Stark, 2012), and even copepods (i.e., *A. tenuiremis*; Belleza et al., 2014). Within individual-based approach, individuals are attributed different properties with respect to their life-history traits; and population dynamics emerge as consequences of individual-level variability (DeAngelis and Mooij, 2005). Life history strategies in the field vary between species, resulting in different vulnerability to toxicants. Additionally, there is a strong life-stage dependent sensitivity in copepods (e.g., Hutchinson et al., 1999; Zafar et al., 2012; Kulkarni et al., 2013b). IBMs, owing to their ability to integrate all life-cycle traits, are perfect tools to approach such issue and quantify emerging properties of vulnerability.

Nevertheless, despite their obvious fit to answer questions of emerging properties, IBMs were often laid aside in favor of other models. This could be associated with the fact that IBMs require consequent computer investment, are more complex (Grimm et al., 1999), and more difficult to communicate that traditional matrix-based or analytical models (expressed through mathematical general language). These issues vanished throughout the last decades with the development of platform and specialized tools (Lorek and Sonnenschein, 1999; Ginot et al., 2002), and the establishment of standardized protocol to describe IBMs (Grimm et al., 2006, 2010). This protocol was applied to describe copepod life-cycle during its development (Grimm et al., 2006) and it has subsequently been used to describe two IBMs developed for studying egg-carrying copepod (Dur et al., 2009, 2013).

The construction of an appropriate framework is essential for the modeling approach. This framework joins together laboratory experiments and mechanistic modeling. Additionally, the combination of ecological and effect models (that consider the toxicokinetics) is needed to calibrate individual-level effects on the species and then to simulate population-level responses (Preuss et al., 2010). IBMs allow the integration of different submodels to simulate the effect of toxicant on organism-level endpoints, e.g., reproduction, survival, heritability (Gasbi et al., 2014). The effect on reproduction can be accounted by stress functions as, for instance, the hermetic model developed by Brain and Cousens (1989) or its modified version (Cedergreen et al., 2005). The former modeled efficiently reproductive effects of a pesticide additive (Dispersogen A) at the population level in *Daphnia* (Gasbi et al., 2014). Nowadays, the General Unified Threshold model for survival (Jager et al., 2011) is commonly used to describe toxic effect based on the survival data (Kulkarni et al., 2013a; Gasbi et al., 2014). Nevertheless, the most adapted models for survival and reproduction may vary with the considered toxicant. For instance, a simple dose–response relationship for survival and for reproduction may suffice for predicting the effect of 3,4-dichloraniline on *Daphnia* (Preuss et al., 2010), but not for the effect of the surfactant-based dispersing agent Dispersogen A (Gasbi et al., 2014). An important point highlighted recently by Gasbi et al. (2014) is that F1 generation effects, which might even control the effects on population more than survival effects, are required to fully explain population level effects.

For IBMs, the model is parameterized using individual-level stressor-response relationship derived from laboratory experiments. The model is then applied to simulate copepod population dynamics on uncontaminated or contaminated conditions. For that purpose measurement data of environmental conditions (i.e., temperature, food availability, salinity, and pollutant concentration) are necessary.

Among the different IBMs developed for copepod so far, some effectively integrate and quantify the impact of both natural and anthropogenic forcing (Belleza et al., 2014; Kulkarni et al., 2014; Korsman et al., 2014) and showed the difference between individual-level and population-level sensitivities (Kulkarni et al., 2014). Those results confirm that models are valuable tools in predicting population-level responses from individual-level data and therefore ecological risk assessment of chemicals. They also represent good examples of the effectiveness of combining experimental work and modeling. Such a combination fits well within the tiered risk assessment scheme proposed by the European commission (EFSA, 2013). Data from laboratory experiments on copepods constitute tier 2. These data can then be implemented into a model constituting tier 3. Validation of the population models should be conducted on data from population experiments and mesocosm studies. After successful testing and validation, the model could serve to predict the population-level effects of toxicants on copepods. Moreover some IBMs are suitable tools to integrate in the future the emerging experimental development in copepod studies aiming to investigate the acclimation and adaptation of copepods through different multigenerational protocols.

12.6 Toxicity Guidelines for Copepods from National and International Organizations

Many regulatory agencies recognize the potential and importance of copepods in marine ecotoxicology. In 1978, the US Environmental Protection Agency published an acute toxicity testing protocol for *A. tonsa* (Sosnowski and Gentile, 1978). This is perhaps the first test protocol for marine copepods. They further published a copper toxicity report based on calanoid copepod *E. affinis* toxicity tests in 1998 (EPA 903/R/98/005). The UK Environmental Agency (2007) has also published a guideline on *T. battagliai* lethality test for direct toxicity assessment.

National standards in Denmark (Danish Standards, 1999) and in Sweden (Swedish Standards Institute, 1991) were based on acute copepod toxicity test performed on *N. spinipes* (Linden et al., 1979).

In 1999, the ISO published a protocol for copepod testing (ISO 14669:1999, determination of acute lethal toxicity to marine copepods). This protocol described test methodology for three copepod species: *T. battagliai, N. spinipes*, and *A. tonsa*. Furthermore, two additional ISO protocols of copepod toxicity testing are under development: ISO/DIS 16778 (Calanoid copepod early-life stage test with *A. tonsa*) and ISO/CD 18220 (Larval development test with the harpacticoid copepod *N. spinipes*). However, no information regarding these two protocols were accessible at the time this chapter was written.

The Organization for Economic Co-operation and Development is also involved in developing guidelines for copepods ecotoxicity testing. In 2005–2006, the Organization for Economic Co-operation and Development identified a number of suitable estuarine and marine arthropod species for performing life-cycle toxicity tests to detect developmental, reproductive, and EDCs (OECD, 2006). Most of the copepod species covered in section 12.3.2 were highlighted as good candidates for life-cycle toxicity tests. The report recommended using all important life-cycle events as toxicity endpoints, including development rates from nauplius to copepodite and adult stages, fertilization success, total viable offspring per mating pair, time to production of first clutch, time between egg clutches, aborted egg sacs, necrotic and infertile eggs, sex ratio, stage-specific mortality, and abnormal behavior. Using the species shortlisted in the report, a process of validating full life-cycle tests with the harpacticoid copepods *N.*

spinipes and *A. tenuiremis* and the calanoid copepod *A. tonsa* was initiated in 2007. Ring tests involving 10 laboratories in Europe using 3,5-dichlorophenol as the reference chemical was organized (OECD, 2007). Although the laboratories were successful in performing the tests, a significant decline of 3,5-dichlorophenol levels in the test medium over time was detected in all laboratories, making it difficult to determine precise effect concentration levels. Based on lowest observed effect concentrations and other observations, the report recommended that partial life-cycle tests may be more appropriate and *A. tenuiremis* should be the best choice for full life-cycle tests, if needed, because of its short generation time. Additional validation work was carried out to evaluate and determine the best test protocol for partial/full life-cycle test with *A. tenuiremis* (OECD, 2011). This interlaboratory calibration include four laboratories, two from the United States and two from Sweden. The assay was carried out in 96-well plates with nauplius individually placed in each well. Life-cycle endpoints such as survival, moulting, development, sex ratio, and reproductive success were assessed. These data were further used to compute population model to examine the impact on population growth rate and projected population size. A draft guideline of *A. tenuiremis* partial life-cycle tests on development and production was produced (OECD, 2013) and population modeling was formally incorporated into the document as one of the required guidelines.

12.7 Research Need to Enhance the Use of Copepods in Ecotoxicology

There are a few key research areas to improve utility of copepods as reference test species in marine and estuarine ecotoxicology.

To facilitate more copepod species use in toxicity testing, a molecular identification technique is a necessary development. Copepods have similar body shape because of morphological conservation and numerous sibling species groups (Bron et al., 2011; Sukhikh et al., 2013; Blanco-Bercial et al., 2014). Identification of species is challenging even for experts, and often involves electron microscopy investigation of appendage morphology. Genetic barcoding using the mitochondrial cytochrome c oxidase subunit I gene has been used to identify many species and will likely be as useful for copepod species identification. A recent study has successfully used the cytochrome c oxidase subunit I gene and generated 800 new barcode sequences for 63 copepod species and reexamined a dataset of 1381 barcode sequences for 195 copepod species (Blanco-Bercial et al., 2014). These new findings will be instrumental in future development of copepod ecology and their application in ecotoxicity.

The next area of development that follows is copepod culturing techniques. Our ability to culture copepod in the laboratory will limit the number of copepod species we can use in toxicity testing. Encouraging developments have been made since an increasing number of researchers are evaluating their native species for ecotoxicity tests (e.g., Ward et al., 2011).

Although many advocate using copepods to understand toxicity of EDCs (see OECD, 2006), endocrine system of copepods is not well-characterized (Kusk and Wollenberger, 2007;

LeBlanc, 2007). Although there are several studies regarding hormones in copepods (e.g., Block et al., 2003), knowledge of the crustacean endocrine system was primarily derived from decapods species, yet noticeable differences exist between copepods and decapods (Kusk and Wollenberger, 2007). For example, a molt-inhibiting hormone is produced in the X organ/sinus gland complex at the eyestalk of decapods (Lachaise et al., 1993; Hopkins, 2012), yet there is no eyestalk in copepods. Nevertheless, a molt-inhibiting hormone is detected in copepod *Caligus rogerresseyi* (Christie, 2014). More research into the copepod endocrine system will be needed to provide mechanistic understanding of EDC toxicity.

12.8 Conclusions

Recent developments in both molecular and modeling techniques for copepods expanded our understanding of chemical toxicity to copepods from the molecular level for mechanism to the population level for ecological relevance. This unique advantage represents an unmatched opportunity to develop molecular biomarkers that can be predictive of population level effects (Forbes et al., 2006). All these developments will ultimately provide scientific basis in ecological risk assessment and regulation of chemicals in the marine and estuarine environments.

Acknowledgment

This work is a contribution to the bilateral French-Korean PHC STAR project "Evaluation of ecotoxicological biomarkers and assays." It is also a contribution to the GIP Seine-Aval as well as the Council of the Région Nord Pas de Calais. This work is also a part of the project (20140342) titled "Development of techniques for assessment and management of hazardous chemicals in the marine environment" funded by the Ministry of Oceans and Fisheries, Korea.

References

Agrone, C., Aluigi, M.G., Fabbri, R., et al., 2011. Exposition biomarkers to organophosphorus pesticides in *Tigriopus fulvus* Fischer (Copepoda, Harpacticoida). J. Biol. Res. 94, 14–17.

ASTM, 2004. Standard Guide for Conducting Renewal Microplate-based Life-cycle Toxicity Tests with a Marine Meiobenthic Copepod. E 2317-04. American Society for Testing and Materials, West Conshohocken, PA. 22 p.

Atkinson, A., 1991. Life cycles of *Calanoides acutus, Calanus simillimus* and *Rhincalanus gigas* (Copepoda: Calanoida) within the Scotia Sea. Mar. Biol. 109, 79–91.

Barka, S., Pavillon, J.F., Amiard, J.C., 2001. Influence of different essential and non-essential metals on MTLP levels in the copepod *Tigriopus brevicornis*. Comp. Biochem. Physiol. 128C, 479–493.

Barnes, H., Stanbury, F.A., 1948. The toxic action of copper and mercury salts both separately and when mixed on the Harpacticid Copepod, *Nitocra Spinipes* (Boeck). J. Exp. Biol. 25, 270–275.

Barreto, F.S., Schoville, S.D., Burton, R.S. Reverse genetics in the tide pool: knockdown of target gene expression via RNA interference in the copepod Tigriopus californicus. Mol. Ecol. Resour. (in press). http://dx.doi.org/10.1111/1755-0998.12359.

Beaugrand, G., Brander, K., Lindley, J., et al., 2003. Plankton effect on cod recruitment in the North Sea. Nature 426, 661–664.

Belleza, E.L., Brinkmann, M., Preuss, T.G., et al., 2014. Population-level effects in *Amphiascus tenuiremis*: contrasting matrix- and individual-based population models. Aquat. Toxicol. 157, 207–214.

Bengtsson, B.E., Bergstrom, B., 1987. A flowthrough fecundity test with *Nitocra spinipes* (Harpacticoidea Crustacea) for aquatic toxicity. Ecotoxicol. Environ. Saf. 14, 260–268.

Bengtsson, B.E., 1978. Use of a harpacticoid copepod in toxicity tests. Mar. Pollut. Bull. 9, 238–241.

Blanco-Bercial, L., Cornils, A., Copley, N., et al., June 23, 2014. DNA barcoding of marine copepods: assessment of analytical approaches to species identification. PLOS Currents Tree of Life Edition 1.

Block, D.S., Bejarano, A.C., Chandler, G.T., 2003. Ecdysteroid concentrations through various life-stages of the meiobenthic harpacticoid copepod, *Amphiascus tenuiremis* and the benthic estuarine amphipod, *Leptocheirus plumulosus*. Gen. Comp. Endocrinol. 132, 151–160.

Brain, P., Cousens, R., 1989. An equation to describe dose–responses where there is stimulation of growth at low doses. Weed Res. 29, 93–96.

Bron, J.E., Frisch, D., Goetze, E., et al., 2011. Observing copepods through genomic lens. Front. Zool 8, 22.

Cailleaud, K., Michalec, F.G., Forget-Leray, J., et al., 2011. Changes in the swimming behavior of *Eurytemora affinis* (Copepoda Calanoida) in response to sub-lethal exposure to nonylphenols. Aquat. Toxicol. 10, 228–231.

Cailleaud, K., Budzinski, H., Le Menach, K., et al., 2009. Uptake and elimination of hydrophobic organic contaminants in estuarine copepods: an experimental study. Environ. Toxicol. Chem. 28, 239–246.

Cailleaud, K., Maillet, G., Budzinski, J., et al., 2007a. Effects of salinity and temperature on the expression of enzymatic biomarkers in *Eurytemora affinis* (Calanoida, Copepoda). Comp. Biochem. Physiol. 147A, 841–849.

Cailleaud, K., Forget-Leray, J., Souissi, S., et al., 2007b. Seasonal variations of hydrophobic organic contaminant concentrations in the water-column of the Seine Estuary and their transfer to a planktonic species *Eurytemora affinis* (Calanoida Copepoda) Part 1: PCBs and PAHs. Chemosphere 70, 270–280.

Cailleaud, K., Forget-Leray, J., Souissi, S., et al., 2007c. Seasonal variations of hydrophobic organic contaminant concentrations in the water-column of the Seine Estuary and their transfer to a planktonic species *Eurytemora affinis* (Calanoida Copepoda) Part 2: Alkylphenol-polyethoxylates. Chemosphere 70, 281–287.

Cedergreen, N., Ritz, C., Streibig, J.C., 2005. Improved empirical models describing hormesis. Environ. Toxicol. Chem. 24, 3166–3172.

Chandler, G.T., Cary, T.L., Volz, D.C., et al., 2004. Fipronil effects on estuarine copepod (*Amphiascus tenuiremis*) development, fertility, and reproduction: a rapid life-cycle assay in 96-well microplate format. Environ. Toxicol. Chem. 23, 117–124.

Chandler, G.T., 1986. High density culture of meiobenthic harpacticoid copepods within a muddy substrate. Can. J. Fish. Aquat. Sci. 43, 53–59.

Chandler, G.T., Coull, B.C., Schizas, N.V., et al., 1997. A culture-based assessment of the effects of chlorpyrifos on multiple meiobenthic copepods using microcosms of intact estuarine sediments. Environ. Toxicol. Chem. 16, 2339–2346.

Chapman, P.M., McDonald, B.G., Kickham, P.E., et al., 2006. Global geographic differences in marine metals toxicity. Mar. Pollut. Bull. 52, 1081–1084.

Christie, A.E., 2014. *In silico* characterization of the peptidome of the sea louse *Caligus rogercresseyi* (Crustacea, Copepoda). Gen. Comp. Endocrinol. 204, 248–260.

Conover, R.J., 1965. Notes on the molting cycle, development of sexual characters and sex ratio in *Calanus hyperboreus*. Crustaceana Leiden 8, 308–320.

Cutts, C.J., 2003. Culture of harpacticoid copepods: potential as live food for rearing marine fish. Adv. Mar. Biol. 44, 295–316.

D'Agostino, A., Finney, C., 1974. The effect of copper and cadmium on the development of *Tigriopus japonicus*. In: Vernberg, F.J., Vernberg, W.B. (Eds.), Pollution and Physiology of Marine Organisms. Academic Press, New York, pp. 445–463.

Danish Standards Association, 1999. Water quality – Determination of acute lethal toxicity to marine copepods (Copepoda, Crustacea). Danish standard DS/ISO 14669:1999 Dansk Standard, Charlottenlund, Denmark. 24 p.

DeAngelis, D.L., Mooij, W.M., 2005. Individual-based modelling of ecological and evolutionary processes. Annu. Rev. Ecol. Evol. Syst. 36, 147–168.

DeAngelis, D.L., Grimm, V., 2014. Individual-based models in ecology after four decades. F1000 Prime Rep. 6, 39.

Devreker, D., Souissi, S., Seuront, L., 2004. Development and mortality of the first naupliar stages of *Eurytemora affinis* (Copepoda, Calanoida) under different conditions of salinity and temperature. J. Exp. Mar. Biol. Ecol. 303, 31–46.

Devreker, D., Souissi, S., Molinero, J.C., et al., 2008. Trade-offs of the copepod *Eurytemora affinis* in mega-tidal estuaries: insight from high frequency sampling in the Seine estuary. J. Plankton Res. 30, 1329–1342.

Devreker, D., Souissi, S., Winkler, G., et al., 2009. Effects of salinity, temperature and individual variability on the reproduction of *Eurytemora affinis* (Copepoda; Calanoida) from the Seine estuary: a laboratory study. J. Exp. Mar. Biol. Ecol. 368, 113–123.

Dur, G., Jiménez-Melero, R., Beyrend-Dur, D., et al., 2013. Individual-based model for the phenology of egg bearing copepods: application to *Eurytemora affinis* from the Seine estuary, France. Ecol. Model 269, 21–36.

Dur, G., Souissi, S., Schmitt, F.G., et al., 2012. Sex ratio and mating behavior in the calanoid copepod *Pseudodiaptomus annandalei*. Zool. Stud. 51, 589–597.

Dur, G., Souissi, S., Devreker, D., et al., 2009. An individual-based model to study the reproduction of egg bearing copepods: application to *Eurytemora affinis* (Copepoda Calanoida) from the Seine estuary, France. Ecol. Model 220, 1073–1089.

EFSA (European Food Safety Authority), 2013. Panel on plant protection products and their residues: guidance on tiered risk assessment for plant protection products for aquatic organisms in edge-of-field surface waters. EFSA J11, 3290.

Ekblom, R., Galindo, J., 2011. Applications of next generation sequencing in molecular ecology of non-model organisms. Heredity 107, 1–15.

Forbes, V.E., Calow, P., Grimm, V., et al., 2011. Adding Value to Ecological Risk Assessment with Population Modelling. Pharmacology, Toxicology and Environmental Health Commons, University of Nebraska, Lincoln http://digitalcommons.unl.edu/biosciforbes/4.

Forbes, V.E., Palmqvist, A., Bach, L., 2006. The use and misuse of biomarkers in ecotoxicology. Environ. Chem. Toxicol. 25, 272–280.

Forget, J., Beliaeff, B., Bocquené, G., 2003. Acetylcholinesterase activity in copepods (*Tigriopus brevicornis*) from the Vilaine River estuary, France, as a biomarker of neurotoxic contaminants. Aquat. Toxicol. 62, 195–204.

Forget, J., Pavillon, J.-F., Beliaeff, B., et al., 1999. Joint action of pollutant combinations (pesticides and metals) on survival (LC50 values) and acetylcholinesterase activity of *Tigriopus brevicornis* (Copepoda, harpacticoida). Environ. Toxicol. Chem. 18, 912–919.

Forget-Leray, J., Landriau, I., Minier, C., et al., 2005. Impact of endocrine toxicants on survival, development, and reproduction of the estuarine copepod *Eurytemora affinis* (Poppe). Ecotox. Env. Saf. 60, 288–294.

Frangoulis, C., Christou, E., Hecq, J., 2005. Comparison of marine copepod outfluxes: nature, rate, fate and role in the carbon and nitrogen cycles. Adv. Mar. Biol. 47, 279–295.

Frangoulis, C., Skliris, N., Lepoint, G., et al., 2011. Importance of copepod carcasses versus faecal pellets in the upper water column of an oligotrophic area. Estuar. Coast. Shelf Sci. 92, 456–463.

Fraser, J.H., 1936. The occurrence, ecology and life history of *Tigriopus fulvus* (Fischer). J. Mar. Biol. Assoc. U.K 20, 523–536.

Gardiner, W.W., Word, J.Q., Word, J.D., et al., 2013. The acute toxicity of chemically and physically dispersed crude oil to key Arctic species under Arctic conditions during the open water season. Environ. Toxicol. Chem. 32, 2284–2300.

Gasbi, F., Hammers-Wirtz, M., Grimm, V., et al., 2014. Coupling different mechanistic effect models for capturing individual- and population-level effect of chemicals: lessons from a case were standard risk assessment failed. Ecol. Model 280, 18–29.

Gerlach, S.A., 1971. On the importance of marine meiofauna for benthos communities. Oecologia 6, 176–190.

Gilbert, J.J., Williamson, C.E., 1983. Sexual dimorphism in zooplankton (Copepoda, Cladocera, and Rotifera). Annu. Rev. Ecol. Syst. 14, 1–33.

Ginot, V., Lepage, C., Souissi, S., 2002. A multi-agents architecture to enhance end user individual based modelling. Ecol. Model 157, 23–41.

Glippa, O., Souissi, S., Denis, L., et al., 2011. Calanoid copepod resting egg abundance and hatching success in the sediment of the Seine estuary (France). Estuar. Coastal Shelf Sci. 92, 255–262.

Gorbi, G., Invidia, M., Savorelli, R., et al., 2012. Standardized methods for acute and semichronic toxicity tests with the copepod *Acartia tonsa*. Environ. Toxicol. Chem. 31, 2023–2028.

Greenstein, D., Bay, S., Anderson, B., et al., 2008. Comparison of methods for evaluating acute and chronic toxicity in marine sediments. Environ. Toxicol. Chem. 27, 933–944.
Grimm, V., Berger, U., DeAngelis, D.L., et al., 2010. The ODD protocol a review and first update. Ecol. Model 221, 2760–2768.
Grimm, V., Berger, U., Bastiansen, F., et al., 2006. A standard protocol for describing individual-based and agent-based models. Ecol. Model 198, 115–126.
Grimm, V., Wyszomirski, T., Aikman, D., et al., 1999. Individual-based modelling and ecological theory: synthesis of a workshop. Ecol. Model 115, 275.
Grimm, V., Railsback, S.F., 2005. Individual-based Modeling and Ecology. Princeton University Press, Princeton.
Gusmao, L.F.M., McKinnon, A.D., 2009. Sex ratios, intersexuality and sex change in copepods. J. Plankton Res. 31, 1101–1117.
Gustafsson, K., Blidberg, E., Elfgren, I.K., et al., 2010. Direct and indirect effects of the fungicide azoxystrobin in outdoor brackish water microcosms. Ecotoxicology 19, 431–444.
Hammervold, S.H., Glud, R.N., Evjemo, J.O., et al., 2015. A new large egg type from the marine live feed calanoid copepod *Acartia tonsa* (Dana)- perspectives for selective breeding of designer feed for hatcheries. Aquaculture 436, 114–120.
Han, J., Won, E.-J., Hwang, D.-S., et al., 2014b. Crude oil exposure results in oxidative stress-mediated dysfunctional development and reproduction in the copepod *Tigriopus japonicus* and modulates expression of *cytochrome P450 (CYP)* genes. Aquat. Toxicol. 152, 308–317.
Han, J., Won, E.-J., Lee, B.-Y., et al., 2014a. Gamma ray induces DNA damage and oxidative stress associated with impaired growth and reproduction in the copepod *Tigriopus japonicus*. Aquat. Toxicol. 152, 264–272.
Hansen, B.H., Altin, D., Booth, A., et al., 2010. Molecular effects of diethanolamine exposure on *Calanus finmarchicus* (Crustacea: Copepoda). Aquat. Toxicol. 99, 212–222.
Hansen, B.H., Altin, D., Nordtug, T., et al., 2007. Suppression subtractive hybridization library prepared from the copepod *Calanus finmarchicus* exposed to a sublethal mixture of environmental stressors. Comp. Biochem. Physiol. 2D, 250–256.
Hansen, B.H., Altin, D., Rorvik, S.F., et al., 2011. Comparative study on acute effects of water accommodated fractions of an artificially weathered crude oil on *Calanus finmarchicus* and *Calanus glacialis* (Crustacea: Copepoda). Sci. Total Environ. 409, 704–709.
Hansen, B.H., Altin, D., Vang, S.H., et al., 2008. Effects of naphthalene on gene transcription in *Calanus finmarchicus* (Crustacea: Copepod). Aquat. Toxicol. 86, 157–165.
Hansen, B.W., Drillet, G., Pedersen, M.F., et al., 2012. Do *Acartia tonsa* (Dana) eggs regulate their volume and osmolality as salinity changes? J. Comp. Physiol. 182B, 613–623.
Hansen, B.H., Nordtug, T., Altin, D., et al., 2009. Gene expression of *GST* and *CYP330A1* in lipid-rich and lipid-poor female *Calanus finmarchicus* (Copepoda: Crustacea) exposed to dispersed oil. J. Toxicol. Environ. Health 72A, 131–139.
Hanson, N., Stark, J., 2012. Comparison of population level and individual level end-points to evaluate ecological risk of chemicals. Environ. Sci. Technol. 46, 5590–5598.
Hatlen, K., Camus, L., Berge, J., et al., 2009. Biological effects of water soluble fraction of crude oil on the Arctic sea ice amphipod *Gammarus wilkitzkii*. Chem. Ecol. 25, 151–162.
Hopkins, P.M., 2012. The eyes have it: a brief history of crustacean neuroendocrinology. Gen. Comp. Endocrinol. 175, 357–366.
Huq, A., Small, E.B., West, P.A., et al., 1983. Ecological relationships between *Vibrio cholerae* and planktonic crustacean copepods. Appl. Environ. Microbiol. 45, 275–283.
Hutchinson, T.H., Williams, T.D., 1989. The use of sheepshead minnow (*Cyprinodon variegates*) and a benthic copepod (*Tisbe battagliai*) in short-term tests for estimating the chronic toxicity of industrial effluents. Hydrobiologia 188/189 567–572.
Hutchinson, T.H., Pounds, N.A., Hampel, M., et al., 1999. Life-cycle studies with marine copepods (*Tisbe battagliai*) exposed to 20-hydroxyecdysone and diethylstilbestrol. Environ. Toxicol. Chem. 18, 2914–2920.
Hwang, D.-S., Lee, J.-S., Rhee, J.-S., et al., 2010a. Modulation of *p53* gene expression in the intertidal copepod *Tigriopus japonicus* exposed to alkylphenols. Mar. Environ. Res. 69, S77–S80.

Hwang, D.-S., Lee, K.-W., Han, J., et al., 2010b. Molecular characterization and expression of *vitellogenin (Vg)* genes from the cyclopoid copepod, *Paracyclopina nana* exposed to heavy metals. Comp. Biochem. Physiol. 151C, 360–368.

ISO (International Organization for Standardization), 1999. Water Quality - Determination of Acute Lethal Toxicity to Marine Copepods (Copepoda, Crustacea). Draft International Standard ISO/DIS 14669, Geneva, Switzerland.

Ito, T., 1970. The biology of a harpacticoid copepod, *Tigriopus japonicus* Mori. J. Fac. Sci. Hokkaido Univ. Ser. VI, Zool. 17, 474–499.

Jager, T., Albert, C., Preuss, T.G., et al., 2011. General unified threshold model of survival – a toxicokinetic-toxicodynamic framework for ecotoxicology. Environ. Sci. Technol. 45, 2529–2540.

Jeong, C.-B., Kim, B.-M., Kim, R.-O., et al., 2014a. Functional characterization of P-glycoprotein in the intertidal copepod *Tigriopus japonicus* and its potential role in remediating metal pollution. Aquat. Toxicol. 156, 135–147.

Jeong, C.-B., Kim, B.-M., Lee, J.-S., et al., 2014b. Genome-wide identification of whole ATP-binding cassette (ABC) transporters in the intertidal copepod *Tigriopus japonicus*. BMC Genomics 15, 651–665.

Jiang, J.L., Wang, G.Z., Mao, M.G., et al., 2013. Differential gene expression profile of the calanoid copepod, *Pseudodiaptomus annandalei*, response to nickel exposure. Comp. Biochem. Physiol. 157C, 203–211.

Ki, J.-S., Raisuddin, S., Lee, K.-W., et al., 2009. Gene expression profiling of copper-induced responses in the intertidal copepod *Tigriopus japonicus* using a 6K oligochip microarray. Aquat. Toxicol. 93, 177–187.

Kim, B.-M., Jeong, C.-B., Han, J., et al., 2013. Role of crustacean hyperglycemic hormone (*CHH*) in the environmental stressor-exposed intertidal copepod *Tigriopus japonicus*. Comp. Biochem. Physiol. 158C, 131–141.

Kim, B.-M., Jeong, C.-B., Rhee, J.-S., et al., 2014a. Transcriptional profiles of Rel/NF-κB, inhibitor of NF-κB (IκB), and lipopolysaccharide-induced TNF-α factor (LITAF) in the lipopolysaccharide (LPS) and two *Vibrio* sp.-exposed intertidal copepod, *Tigriopus japonicus*. Dev. Comp. Immunol. 42, 229–239.

Kim, B.-M., Rhee, J.-S., Jeong, C.-B., et al., 2014b. Heavy metals induce oxidative stress and trigger oxidative stress-mediated heat shock protein (hsp) modulation in the intertidal copepod *Tigriopus japonicus*. Comp. Biochem. Physiol. 166C, 65–74.

Kim, B.-M., Rhee, J.-S., Lee, K.-W., et al., 2015. UV-B radiation-induced oxidative stress and p38 signaling pathway involvement in the benthic copepod *Tigriopus japonicus*. Comp. Biochem. Physiol. 167C, 15–23.

Kim, B.-M., Rhee, J.-S., Park, G.-S., et al., 2011. Cu/Zn- and Mn-superoxide dismutase (*SOD*) from the copepod *Tigriopus japonicus*: molecular cloning and expression in response to environmental pollutants. Chemosphere 84, 1467–1475.

Korsman, J.C., Schipper, A.M., De Hoop, L., et al., 2014. Modeling the impacts of multiple environmental stress factors on estuarine copepod populations. Environ. Sci. Technol. 48, 5709–5717.

Kosobokova, K.N., 1999. The reproductive cycle and life history of the Arctic copepod *Calanus glacialis* in the White Sea. Polar Biol. 22, 254–263.

Kovatch, C.E., Chandler, G.T., Coull, B.C., 1999. Utility of a full life-cycle copepod bioassay approach for assessment of sediment-associated contaminant mixtures. Mar. Pollut. Bull. 38, 692–701.

Kozlowsky-Suzuki, B., Koski, M., Hallberg, E., et al., 2009. Glutathione transferase activity and oocyte development in copepods exposed to toxic phytoplankton. Harmful Algae 8, 395–406.

Kulkarni, D., Hommen, U., Schäffer, A., et al., 2014. Ecological interactions affecting population-level responses to chemical stress in *Mesocyclops leukarti*. Chemosphere 112, 340–347.

Kulkarni, D., Gergs, A., Hommen, U., et al., 2013a. A plea for the use of copepods in freshwater ecotoxicology. Environ. Sci. Pollut. Res. 20, 75–85.

Kulkarni, D., Daniels, B., Preuss, T.G., 2013b. Life-stage-dependent sensitivity of the cyclopoid copepod *Mesocyclops leuckarti* to triphenyltin. Chemosphere 92, 1145–1153.

Kusk, K.O., Wollenberger, L., 2007. Towards an internationally harmonized test method for reproductive and developmental effects of endocrine disrupters in marine copepods. Ecotoxicology 16, 183–195.

Kwok, K.W.H., Leung, K.M.Y., Bao, V.W.W., et al., 2008. Copper toxicity in the marine copepod *Tigropus japonicus*: low variability and high reproducibility of repeated acute and life-cycle tests. Mar. Pollut. Bull. 57, 632–636.

Kwok, K.W.H., Grist, E.P., Leung, K.M., 2009. Acclimation effect and fitness cost of copper resistance in the marine copepod *Tigriopus japonicus*. Ecotoxicol. Environ. Saf. 72, 358–364.

Lachaise, A., Le Roux, A., Hubert, M., et al., 1993. The molting gland of crustaceans: localization, activity, and endocrine control (a review). J. Crustacean Biol. 13, 198–234.

Lauritano, C., Procaccini, G., Ianora, A., 2012. Gene expression patterns and stress response in marine copepods. Mar. Environ. Res. 76, 22–31.

Lauritano, C., Borra, M., Carotenuto, Y., et al., 2011a. First molecular evidence of diatom effects in the copepod *Calanus helgolandicus*. J. Exp. Mar. Biol. Ecol. 404, 79–86.

Lauritano, C., Borra, M., Carotenuto, Y., et al., 2011b. Molecular evidence of the toxic effects of diatom diets on gene expression patterns in copepods. PLOS One 6, e26850.

LeBlanc, G.A., 2007. Crustacean endocrine toxicology: a review. Ecotoxicology 16, 61–81.

Lee, C.E., Petersen, C.H., 2002. Genotype-by-environment interaction for salinity tolerance in the freshwater-invading copepod *Eurytemora affinis*. Physiol. Biochem. Zool. 754, 335–344.

Lee, C.E., 1999. Rapid and repeated invasions of fresh water by the saltwater copepod *Eurytemora affinis*. Evolution 53, 1423–1434.

Lee, J.-S., Rhee, J.-S., Kim, R.-O., et al., 2010. The copepod *Tigriopus japonicus* genomic DNA information (574Mb) and molecular anatomy. Mar. Environ. Res. 69, S21–S23.

Lee, J.-S., Kim, R.-O., Rhee, J.-S., et al., 2011. Sequence analysis of genomic DNA (680 Mb) by GS-FLX-Titanium sequencer in the monogonont rotifer *Brachionus ibericus*. Hydrobiologia 662, 65–75.

Lee, J.-S., Raisuddin, S., 2008. Modulation of expression of oxidative stress genes of the intertidal copepod *Tigriopus japonicus* after exposure to environmental chemicals. In: Murakami, Y., Nakayama, K., Kitamura, S.I., Iwata, H., Tanabe, S. (Eds.), Interdisciplinary Studies on Environmental Chemistry. Biological Responses to Chemical Pollutants, pp. 95–105.

Lee, J.-W., Kim, B.-M., Jeong, C.-B., et al., 2015. Inhibitory effects of biocides on transcription and protein activity of acetylcholinesterase in the intertidal copepod *Tigriopus japonicus*. Comp. Biochem. Physiol. 167C, 147–156.

Lee, K.-W., Hwang, D.-S., Rhee, J.-S., et al., 2008a. Molecular cloning, phylogenetic analysis and developmental expression of a *vitellogenin (Vg)* gene from the intertidal copepod *Tigriopus japonicus*. Comp. Biochem. Physiol. 150B, 395–402.

Lee, K.-W., Rhee, J.-S., Raisuddin, S., et al., 2008b. A corticotropin-releasing hormone binding protein (*CRH-BP*) gene from the intertidal copepod, *Tigriopus japonicus*. Gen. Comp. Endocrinol. 158, 54–60.

Lee, Y.-M., Lee, K.-W., Park, H.G., et al., 2007. Sequence, biochemical characteristics and expression of a novel Sigma-class of glutathione *S*-transferase from the intertidal copepod, *Tigriopus japonicus* with a possible role in antioxidant defense. Chemosphere 69, 893–902.

Lee, Y.-M., Kim, I.-C., Jung, S.-O., et al., 2005. Analysis of 686 expressed sequence tags (ESTs) from the intertidal harpacticoid copepod *Tigriopus japonicus* (Crustacea, Copepoda). Mar. Pollut. Bull. 51, 757–768.

Lee, Y.-M., Park, T.-J., Jung, S.-O., et al., 2006. Cloning and characterization of glutathione *S*-transferase gene in the intertidal copepod *Tigriopus japonicus* and its expression after exposure to endocrine-disrupting chemicals. Mar. Environ. Res. 62, S219–S223.

Lenz, P.H., Unal, E., Hassett, R.P., et al., 2012. Functional genomics resources for the North Atlantic copepod, *Calanus finmarchicus*: EST database and physiological microarray. Comp. Biochem. Physiol. 7D, 110–123.

Lesueur, T., Boulangé-Lecomte, C., Xuereb, B., et al., 2013. Development of a larval bioassay using the calanoid copepod, *Eurytemora affinis* to assess the toxicity of sediment-bound pollutants. Ecotoxicol. Environ. Saf. 94, 60–66.

Liberge, M., Barthélémy, R.M., 2007. Localization of metallothionein, heat shock protein (*Hsp70*), and superoxide dismutase expression in *Hemidiaptomus toubaui* (Copepoda, Crustacea) exposed to cadmium and heat stress. Can. J. Zool. 85, 362–371.

Linden, E., Bengtsson, B.E., Svanberg, O., et al., 1979. The acute toxicity of 78 chemicals and pesticide formulations against two brackish water organisms, the bleak (*Alburnus alburnus*) and the harpacticoid *Nitocra spinipes*. Chemosphere 8, 843–851.

Lorek, H., Sonnenschein, M., 1999. Modelling and simulation software to support individual-based ecological modelling. Ecol. Model 115, 199–216.

Marcus, Z.L., King, C.K., Payne, S.J., et al., 2015. Sensitivity and response time of three common Antarctic marine copepods to metal exposure. Chemosphere 120, 267–272.

McAllen, R., Taylor, A., 2001. The effect of salinity change on the oxygen consumption and swimming activity of the high-shore rockpool copepod *Tigriopus brevicornis*. J. Exp. Mar. Biol. Ecol. 263, 227–240.

Meise, C.J., O'Reilly, J.E., 1996. Spatial and seasonal patterns in abundance and age-composition of *Calanus finmarchicus* in the Gulf of Maine and on Georges Bank: 1977–1987. Deep-Sea Res. II 43, 1473–1501.

Meli, M., Palmqvist, A., Forbes, V.E., et al., 2014. Two pairs of eyes are better than one: combining individual-based and matrix models for ecological risk assessment of chemicals. Ecol. Model 280, 40–52.

Michalec, F.G., Holzner, M., Menu, D., et al., 2013. Behavioral responses of the estuarine calanoid copepod *Eurytemora affinis* to sub-lethal concentration of waterborne pollutants. Aquat. Toxicol. 138-139, 129–138.

Mooij, W., Hülsmann, S., Vijverberg, J., et al., 2003. Modelling *Daphnia* population dynamics and demography under natural conditions. Hydrobiologia 491, 19–34.

Ning, J., Wang, M., Li, C., et al., 2013. Transcriptome sequencing and *de novo* analysis of the copepod *Calanus sinicus* using 454 GS FLX. PLOS One 8, e63741.

Norregaard, R.D., Nielson, T.G., Moller, E.F., et al., 2014. Evaluating pyrene toxicity on Arctic key copepod species *Calanus hyperboreus*. Ecotoxicology 23, 163–174.

OECD, 2006. Detailed review paper on aquatic arthropods in life cycle toxicity tests with an emphasis on developmental, reproductive and endocrine disruptive effects. OECD Series on Testing and Assessment, vol. 55, ENV/JM/MONO(2006)22.

OECD, 2007. Validation report of the full life-cycle test with the harpacticoid copepods *Nitocra spinipes* and *Amphiascus tenuiremis* and the calanoid copepod *Acartia tonsa* – phase 1. OECD Series on Testing and Assessment, vol. 79, ENV/JM/MONO(2007)26.

OECD, 2011. Report of progress on the interlaboratory validation of the OECD harpacticoid copepod development and reproduction test. OECD Series on Testing and Assessment, vol. 158, ENV/JM/MONO(2011)38.

OECD, 2013. Harpacticoid Copepod Development and Reproduction Test with *Amphiascus tenuiremis*. Draft new guidance document http://www.oecd.org/env/ehs/testing/OECD_Harpacticoid_Copepod_Draft_for_REVIEW_Dec2013-clean.pdf.

Overjordet, I.B., Altin, D., Berg, T., et al., 2014. Acute and sub-lethal response to mercury in Arctic and boreal calanoid copepods. Aquat. Toxicol. 155, 160–165.

Perez-Landa, V., Simpson, S.L., 2011. A short life-cycle test with the epibenthic copepod *Nitocra spinipes* for sediment toxicity assessment. Environ. Toxicol. Chem. 30, 1430–1439.

Preuss, T.G., Brinkmann, M., Lundstrom, E., et al., 2011. An individual-based modeling approach for evaluation of endpoint sensitivity in harpacticoid copepod life-cycle tests and optimization of test design. Environ. Toxicol. Chem. 30, 2353–2362.

Preuss, T.G., Hammers-Wirtz, M., Ratte, H.T., 2010. The potential of individual based population models to extrapolate effects measured at standardized test conditions to environmental relevant conditions - an example for 3,4-dichloroaniline on *Daphnia magna*. J. Env. Monit. 12, 2070–2079.

Raisuddin, S., Kwok, K.W.H., Leung, K.M.Y., et al., 2007. The copepod *Tigriopus*: a promising marine model organism for ecotoxicology and environmental genomics. Aquat. Toxicol. 83, 161–173.

Renberg, L., Svanberg, O., Bengtsson, B.E., et al., 1980. Chlorinated guaiacols and catechols bioaccumulation potential in bleaks (*Alburnus alburnus*, Pisces) and reproductive and toxic effects on the harpacticoid *Nitocra spinipes* (Crustacea). Chemosphere 9, 143–150.

Rhee, J.-S., Choi, B.-S., Han, J., et al., 2012. Draft genome database construction from four strains (NIES-298, FCY-26, -27, and -28) of the cyanobacterium, *Microcystis aeruginosa*. J. Microbiol. Biotechnol. 22, 1208–1213.

Rhee, J.-S., Kim, B.-M., Jeong, C.-B., et al., 2013a. Development of enzyme-linked immunosorbent assay (ELISA) for glutathione *S*-transferase (GST-S) protein in the intertidal copepod *Tigriopus japonicus* and its application for environmental monitoring. Chemosphere 93, 2458–2466.

Rhee, J.-S., Raisuddin, S., Lee, K.-W., et al., 2009. Heat shock protein (*Hsp*) gene responses of the intertidal copepod *Tigriopus japonicus* to environmental toxicants. Comp. Biochem. Physiol. 149C, 104–112.

Rhee, J.-S., Yu, I.-T., Kim, B.-M., et al., 2013b. Copper induces apoptotic cell death through reactive oxygen species-triggered oxidative stress in the intertidal copepod *Tigriopus japonicus*. Aquat. Toxicol. 132/133, 182–189.

Roff, J.C., Turner, J.T., Webber, M.K., et al., 1995. Bacterivory by tropical copepod nauplii: extent and possible significance. Aquat. Microb. Ecol. 9, 165–175.

Seo, J.S., Lee, K.-W., Rhee, J.-S., et al., 2006a. Environmental stressors (salinity, heavy metals, H_2O_2) modulate expression of glutathione reductase (GR) gene from the intertidal copepod *Tigriopus japonicus*. Aquat. Toxicol. 80, 281–289.

Seo, J.S., Lee, Y.-M., Park, H.G., et al., 2006b. The intertidal copepod *Tigriopus japonicus* small *heat shock protein 20* gene (*Hsp20*) enhances thermotolerance of transformed *Escherichia coli*. Biochem. Biophys. Res. Comm. 340, 901–908.

Seo, J.S., Park, T.-J., Lee, Y.-M., et al., 2006c. Small heat shock protein 20 gene (*Hsp20*) of the intertidal copepod *Tigriopus japonicus* as a possible biomarker for exposure to endocrine disruptors. Bull. Environ. Contam. Toxicol. 76, 566–572.

Sosnowski, S.O., Gentile, J., 1978. Toxicological Comparison of Natural and Cultured Populations of *Acartia Tonsa* to Cadmium, Copper and Mercury. U.S. Environmental Protection Agency, Washington, D.C. EPA/600/J-78/147 (NTIS PB299467).

Souissi, S., Carlotti, F., Nival, P., 1997. Food and temperature-dependent function of moulting rate in copepods: an example of parameterization for population dynamics models. J. Plankton Res. 19, 1331–1346.

Souissi, A., Souissi, S., Devreker, D., et al., 2010. Occurrence of intersexuality in a laboratory culture of the copepod *Eurytemora affinis* from the Seine estuary (France). Mar. Biol. 157, 851–861.

Souza, M.S., Hansson, L.A., Hylander, S., et al., 2012. Rapid enzymatic response to compensate UV radiation in copepods. PLOS One 7, e32046.

Stottrup, J.G., Richardson, K., Kirkegaard, E., et al., 1986. The cultivation of *Acartia tonsa* Dana for use as live food source for marine fish larvae. Aquaculture 52, 87–96.

Sukhikh, N., Souissi, A., Souissi, S., et al., 2013. Invasion of *Eurytemora* sibling species (Copepoda: Temoridae) from north America into the Baltic Sea and European Atlantic coast estuaries. J. Nat. Hist 47, 753–767.

Sun, P.Y., Foley, H.B., Handschumacher, L., et al., 2014. Acclimation and adaptation to common marine pollutants in the copepod *Tigriopus californicus*. Chemosphere 112, 465–471.

Sverdrup, L.E., Fürst, C.S., Weideborg, M., et al., 2002. Relative sensitivity of one freshwater and two marine acute toxicity tests as determined by testing 30 offshore E & P chemicals. Chemosphere 46, 311–318.

Swedish Institute of Standards, 1991. Determination of Acute Lethal Toxicity of Chemical Substances and Effluents to *Nitocra Spinipes* Boeck - Static Procedure (in Swedish) Swedish Standard SS 02 81 06. SIS - Standardiseringskommissionen i Sverige, Stockholm, Sweden. 17 p.

Tartarotti, B., Torres, J.J., 2009. Sublethal stress: impact of solar UV radiation on protein synthesis in the copepod *Acartia tonsa*. J. Exp. Mar. Biol. Ecol. 375, 106–113.

Topping, C.J., Sibly, R.M., Akcakaya, H.R., et al., 2005. Risk assessment of UK Skylark populations using life-history and individual-based landscape models. Ecotoxicology 14, 925–936.

Tribble, N.D., Burka, J.F., Kibenge, F.S.B., 2007. Evidence for changes in the transcription levels of two putative *P-glycoprotein* genes in sea lice (*Lepeophtheirus salmonis*) in response to emamectin benzoate exposure. Mol. Biochem. Parasitol. 153, 59–65.

U.K. Environmental Agency, 2007. The Direct Toxicity Assessment of Aqueous Environmental Samples Using the Marine Copepod *Tisbe battagliai* Lethality Test. Methods for the Examination of Waters and Associated Materials. U.K. Environmental Agency, Bristol.

Venkateswanran, K., Takai, T., Navarro, I., et al., 1989. Ecology of *Vibrio cholera* non-O1 and *Salmonella* spp. and role of zooplankton in their seasonal distribution in Fukuyama coastal waters. Jpn. Appl. Environ. Microbiol. 55, 1591–1598.

Volz, D.C., Chandler, G.T., 2004. An enzyme-linked immunosorbent assay for lipovitellin quantification in copepods: a screening tool for endocrine toxicity. Environ. Toxicol. Chem. 23, 298–305.

Voznesensky, M., Lenz, P.H., Spanings-Pierrot, C., et al., 2004. Genomic approaches to detecting thermal stress in *Calanus finmarchicus* (Copepoda: Calanoida). J. Exp. Mar. Biol. Ecol. 311, 37–46.

Wang, M.H., Wang, G.Z., 2009. Biochemical response of the copepod *Tigriopus japonicus* Mori experimentally exposed to cadmium. Arch. Environ. Contam. Toxicol. 57, 707–717.

Wang, M., Wang, G., 2010. Oxidative damage effects in the copepod *Tigriopus japonicus* Mori experimentally exposed to nickel. Ecotoxicology 19, 273–284.

Wang, Z., Kwok, K.W.H., Lui, G.C.S., et al., 2014. The difference between temperate and tropical saltwater species' acute sensitivity to chemicals is relatively small. Chemosphere 105, 31–43.

Ward, D.J., Perez-Landa, V., Spadaro, D.A., et al., 2011. An assessment of three harpacticoid copepod species for use in ecotoxicological testing. Arch. Environ. Contam. Toxicol. 61, 414–425.

Willett, C.S., Burton, R.S., 2002. Proline biosynthesis genes and their regulation under salinity stress in the euryhaline copepod Tigriopus californicus. Comp. Biochem. Physiol. B Biochem. & Mol. Biol. 132, 739–750.

Willett, C.S., Burton, R.S., 2003. Characterization of the glutamate dehydrogenase gene and its regulation in a euryhaline copepod. Comp. Biochem. Physiol. 135B, 639–646.

Williams, T.D., 1992. Survival and development of copepod larvae *Tisbe battagliai* in surface microlayer, water and sediment elutriates from the German Bight. Mar. Ecol. Prog. Ser. 91, 221–228.

Wilson, C.B., 1932. The copepods of the Woods Hole region, Massachusetts. U.S. Nat. Mus. Bull. 158, 1–635.

Winkler, G., Souissi, S., Poux, C., et al., 2011. Genetic heterogeneity among several *Eurytemora affinis* populations in Western Europe. Mar. Biol. 158, 1841–1856.

Won, E.-J., Lee, J.-S., 2014. Gamma radiation induces growth retardation, impaired egg production, and oxidative stress in the marine copepod *Paracyclopina nana*. Aquat. Toxicol. 150, 17–26.

Won, E.-J., Lee, Y., Han, J., et al., 2014. Effects of UV radiation on hatching, lipid peroxidation, and fatty acid composition in the copepod *Paracyclopina nana*. Comp. Biochem. Physiol. 165C, 60–66.

Woodhead, P.M.J., Riley, J.D., 1959. Separation of the sexes of *Calanus finmarchicus* (Gunn.) in the fifth copepodite stage, with comments on the sex ratio and the duration in this stage. J. Cons. Cons. Int. Explor. Mer 24, 465–471.

Woods, R.E., Coull, B.C., 1992. Life history responses of *Amphiascus tenuiremis* (Copepoda, Harpacticoida) to mimicked predation. Mar. Ecol. Prog. Ser. 79, 225–234.

Wright, D.A., Savitz, J.D., Dawson, R., et al., 1996. Effect of difluobenzon on the maturation and reproductive success of the copepod *Eurytemora affinis*. Ecotoxicology 5, 47–58.

Xuereb, B., Forget-Leray, J., Souissi, S., et al., 2012. Molecular characterization and mRNA expression of grp78 and hsp90A in the estuarine copepod *Eurytemora affinis*. Cell Stress Chaperones 17, 457–472.

Yi, A.X., Han, J., Lee, J.S., et al., 2014. Ecotoxicity of triphenyltin on the marine copepod *Tigriopus japonicus* at various biological organisations: from molecular to population-level effects. Ecotoxicology 23, 1314–1325.

Zafar, M.I., Belgers, J.D.M., Van Wijngaarden, R.P.A., et al., 2012. Ecological impacts of time-variable exposure regimes to the fungicide azoxystrobin on freshwater communities in outdoor microcosms. Ecotoxicology 21, 1024–1038.

Zhang, Q., Uhlig, G., 1993. Effect of density on larval development and female productivity of *Tisbe holothuriae* (Copepoda, Harpacticoida) under laboratory conditions. Helgoländer Meeresunters 47, 229–241.

Zillioux, E.J., Wilson, D.F., 1966. Culture of a planktonic calanoid copepod through multiple generations. Science 151, 996–998.

CHAPTER 13

Fish as Reference Species in Different Water Masses

Minier Christophe, Amara Rachid, Lepage Mario

Abstract

Assessing water quality is an essential prerequisite for good water body management. In this respect, fish are excellent sentinel species and have proven their usefulness for more than a century. Although the initial use of fish was restricted to acute toxicity testing, recent developments of methodologies, equipment, techniques, and concepts have led to integrated studies that encompass molecular, physiological, and ecological knowledge and tools. Efforts have been made to link these different levels of organization, and the results largely contributed to our understanding of the mechanism of action of chemicals in a complex set of environmental stressors and conditions.

Keywords: Bioindicator; Fish biomarker; Physiological indices.

Chapter Outline

The aquatic world is largely considered as the world of fish. This vision is mainly dictated by the fact that fish is an important source of food for people living near rivers or coastal areas. Capture fisheries supplied the world with about 90 million tons of fish per year in the past decade, with marine production contributing to 86%–89% of the total. More than 120 million

Aquatic Ecotoxicology. http://dx.doi.org/10.1016/B978-0-12-800949-9.00013-9

tons of fish per year, including aquaculture fish, are available for human consumption and fish consumption accounted for 18 kg per capita bringing close to 20% of their intake of animal protein (FAO, 2012). The diversity of fish is second to the vision. They have a wide range of size and forms. They inhabit nearly all aquatic environments and they include more than 32,000 species (Fishbase, 2014) that constitute more than the combined number for all other vertebrates.

Both water quality and fish health have been easily linked together. Observations of fish mortalities have been recognized as indicator of pollution and toxicological testing using fish have been used for a long time (Penny and Adams, 1863; Carpenter, 1925). Because water is an essential component associated with health, including human health, special attention and subsequent development of studies using fish have developed. Rapidly, fish toxicology has moved from a descriptive approach to mechanistic and integrative studies providing information from mechanism of action to ecological risk assessment. This chapter aims at illustrating the recent advances in fish-based environmental studies that, in an integrative perspective, includes measurements at the molecular and biochemical, physiological and ecological levels.

13.1 Molecular and Biochemical Studies Using Fish

13.1.1 From Toxicity Tests to Mechanisms of Toxicity

Aquatic environments have long been recognized as the final sink for many anthropogenic contaminants, and observations of fish mortalities were obvious consequences of the deleterious action of the contaminants. One major example was the disappearance of salmon populations from rivers of the Scotland's Central Lowlands in the late eighteenth century. Reports from the Royal Commission on pollution of rivers made clear that pollution was responsible for this (in Daughty and Gardiner, 2003). As a consequence, fish were easily recognized as surrogates for their environment and used as "sentinel" species for the evaluation of the effects of human activities on ecosystem health. The impacts of mine-tailing effluents were among the first environmental disturbances explored using fish (Carpenter, 1925). Classical toxicology was first applied on fish testing and was designed to evidence specific or nonspecific effects on individuals. In the middle of the twentieth century, efforts were made to standardize techniques for fish acute testing (Hart et al., 1945). These efforts using fish to develop and validate toxicological tests led to some of the most relevant internationally agreed testing methods used by government, industry, and independent laboratories to identify and characterize potential hazards of new and existing chemical substances, chemical preparations, and chemical mixtures. More than 20% of the tests described for effects on biotic systems and recognized by the OECD are using fish (OECD, 2014; Table 13.1).

In the 1960s, concerns about chronic and low dose effects began to grow. Consequently, tests and experiments were more and more directed to long-term exposure of aquatic

Table 13.1: List of OECD Tests Using Fish in Regulatory Safety Testing and Subsequent Chemical and Chemical Product Notification and Chemical Registration*

Test Number	Title of Test	Date of Publication
204	Fish, prolonged toxicity test: 14-day study	April 4, 1984
210	Fish, early life stage toxicity test	July 17, 1992
203	Fish, acute toxicity test	July 17, 1992
212	Fish, short-term toxicity test on embryo and sac-fry stages	September 21, 1998
215	Fish, juvenile growth test	January 21, 2000
230	21-day fish assay	September 8, 2009
229	Fish, short-term reproduction assay	September 8, 2009
234	Fish, sexual development test	July 28, 2011
229	Fish, short-term reproduction assay	October 2, 2012
210	Fish, early life stage toxicity test	July 26, 2013
236	Fish, embryo acute toxicity test	July 26, 2013

*They can also be used for the selection and ranking of candidate chemicals during the development of new chemicals and products and in toxicology research.

organisms to chemicals and flow-through techniques were developed. Early life stages, full reproductive cycles, and full life cycles received more attention. Sublethal endpoints were more thoroughly studied and a variety of biomarkers were developed. Favored by the development of new molecular techniques, research was more focused on the discovery and understanding of the mechanism of toxicity of chemicals combining a diversity of approaches from pharmacokinetics, hematology, metabolism, histology, immunochemistry, or physiology. Studies included a suite of biomarkers combining various effects on a variety of mechanisms and physiologies to be more integrative. By the 2000s, advances in omics technologies further amplified the phenomenon. As the complexity of the system was growing, every effort to conceptualize and deliver a realistic picture of the environmental effects were made, which tended to associate various knowledge and competences into multidisciplinary studies.

Mechanical studies, powered by the rapid development of molecular techniques, led the description of a complex network of events within the cell and the organism. Because any interaction between a xenobiotic and an organism begins at the molecular level, it is only natural that studies using molecular techniques occupy a special place in the understanding of the mechanism of toxicity. Thus, molecular responses were typically the first responses to be detected and quantified. This interaction between a xenobiotic and a biotic ligand is itself a result of particular circumstances that govern bioavailability only eventually leading to chemical-target binding. Then, a suite of events may (or may not) result from this early interaction and led to either pathological alterations or adaptation (Figure 13.1). Molecular studies using fish have largely contributed to the actual knowledge; a few examples will be described to illustrate this in the following paragraphs.

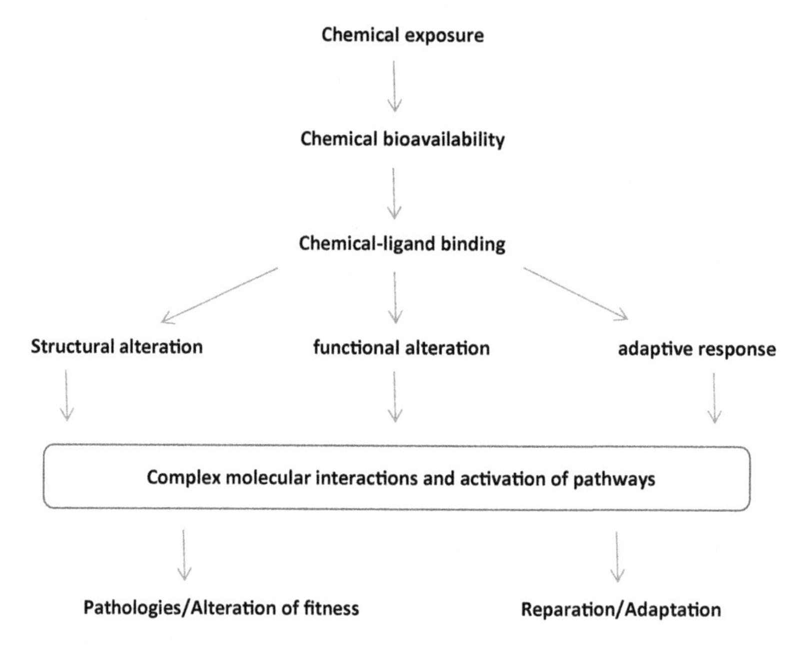

Figure 13.1
Cascade of reactions resulting from chemical exposure in an organism.

13.1.2 From Bioavailability to Modeling

Fish are exposed to a mixture of compounds present in the environment, and toxicological effects depend not on the total concentration of the chemicals in the media but also on how readily the fish can absorb these different compounds at the gill, across the skin, or the digestive tract. Because many chemicals exist in different forms defined as chemical species, these xenobiotics will be more or less bioavailable to a given site of action depending on chemical speciation and various organism attributes.

Gills ought to receive special attention in fish toxicology. Not only are fish living in the aquatic medium and process large volumes of water across a wide surface area, but gills notably serve a variety of physiological functions, including respiratory gas exchange, osmoregulation, nitrogen excretion, and control of acid–base balance (Hoar and Randall, 1984). They are thus a particular site of entry and interactions with chemicals. As reviewed by Paquin et al. (2002), fish toxicologists played a crucial role in elucidating metal toxicity by combining information provided by chemists on one side and fish physiologists on the other and then further prompting the development of a ligand-binding model for metals. This model is particularly focusing on the role of chemical speciation on the resulting toxicity of metals. Early works on fish demonstrated that accumulation of metal in fish tissues is reduced by complexation with organic compounds in exposure water (Muramoto, 1980; Buckley et al., 1984). In turn, binding to ligands at the gills surface are affected (Playle et al., 1993; MacRae et al., 1999; Niyogi and Wood, 2003) and acute toxicological effects are modulated (Zitko et al., 1973; Pagenkopf et al., 1974), suggesting that these complexes are less bioavailable than the free metal. However, metal speciation is dependent on a number of environmental factors. Metals such as copper have high affinity for various ligands found in natural water, including hydroxide, carbonate, sulfide, and various dissolved organic molecules. In marine waters, copper is largely complexed by chloride. Consequently, dissolved copper exist in multiple forms that are constantly associating or dissociating from various ligands. This dynamic speciation has important consequences for bioavailability.

Toxicological studies using fish helped identifying specific targets for metal binding and thus toxicity. Fish physiologists, as early as the 1930s, found out that gills, not kidneys, were the principal route of transfer of ions between the ambient water and the internal fluids of teleost thus playing a determinant role in osmoregulation (Smith, 1930). The disturbance of internal ion balance was found to be the direct cause of acute toxic effects that result from elevated levels of certain metals in freshwater fish (McDonald et al., 1989). This ionoregulatory dysfunction is leading to a redistribution of ions and water between the internal fluid compartments of the fish, and subsequently, to a decrease in levels of plasma sodium, chloride, and other ions, which in turn triggers a sequence of events that potentially leads to cardiovascular collapse and death (Milligan and Wood, 1982; Lauren and McDonald, 1986; Wood, 2001). Specific targets for metals would be the membrane transporters involved in ion regulation. An important component of sodium uptake that can be affected by metals is the active transport of sodium via Na^{+}/K^{+}-adenosinetriphosphatase at the basal membranes of gill epithelial cells. Other ion channels and transporters may be specific sites for metal binding, and environmental cations could compete for these sites, thus explaining results showing that water hardness may decrease toxicity of metals to gills (Chakoumakos et al., 1979; Miller and MacKay, 1980). Hardness and especially cations such as calcium or magnesium may also modify metals speciation by competing for ligands. Furthermore, toxicity resulting from calcium may also involve direct effect on cell permeability through extracellular junctions.

Laboratory toxicity testing using fish as model organisms has provided a wealth of empirical data upon which predictive models, such as the biotic ligand model, are based. In turn, this model can be useful for describing the toxicity of fish and other organisms including animals living in soil (Ardestanti and van Gestel, 2014) and can be viewed as a special achievement of molecular and physiological studies using fish. However, research is still ongoing and is questioning the model. For example, Tao et al. (2002) showed that the microenvironment resulting from the gill activities including mucus secretion and change in pH are affecting metal–gill association constants. Furthermore, there may be different types of biotic ligands (e.g., high- and low-affinity ligands) that may have different contributions to metal uptake and possibly also to metal toxicity (Niyogi et al., 2008).

13.1.3 From Bioavailability to the Chemical Defensome

Bioavailability is the prerequisite for toxicity and special biological processes and structures can alter chemical bioavailability in order to protect the organisms. In this respect, the multixenobiotic resistance (MXR) system appears as the first line of defense that numerous species, including fish species, have developed. Indeed, this system is composed of a number of membrane-bound transporters that can expel chemical compounds out of the cells, thus preventing any intracellular accumulation above a toxic level. In other words, these transporters are bouncers that are meant to exclude unwanted molecules into the cell. These transporters are proteins encoded by the largest gene superfamily in all sequenced genomes to date and correspond to the so-called ABC proteins family. The ABC family is divided into subgroups based on phylogenetic relationships between their nucleotide binding domains and members of each subfamily generally exhibit the same domain structure. Accordingly, fish ABC proteins can be divided into eight families, from ABCA to ABCH, three of which—B, C, and G—are constitutive of the MXR defense system (Annilo et al., 2006).

Research on ABC transporters and the associated concept of multiple resistance to xenobiotics is largely due to investigations conducted on cancer resistance to chemotherapeutic agents. Clinical oncologists were the first to observe that some cancers treated with multiple different anticancer drugs tended to develop cross-resistance to many other cytotoxic agents to which they had never been exposed, effectively leading to failure of the chemotherapy. In vitro studies using mammalian cell lines have help characterize this phenomenon. When exposed to low concentrations of a unique cytotoxic drug, cultured cells spontaneously developed, at low frequency, simultaneous resistance to a large group of structurally and functionally unrelated chemicals (Biedler and Riehm, 1970). Drug transport experiments in multidrug resistant cell lines indicated that the emergence of multidrug resistance was linked to a decreased intracellular drug accumulation and to an increased drug efflux, both of which are strictly adenosine triphosphate–dependent (Danö, 1973). Although this mechanism can be a problem for cancer chemotherapy, the MXR system and its associated cross-resistance to multiple chemical compounds is a crucial defense system for living organisms against pollutants (Minier et al., 2006).

In the 1990s, the first line of evidence supporting the presence and the relevance of the MXR system in fish was provided by studies on transport activity and immunochemical detection revealing that the tissue distribution of the ABC transporter (ABCB1 or P-glycoprotein) in teleosts was similar to its mammalian counterpart, suggesting similar functional properties and physiological roles of ABC pumps in both vertebrate groups. A study in guppy (*Poecilia reticulata*) using the monoclonal antibody C219 reported a strong staining reaction in gill chondrocytes and an absence of signal in filaments and lamellae. The immunohistochemical staining was also present at the lumenal surfaces of the intestinal epithelium (Hemmer et al., 1995). In channel catfish (*Ictalurus punctatus*), reactivity with monoclonal antibody C219 revealed a pronounced expression of reactive protein(s) in the distal and the proximal regions of the gut (Kleinow et al., 2000) and at the apical microvilli of the kidney proximal tubules in winter flounder (*Pleuronectes americanus*, Sussman-Turner and Renfro, 1995). In immunohistochemical investigations of guppy (*P. reticulata*) tissues using antibodies raised against mammalian P-glycoprotein, specifically stained bile canaliculi (Hemmer et al., 1995), whereas ABCB1-like and ABCC2-like proteins were localized to the luminal surface of killifish brain capillaries (Miller et al., 2002). The presence and distribution of the MXR system were then further demonstrated by the development of studies using molecular techniques including quantitative polymerase chain reaction. Tutundjian et al. (2002) cloned a fragment of *abcb1* in turbot (*Scophthalmus maximus*) and demonstrated expression of its mRNA in brain, intestine, kidney, and liver by reverse transcriptase polymerase chain reaction. A cDNA encoding an *abcc2* homologue was cloned in the elasmobranch little skate (*Raja erinacea*) and showed high expression levels in kidney, intestine, and liver, where it located to the canalicular membrane (Cai et al., 2003). Currently available data support the view that ABC transporters are involved in (1) protection of the fish by limiting environmental and food-borne chemicals uptake, (2) in enforced protection of important sites such as the brain, and (3) in transport aimed at discarding either untransformed or metabolites out of the body.

ABC protein-dependent transport activities have been characterized in a number of primary cell cultures from teleosts, employing a range of fluorescent probes and transporter inhibitors of known specificity in mammals and further supporting the role of the MXR system. In killifish (*Fundulus heteroclitus*) and winter flounder (*Pleuronectes americanus*), the transport of the organic base daunomycin into proximal tubule lumen was inhibited by verapamil and cyclosporin A (Miller, 1995; Sussman-Turner and Renfro, 1995) consistent with a mechanism involving *abcb1*. Elevated hepatic and biliary dye accumulation resulted from the presence of verapamil or cyclosporin A in the common carp (*Cyprinus carpio*, Smital and Sauerborn, 2002). Inhibitors of ABC drug transporters also increased the retention of rhodamine 123 in the zebrafish brain (Park et al., 2012). These transport experiments allow the identification of a wide range of compounds that may interact with the transporters thus limiting their toxicological effects or, by overwhelming or blocking the transporters, increasing the effects. Industrial chemicals such as bisphenol A, nonylphenol ethoxylate, or alkylbenzene sulfonate exerted differential effects on xenobiotic transport in killifish kidney tubules (Nickel et al., 2013), rainbow trout hepatocytes (Sturm et al., 2001), or perfused catfish liver (Tan et al., 2010). Several studies provided evidence for a regulation of teleost ABC drug efflux transporters by heavy metals (Long et al., 2011; Della Torre et al., 2012). However, many of these studies have been performed with systems or cells that usually contain a range of drug transporters, which limits their usefulness in studies aiming at the characterization of specific ABC transporters. This can be best identified in cell lines that have been characterized for their ABC transporters expression such as the SAE *Squalus acanthias* shark embryo–derived cell line (Kobayashi et al., 2007) or some rainbow trout cell lines (Fischer et al., 2011). In line with this objective, new systems allowing functional studies of individual ABC drug transporters have recently become available in teleosts. Stepwise selection of the topminnow (*Poeciliopsis lucida*) hepatoma cell line PLHC-1 with increasing levels of doxorubicin was used to generate the subline PLHC-1/dox that shows a 45-fold reduction in doxorubicin sensitivity associated with a 42-fold increase in *abcb1* expression level compared with the parental cell line (Zaja et al., 2008). Other methodologies include an insect cell/baculovirus system suitable for the recombinant expression of ABC pumps and gene knockdown studies (Fischer et al., 2013). These studies allowed the identification of specific substrates or inhibitors of the MXR system that include phenanthrene, arsenic trioxide, pharmaceuticals (pravastatin, atorvastatin, fenofibrate, sildenafil, propranolol, acebutolol, tamoxifen), pesticides (dichlorodiphenyldichloroethylene, endosulfan, diazinon, phosalone, chlorpyrifos, fenoxycarb, malathion), or musks (galaxolide, tonalide) (Caminada et al., 2008; Zaja et al., 2008, 2011; Fischer et al., 2013).

These studies are showing that molecular studies using fish helped characterizing the MXR system and allowed a better understanding of the toxicological effects of identified compounds on fish physiologies. It is now recognized that the MXR is an important component of the defense system in fish and molecular changes can be related to important consequences on the whole organism. Adult zebrafish treated with ivermectin and cyclosporine A were significantly more sensitive to the pesticide compared with fish treated to ivermectin only and inhibition of the ABC transporters was associated with significantly earlier onset and increased mortality in ivermectin-exposed fish. Furthermore, at low doses, neurobehavioral effects could be monitored. These effects included lethargy, slowing of pectoral fin movement, altered posture, and decreased escape response (Bard and Gadbois, 2007). In fish embryos, knocking down the *abcb4* transporter using morpholino in zebrafish significantly enhanced the embryo mortality by 30% in a 48-h exposure experiment to the toxic compound vinblastine (Fischer et al., 2013).

Recent molecular studies have contributed to considerably widen our view on the regulation of gene expression and protein activity. In this framework, receptor-driven mechanisms of toxicity have occupied a particular place as an essential component in a coordinated system. The pregnane xenobiotic receptor (PXR) is involved in the regulation of biotransformation enzymes as well as ABC transporters including ABCB1 and ABCC2 in mammals (Albermann et al., 2005). This might also be true in fish. Exposure of killifish to the metabolically activated form of chlorpyrifos, a known PXR ligand, induced the hepatic expression of P-glycoproteins (Albertus and Laine, 2001). Although, enhanced expression of the MXR system has been observed in areas highly contaminated with polycyclic aromatic hydrocarbons and polychlorobiphenyls, the mechanism of ABCB1 induction does not involve the AhR as exposure of cockscomb blennies or catfish with β-naphthoflavone did not alter hepatic expression of ABCB1-like proteins (Doi et al., 2001; Bard et al., 2002).

The PXR receptor might be a key component in the coordinated signaling pathways of the xenobiotic defense system. The MXR system is reducing the entry of xenobiotic, but is also involved in the transport of metabolites. Indeed, many xenobiotics are metabolized by phase I and phase II enzymes. However, accumulation of metabolites may result in cell injury and their excretion is of particular importance. In this respect, the MXR system is responsible for the phase III (i.e., the export of conjugated compounds out of the cell; Damiens and Minier,

2011). This particularly illustrates the importance of coordinating the different mechanisms dealing with xenobiotics in a given cell. Other studies have also shown that the MXR system is part of a network of pathways related to xenobiotic exposure and, more generally, to stress, endocrine activity or physiological status. Glucocorticoids, either synthetic or natural, are able to stimulate *abcc2*-mediated transport investigated in killifish kidney tubules (Prevoo et al., 2011). During carcinogenesis, a range of changes are occurring in the cells. In the European flounder (*Platichthys flesus*) from polluted sites, preneoplastic basophilic foci, hepatic adenomas, and carcinomas but not early eosinophilic preneoplastic foci showed an increase in P-glycoproteins (Koehler et al., 2004). The MXR system is thus an important component of the chemical defensome (i.e., the coordinated defense against toxic compounds, which includes an array of structures, enzymes, and pathways) (Goldstone et al., 2006). It is obvious that in this view fish molecular studies contributed to better understanding of the consequences and responses of the living animals to toxic compounds which are essential to a scientifically based management of our environment.

13.2 Physiological Condition Indices

Condition is a term widely used to refer to the overall physiological status or health of an individual. Condition indices have been accepted as integrative indicators of general fish fitness and are thought to provide information on energy reserves and the ability of fish to resist pollution or other environmental stresses (Adams, 1999). Sublethal responses such as the measure of physiological condition indices appear to be useful for the ecological risk assessment studies, because of their sensitivity to pollutant and apparent relationships to higher levels (Amara et al., 2007; Kerambrun et al., 2012a). Indeed, individual condition is an important component of performance, reproductive success and survivorship in fish.

Three main types of physiological condition indices can be used to evaluate the effect of environmental stressors (including chemical pollution) in fish: morphometric, biochemical, and growth indices.

13.2.1 Morphometric Condition Indices

Morphometric condition indices are the most easily and widely used types of condition index in fish stress assessment studies. They are based on the assumption that heavier fish of a certain length are in better condition (Jones et al., 1999). Fulton's K condition index (Ricker, 1975) was successfully used to assess the effect of pollution on fish (particularly juveniles) from different aquatic ecosystems (Bervoets and Blust, 2003; Gilliers et al., 2004; Amara et al., 2009; Vasconcelos et al., 2009) and during acute and chronic experimental exposure (Kerambrun et al., 2012a, 2014). Kerambrun et al. (2012b) carried out field caging and laboratory experiments with juvenile sea bass and turbot exposed to polluted harbor and estuarine sediments. They found that Fulton's K and other condition and growth indices were

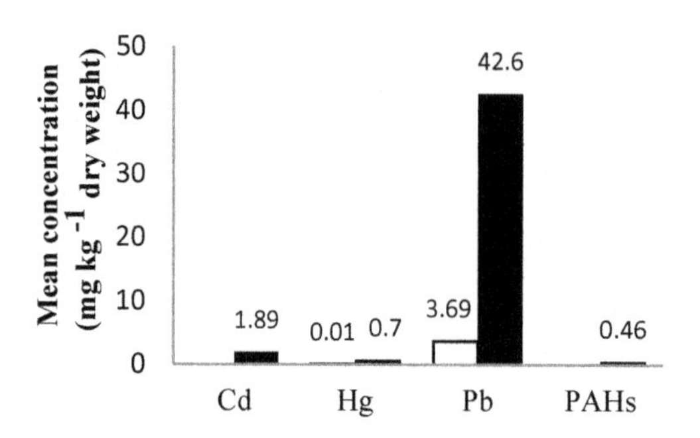

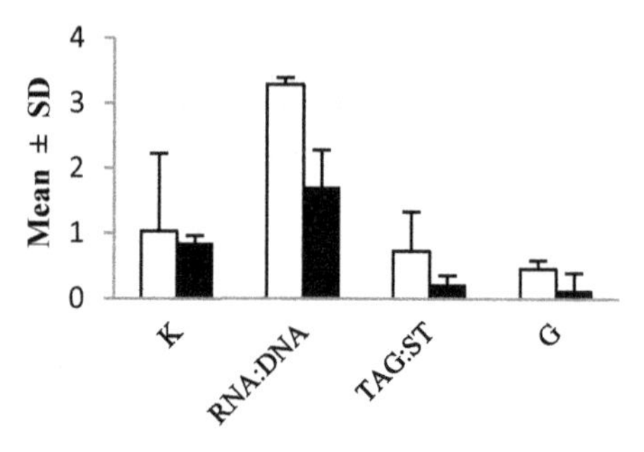

Figure 13.2

Mean sediment metal (Cd, Hg, Pb) and polycyclic aromatic hydrocarbons concentration (mg kg^{-1} dry weight) and different condition indices (Fulton's *K*, RNA:DNA ration, TAG:ST ratio, and growth rate in length (G)) measured on juveniles sea bass caged for 38 days in a harbor at two stations (black or white bars) with different chemical contamination.

significantly higher in the least contaminated station (Figure 13.2). Similarly, juvenile flatfish caught in polluted areas along the eastern English Channel showed lowest condition and growth compared with adjacent low contaminated areas (Amara et al., 2007; Henry et al., 2012; Kerambrun et al., 2013a). Significant negative correlations between *K* and metal concentrations in organs were found in many species (Farkas et al., 2003). These results suggest that environmental chemical contaminants have a negative impact on fish condition.

Fulton's *K* may change for reasons other than pollution stress (Adams, 1999; Lloret et al., 2013). Differences may be caused by sexual maturation and gonad development or by the stomach content. To reduce the effects of feeding condition or gonad development, it is preferable to use the eviscerated weight, so removing the confounding effect of the organ weight. Fulton's *K* condition factor assumes isometric growth (i.e., that the relative proportions of body length, height, and thickness do not change in fish of similar condition as these increase in weight). It has, however, been shown that *K* is influenced by fish length (Figure 13.3). A way to avoid this problem is to compare individual of similar length classes within the same species.

Also easily calculated, accurate morphometric condition assessments are dependent on correct length and weight measurements. It is important to be precise about the weight (total or eviscerated weight) and length measured because *K* may vary according to the type of measurements recorded. For example, *K* are higher when calculated with total weight compared to eviscerated weight and also higher for shorter types of length measurements (i.e., standard length > fork length > total length) (Figure 13.4).

13.2.2 Biochemical and Growth Indices

In comparison with morphometric indices, biochemical and growth indices are time-consuming, expensive, and are often destructive and not feasible for routine field use. However, they offer

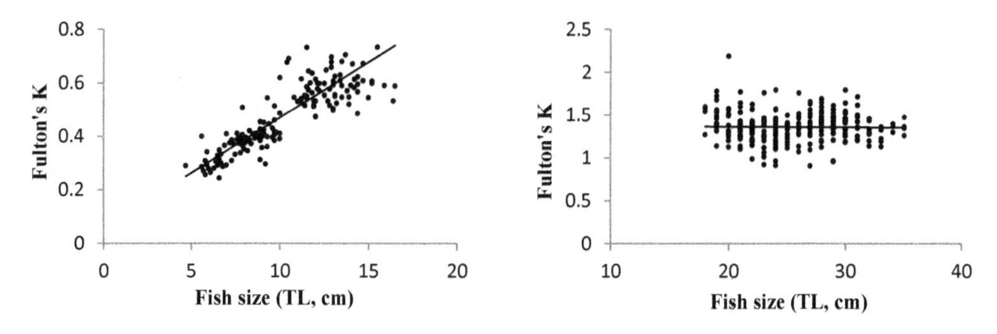

Figure 13.3

Relationship between Fulton's *K* condition factor and fish size in European anchovies (*Engraulis encrasicolus*) from the Algerian coast (W = 0.0006, total length $(TL)^{3.89}$, $r^2 = 0.98$) and red mullet (*Mullus surmuletus*) from the English Channel (W = 0.0139, $TL^{2.99}$, $r^2 = 0.93$). Note that if b-value is very different from "3" Fulton's *K* varies with fish size.

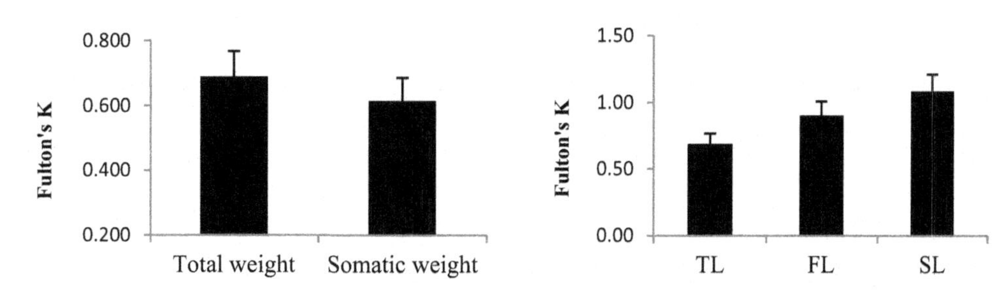

Figure 13.4

Variation in Fulton's condition factor according to the weight (total or eviscerated weight) and length measurements (FL: fork length; SL, standard length; TL, total length) recorded for 150 European anchovy (*Engraulis encrasicolus*) from Algerian coast.

complementary measures to assess, in a more specific way, the condition and underlying health of fish. Lipids are a principal energy reserve in fish and are often the first components to be mobilized during periods of stress (Love, 1974). Many studies have reported depletion in lipid reserves of fish exposed to chemical pollutants both in wild and experimental conditions (Rowe, 2003; Amara et al., 2007; Kerambrun et al., 2012b, 2014). Altered lipid metabolism not only reduces the amount of physiologically available energy for growth, gonad maturation, and repair of damaged tissue, but it also compromises the integrity of the immune system (Arts and Kohler, 2009).

13.2.2.1 Triacylglycerols:Sterols Ratio

Most ecological studies use total body lipids as a measure of energy availability; however, the main form of stored energy in fish are triacylglycerols (TAGs). Other forms of lipids such as cholesterol and phospholipids serve primarily a structural role in cell membranes, and are only metabolized during prolonged periods of stress (e.g., starvation) after TAGs have been depleted (Fraser, 1989; Weber et al., 2003). A lipid storage index based on the ratio of the quantity of TAGs (reserve lipids) to the quantity of sterols (ST; structural lipids) in the fish was developed to study the nutritional status of fish. This index has been showed to be sensitive to pollutant stress. In juveniles sea bass, *Dicentrarchus labrax*, and turbot, *Scophtalmus maximus*, exposed experimentally to harbor polluted sediment for 26 days the value of the TAG:ST ratio diminished respectively by 70% and 64% relative to control fish (Kerambrun et al., 2013a). Similarly, in wild sole TAG:ST ratios, values of individuals caught near polluted harbors were only one-half to one-sixth those of individuals caught in relatively less polluted sites (Amara et al., 2007).

Interpretation of TAG:ST ratio data must be made in the context of ontogenetic changes that occur during the life history. For example, in sole, the TAG content increased during the ontogenetic larval development up to an elevated value corresponding to metamorphosis (Amara and Galois, 2004). In juvenile fish, lipid storage is generally observed before winter and allows juveniles to survive periods of starvation, which are more frequent in winter. TAGs as the major energy storage form in fish have important ecophysiological relevance as indicators of growth potential and survival. TAG:ST ratios are often correlated with growth index such as RNA:DNA ratios or fish growth rates (Figure 13.5).

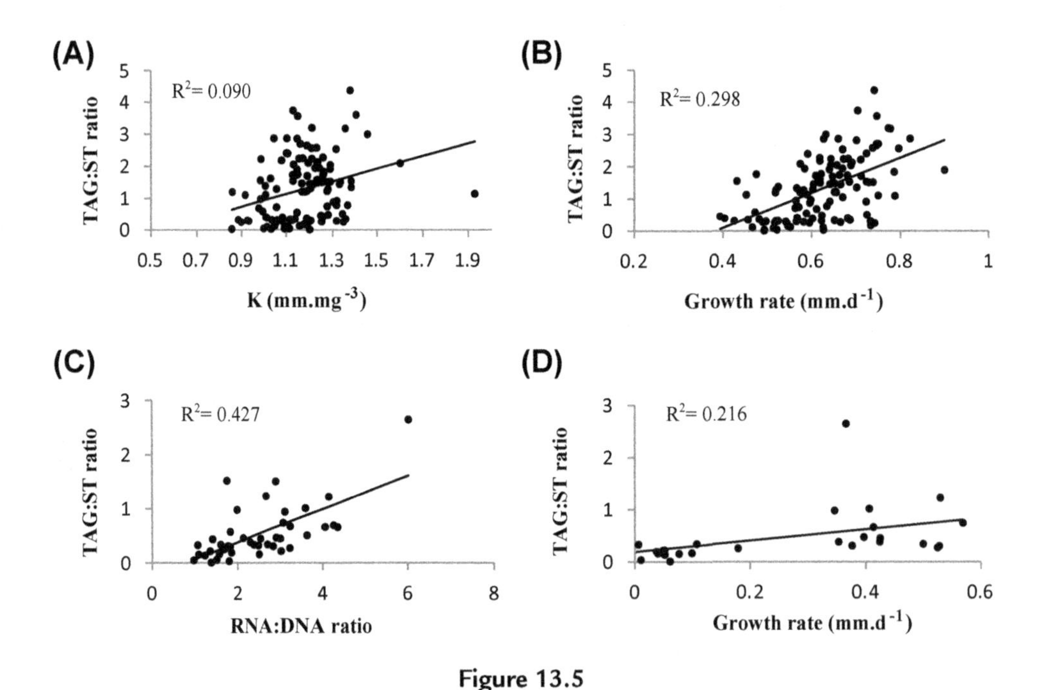

Figure 13.5

Relationship between TAG:ST ratio and different physiological condition indices measured in (A, B) wild juveniles sole (*Solea solea*) from the English channel and (C, D) in juvenile sea bass (*Dicentrarchus labrax*) exposed experimentally to harbor polluted sediment for 26 days.

13.2.2.2 RNA:DNA Ratio

Whole-body and white muscle RNA:DNA ratios have been widely used to assess the condition and growth of fishes in both field and laboratory studies. RNA:DNA ratio reflects variations in growth-related protein synthesis (Chícharo and Chícharo, 2008). Fish in good condition tend to have higher RNA:DNA ratios than those in poor condition. During the past two decades, there has been an increasing number of published studies using RNA:DNA ratio as a biochemical indicator of the physiological and nutritional state of larvae and juvenile fish. However, few have used this index to evaluate the effect of chemical pollutant on fish (Amara et al., 2009; Kerambrun et al., 2012a,b). It was showed that RNA indices respond relatively rapidly to changes in environmental conditions such as starvation (Selleslagh and Amara, 2013). Special caution is advised when comparing RNA:DNA ratio between species or when environmental condition of fish are different.

13.2.2.3 Growth

As a summation of many factors, growth is a useful integrated index of physiological status. Fish growth rates may be a good way to assess habitat quality, provided that fishes grow more quickly in higher-quality habitats, which are those presenting the right set of environmental variables to promote foraging and growth. Both field and laboratory studies have reported reduced growth in fish exposed to pollutants (Burke et al., 1993; Amara et al., 2007; Kerambrun et al., 2013a). Growth is a direct measure of individual fish health, but it is not specific to fish stress responses to pollutants because several other biotic and abiotic stressors can affect it. Experimental studies of fish growth should ideally be carried out using individuals that can be individually identified and remeasured (e.g., Kerambrun et al., 2012b). Fish growth is also commonly estimated from length-at-age data obtained from otoliths. The discovery of daily increments in otoliths has enabled accurate estimates of larva and juvenile growth to be made. When otolith size and body size are strongly related, back-calculation provides a good estimate of the growth trajectory and recent growth index may be determined by measuring the width of the peripheral daily increments of the otoliths (e.g., Henry et al., 2012).

13.2.3 Wild versus Experimental Measure of Fish Physiological Condition

In field studies, because physiological and biochemical responses in fish can be affected by factors other than pollution, it is important to use for comparison valid reference sites with similar environmental conditions. Otherwise, in situ fish-caging studies at contaminated sites can provide information that is more realistic compared to traditional laboratory-based studies. Many recent studies have developed caging method that was used to expose small-bodied fish for up to 6 weeks in duration at field sites (Kerambrun et al., 2011; Miller et al., 2014). The pros and cons of caging in fish experiments are reviewed by Oikari (2006).

Laboratory as well as field experiments with caged animals are likely to overestimate toxicity for species, which would avoid contamination in heterogeneous field settings. On the contrary, they may underestimate toxicity for species given that exposure through the food web is limited or eliminated during experiments. Therefore, if suitable feral species with low risk of migration are available, the use of feral fish is preferable as the results have higher ecological relevance.

13.2.4 Recommendations and Limits of Use of Physiological Conditions Indices

Species-specific differences in condition indices responses to contamination have been observed in studies involving various fish species (Fonseca et al., 2011; Kerambrun et al., 2012b). Physiological condition indices are useful tools to assess the effect of pollution on fish when variation in other environmental factors is limited. They should be evaluated in the context of seasonal metabolic changes that normally occur in fish (Van der Oost et al., 2003). In a recent study, Kerambrun et al. (2014) examined the growth and energetic performance of juvenile turbot after exposure to contaminated sediment and after a recovery period of 35 days in clean seawater with or without food limitation. Juvenile fish affected by chemical pollution can improve their biological performance if pollution events are followed by a period of abundant food. However, if pollution events occur during periods of food scarcity (e.g., in winter), storage of energy reserves will be compromised. In addition, changes in feeding status have significant effects on metal concentrations in fish (Kerambrun et al., 2013b; Ouellet et al., 2013). These authors suggest that starved fish were probably less efficient at eliminating metals from the liver because of energetic deficiencies.

Physiological condition indices present many advantages for ecological risk assessment studies.

- they are sensitive and respond to exposure levels corresponding to encountered environmental levels in polluted aquatic ecosystems
- they have high signification relevance (link between fish health and population impacts)
- results are easily understood by nonscientist (public and user community).

There are, however, several drawbacks with the use of physiological condition indices. They present a lack of specificity because they can be influenced by many biological and environmental factors.

13.3 Fish Population Index and Ecological Quality

Using the wild fish population to assess water quality is nothing new (Karr, 1981; Ramm, 1988). The practice has grown in popularity since the early 2000s, with a marked increase in the development and use of fish indices to determine the ecological status of water bodies (Belliard et al., 1999; Belpaire et al., 2000; Whitfield and Elliott, 2002; Breine et al., 2004; Harrison and Whitfield, 2004; Breine et al., 2007; Coates et al., 2007; Vasconcelos et al., 2007; Henriques et al., 2008; Uriarte and Borja, 2009; Delpech et al., 2010; Argillier et al., 2013;

Laine et al., 2014). The aim of these indices is to detect anthropogenic disturbances and their impact on fish communities in surface waters (lakes, rivers, estuaries, and seas). Heavy metal pollution, organic pollutants, and nutrient enrichment are often identified as having effects on fish communities (Oberdorff and Hughes, 1992; Amara et al., 2007; Courrat et al., 2009; Argillier et al., 2013; Laine et al., 2014), and fish indices demonstrated their ability to detect signs of disturbance.

The creation of bioindicators is based on the hypothesis that pollution may have adverse effects on fish at multiple levels, from cellular level up to entire populations. It is also assumed that there can be a cascade effect (i.e., physiological dysfunction may lead to a total loss of reproduction, causing depletion in fish stocks, which in turn will have a severe impact on the trophic network). The effects of pollution can therefore be seen at a structure-assemblage level (Lima-Junior et al., 2006; Azzurro et al., 2010; Antal et al., 2013) through direct and indirect effects (Fleeger et al., 2003; Iwasaki et al., 2009). Direct effects on a fish population or community will depend on the proportion of the population exposed as well as the intensity and duration of exposure to a specific pollutant (Axiak et al., 2000; Whitfield and Becker, 2014). Effects can be immediate or delayed and will depend on the physicochemical parameters of a habitat's abiotic factors. Temperature, pH, and hardness have previously been shown to influence the bioavailability of metals to fish (Avenant-Oldewage and Marx, 2000). The effects of hypoxia—a phenomenon that is being observed more and more in a range of ecosystem types (Diaz, 2001)—may be exacerbated by exposure to contaminants, leading to long-term consequences in terms of fitness (Gorokhova et al., 2013). The fish community response to these combined impacts is an expected decrease in abundance and species richness.

13.3.1 The Use of Fish Index in the Water Framework Directive in Europe

The implementation of the Water Framework Directive (Directive 2000/60/EC) calls for an evaluation of the status of fish communities in continental water bodies. Thus, a fish index called ELFI (Estuarine and Lagoon Fish Index) able to describe the ecological status of French estuarine water bodies was developed. In compliance with the Water Framework Directive, this index classifies water bodies according to five levels of quality. Seven descriptors (metrics) are used to characterize different aspects of the fish population. These descriptors are rigorously defined based on the outputs of statistical models and the way in which they (the descriptors) respond to a gradient of anthropogenic pressures. The aim of this tool is to attempt to disentangle the effects on fish populations of three elements: physical characteristics of estuaries, natural variation, and human disturbance. The development process of the fish index is depicted in Figure 13.6.

In an initial version of the aforementioned index, Delpech et al. (2010) used four biological metrics to detect the effects of chemical pollution as a proxy of human activities. Biological

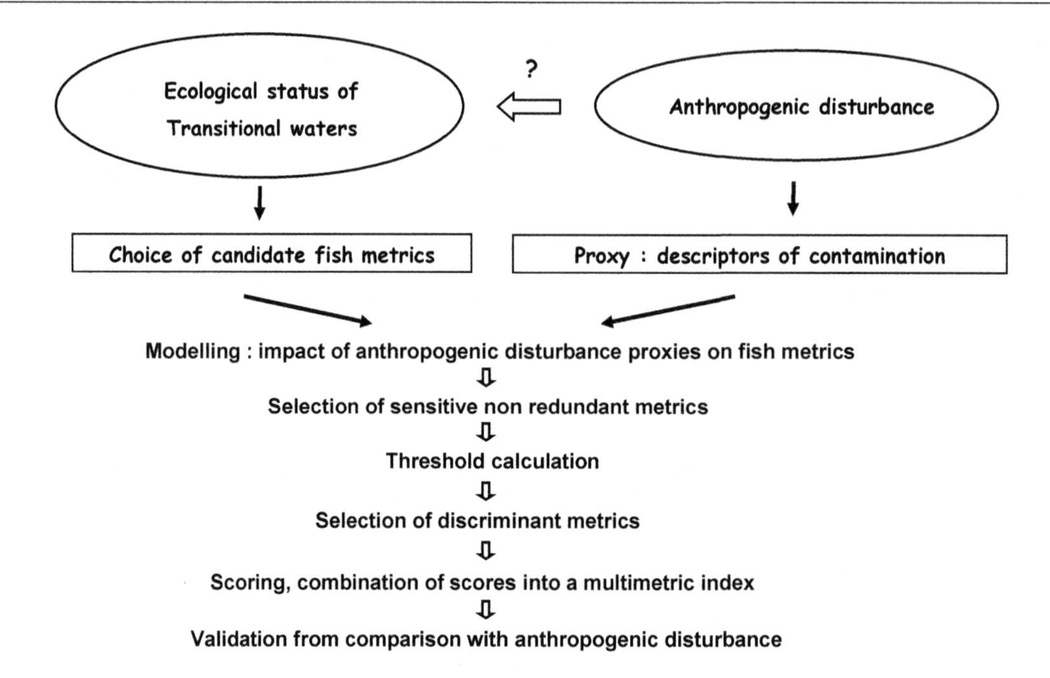

Figure 13.6

General methodology description for the construction of a multimetric fish index. *Adapted from Courrat et al. (2009).*

metrics were selected according to their expected response to an increase in a specific set of human disturbances (degradation). Courrat et al. (2009) demonstrated a correlation between the median level of five heavy metals and two organic pollutants found in mollusc species (oysters *Crassostrea gigas* and mussels *Mytilus edulis*) and the density of marine juvenile fish in 13 French estuaries. To avoid comparison problems in the fish assemblages, ecological guilds were used to describe the fish population (Table 13.2). To achieve disentanglement, generalized linear models were developed for each ecological guild, taking into account the effect of anthropogenic pressure. The most suitable models (i.e., those with the highest level of significance and the highest proportion of explained variance) were selected. This process led to seven metrics being identified as giving significant responses with an increase in human disturbance for use in the final indicator. Because the density of the species is decreasing before a species could disappear from the fish assemblage when facing human disturbance, all metrics are expressed in density to get an early warning. The metrics chosen were the following: (1) diadromous fish density; (2) marine juvenile density; (3) freshwater species density in oligohaline areas; (4) benthic species density; (5) total density; (6) estuarine resident density; and (7) taxonomic richness.

Using ELFI results based on surveillance surveys, it was possible to assess the ecological status of the ecosystems studied. Because ELFI is founded on the pressure–impact relationship, the score generated by each metric also provided an indication of potential causes of

Table 13.2: Ecological Guilds and Their Definitions for the French Fish Index ELFI

Acronym	Name	Definitions
FW	Freshwater species	Freshwater adventitious species, which occasionally enter brackish waters from freshwaters but have no apparent estuarine requirements (definition from (Elliott and Dewailly, 1995)
DIA	Diadromous species	Fish that cross salinity boundaries and are able to maintain stable populations at FW or SW
MA	Marine adventitious species	Main population centers for both adults and juveniles not found in transitional and coastal (TraC) waters. These species may be captured with regularity in TraC waters but numbers are low
ER	Estuarine resident species	>50% population adults and juveniles found in TraC waters. In practical terms, for very small species that are not known to make real migrations (e.g., Gobies, blennies, lepadogaster, tripterigion, hippocampus, syngnathus, stickleback, gambusia). If they are caught, it is probably because they drift in as a larvae and could stay in the estuary/lagoon system for some time if they do not last their entire cycle.
MS	Marine seasonal species	Species that are entering the transitional system only at a certain period of the year and where adults and juveniles are found in numbers.
MJ	Marine juvenile species	Significant shift of the juvenile population to TraC waters because of a distinct migration or larval/juvenile dispersal reaching into TraC waters. In practical terms, for marine species, when the majority of caught fishes are juveniles.

environmental damage (heavy metal and chemical pollution, habitat degradation). These indications would help stakeholders in setting remediation actions. There is still work to be done on the use of pressure–impact models to study fish populations. One key improvement would be to make the descriptors more sensitive, thus providing a more effective early warning when a new disturbance appears.

There are several advantages in developing indices at population level because of high levels of spatiotemporal integration of fish. Such an approach also makes it possible to identify cascade effect, which is otherwise undetectable. Bioindicators in general and fish indices in particular can be very effective in detecting molecules present in water or sediment at concentrations near to the detection threshold with standard chemical methods. Fish indices have proved to be efficient in detecting several types of pollution and their potential combined effect with physicochemical parameters or hydromorphological alterations. Nevertheless, it is never simple or a straightforward process.

13.4 Conclusions

Fish species are obvious sentinel species of the aquatic ecosystems. Finding dead fish in lakes or rivers leads to questions about the quality of the water medium. Consequently, toxicity tests using fish species have become mandatory for the a priori evaluation of the effects of chemicals on the environment. Moreover, fish are often used in biomonitoring studies and are key organisms used to assess the water quality in the Water Framework Directive. Effects can be assessed at different levels of the biological organization. Effects at the population level are highly significant and are generally regarded as the pertinent level to draw conclusions on the environmental status. This level is highly integrative, being the result of toxic action of chemicals (and/or pathogens), combined with molecular responses and physiological regulation in a given physicochemical environment and trophic conditions. Nevertheless, physiological and molecular indices may provide complementary information. They can be more precocious, serving as early warning events and thus contributing to a better management of the environment. However, it is noteworthy that the identified biomarkers for sublethal endpoints are eventually linked to adverse outcomes in terms of survival, reproduction and susceptibility to disease of environmental populations. Concrete steps in this direction have already been taken and were illustrated in this chapter.

The observation of dead fish is also questioning on the reasons that led to its death. Studies using fish may provide valuable information on the causes. Accordingly, population studies are using metrics that are related to identified anthropogenic disturbances. Physiological indices may indicate specific deregulation and altered functions. Molecular investigation can also point out to a specific mode of action. All these studies are important because they may provide evidence for specific danger and may help in identifying chemicals responsible for the observed disturbances.

Knowledge is rapidly growing and new technologies, including omics studies, are offering new opportunities and possibilities to integrate a large amounts of information. Progress in the field has already been significantly facilitated by the completion of different teleost genome projects. In this context, collaboration between fish toxicologists, physiologists, and ecologists is highly relevant and necessary.

References

Adams, S.M., 1999. Ecological role of lipids in the health and success of fish populations. In: Lipids in Freshwater Ecosystems. Springer, New York, pp. 132–160.

Albermann, N., Schmitz-Winnenthal, F.H., Z'graggen, K., et al., 2005. Expression of the drug transporters *MDR1/ABCB1, MRP1/ABCC1, MRP2/ABCC2, BCRP/ABCG2*, and *PXR* in peripheral blood mononuclear cells and their relationship with the expression in intestine and liver. Biochem. Pharmacol. 70, 949–958.

Albertus, J.A., Laine, R.O., 2001. Enhanced xenobiotic transporter expression in normal teleost hepatocytes, response to environmental and chemotherapeutic toxins. J. Exp. Biol. 204, 217–227.

Amara, R., Galois, R., 2004. Nutritional condition of metamorphosis sole, *Solea solea* (L.), spatial and temporal analyses. J. Fish. Biol. 64, 1–17.

Amara, R., Meziane, T., Gilliers, C., et al., 2007. Growth and condition indices in juvenile sole (*Solea solea* L.) measured to assess the quality of essential fish habitat. Mar. Ecol. Progr. Ser. 351, 209–220.

Amara, R., Selleslagh, J., Billon, G., et al., 2009. Growth and condition of 0-group European flounder, *Platichthys flesus* as indicator of estuarine habitat quality. Hydrobiologia 67, 87–98.

Annilo, T., Chen, Z.-Q., Shulenin, S., et al., 2006. Evolution of the vertebrate ABC gene family, analysis of gene birth and death. Genomics 88, 1–11.

Antal, L., Halasi-Kovacs, B., Nagy, S.A., 2013. Changes in fish assemblage in the Hungarian section of River Szamos/Someş after a massive cyanide and heavy metal pollution. North-Western J. Zool. 9, 131–138.

Ardestanti, M.M., van Gestel, C.A.M., 2014. The effect of pH and calcium on copper availability to the springtail *Folsomia candida* in simplified soil solutions. Pedobiologia 57, 53–55.

Argillier, C., Caussé, S., Gevrey, M., et al., 2013. Development of a fish-based index to assess the eutrophication status of European lakes. Hydrobiologia 704, 193–211.

Arts, M.T., Kohler, C.C., 2009. Health and condition in fish, the influence of lipids on membrane competency and immune response. In: Lipids in Aquatic Ecosystems. Springer, New York, pp. 237–256.

Avenant-Oldewage, A., Marx, H.M., 2000. Bioaccumulation of chromium, copper and iron in the organs and tissues of *Clarias gariepinus* in the Olifants River, Kruger National Park. Water SA 26, 569–582.

Axiak, V., Vella, A.J., Agius, D., et al., 2000. Evaluation of environmental levels and biological impact of TBT in Malta (central Mediterranean). Sci. Total Environ. 258, 89–97.

Azzurro, E., Matiddi, M., Fanelli, E., et al., 2010. Sewage pollution impact on Mediterranean rocky-reef fish assemblages. Mar. Environ. Res. 69, 390–397.

Bard, S.M., Gadbois, S., 2007. Assessing neuroprotective P-glycoprotein activity at the blood–brain barrier in killifish (*Fundulus heteroclitus*) using behavioural profiles. Mar. Environ. Res. 64, 679–682.

Bard, S.M., Woodin, B.R., Stegeman, J.J., 2002. Expression of P-glycoprotein and cytochrome p450 1A in intertidal fish (*Anoplarchus purpurescens*) exposed to environmental contaminants. Aquat. Toxicol. 60, 17–32.

Belliard, J., Berrebi dit Thomas, R., Monnier, D., 1999. Fish communities and river alteration in the Seine Basin and nearby coastal streams. Hydrobiologia 400, 155–166.

Belpaire, C., Smolders, R., Auweele, I.V., et al., 2000. An index of biotic integrity characterizing fish populations and the ecological quality of Flandrian water bodies. Hydrobiologia 434, 17–33.

Bervoets, L., Blust, R., 2003. Metal concentrations in water, sediment and gudgeon (*Gobio gobio*) from a pollution gradient, relationship with fish condition factor. Environ. Pollut. 126, 9–19.

Biedler, J.M., Riehm, H., 1970. Cellular resistance to actinomycin D in Chinese hamster cells *in vitro*, cross-resistance, radioautographic and cytogenetic studies. Cancer Res. 30, 1174–1184.

Breine, J., Maes, J., Quataert, P., et al., 2007. A fish-based assessment tool for the ecological quality of the brackish Schelde estuary in Flanders (Belgium). Hydrobiologia 575, 141–159.

Breine, J., Simoens, I., Goethals, P., et al., 2004. A fish-based index of biotic integrity for upstream brooks in Flanders (Belgium). Hydrobiologia 522, 133–148.

Buckley, J.A., Yoshida, G.A., Wells, N.R., 1984. A cupric ion-copper bioaccumulation relationship in coho salmon exposed to copper-containing treated sewage. Comp. Biochem. Physiol. 78C, 105–110.

Burke, J.S., Peters, D.S., Hanson, P.J., 1993. Morphological indices and otolith microstructure of Atlantic croaker, *Micropogonias undulatus*, as indicators of habitat quality along an estuarine pollution gradient. Environ. Biol. Fish. 36, 25–33.

Cai, S.Y., Soroka, C.J., Ballatori, N., et al., 2003. Molecular characterization of a multidrug resistance-associated protein, Mrp2, from the little skate. Am. J. Physiol. Regul. Integr. Comp. Physiol. 284, R125–R130.

Caminada, D., Zaja, R., Smital, T., et al., 2008. Human pharmaceuticals modulate P-gp1 (ABCB1) transport activity in the fish cell line PLHC-1. Aquat. Toxicol. 90, 214–222.

Carpenter, K.E., 1925. On the biological factor involved in the destruction of river fisheries by pollution due to lead mining. Ann. Appl. Biol. 12, 1–13.

Chakoumakos, C., Russo, R.C., Thurston, R.V., 1979. Toxicity of copper to cutthroat trout (*Salmo claki*) under different conditions of alkalinity, pH and hardness. Environ. Sci. Technol. 13, 213–219.

Chícharo, M.A., Chícharo, L., 2008. RNA, DNA ratio and other nucleic acid derived indices in marine ecology. Int. J. Mol. Sci. 9, 1453–1471.

Coates, S., Waugh, A., Anwar, A., et al., 2007. Efficacy of a multi-metric fish index as an analysis tool for the transitional fish component of the Water Framework Directive. Mar. Pollut. Bull. 55, 225–240.

Courrat, A., Lobry, J., Nicolas, D., et al., 2009. Anthropogenic disturbance on nursery function of estuarine areas for marine species. Estuar. Coastal Shelf Sci. 81, 179–190.

Damiens, G., Minier, C., 2011. The multixenobiotic transport system, a system governing intracellular contaminant bioavailability. In: Amiard-Triquet, C., Rainbow, P.S., Roméo, M. (Eds.), Tolerance in Aquatic Organisms. CRC Press, pp. 229–246.
Danö, K., 1973. Active outward transport of daunomycin in resistant Ehrlich ascites tumor cells. Biochim. Biophys. Acta 323, 466–483.
Daughty, R., Gardiner, R., 2003. The return of salmon to cleaner rivers, a Scottish perspective. In: Mills, D. (Ed.), Salmon at the Edge. Blackwell Publishing, Oxford, pp. 175–185.
Delpech, C., Courrat, A., Pasquaud, S., et al., 2010. Development of a fish-based index to assess the ecological quality of transitional waters, the case of French estuaries. Mar. Pollut. Bull. 60, 908–918.
Della Torre, C., Zaja, R., Loncar, J., et al., 2012. Interaction of ABC transport proteins with toxic metals at the level of gene and transport activity in the PLHC-1 fish cell line. Chem. Biol. Interact. 198, 9–17.
Diaz, R.J., 2001. Overview of hypoxia around the world. J. Environ. Qual. 30, 275–281.
Doi, A.M., Holmes, E., Kleinow, K.M., 2001. P-glycoprotein in the catfish intestine, inducibility by xenobiotics and functional properties. Aquat. Toxicol. 55, 157–170.
Elliott, M., Dewailly, F., 1995. The structure and components of European estuarine fish assemblages. Neth. J. Aquat. Ecol. 29, 397–417.
FAO, 2012. In: The State of World Fisheries and Aquaculture. FAO, Rome, p. 230.
Farkas, A., Salánki, J., Specziár, A., 2003. Age-and size-specific patterns of heavy metals in the organs of freshwater fish *Abramis brama* L. populating a low-contaminated site. Water Res. 37, 959–964.
Fishbase. 2014. http://www.fishbase.org/.
Fischer, S., Klüver, N., Burkhardt-Medicke, K., et al., 2013. Abcb4 acts as multixenobiotic transporter and active barrier against chemical uptake in zebrafish (*Danio rerio*) embryos. BMC Biol. 11, 69.
Fischer, S., Loncar, J., Zaja, R., et al., 2011. Constitutive mRNA expression and protein activity levels of nine ABC efflux transporters in seven permanent cell lines derived from different tissues of rainbow trout (*Oncorhynchus mykiss*). Aquat. Toxicol. 101, 438–446.
Fleeger, J.W., Carman, K.R., Nisbet, R.M., 2003. Indirect effects of contaminants in aquatic ecosystems. Sci. Total Environ. 317, 207–233.
Fonseca, V.F., França, S., Serafim, A., et al., 2011. Multi-biomarker responses to estuarine habitat contamination in three fish species, *Dicentrarchus labrax, Solea senegalensis* and *Pomatoschistus microps*. Aquat. Toxicol. 102, 216–227.
Fraser, A.J., 1989. Triacylglycerol content as a condition index for fish, bivalve and crustacean larva. Can. J. Fish. Aquat. Sci. 46, 1868–1873.
Gilliers, C., Amara, R., Bergeron, J.P., 2004. Quality of flatfish nursery habitats in the eastern channel, and southern bight of the North Sea. Environ. Biol. Fish. 71, 189–198.
Goldstone, J.V., Hamdoun, A., Cole, B.J., et al., 2006. The chemical defensome, environmental sensing and response genes in the *Strongylocentrotus purpuratus* genome. Dev. Biol. 300, 366–384.
Gorokhova, E., Lof, M., Reutgard, M., et al., 2013. Exposure to contaminants exacerbates oxidative stress in amphipod *Monoporeia affinis* subjected to fluctuating hypoxia. Aquat. Toxicol. 127, 46–53.
Harrison, T.D., Whitfield, A.K., 2004. A multi-metric fish index to assess the environmental condition of estuaries. J. Fish. Biol. 65, 683–710.
Hart, H., Doudoroff, P., Greenbank, J., 1945. The Evaluation of the Toxicity of Industrial Wastes, Chemicals and Other Substances to Fresh Water Fishes. Waste Water Control Laboratory, Atlantic Refining Co., Philadelphia.
Hemmer, M., Courtney, L., Ortego, L., 1995. Immunohistochemical detection of P-glycoprotein in teleost tissues using mammalian polyclonal and monoclonal antibodies. J. Exp. Zool. 272, 69–77.
Henriques, S., Pais, M.P., Costa, M.J., et al., 2008. Development of a fish-based multimetric index to assess the ecological quality of marine habitats, the marine fish community index. Mar. Pollut. Bull. 56, 1913–1934.
Henry, F., Filipuci, I., Billon, G., et al., 2012. Metal concentrations, growth and condition indices in European juvenile flounder (*Platichthys flesus*) relative to sediment contamination levels in four Eastern English channel estuaries. J. Environ. Monit. 14, 3211–3220.

Hoar, W.S., Randall, D.J., 1984. Fish Physiology. Gills-Anatomy, Gas Transfer, and Acid-base Regulation, vol. X, Academic Press, Orlando, FL. pp. 1–456.

Iwasaki, Y., Kagaya, T., Miyamoto, K-i., et al., 2009. Effects of heavy metals on riverine benthic macroinvertebrate assemblages with reference to potential food availability for drift-feeding fishes. Environ. Toxicol. Chem. 28, 354–363.

Jones, R.E., Petrell, R.J., Pauly, D., 1999. Using modified length-weight relationships to assess the condition of fish. Aquacult. Eng. 20, 261–276.

Karr, J.R., 1981. Assessment of biotic integrity using fish communities. Fisheries 6, 21–27.

Kerambrun, E., Sanchez, W., Henry, F., et al., 2011. Are biochemical biomarker responses related to physiological performance of juvenile sea bass (*Dicentrarchus labrax*) and turbot (*Scophthalmus maximus*) caged in a polluted harbour? Comp. Biochem. Physiol. 154C, 187–195.

Kerambrun, E., Le Floch, S., Thomas-Guyon, H., et al., 2012a. Responses of juvenile sea bass, *Dicentrarchus labrax*, exposed to acute concentrations of crude oil, as assessed by molecular and physiological biomarkers. Chemosphere 87, 692–702.

Kerambrun, E., Henry, F., Sanchez, W., et al., 2012b. Relationships between biochemical and physiological biomarkers responses measured on juvenile marine fish to environmental chemical contamination. Comp. Biochem. Physiol. 63A, 22–23.

Kerambrun, E., Henry, F., Cornille, V., et al., 2013a. A combined measurement of bioaccumulation and condition indices in juvenile European flounders, *Platichthys flesus*, from European estuaries. Chemosphere 91, 498–505.

Kerambrun, E., Amara, R., Henry, F., 2013b. Effects of food limitation on 9 metal concentration in liver and polycyclic aromaitic hydrocarbon metabolites in bile of juvenile turbot (*Scophthalmus maximus*) previously exposed to contaminated sediments. Environ. Toxicol. Chem. 32, 2552–2557.

Kerambrun, E., Henry, F., Rabhi, K., et al., 2014. Effects of chemical stress and food limitation on the energy reserves and growth of turbot, *Scophthalmus maximus*. Environ. Sci. Pollut. Res. http://dx.doi.org/10.1007/s11356-014-3281-1.

Kleinow, K.M., Doi, A.M., Smith, A.A., 2000. Distribution and inducibility of P-glycoprotein in the catfish, immunohistochemical detection using the mammalian C-219 monoclonal. Mar. Environ. Res. 50, 313–317.

Kobayashi, H., Parton, A., Czechanski, A., et al., 2007. Multidrug resistance-associated protein 3 (Mrp3/Abcc3/Moat-D) is expressed in the SAE *Squalus acanthias* shark embryo-derived cell line. Zebrafish 4, 261–275.

Koehler, A., Alpermann, T., Lauritzen, B., et al., 2004. Clonal xenobiotic resistance during pollution-induced toxic injury and hepatocellular carcinogenesis in liver of female flounder (*Platichthys flesus* (L.)). Acta Histochem. 106, 155–170.

Laine, M., Morin, S., Tison-Rosebery, J., 2014. A multicompartment approach – diatoms, macrophytes, benthic macroinvertebrates and fish – to assess the impact of toxic industrial releases on a small french river. PLoS One 9.

Lauren, D.J., McDonald, D.G., 1986. Influence of water hardness, pH, and alkalinity on the mechanisms of copper toxicity in juvenile rainbow trout, *Salmo gairdneri*. Can. J. Fish. Aquat. Sci. 43, 1488–1496.

Lima-Junior, S.E., Cardone, I.B., Goitein, R., 2006. Fish assemblage structure and aquatic pollution in a Brazilian stream, some limitations of diversity indices and models for environmental impact studies. Ecol. Freshwater Fish. 15, 284–290.

Lloret, J., Shulman, G., Love, R.M., 2013. Condition and Health Indicators of Exploited Marine Fishes. John Wiley & Sons, Oxford. http://dx.doi.org/10.1002/9781118752777.

Long, Y., Li, Q., Wang, Y., Cui, Z., 2011. MRP proteins as potential mediators of heavy metal resistance in zebrafish cells. Comp. Biochem. Physiol. 153C, 310–317.

Love, R.M., 1974. The Chemical Biology of Fishes. Academic Press, London.

MacRae, R.K., Smith, D.E., Swoboda-Colberg, N.S., et al., 1999. Copper-binding affinity of rainbow trout (*Oncorhynchus mykiss*) and brook trout (*Salvelinus fontinalis*) gills, implications for assessing bioavailable metal. Environ. Toxicol. Chem. 18, 1180–1189.

McDonald, D.G., Reader, J.P., Dalziel, T.R.K., 1989. The combined effects of pH and trace metals on fish ionoregulation. In: Morris, R., Taylor, E.W., Brown, D.J.A., Brown, J.A. (Eds.), Acid Toxicity and Aquatic Animals, Society for Experimental Biology Seminar Series, vol. 34. Cambridge University Press, Cambridge, UK, pp. 221–242.

Miller, T.G., MacKay, W.C., 1980. The effects of harness, alkalinity and pH of test water on toxicity of copper to rainbow trout (*Salmo gairdneri*). Water Res. 14, 129–133.

Miller, D.S., 1995. Daunomycin secretion by killfish renal proximal tubules. Am. J. Physiol. Regul. Integr. Comp. Physiol. 269, R370–R379.

Miller, D.S., Graeff, C., Droulle, L., et al., 2002. Xenobiotic efflux pumps in isolated fish brain capillaries. Am. J. Physiol. Regul. Integr. Comp. Physiol. 282, R191–R198.

Miller, J.L., Sherry, J., Parrott, J., et al., 2014. A subchronic in situ exposure method for evaluating effects in small–bodied fish at contaminated sites. Environ. Toxicol. 29, 54–63.

Milligan, C.L., Wood, C.M., 1982. Disturbances in haematology, fluid volume distribution and circulatory function associated with low environmental pH in the rainbow trout, *Salmo gairdneri*. J. Exp. Biol. 99, 397–415.

Minier, C., Moore, M.N., Galgani, F., et al., 2006. Multixenobiotic resistance protein expression in *Mytilus edulis, Mytilus galloprovinciallis* and *Crassostrea gigas* from the French coasts. Mar. Ecol. Prog. Ser. 322, 155–168.

Muramoto, S., 1980. Effect of complexants (EDTA, NTA and DTPA) on the exposure of high concentrations of cadmium, copper, zinc and lead. Bull. Environ. Contam. Toxicol. 25, 941–946.

Nickel, S., Bernd, A., Miller, D.S., et al., 2013. Bisphenol - a modulates function of ABC transporters in killifish. MDIBL Bull. 52, 30.

Niyogi, S., Kent, R., Wood, C.M., 2008. Effects of water chemistry variables on gill binding and acute toxicity of cadmium in rainbow trout (*Oncorhynchus mykiss*), a biotic ligand model (BLM) approach. Comp. Biochem. Physiol. 148C, 305–314.

Niyogi, S., Wood, C.M., 2003. Effects of chronic waterborne and dietary metal exposures on gill metal-binding, implications for the biotic ligand model. Hum. Ecol. Risk Assess. 9, 813–846.

Oberdorff, T., Hughes, R.M., 1992. Modification of an index of biotic integrity based on fish assemblages to characterize rivers of the Seine Basin, France. Hydrobiologia 228, 117–130.

OECD, 2014. In: Guidelines for the Testing of Chemicals, Section 2 Effects on Biotic Systems. OECD. http://dx.doi.org/10.1787/20745761.

Oikari, A., 2006. Caging techniques for field exposures of fish to chemical contaminants. Aquat. Toxicol. 78, 370–381.

Ouellet, J.D., Dubé, M.G., Niyogi, S., 2013. The influence of food quantity on metal bioaccumulation and reproduction in fathead minnows (*Pimephales promelas*) during chronic exposures to a metal mine effluent. Ecotox. Environ. Saf. 91, 188–197.

Pagenkopf, G.K., Russo, R.C., Thurston, R.V., 1974. Effect of complexation on toxicity of copper to fishes. J. Fish. Res. Board Can. 31, 462–465.

Paquin, P.R., Gorush, J.W., Apte, S., et al., 2002. The biotic ligand model: a historical overview. Comp. Biochem. Physiol. 133C, 3–35.

Park, D., Haldi, M., Seng, W.L., 2012. Zebrafish, a new in vivo model for identifying P-glycoprotein efflux modulators. In: McGrath, P. (Ed.), Zebrafish, Methods for Assessing Drug Safety and Toxicity. Wiley, Hoboken, NJ, USA, pp. 177–190.

Penny, C., Adams, C., 1863. Fourth Report from the Royal Commission on Pollution of Rivers in Scotland. London.

Playle, R.C., Dixon, D.G., Burnison, K., 1993. Copper and cadmium binding to fish gills, modification by dissolved organic carbon and synthetic lignads. Can. J. Fish. Aquat. Sci. 50, 2678–2687.

Prevoo, B., Miller, D.S., Van DeWater, F.M., et al., 2011. Rapid, nongenomic stimulation of multidrug resistance protein 2 (Mrp2) activity by glucocorticoids in renal proximal tubule. J. Pharmacol. Exp. Ther. 338, 362–371.

Ramm, A.E., 1988. The community degradation index, a new method for assessing the deterioration of aquatic habitats. Water Res. 22, 293–301.

Ricker, W.E., 1975. Computation and interpretation of the biological statistics of fish populations. Bull. Fish. Res. Bd. Can. 191, 1–382.

Rowe, C.L., 2003. Growth responses of an estuarine fish exposed to mixed trace elements in sediments over a full life cycle. Ecotoxicol. Environ. Saf. 54, 229–239.

Selleslagh, J., Amara, R., 2013. Effect of starvation on condition and growth of juvenile plaice *Pleuronectes platessa*, nursery habitat quality assessment during the settlement period. J. Mar. Biol. Ass. UK 93, 479–488

Smital, T., Sauerborn, R., 2002. Measurement of the activity of multixenobiotic resistance mechanism in the common carp *Cyprinus carpio*. Mar. Environ. Res. 54, 449–453.

Smith, H.W., 1930. The absorption and excretion of water and salts by marine teleosts. Am. J. Physiol. 93, 480–505.

Sturm, A., Cravedi, J.P., Segner, H., 2001. Prochloraz and nonylphenol diethoxylate inhibit anmdr1-like activity in vitro, but do not alter hepatic levels of P-glycoprotein in trout exposed *in vivo*. Aquat. Toxicol. 53, 215–228.

Sussman-Turner, C., Renfro, J.L., 1995. Heat-shock stimulated transepithelial daunomycin secretion by flounder renal proximal tubule primary cultures. Am. J. Physiol. 268, F135–F144.

Tan, X., Yim, S.-Y., Uppu, P., et al., 2010. Enhanced bioaccumulation of dietary contaminants in catfish with exposure to the waterborne surfactant linear alkylbenzene sulfonate. Aquat. Toxicol. 99, 300–308.

Tao, S., Liu, G., Xu, F., 2002. Estimation of conditional stability constant for copper binding to fish gill surface with consideration of chemistry of fish gill microenvironment. Comp. Biochem. Physiol. 133C, 219–226.

Tutundjian, R., Cachot, J., Leboulenger, F., et al., 2002. Genetic and immunological characterisation of a multixenobiotic resistance system in the turbot (*Scophthalmus maximus*). Comp. Biochem. Physiol. 132B, 463–471.

Uriarte, A., Borja, A., 2009. Assessing fish quality status in transitional waters, within the European Water Framework Directive, setting boundary classes and responding to anthropogenic pressures. Estuar. Coast. Shelf Sci. 82, 214–224.

Van der Oost, R., Beyer, J., Vermeulen, N.P., 2003. Fish bioaccumulation and biomarkers in environmental risk assessment, a review. Environ. Toxicol. Pharmacol. 13, 57–149.

Vasconcelos, R.P., Reis-Santos, P., Fonseca, V., et al., 2007. Assessing anthropogenic pressures on estuarine fish nurseries along the Portuguese coast, a multi-metric index and conceptual approach. Sci. Total Environ. 374, 199–215.

Vasconcelos, R.P., Reis-Santos, P., Fonseca, V., et al., 2009. Juvenile fish condition in estuarine nurseries along the Portuguese coast. Estuar. Coast. Shelf Sci. 82, 128–138.

Weber, L.P., Higgins, P.S., Carlson, R.I., et al., 2003. Development and validation of methods for measuring multiple biochemical indices of condition in juvenile fishes. J. Fish Biol. 63, 637–658.

Whitfield, A.K., Becker, A., 2014. Impacts of recreational motorboats on fishes, A review. Mar. Pollut. Bull. 83, 24–31.

Whitfield, A.K., Elliott, A., 2002. Fishes as indicators of environmental and ecological changes within estuaries, a review of progress and some suggestions for the future. J. Fish Biol. 61, 229–250.

Wood, C.M., 2001. Toxic responses of the gill. In: Schlenk, D., Benson, W.H. (Eds.), Target Organ Toxicity in Marine and Freshwater Teleosts. Organs, vol. 1. Taylor & Francis, London, pp. 1–89.

Zaja, R., Caminada, D., Lončar, J., et al., 2008. Development and characterization of P-glycoprotein 1 (Pgp1, ABCB1) -mediated doxorubicin-resistant PLHC-1 hepatoma fish cell line. Toxicol. Appl. Pharmacol. 227, 207–218.

Zaja, R., Lončar, J., Popovic, M., et al., 2011. First characterization of fish P-glycoprotein (abcb1) substrate specificity using determinations of its ATPase activity and calcein- AM assay with PLHC-1/dox cell line. Aquat. Toxicol. 103, 53–62.

Zitko, V., Carson, W.V., Carson, W.G., 1973. Prediction of incipient lethal levels of copper to juvenile Atlantic salmon in the presence of humic acid by cupric electrode. Bull. Environ. Contam. Toxicol. 10, 265–271.

CHAPTER 14

Biological Responses at Supraindividual Levels

Angel Borja, Julie Bremner, Iñigo Muxika, J. Germán Rodríguez

Abstract

The effects of pollutants on marine fauna and flora cover many direct and indirect effects at supraindividual levels, from populations to ecosystems. In recent times, hundreds of indicators, metrics, and assessment methods have been developed to determine the impacts of those pollutants on different components of the ecosystem. This development is generally included in the framework of national and international legislation, approved in different continents. Pollution effects on organisms can imply consequences at the population level to different degrees, from changes in population dynamics or genetic diversity, to the local extinction of a population. In turn, ecological integrity assessment requires the study of structure (e.g., richness, diversity), function (e.g., response of sensitive and opportunistic species or biological traits to pollution), and processes at the community level. However, the most important challenge is to understand the response of the complete ecosystem to interactions between multiple stressors (i.e., cumulative, synergistic, antagonistic) and to assess marine health in an integrative way at regional or global scales.

Keywords: Assessment methods; Ecological integrity; Ecosystem health; Function; Integrative assessment; Structure.

Chapter Outline

Aquatic Ecotoxicology. http://dx.doi.org/10.1016/B978-0-12-800949-9.00014-0

The effects of pollutants (metallic, organic, contaminants of emergent concern) on marine fauna and flora cover many direct and indirect effects at supraindividual levels, from populations and communities to ecosystems. In recent times, hundreds of indicators, metrics, and assessment methods have been developed to determine the impacts of those pollutants on different components of the ecosystem (e.g., phytoplankton, zooplankton, macroalgae, sea grasses, macroinvertebrates, fishes) (Mialet et al., 2011; Birk et al., 2012; see also Chapters 10–13). This development is generally included in the framework of national and international legislation, approved in different continents (Borja et al., 2008). As mentioned in Chapter 1, these include the US Clean Water Act, the US and Canada Oceans Act, the European Water Framework Directive (WFD), and the European Marine Strategy Framework Directive (MSFD). All this legislation promotes the assessment of pollutants and other human pressures on marine ecosystems in an integrative way, from an ecosystem approach management view (Borja et al., 2012). This chapter reviews the current status of the knowledge on biological responses to pollution, at the previously mentioned supraindividual levels, taking into account also the approaches that can be used to assess such responses and effects.

14.1 Effects of Pollutants in Marine Populations

Nuss and Eckleman (2014) have recently reviewed the life-cycle of inorganic metallic pollutants. Among them, Cd, Cu, Cr, Pb, Hg, Ni, and Zn are some of the most studied, in terms of their effects (toxicity, bioavailability or bioaccumulation) in different marine ecosystem components. In addition, global sources of organic compounds, such as polycyclic aromatic hydrocarbons (PAHs) and polychlorinated biphenyls (Wolska et al., 2011; Tobiszewski and Namieśnik, 2012), lindane (Vijgen et al., 2011), and other emerging pollutants, such as synthetic musks and benzotriazole ultraviolet stabilizers (Nakata et al., 2012), have also recently been reviewed.

The extent to which these pollutants have toxic effects on marine biota depends upon a variety of factors, such as: (1) their geochemical behavior; (2) the physiology and condition of the target biota; (3) chemical speciation; (4) the presence of other toxicants; and (5) environmental conditions (Ansari et al., 2004; Borja et al., 2011a). Several approaches have been developed to assess the lethal responses of marine biota to metallic and organic pollution. These include laboratory bioassays, mesocosm experiments, and field experiments, among others. The laboratory bioassays have permitted comparisons of toxicity between different pollutants for a variety of species and vice versa: comparisons of toxicity between species for one pollutant. Moreover, they have shown that the first developmental stages of several invertebrate species are highly sensitive to multiple toxicants (Ringwood, 1992; His et al., 1999). Research has also evaluated sublethal responses to pollutants in marine biota: this has included effects on growth, sexual maturity, disease, luminescence, and metabolism (Borja et al., 2011a).

When comparing results obtained in the laboratory with those in the field, some authors (e.g., Mayer-Pinto et al., 2010) have demonstrated that our understanding of the effects of contaminants (metals in this case) on organisms is not as definitive as is often assumed in the literature. Hence, these authors recognize that many laboratory studies have demonstrated the potential for large concentrations of metals to kill almost every type of organism. However, the concern is that, under natural conditions, metals are less likely to be harmful than suggested by such laboratory studies because they may be used in lower concentrations than those tested in the laboratory and because interactions with environmental parameters (biotic and abiotic) may affect their toxicity. For example, dilution in receiving waters will decrease the actual concentrations of contaminants encountered by organisms (Mayer-Pinto et al., 2010). This mismatch between results in the laboratory and the field could pose management and legislative problems because environmental quality standards are mainly based on laboratory bioassays, whereas the assessments are undertaken using field monitoring (e.g., in the case of the WFD, European Commission, 2011). For example, quality standards set for some pollutants in European marine waters (e.g., tributyltin (TBT)) are extremely low, even close to the best available methods of determination, making very difficult to monitor them under an operational basis. Normally this problem occurs because its effects are determined in freshwater biota (usually in *Daphnia*) and then an assessment factor of 1000 is applied to the standards set for freshwater biota to calculate those corresponding to marine waters (Lepper, 2004).

Research undertaken on the biological effects of organometallic pollutants has shown that some of these compounds are substantially more toxic to marine organisms than inorganic metals. As an example, methylmercury is substantially more toxic than inorganic mercury because of its increased ability to cross the blood–gut barrier (Hill, 1997). This contaminant is a potent neurotoxin that bioaccumulates in biota (Chen et al., 2008, 2014). TBT, in common with many other biocides used in antifouling paints, is well-known for its lethal toxicity and effects on growth, reproduction, physiology, and behavior of marine biota (Alzieu et al., 1996; Konstantinou and Albanis, 2004). Some of the negative effects of these biocides (especially TBT) are due to interference with endocrine function as exemplified by the phenomenon of imposex (Chapter 15). In fact, this biomarker is included in biomonitoring networks such as that of Oslo–Paris Convention (OSPAR Commission, 2009).

There are more than 100 different PAHs, providing both natural and anthropogenic inputs into ecosystems (Tobiszewski and Namieśnik, 2012). They generally occur as complex mixtures in the environment, making the evaluation of toxicity difficult (Hylland, 2006). However, toxic (carcinogenic) effects have been noted in marine phytoplankton, zooplankton, invertebrates, and vertebrates because of their ability to form DNA adducts (Hylland, 2006). Persistent organic pollutants, i.e., substances that possess toxic properties and resist degradation, such as pesticides (e.g., DDT, lindane), industrial chemicals (e.g., polychlorinated biphenyls), and dioxins are highly stable and can accumulate in the environment and cause problems for marine biota (Porte et al., 2006; Koenig, 2012).

Pollution effects on organisms can imply consequences at the population level to varying degrees, from changes in population dynamics or genetic diversity (Bickham et al., 2000; Belfiore and Anderson, 2001), to the local extinction of a population (Gibbs and Bryan, 1996). Because biological communities are formed by multiple species interacting with each other, such impacts on populations can have implications for whole communities.

14.2 Effects of Pollutants on Marine Communities

There is general agreement among marine investigators that demonstrating biochemical and physiological responses to pollutants may not be sufficient to assess the health of marine environments (Borja et al., 2009b). Hence, the assessment of ecological integrity requires the study of structure, function, and processes at the community level (Borja et al., 2008, 2009a). Following Clements (2000), the key to predicting the effects of contaminants on communities is to understand the underlying mechanisms, measuring a suite of indicators across levels of biological organization. However, establishing a cause-and-effect relationship between stressors and responses at these higher levels of organization is problematic, because the structure and functioning of communities may be altered for many reasons, other than contaminant exposure. Hence, Clements (2000) suggests that one of the major goals of ecotoxicology is to develop an improved mechanistic understanding of ecologically significant responses to contaminants. Such indicators, termed "ecological biomarkers" have been recently reviewed (Amiard-Triquet et al., 2013). Ecologically relevant biomarkers include behavior, reproduction, growth, energy metabolism, lysosomal integrity, immunotoxicity, and genotoxicity biomarkers (Mouneyrac and Amiard-Triquet, 2013).

Some researchers (e.g., Wolfe, 1992; Islam and Tanaka, 2004; Borja et al., 2011a) have systematized bioindicators of marine pollution at community and ecosystem levels. These approaches include abundance, biomass, richness, dominance, similarity, the ratio of opportunistic to sensitive species, age–size spectra, trophic interactions, energy flow, productivity, and the loss of goods and services, as examined in the following section.

14.2.1 Structural Variables to Detect Pollutants Effects: Richness and Diversity

Various measures of biotic richness and diversity have been used in research related to the assessment of pollution effects (see review by Johnston and Roberts, 2009). Anthropogenic contamination is strongly associated with reductions in species richness and diversity in some biotic components, mainly in macrobenthic communities of soft-bottom habitats (Johnston and Roberts, 2009). These authors found few studies identifying an increase in species richness associated with anthropogenic contamination; such increases were generally associated with nutrient enrichment. But decreasing richness and diversity is not a generality because several exceptions have been found (see e.g., Borja et al., 2011a). As an example, Piola and Johnston (2008) found that heavy metal pollution can reduce the diversity of native species but increase the invader dominance in marine hard-substratum macrobenthic assemblages. Hence, Clements and Rohr (2009) consider that the study of contaminant effects on patterns of species abundance, diversity, and community composition should include the integration of basic ecological principles.

During the twentieth century, marine bacterial and archaeal communities saw relatively little research in relation to pollution. However, the development of genetic methods in recent years has increased the knowledge on ecological diversity of bacterial and archaeal communities (though the suitability of ecological diversity measures for highly diverse bacterial communities was uncertain and seldom considered during the initial development period; Hill et al., 2003). Although in other biotic components, the measures of diversity are mainly based on taxonomic composition, in the bacterial and archaeal ones, the measures are usually based on molecular operational taxonomic units.

Sun et al. (2012, 2014) found that bacterial diversity and evenness in estuaries showed strong associations with sediment contaminant concentrations, particularly with metals. However, the situation, like that of the macrobenthic communities, is complicated and unclear, with previous studies by Ellis et al. (2003) and Gillan et al. (2005) suggesting that bacterial diversity is not related to long-term metal pollution. Research carried out in a polluted harbor by Iannelli et al. (2012) and Chiellini et al. (2013) found that diversity did not diverge significantly among samples with differing pollution levels, and research on the impact of aquaculture (Bissett et al., 2007; Castine et al., 2009; Kawahara et al., 2009) found that sedimentary bacterial communities may change under the organic perturbations, but that does not necessarily translate to a clear change in diversity. Elsewhere, research on the bacterial community response to petroleum contamination has found a decrease in diversity in salt marsh sediments (Ribeiro et al., 2013). In mangroves, experiments with PAHs showed that the diversity of the bacterial community was suppressed for 7 days, but was promoted after 24 days (Zhang et al., 2014). Previously, Wang and Tam (2011) have found a decrease in microbial diversity after 60 days, in experiments with high levels of PAHs. These varying outcomes mean that a global paradigm on the responses of the diversity of bacterial communities to pollution cannot be established.

14.2.2 Functional Variables to Detect Pollutant Effects: Opportunistic and Sensitive Species

The labeling of species in relation to their responses to pollution gradients (i.e., opportunistic, sensitive, tolerant) has acquired great importance in recent years, because species have long been recognized as potential bioindicators of impacted systems (e.g., Pearson and Rosenberg, 1978; Gray, 1979; Hily, 1984; Rygg, 1985). In the case of macroinvertebrates, when experts are required to determine the criteria used to classify the status of an area, the most common criterion across different continents is the presence of indicator species, followed by richness and diversity/dominance (Borja et al., 2014a).

Hence, in some legislation (e.g., WFD), the use of ratios between opportunistic and sensitive (indicator) species is being increasingly used to assess the ecological status of macrobenthic communities, e.g., Borja et al. (2000), for the AZTI's Marine Biotic Index (AMBI) and Rosenberg et al. (2004), for the Benthic Quality Index, macroalgae (e.g., Orfanidis et al. (2001)), for the Ecological Evaluation Index and Juanes et al. (2008), for the Quality of Rocky Bottoms, or fishes (e.g., Borja et al. (2004), for the AZTI's Fish Index). Some reviews in the use of such methods can be found in Díaz et al. (2004), Borja et al. (2009c, 2012), Birk et al. (2012), or Pérez-Domínguez et al. (2012), for different ecosystem components.

The response of these indices to different pollutants in benthic ecosystems has been studied in many publications and showed degradation of soft-bottom benthic communities after an increase in metal concentrations (Marin-Guirao et al., 2005; Simboura et al., 2007; Gray and Delaney, 2008; Josefson et al., 2008), organic compounds, and TBTs (Muxika et al., 2005; Martínez-Lladó et al., 2007; Wetzel et al., 2012). The primary mechanism driving these changes is the elimination of sensitive species and the subsequent dominance by tolerant and opportunistic species (Johnston and Roberts, 2009; Borja et al., 2011a).

In turn, from the relatively few data on hard-bottom substrata, it has been suggested that macroalgal communities are relatively resilient to pollution (Johnston and Roberts, 2009). In a recent review, Chakraborty et al. (2014) stated that though genus-specific macroalgal responses to heavy metal accumulation are significant, species-specific response is insignificant. These authors suggest using some species as indicators, because of their high uptake of metals (e.g., *Ulva lactuca, Enteromorpha (=Ulva) intestinalis, Padina gymnospora*, and *Dictyota bartayresiana*). In general, they are considered to be opportunistic species with rapid growth rates that replace diverse communities of large perennial algae when pollution affects pristine locations (Díez et al., 2009, 2012; Orfanidis et al., 2001; Guinda et al., 2008). However, it should be taken into account that environmental changes affect the physiological status of algae and, thus, their metal binding capacity (Connan and Stengel, 2011). Hence, the response of algae to metallic pollution depends on environmental conditions.

14.2.3 Functional Variables to Detect Pollutant Effects: Trophic Interactions

The importance of trophic interactions in community-level response to disturbance has long been recognized. In his classic ecological experiment, Paine (1966) documented the existence of strong food-web connections in aquatic environments; by removing the top predator, the starfish *Pisaster ochraceus*, from a rocky shore in Washington State, he triggered a series of responses that resulted in the exclusion of most grazers and producers from the food web. A similar phenomenon was described in freshwater lakes, where piscivores controlled populations of zooplanktivores, thereby indirectly controlling herbivorous zooplankton and phytoplankton (Carpenter et al., 1985).

The accumulation of pollutants by producers (e.g., algae, sea grasses) represents a potentially important pathway of contaminant exposure to grazing organisms (herbivores and detritivores), which can be responsible for the transfer of metals to higher trophic levels (see, for example, Roberts et al., 2008). In fact, the algae *Ulva lactuca* and *Enteromorpha (=Ulva) intestinalis*, collected from contaminated sites and used to feed herbivorous gastropods, produced complete mortality of the latter organisms within 1–4 weeks of continuous dietary exposure (Weis and Weis, 1992). Such impacts on aquatic systems can also filter through to terrestrial food webs; for example, contaminant-driven reductions in aquatic insects can remove an important food source for riparian spiders (Kraus et al., 2014).

As well as triggering trophic responses, the same contaminant can act on multiple components of the food web simultaneously. For example, Chen et al. (2014) found that mercury in the Northeastern United States estuaries passed into polychaetes from benthic sediment and into fishes by way of pelagic particulates. These dual contaminant sources (e.g., sediments and the water column) can make prediction of likely food web effects difficult. Such predictions are, however, important as an alternative to empirical studies that can only measure effects once they have occurred. Modeling stressor effects in complex aquatic food webs is an expanding subject and user-friendly model software is becoming available (see, for example, the AQUATOX model at http://www2.epa.gov/exposure-assessment-models/aquatox and the Ecopath model at www.ecopath.org).

14.2.4 Functional Variables to Detect Pollutant Effects: Biological Traits

The trophic/food web concept can also be broadened to incorporate other types of ecological interactions, bridging the gap between populations (Section 14.1) and ecosystems (Section 14.3). This is achieved by examining the biological characteristics, or traits, of species. Biological traits describe species' physiology, morphology, life history, and behavior, capturing both interspecific interactions and the connections between species and their environment. Biological Traits Analysis (BTA) has its roots in classical ecological theory on species–environment relationships such as Southwood's (1977) habitat templet concept, but has been formalized as an approach over the past two decades or so; initially in plant ecology (Diaz and Cabido, 1997), then in freshwater benthos (Charvet et al., 2000) and marine systems (Bremner et al., 2003). The approach utilizes a range of biological traits, combined with species' abundance, density, or biomass to examine changes in the occurrence of traits or groups of traits over space or time. Examples of traits that have been used in BTA are shown in Table 14.1.

BTA can be used in two subtly different, but complementary ways. First, because species' responses to the physical environment are determined by their traits (e.g., the way that a species respires will determine the oxygen conditions in which it can comfortably exist), these traits can be used to understand and predict species' sensitivities to human-derived environmental stressors (Figure 14.1).

Table 14.1: Examples of Traits Used in Aquatic Biological Traits Analysis. Authors Often Use Differing Terminology; the List Provided Here is an Amalgamation of those given in Bremner (2008), Statzner and Beche (2010), and Törnroos and Bonsdorff (2012)

Morphology, Behavior, and Interaction with Environment	
Maximal size	Biogenic habitat provision (type, scale)
Body design/shape	Sediment reworking (e.g., bioturbation)
Body flexibility	Living habit (e.g., free, tube-dweller)
Body protection/armoring/fragility	Feeding method
Defense strategy	Digestive physiology
Flow, drag, or silt adaptations	Food type
Resistance against unfavorable conditions	Migration (e.g., seasonal, reproductive)
Attachment to/dependency on substrate	Adult dispersal (distance, potential)
Environmental position (e.g., within/on substrate)	Locomotion/mobility
Emergence (behavior, location, synchrony, season)	Respiration technique
Sociability	
Life History (Including Reproduction)	
Longevity	Developmental technique
Aquatic stages	Propagule/larval type and dispersal
Adult/egg diapause	Fecundity
Growth rate	Reproductive cycles per year (voltinism)
Time to maturity/development speed	Length of egg phase
Sexual differentiation	
Reproductive strategy/technique	
Fertilization type	

Intrahabitat trait patterns can be remarkably stable across large spatial scales in undisturbed conditions (see Statzner and Beche, 2010), so one of the most common uses of traits in this respect is as the foundation for developing indicators of ecological quality. For example, Borja et al. (2000) developed AMBI as an indicator of disturbance in marine benthic environments, based on ecological groups defined by traits describing species' sensitivity to a specific stressor (organic enrichment; see Section 14.2.2). AMBI is also capable of indicating the presence of other types of stressor (Borja et al., 2003; Muxika et al., 2005; Josefson et al., 2008), giving it utility for large-scale water quality assessments such as those of the WFD, but it is not likely to differentiate between these stressors if they cooccur in an area, which will make it difficult to manage their impacts. Elsewhere in Europe, the newly developed Multi-Metric Invertebrate Index has been developed for streams (Mondy et al., 2012). This is a multistressor indicator combining traditional community metrics (e.g., Shannon diversity and taxon richness) and traits sensitive to disturbance. It has significantly improved detection rates for nitrogen compounds and organic micropollutants. Rodil et al. (2013) have developed a traits-based index specific for intertidal heavy metal contamination in New Zealand, which is so sensitive it can detect the presence of contaminants at levels below international guidelines.

However, when they are viewed in the real world, traits do not always respond in the manner we might expect. For example, Dolédec and Statzner (2008) tried to predict heavy metal

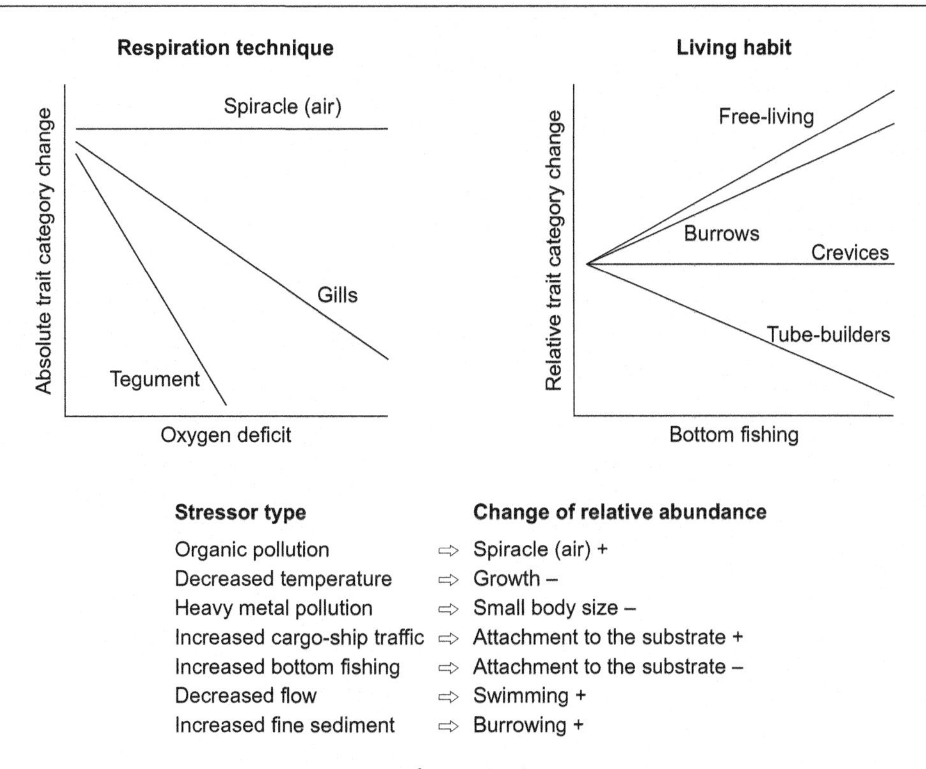

Figure 14.1
Predicting responses of biological traits to environmental stressors. The graphs illustrate how the abundance of species expressing different trait categories might fare across gradients of human impact. Change can be predicted in absolute terms (left) or in relative terms (right). Also shown are further examples of how trait categories might respond to stressors. *Adapted from Statzner and Beche (2010).*

impacts in river reaches using traits they considered to be sensitive to metal pollution. Of 291 impacted river reaches in their validation dataset, they were able to correctly assign only 25–57%, because only one of the three trait categories predicted to decrease in frequency did so in reality. Using traits to develop stressor-specific indicators is a potentially powerful tool, but more work is required before the potential can be fully realized.

Second, a special case of BTA occurs if the traits are linked to—so becoming indicators of—ecosystem functioning. These particular traits are termed functional traits and provide information on how aquatic ecosystem functioning may change in the face of pollution. Traits important for functioning include those that influence trophic interactions, biogeochemical processes, and habitat provision or disruption. Several different parameters are available when using BTA to assess ecosystem function, depending on the questions being asked. These include functional groups, functional diversity indices, functional indicators, and multivariate descriptors. Figure 14.2 shows a generalized framework for BTA.

There are a number of metrics available for describing functional diversity (FD), based on single or multiple traits. Some of the most commonly used are Petchey and Gaston's FD

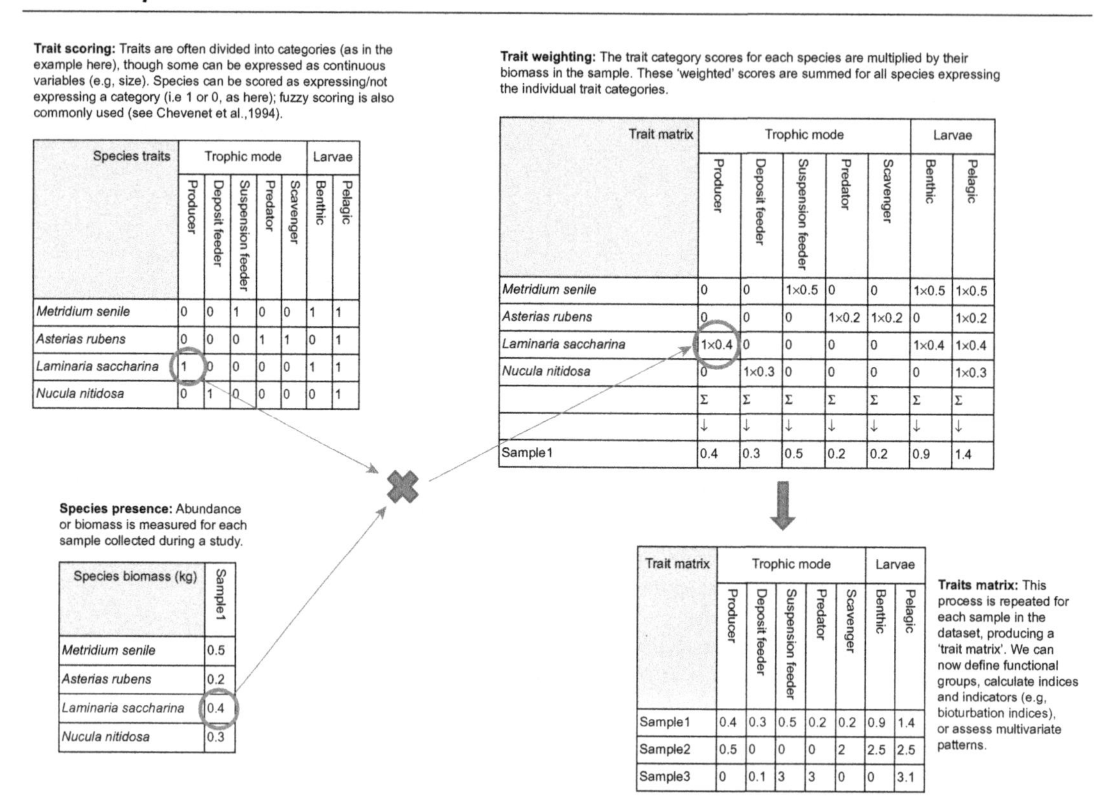

Species traits	Trophic mode					Larvae	
	Producer	Deposit feeder	Suspension feeder	Predator	Scavenger	Benthic	Pelagic
Metridium senile	0	0	1	0	0	1	1
Asterias rubens	0	0	0	1	1	0	1
Laminaria saccharina	1	0	0	0	0	1	1
Nucula nitidosa	0	1	0	0	0	0	1

Species biomass (kg)	Sample1
Metridium senile	0.5
Asterias rubens	0.2
Laminaria saccharina	0.4
Nucula nitidosa	0.3

Trait matrix	Trophic mode					Larvae	
	Producer	Deposit feeder	Suspension feeder	Predator	Scavenger	Benthic	Pelagic
Metridium senile	0	0	1×0.5	0	0	1×0.5	1×0.5
Asterias rubens	0	0	0	1×0.2	1×0.2	0	1×0.2
Laminaria saccharina	1×0.4	0	0	0	0	1×0.4	1×0.4
Nucula nitidosa	0	1×0.3	0	0	0	0	1×0.3
	Σ	Σ	Σ	Σ	Σ	Σ	Σ
	↓	↓	↓	↓	↓	↓	↓
Sample1	0.4	0.3	0.5	0.2	0.2	0.9	1.4

Trait matrix	Trophic mode					Larvae	
	Producer	Deposit feeder	Suspension feeder	Predator	Scavenger	Benthic	Pelagic
Sample1	0.4	0.3	0.5	0.2	0.2	0.9	1.4
Sample2	0.5	0	0	0	2	2.5	2.5
Sample3	0	0.1	3	3	0	0	3.1

Figure 14.2
Generalized framework for the biological traits analysis process, illustrated here using a theoretical example from a study of marine benthic organisms.

family of indices (Petchey and Gaston, 2002); Mason et al.'s functional richness, evenness, and divergence indices (Mason et al., 2005); Rao's Quadratic Entropy (Botta-Dukat, 2005); and Walker et al.'s Functional Attributes Diversity indices (Walker et al., 1999). However, the field is expanding quickly and many others exist. The FDiversity software (Casanoves et al., 2011) has been developed for calculating these indices.

The indices describe the number and variety of functional traits being expressed in communities, which can be helpful when assessing pollution effects against a standard or baseline. However, because they generate a single number, they do not show specifically how function may be changing. For this, approaches that consider trait identity are required. For example, Oug et al. (2012) used multivariate analysis to show that contaminant pollution in the Oslofjord reduced the numbers of deep-dwelling, large and surface deposit-feeding invertebrate taxa, which indicate that contaminants impact on sediment reworking and nutrient cycling. The effects of organic pollution have also been assessed in this way, though outcomes have been inconsistent. Feio and Dolédec (2012) have found changes in feeding strategies and dispersal mechanisms in Portuguese streams (both traits potentially important for carbon cycling and community connectivity), but Marchini et al. (2008) found no clearly defined effects across Italian coastal lagoons varying in disturbance conditions.

One of the major limitations of the biological traits approach is a lack of information on species' biology. For example, a study looking at eight key traits of the fish and invertebrate fauna of the North Sea has concluded that there is complete information for only 9% of the taxa and there is no information at all for 20% of them (Tyler et al., 2012). The North Sea is a well-studied ecosystem, so the problem may be worse elsewhere. The same situation occurs in freshwater systems (see Statzner and Beche, 2010)—it is true that in diverse ecosystems we often know little about the biology of the majority of species.

The approach has other limitations, for example the traits expressed by a species may not be permanently "fixed" and could evolve when it is exposed to stress (see Bremner, 2008). This is problematic because, unless we know whether and how traits could evolve, it will be difficult to accurately predict the effects of pollution on community function. There has not been a great deal of research on the links between biological traits and marine ecosystem function, so deciding which traits should be included in the assessments can be a subjective process based on an educated guess about how biology links to ecosystem function. However, this is an imperfect world and most methods for assessing pollution and other human impacts face difficulties. Biological traits seem to be showing promise for indicating species sensitivity to environmental stressors and assessing the functional effects of pollution.

14.3 Assessing the Effects of Pollutants on Marine Ecosystems

The concept of "health" was extended from its traditional domains of application at the individual and population levels to that of the whole ecosystem as early as 1992 (Rapport et al., 1999). The concept involves the development of methods for assessing the degree to which the functions of complex ecosystems are maintained or impaired by human activity. It also involves formulating new strategies that take account of societal values and biophysical realities to manage human activities so that ecosystem health is enhanced and not compromised further (Rapport et al., 1999).

The study of the interactions of multiple stressors produced by human activities in marine and coastal ecosystems is necessary to understand the responses of the different ecosystem components to disturbance, which will allow a better assessment of ecosystem health (Borja et al., 2013; Tett et al., 2013). Stressor interactions can produce cumulative effects, which can be additive (26%), synergistic (36%), or antagonistic (38%) (Crain et al., 2008). The overall interaction effect across 171 studies reviewed by Crain et al. (2008) was synergistic, but interaction type varied in relation to response level (antagonistic for community, synergistic for population), trophic level (antagonistic for autotrophs, synergistic for heterotrophs), and specific stressor pair (seven pairs additive, three pairs each synergistic and antagonistic).

The addition of a third stressor changed the interaction effects significantly in two-thirds of all of cases, and it doubled the number of synergistic interactions. Hence, pollutants can affect communities and ecosystems and their effects can be reinforced when other pressures or stressors (e.g., nutrient enrichment, habitat loss) are present, so they need to be assessed in an integrative way in order to properly manage marine systems (Borja et al., 2011a,b).

Given the complexity of managing marine ecosystems in the face of potentially interacting stressors, the need to assess the health and resilience of these systems has increased dramatically (Hofmann and Gaines, 2008). Approaches for assessing the ecological integrity of an ecosystem can be divided into two categories (Foden et al., 2008; Borja et al., 2009b): (1) those evaluating risk and state of a particular system (*sensu* the Drivers-Pressures-State-Impacts-Response approach (EEA, 1999)) and (2) those assessing the ecological integrity status of the whole ecosystem, under an ecosystem-based approach.

14.3.1 Evaluating the Risk and State of an Ecosystem

Metrics of ecotoxicological effects must be used, within the context of ecological risk assessment, as a tool to assess the likelihood of harm to ecosystems, or their components, through exposure to a specific concentration of a chemical (Chapman, 2007; van den Brink et al., 2011). Ecotoxicological approaches include multispecies tests or different compartments of the system (i.e., chemical analysis, bioassays, or impacts on benthic communities) in an assessment, representing an integrative assessment that considers multiple lines of evidence (Adams, 2005; Chapman and Hollert, 2006; Chapman et al., 2013).

The weight of evidence concept is the result of combining different measures of environmental quality to establish an overall assessment of environmental health (EPA, 1999; Adams, 2005). The philosophy behind weight of evidence is a preponderance of evidence approach, where the conclusions drawn from individual components are considered not a sum of these components, but relative to one another (Scrimshaw et al., 2007).

The sediment quality triad was one of the first weight of evidence methods developed (Long and Chapman, 1985). It is based on three components: (1) chemistry, to determine the level and extent of pollution and modifying factors (e.g., grain size, total organic carbon) compared with sediment quality guidelines, and to answer the question "are contaminants present at levels of concern?" (Scrimshaw et al., 2007); (2) bioassays, to determine whether the contaminated sediments are affecting the biota (toxicity); and (3) benthic community structure, to determine the status of resident fauna exposed to the sediment contaminants.

Additional lines of evidence can include the following: (1) measures of biomagnification, usually involving measurements of body burdens in sediment-dwelling invertebrates and food chain modeling, to answer the question "are any contaminants of concern capable of biomagnifying and are they likely to do so?"; (2) measures of exposure, such as biomarkers or body burdens (bioaccumulation), to determine which sediment contaminants, if any, are bioavailable and to try to determine causation; (3) toxicity identification evaluations, to attempt to assign causation; and (4) determinations of sediment stability, to determine whether only surficial sediments should be evaluated or whether deeper sediments, which may be exposed during storm or other events, need to be evaluated by answering the question "is the sediment stable or is it liable to erosion resulting in exposure of deeper, more contaminated sediments and/or contamination down-current?" (Borja et al., 2011a).

14.3.2 Assessing the Ecological Integrity of an Ecosystem

The approaches introduced in Section 14.3.1 are useful tools for agency staff and consultants operating in structured contexts for setting and auditing performance criteria for specific aspects of environmental management (Borja et al., 2009b). However, they do not address the multiple scale issues of ecological integrity assessment (the ecosystem-based approach), which indicates the status of the whole ecosystem, with multiple ecosystem components and matrices (Borja et al., 2009a,d; 2013).

Ecological integrity assessment is often undertaken on the basis of univariate or multivariate indices, which address individual components of the ecosystem (Birk et al., 2012). However, considering diversity and richness alone is less powerful in detecting ecological impacts than studies that consider taxonomic relatedness and multivariate community structure (Johnston and Roberts, 2009). Acceptance of this knowledge and that of the interactions and synergistic effects described previously has led managers to use multiple indices and integrative methods to assess ecological integrity in marine waters, combining different ecosystem components together (Borja et al., 2008). These methods are used to determine environmental or ecological status in both Europe and the United States (i.e., in the European MSFD and the WFD, Borja et al., 2010; or in the US Oceans Act, Barnes and McFadden, 2008). One of the difficulties of using such methods is identifying when good environmental status has been achieved. Hence, Borja et al. (2013) propose that good status is achieved when "physico-chemical (including contaminants, litter and noise) and hydrographical conditions are maintained at a level where the structuring components of the ecosystem are present and functioning, enabling the system to be resistant (ability to withstand stress) and resilient (ability to recover after a stressor) to harmful effects of human pressures/activities/impacts, where they maintain and provide the ecosystem services that deliver societal benefits in a sustainable way (i.e., that pressures associated with uses cumulatively do not hinder the ecosystem components in order to retain their natural diversity, productivity and dynamic ecological processes, and where recovery is rapid and sustained if a use ceases)."

This approach is intended to permit an assessment of marine status at the ecosystem level ("ecosystem-based approach" or "holistic approach" methodologies), which is more effective than assessments carried out at a species or chemical level (i.e., quality objectives)

(Borja et al., 2011a). However, there are few examples of assessing marine pollutant impacts at the whole-ecosystem level (Sheehan et al., 1984; Islam and Tanaka, 2004) because these impacts are masked by other human pressures (see the previous section).

Borja et al. (2008) present an overview of integrative tools and methods used to assess ecological integrity in estuarine and coastal systems around the world. The legislative measures do tend to converge in defining environmental water quality in an integrative way. However, the degree of convergence is variable, based generally upon studies carried out in single (or small) systems, which do not permit generalization.

In the United States, the National Oceanic and Atmospheric Administration began to move its management strategy away from traditional marine resource management approaches and toward an Ecosystem Approach to Management in 2005 (Barnes and McFadden, 2008). After conducting surveys in eight different geographic regions across the United States, a qualitative analysis of the attitudes and experiences of participants with respect to implementing Ecosystem Approach to Management identified four major challenges to enhancing cooperation and understanding: (1) integrating the science; (2) encouraging partnership; (3) funding restraints; and (4) decision-making issues.

In practical terms and considering these challenges, managers and decision-makers need simple, but scientifically well-established methodologies that are capable of demonstrating to the general public the status of a marine area, taking into account pollution and other human pressures or recovery processes (Borja and Dauer, 2008). Within this context, the major scientific challenge we face is to develop tools to adequately define the scale and present condition of marine ecosystems, in terms of biological performance as well as to monitor changes through time; similarly, to identify and address, through management, the causes of observed impairments (Borja et al., 2008). This challenge includes developing the methodology to integrate large amounts of information, on different geographical and time scales, from different ecosystem components, to provide that integrative assessment. Many methods are available to undertake such integration (Borja et al., 2014b), which needs appropriate operational indicators and evaluation tools that do not mask or leave out inherent ecosystem properties and dynamics, and that can be selected using empirical and modeling techniques (Rombouts et al., 2013).

One of the first methods for integrating information from different ecosystem components was developed in Europe, within the WFD, using a decision-tree analysis (Borja et al., 2004, 2009d). However, this method was applied within a relatively small part of the Bay of Biscay. In turn, an indicator-based assessment tool-box termed Holistic Assessment of Ecosystem Health Status was developed for use at a regional level in the Baltic Sea (HELCOM, 2010). The indicators used in the thematic assessments within this tool-box included eutrophication, hazardous substances and biodiversity, which are integrated into a holistic assessment of "ecosystem health."

Another example of high-level aggregation at a regional scale, for the southern Bay of Biscay, is the integrative method of Borja et al. (2010, 2011b) that incorporates a weighted scoring method to aggregate multiple indicators from 11 qualitative descriptors to assess environmental status within the MSFD. Tett et al. (2013) have also developed a regional high-level aggregation approach, designed to assess marine ecosystem health within the North Sea. These authors define ecosystem health as follows: "the condition of a system that is self-maintaining, vigorous, resilient to externally imposed pressures, and able to sustain services to humans. It contains healthy organisms and populations, and adequate functional diversity and functional response diversity. All expected trophic levels are present and well interconnected, and there is good spatial connectivity amongst subsystems." This definition is equated by Tett et al. (2013) to good ecological or environmental status (e.g., as referred to by WFD and MSFD, respectively) and links with the definition of good environmental status proposed by Borja et al. (2013) and presented previously. In both definitions, resilience is central to health (i.e., good status). Ecosystems under anthropogenic pressure are at risk of losing resilience and thus of suffering regime shifts and loss of services, but resilience is difficult to measure in an easy way. For monitoring and assessment of whole ecosystems, Tett et al. (2013) propose an approach based on "trajectories in ecosystem state space," which they apply to the North Sea using long-term time series. Change is visualized as a Euclidian distance from an arbitrary reference state. Variability about a trend is used as a proxy for inverse resilience (i.e., variability indicates a lack of resilience).

At this moment, the only method for assessing ecological integrity that is applicable at a global scale is that of Halpern et al. (2012)—the Ocean Health Index; www.oceanhealthindex.org/. The index comprises 10 diverse public goals for a healthy coupled human–ocean system, based upon human activities and pressures. It provides weighted index scores for environmental health, both a global area-weighted average and individual scores by country and region (e.g., in Brazil: Elfes et al., 2014). Globally, the overall index performs better in developed countries than in developing countries. On a scale from 0 to 100, only 5% of countries scored higher than 70, whereas 32% scored lower than 50 (Halpern et al., 2012). The index provides a powerful tool for raising public awareness, directing resource management, improving policy, and prioritizing scientific research.

In conclusion, one of the main challenges in marine research for coming years is to develop approaches to assess the ecosystem health in an integrative way, including as many components of the ecosystem as possible, considering humans as part of the marine ecosystem, and making sustainable the activities they carried out in marine waters (Borja, 2014).

References

Adams, S.M., 2005. Assessing cause and effect of multiple stressors on marine systems. Mar. Poll. Bull. 51, 649–657.

Alzieu, C., 1996. In: de Mora, S.J. (Ed.), Tributyltin: Case Study of an Environmental Contaminant. Cambridge University Press, Cambridge, pp. 167–211.

Amiard-Triquet, C., Amiard, J.C., Rainbow, P.S. (Eds.), 2013. Ecological Biomarkers: Indicators of Ecotoxicological Effects. CRC Press, Boca Raton, p. 464.

Ansari, T.M., Marr, I.L., Tariq, N., 2004. Heavy metals in marine pollution perspective - a mini review. J. Appl. Sci. 4, 1–20.

Barnes, C., McFadden, K.W., 2008. Marine ecosystem approaches to management: challenges and lessons in the United States. Mar. Pol. 32, 387–392.

Belfiore, N.M., Anderson, S.L., 2001. Effects of contaminants on genetic patterns in aquatic organisms: a review. Mut. Res./Rev. Mut. Res. 489, 97–122.

Bickham, J.W., Sandhu, S., Hebert, P.D.N., et al., 2000. Effects of chemical contaminants an genetic diversity in natural populations: implications for biomonitoring and ecotoxicology. Mut. Res./Rev. Mut. Res. 463, 33–51.

Birk, S., Bonne, W., Borja, A., et al., 2012. Three hundred ways to assess Europe's surface waters: an almost complete overview of biological methods to implement the Water Framework Directive. Ecol. Ind. 18, 31–41.

Bissett, A., Burke, C., Cook, P.L.M., et al., 2007. Bacterial community shifts in organically perturbed sediments. Env. Microbiol. 9, 46–60.

Borja, A., 2014. Grand challenges in marine ecosystems ecology. Front. Mar. Sci. 1, 1–6. http://dx.doi.org/10.3389/fmars.2014.00001.

Borja, A., Dauer, D.M., 2008. Assessing the environmental quality status in estuarine and coastal systems: comparing methodologies and indices. Ecol. Ind. 8, 331–337.

Borja, A., Franco, J., Pérez, V., 2000. A marine biotic index to establish the ecological quality of soft-bottom benthos within European estuarine and coastal environments. Mar. Poll. Bull. 40, 1100–1114.

Borja, A., Muxika, I., Franco, J., 2003. The application of a marine biotic index to different impact sources affecting soft-bottom benthic communities along European coasts. Mar. Poll. Bull. 46, 835–845.

Borja, Á., Franco, J., Valencia, V., et al., 2004. Implementation of the European Water Framework Directive from the Basque Country (northern Spain): a methodological approach. Mar. Poll. Bull. 48, 209–218.

Borja, A., Bricker, S.B., Dauer, D.M., et al., 2008. Overview of integrative tools and methods in assessing ecological integrity in estuarine and coastal systems worldwide. Mar. Poll. Bull. 56, 1519–1537.

Borja, A., Ranasinghe, A., Weisberg, S.B., 2009a. Assessing ecological integrity in marine waters, using multiple indices and ecosystem components: challenges for the future. Mar. Poll. Bull. 59, 1–4.

Borja, A., Bricker, S.B., Dauer, D.M., et al., 2009b. Ecological integrity assessment, ecosystem-based approach, and integrative methodologies: are these concepts equivalent? Mar. Poll. Bull. 58, 457–458.

Borja, A., Miles, A., Occhipinti-Ambrogi, A., et al., 2009c. Current status of macroinvertebrate methods used for assessing the quality of European marine waters: implementing the Water Framework Directive. Hydrobiologia 633, 181–196.

Borja, A., Bald, J., Franco, J., et al., 2009d. Using multiple ecosystem components, in assessing ecological status in Spanish (Basque Country) Atlantic marine waters. Mar. Poll. Bull. 59, 54–64.

Borja, Á., Elliott, M., Carstensen, J., et al., 2010. Marine management - towards an integrated implementation of the european marine strategy framework and the Water Framework Directives. Mar. Poll. Bull. 60, 2175–2186.

Borja, A., Belzunce, M.J., Garmendia, J.M., et al., 2011a. Chapter 8. Impact of pollutants on coastal and benthic marine communities. In: Sánchez-Bayo, F., van den Brink, P.J., Mann, R.M. (Eds.), Ecological Impacts of Toxic Chemicals. Bentham Science Publishers Ltd, pp. 165–186.

Borja, Á., Galparsoro, I., Irigoien, X., et al., 2011b. Implementation of the European marine strategy framework directive: a methodological approach for the assessment of environmental status, from the Basque Country (Bay of Biscay). Mar. Poll. Bull. 62, 889–904.

Borja, A., Basset, A., Bricker, S., et al., 2012. Classifying ecological quality and integrity of estuaries. In: Wolanski, E., McLusky, D.S. (Eds.), Treatise on Estuarine and Coastal Science, 1. Academic Press, Waltham, pp. 125–162.

Borja, A., Elliott, M., Andersen, J.H., et al., 2013. Good Environmental Status of marine ecosystems: what is it and how do we know when we have attained it? Mar. Poll. Bull. 76, 16–27.

Borja, Á., Marín, S., Núñez, R., et al., 2014a. Is there a significant relationship between the benthic status of an area, determined by two broadly-used indices, and best professional judgment? Ecol. Ind. 45, 308–312.

Borja, A., Prins, T., Simboura, N., et al., 2014b. Tales from a thousand and one ways to integrate marine ecosystem components when assessing the environmental status. Front. Mar. Sci. 1. http://dx.doi.org/10.3389/fmars.2014.00072.

Botta-Dukat, Z., 2005. Rao's quadratic entropy as a measure of functional diversity based on multiple traits. J. Veg. Sci. 16, 533–540.

Bremner, J., 2008. Species' traits and ecological functioning in marine conservation and management. J. Exp. Mar. Biol. Ecol. 366, 37–47.

Bremner, J., Rogers, S.I., Frid, C.L.J., 2003. Assessing functional diversity in marine benthic ecosystems: a comparison of approaches. Mar. Ecol. Progr. Ser. 254, 11–25.

Carpenter, S.R., Kitchell, J.F., Hodgson, J.R., 1985. Cascading trophic interactions and lake productivity. BioScience 35 (10), 634–639.

Casanoves, F., Pla, L., Di Rienzo, J.A., et al., 2011. FDiversity: a software package for the integrated analysis of functional diversity. Meth. Ecol. Evol. 2 (3), 233–237.

Castine, S.A., Bourne, D.G., Trott, L.A., et al., 2009. Sediment microbial community analysis: establishing impacts of aquaculture on a tropical mangrove ecosystem. Aquaculture 297, 91–98.

Chakraborty, S., Bhattacharya, T., Singh, G., et al., 2014. Benthic macroalgae as biological indicators of heavy metal pollution in the marine environments: a biomonitoring approach for pollution assessment. Ecotoxicol. Environ. Saf. 100, 61–68.

Chapman, P.M., 2007. Determining when contamination is pollution — weight of evidence determinations for sediments and effluents. Environ. Int. 33, 492–501.

Chapman, P.M., Hollert, H., 2006. Should the sediment quality triad become a tetrad, a pentad, or possibly even a hexad? J. Soils Sed. 6 (1), 4–8.

Chapman, P.M., Wang, F., Caeiro, S.S., 2013. Assessing and managing sediment contamination in transitional waters. Environ. Int. 55, 71–91.

Charvet, S., Statzner, B., Usseglio-Polatera, P., et al., 2000. Traits of benthic macroinvertebrates in semi-natural French streams: an initial application to biomonitoring in Europe. Freshwater Biol. 43, 277–296.

Chen, C., Amirbahman, A., Fisher, N., et al., 2008. Methylmercury in marine ecosystems: spatial patterns and processes of production, bioaccumulation, and biomagnification. EcoHealth 5, 399–408.

Chen, C.Y., Bursuk, M.E., Bugge, D.M., et al., 2014. Benthic and pelagic pathways of methylmercury bioaccumulation in estuarine food webs of the Northeast United States. PLoS One 9 (2), e89305.

Chiellini, C., Iannelli, R., Verni, F., et al., 2013. Bacterial communities in polluted seabed sediments: a molecular biology assay in Leghorn harbor. Scient. World J. 165706. http://dx.doi.org/10.1155/2013/165706.

Clements, W.H., 2000. Integrating effects of contaminants across levels of biological organization: an overview. J. Aquat. Ecosys. Stress Recov. 7, 113–116.

Clements, W.H., Rohr, J.R., 2009. Community responses to contaminants: using basic ecological principles to predict ecotoxicological effects. Environ. Toxicol. Chem. 28 (9), 1789–1800.

Connan, S., Stengel, D.B., 2011. Impacts of ambient salinity and copper on brown algae: 1. Interactive effects on photosynthesis, growth and copper accumulation. Aquat. Toxicol. 104, 94–107.

Crain, C.M., Kroeker, K., Halpern, B.S., 2008. Interactive and cumulative effects of multiple human stressors in marine systems. Ecol. Lett. 11, 1304–1315.

Diaz, S., Cabido, M., 1997. Plant functional types and ecosystem function in relation to global change. J. Veg. Sci. 8, 463–474.

Díaz, R.J., Solan, M., Valente, R.M., 2004. A review of approaches for classifying benthic habitats and evaluating habitat quality. J. Environ. Manag. 73, 165–181.

Díez, I., Santolaria, A., Secilla, A., et al., 2009. Recovery stages over long-term monitoring of the intertidal vegetation in the Abra de Bilbao area and on the adjacent coast (N. Spain). Europ. J. Phycol. 44, 1–14.

Díez, I., Bustamante, M., Santolaria, A., et al., 2012. Development of a tool for assessing the ecological quality status of intertidal coastal rocky assemblages, within Atlantic Iberian coasts. Ecol. Ind. 12, 58–71.

Dolédec, S., Statzner, B., 2008. Invertebrate traits for the biomonitoring of large European rivers: an assessment of specific types of human impact. Freshwater Biol. 53 (3), 617–634.

EEA, 1999. State and Pressures of the Marine and Coastal Mediterranean Environment. European Environment Agency, Copenhagen. 44 pp.
Elfes, C.T., Longo, C., Halpern, B.S., et al., 2014. A regional-scale ocean health index for Brazil. PLoS One 9, e92589.
Ellis, R.J., Morgan, P., Weightman, A.J., et al., 2003. Cultivation-dependent and -independent approaches for determining bacterial diversity in heavy-metal-contaminated soil. Appl. Environ. Microbiol. 69 (6), 3223–3230.
EPA, 1999. Targeting Toxics: A Characterization Report. A Tool for Directing Management and Monitoring Actions in the Chesapeake Bay's Tidal Rivers. A Technical Workplan. Chesapeake Bay Program, Annapolis, Maryland. 99 pp.
European Commission, 2011. Common implementation strategy for the Water Framework Directive (2000/60/EC). Guidance Document No: 27 Technical Guidance for Deriving Environmental Quality Standards. Technical Report, 2011-055: 204 p.
Feio, M.J., Dolédec, S., 2012. Integration of invertebrate traits into predictive models for indirect assessment of stream functional integrity: a case study in Portugal. Ecol. Ind. 15, 236–247.
Foden, J., Rogers, S.I., Jones, A.P., 2008. A critical review of approaches to aquatic environmental assessment. Mar. Poll. Bull. 56, 1825–1833.
Gibbs, P.E., Bryan, G.W., 1996. TBT-induced imposex in neogastropod snails: masculinization to mass extinction. In: de Mora, S.J. (Ed.), Tributyltin: Case Study of an Environmental Contaminant. Cambridge Environmental Chemistry Series. Cambridge University Press, Cambridge, pp. 212–236.
Gillan, D.C., Danis, B., Pernet, P., et al., 2005. Structure of sediment associated microbial communities along a heavy-metal contamination gradient in the marine environment. Appl. Env. Microbiol. 71, 679–690.
Gray, J.S., 1979. Pollution-induced changes in populations. Phil. Trans. Roy. Soc. London, B. 286, 545–561.
Gray, N.F., Delaney, E., 2008. Comparison of benthic macroinvertebrate indices for the assessment of the impact of acid mine drainage on an Irish river below an abandoned Cu-S mine. Env. Poll. 155, 31–40.
Guinda, X., Juanes, J.A., Puente, A., et al., 2008. Comparison of two methods for quality assessment of macroalgae assemblages, under different pollution types. Ecol. Ind. 8, 743–753.
Halpern, B.S., Longo, C., Hardy, D., et al., 2012. An index to assess the health and benefits of the global ocean. Nature 488, 615–620.
HELCOM, 2010. Ecosystem health of the Baltic sea 2003–2007: HELCOM initial holistic assessment. Baltic Sea Environ. Proc. 122, 68.
Hill, S.J., 1997. Speciation of trace metals in the environment. Chem. Soc. Rev. 26, 291–298.
Hill, T.C., Walsh, K.A., Harris, J.A., et al., 2003. Using ecological diversity measures with bacterial communities. Microbiol. Ecol. 43 (1), 1–11.
Hily, C., 1984. Variabilité de la macrofaune benthique dans les milieux hypertrophiques de la Rade de Brest (Thèse de Doctorat d'Etat). Université de Bretagne Occidentale.
His, E., Heyvang, I., Geffard, O., et al., 1999. A comparison between oyster (*Crassostrea gigas*) and sea urchin (*Paracentrotus lividus*) larval bioassays for toxicological studies. Water Res. 33, 1706–1718.
Hofmann, G.E., Gaines, S.D., 2008. New tools to meet new challenges: emerging technologies for managing marine ecosystems for resilience. Bioscience 58, 43–52.
Hylland, K., 2006. Polycyclic aromatic hydrocarbon (PAH) ecotoxicology in marine ecosystems. J. Toxicol. Environ. Health Part a 69, 109–123.
Iannelli, R., Bianchi, V., Macci, C., et al., 2012. Assessment of pollution impact on biological activity and structure of seabed bacterial communities in the Port of Livorno (Italy). Sci. Total Environ. 426, 56–64.
Islam, M.S., Tanaka, M., 2004. Impacts of pollution on coastal and marine ecosystem including coastal and marine fisheries and approach for management: a review and synthesis. Mar. Poll. Bull. 48, 624–649.
Johnston, E.L., Roberts, D.A., 2009. Contaminants reduce the richness and evenness of marine communities: a review and meta-analysis. Environ. Poll. 157, 1745–1752.
Josefson, A.B., Hansen, J.L.S., Asmund, G., et al., 2008. Threshold response of benthic macrofauna integrity to metal contamination in West Greenland. Mar. Poll. Bull. 56, 1265–1274.
Juanes, J.A., Guinda, X., Puente, A., et al., 2008. Macroalgae, a suitable indicator of the ecological status of coastal rocky communities in the NE Atlantic. Ecol. Ind. 8, 351–359.

Kawahara, N., Shigematsu, K., Miyadai, T., et al., 2009. Comparison of bacterial communities in fish farm sediments along an organic enrichment gradient. Aquaculture 287, 107–113.

Koenig, S., 2012. Bioaccumulation of Persistent Organic Pollutants (POPs) and Biomarkers of Pollution in Mediterranean Deep-sea Organisms (Ph.D. thesis). University of Barcelona, 227 pp.

Konstantinou, I.K., Albanis, T.A., 2004. Worldwide occurrence and effects of antifouling paint booster biocides in the aquatic environment: a review. Environ. Int. 30, 235–248.

Kraus, J.M., Schmidt, T.S., Walters, D.M., et al., 2014. Cross-ecosystem impacts of stream pollution reduce resource and contaminant flux to riparian food webs. Ecol. Ind. 24, 235–243.

Lepper, P., 2004. Manual of the Methodological Framework Used to Derive Quality Standards for Priority Substances of the Water Framework Directive. Fraunhofer-Institute Molecular Biology and Applied Ecology. 32 pp.

Long, E.R., Chapman, P.M., 1985. A sediment quality triad: measures of sediment contamination, toxicity and infaunal community composition in Puget Sound. Mar. Poll. Bull. 16, 405–415.

Marchini, A., Munari, C., Mistri, M., 2008. Functions and ecological status of eight Italian lagoons examined using biological traits analysis (BTA). Mar. Poll. Bull. 56, 1076–1085.

Marín-Guirao, L., César, A., Marín, A., et al., 2005. Establishing the ecological quality status of soft-bottom mining-impacted coastal water bodies in the scope of the Water Framework Directive. Mar. Poll. Bull. 50, 374–387.

Martínez-Lladó, X., Gibert, O., Martí, V., et al., 2007. Distribution of polycyclic aromatic hydrocarbons (PAHs) and tributyltin (TBT) in Barcelona harbour sediments and their impact on benthic communities. Environ. Poll. 149, 104–113.

Mason, N.W.H., Mouillot, D., Lee, W.G., et al., 2005. Functional richness, functional evenness and functional divergence: the primary components of functional diversity. Oikos 111, 112–118.

Mayer-Pinto, M., Underwood, A.J., Tolhurst, T., et al., 2010. Effects of metals on aquatic assemblages: what do we really know? J. Exp. Mar. Biol. Ecol. 391, 1–9.

Mialet, B., Gouzou, J., Azémar, F., et al., 2011. Response of zooplankton to improving water quality in the Scheldt estuary (Belgium). Est. Coast. Shelf Sci. 93, 47–57.

Mondy, C.P., Villeneuve, B., Archaimbaul, V., et al., 2012. A new macroinvertebrate-based multimetric index (I2M2) to evaluate ecological quality of French wadeable streams fulfilling the WFD demands: a taxonomical and trait approach. Ecol. Ind. 18, 452–467.

Mouneyrac, C., Amiard-Triquet, C., 2013. Biomarkers of ecological relevance. In: Férard, J.F., Blaise, C. (Eds.), Comprehensive Handbook of Ecotoxicological Terms. Springer, pp. 221–236.

Muxika, I., Borja, A., Bonne, W., 2005. The suitability of the marine biotic index (AMBI) to new impact sources along European coasts. Ecol. Ind. 5, 19–31.

Nakata, H., Shinohara, R.-I., Nakazawa, Y., et al., 2012. Asia-Pacific mussel watch for emerging pollutants: distribution of synthetic musks and benzotriazole UV stabilizers in Asian and US coastal waters. Mar. Poll. Bull. 64, 2211–2218.

Nuss, P., Eckelman, M.J., 2014. Life cycle assessment of metals: a scientific synthesis. PLoS One 9, e101298.

Orfanidis, S., Panayotidis, P., Stamatis, N., 2001. Ecological evaluation of transitional and coastal waters: a marine benthic macrophytes-based model. Med. Mar. Sci. 2, 45–65.

OSPAR Commission, 2009. Evaluation of the OSPAR system of Ecological Quality Objectives for the North Sea. Biodiversity Series 102 pp.

Oug, E., Fleddum, A., Rygg, B., et al., 2012. Biological traits analyses in the study of pollution gradients and ecological functioning of marine soft bottom species assemblages in a fjord ecosystem. J. Exp. Mar. Biol. Ecol. 432, 94–105.

Paine, R.T., 1966. Food web complexity and species diversity. Am. Nat. 100 (910), 65–75.

Pearson, T., Rosenberg, R., 1978. Macrobenthic succession in relation to organic enrichment and pollution of the marine environment. Oceanogr. Mar. Biol. Ann. Rev. 16, 229–311.

Pérez-Domínguez, R., Maci, S., Courrat, A., et al., 2012. Current developments on fish-based indices to assess ecological-quality status of estuaries and lagoons. Ecol. Ind. 23, 34–45.

Petchey, O.L., Gaston, K.J., 2002. Functional diversity (FD), species richness and community composition. Ecol. Lett. 5, 402–411.

Piola, R.F., Johnston, E.L., 2008. Pollution reduces native diversity and increases invader dominance in marine hard-substrate communities. Diver. Distribut. 14 (2), 329–342.

Porte, C., Janer, G., Lorusso, L.C., et al., 2006. Endocrine disruptors in marine organisms: approaches and perspectives. Comp. Biochem. Physiol. Part C: Toxicol. Pharmacol. 143, 303–315.

Rapport, D.J., Böhm, G., Buckingham, D., et al., 1999. Ecosystem health: the concept, the ISEH, and the important tasks ahead. Ecosyst. Health 5, 82–90.

Ribeiro, H., Mucha, A.P., Almeida, C.M., et al., 2013. Bacterial community response to petroleum contamination and nutrient addition in sediments from a temperate salt marsh. Sci. Total Environ. 458–460, 568–576.

Ringwood, A.H., 1992. Comparative sensitivity of gametes and early developmental stages of a sea-urchin species (*Echinometra mathaei*) and a bivalve species (*Isognomon californicum*) during metal exposures. Archiv. Env. Cont. Toxicol. 22, 288–295.

Roberts, D.A., Johnston, E.L., Poore, A.G.B., 2008. Contamination of marine biogenic habitats and effects upon associated epifauna. Mar. Poll. Bull. 56, 1057–1065.

Rodil, I.F., Lohrer, A.M., Hewitt, J.E., et al., 2013. Tracking environmental stress gradients using three biotic integrity indices: advantages of a locally-developed traits-based approach. Ecol. Ind. 34, 560–570.

Rombouts, I., Beaugrand, G., Artigas, L.F., et al., 2013. Evaluating marine ecosystem health: case studies of indicators using direct observations and modelling methods. Ecol. Ind. 24, 353–365.

Rosenberg, R., Blomqvist, M., Nilsson, H.C., et al., 2004. Marine quality assessment by use of benthic species-abundance distributions: a proposed new protocol within the European Union Water Framework Directive. Mar. Poll. Bull. 49, 728–739.

Rygg, B., 1985. Distribution of species along pollution-induced diversity gradients in benthic communities in Norwegian fjords. Mar. Poll. Bull. 16, 469–474.

Scrimshaw, M.D., Delvalls, T.A., Blasco, J., et al., 2007. Sediment quality guidelines and weight of evidence assessments. Sust. Manag. Sed. Resour. 1, 295–309.

Sheehan, P.J., Miller, D.R., Butler, G.C., et al. (Eds.), 1984. Effects of Pollutants at the Ecosystem Level, SCOPE 22. John Wiley and Sons, NY.

Simboura, N., Papathanassiou, E., Sakellariou, D., 2007. The use of a biotic index (Bentix) in assessing long-term effects of dumping coarse metalliferous waste on soft bottom benthic communities. Ecol. Ind. 7, 164–180.

Southwood, T.R.E., 1977. Habitat, the templet for ecological strategies? J. Anim. Ecol. 46, 337–365 .

Statzner, B., Beche, L.A., 2010. Can biological invertebrate traits resolve effects of multiple stressors on running water ecosystems? Freshwater Biol. 55 (Supp. 1), 80–199.

Sun, M.Y., Dafforn, K.A., Brown, M.V., et al., 2012. Bacterial communities are sensitive indicators of contaminant stress. Mar. Poll. Bull. 64 (5), 1029–1038.

Sun, M.Y., Dafforn, K.A., Johnston, E.L., et al., 2014. Core sediment bacteria drive community response to anthropogenic contamination over multiple environmental gradients. Environ. Microbiol. 15 (9), 2517–2531.

Tett, P., Gowen, R.J., Painting, S.J., et al., 2013. Framework for understanding marine ecosystem health. Mar. Ecol. Progr. Ser. 494, 1–27.

Tobiszewski, M., Namieśnik, J., 2012. PAH diagnostic ratios for the identification of pollution emission sources. Environ. Poll. 162, 110–119.

Törnroos, A., Bonsdorff, E., 2012. Developing the multitrait concept for functional diversity: lessons from a system rich in functions but poor in species. Ecol. Appl. 22, 2221–2236.

Tyler, E.H.M., Somerfield, P.J., Vanden Berghe, E., et al., 2012. Extensive gaps and biases in our knowledge of a well-known fauna: implications for integrating biological traits into macroecology. Glob. Ecol. Biogeog. 21 (9), 922–934.

Van den Brink, P.J., Alexander, A.C., Desrosiers, M., et al., 2011. Traits-based approaches in bioassessment and ecological risk assessment: strengths, weaknesses, opportunities and threats. Integ. Environ. Assess. Manag. 7, 198–208.

Vijgen, J., Abhilash, P., Li, Y., et al., 2011. Hexachlorocyclohexane (HCH) as new Stockholm Convention POPs - a global perspective on the management of Lindane and its waste isomers. Environ. Sci. Poll. Res. 18, 152–162.

Walker, B., Kinzig, A., Langridge, J., 1999. Plant attribute diversity, resilience, and ecosystem function: the nature and significance of dominant and minor species. Ecosystem 2, 95–113.

Wang, Y.F., Tam, N.F.Y., 2011. Microbial community dynamics and biodegradation of polycyclic aromatic hydrocarbons in polluted marine sediments in Hong Kong. Mar. Poll. Bull. 63 (5–12), SI424–SI430.

Weis, J.S., Weis, P., 1992. Transfer of contaminants from CCA-treated lumber to aquatic biota. J. Exp. Mar. Biol. Ecol. 161, 189–199.

Wetzel, M.A., von der Ohe, P.C., Manz, W., et al., 2012. The ecological quality status of the Elbe estuary. A comparative approach on different benthic biotic indices applied to a highly modified estuary. Ecol. Ind. 19, 118–129.

Wolfe, D.A., 1992. Selection of bioindicators of pollution for marine monitoring programmes. Chem. Ecol. 6, 149–167.

Wolska, L., Mechlińska, A., Rogowska, J., et al., 2011. Sources and fate of PAHs and PCBs in the marine environment. Crit. Rev. Environ. Sci. Technol. 42, 1172–1189.

Zhang, X.Z., Xie, J.J., Sun, F.L., 2014. Effects of three polycyclic aromatic hydrocarbons on sediment bacterial community. Curr. Microbiol. 68 (6), 756–762.

CHAPTER 15

Ecotoxicological Risk of Endocrine Disruptors

Amiard Jean-Claude, Claude Amiard-Triquet

Abstract

Effects of endocrine disrupting chemicals (EDCs) are reported in organisms exposed in the laboratory or in the field mainly for (anti-)estrogens, (anti-)androgens, and thyroid disruptors. Potential or proven EDCs belong to different chemical classes, among which persistent and bioaccumulative substances are particularly at risk. Cascading effects from molecular responses to adverse outcomes at the population level are well-established in a few cases. Conventional risk assessment is hardly applicable to EDCs: because they are active at very low doses, effects vary deeply according to developmental stages, and in many cases the dose–response is nonmonotonic. In silico and in vitro tools should be developed for prioritization of potential EDCs. However, in vivo assays remains indispensable, particularly full life-cycle and multigeneration assays. These bioassays used in global approaches (effects-directed analysis, toxicity identification evaluation) are crucial for assessing environmental health.

Keywords: Developmental stages; Full life cycle; Low doses; Multigeneration assays; Nonmonotonic dose-responses; Prioritization.

Chapter Outline

Aquatic Ecotoxicology. http://dx.doi.org/10.1016/B978-0-12-800949-9.00015-2

Introduction

The endocrine system is involved in the control and regulation of all major functions and processes of the body: energy control, reproduction, immunity, behavior, growth, and development (Amiard et al., in Amiard-Triquet et al., 2013). An endocrine disrupting chemical (EDC) has been defined as "an exogenous substance or mixture that alters function(s) of the endocrine system and consequently causes adverse health effects in an intact organism, or its progeny, or (sub) populations" under the auspices of the World Health Organization/International Programme on Chemical Safety(IPCS, 2002). According to Kortemkamp et al. (2011), the term "consequently causes" has been interpreted as a requirement for detailed information on the relationship between altered function of the endocrine system and the adverse effect. Because such a link is not easily established, a definition for a "possible" endocrine disrupter has recently been launched (OECD, 2011): "A possible endocrine disrupter is a chemical that is able to alter the functioning of the endocrine system but for which information about possible adverse consequences of that alteration in an intact organism is uncertain." Among the modes of action able to "alter the functioning of the endocrine system," estrogenicity/antiestrogenicity, androgenicity/antiandrogenicity, and thyroid disruption are to date the most studied.

According to Bergmann et al. (2013b), close to 800 chemicals are known or suspected to be capable of interfering with different aspects of endocrine system functioning. However, only a small fraction of these chemicals have been investigated in tests capable of identifying overt endocrine effects in intact organisms. Thus the importance of the environmental impact of EDCs and the need to enforce a regulatory framework is still a topic of controversy (Dietrich et al., 2013). However, it is clear that government actions to reduce exposure have proven to be effective in a number of cases (Bergman et al., 2013a). Negative impacts on growth and reproduction of wildlife populations affected by endocrine disruption primarily to form persistent organic pollutants (POPs) have been reported. Bans of these chemicals have reduced exposure and led to recovery in some populations (Bergman et al., 2013a), as exemplified in the case of tributyltin (TBT) (Gubbins et al., in Amiard-Triquet et al., 2013). It is thus crucial to evaluate associations between EDC exposures and health outcomes by further developing methods (Bergman et al., 2013b).

This chapter will provide information about (1) the main sources of EDCs responsible for exposure of wildlife and (2) hazard characterization considering effects at different levels of biological organization (cellular tests, growth, development and reproduction, populational effects). Then environmental risk assessment will be considered, taking into account the specificity of EDCs, particularly that they are efficient at low doses and at specific stages in the life- cycle, and that nonmonotonic dose–response relationships have been documented (Kortenkamp et al., 2011; Vandenberg et al., 2012; Zoeller et al., 2012).

15.1 Chemicals of Concern and Exposure

The chemicals of concern are identified in several reviews (Kortenkamp et al., 2011; Vandenberg et al., 2012; Kidd et al., in Bergman et al., 2013a). They include persistent and bioaccumulative halogenated chemicals (e.g., polychlorinated dibenzo-*p*-dioxin/dibenzofurans, polychlorinated biphenyls (PCBs), polybrominated diphenyl ethers (PBDEs), DDT, perfluorooctane sulfonate (PFOS), perfluorooctanoic acid, hexabromocyclododecane), less persistent and less bioaccumulative chemicals (phthalate esters, polyaromatic hydrocarbons, halogenated phenolic chemicals such as triclosan, nonhalogenated phenolic chemicals such as bisphenol A (BPA), nonylphenol, octylphenol, etc.), current-use pesticides, pharmaceuticals (particularly synthetic hormones such as ethinylestradiol (EE2)) and personal care products, metals and organometallic chemicals (methylmercury, tributyltin, etc.), natural hormones (e.g., 17β-estradiol), and phytoestrogens.

15.1.1 Environmental Distribution and Fate of Endocrine Disrupting Chemicals

Because EDCs belong to different chemical classes with many different uses, their environmental input is also highly diversified (Kidd et al., in Bergman et al., 2013a), but the major sources are the effluents of wastewater treatment plants (Desforges et al., 2010; Gust et al., 2010; Hill et al., 2010; Vajda et al., 2011; Barber et al., 2012; Tanna et al., 2013; Bradley and Journey, 2014) and diffuse sources of pesticides. The general processes involved in the fate of contaminants described in Chapter 2 are at work for different EDCs entering the different compartments of the environment (Figures 2.2 and 2.3). Application of sewage sludges and pesticides in agriculture and subsequent agricultural runoff are responsible for important inputs. Long-range transport via wind and ocean currents can induce EDC exposure for biota in remote environments as illustrated by the presence in polar regions of EDCs initially introduced in the environment of temperate regions (Kidd et al., in Bergman et al., 2013a).

Some EDCs are persistent and will concentrate in soils, sediments, or biota (particularly fatty tissues), whereas others are more soluble in water and rapidly broken down.

15.1.2 The Endocrine Disrupting Chemicals Found in Aquatic Wildlife

Wildlife are exposed to waterborne EDCs and EDCs present in sediment and diet. Wildlife higher in the food chain are particularly at risk for exposure to POPs and other similar chemicals with endocrine disrupting (ED) properties because of their biomagnification through the food web (Chapter 2). In mammalian species (marine mammals), pregnant females are exposed to multiple chemicals, most of which can cross the placenta, leading to fetal exposure. In addition, wildlife can be exposed to EDCs via mothers' milk.

Recent data resulting from increased monitoring of EDC exposures in wildlife show that organisms are exposed to a much greater diversity of chemicals in the environment than was documented 10 years ago (Kidd et al., in Bergman et al., 2013a). In particular, pharmaceutical and personal care product ingredients and halogenated phenolic compounds are now commonly reported in wildlife (Chapter 16). Bioaccumulation and, in some cases, biomagnification of legacy (e.g., chlorinated PCBs, DDTs, chlordanes) and emerging (e.g., brominated flame retardants and in particular PBDEs and perfluorinated compounds, including PFOS and perfluorooctanoic acid) contaminants were found even in Arctic biota (Letcher et al., 2010) because of long-range transport and deposition in polar regions. Unlike other POPs, PBDEs and PFOS are generally highest in wildlife near urban areas around the globe (Kidd et al., in Bergman et al., 2013a).

Recently, a special issue of *Marine Pollution Bulletin* (2014, vol. 88, issue 2) was devoted to the strategy developed in the framework of the National Oceanic and Atmospheric Administration's Mussel Watch Program to incorporate CECs, particularly different EDCs. In addition to being efficient for POPs, mussels are also of interest for less persistent compounds such as BPA, alkyl phenols, and polyaromatic hydrocarbons. To improve aquatic environment monitoring strategies, implementation of multiresidue analytical method for the determination of trace concentrations of polyethylene, BPA, polyaromatic hydrocarbons, organochlorine pesticides, PCBs, PBDEs, and alkylphenols in seawater, river water, wastewater treatment plants effluents, sediments, and mussels must be optimized (Sánchez-Avila et al., 2011).

There are currently very few publications on the distribution and transportation of EDCs in the food web of animals living in the wild, thus analyses in a large array of aquatic organisms are very useful (e.g., Diehl et al., 2012; Staniszewska et al., 2014).

From a review about temporal changes in EDCs, Kidd et al. (in Bergman et al., 2013a) have stated that several POPs (e.g., PCBs, PBDEs, PFOS) have increased and then more recently decreased in most areas where concentrations in wildlife were measured. These trends were attributed to restrictions or bans on their use in many countries. On the other hand, few long-term series are available for less persistent, nonbioaccumulative EDCs in wildlife.

15.2 Biological Effects of Endocrine Disrupting Chemicals

15.2.1 Modes of Action of Endocrine Disrupting Chemicals

It is not in the objective of this chapter to describe in detail the modes of action of EDCs. The mechanisms involved in endocrine physiology are numerous and complex (Brown et al., 2004; Melmed and Williams, 2011; Heindel et al., in Bergman et al., 2013a), and environmental chemicals may interfere at all levels. Environmental chemicals have been shown to exert direct actions on hormone receptors and receptor function, as well as to exert direct actions controlling hormone delivery to the receptor (Figure 15.1).

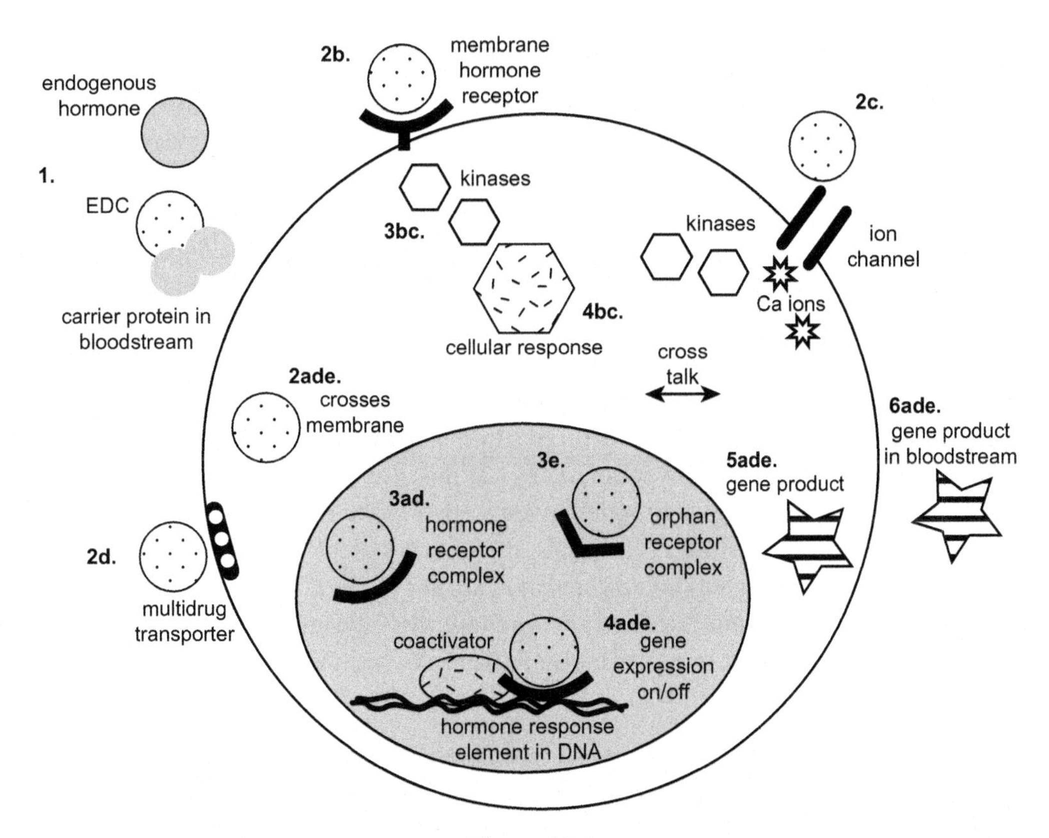

Figure 15.1

Summary of the cellular mechanisms by which hormones and EDCs act. EDCs modulate the function and response of hormones by operating via the same cellular mechanisms as endogenous hormones. EDCs may displace an endogenous hormone from carrier protein outside of the cell (1). (A) An EDC crosses cell and nuclear membranes, binds to a nuclear receptor, which forms a complex that interacts with other transcription factors, and binds to a hormone-response element, which results in gene expression and often the translation of a gene product (protein) (2a to 6a). (B) EDC binds to a membrane hormone receptor and causes a cellular signaling cascade mediated by protein kinases, generating a variety of downstream cellular responses (2b to 4b). (C) An EDC interacts with an ion channel, facilitating the passage of ions (i.e., calcium) into the cell, which causes a cellular signaling cascade as described in b (2c to 4c). (D) An EDC bypasses multidrug transporter, entering the cell and proceeding as described in a (2d to 6d). (E) An EDC crosses cell and nuclear membranes, but binds to orphan receptor (PXR or CAR) instead of nuclear hormone receptor, influencing gene expression and possibly resulting in a gene product (2e to 4e). Cross-talk may occur between any of these pathways. Other mechanisms, not pictured, include alteration of transcription factors (e.g., coactivators or corepressors), altering the rate of nuclear receptor degradation, and binding to neurotransmitter receptors or transporters, resulting in alteration of signaling. *After Brander (2013) with permission.*

The endocrine systems of vertebrates largely share mechanisms common to mammals (including humans) and wildlife (fish, amphibians, and reptiles). For instance, Ankley and Gray (2013) have demonstrated a significant degree of cross-species conservation of the hypothalamic-pituitary-gonadal axis by using fish and rat screening assays with 12 model chemicals. However, the receptors are somewhat different among classes and there is a need to understand the mechanisms underlying species differences because this can influence the ability of exogenous chemicals to interact with them and then exert a noxious effect (Heindel et al., in Bergman et al., 2013a). These authors underline that "despite our lack of knowledge on their fundamental endocrinology, chemicals that affect hormonal activities in vertebrates also appear to affect several invertebrate species." Focusing on the case of environmental estrogens on fish, Segner et al. (2013) underline that the commonality in their biological action is their binding and activation of the endoplasmic reticulum. However, the diversity in estrogen responses between fish species should not be ignored because the commonality in the initial event is not necessarily reflected in a uniformity of the subsequent events (i.e., in the molecular and cellular endoplasmic reticulum signaling cascades, in the physiological target processes, and in the toxicological and ecological consequences) (Segner et al., 2013). In arthropods, reproduction and development are controlled mainly by ecdysteroids. Endocrine functions in crustaceans and their impairment by EDCs have been recently reviewed (Amiard et al., in Amiard-Triquet et al., 2013).

With recent advances in genomic sequencing technology, microarrays have been developed for many fish species, and the ability to evaluate thousands of genes simultaneously has revealed that EDCs influence cellular functions as diverse as stress response, cell division, and apoptosis (Brander, 2013). Segner et al. (2013) emphasize the range of nonreproductive physiological processes in fish that are known to be responsive to environmental estrogens: sensory systems, the brain, the immune system, growth, specifically through the growth hormone/insulin-like growth factor system, and osmoregulation. Omics have also been used to elucidate disruptions of behaviors affecting social hierarchy, and in turn breeding outcome, as a consequence of exposure to EE2 (Filby et al., 2012). This mechanistic analysis has shown complex interconnecting gonadal and neurological control mechanisms that generally conform with those established in mammalian models.

"Evaluating the expression of endocrine-sensitive genes by measuring mRNA transcript levels is an effective and common approach used to discern mechanisms of action of EDCs or EDC mixtures" (Brander, 2013). Changes in gene expression manifest themselves at the level of both the message (mRNA) and the final product (protein), but there is often lack of correlation between mRNA expression and protein amounts or activity, and transcript levels only partially explain levels of protein expression. Thus the impact of EDCs on protein expression may be a better predictor of tissue- and organism-level EDC effects than the mRNA transcripts (Cox and Mann, 2007; Brander, 2013).

15.2.2 Evidence for Endocrine Disruption in Aquatic Wildlife

Endocrine disruption in aquatic wildlife has been recently reviewed (Jobling et al., in Bergman et al., 2013a). The different categories of effects that have been documented are summarized in Table 15.1.

15.2.2.1 Reproductive Health: Imposex and Intersex

Endocrine disruption processes leading to impairment of mechanisms involved in reproduction are the most documented such as the presence of imposex—the development of a penis homologue able to block the emission of sexual products—in prosobranch female snails already mentioned. It has been shown that potent androgen receptor agonists and aromatase inhibitors, as well as TBT were responsible for this impairment. Endocrine disruption from the use of TBT was observed worldwide in a huge number of gastropod species (Titley-O'Neal et al., 2011).

Table 15.1: Evidence for Endocrine Disruption in Aquatic Wildlife after Jobling et al. (in Bergman et al., 2013a)

Female Reproductive Health	Imposex	Prosobranch Female Snails
	Uterine Fibroid	Seals
Male reproductive health	Intersex	Bivalves, crustaceans, fish, *amphibians*,[a] reptiles
Sex ratio	Male or female-biased	Gastropods, bivalves, *crustaceans*[b], fish
Thyroid-related disorders and diseases	Various physiological processes	Fish, marine mammals
Neurodevelopment	Brain sexualization, Reproductive behavior	Mammals, birds, *fish*[c]
Hormone-related cancers	Genital papillomas, genital tract carcinomas	Marine mammals
Immune function	Immunodeficiency, reduction of resistance, *impaired antiviral defenses*	Marine mammals, fish, bivalves, crustaceans, echinoderms, ascidia, *amphibians*[d]
Adrenal disorders	Adrenocortical hyperplasia	Seals
	Cortisol levels, cortisol response to stress	Fish
Bone disorders		Alligators, aquatic birds, seals
Metabolic disorders	Adipogenesis, weight gain	Fish, amphibians
Arthropod-specific EDCs[e]	*Embryo-larval development*	*Insects, crustaceans*
	Molt cycle	

Other sources shown in italics.
[a]Bahamonde et al., 2013.
[b]Forget-Leray et al., 2005; Huang et al., 2006.
[c]Söffker and Tyler, 2012.
[d]Sikarovski et al., 2014.
[e]Amiard et al. (in Amiard-Triquet et al., 2013).

Other examples of failure in reproductive success include the occurrence of uterine leiomyoma in seals exposed to DDT and PCBs in the Baltic sea (Bäcklin et al., 2003).

Studies about male reproductive health in wildlife have revealed symptoms of androgen insufficiency and/or androgen:estrogen imbalance in fish, leading to intersex (developing oocytes within the male testis). In aquatic animals, intersex is often viewed as a signature effect of exposure to EDCs. Links between intersex and vitellogenin (VTG; egg yolk protein) induction have been shown in male fish (Jobling et al., in Bergman et al., 2013a). However, the literature is not definitive on whether field studies are distinguishing between natural intersex and intersex resulting from stressors (Bahamonde et al., 2013). Field-based evidence of endocrine-mediated reproductive disorders in invertebrate males is scarce and solely concerns aquatic mollusks and crustaceans (Jobling et al., in Bergman et al., 2013a). Intersex has been reported in mussels exposed to the oil spill from the *Prestige* oil tanker (Ortiz-Zarragoitia and Cajaraville, 2010) and in the softshell clam (*Scrobicularia plana*) originating from multipolluted estuaries in France, Portugal, and the United Kingdom (Fossi Tankoua et al., 2012 and literature quoted therein). In the latter, experimental exposures to sediment spiked with mixtures of 17β-estradiol (E2), 17α-ethinylestradiol (EE2), octylphenol, and nonylphenol indicated that (xeno)estrogens could be a contributory factor in the induction of intersex (Langston et al., 2007).

Although some parasite-induced intersexuality exists naturally, several authors have shown that crustacean intersexuality was linked to environmental contamination in copepods (Moore and Stevenson, 1991), amphipods (Barbeau and Grecian, 2003; Jungmann et al., 2004; Ford et al., 2006; Yang et al., 2008; Hyne, 2011; Short et al., 2012), and decapods (Sangalang and Jones, 1997; Ayaki et al., 2005; Stentiford, 2012). A proportion of intersexes may be indirectly caused by anthropogenic contamination, as contaminant-exposed crustaceans present significant increases in infection by microsporidia (Short et al., 2014).

15.2.2.2 Reproductive Health: Biased Sex Ratio

According to Jobling et al. (in Bergman et al., 2013a), "EDC-related sex ratio imbalances have been seen in wild fish and mollusks and, in some of these species, are also supported by laboratory evidence." Female-biased sex ratios have been observed in a number of wild fish species exposed to the discharge of estrogenic wastewater effluents, whereas male-biased sex ratios are observed below paper mill effluents, pharmaceutical manufacture discharge (Sanchez et al., 2011), and maybe in TBT-contaminated marinas (Jobling et al., in Bergman et al., 2013a). In roaches (*Rutilus rutilus*), lifelong exposure to environmental concentrations of EE2 ($4\,ng\,L^{-1}$) not only induced intersex but a complete reversal of males, resulting in an all-female population (Lange et al., 2011).

Feminized sex ratios were observed in zebra mussels after a DDT-pollution incident as also in freshwater mussels (*Elliptio complanata*) downstream of two municipal effluent outfalls. Masculinized sex ratios were observed in both softshell clams (*Mya arenaria*) and gastropods (*Thais clavigera*) exposed to TBT (Jobling et al., in Bergman et al., 2013a). In many different populations of the clam *S. plana* collected from different sites in United Kingdom and in France, sex ratios do not differ significantly from unity in most cases but at some sites, either female- or male-biased sex ratios were observed (Fossi Tankoua et al., 2012; Langston et al., 2012).

Skewed sex ratios in favor of females were observed in copepods (*Eurytemora affinis*) experimentally exposed to binary mixtures of 17β-estradiol, 4-nonylphenol, di(ethylhexyl)-phthalate, and atrazine (Forget-Leray et al., 2005). In another species (*Pseudodiaptomus marinus*) exposed to TBT, female-to-male ratio was significantly reduced (Huang et al., 2006).

The mechanisms underlying EDC effects on sex ratios remain unknown for many species. Reversibility of endocrine disruption in zebrafish (*Danio rerio*) after discontinued exposure (0.1–10 ng EE2 L^{-1} up to 60 days posthatch followed by 40 days of depuration in clean water) was shown for the first time at different biological organization levels (brain aromatase mRNA expression, VTG, maturity index, sex ratio) (Baumann et al., 2014).

15.2.2.3 Thyroid Impairments

Relationships between exposure to chemicals and thyroid hormone disruption in wildlife species have increased in the last decade, especially in relation to exposure to PCBs, flame retardants (PBDEs, tetrabromobisphenol A) and other chemicals (perchlorates, alkylphenols). EDCs can induce changes in thyroid histology, levels of thyroxin (T4) and triiodothyroxine (T3), T3/T4 ratios with possible consequences on the different biological processes governed by thyroid hormones (development, reproduction, growth, behavior, etc.). Many studies have reported relationships between individual body burdens of persistent organic pollutants and thyroid-related effects in marine mammals (Jobling et al., in Bergman et al., 2013a). Fish in contaminated locations are known to have impaired thyroid systems, a finding also supported by experimental studies (Amiard et al., in Amiard-Triquet et al., 2013). Thyroid hormone disrupting effects in amphibians are well-established, leading to the development of standardized tests in larval frogs (*Xenopus laevis*) that consider thyroid gland histology as one of the primary endpoints but also developmental stage, hind limb length, snout-vent length, and wet body weight (Miyaka and Ose, 2012). However the statement by Brucker-Davis (1998) that "Real-life exposure of humans and wildlife to a wide spectrum of synthetic chemicals makes it difficult to prove the causative thyroid effect of one single agent. Therefore, it is more prudent to talk of association rather than causation, and the role of potential co-contaminants should always be considered" still remains valid.

15.2.2.4 Other Effects of Endocrine Disrupting Chemicals

Endocrine disruptors affecting neurodevelopment in wildlife include MeHg, PBDEs, PCBs, and DDT (Jobling et al., in Bergman et al., 2013a) as also EE2 and nonylphenol (Vosges et al., 2012). They can cause the impairment of brain sexualization and behavioral disturbances, particularly reproductive behavior, as reviewed by Söffker and Tyler (2012) in fish. In addition to articles clearly establishing a mechanistic link between neuroendocrine effects of exposure to EDCs and behavioral disturbances, many reports exist of behavioral impairments in organisms exposed to EDCs (Sárria et al., 2011; Saaristo et al., 2013; Dzieweczynski et al., 2014; Shenoy, 2014).

Hormone-related cancers are well-documented in humans. In the past several decades, the occurrence of cancers in vertebrate wildlife have consistently increased, particularly in areas submitted to important inputs of anthropogenic contaminants among which the case of beluga whales of the St Lawrence estuary is emblematic. Cancers have been described in different species including marine mammals, sediment-dwelling fish, and also invertebrates, mainly bivalves. Among cancers, endocrine organs are frequently affected, particularly reproductive organs (Jobling et al., in Bergman et al., 2013a). However, the causation cannot be directly attributed to EDCs because the etiology of cancer is multifactorial. In contaminated areas, in addition to endocrine disruptors, pollutants (pharmaceuticals, recombinant biologicals, or environmental and occupational pollutants) can induce immunodeficiency, favoring the reduction of the host's resistance to infections or the development of cancer (Brousseau et al., in Amiard-Triquet et al., 2013). The link between endocrine disruptors and immune function is documented in marine mammals, other vertebrates such as fish and amphibians, and invertebrates (Galloway and Depledge, 2001; Milla et al., 2011; Jobling et al., in Bergman et al., 2013a; Sifkarovski et al., 2014).

Other effects of EDCs (Table 15.1) have been much less studied and the existing knowledge has been reviewed by Jobling et al. (in Bergman et al., 2013a).

15.2.3 Do Endocrine Impairments Represent a Risk for Populations and Communities?

Disturbances at infraindividual and individual levels could have impacts at the scale of populations through breeding success (including egg production and spawning, reduced egg fertility, hatching rate, survival of embryos, etc.) and impaired behaviors (Colman et al., 2009; Coe et al., 2010; Sárria et al., 2011; Filby et al., 2012; Söffker and Tyler, 2012; Saaristo et al., 2013; Elliott et al., 2014; Dzieweczynski et al., 2014; Liu et al., 2014). However, only a few studies have tested these hypothesis in field populations.

15.2.3.1 Invertebrates

Population effects induced by the occurrence of imposex were clearly established in gastropods, particularly *Nucella lapillus* with the decline of this species as a consequence of TBT use and a recovery with the decrease of TBT inputs and the final stop on the use of TBT-based paints (Gubbins et al., in Amiard-Triquet et al., 2013). In bivalves, intersex has not yet been associated with decreased recruitment (Fossi Tankoua et al., 2012).

According to McCurdy et al. (2008), typically, crustacean intersexes are sterile or function as females, but some species escape this general pattern such as the amphipod *Corophium volutator*. Intersexes behave as males by crawling (mate-searching) on mudflat during ebb tides and pairing in burrows with female amphipods but at a lesser degree than males, forming fewer pairs with females, and remaining in tandem less often with receptive females. In addition, because contrary to vertebrates, a crustacean likely requires both feminizing and de-masculinizing agents to cause sexual dysfunction equivalent to that of vertebrates, Short et al. (2014) concluded that "the underlying biology of crustaceans may confer a level of protection for crustaceans against the harmful influences of anthropogenic chemicals."

15.2.3.2 Fish

In fish, intersex is interpreted as feminization of male fish (Lange et al., 2011). Because it has been observed in many fish species (Bahamonde et al., 2013; Blazer et al., 2012, 2014; Yonkos et al., 2014), it could be an important determinant of reproductive performance, particularly in rivers where there is a high prevalence of moderately to severely feminized males (Harris et al., 2011). Yu et al. (2014) have investigated the impact of low doses of PBDEs on the reproduction of zebrafish over two generations. Several reproductive endpoints showed impairments in F0 specimens. Delayed hatching, reduced survival, and decreased growth were observed in the F1 embryos despite them being cultivated in clean water, suggesting that PBDEs might have significant adverse effects on fish population in a highly PBDE-contaminated aquatic environment.

Modeling approaches used to predict the fate of fish populations exposed to EDCs, considering impairments on specimens exposed in the laboratory or in the field, have not allowed a definitive conclusion (Harris et al., 2011). However, Kidd et al. (2007) have shown that experimental exposure of the fish *Pimephales promelas* to realistic concentrations (5–6 ng L^{-1}) of EE2 in a lake led to feminization of males and altered oogenesis in females, thus leading to the collapse of fish population. However, longer-lived fish species (pearl dace, white sucker, and trout) from the experimental lake did not show any effects of EE2 exposure on their population structures (Pelley, 2003; Kidd et al., 2007).

In the fish *Gobio gobio* living downstream from pharmaceutical manufacture discharge, endocrine disruption (VTG concentrations, intersex fish, male-biased sex ratio) was associated to fish population disturbances with a decrease or even the disappearance at the most contaminated site (Sanchez et al., 2011). Considering more widely fish assemblages, a decrease of occurrence of sensitive fish species and fish density was observed. In the planktivorous fish *R. rutilus*, exposure to EE2 in a mesocosm study has induced a reduced foraging performance, leading to a positive indirect effect on zooplankton (Hallgren et al., 2014). Hence, EE2 may have consequences for both the structure and function of freshwater communities.

The literature dealing with the influence of EDCs on behavior is mainly devoted to impairments because of estrogens, but thyroid dysfunction is another source of behavioral disturbances. Links to higher levels of biological organization have been documented by using complementarily laboratory and field approaches as reviewed by Weis and Candelmo (2012).

15.2.3.3 Reptiles

Environmental exposure to anthropogenic contaminants alters the development and functioning of American alligators' reproductive and endocrine systems (Milnes and Guillette, 2008; Woodward et al., 2011; Moore et al., 2012). Thus, repeated episodes of Nile crocodile deaths have been suspected to be due to similar dysfunctioning. As large predators, crocodiles are particularly exposed to biomagnifying chemicals such as DDT and other organic contaminants. However, Bouwman et al. (2014) found it unlikely that pollutants could have caused or contributed substantially to the mortalities as a primary chemical agent. Similarly, in Australian freshwater crocodiles, high concentrations of p,p′-DDE and toxaphene were found in the lipid-rich tissues but no obvious effects on blood chemistry or gonad histology were registered, although the limited number of samples and the variability of the breeding state of the animals examined may have masked possible effects (Yoshikane et al., 2006).

15.2.3.4 Amphibians

If amphibian decline is well-documented in many parts of the world, the role of EDCs (particularly pesticides, including atrazine) in this decline is still a topic of discussion (Gubbins et al., in Amiard-Triquet et al., 2013). For intersex, the lack of full scientific certainty likely results from the general plasticity of amphibians with regard to sexual phenotype (influence of environmental conditions during development, natural hermaphrodism). Meta-analysis of the effect of atrazine (Rohr and McCoy, 2010) or other pesticides (Baker et al., 2013) concluded that these agro-chemicals consistently impact development, survival, and growth of amphibians. It is not a straightforward conclusion that impairments of sexual characteristics represent a risk for populations, but it should not be ignored that pesticides can affect different biological aspects such as immune function and behavior (Rohr and McCoy, 2010) that can reinforce the deleterious effects of reproductive impairments.

15.3 Tools for the Detection and Quantification of Endocrine Disrupting Chemical Effects

15.3.1 Frameworks for Regulatory Testing and Screening

The Organization for Economic Co-operation and Development (OECD) conceptual framework for testing and assessment of endocrine disrupters (OECD, 2012a) lists the OECD test guidelines and standardized test methods available, under development or proposed, that can be used to evaluate chemicals for endocrine disruption. This approach includes five levels:

- Physical and chemical properties, quantitative structure activity relationships (QSARs) and other in silico predictions; absorption, distribution, metabolism, and excretion) model predictions; available data from (eco)toxicological tests
- In vitro assays providing data about selected endocrine mechanism(s)/pathways(s)
- In vivo assays providing data about selected endocrine mechanism(s)/pathway(s)
- In vivo assays providing data on adverse effects on endocrine relevant endpoints (development, growth, reproduction)
- In vivo assays providing more comprehensive data on adverse effects on endocrine relevant endpoints (full life-cycle and multigeneration assays).

Mammalian and nonmammalian methods are included, among which many tests are based on aquatic models (daphnia, chironomids, mysids, oligochaetes, mollusks, fish, amphibians). The use and interpretation of standardized OECD and US Environmental Protection Agency (USEPA) tests that can be carried out in this context have been discussed in detail (OECD, 2012b).

The OECD conceptual framework remains a "tool box," and not a tiered testing strategy, contrary to the two-tiered approach adopted under the auspices of the USEPA Endocrine Disruptor Screening Program (EDSP). It has been approved for use in a regulatory context to screen pesticides, chemicals, and environmental contaminants for effects that are mediated by interactions with estrogen, androgen, or thyroid hormone signaling pathways. In tier 1, chemicals are screened for their potential to interact with the endocrine system using both in vitro and in vivo tests (USEPA, 2009). Automated chemical screening technologies ("high-throughput screening assays") have been developed to expose living cells or isolated proteins to chemicals under the auspices of the USEPA in the framework of the Toxicity Forecaster (ToxCast™). They can be incorporated in the strategy of the USEPA's EDSP tier 1 to prioritize the thousands of chemicals that need to be tested for potential endocrine-related activity. The in vivo assays used in tier 2 are similar to those indicated at level 5 of the OECD framework (see the previous section). Regular updates on the status of tier 1 and 2 tests may be found at: http://www.epa.gov/scipoly/oscpendo/pubs/assayvalidation/status.htm.

Sensitivity to endocrine disruption is highest during tissue development, thus developmental effects will occur at lower doses than are required for effects in adults (Heindel et al., in Bergman et al., 2013a). These authors recommend that testing for endocrine disruption encompass the developmental period and include lifelong follow-up to assess latent effects. In agreement, Kortenkamp et al. (2011) already suggested that in the OECD conceptual framework, the only assays appropriate for dose–response or potency assessment would be those listed in level 5.

For European Community legislation, the problem of endocrine disrupters is taken into account in three legal texts: Plant Protection Product Regulation; Registration, Evaluation,

Authorization and Restriction of Chemicals; and Biocidal Product Regulation. However there are conflicting opinions on how to assess the risk from exposure to EDCs (Kortenkamp et al., 2011, 2012; Rhomberg et al., 2012; Testai et al., 2013). It is necessary to discriminate between adaptive responses vs adverse effects, termed endocrine modulation and endocrine disruption. According to Rhomberg et al. (2012), "many adaptive, compensatory, and even physiologically normal and necessary processes result in measurable endocrine changes, and these cannot be considered endocrine disruption (ED). It is only when there is inappropriate expression of these natural mechanisms to such a degree that adverse effects are induced that ED occurs." When effects are observed on different biological functions of organisms exposed to compounds suspected to be EDCs, the ED basis for those outcomes is often difficult to established conclusively (see above sub-section 17.2.2).

The concept of the Adverse Outcome Pathway first launched by Ankley et al. (2010) can be applied to the effects of EDCs (subindividual responses to EDCs → life history traits and behavior → individual fitness → population fate) (Amiard et al., in Amiard-Triquet et al., 2013). However, many studies fail to integrate that impairments of biological pathways must be sufficiently large to overcome adaptive responses before biological function is compromised. In addition, exposure of wildlife to EDCs can be intermittent (e.g., pesticide treatments) and then the reversibility is important to assess.

In European regulations, substances having ED properties may be included in the list of "substances of very high concern." Thus Testai et al. (2013) considered that this "hazard-based approach" is less relevant than the risk-based approach conducted in the United States and Japan for all potential EDCs integrating both hazard and exposure (Chapter 2). However, Kortenkamp et al. (2011) argue that deriving an exposure dose that may be considered safe is almost a mission impossible because a risk assessment cannot be carried with a satisfactory level of certainty, not least because of the complexity of the issues of low dose, irreversibility, and exposure during critical stages of development.

15.3.2 Research and Development Needs

15.3.2.1 In Silico and In Vitro Tools

Computational methods such as virtual screening, QSARs, and docking can be used to speed up and save costs for screening chemicals for their ED properties (Vuorinena et al., 2013). These authors review a number of case studies, where new active EDCs have been discovered with the help of in silico tools. QSAR models define the relationship between the chemical structure of a compound and its biological or pharmacological activity. QSARs have proved useful for predicting the biological activities of molecules (e.g., their binding affinities to different receptors). Lo Piparo and Worth (2010) have reviewed QSAR models and software tools for predicting developmental and reproductive toxicity. Despite refining these techniques for application in ED, predictions should be clearly encouraged, and it must be kept in mind that they are not suitable for replacing in vitro or in vivo testing. However, computational models are valuable tools for prioritizing potential EDCs for biological evaluation (Vuorinena et al., 2013).

To date, exposure to EDCs in the environment is mainly assessed through an a priori approach with chemical analysis of these substances known or suspected of having ED effects. A global approach can be preferred by adopting unbiased exposure assessment strategies that search for unknowns such as effects-directed analysis (EDA) and toxicity identification evaluation. The global approach combining biotesting, fractionation and chemical analysis, helps to identify hazardous compounds in complex environmental mixtures (Brack, 2011; Kortenkamp et al., 2011; Burgess et al., 2013). For instance, Kinani et al. (2010) have evaluated the contribution of chemically quantified compounds to the biological activity detected by in vitro reporter cell-based bioassays. When the ratio is nearing 100%, it means that all analyzed compounds were explicative for biological results; otherwise, other nonanalyzed compounds may be present in the samples. For estrogenic activity measured in the MELN bioassay, the ratio varied from 28% to 96%, but for antiandrogenic activities, it was as low as 0.01–0.45%.

Toxicity identification evaluation, which was mainly developed in North America in support to the US Clean Water Act, and EDA, which originates from both Europe and North America, differed primarily by the biological endpoints used to reveal toxicity (Figure 15.2). Biological

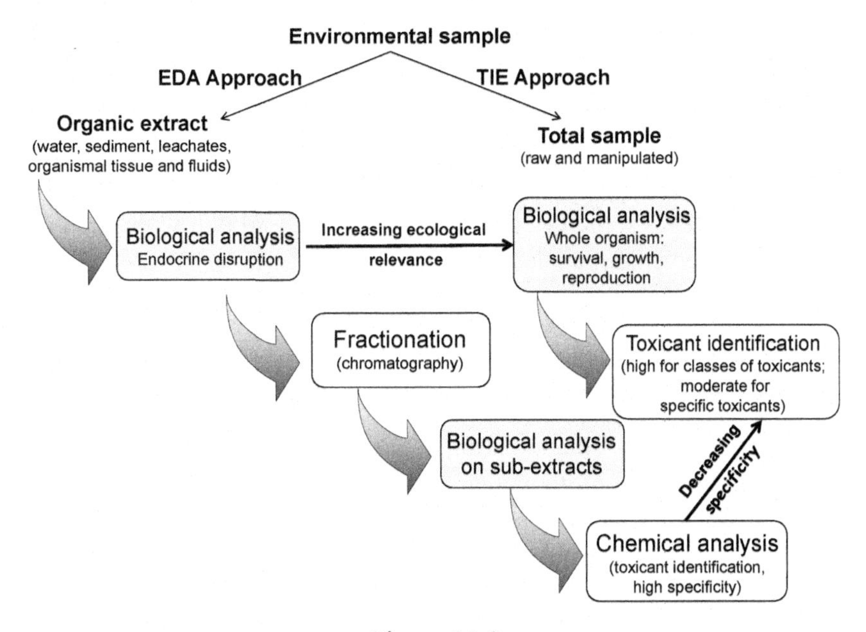

Figure 15.2
Global approach of mixtures: Effects-Directed Analysis (EDA) and Toxicity Identification Evaluation (TIE). *Modified after Burgess et al. (2013) with permission.*

analysis at the level of the whole organism confers an increasing ecological relevance on toxicity identification evaluation, whereas biological analyses pinpointing ED in both total and subextracts confers a better specificity to EDA (Burgess et al., 2013). EDA has been mainly developed to measure (anti)estrogenic and (anti)androgenic activities (Hecker and Giesy, in Brack, 2011), whereas fewer studies have been devoted to thyroid disrupting properties (Houtman et al., in Brack, 2011).

Different in vitro bioassays have been designed and associated to chemical analysis to measure estrogenic activity in different water sources (e.g., Leusch et al., 2010; Creusot et al., 2014) and sediments (e.g., Creusot et al., 2013a, b). According to Dévier et al. (2011), combining bioanalytical approaches with the use of passive sampling devices instead of classical spot water sampling is a promising methodology to improve environmental risk assessment, allowing feasible bioavailability estimations of sediment-bound and waterborne micropollutants (see also Chapter 5).

The mechanisms involved in thyroid homeostasis are numerous and complex as exemplified in the teleost thyroid cascade (Brown et al., 2004), thus thyroid hormone-disrupting chemicals (THDCs) can interfere at many different levels. In vitro assays are available for many of these endpoints, but a review of the state of the science indicates that these assays ranges from those already being used in medium- or high-through put screening to potential targets that lack cost efficient testing methods (Murk et al., 2013). These authors have recommended a battery of test methods to be able to classify chemicals as of less or high concern for further hazard and risk assessment for THDCs. Likewise, in vitro and ex vivo assays for identification of modulators of thyroid hormone signaling have been listed under the auspices of OECD (2013), including central regulation (hypothalamic-pituitary-thyroid axis), thyroid hormone synthesis, secretion and transport, metabolism and excretion, local cellular concentrations, cellular responses, and integrative cellular assays. For each assay, the validation status toward known THDCs is indicated as also if they are used in research and development or already proposed for potential regulatory purpose.

Although relatively less specific compared to analytical methods, bioluminescent bioassays are more cost-effective, more rapid, can be scaled to higher throughput, and can be designed to report not only the presence but also the bioavailability of EDCs, including THDCs (Xu et al., 2014).

In vitro tests have received particular attention as alternatives to animal testing, but improving in vivo validation will be necessary to increase confidence for use of their results in regulatory decisions (Scholz et al., 2013; Murk et al., 2013). In line with these recommendations, Henneberg et al. (2014) have examined the degree at which in vitro methods for the detection of sexual steroid disruptors were consistent with in vivo effects. They have shown that in vitro assays for endocrine potentials in stream water (E-screen and reporter gene assays) reasonably reflect reproduction of gastropods in laboratory tests and ED in different field-exposed fish species. On the other hand, stable cell lines currently used as screening applications to detect EDCs (H295R assay validated under the auspices of the OECD, Hecker et al., 2011) were found to have lower sensitivity than gonad explants from some wild fish species, suggesting that it might not be protective enough (Beitel et al., 2014).

In crustaceans, in addition to EDCs affecting sexual and thyroid hormones, contaminants can interfere with hormones involved in molting. Tools for monitoring crustacean exposure to EDCs includes few in vitro assays (OECD, 2006; Amiard et al., in Amiard-Triquet et al., 2013).

15.3.2.2 In Vivo Approaches

However, as underlined by Fetter et al. (2014), in vitro assays cannot reflect the complexity of whole organisms. These authors proposed the use of embryos of zebrafish as an alternative to experiments with adult animals considering that the use of fish embryos combines the complexity of a full organism with the simplicity and reproducibility of cellular assays. Fluorescent transgenic zebrafish (larvae and adults) have been generated for the in vivo detection of environmental estrogens (Chen et al., 2010) and THDCs (Terrien et al., 2011). In the same way, Fini et al. (2007) have developed an in vivo test for the detection of thyroid receptor agonists and for the determination of effects of thyroid hormone production inhibitors.

Impairments of endocrine functions provide biomarkers as ecotoxicological tools. Their use in case studies based on different taxa (gastropods, fish, amphibians) has been recently described (Gubbins et al., in Amiard-Triquet et al., 2013). However, only a few biomarkers of ED are recognized in the framework of integrated monitoring and assessment of contaminants such as VTG, mRNA transcripts, or protein in male fish plasma, intersex in fish, imposex/intersex in gastropods, and reproductive success in fish (Davies and Vethaak, 2012). The limits for their use are described in Chapter 7, and research and development are needed to enlarge their acceptance in a regulatory context. VTG is widely recognized as a specific biomarker of ED in vertebrates whereas its relevance in crustaceans is a topic of debate (Chapter 11). The development of new potential molecular biomarkers for monitoring the reproductive health of crustaceans is in progress (Short et al., 2014).

In vivo assays providing data on adverse effects on endocrine relevant endpoints (development, growth, reproduction) are already well-developed for regulatory testing, even if not all of them are appropriate for dose–response or potency assessment (see above subsection 17.3.1). In addition, it has been proposed to use biomarkers of contamination by EDCs such as immune parameters (Milla et al., 2011) and behavioral impairments (Sárria et al., 2011; Soffker and Tyler, 2012). These biomarkers have several advantages such as their sensitivity and their ecological relevance (Brousseau et al., Amiard-Triquet and Amiard, both in Amiard-Triquet et al., 2013), the latter of which is important because they are noninvasive (Soffker and Tyler, 2012). On the other hand, their specificity remains questionable since chemicals without ED properties are able to affect these biological parameters and there are also prone to the effects of confounding factors (Chapter 7). Adopting such biomarkers still has many technical and interpretation challenges and extrapolation from laboratory conditions to wild populations is not an easy task (Soffker and Tyler, 2012).

15.3.2.3 Hazard Identification: The Problem of Nonmonotonic Dose–Response Relationships

EDCs can exert effects at very low doses in the experimental media or in the wild. It is questionable if many classical bioassays used as a basis for conventional environmental risk assessment are relevant for EDCs because many studies (reviewed by Vandenberg et al., 2012) have shown nonmonotonic dose–response relationships (NMDRs), in opposition to the dogma of "the dose makes the poison." For many contaminants, dose–response curves are nonlinear but remain monotonic such as the sigmoid curves shown in Figures 2.4 and 2.5 and statistical treatments have been proposed to overcome this difficulty (Chapter 2). Dose–response curves are considered nonmonotonic when the slope changes sign one or more times, i.e., bell-shaped curve, U-shaped curve, or multiphasic curve (Figure 15.3). Such NMDRs are well-documented in both cell culture experiments and animal studies, with most of the examples being obtained with mammalian models but reports also exists for gastropods, fish, and amphibians (Vandenberg et al., 2012; Brodeur et al., 2013; Hass et al., 2013; Vandenberg, 2014). Thus NMDR cannot be considered as a marginal phenomenon when carrying out risk assessment (Birnbaum, 2012; Beausoleil et al., 2013).

The simplest mechanism for NMDR derives from the observation that EDCs can be acutely toxic at high doses, thus hampering the detection of those responses that are mediated by ligand-binding interactions. Consequently, acute toxicity tests that are commonly used in conventional risk assessment (Chapters 2, 6) are irrelevant for the determination of safe doses in the case of EDCs (Vandenberg et al., 2012). In addition to cytotoxicity, several mechanisms that can explain how hormones and EDCs produce nonmonotonic responses in cells, tissues,

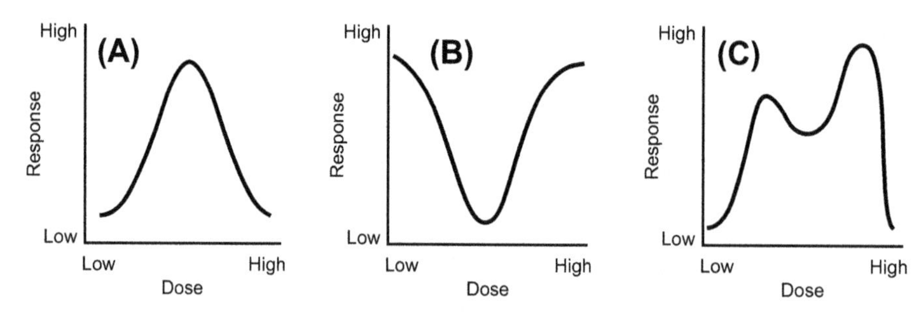

Figure 15.3
Theoretical NMDR curves and articles illustrating these types of relationships. (A) Bell-shaped curve (Oehlmann et al., 2000; Evanson and Van der Kraak, 2001; Duft et al., 2003; Gust et al., 2009; Chaube et al., 2010); (B) U-shaped curve (Freeman et al., 2005; Brodeur et al., 2009; McMahon et al., 2011); and (C) multiphasic curve (Boettcher et al., 2011; McMahon et al., 2011; Sharma and Patiño, 2009).

and animals have been identified and studied (Conolly and Lutz, 2004; Li et al., 2007; Vanderberg et al., 2012; Hass et al., 2013). These mechanisms include receptor selectivity, receptor downregulation and desensitization, receptor competition, and endocrine negative feedback loops. Actions of a toxic agent in an organism (or an organ constituted of different categories of cells) are multifaceted and accordingly, the reaction of the organism is pleiotropic; the dose–response relationship is the result of a superimposition of all interactions that pertain (Conolly and Lutz, 2004; Vanderberg et al., 2012).

Accurate description of NMDRs is a key step for the characterization of hazards associated with the presence of pollutants in the medium, thus needing the development of dedicated models as already proposed in the case of hormesis (Qin et al., 2010; Zhu et al., 2013). Conolly (2012) discussed computational modeling of NMDR, underlining the prerequisites to make this approach efficient such as the dose relevance across the full extent of the nonmonotonic curve ("Are all the exposures and associated doses relevant in the real world?") and the level of biological organization ("Are nonmonotonic responses as likely in vivo as they are in vitro?").

15.4 Conclusions

EDCs mainly enter the aquatic environment with the effluents of wastewater treatment plants and diffuse sources of pesticides. The global transport of EDCs leads to deposition in remote areas such as polar regions (Kidd et al., in Bergman et al., 2013a). Some EDCs are persistent and concentrate in soils, sediments, or biota, whereas others are more soluble in water and rapidly broken down. Thus wildlife are exposed to waterborne EDCs and EDCs present in sediment and diet, with species feeding at the top of the food chain being particularly at risk.

Kidd et al. (in Bergman et al., 2013a) consider that it is crucial to develop the abilities to measure any potential EDCs but this objective seems elusive because of the number of substances of concern (9700 environmental chemicals that should be screened in the framework of the EDSP according to Rotroff et al., 2013). In addition, chemical analyses systematically miss substances with unknown ED potential. Global approaches such as TIE and EDA should be encouraged.

EDCs have many potential mechanisms of action and in vitro assays are available for many of these endpoints. The application of "-omics" technologies, in organisms exposed to EDCs in vivo can help to reveal relevant and possibly new endpoints for inclusion in in vitro test batteries.

Endocrine modes of action and general toxic effects must be clearly distinguished for the interpretation of in vitro and in vivo assays. Almost half of the in vitro examples (45%) evaluated by Hass et al. (2013) showed an NMDR that was not due to ED but to cytotoxicity. Consequently, acute toxicity tests that are commonly used in conventional risk assessment are irrelevant for the determination of safe doses in the case of EDCs

(Vandenberg et al., 2012). That is all the more the case when the studied endpoint is at a high level of biological organization, ecologically relevant (e.g., immunotoxicity, behavior) because the origin of impairment is often multifactorial. Ankley and Jensen (2014) underline that tests historically used in the field of regulatory ecotoxicology mostly focus on apical outcomes with little or no consideration given to the mechanistic basis for the observed effects. Because sensitivity varies deeply at different developmental stages, studies carried out with adults or nonsensitive stages can lead to false "safe" doses of a given chemical (Zoeller et al., 2012).

Data from field and laboratory works support the hypothesis that EDCs in the aquatic environment can impact the reproductive health of various aquatic organisms (Kidd et al., 2007; Sanchez et al., 2011; Gubbins et al., in Amiard-Triquet et al., 2013). On the other hand, a clear decline of biodiversity and abundance of wildlife is documented. However, making direct links between declines in species and endocrine disruption from chemical exposure remains a major challenge.

To characterize potential risks of EDCs, weight of evidence approaches should take into account exposure and pharmacokinetic data, screening assays (identification of potential interactions of EDCs with components of the endocrine system), mode of action studies (identification of toxicological pathways underlying adverse effects), long-term reproductive and developmental tests that define adverse effects, and overall toxicity. Borgert et al. (2011) have developed a quantitative weight of evidence evaluation, proposing specific criteria for evaluating the status of the data used in the weight of evidence approach. In line with this procedure, Van Der Kraak et al. (2014) have evaluated studies used for regulatory purposes as well as those in the open literature that report the effects of the herbicide atrazine on fish (31 species), amphibians (32 species), and reptiles (8 species). Their general conclusion was that atrazine does not adversely affect these aquatic organisms at concentrations that are present in surface waters. However, they have excluded studies on mixtures, because it was not possible to assign causality in these cases but in the real world, contaminants are always present in mixtures (Kortenkamp, 2007), including EDCs with agonistic or antagonistic modes of actions as also different POPs with unknown endocrine disrupting properties (e.g., Zimmer et al., 2011; Lyche et al., 2013).

Challenges posed by EDCs to conventional risk assessment are summarized in Table 15.2 after a review by Futran Fuhrman et al. (2015) devoted to human health but applicable to environmental health for most aspects.

Futran Fuhrman et al. (2015) have listed EDC research needs and priorities, underlying that EDC information gaps covered nearly the whole field of human toxicology and, a fortiori, ecotoxicology:

- Mechanisms and modes of action
- Sources and paths through environment

Table 15.2: Challenges posed by EDCs to conventional risk assessment

Lack of a universal definition of EDC
Incomplete data on most EDCs and potential EDCs
Humans have long-term exposure to a combination of EDCs
EDCs' exposure regimes are fundamentally different from traditional chemicals
Timing of exposure can be as important as dose size for EDCs
EDCs often display latent transgenerational effects
Gaps in knowledge exist regarding mechanism and mode of action for EDCs
EDC effects are not always categorically adverse, and "adversity" is not universally defined
No consensus on what endpoints are best to use in EDC toxicological studies
EDC mixtures can have unknown combination effects (i.e., additive, synergistic)
EDCs can show low dose effects at miniscule concentrations
Some EDCs have nonmonotonic dose-response curves that are less compatible with traditional risk assessment
EDCs in background environment threatens to contaminate study results
EDCs may not have a "safe" threshold for doses linked to detrimental effects
EDC uncertainties/unknowns hinder the accurate calculation of absolute risk (but permit assessment of relative risk)
EDC risk from a given source should be understood in context, in relation to: other sources, improving detection, and what is tolerable by society
Appropriate communication of EDC risk findings needs to balance precaution and alarm

After Futran Fuhrman et al. (2015).

- Response thresholds
- Combination effects
- Low-dose effects
- Dose–response (potentially nonmonotonic) curves in multiple species, ending with humans.

Thus it is crucial to develop approaches that should be used to prioritize EDCs for research on environmental and human health exposure and effects. Computational methods such as virtual screening, QSARs, and docking are valuable tools for prioritization (Vuorinena et al., 2013). Automated chemical screening technologies ("high-throughput screening assays") that expose living cells or isolated proteins to chemicals constitute the next step for prioritization. In vitro tests have received a peculiar attention as alternatives to animal testing but improving in vivo validation is needed to increase confidence for use of their results in regulatory decisions (Scholz et al., 2013; Murk et al., 2013). The specific effects elicited by EDCs are best evaluated by using in vivo tests on whole life-cycle and assays over two generations (EPA's EDSP tier 2 including fish and mysid life cycles, amphibian two-generation toxicity testing, USEPA, 2009; assays listed in level 5 of the OECD conceptual framework, OECD, 2012a).

Publicly available toxicity data resources in Europe and the United States are listed in Zhu et al. (2014). Data on EDCs are only part of these databases, and toxicity data on aquatic organisms are only a small minority. However, the strategies that are developed in this framework provide suggestions to manage the current data pool in ecotoxicity research (in vitro and in vivo assays). Novel techniques include data mining/generation, curation, storage, and management (Zhu et al., 2014). According to Ceger et al. (2014), who have developed a curated database of in vivo estrogenic activity, such a comprehensive database is critical to the success of a holistic approach of ED. It can be used to validate in vitro and in silico models, develop physiologically based pharmacokinetic models, evaluate dose- and duration-specific effects, and link in vivo effects to specific pathway perturbations.

The assessment of hazard/risk (Testai et al., 2013) remains highly controversial, as illustrated by Beausoleil et al. (2013) reporting on a workshop on low doses and nonmonotonic effects of endocrine disruptors. The positions about the existence of NMDRs, the importance of the definition of low doses and adverse effects, the use of all available data (including non-guideline, epidemiological, and biomonitoring data) in risk assessment varied depending on the affiliations of participants (research institutions, public interest, government, industry), highlighting that these controversies are partly fueled by conflicts of interest.

Independently of the improvement of risk assessment, others issues include the reduction of contaminant sources by improving wastewater treatment (Eggen et al., 2014) and developing sound and sustainable agriculture to reduce the input of pesticides. Trends toward green chemistry that intend to eliminate and/or minimize the usage and production of ecologically hazardous reagents and design alternative synthesis pathways must be encouraged. In line with these strategies, green pharmacy must be also envisaged (Chapter 16).

References

Amiard-Triquet, C., Amiard, J.C., Rainbow, P.S., 2013. Ecological Biomarkers: Indicators of Ecotoxicological Effects. CRC Press, Boca Raton.

Ankley, G.T., Bennett, R.S., Erickson, R.J., et al., 2010. Adverse outcome pathways: a conceptual framework to support ecotoxicology research and risk assessment. Environ. Toxicol. Chem. 29, 730–741.

Ankley, G.T., Gray, L.E., 2013. Cross-species conservation of endocrine pathways: a critical analysis of tier 1 fish and rat screening assays with 12 model chemicals. Environ. Toxicol. Chem. 32, 1084–1087.

Ankley, G.T., Jensen, K.M., 2014. A novel framework for interpretation of data from the fish short-term reproduction assay (FSTRA) for the detection of endocrine-disrupting chemicals. Environ. Toxicol. Chem. 33, 2529–2540.

Ayaki, T., Kawauchino, Y., Nishimura, C., et al., 2005. Sexual disruption in the freshwater crab (*Geothelphusa dehaani*). Integr. Comp. Biol. 45, 39–42.

Bäcklin, B.M., Eriksson, L., Olovsson, M., 2003. Histology of uterine leiomyoma and occurrence in relation to reproductive activity in the Baltic gray seal (*Halichoerus grypus*). Vet. Pathol. 40, 175–180.

Bahamonde, P.A., Munkittrick, K.R., Martyniuk, C.J., 2013. Intersex in teleost fish: are we distinguishing endocrine disruption from natural phenomena? Gen. Comp. Endocrinol. 192, 25–35.

Baker, N.J., Bancroft, B.A., Garcia, T.S., 2013. A meta-analysis of the effects of pesticides and fertilizers on survival and growth of amphibians. Sci. Total Environ. 449, 150–156.

Barbeau, M., Grecian, L., 2003. Occurrence of intersexuality in the amphipod *Corophium volutator* (pallas) in the upper Bay of Fundy, Canada. Crustaceana 76, 665–679.

Barber, L.B., Vajda, A.M., Douville, C., et al., 2012. Fish endocrine disruption responses to a major wastewater treatment facility upgrade. Environ. Sci. Technol. 46, 2121–2131.

Baumann, L., Knörr, S., Keiter, S., et al., 2014. Reversibility of endocrine disruption in zebrafish (*Danio rerio*) after discontinued exposure to the estrogen 17α-ethinylestradiol. Toxicol. Appl. Pharmacol. 278, 230–237.

Beausoleil, C., Ormsby, J.N., Gies, A., et al., 2013. Low dose effects and non-monotonic dose responses for endocrine active chemicals: science to practice workshop: workshop summary. Chemosphere 93, 847–856.
Beitel, S.C., Doering, J.A., Patterson, S.E., et al., 2014. Assessment of the sensitivity of three North American fish species to disruptors of steroidogenesis using in vitro tissue explants. Aquat. Toxicol. 152, 273–283.
Bergman, Å., Heindel, J.J., Jobling, S., et al., 2013a. State of the Science of Endocrine Disrupting Chemicals – 2012. UNEP and WHO. 260 p.
Bergman, A., Heindel, J.J., Kasten, T., et al., 2013b. The impact of endocrine disruption: a consensus statement on the state of the science? Environ. Health Perspect. 121, 104–106.
Birnbaum, L.S., 2012. Environmental chemicals: evaluating low-dose effects. Environ. Health Perspect. 120, A143–A144.
Blazer, V.S., Iwanowicz, D.D., Walsh, H.L., et al., 2014. Reproductive health indicators of fishes from Pennsylvania watersheds: association with chemicals of emerging concern. Environ. Monit. Assess. 186, 6471–7491.
Blazer, V.S., Iwanowicz, L.R., Henderson, H., et al., 2012. Reproductive endocrine disruption in smallmouth bass (*Micropterus dolomieu*) in the Potomac river basin: spatial and temporal comparisons of biological effects. Environ. Monit. Assess. 184, 4309–4334.
Boettcher, M., Kosmehl, T., Braunbeck, T., 2011. Low-dose effects and biphasic effect profiles: is trenbolone a genotoxicant? Mutat. Res. 723, 152–157.
Borgert, C.J., Mihaich, E.M., Ortego, L.S., et al., 2011. Hypothesis-driven weight of evidence framework for evaluating data within the USEPA's endocrine disruptor screening program. Regul. Toxicol. Pharmacol. 61, 185–191.
Bouwman, H., Booyens, P., Govender, D., et al., 2014. Chlorinated, brominated, and fluorinated organic pollutants in Nile crocodile eggs from the Kruger National Park, South Africa. Ecotoxicol. Environ. Saf. 104, 393–402.
Brack, W., 2011. Effect-directed analysis of complex environmental contamination. Handb. Environ. Chem. 15, 345.
Bradley, P.M., Journey, C.A., 2014. Assessment of endocrine-disrupting chemicals attenuation in a coastal plain stream prior to wastewater treatment plant closure. J. Am. Water Resour. Assoc. 50, 388–400.
Brander, S.M., 2013. Thinking outside the box: assessing endocrine disruption in aquatic life. In: Ahuja, S. (Ed.), Monitoring Water Quality Pollution Assessment, Analysis, and Remediation. Elsevier, Dordrecht, pp. 103–147.
Brodeur, J.C., Sassone, A., Hermida, G.N., et al., 2013. Environmentally-relevant concentrations of atrazine induce non-monotonic acceleration of developmental rate and increased size at metamorphosis in *Rhinella arenarum* tadpoles. Ecotoxicol. Environ. Saf. 92, 10–17.
Brodeur, J.C., Svartz, G., Perez-Coll, C.S., et al., 2009. Comparative susceptibility to atrazine of three developmental stages of *Rhinella arenarum* and influence on metamorphosis: non-monotonous acceleration of the time to climax and delayed tail resorption. Aquat. Toxicol. 91, 161–170.
Brown, S.B., Adams, B.A., Cyr, D.G., et al., 2004. Contaminant effects on the teleost fish thyroid. Environ. Toxicol. Chem. 23, 1680–1701.
Brucker-Davis, F., 1998. Effects of environmental synthetic chemicals on thyroid function. Thyroid 8, 827–856.
Burgess, R.M., Ho, K.T., Brack, W., Lamoree, M., 2013. Effects-directed analysis (EDA) and toxicity identification evaluation (TIE): complementary but different approaches for diagnosing causes of environmental toxicity. Environ. Toxicol. Chem. 32, 1935–1945.
Ceger, P., Kleinstreuer, N., Chang, X., et al., 2014. Development of a curated database of in vivo estrogenic activity. In: In Vitro Estrogenic Activity for EDSP21, NICEATM FutureTox. http://ntp.niehs.nih.gov/iccvam/meetings/9wc/posters/ceger-edlitrev-fd-text.pdf.
Chaube, R., Mishra, S., Singh, R.K., 2010. In vitro effects of lead nitrate on steroid profiles in the post-vitellogenic ovary of the catfish *Heteropneustes fossilis*. Toxicol. In Vitro 24, 1899–1904.
Chen, H., Hu, J., Yang, J., et al., 2010. Generation of a fluorescent transgenic zebrafish for detection of environmental estrogens. Aquat. Toxicol. 96, 53–61.
Coe, T.S., Söffker, M.K., Filby, A.L., et al., 2010. Impacts of early life exposure to estrogen on subsequent breeding behavior and reproductive success in zebrafish. Environ. Sci. Technol. 44, 6481–6487.
Colman, J.R., Baldwin, D., Johnson, L.L., et al., 2009. Effects of the synthetic estrogen, 17α-ethinylestradiol, on aggression and courtship behavior in male zebrafish (*Danio rerio*). Aquat. Toxicol. 91, 346–354.

Conolly, R., 2012. Computational modeling of nonmonotonic dose-response: Insight into minimum requirements. In: Low Dose Effects and Non-monotonic Dose Responses for Endocrine Active Chemicals: Science to Practice. Intern. Workshop, September 11-13, 2012, Berlin, Germany. https://eurl-ecvam.jrc.ec.europa.eu/laboratories-research/endocrine_disrupters/low-dose-conf-2012-pres/niehs-conference-presentations/conolly.pdf.

Conolly, R.B., Lutz, W.K., 2004. Nonmonotonic dose-response relationships: mechanistic basis, kinetic modeling, and implications for risk assessment. Toxicol. Sci. 77, 151–157.

Cox, J., Mann, M., 2007. Is proteomics the new genomics? Cell 130, 395–398.

Creusot, N., Budzinski, H., Balaguer, P., et al., 2013a. Effect-directed analysis of endocrine-disrupting compounds in multi-contaminated sediment: identification of novel ligands of estrogen and pregnane X receptors. Anal. Bioanal. Chem. 405, 2553–2566.

Creusot, N., Tapie, N., Piccini, B., et al., 2013b. Distribution of steroid- and dioxin-like activities between sediments, POCIS and SPMD in a French river subject to mixed pressures. Environ. Sci. Pollut. Res. 20, 2784–2794.

Creusot, N., Aït-Aïssa, S., Tapie, N., et al., 2014. Identification of synthetic steroids in river water downstream from pharmaceutical manufacture discharges based on a bioanalytical approach and passive sampling. Environ. Sci. Technol. 48, 3649–3657.

Davies, I.M., Vethaak, A.D., 2012. Integrated marine environmental monitoring of chemicals and their effects. In: ICES Cooperative Research Report No. 315, p. 277.

Desforges, J.P.W., Peachey, B.D.L., Sanderson, P.M., et al., 2010. Plasma vitellogenin in male teleost fish from 43 rivers worldwide is correlated with upstream human population size. Environ. Pollut. 158, 3279–3284.

Dévier, M.H., Mazellier, P., Aït-Aïssa, S., et al., 2011. New challenges in environmental analytical chemistry: Identification of toxic compounds in complex mixtures. C. R. Chim. 14, 766–779.

Diehl, J., Johnson, S.E., Xia, K., et al., 2012. The distribution of 4-nonylphenol in marine organisms of North American Pacific Coast estuaries. Chemosphere 87, 490–497.

Dietrich, D.R., von Aulock, S., Marquardt, H., et al., 2013. Scientifically unfounded precaution drives European Commission's recommendations on EDC regulation, while defying common sense, well-established science and risk assessment principles. Chem. Biol. Interact. 205, A1–A5.

Duft, M., Schulte-Oehlmann, U., Weltje, L., et al., 2003. Stimulated embryo production as a parameter of estrogenic exposure via sediments in the freshwater mudsnail *Potamopyrgus antipodarum*. Aquat. Toxicol. 64, 437–449.

Dzieweczynski, T.L., Campbell, B.A., Marks, J.M., et al., 2014. Acute exposure to 17α-ethinylestradiol alters boldness behavioral syndrome in female siamese fighting fish. Horm. Behav. 66, 577–584.

Eggen, R.I.L., Hollender, J., Joss, A., et al., 2014. Reducing the discharge of micropollutants in the aquatic environment: the benefits of upgrading wastewater treatment plants. Environ. Sci. Technol. 48, 7683–7689.

Elliott, S.M., Kiesling, R.L., Jorgenson, Z.G., et al., 2014. Fathead minnow and bluegill sunfish life-stage responses to 17ß-estradiol exposure in outdoor mesocosms. J. Am. Water Res. Ass. 50, 376–387.

Evanson, M., Van Der Kraak, G.J., 2001. Stimulatory effects of selected PAHs on testosterone production in goldfish and rainbow trout and possible mechanisms of action. Comp. Biochem. Physiol. 130C, 249–258.

Fetter, E., Krauss, M., Brion, F., et al., 2014. Effect-directed analysis for estrogenic compounds in a fluvial sediment sample using transgenic cyp19a1b-GFP zebrafish embryos. Aquat. Toxicol. 154, 221–229.

Filby, A.L., Paull, G.C., Searle, F., et al., 2012. Environmental estrogen-induced alterations of male aggression and dominance hierarchies in fish: a mechanistic analysis. Environ. Sci. Technol. 46, 3472–3479.

Fini, J.B., Le Mevel, S., Turque, N., et al., 2007. An in vivo multiwell-based fluorescent screen for monitoring vertebrate thyroid hormone disruption. Environ. Sci. Technol. 41, 5908–5914.

Ford, A.T., Fernandes, T.F., Robinson, C.D., et al., 2006. Can industrial pollution cause intersexuality in the amphipod, *Echinogammarus marinus*? Mar. Pollut. Bull. 53, 100–106.

Forget-Leray, J., Landriau, I., Minier, C., et al., 2005. Impact of endocrine toxicants on survival, development, and reproduction of the estuarine copepod *Eurytemora affinis* (poppe). Ecotoxicol. Environ. Saf. 60, 288–294.

Fossi Tankoua, O., Amiard-Triquet, C., Denis, F., et al., 2012. Physiological status and intersex in the endobenthic bivalve *Scrobicularia plana* from thirteen estuaries in northwest France. Environ. Pollut. 167, 70–77.

Freeman, J.L., Beccue, N., Rayburn, A.L., 2005. Differential metamorphosis alters the endocrine response in anuran larvae exposed to T3 and atrazine. Aquat. Toxicol. 75, 263–276.

Futran Fuhrman, V., Tal, A., Arnon, s., 2015. Why endocrine disrupting chemicals (EDCs) challenge traditional risk assessment and how to respond. J. Hazard. Mater. 286, 589–611.

Galloway, T.S., Depledge, M.H., 2001. Immunotoxicity in invertebrates: measurement and ecotoxicological relevance. Ecotoxicology 10, 5–23.

Gust, M., Buronfosse, T., Geffard, O., et al., 2010. In situ biomonitoring of freshwater quality using the New Zealand mudsnail *Potamopyrgus antipodarum* (gray) exposed to waste water treatment plant (WWTP) effluent discharges. Water Res. 44, 4517–4528.

Gust, M., Buronfosse, T., Giamberini, L., et al., 2009. Effects of fluoxetine on the reproduction of two prosobranch mollusks: *Potamopyrgus antipodarum* and *Valvata piscinalis*. Environ. Pollut. 157, 423–429.

Hallgren, P., Nicolle, A., Hansson, L.A., et al., 2014. Synthetic estrogen directly affects fish biomass and may indirectly disrupt aquatic food webs. Environ. Toxicol. Chem. 33, 930–936.

Harris, C.A., Hamilton, P.B., Runnalls, T.J., et al., 2011. The consequences of feminization in breeding groups of wild fish. Environ. Health Perspect. 119, 306–311.

Hass, U., Christiansen, S., Axelstad, M., et al., 2013. Input for the REACH-review in 2013 on endocrine disrupters. Danish Centre Endocr. Disrupters, 50 p. http://mst.dk/media/mst/9106721/rapport_input_for_the_reach-review.pdf (last accessed 09.01.15).

Hecker, M., Hollert, H., Cooper, R., et al., 2011. The OECD validation program of H295R steroidogen-esis assay: phase 3. Final inter-laboratory validation study. Environ. Sci. Pollut. Res. 18, 503–515.

Henneberg, A., Bender, K., Blaha, L., et al., 2014. Are in vitro methods for the detection of endocrine potentials in the aquatic environment predictive for in vivo effects? Outcomes of the projects SchussenAktiv and SchussenAktiv*plus* in the lake constance area, Germany. PLoS One 9 (6), e98307.

Hill, E.M., Evans, K.L., Horwood, J., et al., 2010. Profiles and some initial identifications of (anti)androgenic compounds in fish exposed to wastewater treatment works effluents. Environ. Sci. Technol. 44, 1137–1143.

Huang, Y., Zhu, L., Liu, G., 2006. The effects of bis(tributyltin) oxide on the development, reproduction and sex ratio of calanoid copepod *Pseudodiaptomus marinus*. Estuarine Coastal Shelf Sci. 69, 147–152.

Hyne, R.V., 2011. Review of the reproductive biology of amphipods and their endocrine regulation: identification of mechanistic pathways for reproductive toxicants. Environ. Toxicol. Chem. 30, 2647–2657.

IPCS (International Programme on Chemical Safety), 2002. Global Assessment of the State-of-the-Science of Endocrine Disruptors. World Health Organization, Geneva, Switzerland.

Jungmann, D., Ladewig, V., Ludwichowski, K.U., et al., 2004. Intersexuality in *Gammarus fossarum* KOCH-a common inducible phenomenon? Archiv für Hydrobiologie 159, 511–529.

Kidd, K.A., Blanchfield, P.J., Mills, K.H., et al., 2007. Collapse of a fish population after exposure to a synthetic estrogen. Proc. Natl. Acad. Sci. 104, 8897–8901.

Kinani, S., Bouchonnet, B., Creusot, N., et al., 2010. Bioanalytical characterisation of multiple endocrine- and dioxin-like activities in sediments from reference and impacted small rivers. Environ. Pollut. 158, 74–83.

Kortenkamp, A., 2007. Ten years of mixing cocktails: a review of combination effects of endocrine-disrupting chemicals. Environ. Health Perspect. 115 (Suppl. 1), 98–105.

Kortenkamp, A., Evans, R., Martin, O., et al., 2011. State of the Art Assessment of Endocrine Disrupters. Final Report. Project Contract Number 070307/2009/550687/SER/D, 135 p.

Kortenkamp, A., Martin, O., Evans, R., et al., 2012. Response to A critique of the European commission document, "state of the art assessment of endocrine disrupters" by Rhomberg and colleagues – letter to the editor. Crit. Rev. Toxicol. 42, 787–789.

Lange, A., Paull, G.C., Hamilton, P.B., et al., 2011. Implications of persistent exposure to treated wastewater effluent for breeding in wild roach (*Rutilus rutilus*) populations. Environ. Sci. Technol. 45, 1673–1679.

Langston, W.J., Burt, G.R., Chesman, B.S., 2007. Feminisation of male clams *Scrobicularia plana* from estuaries in Southwest UK and its induction by endocrine-disrupting chemicals. Mar. Ecol. Prog. Ser. 333, 173–184.

Langston, W.J., Pope, N.D., Davey, M., et al., 2012. Indicators of Endocrine Disruption in Estuaries. Determination of Pertinent Indicators for Environmental Monitoring: A Strategy for Europe (DIESE). http://www.mba.ac.uk/wp-content/uploads/2012/10/DIESE-Poster.pdf.

Letcher, R.J., Bustnes, J.O., Dietz, R., et al., 2010. Exposure and effects assessment of persistent organohalogen contaminants in arctic wildlife and fish. Sci. Total Environ. 408, 2995–3043.
Leusch, F.D.L., de Jager, C., Levi, Y., et al., 2010. Comparison of five in vitro bioassays to measure estrogenic activity in environmental waters. Environ. Sci. Technol. 44, 3853–3860.
Li, L., Andersen, M.E., Heber, S., et al., 2007. Non-monotonic dose–response relationship in steroid hormone receptor-mediated gene expression. J. Mol. Endocrinol. 38, 569–585.
Liu, S.Y., Jin, Q., Huang, X.H., et al., 2014. Disruption of zebrafish (*Danio rerio*) sexualdevelopment after full life-cycle exposure toenvironmental levels of triadimefon. Environ. Toxicol. Pharmacol. 37, 468–475.
Lo Piparo, E., Worth, A., 2010. Review of Qsar Models and Software Tools for Predicting Developmental and Reproductive Toxicity. JRC Sci. Technol. Rep. EUR 24522 EN https://eurl-ecvam.jrc.ec.europa.eu/laboratories-research/predictive_toxicology/doc/EUR_24522_EN.pdf (last accessed 31.12.14).
Lyche, J.L., Grześ, I.M., Karlsson, C., et al., 2013. Parental exposure to natural mixtures of POPs reduced embryo production and altered gene transcription in zebrafish embryos. Aquat. Toxicol. 126, 424–434.
McCurdy, D.G., Painter, D.C., Kopec, M.T., et al., 2008. Reproductive behavior of intersexes of an intertidal amphipod *Corophium volutator*. Invertebr. Biol. 127, 417–425.
McMahon, T., Halstead, N., Johnson, S., et al., 2011. The fungicide chlorothalonil is nonlinearly associated with corticosterone levels, immunity, and mortality in amphibians. Environ. Health Perspect. 119, 1098–1103.
Melmed, S., Williams, R.H., 2011. Williams Textbook of Endocrinology. Elsevier/Saunders, Philadelphia, PA.
Milla, S., Depiereux, S., Kestemont, P., 2011. The effects of estrogenic and androgenic endocrine disruptors on the immune system of fish: a review. Ecotoxicology 20, 305–319.
Milnes, M.R., Guillette Jr., L.J., 2008. Alligator tales: new lessons about environmental contaminants from a sentinel species. BioScience 58, 1027–1036.
Miyata, K., Ose, K., 2012. Thyroid hormone-disrupting effects and the amphibian metamorphosis assay. Rev. J. Toxicol. Pathol. 25, 1–9.
Moore, B.C., Forouhar, S., Kohno, S., et al., 2012. Gonadotropin-induced changes in oviducal mRNA expression levels of sex steroid hormone receptors and activin-related signaling factors in the alligator. Gen. Comp. Endocrinol. 175, 251–258.
Moore, C.G., Stevenson, J.M., 1991. The occurrence of intersexuality in harpacticoid copepods and its relationship with pollution. Mar. Pollut. Bull. 22, 72–74.
Murk, A.T.J., Rijntjes, E., Blaauboer, B.J., et al., 2013. Mechanism-based testing strategy using in vitro approaches for identification of thyroid hormone disrupting chemicals. Toxicol. In Vitro 27, 1320–1346.
OECD, 2006. Detailed Review Paper on Aquatic Arthropods in Life Cycle Toxicity Tests with an Emphasis on Developmental, Reproductive and Endocrine Disruptive Effects. OECD Series on Testing and Assessment. N. 55 http://www.oecd.org/chemicalsafety/testing/seriesontestingandassessmentecotoxicitytesting.htm.
OECD, 2011. Draft Summary Record of the Second Meeting of the Advisory Group on Endocrine Disrupter Testing and Assessment. OECD, Paris, France.
OECD, 2012a. Conceptual Framework for Testing and Assessment of Endocrine Disrupters. http://www.oecd.org/env/ehs/testing/OECD%20Conceptual%20Framework%20for%20Testing%20and%20Assessment%20of%20Endocrine%20Disrupters%20for%20the%20public%20website.pdf.
OECD, 2012b. Guidance Document on Standardised Test Guidelines for Evaluating Chemicals for Endocrine Disruption. Series on Testing and Assessment, No. 150, ENV/JM/MONO(2012)22.
OECD, 2013. *In vitro* and *ex vivo* assays for identification of modulators of thyroid hormone signaling. http://www.oecd.org/env/ehs/testing/CLEAN_Thyroid_scoping_Part%202_WNT_Dec_18%202013.pdf (last accessed 19.12.14).
Oehlmann, J., Schulte-Oehlmann, U., Tillmann, M., et al., 2000. Effects of endocrine disruptors on prosobranch snails (mollusca: gastropoda) in the laboratory. Part I. Bisphenol A and octylphenol as xeno-estrogens. Ecotoxicology 9, 383–397.
Ortiz-Zarragoitia, M., Cajaraville, M.P., 2010. Intersex and oocyte atresia in a mussel population from the biosphere's reserve of Urdaibai (Bay of Biscay). Ecotoxicol. Environ. Saf. 73, 693–701.

Pelley, J., 2003. Estrogen knocks out fish in whole-lake experiment. Environ. Sci. Technol. 37, 313A–314A.
Qin, L.T., Liu, S.S., Liu, H.L., et al., 2010. Support vector regression and least squares support vector regression for hormetic dose–response curves fitting. Chemosphere 78, 327–334.
Rhomberg, L.R., Goodman, J.E., Foster, W.G., et al., 2012. A critique of the European commission document, "state of the art assessment of endocrine disrupters". Crit. Rev. Toxicol. 42, 465–473.
Rohr, J.R., McCoy, K.A., 2010. A qualitative meta-analysis reveals consistent effects of atrazine on freshwater fish and Amphibians. Environ. Health Perspect. 118, 20–32.
Rotroff, D.M., Dix, D.J., Houck, K.A., et al., 2013. Using in vitro high throughput screening assays to identify potential endocrine-disrupting chemicals. Environ. Health Perspect. 121, 7–14.
Saaristo, M., Tomkins, P., Allinson, M., et al., 2013. An androgenic agricultural contaminant impairs female reproductive behaviour in a freshwater fish. PLoS One 8 (5), e62782.
Sanchez, W., Sremski, W., Piccini, B., et al., 2011. Adverse effects in wild fish living downstream from pharmaceutical manufacture discharges. Environ. Int. 37, 1342–1348.
Sánchez-Avila, J., Fernandez-Sanjuan, M., Vice, J., et al., 2011. Development of a multi-residue method for the determination of organic micropollutants in water, sediment and mussels using gas chromatography–tandem mass spectrometry. J. Chromatogr. A 1218, 6799–6811.
Sangalang, G., Jones, G., 1997. Oocytes in testis and intersex in lobsters (*Homarus americanus*) from Nova Scotian sites: natural or site related phenomenon? Can. Tech. l Rep. Fish Aquat. Sci. 2163, 46.
Sárria, M.P., Soares, J., Vieira, M.N., et al., 2011. Rapid-behaviour responses as a reliable indicator of estrogenic chemical toxicity in zebrafish juveniles. Chemosphere 85, 1543–1547.
Scholz, S., Renner, P., Belanger, S.E., et al., 2013. Alternatives to in vivo tests to detect endocrine disrupting chemicals (EDCs) in fish and amphibians – screening for estrogen, androgen and thyroid hormone disruption. Crit. Rev. Toxicol. 43, 45–72.
Segner, H., Casanova-Nakayama, A., Kase, R., et al., 2013. Impact of environmental estrogens on fish considering the diversity of estrogen signaling. Gen. Comp. Endocrinol. 191, 190–201.
Sharma, B., Patiño, R., 2009. Effects of cadmium on growth, metamorphosis and gonadal sex differentiation in tadpoles of the African clawed frog, *Xenopus laevis*. Chemosphere 76, 1048–1055.
Shenoy, K., 2014. Prenatal exposure to low doses of atrazine affects mating behaviors in male guppies. Horm. Behav. 66, 439–448.
Short, S., Yang, G., Guler, Y., et al., 2014. Crustacean intersexuality is feminization without demasculinization: implications for environmental toxicology. Environ. Sci. Technol. 48, 13520–13529.
Short, S., Yang, G., Kille, P., et al., 2012. A widespread and distinctive form of amphipod intersexuality not induced by known feminising parasites. Sex. Dev. 6, 320–324.
Sifkarovski, J., Grayfer, L., De Jesús Andino, F., et al., 2014. Negative effects of low dose atrazine exposure on the development of effective immunity to FV3 in *Xenopus laevis*. Dev. Comp. Immunol. 47, 52–58.
Söffker, M., Tyler, C., 2012. Endocrine disrupting chemicals and sexual behaviors in fish – a critical review on effects and possible consequences. Crit. Rev. Toxicol. 42, 653–668.
Staniszewska, M., Falkowska, L., Grabowski, P., et al., 2014. Bisphenol A, 4-tert-octylphenol, and 4-nonylphenol in the Gulf of Gdan´sk (Southern Baltic). Arch. Environ. Contam. Toxicol. 67, 335–347.
Stentiford, G., 2012. Histological intersex (ovotestis) in the European lobster *Homarus gammarus* and a commentary on its potential mechanistic basis. Dis. Aquat. Org. 100, 185.
Tanna, R.N., Tetreault, G.R., Bennett, C.J., et al., 2013. Occurrence and degree of intersex (testis–ova) in darters (*Etheostoma* spp.) across an urban gradient in the Grand River, Ontario, Canada. Environ. Toxicol. Chem. 32, 1981–1991.
Terrien, X., Fini, J.B., Demeneix, B.A., 2011. Generation of fluorescent zebrafish to study endocrine disruption and potential crosstalk between thyroid hormone and corticosteroids. Aquat. Toxicol. 105, 13–20.
Testai, E., Galli, C.L., Dekant, W., et al., 2013. A plea for risk assessment of endocrine disrupting chemicals. Toxicology 314, 51–59.
Titley-O'Neal, C.P., Munkittrick, K.R., MacDonald, B.A., 2011. The effects of organotin on female gastropods. J. Environ. Monit. 13, 2360–2388.

USEPA, 2009. Endocrine Disruptor Screening Program (EDSP); Announcing the Availability of the Tier 1 Screening Battery and Related Test Guidelines. United States Environmental Protection Agency, Washington (DC) http://docs.regulations.justia.com/entries/2009-10-21/E9-25348.pdf (last accessed 15.01.15).

Vajda, A.M., Barber, L.B., Gray, J.L., et al., 2011. Demasculinization of male fish by wastewater treatment plant effluent. Aquat. Toxicol. 103, 213–221.

Vandenberg, L.N., 2014. Non-monotonic dose responses in studies of endocrine disrupting chemicals: bisphenol A as a case study. Dose-Response 12, 259–276.

Vandenberg, L.N., Colborn, T., Hayes, T.B., et al., 2012. Hormones and endocrine-disrupting chemicals: lowdose effects and nonmonotonic dose responses. Endocr. Rev. 33, 378–455.

Van Der Kraak, G.J., Hosmer, A.J., Hanson, M.L., et al., 2014. Effects of atrazine in fish, amphibians, and reptiles: an analysis based on quantitative weight of evidence. Crit. Rev. Toxicol. 44, 1–66.

Vosges, M., Kah, O., Hinfray, N., et al., 2012. 17α-Ethinylestradiol and nonylphenol affect the development of forebrain GnRH neurons through an estrogen receptors-dependent pathway. Reprod. Toxicol. 33, 198–204.

Vuorinena, A., Odermatt, A., Schuster, D., 2013. In silico methods in the discovery of endocrine disrupting chemicals. J. Steroid. Biochem. Mol. Biol. 137, 18–26.

Weis, J.S., Candelmo, A., 2012. Pollutants and fish predator/prey behavior: a review of laboratory and field approaches. Curr. Zool. 58, 9–20.

Woodward, A.R., Percival, H.F., Rauschenberger, R.H., et al., 2011. Abnormal alligators and organochlorine pesticides in lake Apopka, Florida. In: Elliott, J.E., Bishop, C.A., Morrissey, C.A. (Eds.), Wildlife Ecotoxicology. Emerging Topics in Ecotoxicology, vol. 3. Springer, New York, pp. 153–187.

Xu, T., Close, D., Smartt, A., et al., 2014. Detection of organic compounds with whole-cell bioluminescent bioassays. In: Thouand, G., Marks, R. (Eds.), Bioluminescence: Fundamentals and Applications in Biotechnology, Volume 1. Advances in Biochemical Engineering/Biotechnology, vol. 144. Springer-Verlag, Berlin, pp. 111–151.

Yang, G., Kille, P., Ford, A.T., 2008. Infertility in a marine crustacean: have we been ignoring pollution impacts on male invertebrates? Aquat. Toxicol. 88, 81–87.

Yonkos, L.T., Friedel, E.A., Fisher, D.J., 2014. Intersex (testicular oocytes) in largemouth bass (*Micropterus salmoides*) on the Delmarva peninsula, USA. Environ. Toxicol. Chem. 33, 1163–1169.

Yoshikane, M., Kay, W.R., Shibata, Y., et al., 2006. Very high concentrations of DDE and toxaphene residues in crocodiles from the Ord River, Western Australia: an investigation into possible endocrine disruption. J. Environ. Monit. 8, 649–661.

Yu, L., Liu, C., Chen, Q., et al., 2014. Endocrine disruption and reproduction impairment in zebrafish after long-term exposure to DE-71. Environ. Toxicol. Chem. 33, 1354–1362.

Zhu, X.W., Liu, S.S., Qin, L.T., et al., 2013. Modeling non-monotonic dose-response relationships: model evaluation and hormetic quantities exploration. Ecotoxicol. Environ. Saf. 89, 130–136.

Zhu, H., Zhang, J., Kim, M.T., et al., 2014. Big data in chemical toxicity research: the use of high-throughput screening assays to identify potential toxicants. Chem. Res. Toxicol. 27, 1643–1651.

Zimmer, K.E., Montaño, M., Olsaker, I., et al., 2011. In vitro steroidogenic effects of mixtures of persistent organic pollutants (POPs) extracted from burbot (*Lota lota*) caught in two Norwegian lakes. Sci. Total Environ. 409, 2040–2048.

Zoeller, R.T., Brown, T.R., Doan, L.L., et al., 2012. Endocrine-disrupting chemicals and public health protection: a statement of principles from the endocrine society. Endocrinology 153, 4097–4110.

CHAPTER 16

Ecotoxicological Risk of Personal Care Products and Pharmaceuticals

M.J. Bebianno, M. Gonzalez-Rey

Abstract

Personal care products and pharmaceutical compounds were only detected in the aquatic environment in the past two decades. This chapter critically reviews the current knowledge of the main sources of these compounds and their distribution in the aquatic environment, their impact, and effects (acute, chronic and sublethal) on aquatic organisms, particularly on marine molluscs. In addition, the impact of mixtures and the importance of the use of new technologies such as "omics" are also highlighted along with environmental risk assessment and measures to mitigate their impact in the aquatic environment.

Keywords: Emerging contaminants; Green pharmacy; Hazard assessment; PCPs; Pharmaceutical compounds; Risk assessment; Toxic effects.

Chapter Outline

Aquatic Ecotoxicology. http://dx.doi.org/10.1016/B978-0-12-800949-9.00016-4

Only in the past two decades have environmental exposure to personal care products (PCPs) (moisturizers, lipsticks, shampoos, hair colors, deodorants, and toothpastes) used externally in hygiene and for beautification—and initially thought to be harmless—become a cause of concern because some of these compounds including the disinfectant triclosan, fragrances (musks), preservatives (parabens), and ultraviolet (UV) filters (e.g., methyl benzylidene camphor) provided evidence of endocrine disruptor effects on aquatic organisms. This misconception occurred because their development and application in health care was some of the greatest benefits of modern society, improving substantially the health and quality of individual lifestyle (Roberts and Bersuder, 2006; Boxall et al., 2012). However, the exponential growth of industrial economies and human population resulted in their becoming ubiquitous in the environment, where they are biologically active, persistent, and with the potential for bioaccumulation (Brauch and Rand, 2011). Although PCPs are commonly detected in the aquatic environment at higher concentrations than pharmaceutical compounds, little is known about their toxicity (Brauch and Rand, 2011).

Pharmaceutical products, on the other hand, are a group of chemicals in which the key component is the active pharmaceutical ingredient (API) used for diagnosis, treatment, mitigation, or prevention of diseases or health conditions at low concentrations (Anderson et al., 2004; Kümmerer, 2004). These products are generally categorized according to their therapeutic purpose (e.g., analgesics, antiasthmatics, antibiotics, antidepressants, antidiabetics, antiepileptics, antihypertensives, antilipidemics, antipsychotics, antiulcerants, autoimmune agents, non- and narcotics oncologics), and more than 3000 types of APIs are applied in human and veterinary medicine (Anderson et al., 2004; Kümmerer, 2004). Besides the APIs, these products also include excipients (essential components of a modern drug product that is technically "inactive" in a therapeutic sense).

Modern and more efficient APIs are constantly being developed, and consumers' demand is growing, in particular in emerging countries (IMS Health, 2011). Because the main goal in pharmaceutical development is to discover molecules that are resistant to metabolic degradation processes and therefore persist to exert the desired effect, parts of APIs are excreted unchanged, withstanding wastewater treatment processes and finally entering into aquatic ecosystems (Anderson et al., 2004). WHO (2010, 2011) has warned of the importance of the rational use of medicines: more than 50% of all medicines are prescribed, dispensed, or sold inappropriately, and half of all patients fail to take medicines correctly and the overuse, underuse, or misuse of medicines harm people and produce wastes, whereas others are conjugated or hydroxylated, producing metabolites whose biological activity might be greater than the parent compounds (Bound and Voulvolis, 2006; Fong and Molnar, 2008; Lazarra et al., 2012).

The first report featuring the occurrence of pharmaceuticals in the environment was published by Garrison et al. (1976) but, until recently, the lack of proper analytical techniques resulted in a misreading of their fate, occurrence, and impact on ecosystems (Fent et al., 2006). Because of their chemical properties, more than 150 APIs from all therapeutic classes have been screened at concentrations ranging from $ng\,L^{-1}$ to $\mu g\,L^{-1}$ in raw and treated wastewater, surface, drinking, and seawaters worldwide, including the Arctic environment (see Table 16.1), but concentrations in estuaries and coastal areas are scarce. Moreover, in aquatic and terrestrial organisms, soil, biosolids, and sediments, concentrations range between ng kg^{-1} to mg kg^{-1} (Kümmerer, 2009; Daughton, 2010; Santos et al., 2010; Brausch et al., 2012) (see Tables 16.2 and 16.3). However, ecotoxicological information is scarce and only available for <10% of the currently prescribed compounds and even fewer were subject to environmental risk assessment (Brausch et al., 2012).

As with PCPs, these compounds are considered an emergent class of environmental contaminants (Brausch et al., 2012) because they express certain features: occurrence and effects at very low concentrations, undefined effects of chronic low levels of exposure on ecosystem and human health, being unsuccessfully removed by waste treatment processes, and/or being subject to environmental dispersion and transportation after the release of wastewater treatment plant (WWTP) effluents (Wall, 2004). PCPs and APIs meet all these criteria and although not yet regulated they must be properly managed (Daughton, 2003; Khetan and Collins, 2007). Therefore, the release of PCPs and human pharmaceuticals into aquatic ecosystems is a serious environmental problem (Fong and Ford, 2014).

This chapter summarizes the current knowledge of the main sources of PCPs and pharmaceutical compounds, their distribution in the aquatic environment, and their impact and effects (acute, chronic, and sublethal), in aquatic organisms, particularly in marine mollusks. In addition, the impact of mixtures and the importance of the use of new technologies such as "omics" are highlighted. Finally, environmental risk assessment and measures to mitigate the impact of these compounds in the aquatic environment are reviewed.

16.1 Sources and Routes in the Environment

PCPs enter the aquatic environment by different pathways including emission from production effluents, release to surface waters from WWTP, aquaculture facilities, and soil run off (Boxall et al., 2012). The release of APIs and/or of their metabolites into the aquatic environment occurs through several exposure pathways (Figure 16.1), but the two major point sources (depicted "1" and "2" in the Figure) are domestic and hospital WWTP effluents resulting from the incomplete removal by WWTPs and/or from diffuse sources via out-of-date or unwanted medicines disposal either flushed in the sink and/or toilet or as household waste which ends up in landfill sites leaking into surrounding water compartments. Other APIs pathways are detailed by Halling-Sørensen et al. (1997), Christensen (1998), Ternes (1998), Jones et al. (2001), Andreozzi et al. (2003), Metcalfe et al. (2003a,b), Bendz et al. (2005), Castiglioni et al. (2005), Bound and Voulvoulis (2006), Lishman et al. (2006), Gagné et al. (2007), and Kümmerer (2009), Santos et al. (2010). However, recent studies identified significant impact of pharmaceutical manufacture discharges (Sanchez et al., 2011; Cardoso et al., 2014).

In humans, APIs may be metabolized to nontoxic (detoxification) or more toxic (metabolic activation) compounds (Li, 2001). The transformation and/or degradation of the parent

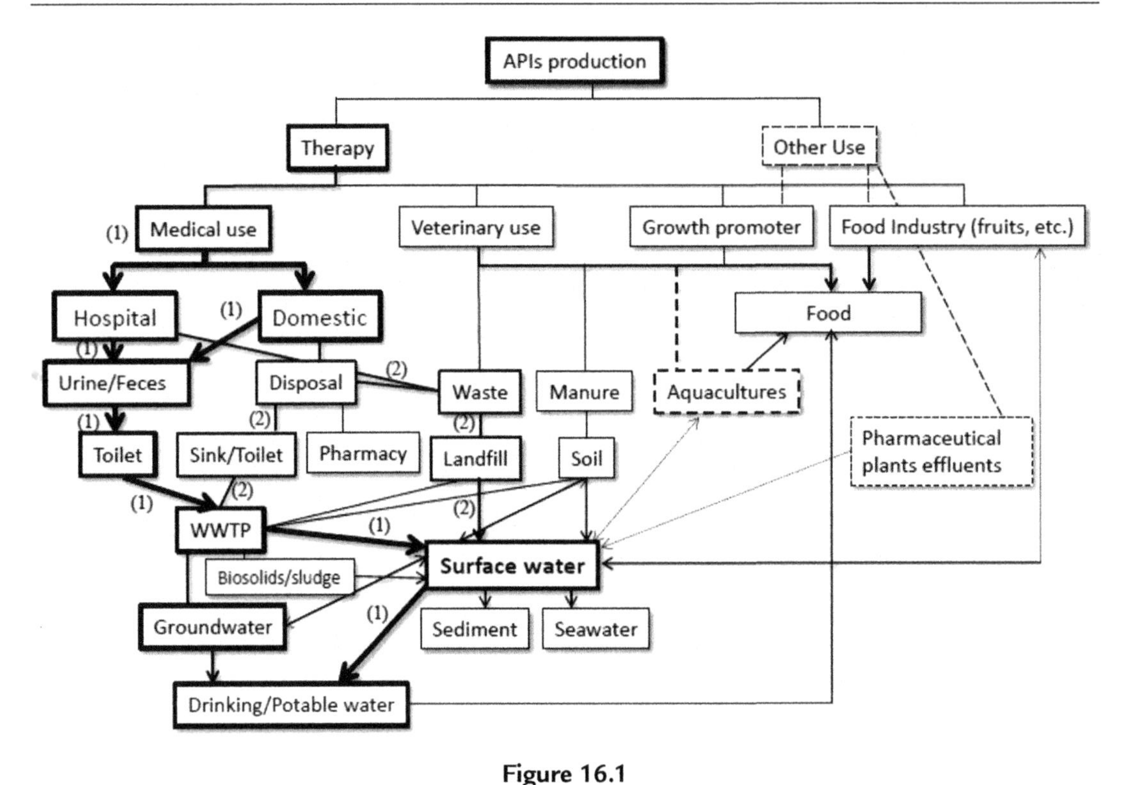

Figure 16.1

Pathways of APIs from production to the impact in the aquatic environment. *Adapted from Heberer (2002), Kümmerer (2004) and Bound and Voulvoulis (2007).*

compound when complete involve mineralization to low molecular weight compounds (i.e., carbon dioxide, sulfate, nitrate) or when the degradation is partial the process stops before mineralization and intermediates are formed (i.e., stable product of biotransformation more stable than the parent compound and with a comparatively higher accumulation potential) (Kümmerer, 2009; Santos et al., 2010). APIs are partially metabolized (biotransformed) by phase I or phase II reactions in the body, being excreted via urine and feces (Halling-Sørensen et al., 1997; Daughton and Ternes, 1999; Kümmerer, 2004). Last, an API parent compound may be excreted unchanged, as a glucuronide or sulfate conjugate, as a metabolite, or as a complex mixture of several metabolites. This metabolic strategy enhances excretion because it creates metabolites successively more polar and more water-soluble than the parent compound (Halling-Sørensen et al., 1997; Daughton and Ternes, 1999; Kümmerer, 2004; Josephy and Mannervik, 2006). API elimination rates vary depending on the individual, drug, and dosage (Bound and Voulvoulis, 2004, 2006).

16.2 Fate of PCPs and APIs in Wastewater Treatment Plants

WWTPs continuously receive wastes that include PCPs and APIs for which conventional treatment technologies were not specifically designed (Carballa et al., 2004; Reif et al., 2008).

A recent study carried out in Portugal revealed that among 62–65 compounds detected out of the 78 screened in effluents from several hospitals, nonsteroidal anti-inflammatory drugs (NSAIDs), analgesics, and antibiotics had the highest inputs, whereas antihypertensive, psychiatric drugs, or lipid regulators sources were attributed to public wastewater. The total mass load of pharmaceuticals into receiving surface waters was estimated between 5 and 14 g d^{-1} per 1000 inhabitants (Santos et al., 2013). Similarly, ibuprofen (IBU) was detected in four hospital effluents (two general, one pediatric, and one maternity) from the north of Portugal that release their effluents into the Douro river with an estimated mass load of 0.162 for maternity to 2.193 g d^{-1} for the general hospital (Paíga et al., 2013).

However, during biological treatment, some PCPs are degraded or removed through sorption to sludge (Ternes et al., 2004; Prasse et al., 2011). Recalcitrant PCPs may also be removed using tertiary treatment processes such as ozonation, activated carbon adsorption, or nanofiltration (Ternes et al., 2004), but in some cases, the WWTPs may increase their environmental risk (Boxall et al., 2012). In addition to the treatment process employed, the removal of APIs in WWTPs is affected by various factors such as chemical properties of specific compounds, age of the sludge, dilution, temperature of the raw sewage, environmental conditions, characteristics of the influent, hydraulics and solid retention time, and plant configuration (Kanda et al., 2003; Carballa et al., 2004; Clara et al., 2004; Kreuzinger et al., 2004; O'Brien and Dietrich, 2004; Clara et al., 2005a,b; Vieno et al., 2005; Tauxe-Wuersch et al., 2005; Castiglioni et al., 2006; Joss et al., 2006; Zuccato et al., 2006). Removal rates vary for each individual API and its elimination in WWTPs is often incomplete (ranging between 50% and 99% of efficiency), which ultimately enables them to reach the aquatic environment at continuous influx (Lindqvist et al., 2005; Castiglioni et al., 2006; De Lange et al., 2006; Han et al., 2006; Lishman et al., 2006; Gagné et al., 2007; Gros et al., 2010; Santos et al., 2013). Depending on the chemical properties of specific compounds, three different behaviors linking removal rates to therapeutic classes in WWTPs have been verified (Gros et al., 2010):

- an increase in APIs concentrations through its passage in the WWTP (macrolide antibiotics, antiepileptic carbamazepine, benzodiazepines, and selective serotonin reuptake inhibitors (SSRIs), such as fluoxetine presented poor or no elimination at WWTPs);
- no significant to medium removal (lipid regulators, fluoroquinolone, tetracycline antibiotics, anticholesterol, antihistaminics, β-blockers, β-agonists. and antidiabetic glibenclamide were partially degraded with either average removal efficiencies—40 to 60–70%—or in other situations were not eliminated at all);
- high removal efficiency (NSAIDs and the antihypertensive enalapril showed consistently high removal rates, with the exception of diclofenac).

API therapeutic classes detected in the environment described thoroughly by Santos et al. (2010) are ranked in the decreasing order: NSAIDs (16%), antibiotics (15%), antilipidemics (12%), sexual hormones (9%), antiepileptics (8%), β-blockers (8%), anxiolytics (4%), antidepressants (4%), antihypertensives (4%), antineoplastics (4%), antacids (3%), X-ray contrast media (3%), antiasthmatics (3%), veterinary products (3%), oral antidiabetics (3%), and antipsychotics (1%).

16.3 API Transformation and Degradation in the Environment

APIs can undergo different structural changes such as biotic (biotransformation, biodegradation) and nonbiotic (oxidation, hydrolysis, and photolysis) transformation processes after their introduction in different environmental compartments such as surface waters, soil, or sewage, which alter their physicochemical and pharmaceutical properties (Figure 16.2). Bacteria and fungi are the organisms most capable of degrading organic compounds. Bacteria are particularly efficient in the aquatic environment and fungi in soils. For this reason, microbial degradation will be slower in surface water than in the sewage system because of its lower bacterial density and lower diversity (Kümmerer, 2009). Nevertheless, Sammartino et al. (2008) report that APIs are not subjected to microbial degradation; on the contrary, they can reduce the bacterial flora present in the receiving medium and therefore the nonbiotic process of photodegradation is considered the main way to eliminate APIs and their residues from the environment. Photodegradation depends on UV radiation energy and intensity, latitude (seasonal and geographical variation), physical state of APIs, chemical parameters such as pH, and presence of photosensitizers (e.g., nitrates, humic acids) (Andreozzi et al.,

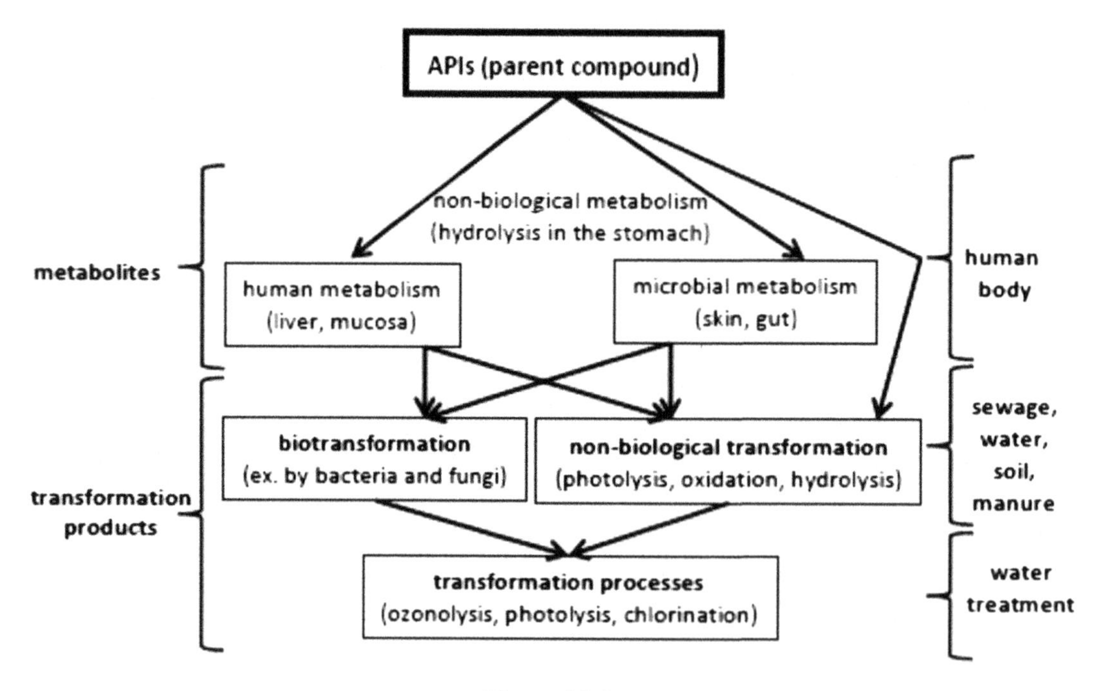

Figure 16.2
APIs metabolites and transformation products. *Adapted from Kümmerer (2009).*

2003; Sammartino et al., 2008; Santos et al., 2010). Even if these transformation processes usually lead to decreased API toxicity in the environment, the inverse also occurs for instance in the case of pro-drugs (Kümmerer, 2009).

16.4 Personal Care Products and Active Pharmaceutical Ingredients in the Aquatic Environment

The environmental fate of PCPs and APIs after effluent release into surface waters is largely unknown, particularly in the marine environment (McEneff et al., 2014). APIs need to be quantified in situ to assess pharmaceutical residues in marine environmental matrices. It is anticipated that their physicochemical properties play an important role in the fate and transport kinetics in coastal marine systems and that hydrodynamics have also a great influence on the magnitude of API concentrations (Bayen et al., 2013). In addition, APIs having a long half-life in water are distributed more widely in coastal areas. For instance, carbamazepine with a half-life of 900 h in surface waters is one of the most frequently detected compounds in coastal areas (Benotti and Brownawell, 2009; Matamoros et al., 2009). Even molecules that have a short half-life are present at relatively high concentrations due to the continuous emissions. Moreover, APIs with a relatively high partition coefficient such as simvastatin, sertraline, fluoxetine, paroxetine, estrogens (estrone (E1), estradiol (E2), ethinyl-estradiol (EE2)), and PCPs, such as triclosan, are likely to bind to suspended particles and subsequently settle and can be resuspended. In general, the APIs more frequently detected in receiving waters match with those more ubiquitous in wastewater effluents, but in most cases after great dilution. Yet, the highly consumed analgesics and NSAIDs (such as IBU) are an exception. Even after removal by WWTPs and although not persistent in surface waters because of its short half-life (360 h), IBU is the NSAID most frequently detected in seawater, as is carboxy-ibuprofen as the dominant IBU metabolite (Weigel et al., 2004), indicating the importance of emission sources (Gros et al., 2010; Bayen et al., 2013).

Table 16.1 shows data on the occurrence of PCPs and APIs and their metabolites in surface and seawaters in different world regions. In most cases, there is no relationship between the amount of APIs sold and the levels detected in the aquatic environment, indicating that other factors such as dosage, pharmacokinetics, transformation, or removal in WWTP and environmental fate (dilution) are critical on the occurrence of these compounds in estuarine and marine environments (Klosterhaus et al., 2013). Extremely high levels of APIs were detected in India and China (two countries that are responsible for half of the world production of APIs) in areas directly impacted by pharmaceutical plant effluents (Corcoran et al., 2010; Sanchez et al., 2011; Cardoso et al., 2014). Also, in France, severe signs of endocrine disruption were detected in fish living downstream from a pharmaceutical discharge as also fish population disturbances (Sanchez et al., 2011).

Table 16.1: Concentration of PCPs and APIs in Surface Waters

Compounds	Classes	Country	Environment	Concentration ng L^{-1}	References
a-Hydroxy metoprolol	β-blocker	Norway	Tromsø Sound	25–108	Langford and Thomas (2011)
Amitriptyline	Antidepressant	Portugal	Arade Estuary	<2	Gonzalez-Rey et al. (in press)
Atenolol	β-blocker	Belgium	North Sea	293	Wille et al. (2010)
Atenolol	β-blocker	United States	San Francisco Bay	3.8–37.0	Klosterhaus et al. (2013)
Atenolol	β-blocker	United States	Seawater	2.3	Vidal-Dorsch et al. (2012)
Benzoylecgonine	Cocaine metabolite	United States	San Francisco Bay	2.8–7.2	Klosterhaus et al. (2013)
Bezafibrate	Lipid regulator	Belgium	North Sea	<18	Wille et al. (2010)
Bezafibrate	Lipid regulator	Spain	Turia River	4	Carmona et al. (2014)
Bisphenol A	PCP	Spain	Turia River	57	Carmona et al. (2014)
BPA	PCP	Singapore	Seawater	<96–694	Bayen et al. (2013)
Butylparaben	PCP	Spain	Turia River	14	Carmona et al. (2014)
Caffeine	Stimulant	Norway	Tromsø	7–87	Weigel et al. (2004)
Caffeine	Stimulant	United States	San Francisco Bay	2.8–7.2	Klosterhaus et al. (2013)
Carbamazepine	Antiepileptic	Belgium	Belgium Coastal region	Up to 732	Wille et al. (2010)
Carbamazepine	Antiepileptic	Belgium	North Sea	321	Wille et al. (2010)
Carbamazepine	Antiepileptic	France	Mediterranean Sea	10–40	Togola and Budzinski (2008)
Carbamazepine	Antiepileptic	Ireland	Irish Coast	0.21–1.56[a]	McEneff et al. (2014)
Carbamazepine	Antiepileptic	Norway	Tromsø Sound	ND–10	Langford and Thomas (2011)
Carbamazepine	Antiepileptic	Portugal	Arade Estuary	31	Gonzalez-Rey et al. (in press)
Carbamazepine	Antiepileptic	Singapore	Singapore marine environment	<0.28–10.9	Bayen et al. (2013)
Carbamazepine	Antiepileptic	United States	San Francisco Bay	5.2–44.2	Klosterhaus et al. (2013)
Carbamazepine-10,11-epoxide	Antiepileptic	Norway	Tromsø Sound	14–77	Langford and Thomas (2011)

Table 16.1: Concentration of PCPs and APIs in Surface Waters—cont'd

Compounds	Classes	Country	Environment	Concentration ng L^{-1}	References
Chloramphenicol	Antibiotic	Spain	Turia River	3	Carmona et al. (2014)
Citalopram	SSRI	Canada	Grand River	4–206	Metcalfe et al. (2010)
Citalopram	SSRI	India	Isakavagu-Nakkavagu Rivers	40–7600	Fick et al. (2009)
Citalopram	SSRI	Spain	Guadarrama River	13–120	Alonso et al. (2010)
Citalopram	SSRI	Spain	Jarama River	3–43	Alonso et al. (2010)
Citalopram	SSRI	Spain	Manzanares River	58	Alonso et al. (2010)
Citalopram	SSRI	United States	Surface water	40–90	Schultz and Furlong (2008)
Citalopram	SSRI	United States	Surface water	5.7–63.7	USEPA (2007)
Citalopram	SSRI		Receiving water	0.011	Lajeunesse et al. (2008)
Citalopram	SSRI	Canada	St. Lawrence	3.4–11.5	Lajeunesse et al. (2008)
Citalopram	SSRI	India	Surface Water	2000–8000	Fick et al. (2009)
Citalopram	SSRI	United States	Surface Water	<0.5–219	Schultz et al. (2010)
Clofibric acid	Lipid regulator	United Kingdom	UK estuaries	<20–98.4	Thomas and Hilton (2004)
Clofibric acid	Lipid regulator	Spain	Turia River	17	Carmona et al. (2014)
Clotrirr	PCP	United Kingdom	UK estuaries	<1–22	Thomas and Hilton (2004)
DEET	PCP	United States	San Francisco Bay	4.9–21.0	Klosterhaus et al. (2013)
Dextropropoxyphene	PCP	United Kingdom	UK Estuaries	<8–33	Thomas and Hilton (2004)
Diclofenac	NSAID	Ireland	Irish Coast	0.02–0.60[a]	McEneff et al. (2014)
Diclofenac	NSAID	Portugal	Arade Estuary	31	Gonzalez-Rey et al. (in press)
Diclofenac	NSAID	Singapore	Marine environment	<1.5–11.6	Bayen et al. (2013)
Diclofenac	NSAID	Spain	Turia River	49	Carmona et al. (2014)
Diclofenac	NSAID	United Kingdom	UK Estuaries	<8–95	Thomas and Hilton (2004)
Diltiazem	Antihypertensive	Singapore	Seawater	<0.9–1.7	Bayen et al. (2013)
Diltiazem	Antihypertensive	United States	San Francisco Bay	0.4–3.5	Klosterhaus et al. (2013)

Continued

Table 16.1: Concentration of PCPs and APIs in Surface Waters—cont'd

Compounds	Classes	Country	Environment	Concentration $ng\,L^{-1}$	References
Diphenhydr-amine	Antihistamine	Singapore	Seawater	<0.3–4.6	Bayen et al. (2013).
E1	Contraceptive	Singapore	Seawater	<0.8–11	Bayen et al. (2013)
Erythromycin-H_2O	Antibiotic	United States	San Francisco Bay	1.0–12.1	Klosterhaus et al. (2013)
Ethylparaben	PCP	Spain	Turia River	16	Carmona et al. (2014)
Flufenamic acid	PCP	Spain	Turia River	21	Carmona et al. (2014)
Fluoxetine	SSRI	Canada	Grand River	4–141	Metcalfe et al. (2010)
Fluoxetine	SSRI	Canada	Hamilton Harbor Little River	13–46	Lajeunesse et al. (2008)
Fluoxetine	SSRI	Canada	St. Lawrence	0.42–1.3	Lajeunesse et al. (2008)
Fluoxetine	SSRI	Norway	Thromso	3	Vasskog et al. (2008)
Fluoxetine	SSRI	Portugal	Arade Estuary	<2	Gonzalez-Rey et al. (in press)
Fluoxetine	SSRI	Spain	Ebro River	3	Gros et al. (2010)
Fluoxetine	SSRI	Spain	Guadarrama River	8–44	Alonso et al. (2010)
Fluoxetine	SSRI	Spain	Henares River	11	Alonso et al. (2010)
Fluoxetine	SSRI	Spain	Jarama River	13–16	Alonso et al. (2010)
Fluoxetine	SSRI	Spain	Manzanares River	22	Alonso et al. (2010).
Fluoxetine	SSRI	Spain	Tajo River	12	Alonso et al. (2010)
Fluoxetine	SSRI	United States	Chesapeake Bay	3	Pait et al. (2006)
Fluoxetine	SSRI	United States	Surface water	12	Kolpin et al. (2002)
Fluoxetine	SSRI	United States	Surface water	12–20	Schultz and Furlong (2008)
Fluoxetine	SSRI	United States	Surface water	<0.5–43.2	Schultz et al. (2010)
Fluoxetine	SSRI	United States	Surface water	5.6–48	USEPA (2007)
Fluvoxamine	SSRI	Norway	Seawater	0.5–0.8	Vasskog et al. (2008)
Fluvoxamine	SSRI	United States	Surface water	<0.5–4.6	Schultz et al. (2010)
Fluvoxamine	SSRI	United States	Surface water	3.7	USEPA (2007)

Table 16.1: Concentration of PCPs and APIs in Surface Waters—cont'd

Compounds	Classes	Country	Environment	Concentration ng L^{-1}	References
Gemfibrozil	Lipid regulator	Ireland	Irish Coast	0.04–0.78[a]	McEneff et al. (2014)
Gemfibrozil	Lipid regulator	Portugal	Arade Estuary	10	Gonzalez-Rey et al. (in press)
Gemfibrozil	Lipid regulator	Singapore	Marine environment	<0.09–19.8	Bayen et al. (2013)
Gemfibrozil	Lipid regulator	Singapore	Seawater	<0.09–19.8	Bayen et al. (2013)
Gemfibrozil	Lipid regulator	Spain	Turia River	77	Carmona et al. (2014)
Gemfibrozil	Lipid regulator	United States	San Francisco Bay	12–38.2	Klosterhaus et al. (2013)
Gemfibrozil	Lipid regulator	United States	Seawater	3	Vidal-Dorsch et al. (2012)
Ibuprofen	NSAID	Norway	Tromsø	2	Weigel et al. (2002)
Ibuprofen	NSAID	Norway	Tromsø	0.01	Weigel et al. (2004)
Ibuprofen	NSAID	Norway	Tromsø	0.8–7.8	Weigel et al. (2004)
Ibuprofen	NSAID	Portugal	Ave River	ND–362	Paíga et al. (2013)
Ibuprofen	NSAID	Portugal	Aveiro Lagoon	242 ± 9	Paíga et al. (2013)
Ibuprofen	NSAID	Portugal	Douro River	ND–239	Paíga et al. (2013)
Ibuprofen	NSAID	Portugal	Leça River	ND–265	Paíga et al. (2013)
Ibuprofen	NSAID	Portugal	Lima River	42–739	Paíga et al. (2013)
Ibuprofen	NSAID	Portugal	Minho River	204 ± 10	Paíga et al. (2013)
Ibuprofen	NSAID	Portugal	Tamega River	359	Paíga et al. (2013)
Ibuprofen	NSAID	Portugal	Ulma River	173	Paíga et al. (2013)
Ibuprofen	NSAID	Singapore	Marine environment	<2.2–9.1	Bayen et al. (2013)
Ibuprofen	NSAID	Spain	Turia River	830	Carmona et al. (2014)
Ibuprofen	NSAID	United Kingdom	UK estuaries	<8–928	Thomas and Hilton (2004)
Ibuprofen	NSAID	Portugal	Arade Estuary	28	Gonzalez-Rey et al. (in press)
Indomethacin	NSAID	Spain	Turia River	3	Carmona et al. (2014)

Continued

Table 16.1: Concentration of PCPs and APIs in Surface Waters—cont'd

Compounds	Classes	Country	Environment	Concentration ng L^{-1}	References
Mefenamic acid	NSAID	Ireland	Irish Coast	0.03–0.75[a]	McEneff et al. (2014)
Mefenamic acid	NSAID	United Kingdom	UK estuaries	<20–196	Thomas and Hilton (2004)
Meprobamate	Anxiolytic	United States	San Francisco Bay	6.2–36.1	Klosterhaus et al. (2013)
Meprobamate	Anxiolytic	United States	Seawater	0.4	Vidal-Dorsch et al. (2012)
Methylparaben	PCP	Spain	Turia River	119	Carmona et al. (2014)
N,N-diethyl-3-toluamide (DEET)	Insecticide	Norway	Tromsø	0.4–13	Weigel et al. (2004)
N,N-diethyl-*m*-toluamide (DEET)	Insecticide	USA	San Francisco Bay	4.9–21	Klosterhaus et al. (2013)
Naproxen	NSAID	Norway	Turia River	278	Carmona et al. (2014)
Naproxen	NSAID	Portugal	Arade Estuary	17	Gonzalez-Rey et al. (in press)
Naproxen	NSAID	Singapore	Marine environment	<0.9–7.3	Bayen et al., 2013.
Naproxen	NSAID	Singapore	Seawater	<0.9–7.3	Bayen et al. (2013).
Naproxen	NSAID	USA	Seawater	3.5	Vidal-Dorsch et al. (2012)
Nordazepam	Anxiolytic	Portugal	Arade Estuary	3	Gonzalez-Rey et al. (in press)
Paracetamol	NSAID	Portugal	Arade Estuary	88	Gonzalez-Rey et al. (in press)
Paroxetine	SSRI	Canada	St. Lawrence	1.3–3	Lajeunesse et al. (2008)
Paroxetine	SSRI	Norway	Seawater	0.6–1.4	Vasskog et al. (2008)
Paroxetine	SSRI	United States	Surface water	1.6	USEPA (2007)
Propranolol	β-blocker	Belgium	North Sea	24	Wille et al. (2010)
Propranolol	β-blocker	United Kingdom	UK estuaries	<4–56	Thomas and Hilton, (2004)
Propylparaben	β-blocker	Spain	Turia River	145	Carmona et al. (2014)
Salicylic acid	NSAID	Belgium	North Sea	855	Wille et al. (2010)
Salicylic acid	NSAID	Spain	Turia river	70	Carmona et al. (2014).

Table 16.1: Concentration of PCPs and APIs in Surface Waters—cont'd

Compounds	Classes	Country	Environment	Concentration ng L^{-1}	References
Sertraline	SSRI	Canada	Grand River	6–17	Metcalfe et al. (2010)
Sertraline	SSRI	Canada	St. Lawrence	0.84–2.4	Lajeunesse et al. (2008)
Sertraline	SSRI	Norway	Seawater	<0.16	Vasskog et al. (2008)
Sertraline	SSRI	Spain	Llobregat River	11	Huerta-Fontela et al. (2011)
Sertraline	SSRI	United States	Surface water	33–49	Schultz and Furlong (2008)
Sertraline	SSRI	United States	Surface water	<0.5–37.5	Schultz et al. (2010)
Sertraline	SSRI	United States	Surface water	1.1–34.9	USEPA (2007)
Sulfamethoxazole	Antibiotic	Belgium	North Sea	96	Wille et al. (2010)
Sulfamethoxazole	Antibiotic	United States	Seawater	0.8	Vidal-Dorsch et al. (2012)
Tamoxifen	Anticancer drug	United Kingdom	UK estuaries	<4–71	Thomas and Hilton (2004)
THCCOOH	PCP	Spain	Turia River	7	Carmona et al. (2014)
Thiamphenicol	Antibiotic	Spain	Turia River	7	Carmona et al. (2014)
Triamterene	Diuretic	United States	San Francisco Bay	1.1–9.6	Klosterhaus et al. (2013)
Triclocarban	PCP	Spain	Turia River	4	Carmona et al. (2014)
Triclosan	PCP	Germany	Seawater	0.008–6.87	Xie et al. (2008)
Triclosan	PCP	Hong Kong	Seawater	15–110	Wu et al. (2007)
Triclosan	PCP	Singapore	Marine environment	<0.55–10.5	Bayen et al. (2013)
Triclosan	PCP	Singapore	Seawater	<0.55–10.5	Bayen et al. (2013)
Triclosan	PCP	Spain	Turia River	1	Carmona et al. (2014)
Triclosan	PCP	United States	Seawater	1.7	Vidal-Dorsch et al. (2012)
Trimethoprim	Antibiotic	Belgium	North Sea	29	Wille et al. (2010)
Trimethoprim	Antibiotic	Ireland	Irish Coast	0.10–1.01[a]	McEneff et al. (2014)
Trimethoprim	Antibiotic	United Kingdom	UK estuaries	<4–569	Thomas and Hilton (2004)
Trimethoprim	Antibiotic	United States	Seawater	0.5	Vidal-Dorsch et al. (2012)

Continued

Table 16.1: Concentration of PCPs and APIs in Surface Waters—cont'd

Compounds	Classes	Country	Environment	Concentration ng L^{-1}	References
Valsartan	High Blood Pressure	United States	San Francisco Bay	18–92.1	Klosterhaus et al. (2013)
Venlafaxine	SSRI	Canada	St. Lawrence River	0.013–0.045	Lajeunesse et al. (2008)
Venlafaxine	SSRI	United States	Grand River Southern Ontario	0.5	Metcalfe et al. (2010)
Venlafaxine	SSRI	United States	Colorado	0.22	Schultz et al. (2010)
Venlafaxine	SSRI	United States	Texas	1.31	Schultz and Furlong (2008)
Warfarin	Anticoagulant	Spain	Turia River	1	Carmona et al. (2014)

ND, no data.
[a]μg L^{-1}.

Table 16.2 shows data on the occurrence of PCPs and APIs in sediments. Chen et al. (2013) identified 330 pharmaceuticals from nine therapeutic groups in surface sediments from three Chinese rivers, 291 of which were common to all rivers, but levels were not quantified. Moreover, in most cases the compounds most frequently detected in sediments are not the ones detected at highest concentrations in water. Triclocarban, for instance, was not detected in San Francisco Bay surface waters but was strongly adsorbed to sediments (Klosterhaus et al., 2013).

Table 16.2: PCP and API concentrations in sediments

Compounds	Country	Environment	Concentration ng g^{-1} dry weight	References
a-Hydroxy metoprolol	Norway	Tromsø Sound	1–3	Langford and Thomas (2011)
Atenolol	Brazil	Todos os Santos Bay	0.49–9.84	Beretta et al. (2014)
Bisphenol A	Spain	Turia River	7	Carmona et al. (2014)
Butylparaben	Spain	Turia River	3	Carmona et al. (2014)
Caffeine	Brazil	Todos os Santos Bay	0.28–23.4	Beretta et al. (2014)
Caffeine	United States	San Francisco Bay	<1.5–29.7	Klosterhaus et al. (2013)
Carbamazepine	Brazil	Todos os Santos Bay	<0.10–4.4	Beretta et al. (2014)
Carbamazepine	France	Mediterranean Sea	3.5	Martinez Bueno et al. (2013)
Carbamazepine	Portugal	Arade River	<0.48	Blasco et al. (2013)
Carbamazepine	Spain	Chiclana	1.20	Blasco et al. (2013)
Carbamazepine	Spain	Puente Suazo	<0.48	Blasco et al. (2013)
Carbamazepine	Spain	Rio San Pedro	5.09	Blasco et al. (2013)

Table 16.2: PCP and API concentrations in sediments—cont'd

Compounds	Country	Environment	Concentration $ng\,g^{-1}$ dry weight	References
Carbamazepine	Spain	Trocadero	1.38	Blasco et al. (2013)
Chloramphenicol	Spain	Turia River	6	Carmona et al. (2014)
Clofibric acid	Spain	Turia River	10	Carmona et al. (2014)
Citalopram	United States	Chesapeake Bay	0.40–2.88	USEPA (2007)
Diazepam	Brazil	Todos os Santos Bay	<0.10–0.71	Beretta et al. (2014)
Diclofenac	Brazil	Todos os Santos Bay	<0.10–1.06	Beretta et al. (2014)
Diclofenac	Portugal	Ria Formosa	<1.19	Blasco et al. (2013)
Diclofenac	Spain	Chiclana	1.25	Blasco et al. (2013)
Diclofenac	Spain	Puente Suazo	2.06	Blasco et al. (2013)
Diclofenac	Spain	Rio San Pedro	4.00	Blasco et al.(2013)
Diclofenac	Spain	Trocadero	3.17	Blasco et al. (2013)
Diclofenac	Spain	Turia river	15	Carmona et al. (2014)
Erythromycin	Brazil	Todos os Santos Bay	<0.10–2.29	Beretta et al. (2014)
Ethylparaben	Spain	Turia River	23	Carmona et al. (2014)
Flufenamic acid	Spain	Turia River	7	Carmona et al. (2014)
Fluoxetine	United States	Chesapeake Bay	1.26	USEPA (2007)
Galaxolide	Brazil	Todos os Santos Bay	2.39–14.54	Beretta et al. (2014)
Gemfibrozil	Spain	Turia River	6	Carmona et al. (2014)
Ibuprofen	Brazil	Todos os Santos Bay	0.77–18.8	Beretta et al. (2014)
Ibuprofen	Portugal	Arade River	<0.41–0.62	Blasco et al. (2013)
Ibuprofen	Spain	Chiclana	1.23	Blasco et al. (2013)
Ibuprofen	Spain	Puente Suazo	1.01	Blasco et al. (2013)
Ibuprofen	Spain	Rio San Pedro	4.58	Blasco et al. (2013)
Ibuprofen	Spain	Trocadero	1.10	Blasco et al. (2013)
Ibuprofen	Spain	Turia River	30	Carmona et al. (2014)
Indomethacin	Spain	Turia River	4	Carmona et al. (2014)
Methylparaben	Spain	Turia River	152	Carmona et al. (2014)
Naproxen	Spain	Turia River	13	Carmona et al. (2014)
Propylparaben	Spain	Turia River	9	Carmona et al. (2014)
Salicylic acid	Spain	Turia River	318	Carmona et al. (2014)
Sertraline	United States		0.09–0.67	USEPA (2007)
Simvastatin hydroxy carboxylic acid	Normay	Tromsø Sound	2	Langford and Thomas (2011)
THC	Spain	Turia River	42	Carmona et al. (2014)
THCCOOH	Spain	Turia River	5	Carmona et al. (2014)
Thiamphenicol	Spain	Turia River	2	Carmona et al. (2014)
Tonalide	Brazil	Todos os Santos Bay	2.81–27.9	Beretta et al. (2014)
Triamterene	United States	San Francisco Bay	0.3–10.8	Klosterhaus et al. (2013)
Triclocarban	United States	San Francisco Bay	<3–32.7	Klosterhaus et al. (2013)
Triclosan	Spain	Turia River	6	Carmona et al. (2014)
Warfarin	Spain	Turia River	9	Carmona et al. (2014)

One of the main concerns arising from APIs released into surface waters is their potential for bioaccumulation that is age-, species-, diet-, habitat-, and reproductive cycle–dependent (Meredith-Williams et al., 2012). Their polar nature makes them directly bioavailable to filter-feeding organisms such as bivalves so there is a need to quantify pharmaceutical residues in marine organisms collected from the wild. Table 16.3 shows levels of PCPs and APIs in freshwater and seawater mussels (wild or caged). In the San Francisco Bay, several compounds (atenolol, caffeine, gemfibrozil, sulfamethoxazole, meprobamate, valsartan) were detected in surface waters but not in benthic mussels *Geukensia demissa*, suggesting a low bioaccumulation potential for these compounds. However, in this species, lipid partitioning does not explain API bioaccumulation pattern. Moreover, the concentrations of diphenhydramine, carbamazepine, and diltiazem in *G. demissa* were similar to those detected in fish and snails (Klosterhaus et al., 2013). In the mussel *Mytilus galloprovincialis*, the exposure to two UV filters via spiked food, and two benzodiazepines in water, have revealed that benzodiazepines were bioaccumulated following a first-order kinetic model, whereas UV filters were rapidly accumulated, followed by an elimination within 24 h (Gómez et al., 2012).

Table 16.3: PCPs and APIs Concentrations in Bivalves

Compound	Species	Country	Environment	Concentration $ng\,g^{-1}$ wet weight	References
Amphetamine	*Geukensia demissa*	United States	San Francisco Bay	<0.292–4.2	Klosterhaus et al. (2013)
Atenolol	*Geukensia demissa*	United States	San Francisco Bay	<0.1–0.3	Klosterhaus et al. (2013)
Carbamazepine	*Geukensia demissa*	United States	San Francisco Bay	1.3–5.3	Klosterhaus et al. (2013)
Carbamazepine	*Mytilus galloprovincialis*	France	Mediterranean Sea	0.5–3.5	Martinez Bueno et al. (2013)
Carbamazepine	*Mytilus* spp.	Ireland	Irish Coast	<0.6–0.50	McEneff et al. (2014)
Cashmeran	*M. galloprovincialis*	Portugal	Southern Coast	ND	Groz et al. (2014)
Celestolide	*M. galloprovincialis*	Portugal	Southern Coast	ND	Groz et al. (2014)
Diclofenac	*Mytilus* spp.	Ireland	Irish Coast	ND–0.25	McEneff et al. (2014)
EHMC (2-ethylhexyl-4-trimethoxy-cinnamate)	*M. edulis* and *M. galloprovincialis*	France	Mediterranean Sea	3–256	Bachelot et al. (2012)
EHMC (2-ethylhexyl-4-trimethoxy-cinnamate)	*M. galloprovincialis*	Portugal	Southern Coast	1765[a]	Groz et al. (2014)

Table 16.3: PCPs and APIs Concentrations in Bivalves—cont'd

Compound	Species	Country	Environment	Concentration $ng\,g^{-1}$ wet weight	References
Fluoxetine	*Mytilus* spp.	United States	North Carolina	79	Bringolf et al. (2010)
Galaxolide	*M. galloprovincialis*	Portugal	Southern Coast	10[a]	Groz et al. (2014)
Gemfibrozil	*Mytilus* spp.	Ireland	Irish Coast	ND	McEneff et al. (2014)
Mefenamic acid	*Mytilus* spp.	Ireland	Irish Coast	ND–0.36	McEneff et al. (2014)
Musk ketone	*M. galloprovincialis*	Portugal	Southern Coast	<50	Groz et al. (2014)
N,N-diethyl-*m*-toluamide (DEET)	*Geukensia demissa*	United States	San Francisco Bay	3.7–13.7	Klosterhaus et al. (2013)
OCT (octocrylene)	*M. edulis* and *M. galloprovincialis*	France	Mediterranean Sea	2–7112	Bachelot et al. (2012)
OCT (octocrylene)	*M. galloprovincialis*	Portugal	Southern Coast	3992[a]	Groz et al. (2014)
OD-PABA (octyldimethyl *p*-amino benzoic acid)	*M. galloprovincialis*	Portugal	Southern Coast	833[a]	Groz et al. (2014)
Salicylic acid	*Mytilus* spp	Belgium	Belgium coast	490	Wille et al. (2010)
Sertraline	*Geukensia demissa*	United States	San Francisco Bay	0.1–1.4	Klosterhaus et al. (2013)
Triamterene	*Geukensia demissa*	United States	San Francisco Bay	<0.06–0.6	Klosterhaus et al. (2013)
Triclocarban	*Geukensia demissa*	United States	San Francisco Bay	<0.6–1.5	Klosterhaus et al. (2013)
Trimethoprim	*Mytilus* spp	Ireland	Irish Coast	<4–1.08	McEneff et al. (2014)
UV-326	*M. galloprovincialis*	Portugal	Southern Coast	13–59	Groz et al. (2014)
UV-P (2-(2-hydroxy-5-methylphenyl) benzotriazole)	*M. galloprovincialis*	Portugal	Southern Coast	ND	Groz et al. (2014)

ND, not detected.
[a]Maximum concentration.

The first field-derived bioaccumulation factors (BAFs) for mussels are shown in Table 16.4. BAFs were <1500, which suggests low bioaccumulation potential for these compounds that have low octanol–water partition coefficients (log $K_{ow} < 3$). However, for certain compounds, biological effects may occur even without their accumulation in aquatic species.

Table 16.4: PCP and API BAFs in Benthic Mussels (Klosterhaus et al., 2013)

Compound	Species	Environment	BAF ($L kg^{-1}$)
Carbamazepine	*Geukensia demissa*	San Francisco Bay	208 ± 102
Dehydronifedipine	*Geukensia demissa*	San Francisco Bay	511 ± 196
Diphenhydramine	*Geukensia demissa*	San Francisco Bay	164 ± 50
Erythromycin	*Geukensia demissa*	San Francisco Bay	40 ± 20
N,N-diethyl-*m*-toluamide (DEET)	*Geukensia demissa*	San Francisco Bay	779 ± 558
Triamterene	*Geukensia demissa*	San Francisco Bay	65 ± 7

16.5 Hazards to Aquatic Organisms

It is important to understand the effects that PCPs and APIs may inflict based on pharmacological mode of action (MoA) on human and environmental health (Boxall et al., 2012; Klosterhaus et al., 2013). Among APIs, anticancer or antineoplastic drugs are suspected to represent a specific risk for aquatic species. However, because of their specific MoA, new approaches are needed for standardized test and risk assessment to assess the presence and effects of parent compounds and their metabolites after long-term exposure to their mixtures in aquatic systems. Ecotoxicity, particularly genotoxicity and endocrine disrupting effects in aquatic species at realistic environmental concentrations, are relevant endpoints (Besse et al., 2012).

To support an Environmental Risk Assessment (ERA), a growing number of ecotoxicological acute (short-time exposure) and chronic (long-time exposure) studies are available. The effects of APIs on aquatic organisms (i.e., algae, zooplankton, invertebrates, and fish) are addressed using endpoints such as survival rate or reproductive, feeding, and growth rate alterations through the exposure to individual pharmaceutical compounds or mixtures (see reviews by Kümmerer, 2009; Boxall et al., 2012; Brausch et al., 2012; Fong and Ford, 2014). However, there is a big gap of information on ecotoxicological effects of APIs in estuarine and marine species.

16.5.1 Acute Toxicity Data

Most of the published data deal only with acute toxic effects available for more than 150 compounds belonging to 35 pharmaceutical classes. It is estimated that 10%–15% of these compounds are acutely or chronically toxic. Moreover, the information available is only for single pharmaceutical compounds on bacteria, protozoa, algae, invertebrates, fish, and amphibians (see Brausch et al., 2012). Acute toxicity data for 12 NSAIDs indicated that invertebrates and phytoplankton are the most sensitive groups, especially to IBU, paracetamol (or acetaminophen), and the narcotic analgesic dextropropoxyphene, whereas bacteria, fish, and amphibians were less sensitive (Brausch et al., 2012). Acute toxicity end points for NSAIDs on *Daphnia magna* ranked as follows: dextropropoxyphene (opioid) > paracetamol (nonnarcotic) > tramadol (nonnarcotic) > IBU (NSAID) > naproxen sodium (NSAID) > diclofenac ((DCF) NSAID) > salicylic acid (NSAID) (Brausch et al., 2012). Moreover, the marine microalgae *Phaeodactylum tricornutum* did not show any inhibition of growth after 72 h of exposure to carbamazepine, acetaminophen and IBU at levels below $2\,mg\,L^{-1}$ (Blasco and Del Valls, 2008). In mussels injected with DCF and gemfibrozil, the effective concentration, inhibitory concentration, and lethal concentration were in the $mg\,L^{-1}$ range for both pharmaceuticals, with DCF showing an order of magnitude of toxicity higher than gemfibrozil (Schmidt et al., 2011).

According to Table 16.1, these acute doses, generally in $mg\,L^{-1}$, are unlikely to be found in the aquatic environment (Fent et al., 2006). Therefore, standard acute toxicity tests are not appropriate to predict adverse biological effects at relevant environmental concentrations (Blasco and Del Valls, 2008) and given the continuous release of APIs to the aquatic environment, these tests are only relevant when accidental discharges or spills occurs (Santos et al., 2010) but sublethal effects remain a concern (Klosterhaus et al., 2013).

16.5.2 Chronic Effects and Biomarker Responses

With the huge number of human pharmaceuticals entering the aquatic environment, it is important to identify drivers of chronic stress on the health of aquatic species. Although it is unrealistic to evaluate every compound and their potential adverse effects, the chronic effects of 65 APIs of more than 20 classes were assessed in several aquatic species including invertebrates, algae, fish, amphibians and plants (for details, see Brausch et al., 2012). In addition, DCF and gemfibrozil cause toxic effects in nontarget organisms at a subcellular tissue level at an environment relevant concentration ($1\,\mu g\,L^{-1}$) (Schmidt et al., 2011). Adverse effects of DCF were reported in the liver, kidney, and gills of rainbow trout, resulting in pathological effects on renal and gill functionality (Schwaiger et al., 2004; Triebskorn et al., 2004; Hoeger et al., 2008). Pharmaceutical effects are well known in humans and/or mammals, and it is expected that APIs have the potential to affect analogous pathways in aquatic organisms, presenting similar MoAs on target organs, tissues, cells, or biomolecules (Fent et al., 2006; Boxall et al., 2012). However, biochemical, metabolic and physiological responses that can be used as biomarkers (Chapters 7 and 8) in aquatic organisms chronically exposed to PCPs and APIs at realistic concentrations remained poorly documented (Fent et al., 2006; Kümmerer, 2009; Santos et al., 2010), particularly in the framework of multibiomarker approaches. As stated by Van der Oost et al. (2003), the knowledge and understanding of the relationships between biomarkers and bioassays is essential for a more reliable and holistic ERA of contaminants.

IBU, for instance, is one of the most applied NSAIDs in the world and is defined as a class IA compound, which means it poses a "high environmental risk" (Besse and Garric, 2008). Because most of the studies concerning exposure to IBU focus on acute and chronic tests (e.g., reproductive, behavior, growth alterations) (De Lange et al., 2006; Flippin et al., 2007), it is important to address IBU impact on the antioxidant system of bivalves as most pharmaceutical therapeutic action is related to specific redox reactivity (Martín-Díaz et al., 2009). Mussels are suitable models for the determination of biomarkers of exposure to APIs using environmentally realistic concentrations.

Tissue- and time-specific alterations of antioxidant enzyme activities were observed in *M. galloprovincialis* exposed to IBU (250 ng L^{-1}) for 2 weeks. An increase of superoxide dismutase (SOD) activity was evident in gills on the first week, followed by its recovery by the end of the exposure period. The activities of catalase (CAT), glutathione reductase (GR), and glutathione *S*-transferase (GST) concomitantly decreased, whereas in digestive glands, all the enzymes (except GST) were significantly enhanced after the third (except SOD) and seventh days, recovering to basal levels afterwards. Nevertheless, lipid peroxidation (LPO) levels were elevated in both tissues during the same period, indicating that IBU induces damage in structural membrane integrity. Moreover, the constant decrease of GR activity in gills suggested that IBU seriously affects cell ability to maintain a homeostatic level between glutathione disulfideglutathione (GSH) in this tissue (Gonzalez-Rey and Bebianno, 2011, 2012). This is further supported by the concomitant inhibition trend of phase II GST activity because of a lesser presence of its substrate GSH (Regoli et al., 2004). These results are in agreement with LPO levels in *M. galloprovincialis* gills after 1-week exposure to anticonvulsant carbamazepine (100 ng L^{-1}) and with LPO levels in digestive glands of the same species after exposure to 23 μg L^{-1} of paracetamol and of *Elliptio complanata* to 0.4, 2, and 10 mM carbamazepine (Martín-Díaz et al., 2009; Solé et al., 2010). Therefore, IBU exerts transient oxidative stress in mussel tissues (Gonzalez-Rey and Bebianno, 2011, 2012).

Similarly, for fluoxetine (FLX) a transient antioxidant status alteration was observed in both *M. galloprovincialis* tissues after exposure to 75 ng FLX L^{-1}, although with totally different activity profiles (Gonzalez-Rey and Bebianno, 2013). In both tissues, SOD activity was downregulated over time, although the difference was significant in gills only. In the digestive gland, SOD activity was inversely related to CAT activity. Higher CAT activities were also reported in digestive glands than in gills of the same species after exposure to IBU or carbamazepine (Martín-Díaz et al., 2009; Gonzalez-Rey and Bebianno, 2011, 2012). Again, but to a lesser extent than that observed in IBU exposure, FLX-exposed gills were more prone to LPO damage and antioxidant system breakdown than digestive gland (Gonzalez-Rey and Bebianno, 2012, 2013). In addition, *M. galloprovincialis* exposed to FLX (75 ng L^{-1}) showed an inhibition of acetylcholinesterase (AChE) activity only after 2 weeks (Gonzalez-Rey and Bebianno, 2013). It was hypothesized that the MoA of SSRIs involves the increase of serotonin (5-HT) concentration and that AChE function was also affected by derived 5-HT induction. Similarly, AChE activity was inhibited in *M. galloprovincialis* gills exposed to paracetamol (23 and 403 μg L^{-1}) and *E. complanata* visceral mass exposed to diazepam (a common benzodiazepine at 4, 20, and 100 nmol per mussel) (Solé et al., 2010; Gagné et al., 2011).

Along with IBU, which has been ubiquitously detected in surface waters worldwide (Table 16.1), DCF—the most highly consumed anti-inflammatory drug—was recently included in the list of priority substances under the European Commission (European Commission, 2012a,b). Exposure of *M. galloprovincialis* to DCF (250 ng L^{-1}) suggested different tissue functions and transport to counteract DCF-induced reactive oxygen species (ROS) increase. However, the inhibition of antioxidant system activities and LPO levels showed an antioxidant system recovery after the first days of DCF-induced oxidative stress effects. Moreover, even though GSH was available for the eventual phase II conjugation reaction by GST, this enzyme activity was not altered showing a direct relationship with digestive gland SOD rather than with GR or CAT activity. Therefore, at this concentration, DCF does not promote phase II detoxification contrasting with that showed by the concomitant induction of GST and LPO in *Dreissena polymorpha* and *Mytilus* spp. exposed to a four-fold higher DCF concentration (Quinn et al., 2011; Schmidt et al., 2011). Therefore, mussel tissues showed significant alterations in antioxidant enzyme responses after DCF exposure (Gonzalez-Rey and Bebianno, 2014). In addition, DCF clearly induced AChE activity, which was not expected and contrasted to observations in mussels exposed to paracetamol and diazepam (Solé et al., 2010; Gagné et al., 2011). It is extremely difficult to ascertain the reason why AChE increased in the presence of DCF, but it has been hypothesized that it could be related to cell apoptosis as suggested by Zhang et al. (2002) or to an inverse relationship with endogenous E2 levels (which in turn are related to vitellogenesis) (Matozzo et al., 2008). In *Mytilus* spp. injected with a lipid-lowering fibrate gemfibrozil (at environmentally relevant and elevated concentrations 1 and 1000 $\mu g L^{-1}$), gemfibrozil affected mussel detoxification and defense systems by inducing GST and metallothioneins at both concentrations. LPO, DNA damage, and alkali-labile phosphate assay (biomarker for reproduction) were also affected indicating that gemfibrozil has the potential to induce oxidative stress and endocrine disruption in these species. On the other hand, DCF significantly induced oxidative stress and tissue damage (Schmidt et al., 2011). In the freshwater mussel *E. complanata* exposed to morphine and diazepam, morphine reduced serotonin and increased dopamine in mussel tissues and reduced AChE activity and increased GABA levels, suggesting the induction of a relaxation state in mussels, whereas diazepam also reduced serotonin levels but produced no change in dopamine levels. However, dopamine-sensitive ADC activity was readily activated, indicating the potential effect on opiate signaling. Diazepam increased glutamate levels slightly, but AChE remained stable. The increase in both dopamine ADC activity and glutamate concentrations was associated with greater oxidative stress on the mitochondrial and postmitochondrial fractions in cells (Ericson et al., 2010).

16.5.3 Mixture Toxicity

From an ecotoxicological point of view, a realistic scenario should consider that APIs are often applied jointly and the overall input in surface waters is related to a rather complex mixture. Even if mixture components are present below their individual ERA threshold concentrations, as a mixture they can eventually contribute to an increased toxicity through combined MoA. The concepts of concentration addition and independent action and the phenomena of synergism or antagonism (described in detail in Chapter 18) are also relevant for APIs (Cleuvers, 2003, 2004; Schnell et al., 2009).

Very few studies have addressed APIs mixture effects in nontarget organisms (see Gonzalez-Rey et al., 2014). In Baltic Sea blue mussels, *Mytilus edulis trossulus* exposed to DCF, IBU, and propranolol (concentrations ranging from 1 to 10,000 $\mu g\,L^{-1}$ separately or in combination), APIs were taken up by mussel tissues and affect their physiology by decreasing the scope for growth as well as their ability to attach to the substrate for the highest concentration (Ericson et al., 2010). Although it is difficult to mimic a totally realistic scenario, mussels were exposed to a mixture of IBU, FLX, and DCF to assess the possible interactions between these compounds (MIX1), or coupled with a common environmental contaminant Cu (MIX2) using a multibiomarker approach (Gonzalez-Rey et al., 2014). Antioxidant enzyme activities and LPO levels were mixture-, tissue-, and time-specific. MIX1-exposed gills showed the same antioxidant enzymes as well as phase II GST enhancement rendering a more effective protection against ROS-induced LPO. On the other hand, in MIX1-exposed digestive gland there was a swifter antioxidant enzyme response than in gills preventing this tissue from LPO damage, and confirming that gills are better "equipped" to cope with the oxidative stress induced by this mixture. Compared with MIX1, Cu in MIX2 generally induced higher and more progressive LPO levels in gills by the concomitant inactivation or inhibition of SOD, CAT, and GR activities, whereas in the digestive gland Cu in MIX2 promoted a general inactivation of these enzymes (except GR activity). However, GR activity enhancement, freeing ROS-scavenger GSH in digestive gland cells, was not enough to prevent the significant induction of LPO. Induction of alkyl-labile phosphates ((ALPs), released by vitellogenin after alkali hydrolysis) occurred in females exposed to both mixtures. Nevertheless, both sex-differentiated gonads showed at some point the induction of ALP levels, giving evidence of these mixtures ability (particularly MIX1) to induce endocrine disruption in exposed mussels (Gonzalez-Rey et al., 2014). Moreover, gene expression alterations at transcriptional level encoding antioxidants and biotransformation genes (SOD, CAT, GST, and CYP4YA of the CYP450 subfamily), associated with fatty acid, prostaglandin metabolism, and xenobiotic response, revealed that alterations were only detected in mussel gills. The downregulation of CAT gene expression after MIX1 exposure and the upregulation of the gene encoding CYP4Y1 after MIX2 exposure highlight different genotoxic effects of these mixtures in *M. galloprovincialis* (Gonzalez-Rey et al., 2014). Variable tissue and time-specific biomarker responses and gene expression alterations, along with several interactions between each mixture component on each biomarker confirm the susceptibility of mussels to API mixtures. Additionally, there were several differences between each mixture and single API exposure effects on oxidative stress responses (antioxidant enzyme activities and LPO levels), neurotoxic effects, and endocrine disruption (Gonzalez-Rey and Bebianno, 2011, 2012, 2013, 2014; Gonzalez-Rey et al., 2014).

16.5.4 Endocrine Disruption

In the clam *Ruditapes philippinarum* exposed to triclosan (300, 600, and 900 $ng\,L^{-1}$) results indicated that this compound was not estrogenic but can act as an estrogen receptor antagonist in male clams. Although not promoting oxidative stress, results indicated that triclosan might act as a neurotoxicant in this species (Matozzo et al., 2012).

One of the very first effects associated with exposure to SSRIs in aquatic organisms was the induction and potentiation of parturition/spawning (Fong, 1998; Fong and Molnar, 2008). IBU

(250 ng L^{-1}) induced significant increment of ALP levels in both males and females' gonads *M. galloprovincialis* (two-fold higher in exposed males after 1 week). The induction of these proteins was also an effective biomarker of feminization in *E. complanata* exposed for 30 days to primary-treated municipal extract containing NSAIDs (including IBU at μg L^{-1} concentrations) (Gagné et al., 2005). Results clearly indicate that IBU exposure effectively induces endocrine disruption effects that potentially lead to mussels' reproductive fitness impairment or ultimately to its feminization (Gonzalez-Rey and Bebianno, 2011, 2012). On the contrary, ALP levels in *M. galloprovincialis* exposed to FLX were downregulated in both sex-differentiated gonads. These results also attest an FLX impact in mussels' reproductive fitness because this neurotransmitter:

- is known to control gonad development and reproduction (Gagné and Blaise, 2003);
- is associated to the inverse relationships between vitellogenin synthesis and serotonin (5-HT) levels (present in bivalves that mediate endocrine functions) and/or between 5-HT and E2 levels as well as to the increase of 5-HT during the active spawning period as observed in *E. complanata* (Gagné and Blaise, 2003; Matozzo et al., 2008);
- decreases the E2 levels between spawning and after-spawning phases in mussel *D. polymorpha* after exposure (20 and 200 FLXng L^{-1}) (Lazzara et al., 2012).

A link was also hypothesized for AChE activity increase with the concomitant Vtg-like proteins decrease (implicating lower E2 levels). The exposure to sublethal FLX concentrations (0, 1, 5, 25, 125, 625 mg L^{-1}) of *Venerupis phillipinarum* for 7 days demonstrated that FLX markedly affects the immune parameters and AChE activity (Munari et al., 2014). Finally, even though FLX induces ecotoxicological effects in *M. galloprovincialis*, affecting its reproduction fitness, the mechanisms of 5-HT receptor activation are still unknown (Gagné and Blaise, 2003) and should be further assessed (Gonzalez-Rey and Bebianno, 2012). The induction of ALP levels in DCF-exposed females and in males occurred at the beginning of DCF exposure period (Gonzalez-Rey and Bebianno, 2014). These results are not in agreement to that of the exposure to a four-fold higher concentration of DCF, as no alterations of ALP levels were detected in exposed *D. polymorpha* and *Mytilus* spp. (Quinn et al., 2011; Schmidt et al., 2011). The direct relationship between ALP levels in females and AChE induction also gave evidence to an unknown and possible DCF-derived interference with mussels' estrogen receptors. This statement, however, can only be confirmed through the concomitant quantification and interaction assessment of E2, prostaglandins, and cyclooxygenase activity in mussel tissues (Gonzalez-Rey and Bebianno, 2014).

16.6 Omics

New technologies such as proteomics, genomics, and metabolomics (see Chapter 8) are potential tools for identifying the effects of PCPs and APIs of potential concern as well as in the most sensitive species (Boxall et al., 2012). These responses are generally not included in current ERA framework (Blasco and Del Valls, 2008), but can occur at concentrations that are orders of magnitude lower than those at which effects are not observed in regulatory tests, such as acute or chronic studies (Boxall et al., 2012). Differential protein expression signatures in the digestive gland of *Mytilus* spp. obtained by two-dimensional gel electrophoresis after exposure to gemfibrozil and DCF (1 and 1000 $\mu g L^{-1}$), identified 12 proteins that were upregulated or downregulated. Seven were identified by liquid chromatography-tandem mass spectrometry and indicated changes in proteins involved in energy metabolism, oxidative stress response, protein folding, and immune responses (Schmidt et al., 2013). Moreover exposure of *D. polymorpha* to benzoylecgonine (main metabolite of cocaine) (0.5 and 1 $\mu g L^{-1}$) for 2 weeks revealed an increase in protein carbonylation and oxidative stress translated by modifications in cytoskeleton, energetic metabolism, and stress response proteins in gills (Pedriali et al., 2013). Therefore, the application of an environmental proteomic approach supports its potential to assess the effects of PCPs and APIs in bivalves and highlights its potential for new biomarker discovery.

16.7 Environmental Risk Assessment of Personal Care Products and Active Pharmaceutical Ingredients

Since 1993, the significant impact of PCPs and APIs on ecosystems was acknowledged by the European Medicine Evaluation Agency (EMEA) and the US Food and Drug Administration (FDA), leading to the establishment by the European Union of mandatory guidelines for the elaboration of an ERA to authorize any new API commercialization (FDA, 1998; EMEA, 2006; Boxall et al., 2012). Drug substances with a log octanol–water partition coefficient <4.5 are screened according to the European Union technical guidance document for their persistence, bioaccumulation, and toxicity (EMEA, 2006). The ERA of APIs involves a two-phase process. Phase 1 is limited to the estimation of environmental exposure because of a parental API or its metabolites. If the predicted environmental concentration (PEC) exceeds a threshold value that poses environmental concern, additional studies on environment fate and effects are necessary in a second phase (phase 2). In Europe, that happens when the API's PEC exceeds 10 $ng L^{-1}$ in surface waters (in the United States, it is 1000 $ng L^{-1}$). A hazard quotient is estimated as the ratio between PEC and predicted no effect concentration (PNEC) for various compartments, that has to be <1 for environmental security (FDA, 1998; EMEA, 2006, 2007; EEA, 2010; Santos et al., 2010; Buchberger, 2011). If PEC/PNEC >1, other appropriate measures should be taken (for details, see Chapter 2). Approximately 200–300 new APIs have been tested and none of them present any risk to the environment (EMEA, 2006). However, once the authorization is given, there are no requirements for the ERA to be updated or reviewed (Holm et al., 2013). Because the EMEA guidelines only require environmental assessment for newly authorized pharmaceuticals, more than 3000 to 4000 APIs, authorized before 2006, including most of the anticancer drugs were not submitted to such a procedure (Besse et al., 2012). Therefore, there is a need to prioritize those compounds that are likely to pose an environmental risk.

A good portrait of this process complexity is the one applied to the French situation: this prioritization methodology based itself on the API threshold values of PECs established by the EMEA ranking APIs from class IA—"highest risk compound" to IV—"very low risk for the environment" (Figures 16.3 and 16.4).

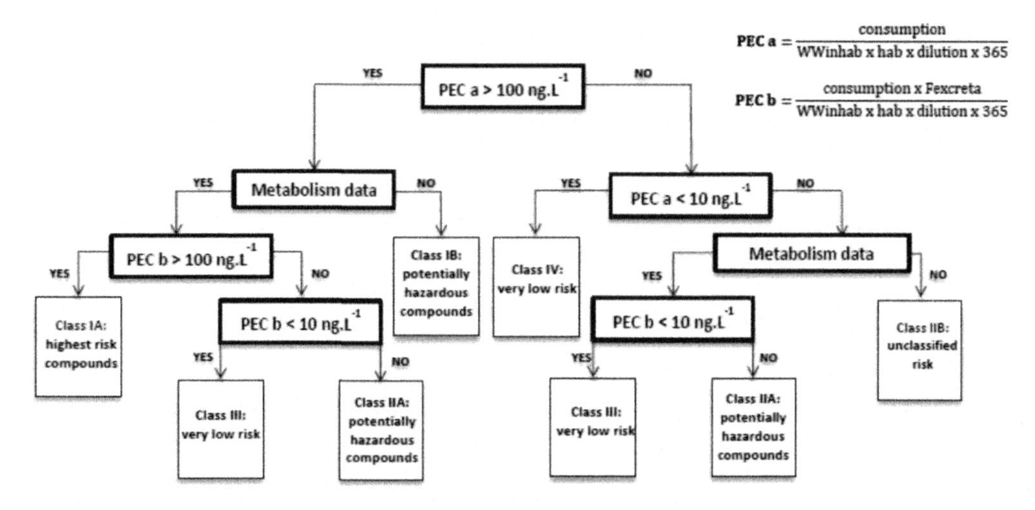

Figure 16.3
Phase I of the prioritization methodology proposal based on EMEA guidelines applied to French situation. PECa assumes the excretion of the API 100% as parental compound and 0% removal rate at WWTPs; PECb takes in consideration the real amount of parental compound excreted. *Adapted from Besse and Garric (2008).*

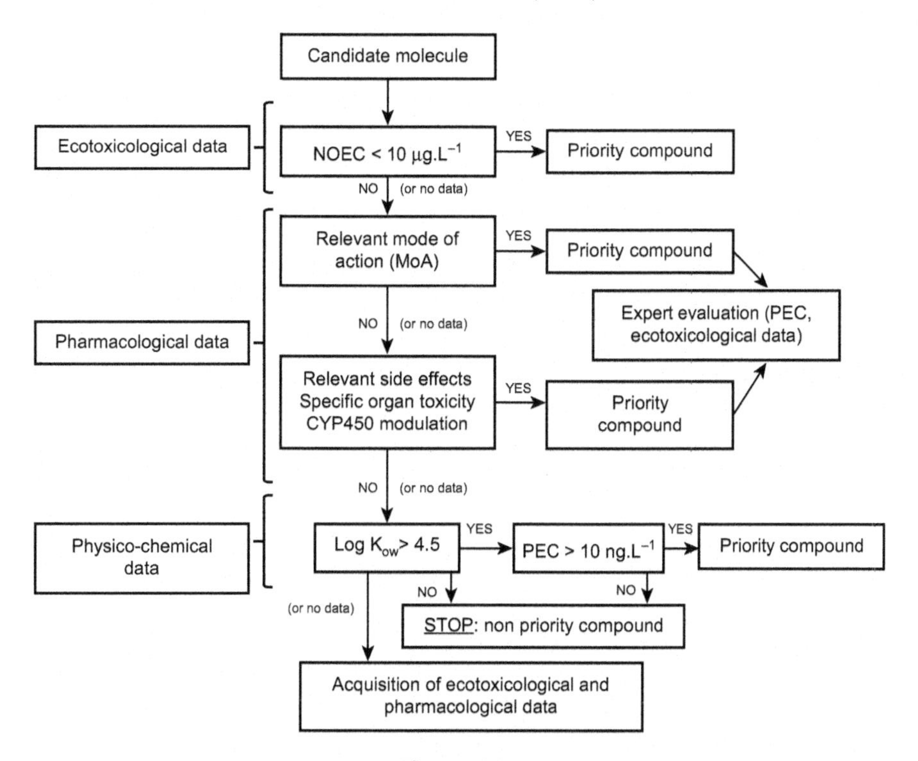

Figure 16.4
Phase II of the prioritization methodology proposal based on ecotoxicological, pharmacological, and physicochemical data. The YES is chosen only in the case of agreement with two criteria. *Adapted from Besse and Garric (2008).*

Although there is the possibility to estimate the PEC for several APIs, it is difficult to calculate PNEC for most APIs because their ecotoxicological data are still too scarce (sufficient data only exist for about 10% of APIs to enable PEC/PNEC ratio to be calculated) and is only based on results of acute and chronic toxicity tests. Therefore, a read-across hypothesis was proposed by Huggett et al. (2003) to assess an effect in nontarget organisms. This effect can only be established if plasma concentrations are similar to human plasma therapeutic concentrations. According to several authors, without this information, incorrect conclusions can be drawn regarding the effects and threats posed by APIs to no-target organisms. Because APIs are designed to have highly specific interactions with their biochemical target, their biological effect endpoints need to be assessed at the same time in relevant tissues. Another possible option is the use of the Fish Plasma Model to prioritize APIs (Rand-Weaver et al., 2013). Recently, a new concept of ecopharmacovigilance based on the detection, evaluation, understanding, and prevention of adverse effects of APIs in the environment but also in humans through indirect nontherapeutic exposure is being developed (Holm et al., 2013).

Nonetheless, DCF, whose veterinary use has been linked to the extinction of vulture species (*Gyps* sp.) in Asia (Oaks et al., 2004; Saini et al., 2012), 17α-ethinylestradiol, and 17β-estradiol were the first APIs to be included in the new list of priority substances adopted by the European Union (European Commission, 2012a,b) and previously classified as a class IIA "potentially hazardous compound." Johnson et al. (2013), using a geographically based water model, found that the concentration of 17α-ethinylestradiol exceeds the environmental quality standard established by the European Union ($0.035\,ng\,L^{-1}$) in 12% of European rivers, whereas concentrations of 17β-estradiol and DCF exceed 1% and 2%, the threshold levels of $0.4\,ng\,L^{-1}$ and $100\,ng\,L^{-1}$, respectively.

Still, based on the ERA process, an API applied in human medicine cannot be refused to be marketed, whereas a veterinary use one will be refused or need further risk mitigation measures (FDA, 1998; EMEA, 2006, 2007; EEA, 2010; Santos et al., 2010; Buchberger, 2011). Finally, concerning known ERA on APIs results, EEA (2010) refers that 95% of the applied APIs were not readily biodegradable according to OECD Test 301 (OECD, 1992) and 15% of the human pharmaceuticals were persistent in water and 50% in sediments according to OECD test 308 (OECD, 2002).

Moreover, ERA of these compounds in marine systems is largely dependent on mixing characteristics and residence times that can influence the transport, kinetics, pathways, and site-specific concentrations of these contaminants (Bayen et al., 2013).

An ERA carried out for IBU from surface waters, WWTP effluents, and landfill leachates, indicated that landfill leachates represented a potential risk for aquatic organisms, algae being the most sensitive species (Paíga et al., 2013). In California, a similar approach for the synthetic musk galaxolide and DCF revealed that the risk quotient was >1, indicating that environment concentrations need special attention because galaxolide is suspected of endocrine disruption. However, for IBU, sulfamethoxazole, carbamazepine, antibiotics triclocarban, sulfamethoxazole, and triclosan, the risk quotient was between 0.1 and 1 (Sengputa et al., 2014). Along the Irish coast, carbamazepine and DCF presented a potential risk for aquatic organisms (McEneff et al., 2014).

Because no regulatory program for ERA of PCPs and APIs exist and taking into account the long-term combined toxicity of mixtures of compounds and drug–drug interactions, there is a priority to develop new approaches for assessing the risks arising from the long-term exposure to these mixtures (Boxall et al., 2012; Bayen et al., 2013). In addition, by better understanding the drivers for PCPs and APIs exposure in different regions, it may be possible to identify areas at great risk and implement most effective control options (Boxall et al., 2012). Therefore, a systematic assessment is needed involving all stake holders to allow decisions to be made on the best mitigation strategies for an adequate environmental protection (Boxall et al., 2012).

16.8 "Green Pharmacy"

Following the same logical principles of "green chemistry," which intend to eliminate and/or minimize the usage and production of ecologically hazardous reagents and design alternative synthesis pathways, "green pharmacy" involves, as proposed by Daughton (2003), the application of several strategies and measures to reduce the impact of APIs in the environment. This includes developments on: (1) drug design—i.e., increasing the specificity of APIs MoA at the target receptor would greatly favor the evaluation of potential effects on nontarget species; (2) drug delivery—i.e., provide prescriptions at lower doses and/or promote individualization of therapy; (3) packaging—i.e., reduce packages size, offer unit doses alternative; and (4) disposal—providing consumer-oriented guidance on the package for how to dispose unused medicines. Furthermore, "green" APIs would have the advantage of—while enhancing and/or maintaining therapeutic efficacy—improving API susceptibility to biodegradation, photolysis, or other physicochemical alterations to convert them to harmless end products (Daughton, 2003). Green and sustainable pharmacy is an emerging topic and is in early stages of implementation, with it not being a high priority for the pharmaceutical industry (EEA, 2010). Most of all, because very little is yet known concerning environmental effects, fate, and behavior of APIs, it is crucial to create awareness and inform the general population, health care practitioners, and manufacturers on these matters (Daughton, 2003; EEA, 2010).

16.9 Conclusions

This review collectively shows that PCPs and APIs are ubiquitous in all types of surface waters and promote significant impact on no-target aquatic species. Therefore, it is evident that to prevent further environmental implications in wildlife, it is absolutely necessary to review and update current WWTP processes, to increase the knowledge of biological effects, and to further implement and abide current EMEA and FDA environmental guidelines, but also to inform the general public, scientists, medical doctors, and manufacturers about how PCPs and APIs are problematic in the environment to ensure an easier adoption of "green pharmacy" strategies to reduce the impact of these compounds in the aquatic environment.

References

Alonso, S.G., Catalá, M., Maroto, R.R., 2010. Pollution by psychoactive pharmaceuticals in the Rivers of Madrid metropolitan area (Spain). Environ. Intern. 36, 195–201.

Anderson, P.D., D´Aco, V.J., Shanahan, P., et al., 2004. Screening analysis of human pharmaceutical compounds in U.S. Surface waters. Environ. Sci. Technol. 38, 838–849.

Andreozzi, R., Raffaele, M., Nicklas, P., 2003. Pharmaceuticals in STP effluents and their solar photodegradation in aquatic environment. Chemosphere 50, 1319–1330.

Bachelot, M., Li, Z., Munaron, D., et al., 2012. Organic UV filter concentrations in marine mussels from French coastal regions. Sci. Total Environ. 420, 273–279.

Bayen, S., Zhang, H., Desai, M.M., et al., 2013. Occurrence and distribution of pharmaceutically active and endocrine disrupting compounds in Singapore's marine environment: Influence of hydrodynamics and physical-chemical properties. Environ. Pollut. 182, 1–8.

Bendz, D., Paxeus, N.A., Ginn, T.R., et al., 2005. Occurrence and fate of pharmaceutically active compounds in the environment, a case study: Hoje River in Sweden. J. Hazard. Mater. 122, 195–204.

Benotti, M.J., Brownawell, B.J., 2009. Microbial degradation of pharmaceuticals in estuarine and coastal seawater. Environ. Pollut. 157, 994–1002.

Beretta, M., Britto, V., Tavares, T.M., et al., 2014. Occurrence of pharmaceutical and personal care products (PPCPs) in marine sediments in the Todos os Santos Bay and the north coast of Salvador, Bahia, Brazil. J. Soils Sed. 14, 1278–1286.

Besse, J.-P., Garric, J., 2008. Human pharmaceuticals in surface waters. Implementation of a prioritization methodology and application to the French situation. Toxicol. Lett. 176, 104–123.

Besse, J.-P., Latour, J.-F., Garric, J., 2012. Anticancer drugs in surface waters what can we say about the occurrence and environmental significance of cytotoxic, cytostatic and endocrine therapy drugs? Environ. Int. 39, 73–86.

Blasco, J., Del Valls, A., 2008. Impact of emergent contaminants in the environment: environmental risk assessment. Environ. Chem. 5 (Part S/1), 169–188.

Blasco, J., Hampel M., Pereira, C., et al., 2013. Occurrence pharmaceuticals in South-West Iberian Peninsula: a preliminary risk assessment. 17th Pollutant Responses in Marine Organisms Congress, PRIMO 17.

Bound, J.P., Voulvoulis, N., 2004. Pharmaceuticals in the aquatic environment - a comparison of risk assessment strategies. Chemosphere 56, 1143–1155.

Bound, J.P., Voulvoulis, N., 2006. Predicted and measured concentrations for selected pharmaceuticals in UK rivers: Implications for risk assessment. Water Res. 40, 2885–2892.

Bound, J.P., Voulvoulis, N., 2007. Household disposal of pharmaceuticals as a pathway for aquatic contamination in the United Kingdom. Environ. Health Perspect. 113, 1705–1711.

Boxall, A.B.A., Rudd, M.A., Brooks, B.W., et al., 2012. Pharmaceuticals and personal care products in the environment: what are the big questions? Environ. Health Perspect. 120, 1221–1229.

Brausch, J.M., Connors, K.A., Brooks, B.W., et al., 2012. Human pharmaceuticals in the aquatic environment: a review of recent toxicological studies and considerations for toxicity testing. Rev. Environ. Contam. Toxicol. 218, 1–99.

Brausch, J.M., Rand, G.M., 2011. A review of personal care products in the aquatic environment: environmental concentrations and toxicity. Chemosphere 82, 1518–1532.

Bringolf, R.B., Heltsley, R.M., Newton, T.J., et al., 2010. Environmental occurrence and reproductive effects of the pharmaceutical fluoxetine in native freshwater mussels. Environ. Toxicol. Chem. 29, 13111318.

Buchberger, W.W., 2011. Current approaches to trace analysis of pharmaceuticals and personal care products in the environment. J. Chromatogr. A 1218, 603–618.

Carballa, M., Omil, F., Lema, J.M., et al., 2004. Behavior of pharmaceuticals, cosmetics and hormones in a sewage treatment plant. Water Res. 38, 2918–2926.

Cardoso, O., Porcher, J.-M., Sanchez, W., 2014. Factory-discharged pharmaceuticals could be a relevant source of aquatic environment contamination: review of evidence and need for knowledge. Chemosphere 115, 20–30.

Carmona, E., Andreu, V., Picó, Y., 2014. Occurrence of acidic pharmaceuticals and personal care products in Turia River Basin: from waste to drinking water. Sci. Total Environ. 484, 53–63.

Castiglioni, S., Bagnati, R., Calamari, D., et al., 2005. A multiresidue analytical method using solid-phase extraction and high-pressure liquid chromatography tandem mass spectrometry to measure pharmaceuticals of different therapeutic classes in urban wastewaters. J. Chromatogr. A 1092, 206–215.

Castiglioni, S., Bagnati, R., Fanelli, R., et al., 2006. Removal of pharmaceuticals in sewage treatment plants in Italy. Environ. Sci. Technol. 40, 357–363.

Chen, Y.S., Yu, S., Hong, Y.W., et al., 2013. Pharmaceutical residues in tidal surface sediments of three rivers in southeastern China at detectable and measurable levels. Environ. Sci. Pollut. Res. 20, 8391–8403.

Christensen, F.M., 1998. Pharmaceuticals in the environment–a human risk? Reg. Toxicol. Pharmacol. 28, 212–221.

Clara, M., Strenn, B., Ausserleitner, M., et al., 2004. Comparison of the behaviour of selected micropollutants in a membrane bioreactor and a conventional wastewater treatment plant. Water Sci. Technol. 50, 29–36.

Clara, M., Kreuzinger, N., Strenn, B., et al., 2005a. The solids retention time - a suitable design parameter to evaluate the capacity of wastewater treatment plants to remove micropollutants. Water Res. 39, 97–106.

Clara, M., Strenn, B., Gans, O., et al., 2005b. Removal of selected pharmaceuticals, fragrances and endocrine disrupting compounds in a membrane bioreactor and conventional wastewater treatment plants. Water Res. 39, 4797–4807.

Cleuvers, M., 2003. Aquatic ecotoxicity of pharmaceuticals including the assessment of combination effects. Toxicol. Lett. 142, 185–194.

Cleuvers, M., 2004. Mixture toxicity of the anti-inflammatory drugs diclofenac, ibuprofen, naproxen, and acetylsalicylic acid. Ecotoxicol. Environ. Saf. 59, 309–315.

Corcoran, J., Winter, C.J., Tyler, C.R., 2010. Pharmaceuticals in the aquatic environment: a critical review of the evidence for health effects in fish. Crit. Rev. Toxicol. 40, 287–304.

Daughton, C.G., 2003. Cradle-to-cradle stewardship of drugs for minimizing their environmental disposition while promoting human health: I. Rationale for and avenues toward a Green Pharmacy. Environ. Health Perspect. 111, 775–785.

Daughton, C.G., 2010. Pharmaceutical ingredients in drinking water: overview of occurrence and significance of human exposure. In: Halden, R. (Ed.), Contaminants of Emerging Concern in the Environment: Ecological and Human Health Consideration. ACS Symposium, Series, vol. 1048. American Chemical Society, Washington, D.C, pp. 9–68 (Chapter 2).

Daughton, C.G., Ternes, T.A., 1999. Pharmaceuticals and health care products in the environment: agents of subtle change? Environ. Health Perspect. 107, 907–938.

De Lange, H.J., Noordoven, W., Murkc, A.J., et al., 2006. Behavioural responses of *Gammarus pulex* (Crustacea, Amphipoda) to low concentrations of pharmaceuticals. Aquat. Toxicol. 78, 209–216.

EEA Technical Report No 1/2010 EEA (European Environment Agency), 2010. Pharmaceuticals in the Environment: Results of an EEA Workshop. ISSN:1725–2237.

EMEA (European Medicines Agency), 2006. Guideline on the Environmental Risk Assessment of Medicinal Products for Human Use. EMEA/CHMP/SWP4447/00. London, UK.

EMEA (European Medicines Agency), 2007. Guideline on Environmental Impact Assessment for Veterinary Medicinal Products in Support of the VICH Guidelines GL6 and GL38. European Medicines Agency. Committee for Medicinal Products for Veterinary Use (CVMP).

Ericson, H., Thorsén, G., Kumblad, L., 2010. Physiological effects of diclofenac, ibuprofen and propranolol on Baltic sea blue mussels. Aquat. Toxicol. 99, 223–231.

European Commission, 2012a. Proposal for a Directive of the European Parliament and of the Council Amending Directives 2000/60/EC and 2008/105/EC as Regards Priority Substances in the Field of Water Policy. European Environment Agency, Brussels. 35 p.

European Commission, 2012b. Report from the Commission to the European Parliament and the Council on the Outcome of the Review of Annex X to Directive 2000/60/EC of the European Parliament and of the Council on Priority Substances in the Field of Water Policy. European Environment Agency, Brussels. 6 p.

FDA–Food and Drug Administration, 1998. Guidance for Industry: Environmental Assessment of Human Drug and Biologics Applications, vol. 6,Food and Drug Administration (Center for Drug Evaluation and Research), CMC. Revision 1.

Fent, K., Weston, A.A., Caminada, D., 2006. Ecotoxicology of human pharmaceuticals. Aquat. Toxicol. 76, 122–159.

Fick, J., Söderstrom, H., Lindberg, R.H., et al., 2009. Contamination of surface, ground and drinking water from pharmaceutical production. Environ. Toxicol. Chem. 28, 2522–2527.

Flippin, J.L., Huggett, D., Foran, C.M., 2007. Changes in the timing of reproduction following chronic exposure to ibuprofen in Japanese medaka, *Oryzias latipes*. Aquat. Toxicol. 81, 73–78.

Fong, P., 1998. Zebra mussel spawning is induced in low concentrations of putative serotonin reuptake inhibitors. Biol. Bull. 194, 143–148.

Fong, P.P., Ford, A.T., 2014. The biological effects of antidepressants on the molluscs andcrustaceans: a review. Aquat. Toxicol 151, 4–13.

Fong, P.P., Molnar, N., 2008. Norfluoxetine induces spawning and parturition in estuarine and freshwater bivalves. Bull. Environ. Contam. Toxicol. 81, 535–538.

Gagné, F., Blaise, C., 2003. Effects of municipal effluents on serotonin and dopamine levels in the freshwater mussel *Elliptio complanata*. Comp. Biochem. Physiol. C 136, 117–125.

Gagné, F., Bérubé, E., Fournier, M., et al., 2005. Inflammatory properties of municipal effluents to *Elliptio complanata* mussels – lack of effects from anti-inflammatory drugs. Comp. Biochem. Physiol. C 141, 332–337.

Gagné, F., Blaise, C., André, C., et al., 2007. Neuroendocrine disruption and health effects in *Elliptio complanata* mussels exposed to aeration lagoons for wastewater treatment. Chemosphere 68, 731–743.

Gagné, F., Bouchard, B., André, C., et al., 2011. Evidence of feminization in wild Elliptio complanata mussels in the receiving waters downstream of a municipal effluent outfall. Comp. Biochem. Physiol. C 153, 99–106.

Garrison, A.W., Pope, J.D., Allen, F.R., 1976. GC/MS analysis of organic compounds in domestic wastewaters. In: Keith, C.H. (Ed.), Identification and Analysis of Organic Pollutants in Water. Ann Arbor Science Publishers, Ann Arbor, MI, pp. 517–556.

Gomez, E., Bachelot, M., Boillot, C., et al., 2012. Bioconcentration of two pharmaceuticals (benzodiazepines) and two personal care products (UV filters) in marine mussels (*Mytilus galloprovincialis*) under controlled laboratory conditions. Environ. Sci. Pollut. Res. 19, 2561–2569.

Gonzalez-Rey, M., Bebianno, M.J., 2011. Non-steroidal anti-inflammatory drug (NSAID) Ibuprofen distresses antioxidant defense system in mussel *Mytilus galloprovincialis* gills. Aquat. Toxicol. 105, 264–269.

Gonzalez-Rey, M., Bebianno, M.J., 2012. Does non-steroidal anti-inflammatory (NSAID) Ibuprofen induces antioxidant stress and endocrine disruption in mussel *Mytilus galloprovincialis*? Environ. Toxicol. Pharmacol. 33, 361–371.

Gonzalez-Rey, M., Bebianno, M.J., 2013. Does selective serotonin reuptake inhibitor (SSRI) fluoxetine affect mussel *Mytilus galloprovincialis*? Environ. Pollut. 174, 200–209.

Gonzalez-Rey, M., Bebianno, M.J., 2014. Effect of non-steroidal anti-inflammatory drug (NSAID) diclofenac exposure in mussel *Mytilus galloprovincialis*. Aquat. Toxicol. 148, 221–230.

Gonzalez-Rey, M., Mattos, J.J., Piazza, C.E., et al., 2014. Effects of active pharmaceutical ingredients mixtures in mussel *Mytilus galloprovincialis*. Aquat. Toxicol. 153, 12–26.

Gonzalez-Rey, M., Tapie, N., Le Menach, K., et al. Occurrence of APIS and pesticides in aquatic systems. Mar. Pollut. Bull. (in press)

Gros, M., Petrovic, M., Ginebreda, A., et al., 2010. Removal of pharmaceuticals during wastewater treatment and environmental risk assessment using hazard indexes. Environ. Int. 36, 15–26.

Groz, M.C., Martinez Bueno, M.J., Rosain, D., et al., 2014. Detection of emerging contaminants (UV filters, UV stabilizers and musks) in marine mussels from Portuguese coast by QuEChERSextraction and GC–MS/MS. Sci. Total Environ. 493, 162–169.

Halling-Sørensen, B., Nielsen, S.N., Lanzky, P.F., et al., 1997. Occurrence, fate and effects of pharmaceutical substances in the environment - a review. Chemosphere 36, 357–394.

Han, G.H., Hur, H.G., Kim, S.D., 2006. Ecotoxicological risk of pharmaceuticals from wastewater treatment plants in Korea: occurrence and toxicity to *Daphnia magna*. Environ. Toxicol. Chem. 25, 265–271.
Heberer, T., 2002. Occurrence, fate, and removal of pharmaceutical residues in the aquatic environment: a review of recent research data. Toxicol. Lett. 131, 5–17.
Hoeger, B., Dietrich, D.R., Schmid, D., Hartmann, A., Hitzfeld, B., 2008. Distribution of intraperitoneally injected diclofenac in brown trout (*Salmo trutta f. fario*). Ecotoxicol. Environ. Saf. 71, 412–418.
Holm, G., Snape, J.R., Murray-Smith, R., et al., 2013. Implementing ecopharmacovigilance in practice: challenges and potential opportunities. Drug Saf. 36, 533–546.
Huerta-Fontela, M., Galceran, M.T., Ventura, F., 2011. Occurrence and removal of pharmaceuticals and hormones through drinking water treatment. Water Res. 45, 1432–1442.
Huggett, D.B., Cook, J.C., Ericson, J.E., et al., 2003. Theoretical model for utilizing mammalian pharmacology and safety data to prioritise potential impacts of human pharmaceuticals to fish. J. Hum. Ecol. Risk Assess. 9, 1789–1799.
IMS Institute for Healthcare informatics, 2011. The Global Use of Medicines: Outlook through 2015. 27 p.
Johnson, A.C., Dumont, E., Williams, R.J., et al., 2013. Do concentrations of ethinylestradiol, estradiol, and diclofenac in European rivers exceed proposed EU environmental quality standards? Environ. Sci. Technol. 47, 12297–12304.
Jones, O.A.H., Voulvoulis, N., Lester, J.N., 2001. Human pharmaceuticals in the aquatic environment: a review. Environ. Technol. 22, 1383–1395.
Josephy, P.D., Mannervik, B., 2006. Molecular Toxicology, second ed. Oxford University Press, New York, U.S.A. 589 p.
Joss, A., Zabczynski, S., Gobel, A., et al., 2006. Biological degradation of pharmaceuticals in municipal wastewater treatment: proposing a classification scheme. Water Res. 40, 1686–1696.
Kanda, R., Griffin, P., James, H.A., et al., 2003. Pharmaceutical and personal care products in sewage treatment works. J. Environ. Monit. 5, 823–830.
Khetan, S.K., Collins, T.J., 2007. Human pharmaceuticals in the aquatic environment: a challenge to green chemistry. Chem. Rev. 107, 2319–2364.
Klosterhaus, S.L., Grace, R., Hamilton, M.C., et al., 2013. Method validation and reconnaissance of pharmaceuticals, personal care products, and alkylphenols in surface waters, sediments, and mussels in an urban estuary. Environ. Int. 54, 92–99.
Kolpin, D.W., Furlong, E.T., Meyer, M.T., et al., 2002. Pharmaceuticals, hormones and other organic wastewater contaminants in US streams, 1999–2000: a national reconnaissance. Environ. Sci. Technol. 36, 1202–1211.
Kreuzinger, N., Clara, M., Strenn, B., et al., 2004. Investigation on the behaviour of selected pharmaceuticals in the groundwater after infiltration of treated wastewater. Water Sci. Technol. 50, 221–228.
Kümmerer, K., 2004. Pharmaceuticals in the environment - scope of the book and introduction. In: Kümmerer, K. (Ed.), Pharmaceuticals in the Environment - Sources, Fate, Effects and Risks, second ed. Springer, Berlin, Germany, pp. 3–11 (Chapter 1).
Kümmerer, K., 2009. The presence of pharmaceuticals in the environment due to human use - present knowledge and future challenges. J. Environ. Manag. 90, 2354–2366.
Lajeunesse, A., Gagnon, C., Sauve, S., 2008. Determination of basic antidepres-sants and their N-desmethyl metabolites in raw sewage and wastewater usingsolid-phase extraction and liquid chromatography-tandem mass spectrometry. Anal. Chem. 80, 5325–5333.
Langford, K., Thomas, K.V., 2011. Input of selected human pharmaceutical metabolites into the Norwegian aquatic environment. J. Environ. Monit. 13, 416–421.
Lazzara, R., Blázquez, M., Porte, C., et al., 2012. Low environmental levels of fluoxetine induce spawning and changes in endogenous estradiol levels in the zebra mussel *Dreissena polymorpha*. Aquat. Toxicol. 106-107, 123–130.
Li, A.P., 2001. Screening for human ADME/Tox drug properties in drug discovery. Drug Disc. Today 6, 357–366.
Lindqvist, N., Tuhkanen, T., Kronberg, L., 2005. Occurrence of acidic pharmaceuticals in raw and treated sewages and in receiving waters. Water Res. 39, 2219–2228.
Lishman, L., Smyth, S.A., Sarafin, K., et al., 2006. Occurrence and reductions of pharmaceuticals and personal care products and estrogens by municipal wastewater treatment plants in Ontario. Sci. Total Environ. 367, 544–558.

Martín-Díaz, M.L., Gagné, F., Blaise, C., 2009. The use of biochemical responses to assess ecotoxicological effects of pharmaceutical and personal care products (PPCPs) after injection in the mussel *Elliptio complanata*. Environ. Toxicol. Pharmacol. 28, 237–242.

Martinez Bueno, M.J., Boillot, C., Fenet, H., et al., 2013. Fast and easy extraction combined with high resolution-mass spectrometry for residue analysis of two anticonvulsants and their transformation products in marine mussels. J. Chromatogr. A 1305, 27–34.

Matamoros, V., Duhec, A., Albaiges, J., et al., 2009. Photodegradation of carbamazepine, ibuprofen, ketoprofen and 17 alpha-Ethinylestradiol in fresh and seawater. Water Air Soil Pollut. 196, 161–168.

Matozzo, V., Gagné, F., Marin, M.G., et al., 2008. Vitellogenin as a biomarker of exposure to estrogenic compounds in aquatic invertebrates: a review. Environ. Int. 34, 531–545.

Matozzo, V., Costa Devoti, A., Marin, M.G., 2012. Immunotoxic effects of triclosan in the clam *Ruditapes philippinarum*. Ecotoxicology 21, 66–74.

McEneff, G., Barron, L., Kelleher, B., et al., 2014. A year-long study of the spatial occurrence and relative distribution of pharmaceutical residues in sewage effluent, receiving marine waters and marine bivalves. Sci. Total Environ. 476–477, 317–326.

Metcalfe, C.D., Koenig, B.G., Bennie, D.T., et al., 2003a. Occurrence of neutral and acidic drugs in the effluents of Canadian sewage treatment plants. Environ. Toxicol. Chem. 22, 2872–2880.

Metcalfe, C.D., Miao, X.-S., Koenig, B.G., et al., 2003b. Distribution of acidic and neutral drugs in surface waters near sewage treatment plants in the lower great lakes, Canada. Environ. Toxicol. Chem. 22, 2881–2889.

Metcalfe, C.D., Chu, S.G., Judt, C., et al., 2010. Antidepressants and their metabolites in municipal wastewater, and downstream exposure in an urban watershed. Environ. Toxicol. Chem. 29, 79–89.

Meredith-Williams, M., Carter, L.J., Fussell, R., et al., 2012. Uptake and depuration of pharmaceuticals in aquatic invertebrates. Environ. Pollut. 165, 250–258.

Munari, M., Marin, M.G., Matozzo, V., 2014. Effects of the antidepressant fluoxetine on the immune parameters and acetylcholinesterase activity of the clam *Venerupis philippinarum*. Mar. Environ. Res. 94, 32–37.

Oaks, J.L., Gilbert, M., Virani, M.Z., et al., 2004. Diclofenac residues as the cause of vulture population decline in Pakistan. Nature 427, 630–633.

O'Brien, E., Dietrich, D.R., 2004. Hindsight rather than foresight: reality versus the EU draft guideline on pharmaceuticals in the environment. Trends Biotechnol. 22, 326–330.

OECD, 1992. OECD Guidelines for the Testing of Chemicals, Test 301: Ready Biodegradability. OECD/OCDE. 62 p.

OECD, 2002. OECD Guidelines for the Testing of Chemicals, Test 308: Aerobic and Anaerobic Transformation in Aquatic Sediment Systems. OECD/OCDE. 19 p.

Paíga, P., Santos, H.M.L.M., Amorim, C.G., et al., 2013. Pilot monitoring study of ibuprofen in surface waters of north of Portugal. Environ. Sci. Pollut. Res. 20, 2410–2420.

Pait, A.S., Warner, R.A., Hartwell, S.I., et al., 2006. Human use pharmaceuticals in the estuarine environment: a Survey of the Chesapeake Bay, Biscayne Bay and Gulf of the Farallones. NOS NCCOS 7. Silver Spring, MD, NOAA/NOS/NCCOS/Center for Coastal Monitoring and Assessment, 21 p.

Pedriali, A., Riva, C., Parolini, M., et al., 2013. A redox proteomic investigation of oxidative stress caused by benzoylecgonine in the freshwater bivalve *Dreissena polymorpha*. Drug Test. Anal. 5, 646–656.

Prasse, C., Wagner, M., Schulz, R., et al., 2011. Biotransformation of the antiviral drugs acyclovir and penciclovir in activated sludge treatment. Environ. Sci. Technol. 45, 2761–2769.

Quinn, B., Schmidt, W., O'Rourke, K., et al., 2011. Effects of the pharmaceuticals gemfibrozil and diclofenac on biomarker expression in the zebra mussel (*Dreissena polymorpha*) and their comparison with standardised toxicity tests. Chemosphere 84, 657–663.

Rand-Weaver, M., Margiotta-Casaluci, L., Patel, A., et al., 2013. The read-across hypothesis and environmental risk assessment of pharmaceuticals. Environ. Sci. Technol. 47, 11384–11395.

Regoli, F., Frenzillib, G., Bocchettia, R., et al., 2004. Time-course variations of oxyradical metabolism, DNA integrity and lysosomal stability in mussels, *Mytilus galloprovincialis*, during a field translocation experiment. Aquat. Toxicol. 68, 167–178.

Reif, R., Suárez, S., Omil, F., et al., 2008. Fate of pharmaceuticals and cosmetic ingredients during the operation of a MBR treating sewage. Desalination 221, 511–517.

Roberts, P.H., Bersuder, P., 2006. Analysis of OSPAR priority pharmaceuticals using high-performance liquid chromatography-electrospray ionization tandem mass spectrometry. J. Chromatogr. A 1134, 143–150.

Saini, M., Taggart, M.A., Knopp, D., et al., 2012. Detecting diclofenac in livestock carcasses in India with an ELISA: a tool to prevent widespread vulture poisoning. Environ. Pollut. 160, 11–16.

Sammartino, M.P., Bellanti, F., Castrucci, M., et al., 2008. Ecopharmacology: deliberated or casual dispersion of pharmaceutical principles, phytosanitary, personal health care and veterinary products in environment needs a multivariate analysis or an expert systems for the control, the measure and the remediation. Microchem. J. 88, 201–209.

Sanchez, W., Sremski, W., Piccini, B., et al., 2011. Adverse effects in wild fish living downstream from pharmaceutical manufacture discharges. Environ. Intern. 37, 1342–1348.

Santos, L.H.M.L.M., Araújo, A.N., Fachini, A., et al., 2010. Ecotoxicological aspects related to the presence of pharmaceuticals in the aquatic environment. J. Hazard. Mater. 175, 45–95.

Santos, L.H.M.L.M., Gros, M., Rodriguez-Mozaz, S., et al., 2013. Contribution of hospital effluents to the load of pharmaceuticals in urban wastewaters: Identification of ecologically relevant pharmaceuticals. Sci. Total Environ. 461–462, 302–316.

Schnell, S., Bols, N.C., Barata, C., et al., 2009. Single and combined toxicity of pharmaceuticals and personal care products (PPCPs) on the rainbow trout liver cell line RTL-W1. Aquat. Toxicol. 93, 244–252.

Schmidt, W., O'Rourke, K., Hernan, R., et al., 2011. Effect of the pharmaceuticals gemfibrozil and diclofenac on the marine mussel (*Mytilus* spp.) and their comparison with standardized toxicity tests. Mar. Pollut. Bull. 62, 1389–1395.

Schmidt, W., Rainville, L.-C., McEneff, G., et al., 2013. A proteomic evaluation of the effects of the pharmaceuticals diclofenac and gemfibrozil on marine mussels (Mytilus spp.): evidence for chronic sublethal effects on stress-response proteins. Drug Test. Anal. 5, 210–219.

Schwaiger, J., Ferling, H., Mallow, U., et al., 2004. Toxic effects of the non-steroidal anti-inflammatory drug diclofenac. Part 1: histopathological alterations and bioaccumulation in rainbow trout. Aquat. Toxicol. 68, 141–150.

Schultz, M.M., Furlong, E.T., 2008. Trace analysis of antidepressant pharmaceuticals and their select degradates in aquatic matrixes by LC/ESI/MS/MS. Anal. Chem. 80, 1756–1762.

Schultz, M.M., Furlong, E.T., Kolpin, D.W., et al., 2010. Antidepressant pharmaceuticals in two U.S. effluent-impacted streams: occurrence, fate in water and sediment, and selective uptake in fish neural tissue. Environ. Sci. Technol. 44, 1918–1925.

Sengputa, A., Lyons, J.M., Smith, D.J., et al., 2014. The occurrence and fate of chemicals of emerging concern in coastal urban rivers receiving discharge of treated municipal wastewater effluent. Environ. Toxicol. Chem. 33, 350–358.

Solé, M., Shaw, J.P., Frickers, P.E., et al., 2010. Effects on feeding rate and biomarker responses of marine mussels experimentally exposed to propranolol and acetaminophen. Anal. Bioanal. Chem. 396, 649–656.

Tauxe-Wuersch, A., De Alencastro, L.F., Grandjean, D., et al., 2005. Occurrence of several acidic drugs in sewage treatment plants in Switzerland and risk assessment. Water Res. 39, 1761–1772.

Ternes, T.A., 1998. Occurrence of drugs in German sewage treatment plants and rivers. Water Res. 32, 3245–3260.

Ternes, T.A., Joss, A., Siegrist, H., 2004. Scrutinizing pharmaceuticals and personal care products in wastewater treatment. Environ. Sci. Technol. 38, 392A–399A.

Thomas, K.V., Hilton, M.J., 2004. The occurrence of selected human pharmaceutical compounds in UK estuaries. Mar. Pollut. Bull. 49, 436–444.

Togola, A., Budzinski, H., 2008. Multi-residue analysis of pharmaceutical compounds in aqueous samples. J. Chromatogr. A 1177, 150–158.

Triebskorn, R., Casper, H., Heyd, A., et al., 2004. Toxic effects of the non-steroidal anti-inflammatory drug diclofenac. Part II: cytological effects in liver, kidney, gills and intestine of rainbow trout (*Oncorhynchus mykiss*). Aquat. Toxicol. 68, 151–166.

USEPA, 2007. Final report: the environmental occurrence, fate, and ecotoxicity of selective serotonin reuptake inhibitors(SSRIs) in aquatic environments.

Van der Oost, R., Beyer, J., Vermeulen, N.P.E., 2003. Review article fish bioaccumulation and biomarkers in environmental risk assessment: a review. Environ. Toxicol. Pharmacol. 13, 57–149.

Vasskog, T., Anderssen, T., Pedersen-Bjergaard, S., et al., 2008. Occurrence of selective serotonin reuptake inhibitors in sewage and receiving waters at Spitsbergen and in Norway. J. Chromatogr. 1185A, 194–205.
Vidal-Dorsch, D.E., Bay, S.M., Maruya, K., et al., 2012. Contaminants of emerging concern in municipal wastewater effluents and marine receiving water. Environ. Toxicol. Chem. 31, 2674–2682.
Vieno, N.M., Tuhkanen, T., Kronberg, L., 2005. Seasonal variation in the occurrence of pharmaceuticals in effluents from a sewage treatment plant and in the recipient water. Environ. Sci. Technol. 39, 8220–8226.
Wall, R., 2004. "Emerging" Contaminants in U.S. Water Supplies Part 1 – a New Kind of Pollution? (Online) http://www.acnatsci.org/education/kye/hi/kye5152004.html.
Weigel, S., Kuhlmann, J., Huhnerfuss, H., 2002. Drugs and personal care products as ubiquitous pollutants: occurrence and distribution of clofibric acid, caffeine and DEET in the North Sea. Sci. Total Environ. 295, 131–141.
Weigel, S., Berger, U., Jensen, E., Kallenborn, R., Thoresen, H., Huhnerfuss, H., 2004. Determination of selected pharmaceuticals and caffeine in sewage and seawater from Tromso/Norway with emphasis on ibuprofen and its metabolites. Chemosphere 56, 583–592.
WHO - World Health Organization, May 2010. Medicines: Rational Use of Medicines, Fact Sheet No 338. (Retrieved).
WHO - World Health Organization, 2011. The World Medicines Situation 2011, Rational Use of Medicines. WHO/EMP/MIE/2011.2.2, third ed. 22 p.
Wille, K., Noppe, H., Verheyden, K., et al., 2010. Validation and application of an LC-MS/MS method for the simultaneous quantification of 13 pharmaceuticals in seawater. Anal. Bioanal. Chem. 397, 1797–1808.
Wu, J.L., Lam, N.P., Marten, D., et al., 2007. Triclosan determination in water related to wastewater treatment. Talanta 72, 1650–1654.
Xie, Z., Ebinghaus, R., Floser, G., et al., 2008. Occurrence and distribution of triclosan in the German Bight (North Sea). Environ. Pollut. 156, 1190–1195.
Zhang, X.J., Yang, L., Zhao, Q., et al., 2002. Induction of acetylcholinesterase expression during apoptosis in various cell types. Cell Death Different 9, 790–800.
Zuccato, E., Castiglioni, S., Fanelli, R., et al., 2006. Pharmaceuticals in the environment in Italy: causes, occurrence, effects and control. Environ. Sci. Pollut. Res. 13, 15–21.

CHAPTER 17

Ecotoxicological Risk of Nanomaterials

Catherine Mouneyrac, Kristian Syberg, Henriette Selck

Abstract

A great variety of engineered nanomaterials (ENMs) is being produced and eventually have been introduced to the aquatic environment. ENMs are assumed to aggregate/agglomerate, undergo transformation in the water column, and subsequently accumulate in the sediment. ENMs have been found to be available for uptake both in pelagic and benthic organisms. ENM environmental concentrations can be estimated from models but few quantitative data exist at present. Consequently, there is a great need for developing techniques that are readily implementable in a standard laboratory and at the same time is robust. Concerning toxic effects, biochemical biomarkers at the subindividual level (e.g., oxidative stress, genotoxicity, apoptosis) and omic approaches (genomic, transcriptomic, proteomic) are useful tools to gain insight into mechanisms whereas endpoints at individual level (behavior, feeding rate, reproduction) are relevant for integration in ecological risk assessment. Further investigations are needed to better understand chronic and delayed ENM effects.

Keywords: Aquatic species; Behavior; Bioaccumulation; Biomarkers; Engineered nanomaterials/nanoparticles; Fate; Test-systems; Toxicity.

Chapter Outline

Nanotechnology has increased rapidly during the past decade and the same is observed for the scientific body of evidence. The International Organization for Standardization (ISO) defines engineered nanomaterials (ENMs) as materials with any external dimension in the nanoscale or having an internal surface structure at those dimensions (between 1 and 100nm) that are designed for a specific purpose or function (ISO/TS 27687:2008; ISO/TS 80004-4: 2011).

Aquatic Ecotoxicology. http://dx.doi.org/10.1016/B978-0-12-800949-9.00017-6

However, the definition is much debated. One of the challenges faced is that ENMs and engineered nanoparticles (ENPs) are produced to have many different applications (e.g., electronics, optics, textiles, cosmetics, medical devices, water treatment technology), which mean that they have many different chemical-physical properties. Thus, there are many types of ENMs (carbon nanotubes, metal-based nanoparticles (NPs), composite nanomaterials, or multilayer ENPs, functionalized NPs) being introduced to the aquatic environment, and the scientific community has obviously only managed to focus on a selection of these with main bias toward ENPs. Once introduced into the aquatic environment, ENPs will behave differently between freshwater and marine systems and, for instance, aggregation chemistry and dissolution kinetics will depend on a number of environmental factors as well as ENP characteristics such as shape (spherical, rods, platelets), size, surface area, coating, and presence of functional groups (Handy et al., 2008). Regarding ecotoxicology, a vast number of different methods have been employed to assess the potential ENP effects and the importance of ENP characteristics for toxicity (Handy et al., 2012). These range from standardized water-only test methods to more sophisticated test methods addressing ENP toxicity in semifield setups. Based on these primarily short-term water tests with pristine ENPs, the ecotoxicology community is aiming to understand the potential risk to the ecosystem. However, environmental fate, behavior, and impact of ENPs, ENMs, or commercial products containing ENMs in realistic environmental conditions using outdoor mesocosms have only recently been employed (Ferry et al., 2009; Lowry et al., 2012; Cleveland et al., 2012; Colman et al., 2013; Buffet et al., 2013, 2014). To determine adverse effects of ENMs toward organisms belonging to different trophic levels, a combination of endpoints at the subindividual level (e.g., molecular and biochemical biomarkers) and at the individual level (e.g., survival/mortality, development, reproduction, behavior, feeding rate) are frequently employed.

This chapter provides insight into our current understanding of exposure, hazard, and risk of ENPs in the aquatic environment and provides tools and guidance on future focus. Because most existent aquatic ENP literature concern invertebrates, the chapter is biased toward this group thus excluding plants and mammals.

17.1 Environmental Fate and Behavior

Once introduced to the aquatic environment, ENPs will undergo transformation via a number of processes including aggregation/agglomeration and subsequent sedimentation as well as interactions with abiotic and biotic components in the aquatic system (Handy et al., 2008; Klaine et al., 2008; Petersen et al., 2014). The transformation will provide the ENP with a new “environmental identity.” However, this transformation is rather unpredictable both because of limitations regarding characterization methods and because ENP behavior may differ even in similar experimental setups (Petersen et al., 2014). Consequently, ENP behavior in the environment is complex and not fully understood and the link to the new environmental identity, and thus to ecotoxicology, is therefore not clear.

17.1.1 Engineered Nanoparticle Entry and Transformation in the Aquatic Environment

ENPs enter the environment through intentional releases resulting from ENM utilization as well as unintentional releases such as atmospheric emissions and solid or liquid waste streams from production facilities. In addition, some ENMs (http://ec.europa.eu/health/scientific_committees/opinions_layman/nanomaterials/en/glossary/mno/nanomaterial.htm) are used in environmental remediation applications and, as such, they are applied as primary ENMs to the environment (review by Farré et al., 2009). Considering the growing variety of ENMs and the diversity of aquatic species and environments, we need to fully understand ENM fate processes in the aquatic environment and their behavior at the ENP-biota interface to be able to establish sound ecotoxicological risk assessment. Fate processes refers to the transport, transformation, and subsequent accumulation of the original or transformed ENMs. Compared with conventional chemical contaminants, transformation processes of ENPs are more complex because the timeframe for achieving thermodynamic equilibrium is much larger for ENPs. ENP fate and behavior is expected to depend not only on the physical and chemical properties of the ENMs, but also on the characteristics of the receiving aquatic environment such as pH, temperature, ionic strength, and natural organic matter. The major processes that govern NP fate and behavior in the aquatic environment are aggregation (discrete assemblages of primary particles that are strongly bonded) and agglomeration (assemblages of particles held together by relatively weak forces), sedimentation, and dissolution (Christian et al., 2008; Handy et al., 2008; Klaine et al., 2008; Petersen et al., 2014). Differences between pristine ENPs and materials actually released or altered after release are also highly significant for fate and behavior processes in the environment but little information is available. Scientific knowledge of fate and behavior of environmental colloids significantly aids the understanding of ENP fate in aquatic environments (Klaine et al., 2008).

17.1.1.1 Impact of Natural Water Components for Engineered Nanoparticle Fate

Components of natural waters induce complex effects on ENP stability, aggregation, and sedimentation through different mechanisms of nanoscale surface film formation, charge enhancement, and steric stabilization by natural organic matter, charge neutralization by ionic strength or specifically by binding cations such as calcium, bridging by fibrils, and bridging by aggregated natural organic matter (Christian et al., 2008). Homoaggregation (aggregation of ENPs with each other) is the dominant process when relatively high initial nanoparticle concentrations are present in the medium, whereas in the case of low initial concentrations of NPs, aggregation of ENPs with other particles (i.e., heteroaggregation) such as natural colloids is likely to be the main route of ENP removal from the water column (Quik et al., 2012). Disaggregation of NPs is as important a process as aggregation because the resulting released small aggregates can be resuspended and thus become mobile in the water column and beneficiate the cotransport of pollutants as well as nutrients (Baalousha et al., 2008; Christian et al., 2008). For example, humic and fulvic substances have been shown to inhibit the aggregation of carbon nanotubes (CNTs) (Hyung et al., 2007), but also to enhance the aggregation and stability of FeO ENPs (Baalousha et al., 2008). These changes in aggregation/agglomeration/disaggregation because interactions between NPs (engineered or natural) and between ENPs and other particles will affect the potential transport of ENPs in the aquatic environment. Contaminant sorption to ENPs is also an important process that can lead to enhanced transport or changed bioavailability of the contaminant.

In oxidation and reduction reactions, sulfidation may be of concern, particularly for metal-based ENPs (Levard et al., 2012). Dissolution (release of metal ions) is a significant factor in determining ENP toxicity, particularly for ENPs composed of elements, which in solution are known to be toxic (e.g., Ag, Cu, Zn, Cd). In nanotoxicological studies, it is necessary to determine if the toxicity observed is from the nanoparticulate form or from the release of soluble ions, or a combination. Literature data seem to favor a complex interplay between ENPs and dissolved species, both potentially contributing to the biological response. The thermodynamic parameter controlling dissolution is described as solubility. Both solubility and dissolution kinetics are dependent on ENP physicochemical properties (chemical composition, size, surface) and are further influenced by the environmental media (temperature, pH, ionic strength, organic components). For example, two nominally identical ENPs, in terms of size and composition, could have totally different dissolution behavior because of different surface modifications.

Dialysis, filtration, or centrifugal ultrafiltration methods are usually employed to separate the dissolved fraction from the NP suspension (Misra et al., 2012). The dissolved fraction of metals in the aquatic media (water column, sediment) during the time of exposure can be determined using diffusive gradients in thin film tools (Mouneyrac et al., 2014). There is a range of analytical techniques (e.g., inductively coupled plasma [ICP]–mass spectrometry, ICP–atomic emission spectroscopy, ion selective electrodes) that can be used to measure dissolution.

Although the impacts of ENMs in freshwater media have largely been studied in the past decade, data covering a salinity gradient (estuaries) or regarding salted aquatic media (seawater) are lacking. However, chemical composition of the aqueous media drastically govern metal speciation (under ionic or nanoparticulate forms), their physicochemical behavior and consequently their ecotoxicity (Klaine et al., 2008). Regarding the influence of temperature, Wong and Leung (2014) reported that the Zn ion dissolution rate from ZnO NPs significantly increased with decreasing temperature, whereas aggregation size was only slightly influenced by temperature, and toxicity toward marine diatom, amphipod, and fish was temperature-dependent and species-specific. Nonetheless, the combined effect of temperature and salinity on the physicochemical properties and toxicities to marine organisms is still largely unknown.

17.1.1.2 Fate Models

Currently, environmental monitoring data of ENMs are lacking as a consequence of the difficulty to detect and quantify ENPs in complex matrices such as surface waters, wastewater treatment plant effluents, biosolids, sediments, soils, and air as well as organisms and their susceptible tissues. Consequently, estimation of fate and quantities of ENMs (http://ec.europa.eu/health/scientific_committees/opinions_layman/nanomaterials/en/glossary/mno/nanomaterial.htm) present in aquatic environments generally derive from calculated exposure scenarios based on predicted ENMs (http://ec.europa.eu/health/scientific_committees/opinions_layman/nanomaterials/en/glossary/mno/nanomaterial.htm) use and not from actual measurements. Boxall et al. (2007) estimated ENP (Ag, Al_2O_3, Au, CeO_2, fullerenes, SiO_2, TiO_2, ZnO) emissions from different sources and their distribution among water, soil, and air, but did not include ENP transport between the compartments. In contrast, the model proposed by Gottschalk et al. (2009) allows for transfer of various ENPs (TiO_2, ZnO, Ag) among different compartments. Praetorius et al. (2012) proposed a mechanistic model for TiO_2 NPs including fate mechanisms such as heteroagglomeration, sedimentation, advection with the flow of the water body, bed load transport within the sediment, burial in the sediment, and resuspension from sediment to water. Recently, Gottschalk et al. (2013) reviewed all available literature with the aim to determine environmental concentrations of ENMs (TiO_2, ZnO, Ag, fullerenes, CNT, and CeO_2). One conclusion from this review was that environmentally relevant ENM exposure levels are significantly lower compared with concentrations typically employed in ecotoxicity testing. Thus, results of many of the published studies may not represent realistic scenario impacts in the environment.

17.1.1.3 Impact by Biota

The tendency of ENPs to aggregate/agglomerate in the aquatic environment inevitably leads to their sedimentation and accumulation in bottom sediments where sediment-dwelling organisms may be at particular risk. Bioturbation processes (i.e., burrowing or irrigation activity by benthic species) may lead to the remobilization of sediment-associated ENPs (Buffet et al., 2012).

17.2 Hazard: Consideration for Test Protocols

As illustrated in section 17.1, even when the basic characteristics of the physicochemistry of an ENP have been established, there remain many challenges before we understand the environmental behavior of ENPs in the aquatic environment. These differences need to be taken into account when designing experimental approaches because they will have implications for bioavailability, accumulation, and thus toxicity. This is in contrast to tests with conventional chemicals, such as metals and organic compounds, in which we have a good understanding of their fate and are able to keep a fairly constant exposure concentration in standardized water-only toxicity test and, in addition, have the possibility to measure actual exposure concentrations at experimental beginning and end as well as during exposure.

ENPs transformation in test media (e.g., water, in vitro media) is fast and rather unpredictable (see Petersen et al., 2014 for a detailed discussion). Thus, exposure is not assumed to

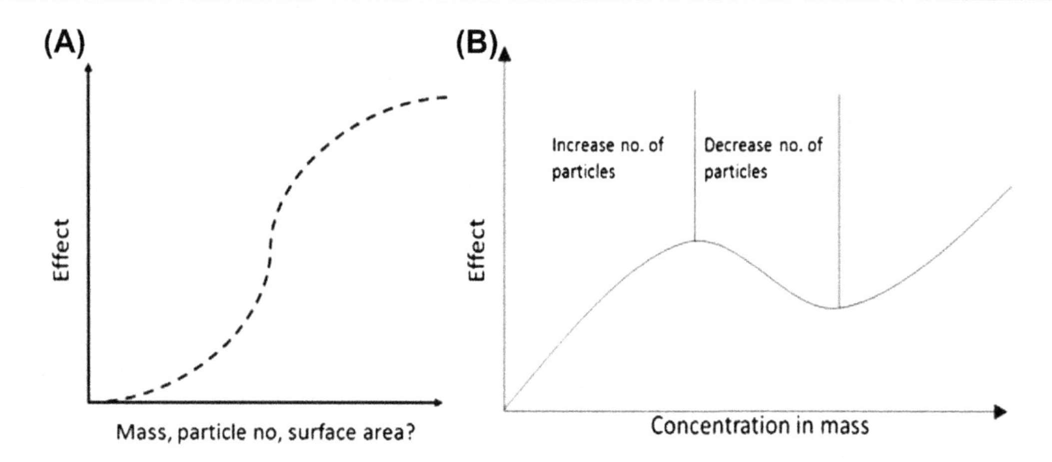

Figure 17.1
Conceptual dose-response relation of ENPs. (A) A classical sigmoid dose-response curve, where the effect (e.g., median effect concentration) is dependent on the dose metric used. (B) How agglomeration and/or aggregation might result in a nonmonotonic dose-response curve because of, for example, decreasing the number of particles and/or surface area.

be constant. The challenge is that the current ability to characterize ENPs and their new environmental identity in experimental samples is limited. In addition, aggregation is considered concentration-dependent such that aggregation likely increases with increasing concentrations, making the dose–response not just two- (increased effect with increasing concentration) but three-dimensional (Baun et al., 2009) (Figure 17.1). The implication is that the relationship between dose and effect is hard to establish and therefore we need to be very careful in interpreting results (Petersen et al., 2014). Acknowledging that exposure concentration is difficult, if at all possible, to keep constant, other experimental approaches including modifications to standard tests are evolving for both water and sediment/diet exposures. This section focus on ecotoxicological test systems and the challenges we are facing regarding ENP hazard assessment (test protocols, uptake routes, target organisms, bioaccumulation, and toxicity).

17.2.1 Experimental Considerations

When designing experiments, both practical experimental issues and indirect effects impacting accuracy of the obtained experimental results must be considered. Practical issues include cleaning of glassware before use, choice of glass versus plastic exposure chambers, and assessment of if or when to change test media during the test. Furthermore, decisions regarding the use of reference/control treatments (e.g., ionic metal form, coating control) and particle size controls (e.g., inert particles, bulk form) also need to be addressed. The indirect effects relate to experimental artifacts, which need to be assessed both in designing and analyzing experimental results (Petersen et al., 2014).

17.2.1.1 Practical Experimental Issues

The choice of cleaning type will depend on the specific ENP to be tested, but, in general, nitric acid or aqua regia is adequate for most ENPs (Handy et al., 2012). Regarding the choice of glass versus plastic exposure chambers, a general rule is to use high-quality glassware but other practical factors may require that plastic ware is used instead of glass. In this case, care must be taken to select the most suitable kind and not to reuse but dispose these after use. For sediment exposures, hard plastic ware (i.e., without softeners) is appropriate as ENPs are thought to have higher affinity for sediment and sediment components than for plastic. Because of surface tension and/or release of chemicals potentially sorbed to plastic surfaces, it is recommended to soak plastic ware (e.g., multiwells) in water overnight before experimental use.

Assessment of if or when to change test media during the test depends on whether an acute or chronic test is to be conducted. The overall idea is to keep the exposure concentration constant. For acute tests, either static (i.e., no change of test media) or semistatic (i.e., regular renewal of all or fraction of test media) systems are the most common approaches but variation in exposure concentration is likely in any case. In chronic tests, the test media is often required either to be changed (e.g., water renewal, flow-through) or rinsed (e.g., filtration) to ensure a good water quality, constant chemical concentration, and adequate food levels. Because of ENP aging and transformation in all test systems, evaluating the pro and cons of using static, (partly) renewal, flow-through, or filtration test setups (for a detailed discussion, see Handy et al., 2012) is required. For nonstandard tests in which the focus may be on the impact of ecological parameters (e.g., pH, salinity, temperature, water flow, organic matter) and their importance for ENP availability and toxicity, researchers might want to avoid renewing the test media to be able to assess the potential impact of ENM transformation and aggregation.

17.2.1.2 Preparation of ENPs for Experimental Protocols

ENPs may either be purchased or synthesized using different methods, which will determine size, shape, and coating. Because discrepancies between the size of the ENPs reported for purchased particles and actual size distribution have been reported, it is important to always characterize purchased ENPs before using them experimentally. Information on storage condition of newly synthesized ENPs and time to actual experimental use are important to report as physicochemical transformations may appear during storage (e.g., particle dissolution). The effect of storage on AgNP toxicity was illustrated by Kennedy et al. (2012), who showed that toxicity to zooplankton, *Ceriodaphnia dubia*, increased with increasing storage time, probably as a function of increased Ag ion release. Gorham et al. (2014) suggest to store Ag NPs in the dark (reduce photooxidation) in dispersions bubbled with nitrogen and containing the highest Ag and coating (in this case, citrate) concentrations, and Petersen et al. (2014) recommend to add a purification step to remove impurities from the ENP dispersion before testing.

Differences in experimental test media (freshwater, marine, cell media, etc.), and exposure route (water, sediment/food) vary greatly and will have implication for ENP fate (e.g., aggregation, interference with assay reagents, particle dissolution to ions, creation of smaller-sized particles) and thus for the test results (bioavailability, uptake/depuration kinetics, internal fate, toxicity) (Figure 17.2).

Hence, comparisons among experiments are difficult even when comparing the same particle type. Preparation of ENM/ENP dispersions before addition to test media by sonication (high-frequency sound methods enabling disruption of agglomerates) or stirring, may be advantageous and necessary in cases where this will lead to higher dispersion and increased homogeneity of the exposure. Another advantage is that sonication and stirring do not require

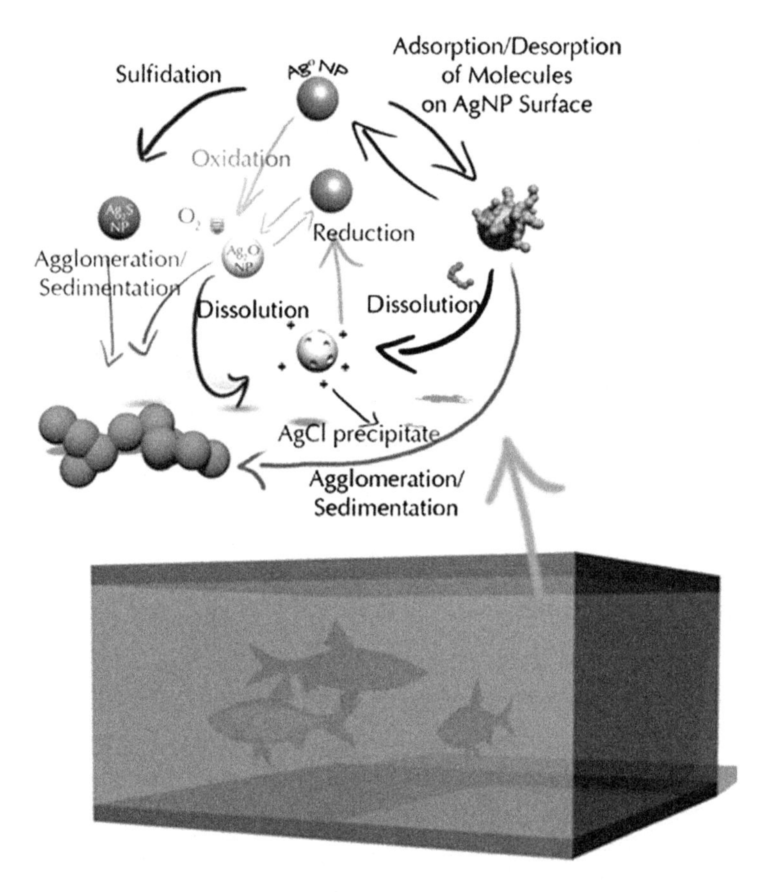

Figure 17.2
Possible physicochemical transformations of silver nanoparticles during storage or ecotoxicology testing with aquatic organisms. Red lines (dark gray in print) indicate transformations that remove the AgNPs from the aqueous phase. Yellow lines (light gray in print) indicate transformations that can occur as a result of laboratory light. Black lines describe transformations that can occur in the aqueous phase in the dark. *After Petersen et al., 2014, with permission.*

addition of chemicals. However, disadvantages have also been observed. For example, fullerene nanocrystal was found to be more toxic to fish (fathead minnow, zebrafish larvae) and crustaceans (e.g., *Daphnia*) when prepared with stirring and use of the solvent tetrahydrofuran (THF) when used without THF. However, the difference in toxicity seemed not to be related to fullerene toxicity but rather to the formation of byproducts in the presence of THF (see discussions in Baun et al., 2009; Petersen et al., 2014). Artifacts arising from other methods such as ultracentrifugation are discussed in Crane et al. (2008), Handy et al. (2012), and critically reviewed in Petersen et al. (2014), who recommend limiting the intensity and time of sonication to what is minimally needed to obtain the required ENP size. Standard operating procedures concerning batch dispersion of ENPs and sonication (duration, intensity) are being implemented for regulatory purposes.

17.2.1.3 ENP Characterization

ENPs differ from most conventional "dissolved" chemicals in terms of their heterogeneous distributions in size, shape, surface charge, composition, and agglomeration state in environmental media, etc. Therefore, it is necessary to determine and present several ENP matrices, along with mass doses of dissolved chemicals. The appropriate physicochemical characterization parameters identified are measures of particle size (including particle-size distribution), particle shape, number of particles, aggregation/agglomeration, dispersability, dissolution (solubility), surface area, surface charge, surface chemistry/composition (e.g., coating), and mass concentration (mg L^{-1} or μg g^{-1}) (Stone et al., 2010). Ideally, these parameters should be determined in three steps including characterization of particles (1) as synthesized or as received in solution or in its dry native state, (2) in suspension in aqueous medium (e.g., ultrapure water), and (3) in the exposure medium including measurement of dissolved and particulate forms of metal ENPs. In addition, characterization data should include information concerning ENP synthesis and storage condition (see the previous section). Furthermore, it is recommended that sufficient information is provided concerning the test media (e.g., water, sediment, food) to enable reassessment of the results once we have a better understanding of ENP interactions with abiotic and biotic components. This includes, as a minimum, information on temperature, oxygen level, pH, salinity, light cycle, composition of water/media (e.g., for artificial water), and sediment (e.g., sediment particle size fraction, organic matter content, collection site).

A wide range of analytical tools is currently available for initial material characterization whereas complex media pose more challenges. The first challenge is that, for low concentrations (e.g., ng g^{-1}), the detection limit for most available techniques are not sufficiently sensitive. Second, there is a high natural background of NPs in environmental media. Characterization techniques of ENPs reviewed by von den Kammer et al. (2012) include microscopy-based approaches (transmission or scanning electron microscopy), centrifugation, dynamic light scattering, voltammetry, and size separation methods (e.g., hydrodynamic chromatography such as field flow fractionation techniques) coupled to analytical instruments (ICP-mass spectometry or ICP-atomic emission spectroscopy). Transmission electron microscopy allows for the highest magnification of nano-sized structures but solely from a two-dimensional perspective, whereas scanning electron microscopy is particularly useful for its ability to create three-dimensional images. Dynamic light scattering currently represents the most common technique for spherical particle size analysis. Outfitting a dynamic light scattering instrument with a special cell that allows for an applied voltage provides measures of particle charge or zeta potential. However, all these techniques have inherent limitations in terms of counting statistics or sensitivity. Thus, to confirm the primary particle size and the size distribution of NPs in the exposure medium, it is recommended to use more than one technique (Handy et al., 2012). Internal distribution of ENMs in the different organs of organisms and their cellular localization can be assessed using advanced microscopic and imaging techniques such as X-ray absorption spectroscopy. ICP-MS coupled to laser ablation can be used in addition to X-ray based two-dimensional images to enhance the limit of detection. Techniques allowing the discrimination between ENPs and other particles (natural NPs or colloids) need considerable development. However, application of tracer techniques such as stable isotopes labeling are well suited to distinguish metal added with NPs from metals naturally present in sediment and biota (Luoma et al., 2014; Mouneyrac et al., 2014). More recently, synchrotron-based X-ray absorption techniques have been used to probe chemical speciation and the local electronic structure of elements. However, sensitivity is in the milligram-per-kilogram or submilligram-per-kilogram range. Development and adaption of single-particle analysis methods seem promising because they are limited by the smallest detectable particle size and volume of a sample rather than by total mass (von den Kammer et al., 2012).

17.2.1.4 Use of Reference (Control) Treatments

To minimize artifacts of ENP behavior and thus misinterpretations related to the unpredictable transformation in the test system, control treatments have been implemented in many ecotoxicological tests. For metal ENPs, micronsized metal particles and/or the ionic metal form have been implemented as reference treatment (e.g., Heinlaan et al., 2008; Croteau et al., 2011a; Dai et al., 2013). For organic ENM, like fullerenes, graphite has been used as a comparison. Preparation of aqueous C_{60} dispersions has included dissolution in organic solvents (C_{60} is insoluble in water), transfer of the dispersion to water, and subsequent evaporation of the solvent. Unfortunately, this procedure leads to the formation of THF decomposition products, which are toxic. Consequently, results have been interpreted as C_{60} toxicity when, in fact, it was the decomposition products that produced toxicity (Henry et al., 2007). Petersen et al. (2014) suggest including an extra control consisting of the decomposition products (i.e., the filtrate passing through a filter size less than the ENM size) to assess toxicity of these products.

Regarding sediment systems, the biggest challenge we are facing is associated with characterizing and quantifying ENMs/ENPs in sediment. There is a severe lack of understanding of what happens once the ENMs have been introduced to the sediment compartment. It is still an open question if ENMs distribute homogenously (e.g., such as single ENPs) or heterogeneously (e.g., patchy distribution of agglomerates) when introduced to sediment and whether this is concentration dependent as suggested for water tests and/or dependent on ENP size, shape, surface charge, and coating. Consequently, it is not possible to relate toxicity to actual exposure concentration. To progress ecotoxicology of ENPs, reference treatments have been and should be included in sediment tests (e.g., Pang et al., 2012; Dai et al., 2013; Ramskov et al., 2014). Certified reference ENMs have been proposed by international organizations (ISO, Organization for Economic Co-operation and Development [OECD]), government agencies (e.g., US National Institute of Standards and Technology), and research centers (Joint Research Center–Institute for Reference Materials and Measurements, European Commission). Depending on which reference treatment is included, there are several interpretations that can be made, and if evidence suggests the nanoscale material has different toxicological properties, then this might warrant a full series of ecotoxicity tests (Handy et al., 2008).

17.2.1.5 Experimental Test Systems

The use of standardized test systems to test ENM toxicity is questionable. Handy et al. (2012) provide a brief discussion regarding the *Daphnia* immobilization test and the acute fish test and a detailed discussion for the algal test (OECD 201, USEPA, 1996). Briefly, methodological factors that may vary somewhat among tests, such as shaking the test vessels, light regime, and sonication may have a critical impact on toxicity toward algae and it is therefore recommended, among other factors, to standardize such parameters especially for ENMs that absorb light or with photoreactive properties. Furthermore, cell-counting methods such as flow cytometry counting may be interfered with by ENPs. ENM toxicity endpoints seem not to be limited to traditional toxicity endpoints. For example, ENM sorption to organism surfaces in *Daphnia* may potentially impact swimming and feeding ability as well as respiration. Fish tend to produce mucus, which may interact with ENMs, making them agglomerate or aggregate. ENM-caused shading of algae and nutrient depletion caused by sorption of nutrients to ENM surfaces can reduce growth (Petersen et al., 2014). Nutrient availability will theoretically decrease with increasing ENM concentration in a dose–response setup. Behavioral change concerning fin nipping and aggression has been observed in trout exposed to single-walled carbon nanotubes (Smith et al., 2007). Whether the aggressive behavior stemmed from single-walled carbon nanotubes exposure or from related reduced food levels or a combination still needs addressing.

Sediments are recognized as accumulation sites for ENMs and benthic organisms may therefore be at particular risk of ENM exposure and will constitute an important path for trophic transfer. Sediment organic matter quantity is low (usually $<2\%$ weight), and the quality differ greatly covering organic detritus, refractory to labile (easily digestible) organic matter as well as microorganisms (e.g., bacteria, microphytobenthos) that can create biofilms on most sediment surfaces. ENMs likely interact with sediment organic matter and, depending on the type (quality) of organic matter, ENMs will be more or less available for dietary uptake in sediment-dwelling organisms. A general approach to test sediment-associated ENMs is initiated by adding (dispersed) ENMs to wet processed natural sediment, then mixing the sediment slurry on a shaking table for days to achieve a homogeneous ENM distribution. The spiked wet sediment is subsequently transferred to the test vessels, left to equilibrate with overlying water, and finally test organisms are added and left in the system for a given amount of time (hours to weeks) (Cong et al., 2011; Pang et al., 2012; Dai et al., 2013; Ramskov et al., 2014).

17.2.2 Uptake Routes and Biological Targets

Three main biological targets can be identified as of particular importance regarding ENM exposure: (1) filter-feeders, which will be exposed to high ENM concentrations including single ENM particles and ENM agglomerates in overlying water, and via food particles; (2) pelagic organisms, including algae, invertebrates, and fish, which will be exposed to ENMs and their transformation products through the entire water column; and (3) benthic species, covering epi- and infaunal organisms, which will be exposed to ENMs associated with sediment, biofilm, and other food items. These three groups should in principle be covered when assessing effects in both freshwater and marine systems. Some of the most applied tests regarding ENMs include the three pelagic organism groups recommended for risk assessment purposes: algae, daphnia, and fish. Here the algae growth test (OECD 221), *Daphnia* immobilization test (OECD 202), and the fish toxicity test (OECD 203) are widely used (OECD, 2004, 2006, 2009). However, because of the tendency of ENMs to aggregate and subsequently precipitate to sediment, it is argued that benthic organism (e.g., Pang et al., 2012; Handy et al., 2012; Mouneyrac et al., 2014; Ramskov et al., 2014) and filter-feeder tests (Canesi et al., 2012) are needed. Which specific species should be included from each group is not trivial and still to be decided. For example, benthic organisms vary widely regarding behavior (some are sessile, some have limited movement, and yet others are mobile) and feeding mode (grazers, filter-feeders, and deposit-feeders). Some feed at the sediment surface, others in depth defecating on the sediment surface, some are particle-selective, and some have internal particle-sorting. It is therefore of utmost importance that the biology of the test organisms is taken into account when designing sediment-/food-exposure tests to ensure that exposure resembles the feeding mode. For example, a deposit-feeder that processes huge amounts of sediment daily (increasing exposure through the gut) should be chosen when assessing uptake and toxicity of sediment-associated ENMs; a grazer should be selected for examination of uptake and toxicity of ENMs in biofilms; an epi-faunal filter-feeder when assessing ENM uptake at the interface between sediment and overlying water.

17.2.3 Engineered Nanomaterial Accumulation

Although bioaccumulation is an important part of hazard assessment, there is surprisingly little information in the literature. Most accumulation studies have focused on water exposure and only a limited number exists concerning sediment/food exposure. Net accumulation, and subsequent toxicity, is a function of processes related to absorption, distribution, metabolism, and elimination. Evidently, for ENMs to be available for uptake, they have to be in contact with the surfaces of biota (outer surface or gut surface) and this will depend on the physicochemical properties of the ENM. Considering uptake across membranes, endocytosis is considered the main route over diffusion across the cell membrane (Figure 17.3).

Most emphasis has been placed on body burden of bulk ENPs (e.g., total concentration of metal in organisms exposed to metal-based ENPs) and less emphasis has been directed toward uptake and elimination kinetics and even less on internal distribution, likely because of methodological and analytical limitations. It is not known whether steady state is achievable (Handy et al., 2012). During the past couple of years, more focus has been directed toward the need for assessment of ENP bioaccumulation and results, primarily on whole-body burden, are beginning to appear in the literature. However, routine methods for detecting ENPs in tissues are in the developmental phase and needed to increase our understanding. Therefore, the pathway for ENM uptake and elimination is one of the areas we need to focus future research (Kettler et al., 2014).

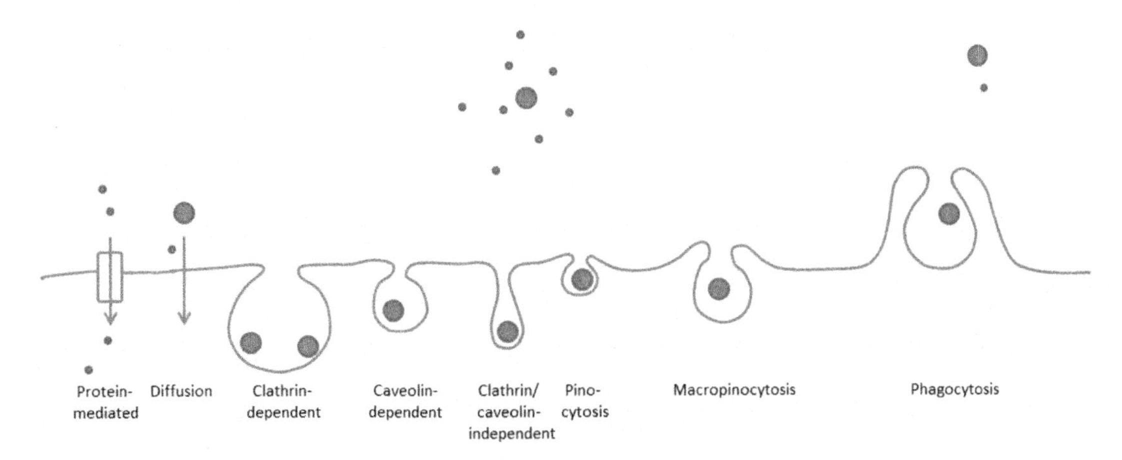

Figure 17.3

Schematic overview of potential intracellular uptake pathways for nanoparticles. Nanoparticles are most likely taken up via endocytosis. Endocytosis occurs in all eukaryotic cells, but the forms of endocytosis (size of formed vesicles and molecular process) may be organism- and cell type–specific. Clathrin- and caveolin-dependent endocytosis are initiated at clathrin coated pits and caveolae in the plasma membrane, respectively, occurs in many different varieties, and can be vesicular or tubular, as illustrated. Pinocytosis and macropinocytosis concern uptake of fluids and solutes. Phagocytosis only occurs in free-living unicellular organisms and specialized cells of higher organisms, such as neutrophils and macrophages, and results in the formation of larger vesicles. Alternatively to endocytotic uptake, NPs may diffuse through or disrupt the plasma membrane, and metal nanoparticles may dissolve into ions, which via protein-mediated transport may be taken up over channels or transporters (including metal specific transporters, such as the Cu transporter). *Figure and text by Amalie Thit, Roskilde University, DK.*

Only few studies have investigated ENM bioaccumulation through the aquatic food chain. Usually, results showed that dietary exposure and bioaccumulation occur although biomagnification was not detected (Wong et al., 2013). Dietary uptake is increasingly recognized as a pathway of ENP uptake, especially in sediment-dwelling organisms. So far, most studies have focused on comparing metal ENPs of different sizes using the ionic/dissolved, and potentially also the bulk, metal form as reference. The dependence of form and shape of ENPs for uptake, elimination, and toxicity has been addressed in a few studies focusing on the sediment compartment and metals (e.g., Cong et al., 2011; Dai et al., 2013; Pang et al., 2013; Ramskov et al., 2014). However, the degree to which metal ENPs in different sizes and shapes are more or less available than the dissolved metal form, bulk, or micro-sized metal particles are not straightforward. ENP shape effect on accumulation may be species-dependent and potentially related to aspects like digestive complexity and feeding mode. On the other hand, bioaccumulation of similar-sized CuO NPs (6 nm) seems to differ among species with similar digestive complexity (Dai et al., 2013; Ramskov et al., 2014). In contrast, Buffet et al. (2011) reported that the polychaete *Nereis diversicolor*, and the clam *Scrobicularia plana*, accumulated sediment-associated Cu equally whether Cu was added in the ionic or particulate form. In conclusion, different species are able to accumulate similar-sized/shaped metal NPs to different degrees. Regarding intracellular compartmentalization, García-Alonso et al. (2011) reported that dissolved Ag was accumulated in the metallothionein fraction in *N. diversicolor*, whereas Ag NP was found in the inorganic granules, organelles, and heat-denatured proteins. There is, however, limited literature on this topic, which is mostly because of the lack of standardized methods to characterize ENPs in tissues. During the past couple of years more focus has been directed toward this research area and recent development in methods to analyze internal distribution has been implemented in a number of laboratories (see section 17.1.).

17.2.4 ENM Toxicity and Endpoints

The literature on ENM ecotoxicity is still emerging, and reviews have been published (e.g., Handy et al., 2008, 2012; Klaine et al., 2008; Wong et al., 2013). Most of the currently available ecotoxicological data are limited to species used in regulatory testing of freshwater species (algae, daphnia, fish), using short-term water-only tests, high concentrations, and pristine ENPs.

17.2.4.1 Whole Body Endpoints

There are studies suggesting that metal ENPs are more toxic than bulk forms and less toxic than ionic metal forms. For example, Heinlaan et al. (2008) reported that CuO NP toxicity was 44–51 times higher than for micron-sized CuO in the bacteria *Vibrio fischeria* and the crustaceans *Daphnia magna* and *Thamnocephalus platyurus*. These authors also reported that Cu toxicity was higher when Cu was added as the ionic metal form compared with CuO NPs. Recently, 317 toxicity values (median effective concentration and lethal concentration or minimal inhibitory concentrations values) were compared between algae, crustaceans, fish, bacteria, yeast, nematodes, protozoa, and mammalian cell lines exposed to Ag, CuO, and ZnO NPs. As a general rule, crustaceans, algae, and fish proved most sensitive to the studied ENPs (Bondarenko et al., 2013). Effects of sediment-associated Cu in *Potamopyrgus antipodarum* have been shown to depend both on the form and shape of Cu added to sediment (Pang et al., 2012; Ramskov et al., 2014), as exemplified for Cu, which was generally more toxic when added as CuO ENP compared with the soluble form ($CuCl_2$). For silver, endpoints such as mortality and growth do not generally show acute toxicity at concentrations up to 200 $\mu g\ g^{-1}$ Ag dry weight sediment regardless of form (ionic, particulate) and particle size added to the sediment, whereas *N. diversicolor* burrowing time increased in Ag ENP treatments (Cong et al., 2011, 2014). Studies focusing on behavior, reproduction, development (early life stages, embryos, and larvae), and transfer to progeny are needed to clarify the overall potential adverse effects of ENPs on aquatic organisms. Behavioral impairments and decreases of feeding rate have been frequently observed in endobenthic organisms (*N. diversicolor*, *S. plana*) exposed to different metal-based ENPs compared with the soluble forms (Mouneyrac et al., 2014). Adverse effects on feeding and digestive processes have been shown in the freshwater snail (*Lymnaea stagnalis*) exposed to ZnO ENPs with implications for growth and reproduction (Croteau et al., 2011b). In zebrafish (*Danio rerio*) exposed to quantum dots (QDs), developmental and behavioral abnormalities have been shown (Zhang et al., 2012). Recently, Blickley et al. (2014) showed a trend of declining fecundity in the reproductively active adult fish (*Fundulus heteroclitus*) exposed to dietary CdSe/ZnS QDs, suggesting that chronic QD exposure could have consequences at the population level. QD uptake and maternal transfer of QDs or their degradation products to developing progeny may pose a threat to future generations of aquatic organisms.

17.2.4.2 Subcellular Effects

There are indications that subcellular effects may be more sensitive for ENPs than other forms. For example, even though whole body effects were not observed in *N. diversicolor* exposed to sediment contaminated with different Ag forms (Ag ion, Ag ENPs), subcellular endpoints were more responsive for ENPs (Cong et al., 2014). Changes in structural and physicochemical properties of ENPs can lead to changes in biological activities including reactive oxygen species generation, one of the most frequently reported NP-associated toxicities. Oxidative stress is a parameter that is convenient to measure because cells respond to oxidative stress by induction of protective or damage responses that can easily be measured as enzymatic activity responses. The well-known sensitivity of DNA to oxidizing agents, suggest that genotoxicity (revealed by comet assay, micronucleus induction) may be a potential mechanism of toxicity. Among biochemical tools, glutathione S-transferase and catalase were the most sensitive revealing the enhancement of anti-oxidant defenses in animals exposed to sublethal concentrations of ENPs (Mouneyrac et al., 2014). Lysosomal damage, alterations in immune parameters, apoptosis, and genotoxicity have also been frequently observed (Cong et al., 2014; Mouneyrac et al., 2014). Moreover, to unravel molecular mechanisms associated with ENP toxicity, omic (genomic, proteomic, transcriptomic) approaches are recommended (Matranga and Corsi, 2012; see Chapter 8).

17.3 Risk Assessment

Despite the challenges concerning the quantification of exposure and hazard as described above, the OECD (and other organizations) believes that the existing risk assessment (RA) framework for conventional chemicals is essentially applicable for ENPs (OECD, 2012). It is acknowledged however that there are nano-specific challenges that have to be addressed. In 2009, the US National Research Council (NRC) reevaluated its 1983 RA strategy with the aim to improve the "utility" of the risk assessment paradigm and thereby reduce the "paralysis by analysis[1]" problem (NRC, 2009). The improved utility refers to a more flexible and thus more specific RA where the problem formulation phase is set up to identify the exact scope of the specific RA at hand as well as possible risk management options. The RA scope is therefore both to provide assessments of the risk related to the different risk management options as well as to answer the problems formulated. For example, in evaluating whether the use of an ENP in a specific production poses new risk compared with using a non-nano counterpart, it would be relevant to identify whether there are critical data gaps regarding nano-specific effects, and if so, what steps should be taken. This type of "utility" is especially important considering "new and emerging risks," because such risks are often relatively poorly understood and uncertainties concerning actual risk thus play a dominant role. ENP still classifies as "new and emerging" and the many uncertainties described in this chapter makes it important to apply an RA approach where the utility is sufficient to allow for risk assessment to account for all relevant nano-specific aspects, even when standard risk assessment tools might not provide adequate estimates of the risk. The NRC improved risk assessment strategy is recommendable, since it, in principle, identifies all relevant nano-specific aspects, and either address them with measurements or in-build uncertainty analyses. The improved framework consists of three phases (NRC, 2009, Figure 17.4). In this chapter, we focus on the nano-specific aspects that should be considered when conducting a risk assessment based on the NRC approach.

17.3.1 Phase I: Problem Formulation

In the first phase, relevant stakeholders are involved in problem formulation and scoping aiming at identifying relevant risk management options. Technical options (including RA

[1] Paralysis by analysis is a term that refers to a situation in which the lack of scientific understanding blocks any regulatory action, because the system awaits more data. The system is thus paralyzed while more and more analyses are being conducted, even though regulatory action is required.

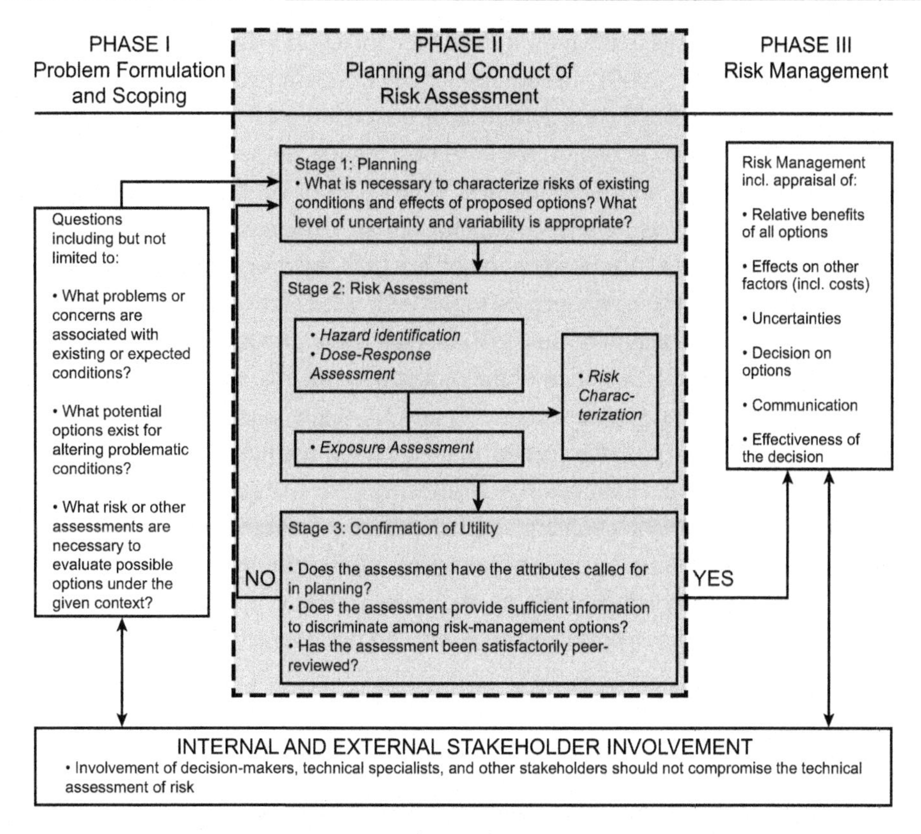

Figure 17.4

The improved risk assessment (RA) framework suggested by NRC in 2009. Phases I-III aim at improving the technical analysis supporting decision-making and enhancing utility by enabling more focused RA. Phase I: aims at focusing RA by defining the problems to be addressed; phase II (gray box): the traditional RA part; phase III: risk management (RM) phase emphasizing decision-making *Modified after NRC, 2009.*

analyses) needed to conduct risk management are identified and their appropriateness evaluated to focus the RA (NRC, 2009; see Figure 17.4 for examples of process facilitating questions). The relevant risk management option(s) vary depending on parameters, such as production type, determining the relevant regulatory legislation as well as the use and disposal-to-waste phase determining who is at risk (e.g., workers, consumers, aquatic environment).

17.3.2 Phase II: Planning and Conducting Risk Assessment

Phase II is where the actual risk assessment is conducted (NRC, 2009). It initiates with the important planning phase (stage 1), where the goal and aim of the risk assessment are identified and execution is planned. The planning phase mainly address two questions: (1) What information is necessary to characterize risks in the relevant risk scenarios? (2) What level of uncertainty and variability is acceptable (NRC, 2009)? In addition, other considerations include whether model data can substitute actual measured data, definition of protection goal(s), selection of appropriate endpoint(s), and finally analyses of achievable level of scientific certainty including the current scientific understanding and technical capabilities. This is paramount because nonachievable data demands can lead to the paralysis-by-analysis dilemma (e.g., demands for characterization of ENP in all relevant compartments) or misclassification of risk (e.g., by only accepting data produced using standard test systems). The nano-specific topics addressed earlier should be considered in this light because the complex nature of the ENP can have great importance for both exposure and hazard scenarios (OECD, 2012). Selection of the proper dose metric is another important consideration, which will depend on exposure scenarios, particles nature, and protection goal(s), as described in section 17.2. Extrapolation scenarios are another important topic to consider. For example, is it feasible to consider extrapolation from other ENP risk scenarios and does data support read-across from non-nano counterparts such as larger particles with similar composition? Based on the output of this discussion and once the areas of uncertainty have been identified, the actual risk assessment can be conducted, by the traditional four step approach: hazard identification, dose–response assessment, exposure assessment, and risk characterization (NRC, 2009) as described in Chapter 2.

17.3.2.1 Hazard Identification

The toxicity of any ENP depends on factors such as the physicochemical properties of the surface coatings, ENP composition and the extent to which the ENP interact with biological systems and relevant environmental components (OECD, 2012). As described previously, many of these aspects are poorly understood and are thus governed by scientific uncertainty (SCENIHR, 2006). Furthermore, standard toxicity tests, such as *Daphnia* immobilization test, might be inadequate for identification of nano-specific effects (subsection 17.2.1). Subcellular biomarkers seem generally to be more sensitive to ENP than standard in vivo tests and may provide valuable insight in mechanisms of toxicity. However, because the link between these subcellular effects and effects at organism or population level are poorly understood, even for conventional chemicals, care should be taken interpreting results in a risk assessment context. To achieve the highest degree of transparency, such uncertainties in hazard identification must be properly communicated to decision-makers. Uncertainty analysis can further inform on the most important uncertainties and thus help guide further research. Nano-specific properties can also lead to an increase in testing variability (i.e., fate may differ even among similar test vessels thus affecting toxicity), which can further complicate hazard identification (OECD, 2012).

17.3.2.2 Dose–Response Assessment

The aim of this stage is to quantify the hazardous potential of the ENP and thereby determine a threshold below which no toxicity is expected (e.g., predicted no effect concentration) based on classical toxicity parameters such as median lethal (effective) concentration ($L(E)C_{50}$). Several nano-specific aspects complicate this process, including selection of a dose metric (see the previous section). This is a critical issue, because the quantification of ENP toxicity, and thus the quality of the subsequent threshold values (i.e., predicted no effect concentrations) depend on the selection of the appropriate metric (Oberdörster et al., 2005) (Figure 17.1). Suggestions to use more than one dose metric have been proposed (Baun et al., 2009). Differences in output (threshold value) in relation to dose metric should be properly communicated. Another important challenge concerns ENP agglomeration and aggregation, which inevitably will complicate dose–response assessment (both dose and toxicity) (Bae et al., 2013).

17.3.2.3 Exposure Assessment

Exposure assessment should take the entire life-cycle of the ENP into consideration, including production patterns, use, and disposal to the waste phase as well as environmental distribution in ecological compartments (ECHA, 2013). This requirement poses several challenges to ENP exposure assessment, largely because of our limited ability to analyze ENM in environmental samples, thus impacting our ability to predict ENM concentrations in the environment (SCENIHR, 2009). Nanotechnology development is largely observed for small- and medium-sized enterprises as illustrated in Germany where about 80% of nanotechnology companies are small- and medium-sized enterprises (Research in Germany (2014)). Compiling life-cycle assessment data from the great number of small- and medium-sized enterprises makes estimation on actual production patterns and subsequent release (i.e., to wastewater treatment plants and the environment) complex and challenging.

17.3.2.4 Risk Characterization

The last step in stage 2 is risk characterization. Because risk characterization depends on the quality of scientific data on hazard and exposure, it is obviously also governed by the uncertainties mentioned previously. Because standard tests for ENP hazard and exposure assessment is currently lacking, risk characterization depends largely on expert evaluations. It is therefore important that any consideration and uncertainties are thoroughly reported to decision-makers.

17.3.2.5 Stage 3: Confirmation of Utility

This final stage of phase 2 serves as a quality check of the risk assessment using questions as illustrated in Figure 17.4. The purpose of this stage is to ensure that the risk assessment and risk characterization provide the answers needed to inform decision making, regarding the scope identified in the problem formulation phase.

17.3.3 Phase III: Risk Management

Upon having finalized phase II, any relevant risk management option is considered by risk managers. Risk management is informed by risk assessment but include other aspects in addition to risk assessment, such as cost (including socioeconomic interests), effectiveness of management, and possible benefits during this phase. Risk communication is also considered under phase III.

17.4 Conclusions and Recommendations

Determining the effects of ENMs in environmental and biological systems is complex and requires sound characterization of ENMs. However, scientific knowledge is challenged by a lack of adequate techniques for the detection and quantification of ENPs at environmentally relevant concentrations in complex media. At present, methods and techniques are being developed to characterize ENPs in complex experimental compartments (e.g., water, sediment, food, tissue) as well as in environmental samples. However, these methods are not yet implemented in regular laboratories and are challenged regarding distinguishing between natural and engineered particles and with respect to detection limit (Baun et al., 2009; von den Kammer et al., 2012). Another challenge concerns the role of environmental factors such as fluctuations in, for example, salinity, temperature, organic matter content for ENP transformation (i.e., creation of new environmental identity), and thus for environmental fate.

Once characterization methods have been implemented in regular laboratories, we will be able to relate exact dose to observed effects and to monitor environmental ENM concentrations and thus to perform sound risk assessment. Until such methods are available, we recommend that efforts should be taken to carefully characterize the particles before addition to experimental systems and that care should be taken to report observations on settling, agglomeration, and dissolution in the experimental media (e.g., using diffusive gradients in thin film tools, dialyze bags). Information should be provided on (1) ENM storage time, storage condition, processing history (from synthesis to actual use); (2) experimental system, including environmental parameters such as pH, temperature, light regime, water composition, sediment composition; and (3) biological parameters like feeding rate that will allow calculation on amount ENMs passing through the gut. Such a detailed information level will enable comparisons between datasets on different species, different systems and different levels of biological organization both with the same and different nanomaterials. Furthermore, and very importantly, it will enable retrospective assessment of datasets to normalize to, for example, organic matter content and to sort datasets for particle size effects, surface area effects, or confounding effects of solvents or impurities. In any cases, the artifacts of the test system and their implication for hazard assessment should be critically evaluated before performing the experiment and be presented and discussed when analyzing the results. For example, when characterization is not possible, the assessment of results should keep this limitation in mind and explicitly state that it is recognized that the ENM inevitably will undergo transformation once introduced to the test media. Furthermore, it is recommended to include relevant reference materials (e.g., as suggested by OECD) and potential standards in ecotoxicological testing. For instance, relevant reference systems for metal ENPs may encompass: (1) ionic metal form; (2) bulk form; and (3) an inert form like nanodiamonds to mimic particle effects (Paget et al., 2014).

The choice of biological models (e.g., filter-feeders, pelagic and benthic organisms) for investigating the effects and mechanisms of action underlying the potential toxicity of ENMs still deserves careful thought. Regarding toxic effects, the application of lysosomal, oxidative stress, immunotoxicity, and physiological biomarkers represents a sensitive tool for evaluating the mechanisms causing effects of different ENPs in aquatic invertebrates. Omic approaches need to be investigated to unravel molecular mechanisms associated with ENP toxicity. To fill the gap existing between (sub)organismal responses to NPs and effects occurring at higher levels of biological organization, ecologically relevant biomarkers such as behavior and genotoxicity appear as promising candidates because their responses could impair development, reproduction, and survival of individuals and finally contribute in endangering populations.

We recommend the following immediate research areas to progress ENM ecotoxicology:

- There is limited information on chronic effects, bioavailability, bioaccumulation, and internal deposition and especially so for diet/sediment exposure. Therefore, fundamental research covering these research areas is needed to better understand uptake mechanisms and subsequent internal fate and how this links to toxicity.
- The complex chemical and physical interactions and behavior of ENPs with other components in natural environments need further development to better understand fate, behavior, bioavailability, and toxicity of these particles in aquatic systems. This will furthermore allow us to assess consequences of ENM transformation in natural environments including impact of seasonal and yearly fluctuation for ENM hazard assessment.
- We need to get a better understanding of the relation between short- and long-term effects, including observations of delayed effects as well as trophic transfer.

References

Baalousha, M., Manciulea, A., Cumberland, S., et al., 2008. Aggregation and surface properties of iron oxide nanoparticles: Influence of pH and natural organic matter. Environ. Toxicol. Chem. 27, 1875–1882.

Bae, J., Lee, B.C., Kim, Y., et al., 2013. Effect of agglomeration of silver nanoparticle on nanotoxicity depression. Korean J. Chem. Eng. 30, 364–368.

Baun, A., Hartmann, N.B., Grieger, K., et al., 2009. Setting the limits for engineered nanoparticles in European surface waters – are current approaches appropriate. J. Environ. Monit. 11, 1774–1781.

Blickley, M., Matson, C.W., Vreeland, W.N., et al., 2014. Dietary CdSe/ZnS quantum dot exposure in estuarine fish: bioavailability, oxidative stress responses, reproduction, and maternal transfer. Aquat. Toxicol. 148, 27–39.

Bondarenko, O., Juganson, K., Ivask, A., et al., 2013. Toxicity of Ag, CuO and ZnO nanoparticles to selected environmentally relevant test organisms and mammalian cells in vitro: a critical review. Arch. Toxicol. 87, 1181–1200.

Boxall, A.B.A., Chaudhry, Q., Jones, A., et al., 2007. Current and Future Predicted Environmental Exposure to Engineered Nanoparticles. Central Science Laboratory, Sand Hutton, UK.

Buffet, P.-E., Fossi Tankoua, O., Pan, J.-F., et al., 2011. Behavioral and biochemical responses of two marine invertebrates *Scrobicularia plana* and *Hediste diversicolor* to copper oxide nanoparticles. Chemosphere 84, 166–174.

Buffet, P.-E., Amiard-Triquet, C., Dybowska, A., et al., 2012. Fate of isotopically labeled zinc oxide nanoparticles in sediment and effects on two endobenthic species the clam *Scrobicularia plana* and the ragworm *Hediste diversicolor*. Ecotoxicol. Environ. Saf. 84, 191–198.

Buffet, P.E., Richard, M., Caupos, F., et al., 2013. A mesocosm study of fate and effects of CuO nanoparticles on endobenthic species (*Scrobicularia plana, Hediste diversicolor*). Environ. Sci. Technol. 47, 1220–1228.

Buffet, P.-E., Zalouk-Vergnoux, A., Châtel, A., et al., 2014. A marine mesocosm study on the environmental fate of silver nanoparticles and toxicity effects on two endobenthic species: the ragworm Hediste diversicolor and the bivalve mollusc *Scrobicularia plana*. Sci. Total. Environ. 470–471C, 1151–1159.

Canesi, L., Ciacci, C., Fabbri, R., et al., 2012. Bivalve molluscs as a unique target group for nanoparticle toxicity. Mar. Environ. Res. 76, 16–21.

Christian, P., von der Kammer, F., Baalousha, M., et al., 2008. Nanoparticles: structure, properties, preparation and behavior in environmental media. Ecotoxicology 17, 326–343.

Cleveland, D., Long, S.E., Pennington, P.L., et al., 2012. Pilot estuarine mesocosm study on the environmental fate of silver nanomaterials leached from consumer products. Sci. Total Environ. 421–422, 267–272.

Colman, B.P., Arnaout, C.L., Anciaux, S., et al., 2013. Low concentrations of silver nanoparticles in biosolids cause adverse ecosystem responses under realistic field scenario. PLoS One 8 (2), e57189.

Cong, Y., Banta, G.T., Selck, H., et al., 2011. Toxic effects and bioaccumulation of nano-, micron- and ionic-Ag in the polychaete, *Nereis diversicolor*. Aquat. Toxicol. 105, 403–411.

Cong, Y., Banta, G.T., Selck, H., et al., 2014. Toxicity and bioaccumulation of sediment –associated silver nanoparticles in the estuarine polychaete, *Nereis (Hediste) diversicolor*. Aquat. Toxicol. 156, 106–115.

Crane, M., Handy, R., Garrod, J., et al., 2008. Ecotoxicity test methods and environmental hazard assessment for engineered nanoparticles. Ecotoxicology 17, 421–437.

Croteau, M.N., Misra, S.K., Luoma, S.N., et al., 2011a. Silver bioaccumulation dynamics in a freshwater invertebrate after aqueous and dietary exposures to nanosized and ionic Ag. Environ. Sci. Technol. 45, 6600–6607.

Croteau, M.N., Dybowska, A.D., Luoma, S.N., et al., 2011b. A novel approach reveals that zinc oxide nanoparticles are bioavailable and toxic after dietary exposures. Nanotoxicology 5, 79–90.

Dai, L., Syberg, K., Banta, G.T., et al., 2013. Effects, uptake and depuration kinetics of silver- and copper nanoparticles in a marine deposit feeder, *Macoma balthica*. ACS Sustainable Chem. Eng. 1, 760–767.

ECHA, 2013. Human Health and Environmental Exposure Assessment and Risk Characterisation of Nanomaterials. Best Practice for REACH Registrants. ECHA, Helsinki, Finland.

Farré, M., Gajda-Schrantz, K., Kantiani, L., et al., 2009. Ecotoxicity and analysis of nanomaterials in the aquatic environment. Anal. Bioanal. Chem. 393, 81–95.

Ferry, J.L., Craig, P., Hexel, C., et al., 2009. Transfer of gold nanoparticles from the water column to the estuarine food web. Nat. Nano 4, 441–444.

Garcia-Alonso, J., Khan, F.R., Misra, S.K., et al., 2011. Cellular internalization of silver nanoparticles in gut epithelia of the estuarine polychaete *Nereis diversicolor*. Environ. Sci. Technol. 45, 4630–4636.

Gorham, J.M., Rohlfing, A.B., Lippa, K.A., et al., 2014. Storage wars: how citrate capped silver nanoparticle suspensions are affected by not-so-trivial decisions. J. Nanopart. Res. 16, 1–14.

Gottschalk, F., Sonderer, T., Scholz, R.W., et al., 2009. Modeled environmental concentrations of engineered nanomaterials (TiO_2, ZnO, Ag, CNT, fullerenes) for different regions. Environ. Sci. Technol. 43, 9216–9222.

Gottschalk, F., Sun, T.Y., Nowack, B., 2013. Environmental concentrations of engineered nanomaterials: review of modeling and analytical studies. Environ. Pollut. 181, 287–330.

Handy, R.D., von der Kammer, F., Lead, J.R., et al., 2008. The ecotoxicology and chemistry of manufactured nanoparticles. Ecotoxicology 17, 287–314.

Handy, R.D., Van den Brink, N., Chappell, M., et al., 2012. Practical considerations for conducting ecotoxicity test methods with manufactured nanomaterials: what have we learnt so far? Ecotoxicology 21, 933–972.

Henry, T.B., Menn, F.M., Fleming, J.T., et al., 2007. Attributing effects of aqueous C60 nano-aggregates to tetrahydrofuran decomposition products in larval zebrafish by assessment of gene expression. Environ. Health Perspect. 115, 1059–1065.

Heinlaan, M., Ivask, A., Blinova, I., et al., 2008. Toxicity of nanosized and bulk ZnO, CuO and TiO_2 to bacteria *Vibrio fischeri* and crustaceans *Daphnia magna* and *Thamnocephalus platyurus*. Chemosphere 71, 1308–1316.

Hyung, H., Fortner, J.D., Hughes, J.B., et al., 2007. Natural organic matter stabilizes carbon nanotubes in the aqueous phase. Environ. Sci. Technol. 41, 179–184.

ISO/TS 27687, 2008. Nanotechnologies – Terminology and Definitions for Nano-objects – Nanoparticle, Nanofibre and Nanoplate. 7pp.

ISO/TS 80004–4, 2011. Nanotechnologies – Vocabulary – Part 4: Nanostructured Materials. 7pp.

Kennedy, A.J., Chappell, M.A., Bednar, A.J., et al., 2012. Impact of organic carbon on the stability and toxicity of fresh and stored silver nanoparticles. Environ. Sci. Technol. 46, 10772–10780.

Kettler, A., Veltman, K., van de Meent, D., et al., 2014. Cellular uptake on nanoparticles as determined by particle properties, experimental conditions, and cell type. Environ. Toxicol. Chem. 33, 481–492.

Klaine, S.J., Alvarez, P.J.J., Batley, G.E., et al., 2008. Nanomaterials in the environment: behavior, fate, bioavailability, and effects. Environ. Toxicol. Chem. 27, 1825–1851.

Levard, C., Hotze, E.M., Lowry, G.V., et al., 2012. Environmental transformations of silver nanoparticles: impact on stability and toxicity. Environ. Sci. Technol. 46, 6900–6914.

Lowry, G.V., Espinasse, B.P., Badireddy, A.R., et al., 2012. Long-term transformation and fate of manufactured Ag nanoparticles in a simulated large scale freshwater emergent wetland. Environ. Sci.Technol. 46, 7027–7036.

Luoma, S.N., Khan, F.R., Croteau, M.N., 2014. Bioavailability and bioaccumulation of metal-based engineered nanomaterials in aquatic environments: concepts and processes. In: Lead, J.R., Valsami-Jones, E. (Eds.), Frontiers of Nanoscience, vol. 7. Elsevier, Oxford, pp. 157–193.

Matranga, V., Corsi, I., 2012. Toxic effect of engineered nanoparticles in the marine environment: model organisms and molecular approaches. Mar. Environ. Res. 76, 32–40.

Misra, S.K., Dybowska, A., Berhanu, D., et al., 2012. The complexity of nanoparticle dissolution and its importance in nanotoxicological studies. Sci. Total Environ. 438, 225–232.

Mouneyrac, C., Buffet, P.-E., Poirier, L., et al., 2014. Fate and effects of metal-based nanoparticles in two marine invertebrates, the bivalve mollusc *Scrobicularia plana* and the annelid polychaete *Hediste diversicolor*. Environ. Sci. Pollut. Res. 21, 7899–7912.

NRC, 2009. Committee on Improving Risk Analysis Approaches Used by the U.S. EPA, National Research Council. Science and Decisions: Advancing Risk Assessment, ISBN: 978-0-309-12046-3.

OECD, 2004. Guidelines for Testing of Chemicals. *Daphnia* Sp. Acute Immobilisation Test Guideline 202. Organization for Economic Cooperation and Development, Paris.

OECD, 2006. Guidelines for Testing of Chemicals. *Lemna* Sp. Growth Inhibition Test Guideline 221. Organization for Economic Cooperation and Development, Paris.

OECD, 2009. Guidelines for Testing of Chemicals. Fish, Acute Toxicity Test Guideline 203. Organization for Economic Cooperation and Development, Paris.

OECD, 2012. Chemicals Committee of the OECD. Important Issues on Risk Assessment of Manufactured Nanomaterials. OECD Environment Directorate, Environment, Health and Safety Division (France).

Oberdörster, G., Oberdörster, E., Oberdörster, J., 2005. Nanotoxicology: an emerging discipline evolving from studies of ultrafine particles. Environ. Health Perspect. 113, 823–839.

Paget, V., Sergent, J.A., Grall, R., et al., 2014. Carboxylated nanodiamonds are neither cytotoxic nor genotoxic on liver, kidney, intestine and lung human cell lines. Nanotoxicology 8, 46–56.

Pang, C., Selck, H., Misra, S.K., et al., 2012. Effects of sediment-associated copper to the deposit-feeding snail, *Potamopyrgus antipodarum*: a comparison of Cu added in aqueous form or as nano- and micro-CuO particles. Aquat. Toxicol. 106-107, 114–122.

Pang, C., Selck, H., Banta, G.B., et al., 2013. Bioaccumulation, toxicokinetics, and effects of copper from sediment spiked with aqueous Cu, nano-CuO or micro-CuO in the deposit-feeding snail, *Potamopyrgus antipodarum*. Environ. Toxicol. Chem. 32, 1561–1573.

Petersen, E.J., Henry, T.B., Zhao, J., et al., 2014. Identification and avoidance of potential artifacts and misinterpretations in nanomaterial ecotoxicity measurements. Environ. Sci. Technol. 48, 4226–4246.

Praetorius, A., Scheringer, M., Hungerbühler, K., 2012. Development of environmental fate models for engineered nanoparticles – a case study of TiO_2 nanoparticles in the Rhine River. Environ. Sci. Technol. 46, 6705–6713.

Quik, J.T.K., Stuart, M.C., Wouterse, M., et al., 2012. Natural colloids are the dominant factor in the sedimentation of nanoparticles. Environ. Toxicol. Chem. 31, 1019–1022.

Ramskov, T., Selck, H., Banta, G.T., et al., 2014. Bioaccumulation and effects of different-shaped copper oxide nanoparticles in the deposit-feeding snail *Potamopyrgus antipodarum*. Environ. Toxicol. Chem. 33, 1976–1987.

Research in Germany, 2014. Federal Ministry of Education and Research. Nanotechnology, Companies and Institutes. Can be accessed at: http://www.research-in-germany.de/dachportal/en/Research-Areas-A-Z/Nanotechnology/Companies-and-Institutes.html.

SCENIHR, 2006. Scientific Committee on Emerging and Newly Identified Health Risks (SCENIHR) Opinion on: The Appropriateness of Existing Methodologies to Assess the Potential Risks Associated with Engineered and Adventitious Products of Nanotechnologies. European Commission, Health & comsumer protection, Directorate-General.

SCENIHR, 2009. Scientific Committee on Emerging and Newly Identified Health Risks (SCENIHR) Opinion on: Risk Assessment of Products of Nanotechnologies. European Commission, Health & comsumer protection, Directorate-General.

Smith, C.J., Shaw, B.J., Handy, R.D., 2007. Toxicity of single walled carbon nanotubes to rainbow trout, (*Oncorhynchus mykiss*): respiratory toxicity, organ pathologies, and other physiological effects. Aquat. Toxicol 82, 94–109.

Stone, V., Nowack, B., Baun, A., et al., 2010. Nanomaterials for environmental studies: classification, reference material issues, and strategies for physico-chemical characterization. Sci. Total Environ. 408, 1745–1754.

US EPA, 1996. Ecological Effects Test Guidelines. OPPTS 850.5400. Algal Toxicity, Tiers I and II. EPA 712-C-96-164. US Environmental Protection Agency, Washington.

von der Kammer, F., Ferguson, P.L., Holden, P., et al., 2012. Analysis of nanomaterials in complex matrices (environment and biota): general considerations and conceptual case studies. Environ. Toxicol. Chem. 31, 32–49.

Wong, S.W.Y., Leung, K.M.Y., 2014. Temperature-dependent toxicities of nano zinc oxide to marine diatom, amphipod and fish in relation to its aggregation size and ion dissolution. Nanotoxicology 8, 24–35.

Wong, S.W.Y., Leung, K.M.Y., Djurišic, A.B., 2013. A comprehensive review on the aquatic toxicity of engineered nanomaterials. Rev. Nanosci. Nanotechnol 2, 79–105.

Zhang, W., Lin, K., Miao, Y., et al., 2012. Toxicity assessment of zebrafish following exposure to CdTe QDs. J. Hazard Mater. 213–214, 413–420.

CHAPTER 18

Ecotoxicological Risk of Mixtures

Cathy A. Laetz, Scott A. Hecht, John P. Incardona, Tracy K. Collier, Nathaniel L. Scholz

Abstract

Water pollution is a global environmental challenge that nearly always involves the degradation of aquatic habitats by mixtures of chemical contaminants. Despite this practical reality, environmental regulations and resource management institutions in most countries are inadequate to the task of addressing complex and dynamic combinations of chemicals. Moreover, our scientific understanding of mixture toxicity and the assessment of corresponding risks to aquatic species and communities have not kept pace with worldwide declines in biodiversity or the introduction of thousands of new chemicals into societal use. In this chapter, we review recent research specific to mixtures in three contexts that are broadly applicable to freshwater and marine ecosystems. These include oil spills, urban non-point source pollution, and the agricultural use of modern pesticides. Each of these familiar and geographically extensive forcing pressures is threaded with uncertainty about interactions between contaminants in mixtures. We also briefly consider relevant and often overlapping environmental regulations in the United States and Europe to illustrate why a proactive consideration of chemical mixtures remains elusive in institutional ecological risk assessment. As the case examples show, however, the problem of mixtures is not intractable and targeted research can guide effective conservation and restoration strategies in a chemically complex world.

Keywords: Ecological risk assessment; Environmental monitoring; Habitat; High-throughput screening; Non-point source pollution; Oil spills; Pesticides; Stormwater; Synergism.

Chapter Outline

Introduction

In aquatic ecosystems, exposure to mixtures of toxic chemicals resulting from human activities is the norm, and assessing the biological effects of chemical mixtures remains an enduring challenge in the field of ecotoxicology (Lydy et al., 2004; Villanueva et al., 2014). Examples of anthropogenic sources of chemical mixtures include oil spills, agricultural uses of pesticides,

Aquatic Ecotoxicology. http://dx.doi.org/10.1016/B978-0-12-800949-9.00018-8

wastewater discharges containing pharmaceuticals and personal care products, and urban stormwater runoff. Monitoring provides ample evidence that chemical mixtures contaminate surface water (Gilliom, 2007), ground water (Toccalino et al., 2012), and drinking water (Donald et al., 2007) in urban, agricultural, and undeveloped areas. Likewise, numerous studies document adverse effects of exposure to chemical mixtures in fish (e.g., Incardona et al., 2013), invertebrates (e.g., Bjergager et al., 2012), amphibians (e.g., Hayes et al., 2006), and aquatic communities (e.g., Relyea, 2008), even when mixture constituents are present at very low concentrations.

Given that thousands of synthetic chemicals are currently in societal use worldwide, it would be logistically impossible to empirically assess the toxicity of each possible combination to aquatic species. Moreover, the co-occurrence of chemical and nonchemical stressors such as pathogens, thermal extremes, and salinity gradients further complicates the ecotoxicology of mixtures in natural systems. This environmental complexity is increasingly necessitating the adaptation of ecological theory, particularly in the design of field studies (Relyea and Hoverman, 2006; Rohr et al., 2006).

Despite the inherent difficulties of measuring the toxicity of complex chemical mixtures, several methods have been developed including concentration-addition and independent-action models, toxic unit approaches, and probabilistic methods (Backhaus and Faust, 2012; Nowell et al., 2014). These methods are mostly limited to cases of noninteracting chemicals where the individual constituents have been identified and the concentration–response relationship of single chemicals is known. This includes, for example, toxicity estimates for pesticides with similar and dissimilar modes of action using concentration-addition and response-addition models, respectively (Belden et al., 2007; Backhaus and Faust, 2012). The accurate prediction of chemical interactions in mixtures, particularly those that result in synergistic toxicity, remains a priority in aquatic ecotoxicology (Cedergreen, 2014). Although some attempt has been made to model complex interactions (see Rodney et al., 2013), the ability to predict chemical interactions is hindered by the lack of a generally acceptable conceptual framework.

A longstanding goal in ecotoxicology is the greater inclusion of environmental realism in the design of scientific studies, in both the laboratory and the field (Chapter 6). On the "eco" side, this has included a push for research on native species (vs laboratory models) (Chapter 9), the scaling of toxicological effects from individuals to populations and communities, the use of ecological concepts to guide study design, and hypothesis-driven experimental field manipulations. On the "tox" side, there has been a corresponding push to embrace environmental complexity, including chemical mixtures, interactions between chemical and nonchemical stressors, indirect (i.e., food web-based) impacts, and harmful toxicological effect that are delayed in time. In this way, our evolving understanding of chemical mixtures has been, and will continue to be, at the forefront of advancements in aquatic ecotoxicology.

In the United States and Europe, mixtures toxicity is not routinely assessed under the regulatory authority of the federal and state governments that administer environmental laws and statutes. As such, most chemicals are regulated on a single-chemical basis, which may lead to an underestimation of actual risk facing aquatic communities exposed to chemical mixtures. A nonsystematic approach to assessing chemical mixtures toxicity may lead to different toxicity predictions, varying degrees of chemical regulation, and ultimately different interpretations of risk. In this chapter, we discuss recent research specific to mixtures in three contexts (i.e., oil spills, urban non-point source pollution, and current-use agricultural pesticides) that are reflective of anthropogenic stressors in aquatic ecosystems worldwide. Additionally, we discuss the importance and challenges of assessing chemical mixtures under the authority of certain overlapping environmental statutes.

18.1 Worldwide Oil Spills

Over the past few decades, technological improvements have generally led to a decline in the total volume of petroleum spilled from production and transportation processes into the world's oceans (Council, 2003). Nevertheless, large marine oil spills still occur with regularity. These include, for example, the 2002 *Prestige* tanker spill off the coast of Spain, the 2007 *Hebei Spirit* tanker spill off the coast of South Korea, the 2009 Montara/*West Atlas* wellhead blowout in the Timor Sea, and the 2010 Macondo/*Deepwater Horizon* disaster in the Gulf of Mexico. Moreover, with increasing overland pipeline transport of products such as diluted bitumen from oil sands extraction, accidents near freshwater habitats are likely to increase in frequency such as the pipeline spills into the Yellowstone River in 2011 and 2015. Although larger oil spills are highly publicized and intensively studied, the dramatic increase in global shipping traffic will likely lead to an increase of smaller fuel oil spills stemming from container ship accidents. There has yet to be a formal quantification of the frequency of these smaller spills, but this number clearly has increased along with container ship traffic, based on the frequency of internet news articles on container ship accidents over the past 15 years. Although much smaller in scale, these spills can have significant localized impacts on marine resources, as indicated by the effects of the 2007 Cosco Busan fuel oil spill on Pacific herring (*Clupea pallasi*) spawning grounds in San Francisco Bay (Incardona et al., 2012a). An improved understanding of the toxicity of petroleum-derived mixtures is critical for guiding spill response activities, assessing natural resource injury in the aftermath of spills, and to better support the reparation and restoration processes.

Crude oils are complex mixtures that contain thousands of different compounds as well as elemental metals. Geologically distinct crude oils differ primarily in the ratios of large and interrelated families of compounds (Stout and Wang, 2007), including a range of aliphatic and aromatic hydrocarbon classes, resins, asphaltenes, and polar compounds containing nitrogen, sulfur, or oxygen atoms. The relative composition of these compounds determines the general physicochemical properties of a given crude oil. For example, the lighter, low-sulfur crude oils from the Northern Gulf of Mexico have proportionally less sulfur-containing aromatic heterocycles than heavier, sulfur-rich crude oils from Alaska or the Middle East (Wang et al., 2003). A major challenge for understanding mixtures toxicity relating to crude oils and their products is that mixtures are changed by the refining process, and mixtures change over time after petroleum products are released into marine and aquatic environments.

The smaller but more frequent spills of marine fuel oils present a challenge with a distinct suite of characteristics. Bunker fuels, the generic term applied to fuel stored on board ships, typically consist of a highly viscous heavy residual oil mixed with a lighter fuel (typically diesel), to facilitate pumping and flow. Because of the nature of modern refinery practices, these residual oils have chemical compositions that are distinct from unrefined (parent) crude oils (Uhler et al., 2007). By definition, residual oils are the dregs of the refining process, and have often been subjected to conversion processes such as catalytic cracking (Matar and Hatch, 2001). Consequently, many chemical and elemental components of the parent crude oil are much more highly concentrated in the remaining residual oil (Clark and Brown, 1977; Wang et al., 2003; Uhler et al., 2007). Residual oils and mixed fuel oil products have a higher percentage and total mass of relatively more water-soluble aromatic compounds and larger fractions of uncharacterized polar compounds (an "unresolved complex mixture") that can approach 30% of the mass (Clark and Brown, 1977). Residual fuel oils also often contain a higher metals content, typically nickel and vanadium (Matar and Hatch, 2001). Generally, there is a higher degree of chemical heterogeneity among modern residual fuel oils (Uhler et al., 2007).

In the case of any marine or aquatic oil spill, be it crude or residual fuel, different taxa can be exposed to different mixtures from the same oil depending on the exposure pathway. Although there are likely tens of thousands of unique compounds in crude oil, only a small fraction of these have appreciable water solubility, with less than 1% becoming dissolved in water (Gros et al., 2014). Thus, mixtures can be extremely different if organisms are exposed to either whole particulate oil or water-dissolved components, or both. If oil is dispersed into small droplets, filter-feeding organisms can be exposed to whole oil through ingestion of droplets, for example. "Weathering" of petroleum released into the environment changes both the composition of oil that remains and the composition of dissolved mixtures derived from that oil. Weathering includes both biological processes (e.g., microbial degradation) and abiotic (physicochemical) factors such as volatilization, photodegradation, and dissolution. Dissolution weathering determines the composition of water-soluble mixtures that are available for uptake by lipid-rich organisms such as fish embryos.

Dissolution weathering represents a major challenge for understanding oil spill toxicity because it leads to a virtually constant change in mixture composition over time. This process has been best characterized in the context of a gravel/cobble shoreline contaminated by stranded oil, emulating the shoreline oiling in Prince William Sound following the 1989 *Exxon Valdez* spill (Short and Heintz, 1997). However, a similar process governs the dissolution of water-soluble compounds from an oil slick or dispersed oil droplets in open water

(Gros et al., 2014). As discussed in detail below, toxicity studies on petroleum products have focused extensively on polycyclic aromatic hydrocarbons (PAHs) because of previously known toxicity of these compounds and their general, albeit low, water solubility and bioavailability. Solubility of PAHs decreases with increasing molecular weight, as determined by both the number of rings and degree of alkyl substitution. For example, two-ringed naphthalene (molecular weight 128) has higher solubility than three-ringed phenanthrene (molecular weight 178); parent phenanthrene in turn has higher solubility than a multiply methylated C4-phenanthrene (molecular weight 234). Dissolution of PAHs as water percolates through oiled gravel follows first-order loss-rate kinetics (Short and Heintz, 1997). Consequently, oil-derived PAH mixtures are initially dominated by the most water-soluble compounds (e.g., naphthalene, parent three-ringed compounds). Over time, the mixture becomes dominated by the less water-soluble alkylated compounds (Figure 18.1).

Petroleum toxicity represents an exceptionally complicated example of the challenge of mixtures ecotoxicology. This is due to the dynamic complexity of the mixture over time, even for situations in which the toxic properties of many of the component PAHs are known. For organisms exposed to whole oil, the uncertainty is magnified by the large number of unidentified compounds in the mixture, together with known compounds with poorly characterized toxicity. Nevertheless, there has been a considerable effort in recent years to break down complex PAH mixtures into component parts to identify which compounds are driving harmful impacts to aquatic species, particularly developing fish exposed to water-soluble components of oil.

A major lesson learned from the *Exxon Valdez* oil spill is that developing fish embryos are very sensitive to low-level exposure to crude oil, which leads to a syndrome of developmental defects related to heart failure (Incardona et al., 2011a). This form of crude oil developmental toxicity was attributed primarily to the proportional content of PAHs and related families of heterocyclic nitrogen, sulfur, or oxygen compounds, such as dibenzothiophenes. Uptake by oil-exposed organisms is more likely given their higher water solubility relative to aliphatic hydrocarbons and polycyclic alkanes (Council, 2003; Carls and Meador, 2009). Early studies with embryos of pink salmon (*Oncorhynchus gorbuscha*) and Pacific herring showed that toxicity per unit mass of total dissolved PAHs increased as the mixture shifted over time from compositions dominated by two-ringed naphthalenes to three-ringed fluorenes, dibenzothiophenes, and phenanthrenes (Marty et al., 1997; Carls et al., 1999; Heintz et al., 1999). In the past decade, studies of single PAH compounds, primarily using the zebrafish model, and isolated petroleum fractions further linked developmental toxicity in fish to three-ringed compounds (Incardona et al., 2004, 2005; Adams et al., 2014a). The zebrafish studies also showed that the etiology of the petroleum developmental toxicity syndrome was disruption of embryonic cardiac function and morphogenesis (Incardona et al., 2005; Adams et al., 2014a). It is now well-established that both crude and residual oils from different geological sources disrupt heart development in a diversity of fish species (Incardona et al., 2009; Hatlen et al.,

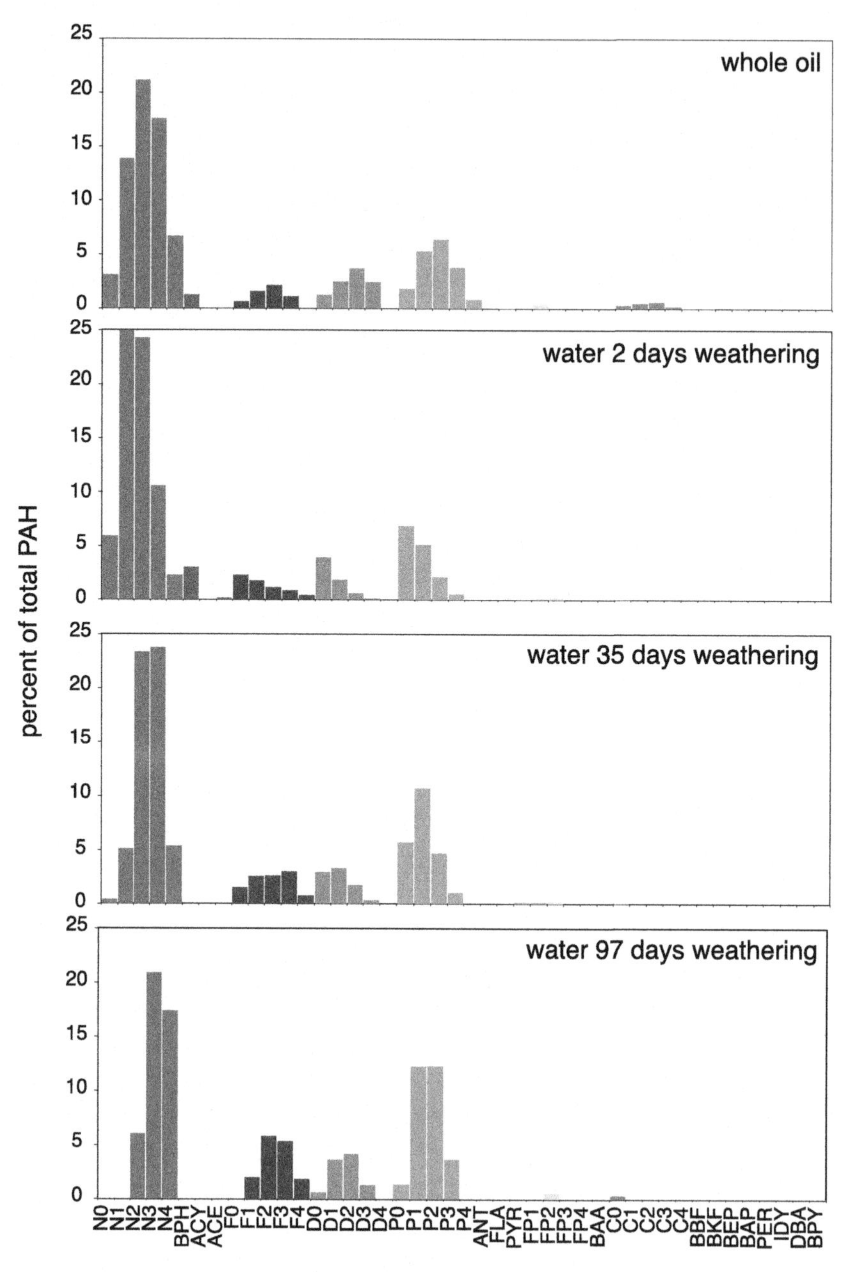

Figure 18.1

Gravel coated with Alaska North Slope Crude oil was weathered with freshwater for zebrafish embryo exposures (see Hicken et al., 2011). PAH compositions are shown for the source oil (top) and in extracts of water at the indicated days of water flow through the column (weathering). ACE, acenaphthene; ANT, anthracene; ACY, acenaphthylene; BAA, benz[a]anthracene; BAP, benzo[a]pyrene; BBF, benzo[b]fluoranthene; BEP, benzo[e]pyrene; BKF, benzo[j]fluoranthene/benzo[k]fluoranthene; BPH, biphenyl; BPY, benzo[ghi]perylene; C, chrysene; D, dibenzothiophene; DBA, dibenz[a,h]anthracene/dibenz[a,c]anthracene; F, fluorene; FLA, fluoranthene; FP, fluoranthenes/pyrenes; IDY, indeno[1,2,3-cd]pyrene; N, naphthalenes; P, phenanthrene; PER, perylene; PYR, pyrene. Parent compound is indicated by a 0 (e.g., N0), whereas numbers of additional carbons (e.g., methyl groups) for alkylated homologs are indicated as N1, N2, etc.

2010; Incardona et al., 2012a, 2013; Jung et al., 2013; Incardona et al., 2014), by a mechanism involving the blockade of potassium and calcium ion channels essential for excitation-contraction coupling in heart muscle cells (Brette et al., 2013). These studies all implicate the three-ringed PAHs (and sulfur heterocycles), although direct evidence remains to be demonstrated (e.g., receptor binding).

Studies of crude oil developmental cardiotoxicity have thus been simplified to a degree because of the common biological response to an array of different oil types with seemingly different chemistries. This is consistent with a subset of the PAHs causing most (but not all) of the toxicity. However, there are still many open questions and unresolved issues relating to crude oil toxicity and other complex PAH mixtures. First, the cardiotoxic potencies of single PAH compounds are far less than the potency of oil-derived mixtures, and there is currently no clear explanation for this. Simple models based on additivity or classical metabolic synergism are unlikely to account for the potency of PAH mixtures.

Second, the developmental toxicity of PAH mixtures becomes yet more complicated when considering habitats affected by industrialization and urbanization in which PAH mixtures include the typical petrogenic compounds plus an array of higher molecular four- to six-ringed compounds typically derived from combustion (pyrogenic PAHs). Single-compound studies have indicated a variety of mechanisms for the toxicity of the higher molecular weight PAHs, including dioxin-like cardiotoxicity mediated by the aryl hydrocarbon receptor (AHR), AHR-independent cardiotoxicity, and metabolism-dependent toxicity (Incardona et al., 2006, 2011b; Van Tiem and Di Giulio, 2011; Jayasundara et al., 2015). Thus the potency and specific effects of different petrogenic and pyrogenic PAH mixtures are likely to arise from much more complicated interactions and include contributions from not only additivity and classical metabolic synergism, but also from distinct synergies between independent toxic mechanisms converging on the cardiac system (e.g., disruption of excitation-contraction coupling by ion channel blockade combined with AHR-dependent disruption of normal cardiac gene expression).

Finally, there is very little information on the interaction between the toxicity of PAH mixtures and other environmental stressors. The potential importance of these interactions is highlighted by the markedly enhanced toxicity of PAHs (and/or other petroleum-derived compounds) by photosensitization reactions with ultraviolet radiation in sunlight (Arfsten et al., 1996; Yu, 2002; Diamond, 2003). This represents an entirely different form of lethal toxicity in transparent organisms such as fish embryos and can occur at tissue concentrations below that causing serious cardiotoxicity (Barron et al., 2003; Hatlen et al., 2010; Incardona et al., 2012b). Interactions with sunlight are particularly important for residual oils, which have a very potent phototoxicity associated with uncharacterized compounds rather than the conventionally measured PAHs (Hatlen et al., 2010; Incardona et al., 2012b). The large-scale uses of chemical dispersants in the response to the *Deepwater Horizon*—MC252 oil spill prompted a renewed focus on oil-dispersant interactions. However, most studies indicate that dispersants primarily increase the bioavailability of oil compounds such as PAHs, rather than specifically enhancing toxicity or acting synergistically (e.g., Adams et al., 2014b). Nevertheless, few studies have addressed this issue with a high degree of rigor using physical biochemical methods appropriate for assessing interactions between hydrocarbons, dispersant surfactants, and biological membranes.

Despite major advances in the past decade described previously, much about the ecotoxicology of petroleum-derived mixtures remains unknown. This is understandable given the considerable chemical complexity of petroleum, but progress is hampered by lack of consistent policy or funding for proactive research. Almost all of our understanding of oil spills is based on assessment activities developed in response to a spill—after the fact. Future research needs to continue along the lines described here, to dissociate the biological effects of individual oil hydrocarbons as drivers for the toxic potency of dynamic mixtures. Moreover, whereas traditional toxicity testing generally relies heavily on model organisms and standardized bioassays, natural resource injury assessments usually involve wild species. At a minimum, future work should focus on (1) characterization of mixture toxicity in species native to areas where future spills are likely to occur or recur, such as the Arctic and the Gulf of Mexico; (2) interactions between toxic oil and other environmental stressors such as temperature (particularly the extreme cold of the Arctic) and sunlight; and (3) understanding how toxic effects at the level of target organs (e.g., the heart) or processes (e.g., development, growth) in individual organisms translates into larger scale ecosystem-level impacts.

18.2 Urban Stormwater Runoff

At present, one of the most important and complicated aquatic conservation challenges worldwide is urban stormwater runoff. For the latter part of the twentieth century, most developed countries were successful at controlling end-of-pipe or point source pollution discharges from municipal and industrial sources. However, these efforts have been less effective at reducing diffuse land-based runoff to aquatic habitats. In urban watersheds, rainfall that would normally infiltrate to soils falls instead on impervious surfaces, including the roofs of residential and commercial buildings, parking lots, streets and roads, highways, and similar types of land cover. The corresponding increase in surface runoff creates significant problems related to both water quantity (i.e., urban flooding) and water quality, the latter a consequence of stormwater mobilizing complex mixtures of toxic chemicals.

Waterways near urban areas throughout the world are almost universally afflicted with the so-called "urban stream syndrome" (Walsh et al., 2005). The syndrome is characterized by poor water quality, declines in aquatic species abundance and diversity, and the proliferation of nonnative, pollution-tolerant taxa. However, urban streams are also negatively affected by a variety of physical habitat-forming processes, including extreme flow regimes, erosion and sedimentation, loss of stream channel complexity, and loss of substrate. The relative contributions of physical and chemical drivers to species declines are not well understood. This creates a frequent dilemma in the arena of urban habitat restoration (i.e., whether and to what extent to expend resources to improve water quality) as opposed to more conventional physical habitat restoration.

Urban stormwater is chemically complex, and numerous environmental monitoring studies have shown that motor vehicles are major sources of toxic contaminants. These originate from vehicle exhaust, the wearing of friction materials (brake pads), leachate from tires, and the leaking of lubricants, fuel additives, and other materials. Untreated runoff from roadways and parking areas therefore contains a highly diverse mixture of chemicals, only a fraction of which has been assessed for toxicity to aquatic species. These include the previously discussed PAHs from internal combustion engines, incomplete fuel combustion, oil, and grease. Urban runoff also contains metals from brake pad wear and other sources (e.g., metal roofs, other building materials), most notably copper, zinc, nickel, lead, and cadmium. For the most part, studies of stormwater quality in the built environment have focused on fluctuating concentrations of PAHs and metals (e.g., Shinya et al., 2000) as a consequence of varying rainfall patterns, antecedent dry intervals, traffic density, and various mitigation measures.

The toxicity of PAHs and metals to fish and other aquatic organisms, alone and in combination, are reasonably well understood. This is due, in part, to the fact that these contaminants pose ecological hazards in many other contexts. In the case of PAHs, for example, this includes oil spills and legacy industrial contamination from the past century. Certain PAHs known to be cardiotoxic to fish are prevalent in both crude oil and urban runoff (Incardona et al., 2011a), and decades of research on the former have informed more recent science on the latter. For example, developing zebrafish embryos and larvae exposed to stormwater-derived mixtures of PAHs (McIntyre et al., 2014) show nearly the same symptoms of cardiac dysregulation and heart failure evident in zebrafish exposed to PAHs from crude oil (Incardona et al., 2013). The ecotoxicology of metals have also been studied for many years in association with the impacts of mining, wood preservatives, antifoulant paints, and common pesticides.

Among toxics, the predominant focus on PAHs and metals in urban stormwater science (e.g., Brown and Peake, 2006) has captured only part of the picture, however. There are hundreds of additional chemicals in urban runoff, most of which are very poorly understood in terms of toxicity to aquatic species. They include chemicals used in the manufacture of tires, uncharacterized components of refined fuels and fuel additives, oxygenated hydrocarbons, surfactants, plasticizers, nanomaterials, and the like. Furthermore, the contaminant exposure scenario gets considerably more complicated when stormwater conveyance systems are connected to sewerage systems, leading to combined sewer overflow events during intense rainstorms (see example, in Figure 18.2). Where this occurs, the chemicals in impervious surface runoff are discharged to receiving waters in combination with a broader mixture of pharmaceuticals, personal care products, flame retardants, and other chemicals used in everyday household life.

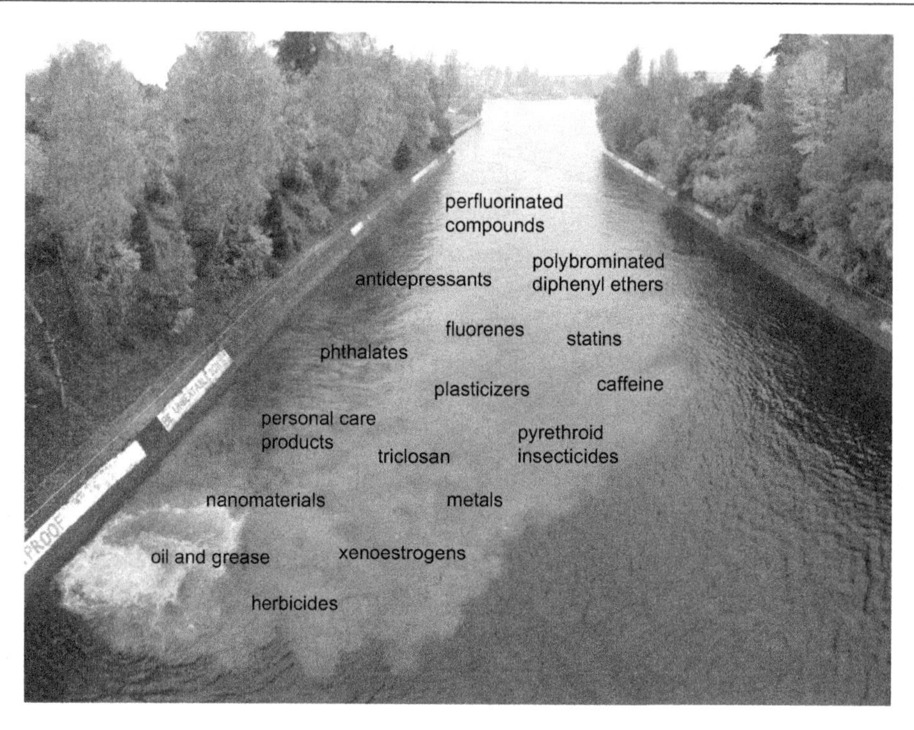

Figure 18.2

Combined inputs of stormwater and untreated sewage to an urban waterway in Seattle, Washington, during a period of high rainfall. Subsurface discharge from a combined sewer overflow (CSO) is evident in the lower left. *Photo by Blake Feist, National Oceanic and Atmospheric Administration (NOAA).*

Despite the ubiquity of urban runoff around the world, the toxic impacts to aquatic biota remain, surprisingly, poorly understood. In the United States, researchers began turning their attention to "unrecorded pollution" in the years following the passage of the 1972 Clean Water Act (e.g., Pratt et al., 1981). In the ensuing decades, urban runoff has been tested primarily using a range of different bioassays, with various toxicity identification and evaluation procedures to correlate measured contaminants with observed effects on species growth or survival (e.g., Kayhanian et al., 2008). In general, urban runoff is more acutely toxic to aquatic invertebrates than to fish, and the initial phase (the first flush) is usually more toxic than runoff collected subsequently during a storm event. More recently, modern molecular tools available for zebrafish and other species have proven useful for confirming contaminant exposure (e.g., upregulation of *cyp1a* gene expression) and for shedding light on which components of complex chemical mixtures in stormwater are driving more nuanced, sublethal toxicity (McIntyre et al., 2014).

Nevertheless, conventional bioassays have remained the state of the science in urban runoff toxicology, often using nonnative model species (but see Snodgrass et al., 2008; McIntyre et al., in press). Thus, although freshwater ecologists have widely documented the urban stream syndrome ("eco") and civil engineers and their colleagues have shown that urban runoff is usually toxic ("tox"), the two areas of science have not converged in recent decades to the extent that we know which chemical components are driving real-world biological decline in receiving waters. Moreover, as noted earlier, this may include uncharacterized contaminants that are simply not being monitored in the laboratory or the field.

As a case example, a research effort focused on adult coho salmon has been ongoing for more than a decade in the urban streams of Puget Sound, in the vicinity of Seattle in the northwestern United States. Coho salmon return from the ocean to coastal urban watersheds in the fall of each year, when autumn rains increase surface flows in spawning habitats. In the late 1990s, spawner surveys took note of fish behaving erratically, including surface swimming, gaping, loss of orientation, and loss of equilibrium. Affected fish died within a few hours, and in most cases females did not survive long enough to spawn (Scholz et al., 2012). Overall mortality rates have been high over the years (>50% of the total run within an urban watershed each fall), at levels very likely to limit the long-term viability of wild coho populations in Puget Sound (Spromberg and Scholz, 2011). This remains an important resource management challenge because coho are commercially important, are culturally iconic, and are designated as a species of concern under the US Endangered Species Act.

The evidence now suggests that coho spawners are succumbing to urban runoff, but the causative agents remain unidentified (Scholz et al., 2011). The severity of the recurring fish kills across urban watersheds is proportional to the amount of impervious surface and, in particular, the density of roads (Feist et al., 2011). This suggests a role for contaminants originating from motor vehicles. However, controlled exposures to artificial stormwater containing realistic mixtures of metals and PAHs do not produce the mortality syndrome in otherwise healthy adult coho (Spromberg et al., submitted). Rather, this requires exposure to actual highway runoff, which rapidly (within 4 h) evokes the same suite of symptoms as observed in field surveys. As further evidence for a toxic chemical component, pretreating highway runoff with conventional bioinfiltration (a form of green stormwater infrastructure) completely reverses the lethal impacts on coho spawners (Spromberg et al., submitted). Therefore, stormwater is very likely conveying an as-yet unidentified contaminant(s) from motor vehicles to coastal drainages where it consistently kills an ecologically important migratory fish species. Importantly, it would be impossible to infer this phenomenon from conventional toxicity bioassays.

18.3 Pesticide Mixtures

According to recent estimates, approximately 5.2 billion pounds of pesticides were used globally in 2006 and 2007, with nearly 1.1 billion of those pounds applied in the United States alone (EPA, 2011). Pesticides are grouped into classes and defined by their intended biological target, chemical structure, and mode of toxic action (Fulton et al., 2014; Solomon et al., 2014). For example, organophosphates and pyrethroids are classes of neurotoxic insecticides that are widely used to kill insect pests (Table 18.1). Pesticides are used in numerous residential and urban applications, but by far the largest volumes are used on agricultural lands (EPA, 2011) to control biological pests and weeds that threaten crop production. Pesticide mixtures, whether as coapplied individual compounds or formulated products, contaminate waterways adjacent to agricultural areas following application spray drift, surface runoff, and accidental direct applications. Agricultural pesticide products are typically sold as formulations, or mixtures of other chemicals in addition to the pesticide active ingredient eliciting the intended toxic effect. These additional ingredients serve to enhance the application or effectiveness of the pesticide, but may themselves be toxic to nontarget organisms (Stark and Walthall, 2003). In the United States, pesticide formulations include other toxic ingredients that are frequently not disclosed on product labels because they are proprietary information. Likewise, toxicity data on pesticide formulations are frequently lacking because even the most basic toxicity tests (e.g., LC_{50}s) of the formulated product are often not required during product registration.

Surface water and sediment monitoring confirms that pesticides commonly cooccur as mixtures in aquatic habitats, in both urban and agricultural areas (Kolpin et al., 2002; Stone et al., 2014). Some of the more commonly detected pesticide classes are shown in Table 18.1. Large-scale water quality syntheses across the United States (Gilliom, 2007) and Europe (Loos et al., 2009, 2013) have shown that these and other pesticides are frequently detected as complex mixtures of many chemicals. For example, aquatic monitoring by the US Geological

Table 18.1: Commonly Detected Current-Use Pesticides in Aquatic Habitats

Group	Class	Mode of Action	Example Pesticides
Insecticide	Organophosphate	Irreversible inhibition of AChE enzymes	Malathion Chlorpyrifos Diazinon
	Carbamate	Reversible inhibition of AChE enzymes	Carbaryl Methomyl
	Pyrethroid	Blocks sodium channel inactivation	Bifenthrin Permethrin Cypermethrin
	Neonicotinoid	Nicotinic ACh receptor agonist	Imidacloprid Thiametoxam
	Phenylpyrazole	Blocks GABA-regulated chloride channel	Fipronil
Herbicide	Triazine	Inhibits photosystem II	Atrazine Cyanazine Simazine
	Phosphonoglycine	Inhibits photosystem II	Glyphosate
	Chlorophenoxy acid	Auxin growth regulator	2,4-D
Fungicide	Azole	Inhibits ergosterol biosynthesis; Inhibits P450 monooxygenases	Prochloraz

AChE, acetylcholinesterase; GABA, gamma-aminobutyric acid

Survey has shown that urban streams were contaminated with at least four pesticides 75% of the time and agricultural streams contained seven or more pesticides 50% of the time (Gilliom, 2007). Chemical concentrations in aquatic systems are inherently variable and influenced by differences in the timing, intensity, and duration of pesticide applications, as well as by site-specific meteorological and geological conditions. Nevertheless, a recent global synthesis estimates that more than 40% of agricultural land area is predicted to be at risk of contaminating nearby streams via runoff (Ippolito et al., 2015).

Considering that the United States alone has nearly 17,000 pesticide products currently registered for use, it is generally impracticable to measure the toxicity of all possible mixture combinations for different aquatic species under different exposure conditions. However, progress has been made using known principles of toxic action. The two most widely accepted mixture models are concentration-addition and response-addition, for use with mixtures of pesticides having the same and different modes of action, respectively. These methodologies rely on the assumption of no interaction (Hertzberg and MacDonell, 2002). Notably, an investigation of large mixture datasets has shown that concentration-addition is a conservative predictor of mixture toxicity, regardless of the modes of action of the mixture constituents (Belden et al., 2007). Additionally, concentration-addition has been suggested as a first-tier model of potential mixture effects (Backhaus and Faust, 2012; Beyer et al., 2014). Derivations of concentration-addition methods have been used to develop mixture assessment approaches such as toxic units (Schafer et al., 2013; Ginebreda et al., 2014) and the pesticide toxicity index (Nowell et al., 2014). For dissimilarly acting pesticides, response-addition is an appropriate model to predict mixture toxicity (Cedergreen et al., 2008). This model assumes that, although the mixture components do not influence each other's uptake or metabolism, they still affect the same toxicological endpoint (e.g., mortality, swimming behavior).

Concentration-additive models generally predict that exposure to pesticide mixtures will produce greater toxicity than exposure to individual chemicals. This depends on the toxicity of individual pesticides, and is especially apparent in mixtures that produce toxicity when the constituent pesticides are present at concentrations below their single-chemical no-effect levels. The mode of toxic action is also an important factor when considering combined biological effects. For example, the organophosphate, pyrethroid, and neonicotinoid classes of insecticides have different modes of action—acetylcholinesterase inhibition, sodium channel inactivation blockade, and nicotinic acetylcholine receptor agonism, respectively. However, all three of these neurotoxicants increase postsynaptic activity in the nervous system, and mixtures are expected to produce enhanced (additive) neurotoxicity in aquatic species (Figure 18.3). Therefore, the biological underpinnings of toxicity are important considerations when assessing pesticide mixtures of potential concern.

For fish, environmental surveillance data have shown that most environmental exposures to pesticide are likely to be sublethal. Also, because many of the more toxic modern pesticides target the nervous system, there has been a focus on fish behavior as a toxicological endpoint.

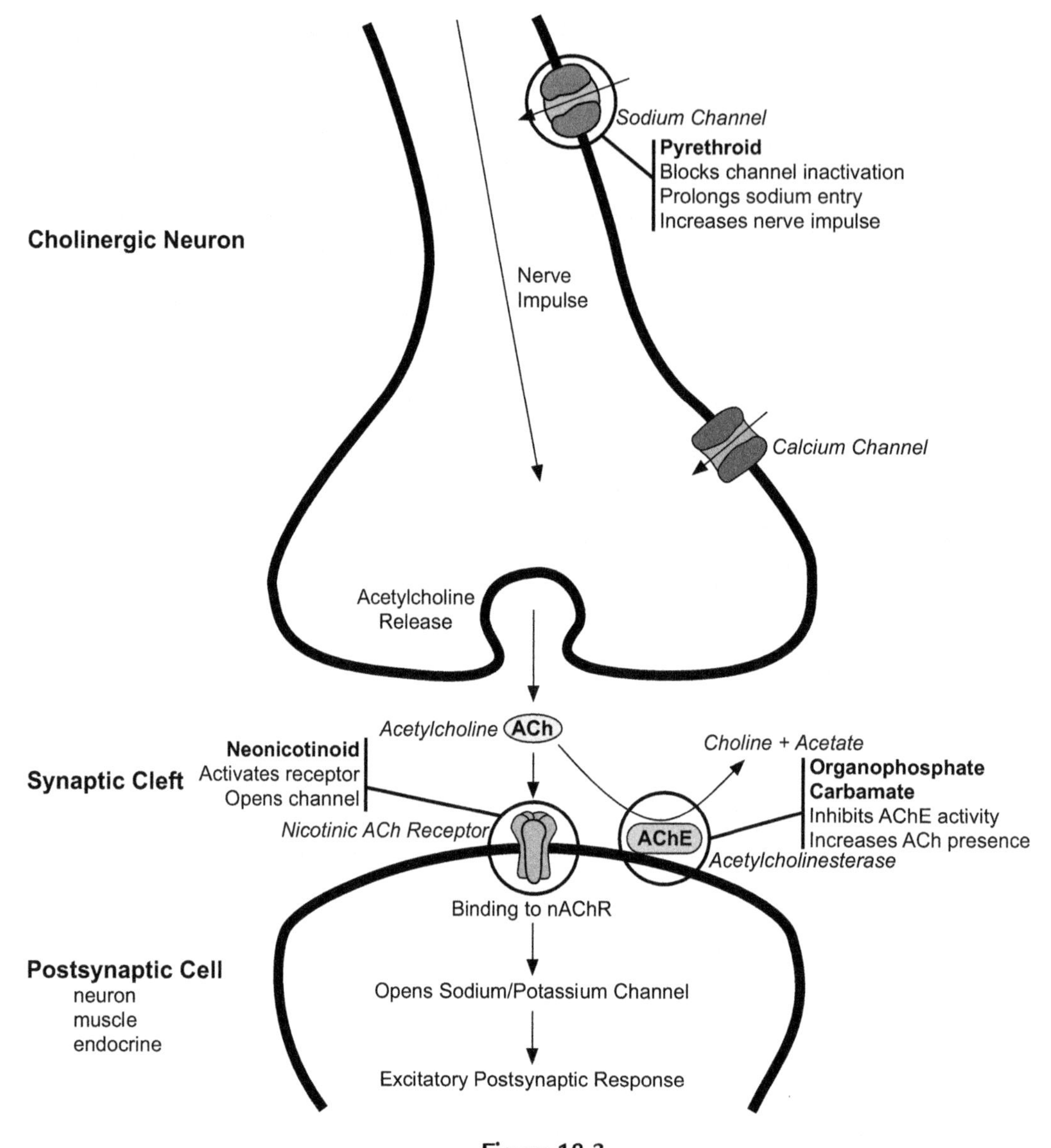

Figure 18.3
Diagram of a cholinergic nerve synapse highlighting the mode of action of organophosphate, pyrethroid, and neonicotinoid pesticides.

For example, Laetz et al. (2013) demonstrated reduced swimming speeds and decreased food strikes in juvenile salmon exposed to binary mixtures of organophosphates. Behavioral impairment was evident at low, environmentally relevant concentrations (1 ppb cumulative concentration) and was correlated with acetylcholinesterase inhibition.

From a natural resource management perspective, reducing or eliminating exposures to combinations of pesticides that interact synergistically in mixtures remains a priority. There are currently no reliable models to predict synergism from single-chemical toxicity data, making direct experimentation the only available tool (Altenburger et al., 2012). Among mammals, coexposure to the organophosphate malathion has long been known to increase the toxicity of other organophosphates (Frawley et al., 1957; Casida et al., 1963). Using juvenile salmon, Laetz et al. (2009) showed that binary mixtures of organophosphate and carbamate insecticides are synergistic, producing greater acetylcholinesterase inhibition in vivo than predicted from concentration-addition. Similarly, the azole fungicide prochloraz was found to be synergistic with other fungicides (Norgaard and Cedergreen, 2010) in experiments with *Daphnia magna*. Additionally, prochloraz and esfenvalerate produced synergistic toxicity in *D. magna* (Bjergager et al., 2012). Looking at the occurrence of synergy more holistically, Cedergreen (2014) analyzed datasets from numerous mixtures experiments and showed that synergy occurred in about 7% of the cases, with organophosphates and azole fungicides almost always involved in synergistic mixtures. Therefore, data showing synergy of specific pesticide combinations, or in certain pesticide classes (i.e., organophosphates), can be valuable in predicting the ecological risk of greater-than-additive toxic impacts.

Looking to the future, there will be an increasing need to move beyond the traditional reductionist approach to estimating toxicity (i.e., single-chemical, single-species laboratory tests) to better anticipate pesticide risks to wild populations, species, and communities (Macneale et al., 2010; Scholz et al., 2012). Ecological theory has been useful in this regard (Clements and Rohr, 2009). For example, mesocosms composed of organisms in multiple trophic levels were exposed to a complex pesticide mixture and followed for 18 weeks (Hua and Relyea, 2014). The direct mortality that occurred in some taxa created complex trophic interactions. Additionally, food web theory may offer a promising framework for predicting ecosystem-level effects of pesticide mixtures (Halstead et al., 2014). Knowledge of ecological parameters such as food web relationships and species interactions may allow us to make reliable predictions of the responses of complex aquatic systems to the toxic effects of pesticide mixtures.

18.4 Regulatory Provisions for Mixtures

Most US and European environmental regulations do not explicitly provide for the evaluation of mixtures toxicity. One notable exception is the US Comprehensive Environmental Response Compensation and Liability Act of 1990, which mandates that mixtures are assessed when determining the potential risks from chemical spills and certain legacy hazardous waste sites (EPA, 2015a). Other laws that do not explicitly call for mixtures assessments may nonetheless still incorporate mixtures analyses within the scope and intent of the legislation. For example, in 2006, the European Union issued the Registration, Evaluation, Authorization, and Restriction of Chemicals (REACH) legislation in an effort to enhance the regulation of chemical contaminants (European Union, 2006). This legislation was primarily based on assessing potential adverse effects from single-chemical toxicity assays (Syberg et al., 2009).

Accordingly, the assessment of mixture effects was not included except in certain specific cases (e.g., petroleum products). For the ensuing decade, efforts have been made to broaden REACH by incorporating mixture assessment methods (Syberg et al., 2009; Swedish Chemicals Industry, 2010; Cedergreen, 2014).

In the United States, there is currently no regulatory or policy framework that calls for an increased understanding of toxicity and environmental effects of PAHs or other petroleum hydrocarbon mixtures. Oil spills may trigger potential violations of the Clean Water Act, but violations are determined on the basis of potential for harm on total volumes of released oil, and there is no provision for different levels of harm based on mixture composition. Following the Exxon Valdez oil spill, the Oil Pollution Act of 1990 and subsequent amendments improved the natural resource damage assessment process and provided a framework for regulating the response, reparation, and restoration activities surrounding the release of a complex mixture (petroleum) into the environment. However, the activities regulated by the Oil Pollution Act of 1990 are not contingent on an understanding of which components of the spilled mixture cause environmental injury. The only water quality criteria for either individual PAHs or PAH mixtures in the United States are in the state of Alaska, and this criterion is based on a total PAH value of 10 μg/L irrespective of the mixture composition. Regulation of petroleum-derived mixtures is further challenged by the fact that some components of oils (and fuel oils in particular) have never been toxicologically characterized.

The pesticide registration process in the United States provides a prominent example of the discrepancy between single chemical and mixture toxicity assessments. Pesticides are registered or re-registered under the Federal Insecticide, Fungicide, and Rodenticide Act (FIFRA), which is an economic cost-benefit statute that has no explicit mandate to assess mixture effects to nontarget organisms (EPA, 2015b). Under FIFRA, registrations of agricultural and commercial pesticides are supported by single-chemical toxicity tests. The goal is to ensure, to the extent practicable, that no unacceptable adverse ecological effects will occur based on data produced by these standardized, single-chemical assays. Unfortunately, traditional toxicity tests do not incorporate simultaneous exposure to multiple pesticides or other natural stressors. Furthermore, guidance for conducting cumulative risk assessments of pesticide has been developed for only select circumstances (i.e., organophosphate pesticides that have the same mechanism of toxic action), and for assessments of potential impacts to human health (EPA, 2002).

In the decades following the publication of *Silent Spring* (Carson, 1962), pesticides have been a societal concern for vulnerable fish and wildlife throughout the world. In the United States, the risks that the federal registration process may pose are evaluated under the Endangered Species Act (ESA). The aim of the ESA is to ensure that a federal action (i.e., registration of an individual pesticide) does not jeopardize a listed species or adversely modify their designated critical habitats. The ESA is broadly intended to consider all risks associated with a proposed federal action. Therefore, real-world environmental exposures to mixtures may be incorporated into an ecological risk analysis. Cumulative assessment methodologies developed by the EPA (EPA, 2002) have been applied in some cases, such as threatened or endangered Pacific salmonids (NMFS, 2008, 2009, 2010).

In the United States, the National Research Council recently reviewed current methodologies used by federal agencies for conducting pesticide risk assessments under FIFRA and the ESA (National Research Council, 2013). The importance of assessing the risk of pesticide mixtures was a primary component of the report. Since the release of the report, several US federal agencies, including the National Marine Fisheries Service, the Fish and Wildlife Service, and the Environmental Protection Agency have been coordinating the development of new procedures to incorporate pesticide mixtures into the ecological risk assessment framework that underpins pesticide registrations (or re-registrations). An explicit consideration of mixtures would represent a key advancement for natural resource management because (1) exposures to pesticide mixtures are the rule rather than the exception in most aquatic habitats and (2) assessments based on individual chemicals are likely to underestimate actual risk where mixtures occur. The inclusion of pesticide mixtures analyses in both ESA and FIFRA regulatory decision-making is expected to yield more consistent and ecologically relevant determinations of risk to threatened and endangered species.

18.5 Conclusions

Chemical mixtures have been a longstanding issue in aquatic ecotoxicology. This is unlikely to change, given the ever-increasing number of anthropogenic chemicals in global societal use. Other important environmental stressors, such as shifting thermal regimes in aquatic systems related to climate change, are expected to have important modulating effects on the direct and indirect toxicity of mixtures to aquatic organisms (Schiedek et al., 2007; Hooper et al., 2013). Some fields of research, including epigenetics and genomics, offer promising advances in the detection of toxic effects in a host of nontarget aquatic species (Altenburger et al., 2012). Such advances will improve the tractability of performing more robust and systematic assessments of complex chemical mixtures. Therefore, incorporating the toxic effects of chemical mixtures into ecological risk assessments and regulatory decision-making processes will remain vitally important for the conservation and protection of imperiled aquatic species and the habitats upon which they depend.

References

Adams, J., Bornstein, J.M., Munno, K., et al., 2014a. Identification of compounds in heavy fuel oil that are chronically toxic to rainbow trout embryos by effects-driven chemical fractionation. Environ. Toxicol. Chem. 33, 825–835.

Adams, J., Sweezey, M., Hodson, P.V., 2014b. Oil and oil dispersant do not cause synergistic toxicity to fish embryos. Environ. Toxicol. Chem. 33, 107–114.

Altenburger, R., Scholz, S., Schmitt-Jansen, M., et al., 2012. Mixture toxicity revisited from a toxicogenomic perspective. Environ. Sci. Technol. 46, 2508–2522.

Arfsten, D.P., Schaeffer, D.J., Mulveny, D.C., 1996. The effects of near ultraviolet radiation on the toxic effects of polycyclic aromatic hydrocarbons in animals and plants: a review. Ecotoxicol. Environ. Saf. 33, 1–24.

Backhaus, T., Faust, M., 2012. Predictive environmental risk assessment of chemical mixtures: a conceptual framework. Environ. Sci. Technol. 46, 2564–2573.

Barron, M.G., Carls, M.G., Short, J.W., et al., 2003. Photoenhanced toxicity of aqueous phase and chemically dispersed weathered Alaska North Slope crude oil to pacific herring eggs and larvae. Environ. Toxicol. Chem. 22, 650–660.

Belden, J.B., Gilliom, R.J., Lydy, M.J., 2007. How well can we predict the toxicity of pesticide mixtures to aquatic life? Int. Environ. Assess. Manag. 3, 364–372.

Beyer, J., Petersen, K., Song, Y., et al., 2014. Environmental risk assessment of combined effects in aquatic ecotoxicology: a discussion paper. Mar. Environ. Res. 96, 81–91.

Bjergager, M.B.A., Hanson, M.L., Solomon, K.R., et al., 2012. Synergy between prochloraz and esfenvalerate in *Daphnia magna* from acute and subchronic exposures in the laboratory and microcosms. Aquat. Toxicol. 110–111, 17–24.

Brette, F., Machado, B., Cros, C., et al., 2013. Crude oil impairs cardiac excitation-contraction coupling in fish. Science 343, 772–776.

Brown, J.N., Peake, B.M., 2006. Sources of heavy metals and polycyclic aromatic hydrocarbons in urban stormwater runoff. Sci. Total Environ. 359, 145–155.

Carls, M.G., Meador, J.P., 2009. A perspective on the toxicity of petrogenic PAHs to developing fish embryos related to environmental chemistry. Hum. Ecol. Risk Assess. 15, 1084–1098.

Carls, M.G., Rice, S.D., Hose, J.E., 1999. Sensitivity of fish embryos to weathered crude oil: part i. Low-level exposure during incubation causes malformations, genetic damage, and mortality in larval pacific herring (*Clupea pallasi*). Environ. Toxicol. Chem. 18, 481–493.

Carson, R., 1962. Silent Spring. Houghton Mifflin, Cambridge, MA. 368 p.

Casida, J.E., Baron, R.L., Eto, M., et al., 1963. Potentiation and neurotoxicity induced by certain organophosphates. Biochem. Pharmacol. 12, 73–83.

Cedergreen, N., Christensen, A.M., Kamper, A., et al., 2008. A review of independent action compared to concentration addition as reference models for mixtures of compounds with different molecular target sites. Environ. Toxicol. Chem. 27, 1621–1632.

Cedergreen, N., 2014. Quantifying synergy: a systematic review of mixture toxicity studies within environmental toxicology. PLoS One 9, 1–11. http://dx.doi.org/10.1371/journal.pone.0096580.

Clark, R.C.J., Brown, D.W., 1977. Petroleum: properties and analyses in biotic and abiotic systems. In: Malins, D.C. (Ed.), Effects of Petroleum on Arctic and Subarctic Marine Environments and Organisms, vol. I. Academic Press, Inc., New York, pp. 1–89.

Clements, W.H., Rohr, J.R., 2009. Community responses to contaminants: using basic ecological principles to predict ecotoxicological effects. Environ. Toxicol. Chem. 28, 1789–1800.

Council, N.R., 2003. Oil in the Sea III: Inputs, Fates, and Effects. National Research Council, Washington, DC. pp. 446.

Diamond, S.A., 2003. Photoactivated toxicity in aquatic environments. In: Hader, D., Joir, G. (Eds.), UV Effects in Aquatic Organisms and Ecosystems. John Wiley & Sons, Inc., New York, pp. 219–250.

Donald, D.B., Cessna, A.J., Sverko, E., et al., 2007. Pesticides in surface drinking-water supplies of the northern great plains. Environ. Health Perspect. 115, 1183–1191.

EPA (Environmental Protection Agency), 2002. Guidance on Cumulative Risk Assessment of Pesticide Chemicals that Have a Common Mechanism of Toxicity. Washington, DC. 81 pp.

EPA (Environmental Protection Agency), 2011. Pesticide Industry Sales and Usage: 2006 and 2007 Market Estimates. Washington, DC. 41 pp.

EPA (Environmental Protection Agency), 2015a. Federal Facilities and the Comprehensive Environmental Response, Compensation, and Liability Act (CERCLA)/Superfund of 1980. U.S. Protection Agency, Washington, D.C., USA. Available online at: http://www.epa.gov/oecaerth/federalfacilities/enforcement/civil/cercla.html.

EPA (Environmental Protection Agency), 2015b. Federal Insecticide, Fungicide, and Rodenticide Act (FIFRA) of 1972 Including Amendments from the Food Quality Protection Act (FQPA) of 1996. Available online at: http://www.epa.gov/agriculture/lfra.html.

European Union, 2006. Regulation (EC) No 1907/2006 of the European Parliament and of the Council of 18 December 2006 Concerning the Registration, Evaluation, Authorization and Restriction of Chemicals (REACH), Establishing a European Chemicals Agency.

Feist, B.E., Buhle, E.R., Arnold, P., et al., 2011. Landscape ecotoxicology of coho salmon spawner mortality in urban streams. PLoS One 6 (8), e23424.

Frawley, J.P., Fuyat, H.N., Hagan, E.C., et al., 1957. Marked potentiation in mammalian toxicity from simultaneous administration of two anticholinesterase compounds. J. Pharmacol. Exp. Ther. 121, 96–106.

Fulton, M.H., Key, P.B., DeLorenzo, M.E., 2014. Insecticide toxicity in fish. In: Tierney, K.B., Farrell, A.P., Brauner, C.J. (Eds.), Organic Chemical Toxicology of Fishes. Elsevier, pp. 309–368.

Gilliom, R.J., 2007. Pesticides in U.S. streams and groundwater. Environ. Sci. Technol. 41, 3407–3413.

Ginebreda, A., Kuzmanovic, M., Guasch, H., et al., 2014. Assessment of multi-chemical pollution in aquatic ecosystems using toxic units: compound prioritization, mixture characterization and relationships with biological descriptors. Sci. Total Environ. 468–469, 715–723.

Gros, J., Nabi, D., Würz, B., et al., 2014. First day of an oil spill on the open sea: early mass transfers of hydrocarbons to air and water. Environ. Sci. Technol. 48, 9400–9411.

Halstead, N.T., McMahon, T.A., Johnson, S.A., et al., 2014. Community ecology theory predicts the effects of agrochemical mixtures on aquatic biodiversity and ecosystem properties. Ecol. Lett. 17, 932–941.

Hatlen, K., Sloan, C.A., Burrows, D.G., et al., 2010. Natural sunlight and residual fuel oils are an acutely lethal combination for fish embryos. Aquat. Toxicol. 99, 56–64.

Hayes, T.B., Case, P., Chui, S., et al., 2006. Pesticide mixtures, endocrine disruption, and amphibian declines: are we underestimating the impact? Environ. Health Perspect. 114, 40–50.

Heintz, R.A., Short, J.W., Rice, S.D., 1999. Sensitivity of fish embryos to weathered crude oil: part ii. Increased mortality of pink salmon (*Oncorhynchus gorbuscha*) embryos incubating downstream from weathered Exxon Valdez crude oil. Environ. Toxicol. Chem. 18, 494–503.

Hertzberg, R.C., MacDonell, M.M., 2002. Synergy and other ineffective mixture risk definitions. Sci. Total Environ. 288, 31–42.

Hicken, C.E., Linbo, T.L., Baldwin, D.H., et al., 2011. Sublethal exposure to crude oil during embryonic development alters cardiac morphology and reduces aerobic capacity in adult fish. PNAS 108, 7086–7090.

Hooper, M.J., Ankley, G.T., Cristol, D.A., et al., 2013. Interactions between chemical and climate stressors: a role for mechanistic toxicology in assessing climate change risks. Environ. Toxicol. Chem. 32, 32–48.

Hua, J., Relyea, R., 2014. Chemical cocktails in aquatic systems: pesticide effects on the response and recovery of >20 animal taxa. Environ. Pollut. 189, 18–26.

Incardona, J.P., Collier, T.K., Scholz, N.L., 2004. Defects in cardiac function precede morphological abnormalities in fish embryos exposed to polycyclic aromatic hydrocarbons. Toxicol. Appl. Pharmacol. 196, 191–205.

Incardona, J.P., Carls, M.G., Teraoka, H., et al., 2005. Aryl hydrocarbon receptor-independent toxicity of weathered crude oil during fish development. Environ. Health Perspect. 113, 1755–1762.

Incardona, J.P., Day, H.L., Collier, T.K., et al., 2006. Developmental toxicity of 4-ring polycyclic aromatic hydrocarbons in zebrafish is differentially dependent on AH receptor isoforms and hepatic cytochrome P450 1A metabolism. Toxicol. Appl. Pharmacol. 217, 308–321.

Incardona, J.P., Carls, M.G., Day, H.L., et al., 2009. Cardiac arrhythmia is the primary response of embryonic pacific herring (*Clupea pallasi*) exposed to crude oil during weathering. Environ. Sci. Technol. 43, 201–207.

Incardona, J.P., Collier, T.K., Scholz, N.L., 2011a. Oil spills and fish health: exposing the heart of the matter. J. Expo. Sci. Environ. Epidemiol. 21, 3–4.

Incardona, J.P., Linbo, T.L., Scholz, N.L., 2011b. Cardiac toxicity of 5-ring polycyclic aromatic hydrocarbons is differentially dependent on the aryl hydrocarbon receptor 2 isoform during zebrafish development. Toxicol. Appl. Pharmacol. 257, 242–249.

Incardona, J.P., Vines, C.A., Anulacion, B.F., et al., 2012a. Unexpectedly high mortality in pacific herring embryos exposed to the 2007 Cosco Busan oil spill in San Francisco Bay. Proc. Natl. Acad. Sci. USA 109, E51–E58.

Incardona, J.P., Vines, C.A., Linbo, T.L., et al., 2012b. Potent phototoxicity of marine bunker oil to translucent herring embryos after prolonged weathering. PLoS One 7, e30116.

Incardona, J.P., Swarts, T.L., Edmunds, R.C., et al., 2013. Exxon Valdez to Deepwater Horizon: comparable toxicity of both crude oils to fish early life stages. Aquat. Toxicol. 142–143, 303–316.

Incardona, J.P., Gardner, L.D., Linbo, T.L., et al., 2014. Deepwater Horizon crude oil impacts the developing hearts of large predatory pelagic fish. Proc. Natl. Acad. Sci. USA 111, E1510–E1518.
Ippolito, A., Kattwinkel, M., Rasmussen, J.J., et al., 2015. Modeling global distribution of agricultural insecticides in surface waters. Environ. Pollut. 198, 54–60.
Jayasundara, N., Van Tiem Garner, L., Meyer, J.N., et al., 2015. AHR2-mediated transcriptomic responses underlying the synergistic cardiac developmental toxicity of PAHs. Toxicol. Sci. 143, 469–481.
Jung, J.H., Hicken, C.E., Boyd, D., et al., 2013. Geologically distinct crude oils cause a common cardiotoxicity syndrome in developing zebrafish. Chemosphere 91, 1146–1155.
Kayhanian, C., Stransky, S., Bay, S.L., et al., 2008. Toxicity of urban highway runoff with respect to storm duration. Sci. Total Environ. 389, 386–406.
Kolpin, D.W., Furlong, E.T., Meyer, M.T., et al., 2002. Pharmaceuticals, hormones, and other organic wastewater contaminants in U.S. streams, 1999–2000: a national reconnaissance. Environ. Sci. Technol. 36, 1202–1211.
Laetz, C.A., Baldwin, D.H., Hebert, V., et al., 2013. Interactive neurobehavioral toxicity of diazinon, malathion, and ethoprop to juvenile coho salmon. Environ. Sci. Technol. 47, 2925–2931.
Laetz, C.A., Baldwin, D.H., Hebert, V., et al., 2009. The synergistic toxicity of pesticide mixtures: implications for risk assessment and the conservation of endangered pacific salmon. Environ. Health Perspect. 117, 348–353.
Loos, R., Carvalho, R., Antonio, D.C., et al., 2013. EU-wide monitoring survey on emerging polar organic contaminants in wastewater treatment plant effluents. Water Res. 47, 6375–6487.
Loos, R., Gawlik, B.M., Locoro, G., et al., 2009. EU-wide survey of polar organic persistent pollutants in European river waters. Environ. Pollut. 157, 561–568.
Lydy, M., Belden, J., Wheelock, C., et al., 2004. Challenges in regulating pesticide mixtures. Ecol. Soc. 9 (1). http://www.ecologyandsociety.org/vol9/iss6/art1/.
Macneale, K.H., Kiffney, P.M., Scholz, N.L., 2010. Pesticides, aquatic food webs, and the conservation of Pacific salmon. Front. Ecol. Environ. 8, 475–482.
Marty, G.D., Short, J.W., Dambach, D.M., et al., 1997. Ascites, premature emergence, increased gonadal cell apoptosis, and cytochrome P4501A induction in pink salmon larvae continuously exposed to oil-contaminated gravel during development. Can. J. Zool.-Rev. Can. Zool. 75, 989–1007.
Matar, S., Hatch, L.F., 2001. Chemistry of Petrochemical Processes. Gulf Professional Publishing, Boston. pp. 356.
McIntyre, J.K., Davis, J.W., Incardona, J.P., et al., 2014. Zebrafish and clean water technology: assessing the protective effects of bioinfiltration as a treatment for toxic urban runoff. Sci. Total Environ. 500, 173–180.
McIntyre, J.K., Davis, J.W., Hinman, C., et al. Soil bioretention protects juvenile salmon and their prey from the toxic impacts of urban stormwater runoff. Chemosphere, in press. http://dx.doi.org/10.1016/j.chemosphere.2014.12.052.
National Research Council, 2013. Assessing Risks to Endangered and Threatened Species from Pesticides. National Academies Press, Washington DC.
NMFS (National Marine Fisheries Service), 2008. Endangered Species Act Section 7 Consultation: Biological Opinion on Environmental Protection Agency Registration of Pesticides Containing Chlorpyrifos, Diazinon, and Malathion. U.S. Department of Commerce, Silver Spring, Maryland. 482 p.
NMFS (National Marine Fisheries Service), 2009. Endangered Species Act Section 7 Consultation: Biological Opinion on Environmental Protection Agency Registration of Pesticides Containing Carbaryl, Carbofuran, and Methomyl. U.S. Department of Commerce, Silver Spring, Maryland. 607 p.
NMFS (National Marine Fisheries Service), 2010. NMFS Endangered Species Act Section 7 Consultation, Biological Opinion: Environmental Protection Agency Registration of Pesticides Containing Azinphos Methyl, Bensulide, Dimethoate, Disulfoton, Ethoprop, Fenamiphos, Naled, Methamidophos, Methidathion, Methyl Parathion, Phorate, and Phosmet. Department of Commerce, Silver Spring, MD. U.S. 1010 p.
Norgaard, K.B., Cedergreen, N., 2010. Pesticide cocktails can interact synergistically on aquatic crustaceans. Environ. Sci. Pollut. Res. 17, 957–967.
Nowell, L.H., Norman, J.E., Moran, P.W., et al., 2014. Pesticide toxicity index – a tool for assessing potential toxicity of pesticide mixtures to freshwater aquatic organisms. Sci. Total Environ. 476–477, 144–157.

Pratt, J.M., Coler, R.A., Godfrey, P.J., 1981. Ecological effects of urban stormwater runoff on benthic macroinvertebrates inhabiting the Green River, Massachusetts. Hydrobiologia 83, 29–42.

Relyea, R.A., 2008. A cocktail of contaminants: how mixtures of pesticides at low concentrations affect aquatic commnities. Oecologia 159, 363–376.

Relyea, R., Hoverman, J., 2006. Assessing the ecology in ecotoxicology: a review and synthesis in freshwater systems. Ecol. Lett. 9, 1157–1171.

Rodney, S.I., Teed, R.S., Moore, D.R.J., 2013. Estimating the toxicity of pesticide mixtures to aquatic organisms: a review. Hum. Ecol. Risk Assess. 19, 1557–1575.

Rohr, J.R., Kerby, J.L., Sih, A., 2006. Community ecology as a framework for predicting contaminant effects. Trends Ecol. Evol. 21, 606–613.

Schäfer, R.B., Bemer, N., Kefford, B., et al., 2013. How to characterize chemical exposure to predict ecologic effects on aquatic communities? Environ. Sci. Technol. 47, 7996–8004.

Schiedek, D., Sundelin, B., Readman, J.W., et al., 2007. Interactions between climate change and contaminants. Mar. Pollut. Bull. 54, 1845–1856.

Scholz, N.L., Myers, M.S., McCarthy, S.G., et al., 2011. Recurrent die-offs of adult coho salmon returning to spawn in Puget Sound lowland urban streams. PLoS One 6 (12), e28013.

Scholz, N.L., Fleishman, E., Brooks, M.L., et al., 2012. A perspective on modern pesticides, pelagic fish declines, and unknown ecological resiliency in highly managed ecosystems. BioScience 62, 428–434.

Shinya, M., Tsuchinaga, T., Kitano, M., et al., 2000. Characterization of heavy metals and polycyclic aromatic hydrocarbons in urban highway runoff. Water Sci. Technol. 42, 201–208.

Short, J.W., Heintz, R.A., 1997. Identification of *Exxon Valdez* oil in sediments and tissues from Prince William Sound and the Northwestern Gulf of Alaska based on a PAH weathering model. Environ. Sci. Technol. 31, 2375–2384.

Snodgrass, J.W., Casey, R.E., Joseph, D., et al., 2008. Microcosm investigations of stormwater pond sediment toxicity to embryonic and larval amphibians: variation in sensitivity among species. Environ. Pollut. 154, 291–297.

Solomon, K.R., Dalhoff, K., Volz, D., et al., 2014. Effects of herbicides on fish. In: Tierney, K.B., Farrell, A.P., Brauner, C.J. (Eds.), Organic Chemical Toxicology of Fishes. Elsevier, pp. 369–409.

Spromberg, J.A., Scholz, N.L., 2011. Estimating the future decline of wild coho salmon populations resulting from early spawner die-offs in urbanizing watersheds of the Pacific Northwest, USA. Integr. Environ. Assess. Manag. 7, 648–656.

Spromberg, J.A., Baldwin, D.H., Damm, S.E., et al. Widespread Adult Coho Salmon Spawner Mortality in Western U.S. Urban Watersheds: Lethal Impacts of Stormwater Runoff Are Reversed by Soil Bioinfiltration, submitted for publication.

Stark, J.D., Walthall, W.K., 2003. Agricultural adjuvants: acute mortality and effects on population growth rate of *Daphnia pulex* after chronic exposure. Environ. Toxicol. Chem. 22, 3056–3061.

Stone, W.W., Gilliom, R.J., Ryberg, K.R., 2014. Pesticides in U.S. streams and rivers: occurrence and trends during 1992-2011. Environ. Sci. Technol. 48, 11025–11030.

Stout, S.A., Wang, Z., 2007. Chemical fingerprinting of spilled or discharged petroleum - methods and factors affecting petroleum fingerprints in the environment. In: Wang, Z., Stout, S.A. (Eds.), Oil Spill Environmental Forensics. Academic Press, London, pp. 1–53.

Swedish Chemicals Agency, 2010. Hazard and Risk Assessment of Chemical Mixtures under REACH: Stat of the Art, Gaps, and Options for Improvement. http://www.kemi.se/Documents/Publikationer/Trycksaker/PM/PM3_10.pdf?epslanguage=en.

Syberg, K., Jensen, T.S., Cedergreen, N., et al., 2009. On the use of mixture toxicity assessment in REACH and the water framework directive: a review. Hum. Ecol. Risk Assess. 15, 1257–1272.

Toccalino, P.L., Norman, J.E., Scott, J.C., 2012. Chemical mixtures in untreated water from public-supply wells in the U.S. – occurrence, composition, and potential toxicity. Sci. Total Environ. 431, 262–270.

Uhler, A.D., Stout, S.A., Douglas, G.S., 2007. Chemical heterogeneity in modern marine residual fuel oils. In: Wang, Z., Stout, S.A. (Eds.), Oil Spill Environmental Forensics. Academic Press, London, pp. 327–348.

Van Tiem, L.A., Di Giulio, R.T., 2011. AHR2 knockdown prevents PAH-mediated cardiac toxicity and XRE- and ARE-associated gene induction in zebrafish (*Danio rerio*). Toxicol. Appl. Pharmacol. 254, 280–287.

Villanueva, C.M., Kogevinas, M., Cordier, S., et al., 2014. Assessing exposure and health consequences of chemicals in drinking water: current state of knowledge and research needs. Environ. Health Perspect. 122, 213–221.

Walsh, C.J., Roy, A.H., Feminella, J.W., et al., 2005. The urban stream syndrome: current knowledge and the search for a cure. J. N. Am. Benthol. Soc. 24, 706–723.

Wang, Z., Hollebone, B.P., Fingas, M., et al., 2003. Characteristics of Spilled Oils, Fuels, and Petroleum Products: 1. Composition and Properties of Selected Oils. U.S. Environmental Protection Agency, Washington, D.C.

Yu, H., 2002. Environmental carcinogenic polycyclic aromatic hydrocarbons: photochemistry and phototoxicity. J. Environ. Sci. Health C Environ. Carcinog. Ecotoxicol. Rev. 20, 149–183.

CHAPTER 19

Predictive Ecotoxicology and Environmental Assessment

Claude Amiard-Triquet, Jean-Claude Amiard, Catherine Mouneyrac

Abstract

The content of this book about tools, methodologies, and concepts, coauthored by scientific experts from many different institutes worldwide, highlights how assessing environmental risks of chemicals entering the aquatic environment as a consequence of human activities is a complicated task. Taking into account the lessons of the past, this chapter summarizes relevant strategies for emerging risks (e.g., innovative chemical methods, issue of mixtures, prioritization of research questions, investigations about modes of action for integration in adverse outcome pathways, ecotoxicological modeling). Important related issues have been taken into account, including global climate change, the emerging concern about plastic debris and the design of alternative methodologies to spare animal lives. Beyond the field of ecotoxicology, strategies for environmental protection (remediation, reduction, or replacing of harmful chemicals) are evoked. The conclusion insists on the necessary integration of environmental studies in the social, economic, and political context.

Keywords: Adverse outcome pathways; Alternative methods; Bioaccumulation models; Bioanalytical tools; Bioavailability; Biomarkers; Ecotoxicological models; Interdisciplinarity; Mixtures.

Chapter Outline

Since disasters, such as the one in Minamata, Japan (mid-twentieth century), in which hundreds of villagers were exposed to lethal levels of mercury from eating fish (Yorifuji et al., in EEA, 2013); the book *Silent Spring* by Rachel Carson (1962) documenting the detrimental effects on the environment of the indiscriminate use of pesticides (updated by Bouwman et al., in EEA, 2013; and the first massive oil spill of the Torrey Canyon (1967) that ran aground off Cornwall, UK, spilling 80,000 tons (919,000 barrels) of crude oil, we have been clearly aware of environmental deteriorations induced by human activities, thus leading to the

Aquatic Ecotoxicology. http://dx.doi.org/10.1016/B978-0-12-800949-9.00019-X

development of ecotoxicology defined first by René Truhaut in 1969 (in Truhaut, 1977). Despite much having been made in this research area for nearly half a century, chemical contaminants still represent serious threats to environmental and human health. For instance, legacy contaminants such as mercury or polychlorinated biphenyl (PCBs) still pose a wicked problem (Amiard et al., 2015), leading to fish consumption advisories in many countries, with a US Environmental Protection Agency (USEPA) dedicated Website http://fishadvisoryonline.epa.gov/General.aspx.

19.1 Lessons of the Past

19.1.1 Standard Quality Values

For legacy contaminants, risk assessment procedures are well established (Chapters 2 and 3), and it would be expected that multiple indices and guidelines can help to assess the degree of contamination of environmental matrices (water, sediment, food) and resulting ecological risks. However, even in the case of metals that have been among the first studied contaminants in the 1970s, recent studies have highlighted that contradictory conclusions could be obtained when different indices and sediment quality guidelines are used to assess metal pollution in surface sediments (Zhuang and Gao, 2014). Similarly, Conder et al. (2015), reviewing biologically based standard quality values that have been developed for characterizing mercury risks to benthic invertebrates, concluded, to their lack of predictive ability, quoting many examples for mercury and also some other legacy contaminants (nickel, total PCBs, chlordane, lindane, total DDT, and 1,1-dichloro-2,2-bis(4-chlorophenyl)ethylene (p,p′-DDE)). Such discrepancies are often the concentrations of contaminants may be much higher than guidelines (both expressed as total concentrations) whereas bioavailable concentrations are consistently lower. This pattern has been exemplified with some detail in Chapter 10 (Table 10.1).

The only water quality criteria for either individual polyaromatic hydrocarbons (PAHs) or PAH mixtures in the United States are in the state of Alaska, and this criterion is based on a total PAH value of $10\,\mu g\,L^{-1}$, irrespective of the mixture composition. Such a total concentration has no ecotoxicological value because the capacity of a mixture to cause environmental impairment depends on the relative abundance of the most harmful constituents (Chapter 18).

The limited utility of standard quality values in predicting environmental toxicity may be also linked in a number of cases to the acquisition of tolerance—through physiological acclimation or genetic adaptation—by local populations chronically exposed to contaminants (to learn more, the reader can refer to the book by Amiard-Triquet et al., 2011). Thus for more realistic environmental risk assessment, the use of local species may be attractive (Chapter 9). This procedure has been tested for the determination of the local predicted no effect concentration (PNEC) in a highly contaminated estuary in Europe, the Seine estuary, with comparison with the generic PNEC recognized in Europe (Bocquené et al., 2012). The problem is that the number of data available for local species is limited thus leading to the application of high assessment factors (Chapter 2). In the case of atrazine, the local PNEC varied between 0.28 and 0.30 $\mu g L^{-1}$ in different sections of the estuary and remained consistent with the generic PNEC (60 $\mu g L^{-1}$). In other cases, the bias of using high assessment factors can lead to an overvaluation of the toxic risk as shown for tributyltin, the PNEC of which (0.2 $ng L^{-1}$) is lower than the detection limit of analytical methods (Bocquené et al., 2012).

However, these standards remain useful for initial screening steps or as part of a weight of evidence approach (Zhuang and Gao, 2014; Conder et al., 2015) as implemented by Benedetti et al. (2012) for the classification of polluted sediments integrating sediment chemistry, bioavailability, biomarker responses, and bioassays. Dagnino and Variengo (2014) have developed an expert decision support system for the management of contamination in marine coastal ecosystems, the first step of which consists in a comparison of concentrations to effect-based thresholds (threshold effect levels and probable effect levels developed by MacDonald et al. (1996)), ignoring the problem of bioavailability. Sites are then placed into three categories of contamination: "uncontaminated," mildly contaminated, and highly contaminated and depending on their classification, different ecotoxicological endpoints are considered, up to and including alterations in community structure and ecosystem functions as also recommended by Chapman et al. (2013) in a review dealing with the assessment and management of sediment contamination in transitional waters.

Integrated approaches are absolutely needed because different indices do not provide the same information, and considered individually could lead to misinterpretations in terms of environmental assessment. Hazard quotients (HQs) and normative limits of Venice protocol (1993, in Benedetti et al., 2012) for sediments of the Venice lagoon are shown in Figure 19.1. Considering sediment contamination, bioavailability, biomarkers, or bioassays, sediment toxicity ranking clearly differs: for instance, severe chemical hazard are depicted for sites S4 and S5 considering sediment contamination and biomarkers; bioassays per se would conclude only to slight hazard for sites S1, S2, and S3; whereas the global assessment indicates a moderate hazard at sites S1 and S2 and a major hazard at sites S3, S4, and S5.

The need for integrated environmental monitoring of chemicals and their effects is generally agreed (Chapter 4) as exemplified by reports of expert groups under the auspices of the French Ministry of Environment (Amiard-Triquet and Rainbow, 2009) or the International Council for the Exploration of the Sea (ICES, 2012). They underline that, in addition to the simultaneous measurement of contaminant concentrations and biological responses at

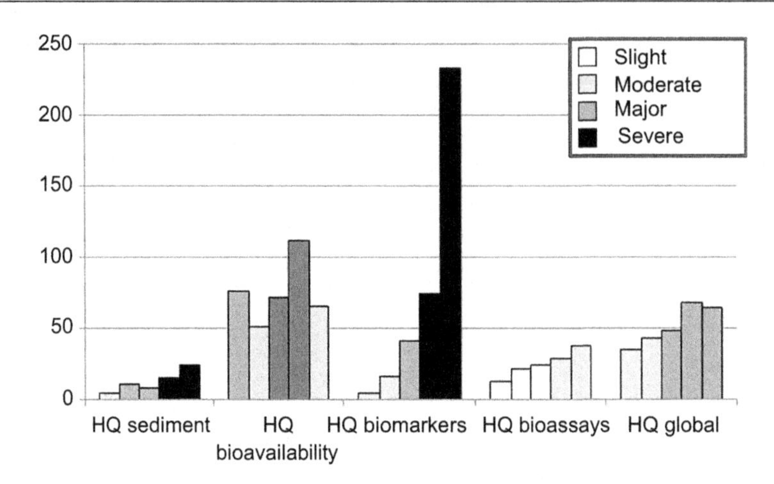

Figure 19.1
Hazard quotients determined using metal and PAH concentrations in sediments (HQ sediment) or eel livers (HQ bioavailability), from a set of many different biomarkers in eels (HQ biomarkers), from seven bioassays in different species (HQ bioassays) and hazard elaboration obtained from various lines of evidence (HQ global). The five bars for each HQ correspond to sediments from five stations S1 to S5. *Drawn after the data published by Benedetti et al. (2012).*

different levels of biological organization, a range of physical and other chemical measurements are indispensable to permit normalization (for examples, see Figures 3.2 and 3.3 in this book) and appropriate assessment. In addition, they should be used to take into account the influence of extrinsic confounding factors on both bioaccumulation (NAS, 1980) and effects (Chapters 7, 10 and 11) in complementarity with the inclusion of intrinsic factors. More widely, it is recommended to better consider the role of chemical contamination in the context of multiple stressors such as alterations to habitat and flow, physical disturbance, nutrient addition, and food availability (Amiard-Triquet and Rainbow, 2009; Burton and Johnston, 2010; Kapo et al., 2014) with a peculiar attention to the global climate change (Stahl et al., 2013).

19.1.2 Bioavailability

The importance of taking into account bioavailability has been highlighted in Chapters 4 and 5. Incorporating bioavailability considerations into the evaluation of contaminated sediments is already incorporated in regulatory guidance (e.g., ITRC, 2011).

Among approaches to environmental exposure assessments of metals, the free ion activity model is an equilibrium-based geochemical speciation model used to predict free metal ion concentrations in waters, which are mainly associated with toxic effects. Working either at equilibrium or in the kinetic phase, passive samplers are increasingly used to determine many classes of contaminants (hydrophobic organics, hydrophilic compounds, metals) in the environment (Chapter 4). For sediments, Lu et al. (2014) propose to assess bioavailability of hydrophobic organic compounds and metals using freely available porewater concentrations. Strategies to assess bioavailability also include partial extraction methods and in vitro digestion (Chapter 5).

Biotic ligand models expand the free ion activity model and predict the short-term binding of free ions to a biotic ligand, also based on equilibrium processes (Chapter 4). In addition, a direct assessment of bioavailability may be based on the determination of bioaccumulation from sediments (in endobenthic organisms, see Chapter 10, or in fish models, e.g., Gaillard et al., 2014). Conventional assessment of chemical hazards to aquatic life is mainly based on the relationship between external exposure (contaminant concentrations in water, much less frequently sediment or food) and biological effects in organisms (Chapter 2). To short circuit the question of bioavailability, it has been proposed to assess the toxicity of environmental contaminants by using concentrations measured in either whole-body or organ-specific rather than external exposure. The relevance of this so-called Tissue Residue Approach for toxicity assessment differs depending on various chemical classes of contaminants (Chapter 5).

Last, bioavailability can be assessed indirectly through effects observed in exposed organisms or communities (Chapter 5).

19.1.3 Biomarkers

Since biomarkers were defined by Depledge in 1994, they have gain credibility, particularly the so-called "core" biomarkers that are in current use for biomonitoring in many countries (Chapter 7). From an operational point of view, it is important to have background assessment criteria (BAC) and environmental assessment criteria (EAC) for each of these biomarkers (ICES, 2012). Briefly, BAC are estimated from data typical of remote areas whereas EAC are usually derived from toxicological data and indicate a sublethal effect. In-depth literature reviews challenge the recurrent criticism about the lack of ecological relevance of a series of biomarkers including acetylcholinesterase (AChE), DNA damage, lysosomal stability, stress on stress, scope for growth, and behavior (ICES, 2012; Amiard-Triquet et al., 2013). However, if cascading effects from molecular to population/community levels—conceptualized by Ankley et al. (2010) as adverse outcome pathways (AOPs) and now widely adopted (OECD, 2013)—have been described for some contaminants, quantitative values are even scarcer and EACs are often provisional values, available for a restricted number of species and contaminants. Even in the case of AChE inhibition, which is one of the most studied biomarker of ecological relevance, it is only recently that an AOP applicable to a wide range of species at multiple life stages with a diversity of organophosphate and carbamate insecticides has been developed (Russom et al., 2014). It was based on information captured from acute mortality studies at very high doses and generally ignored more subtle effects. In general, it has been accepted that 20% reduction in AChE activity in fish and invertebrates indicates exposure to neurotoxic compounds, whereas depression of AChE activity by 20–50% indicates sublethal impact (quoted in ICES, 2012). Several studies with estuarine fish have suggested that brain AChE inhibition levels of 70% are associated with mortality in most species. Selected species, however, appear capable of tolerating much higher levels (90%) of brain inhibition. In addition, sublethal effects on stamina have been reported for some estuarine fish in association with brain AChE inhibition levels as low as 50% (Fulton and Key, 2001). In *Oncorhynchus mykiss* (rainbow trout) exposed to acephate, a dose-dependent relationship between inhibition of AChE, respiratory activity, and mortality, with 76–89% inhibition of AChE in dead fish was found. Survival of *Daphnia magna* to chlorpyrifos, malathion, and carbofuran was impacted 10–50% when AChE was inhibited between 56% and 80%, respectively (quoted in Russom et al., 2014).

Improving the determination of BAC for "core" biomarkers as well as for innovative biomarkers (Chapter 11) is crucial to counteract the effects of confounding factors (Chapter 7) that so far limit a more general implementation of the methodology of biomarkers. Enzyme responses and biomarkers based on different aspects of the reproductive function are particularly sensitive to temperature and then latitude. In such cases, it may be better to establish these criteria at a local rather than at a global scale. Examples are provided for biomarkers recommended under the auspices of the ICES (2012).

It is always possible to search for new biomarkers such as biomarkers of neurotoxicity proposed by Basu et al. (2015) or biomarkers of reproduction in crustaceans (Short et al., 2014). In particular, proteomic approaches are of great interest to serve as a biomarker discovery tool (Chapter 8). However, considering the limited budgets available for ecotoxicology, researches in this domain must be very well targeted because otherwise, it is possible to propose as much biomarkers as biological responses to chemical stress, either specific or more generalist. Another way is to test the validity of well validated biomarkers to assess the potential toxicity of new chemicals of interest as exemplified for pharmaceuticals and personal care products (PPCPs, Chapter 16) and nanomaterials (Chapter 17). Plagiarizing Payne et al. (1996) for AChE, one may wonder if behavioral impairment is "an old biomarker with a new future?" Behavior is a sensitive biomarker, responding at low doses of different classes of contaminants in many different species (Chapters 6, 10, 11, 12, 15 and 18) and because the origin and consequences of behavioral impairments (Figure 7.1, this book) are well established, it has a good potential for use in AOPs. In addition, the application of biological early warning systems based on behavioral responses and the computational methods used to process the behavioral monitoring data have been developed (Gerhardt et al., 2006; Bae and Park, 2014). A standard guide for behavioral testing launched recently by the ASTM (2012) enables effective operationalization of behavioral biomarkers. To assess the health status of ecosystems, it is indispensable to be aware of the toxicological significance of changes observed in biomarker values. A change by itself is a warning sign that organisms are submitted to stress, but for risk management, we need to know what the nature and the degree of the stress syndrome are: acclimation activating biomarkers of defense, deleterious effects activating biomarkers of damage, with which consequences on individual survival, population sustainability, ecosystem functioning? Improving the determination of EAC and developing AOPs should favor the use of ecological biomarkers for risk assessment and management.

19.1.4 The Issue of Low Doses, Nonmonotonic Dose–Response Relationships and Sensitive Life Stages

This book provides critical reviews of the literature dealing with emerging contaminants (PPCPs, Chapter 16; nanomaterials, Chapter 17) or legacy contaminants with emerging risks (endocrine disrupting chemicals (EDCs), Chapter 15), whereas aquatic toxicology of poly- and perfluorinated compounds has been recently reviewed in full by Giesy et al. (2010) and Ding and Peijnenburg (2013). Strikingly, many toxicological studies have been carried out according to conventional procedures of acute toxicity tests ignoring the issue of low doses.

Many standardized international protocols have been launched for acute toxicity tests with different taxonomic groups and in addition, toxicologists interested in mechanisms of toxicity often use high experimental doses of exposure, expecting “clearer” results. In fact, using acute doses immediately leads to the saturation of defense mechanisms (Chapter 7) and then does not allow organisms to cope with pollutants as they might do in the field. At sublethal, more realistic doses, detoxificatory processes are functional and ecotoxicity studies carried out under these conditions are more representative of effects of chronic exposure in the field (Amiard-Triquet and Roméo, in Amiard-Triquet et al., 2011), even if the duration of exposure in the so-called chronic toxicity tests (often 2 or 3 weeks) remains short compared with the lifespan of test organisms.

On the other hand, standardized acute toxicity tests not only used high doses of contaminants, but also effects are observed on the short term. Under these conditions, mortality occurs at very high doses only, leading prima facie to the feeling that effects occur at doses that cannot be encountered in the environment. In fact, more subtle effects occur at low doses that can be deleterious in the case of chronic exposure through AOPs. In the case of PPCPs (Chapter 16) and engineered nanomaterials (Chapter 17), acute doses, generally in the $mg\,L^{-1}$ range, are unlikely to be found in the aquatic environment and standard acute toxicity tests are not appropriate. On the other hand, at more realistic doses, ecologically relevant biomarkers (oxidative stress, neurotoxicity, DNA damage, behavioral impairments) are affected.

Nonmonotonic dose–response relationships may be observed in the case of hormesis for many environmental contaminants (Calabrese and Blain, 2011) and it is also well-documented in the case of EDCs (Chapter 15). Conventional risk assessment (Chapter 2) based on the traditional assumption that "the dose makes the poison" can fail for nonmonotonic data. Accurate description of nonmonotonic dose–response curves is a key step for the characterization of hazards associated to the presence of pollutants in the medium, thus needing the development of dedicated models (Qin et al., 2010; Zhu et al., 2013).

It is commonly recognized that early life stages of organisms are the most sensitive and they are already in current use for ecotoxicological bioassays, including many standardized ones (Chapter 6). This is particularly crucial in the case of EDCs because sensitivity varies deeply at different developmental stages, thus studies carried out with adults or nonsensitive stages can lead to false "safe" doses of a given chemical (Chapter 15).

19.1.5 Competition Matters

Interspecific variability of sensitivity to chemical stress is well-documented and the primary mechanism driving degradation of communities exposed to chemical stress is the elimination of sensitive species and the subsequent dominance by tolerant and opportunistic species (Chapters 9 and 14). As soon as 1967, Moore has established the typology of ecological consequences of this interspecific variability in terms of population effects and food chain contamination (Figure 19.2). The message is clearly that competition matters and

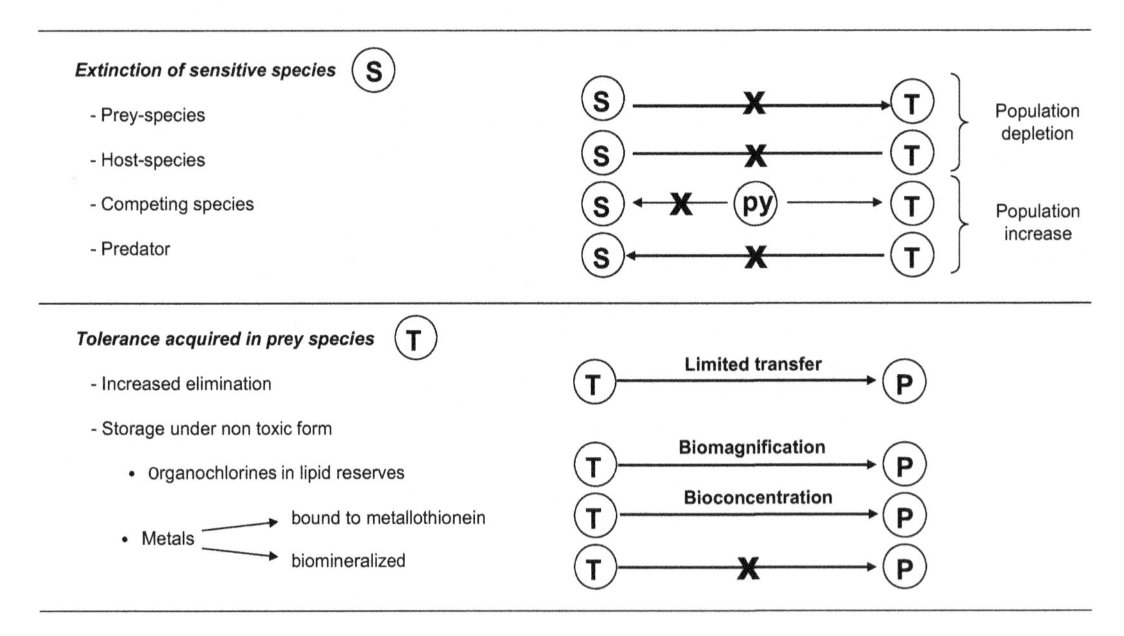

Figure 19.2

How sensitivity/tolerance can affect food web contamination and population fate. P, predator; py, prey; X, disruption of the relationship. *After Amiard and Amiard-Triquet (2008) with permission.*

consequently, single-species assays (the most commonly used among internationally standardized bioassays; see Chapter 2) are poorly informative about ecological consequences.

In line with these assumptions, Kattwinkel and Liess (2014) applied a generic individual-based simulation model of two competing species in a zooplankton community to analyze the consequences of interspecific competition for population dynamics of *Daphnia* spp. under pulsed acute contamination, a realistic scenario for pesticide stress. They conclude that toxicant concentrations derived from conventional risk assessments that do not consider competition might be underprotective for wild populations, particularly for species with low reproductive capacities.

19.2 Which Strategies for Emerging Risks?

It is generally accepted that about 100,000 substances are placed on the global market among which 30,000 would be neurotoxicants (quoted in Basu, 2015), 9700 environmental chemicals should be screened in the framework of the USEPA Endocrine Disruptor Screening Program, etc. It is evident that a complete ecotoxicological assessment of all these substances using the best available procedures (Chapter 6) is completely unrealistic. Indeed, the analysis of the ecotoxicity data submitted within the framework of the European Registration, Evaluation and Authorization of Chemicals (REACH) Regulation revealed that innovative techniques are poorly used (Cesnaitis et al., 2014; Tarazona et al., 2014). Thus it is expected that using high assessment factors will remain the rule, continuing to drive a high degree of uncertainty in the determination of PNECs and subsequently environmental quality standards in water, sediment, or biota (Chapters 2 and 3). However, the interpretation of statistical tests in ecotoxicology can help "making the right conclusions based on wrong results and small sample sizes" (Wang and Riffel, 2011). The recommendations of regulatory bodies (EFSA, 2009; USEPA, 2012a) could lead to a better determination of toxicological parameters through the benchmark dose approach (Chapter 2).

19.2.1 Prioritization of Research Questions

Given the reality of limited resources with which to engage large numbers of subject matter experts with diverse sector perspectives (e.g., industry, academia, nongovernmental organizations), prioritization of research questions is of paramount importance. To narrow down which substances to look for, US and EU chemical regulations (Toxic Substances Control Act; high production volume challenge program, REACH) link the chemical information requirements to the tonnage of the substance, produced, and imported per year (USGAO, 2007). However, independent of any tonnage produced, specific obligations may also depend on the classification as substance of very high concern, including carcinogenic, mutagenic, and reprotoxic; persistent, bioaccumulative, and toxic (PBT), and very persistent and very bioaccumulative substances as well as EDCs (REACH Article 57 ff.).

Methods and techniques that can be used to prioritize the thousands of chemicals that need to be tested for potential endocrine-related activity have been reviewed in Chapter 15. Computational methods such as virtual screening, quantitative structure–activity relationships (QSARs) and docking, are valuable tools for prioritization. Automated chemical screening technologies ("high-throughput screening assays") that expose living cells or isolated proteins to chemicals constitute the next step for prioritization. It is quite appropriate to apply similar techniques when attempting to understand the properties of many other classes of contaminants that may lead to chemical hazard.

The limited availability of experimental data necessary for the hazard/risk assessment of PBT substances under REACH and the expected high costs, have increased the interest for alternative predictive in silico methods such as quantitative structure–activity (property) relationships (Papa and Gramatica, 2010). The Organization for Economic Co-operation and Development (OECD) QSAR project aims at facilitating practical application of QSAR approaches in regulatory contexts by governments and industry and to improve their regulatory acceptance (http://www.oecd.org/env/ehs/risk-assessment/oecdquantitativestructure-activityrelationshipsprojectqsars.htm). It provides a software application, the "QSAR Toolbox," the most recent version of which was released in November 2014. To support computational toxicology, Goldsmith et al. (2014) have developed a database of in silico biomolecular interactions (DockScreen) composed of chemical/target (receptor and enzyme) binding scores calculated by molecular docking of more than 1000 environmental and therapeutic chemicals into 150 protein targets. In addition to providing a decision to support in chemical screening and prioritization, DockScreen may improve linking chemical structures with in vitro and in vivo effects. In the framework of the Toxicity Forecaster (ToxCast™) program developed under the auspices of the USEPA, the assessment encompasses a wide range of structural classes and phenotypic outcomes, including tumorigens, developmental and reproductive toxicants, neurotoxicants, and immunotoxicants. Data about physicochemical, predicted biological activities based on existing structure-activity models, biochemical properties based on high-throughput screening assays, cell-based phenotypic assays, and genomic and metabolomic analyses of cells may be used for a tiered testing approach (Dix et al., 2007). Such screening programs produced enormous amounts of toxicity data that overwhelmed available database management tools or traditional data processing applications. Recently, Zhu et al. (2014) have reviewed existing structured in vitro data as response profiles for compounds of environmental interest and have described potential modeling and mining tools relevant for a big data approach when conducting modern chemical toxicity research.

However, in vitro assays cannot reflect the complexity of whole organisms. When in silico then in vitro tests suggest that a contaminant is at risk, in vivo validation remain necessary—in the light of current scientific knowledge—to increase confidence for use of their results in regulatory decisions.

Finally, Powers et al. (2014) described a web-based tool allowing the implementation of USEPA's comprehensive environmental assessment approach to prioritize research gaps in the field of emerging materials. Briefly, a prioritization matrix was built taking into account expert judgments about the "importance" of research areas to risk assessment efforts and the "confidence" in the availability and utility of current information to support risk management decisions. The areas rated as most important and that had the least confidence were then identified as "high-priority research areas." This web-based tool as the potential to help developing research strategies for other emerging contaminants.

19.2.2 Exposure Assessment

In the laboratory, the discrepancy between nominal concentrations added by the experimenter and the actual concentrations that test organisms have to cope with all along the experiment is not a new topic of debate. The use of semistatic or flow-through bioassays (Chapter 2) can limit this gap, but providing measured concentrations in addition to nominal concentrations remains useful to improve the confidence in the results of the bioassay. In addition, we will see in the following section (Section 19.2.4) that toxicokinetic-toxicodynamic models are very useful to account for time-varying conditions of exposure.

The changes of total concentrations in the experimental medium are not the only important point. The initial physicochemical characteristics and the fate of contaminants added in the experimental medium can deeply influence the results of bioassays. Mouneyrac et al. (Chapter 17) have underlined the importance of aggregation chemistry and dissolution kinetics in ecotoxicological studies with nanomaterials. Developing methods for testing of poorly water-soluble substances is still a topic of research because solvents in current use can exhibit toxicity at lower concentrations than expected by regulatory bodies (Chapter 6). For common aquatic toxicity tests of hydrophobic compounds, passive dosing can help to face the challenge of producing exposure concentrations close to aqueous solubility and keeping these constant for the duration of the test (Chapter 4).

The total concentration in the bioassay medium may be insufficient when certain constituents (e.g., of an effluent, an environmental sample) are mainly responsible of the observed toxicity. For instance, PAH monitoring is key to accurate risk assessment of petroleum (Chapter 18). Laboratory experiments are useful in showing the relative toxicity of various oil and dispersed oil test solutions, but the preparation and characterization of laboratory exposure solutions remain a topic of discussion (Bejarano et al., 2014; Martin et al., 2014) as illustrated by the controversy about the "real world" assessment of oil dispersed as a consequence of the Deepwater Horizon accident (Coelho et al., 2013; Rico-Martínez et al., 2013).

Currently, exposure of biota in the field is mainly assessed using a priori analysis of an important number of well-identified pollutants (however, far from the 100,000 substances potentially present in the environment) in the framework of monitoring programs managed by regulatory bodies (Chapter 3). In this context, the determination of total concentrations is usually preferred. However, taking into account the physicochemical characteristics of contaminant and the resulting bioavailability should be based on the full implementation of novel tools. Passive sampling devices have a potential for providing partial characterization of the physicochemical characteristics that govern fate and effects of contaminants; ensure the preconcentration of contaminants, thus allowing a lowering of the detection and quantification limits; and have significant advantages over the spot sampling techniques, allowing the measurement of time-weighted average concentrations in fluctuating environments (Chapters 3 and 4).

A global approach can be preferred to a priori approach by adopting unbiased exposure assessment strategies that search for unknowns such as effects-directed analysis (EDA) and toxicity identification evaluation (TIE). The global approach combining biotesting, fractionation and chemical analysis, helps to identify hazardous compounds in complex environmental mixtures (Chapters 4 and 15).

19.2.3 Modes of Action

The first step of chemical-induced perturbations of biological systems is at the molecular level. This "molecular initiating event" (MIE) (OECD, 2013) also termed primary mechanism of action (Escher et al., 2011) is at the beginning of the cascade leading to adverse effects at higher levels of biological organization (Figure 12.1, this book). The mode of action describes the key events and processes, starting with the MIE/mechanism of action and proceeding through functional and anatomical changes in the organism. In fish, efforts that have been made to link molecular, physiological and ecological levels of organization have largely contributed to our understanding of the mode of action of chemicals in a complex set of environmental stressors and conditions (Chapter 13).

Most chemicals can interact with more than one, the different molecular targets including biological membranes, proteins and peptides, DNA and RNA, thus leading to different modes of action (Table 19.1).

Several in silico tools are dedicated to providing more mechanistic understanding of the toxicological process, for example, development of structural alerts and profilers (Madden et al., 2014). These authors underline the interest in applying these tools to the development of AOPs that are the central element of a toxicological knowledge framework being built to support chemical risk assessment based on mechanistic reasoning (OECD, 2013). A greater understanding of the mechanisms of toxicity and, in particular, where these mechanisms may be conserved across taxa, such as between model animals and related wild species would help to resolve the issue of extrapolation to diverse species (Madden et al., 2014).

In recent decades, genomic and proteomic have been described as able to help biologists to unravel all the mechanisms used by organisms to respond to changes in their environment, increasing our knowledge for understanding the cellular mechanisms involved in stress response (Chapter 8). Toxicogenomics uses the three major "omics" technologies

Table 19.1: Chemical-Induced Perturbations of Biological Systems at the Molecular Level: Primary (Molecular Initiating Event (MIE) or Primary Mechanism of Action (MeOA)) and Secondary Actions at the Basis of the Mode of Action

Target Site	MIE/MeOA	Mode of Action
All biological membranes	Nonspecific disturbance of membrane structure and functioning	Baseline toxicity
	Formation of reactive intermediates (e.g., reactive oxygen species) causing peroxidation of membrane lipids and proteins	Degradation of membrane lipids and membrane proteins
Energy-transducing membranes	Ionophoric shuttle mechanisms	Uncoupling
	Blocking of quinone and other binding sites, etc.	Inhibition of the electron transport chain
	Blocking of proton channels and other transport channels	Inhibition of adenosine triphosphate synthesis/depletion of adenosine triphosphate
Photosynthetic membranes	Blocking of photosynthetic electron transport	Photosynthesis inhibition
Proteins, peptides	Electrophilic reactivity, alkylation and oxidation of proteins and glutathione	Damage and depletion of biomolecules
Specific enzymes and receptor	Noncovalent or covalent binding to enzymes and receptors	Inhibition or competition (e.g., acetylcholine esterase, estrogen receptor, Ah receptor)
	Noncovalent or covalent binding to enzymes of the nucleic acid metabolism, effect on replication or repair	Indirect mutagenicity (DNA repair, recombination, regulation
DNA, RNA	Base modification and damage: Electrophilic (alkylation) and oxidative damage, bulky adducts	Direct mutagenicity (e.g., frameshift, cross-links, strand breaks, deletion)

After Escher et al. (2011).

(transcriptomics, proteomics. and metabolomics) to study the adverse effects of environmental chemicals on organisms, integrating toxicant specific alterations in gene, protein, and metabolite expression patterns with phenotypic responses of cells, tissues, and organisms (OECD, 2013). Application of omics in the field of in vitro toxicology and the degree of maturity of omics technologies and omics data for incorporation in the regulatory decision-making process have also been recently reviewed (Balmer et al., 2014).

Key events at higher levels of biological organization include organ responses (altered physiology, disrupted homeostasis, altered tissue development/function), organism responses (lethality, impaired development and reproduction), and population responses (structure, extinction) (OECD, 2013).

Knowing both MIE and mode of action is important in the assessment of ecotoxicity of mixtures. The two most widely accepted mixture models are concentration-addition and response-addition for use with mixtures having the same and different modes of action, respectively, provided that no interaction occurs between the different constituents of the mixture under examination (Chapter 18).

19.2.4 Modeling

We have previously evoked (Section 19.2.1) the potential of molecular modeling in the prioritization of research questions. Once again because assessing the fate and effects of the huge number of substances entering the environment is too costly and labor-intensive, modeling has been proposed in many domains of ecotoxicological research, including fate and effects of contaminants (Devillers, 2009; Artigas et al., 2012).

This includes the prediction of chemical fate within the environment with the determination of predicted environmental concentrations that is so important in risk assessment (Chapter 2). The modeling strategies in current use for conventional risk assessment are still in progress, considering the contrasted behaviors of conservative contaminants versus those that have more complex behaviors, including oil spills (James, 2002). Recently, Hritonenko and Yatsenko (2013) have reviewed different models of water pollution propagation, including (1) a general three-dimensional model of water pollution, which takes into account transport and diffusion of pollutants in dissolved state, suspension, and bottom deposits, adsorption and sedimentation of pollutants, subjected to water dynamics (see Figure 2.3, this book); (2) a two-dimensional horizontal model of water pollution dissemination; (3) a simple one-dimensional model relevant in case of a short distance between a point source of pollution and the exposure location; and (4) multimedia modeling allowing investigations about exchanges between environmental compartments. Linking a geographical information system to a chemical exposure model can allow the integration of chemical levels with spatially explicit information on habitat properties and species specific traits (Artigas et al., 2012).

Bioaccumulation models predict contaminant concentrations in biota using chemical and biological parameters (Abarnou, in Amiard-Triquet and Rainbow, 2009; De Laender et al., 2010). They are in current use for both organic and inorganic substances (e.g., Rainbow and Luoma, in Amiard-Triquet et al., 2011; De Hoop et al., 2013; Dutton and Fisher, 2014; Nyman et al., 2014). Briefly, the basic equation of bioaccumulation establishes a mass balance of contaminants within an organism: the inputs from water and from feeding on contaminated preys equal their output because of the elimination, transformation of compounds, and their redistribution within the body (Figure 19.3). Growth, which can be considered as dilution of contaminant, and transfer to the next generation during reproduction also contribute to a decrease of contaminant concentration in biota. The dietary uptake of contaminants may become quite complicated when several types of prey from different trophic levels are considered; that is the case for biomagnification (Abarnou, in Amiard-Triquet and Rainbow, 2009). In line with this concern, De Laender et al. (2010) concluded that food web structure but also chemical parameters (log K_{ow}, the organic carbon–octanol proportionality constant,

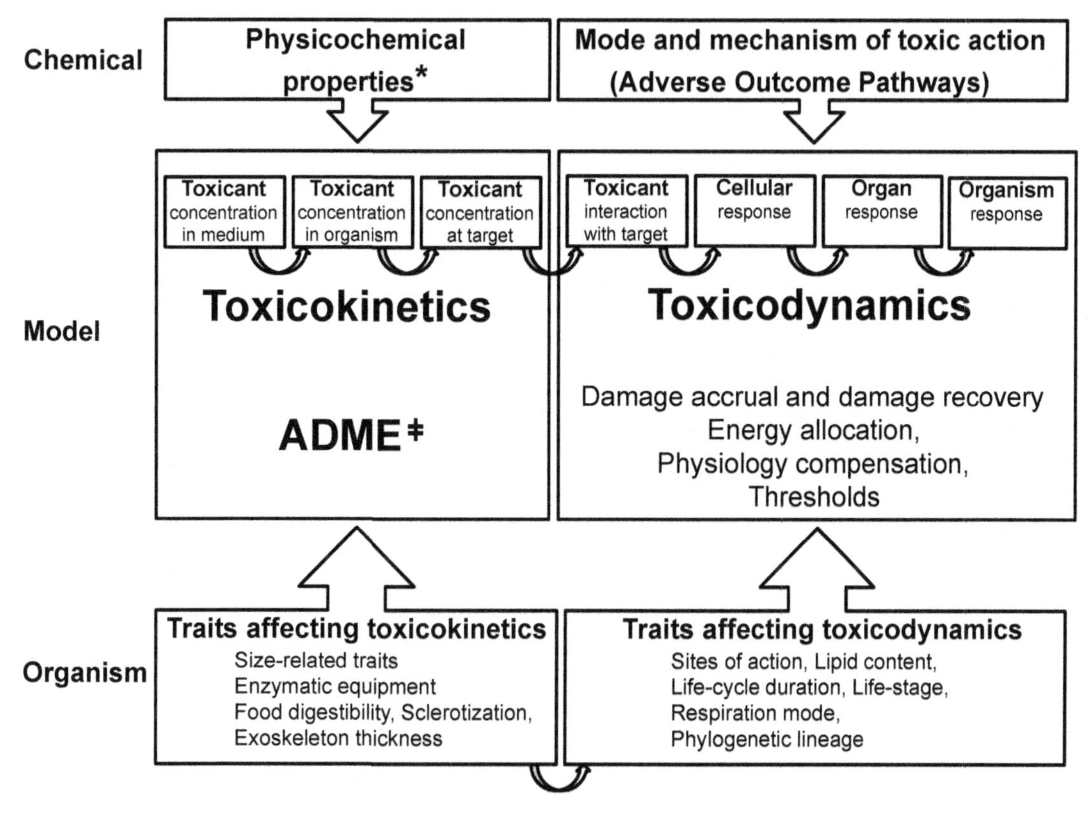

Figure 19.3

Toxicokinetic-toxicodynamic (TK-TD) models. *Log K_{ow}, the organic carbon–octanol proportionality constant, structure, functional groups present, hydrophobicity. ‡ADME, absorption, distribution, metabolism, excretion. Arrows indicate causal relationships. *Modified after Ashauer lab Website:* http://www.ecotoxmodels.org/toxicokinetic-toxicodynamic-models/.

dissolved aqueous contaminant concentration) were the main contributors to uncertainty ranges of internal concentrations. Despite these uncertainties, De Laender et al. (2010) claim that the tissue-based ecological risk assessment reduces variability among species and contaminants, as tissue-based effect thresholds compare better among trophic levels and within contaminant classes than exposure-based metrics. Prompted by the problems encountered with the latter (particularly the question of bioavailability), the tissue residue approach for toxicity assessment has made significant progress in theory and application (Chapter 5).

Inter- and intraspecies variations in sensitivity to contaminants can be quite large (Chapters 9 and 14). Toxicokinetics models (Figure 19.3) can help understanding both of these sources of variation. For instance, toxicokinetics (i.e., biotransformation and distribution) explain the higher tolerance of *L. stagnalis* to the organothiophosphate insecticide diazinon when compared with gammarids (Nyman et al., 2014). In the case of inorganic contaminants, Rainbow and

Luoma (in Amiard-Triquet et al., 2011) conclude that is an enhanced rate of detoxification that is often the result of selection in the field to bring about metal tolerance in a particular animal population.

Toxicodynamic parameters can be chosen at different levels of biological organization involved in AOPs (Figure 19.3). For instance, Kretschmann et al. (2012), comparing the relative sensitivity of two crustacean species (*Gammarus pulex* and *Daphnia magna*) to diazinon, have selected molecular (inhibition and recovery of AChE activity) and individual (immobilization, mortality) parameters. However, it remains unclear whether and in what ways traits can be linked to mechanistic effects to predict intrinsic sensitivity of organisms (Rubach et al., 2012 and literature cited therein). In freshwater arthropods exposed to chlorpyrifos, these authors have tested a large range of biological traits: size-related traits, water content, thickness of exoskeleton, lipid content, respiratory regulation, source of oxygen, mode of respiration, trophic relations, degree of sclerotization, shape of the organisms, life stage, and phylogeny. Several traits are relevant for inclusion in the toxicokinetics part of the model, being predictive of uptake and elimination, whereas a lack of relationship was observed between traits and routine ecotoxicity endpoints (24–96 h $L(E)C_{50}$). In addition, it remains unsure that traits statistically linked to ecotoxicity can exert their influence directly or only indirectly through intermediate effects on toxicokinetics.

Coupling toxicokinetics ("what the organism does with the chemical") and toxicodynamics ("what the chemical does to the organism") in toxicokinetic-toxicodynamic models may help to extrapolate toxic effects between different exposure regimes. This includes static versus flow-through tests (Billoir et al., 2012) and fluctuating or pulse exposures encountered, for example, in risk assessment of pesticides. Other potential uses have been reviewed recently (Ashauer and Escher, 2010; Ashauer and Brown, 2013). However, more research is needed to go beyond the current limitations hampering a wide use, particularly development of more mechanistic approaches for modeling sublethal toxicity.

Monitoring the health status of ecosystems by using indicators at low levels of biological organization (Omics, Chapter 8; Biochemical and individual biomarkers, Chapter 7) aims to establish a link between the presence of chemical compounds and their sublethal effects on biota. However, this makes sense only if these indicators may be linked to effects at higher levels of biological organization. Kramer et al. (2011) have described how mechanistic data, as captured in AOPs, can be translated into modeling focused on population-level risk assessments. Population models are currently increasingly used in predictive approaches. For example, gammarids are good candidates to illustrate how one could address questions concerning the ecological relevance of molecular biomarkers and suborganism responses to stressors (Chapter 11). The biological traits that have been well developed in the field of ecological assessment (Biological Traits Analysis, Chapter 14) are also relevant for ecotoxicological modeling as exemplified in copepods (Chapter 12).

A review by Galic et al. (2010) based upon 148 publications revealed that most models were developed to estimate population-level responses on the basis of individual effects. In addition, it is very important to take into account interactions between species because population responses to chemicals are highly sensitive to predation or competition and food availability (De Laender et al., 2009; Gabsi et al., 2014; Kattwinkel and Liess, 2014).

Assessing biological responses at supraindividual levels (from population to ecosystem) is a complicated task. Recently hundreds of indicators, metrics, and assessment methods have been developed to meet the demands of regulatory bodies in agreement with national and international regulations (US Clean Water Act, the US and Canada Oceans Act, the European Water Framework Directive, and the European Marine Strategy Framework Directive) (Chapter 14). Because it is so difficult, ecological models for ecotoxicology and chemical risk assessment appear attractive substitutes. Several authors have assessed their suitability for use in a regulatory context (Galic et al., 2010; Schmolke et al., 2010; Forbes et al., 2011). They conclude that, in spite of meeting high scientific standards, most currently available ecological models would need modifications before being suitable for regulatory risk assessments. It is now a question of taking a further step toward improving their effectiveness because it has become obvious that the extrapolative and integrative power of such models can add value to environmental risk assessment (Galic and Forbes, 2014; introducing a special section of the Society of Environmental Toxicology and Chemistry (*SETAC*) *Journal*, Vol. 33, No. 7). Several contributions in this issue are outputs from a European project on mechanistic effect models (Grimm et al., 2009) funded within the EU 7th Framework Program, underlying the strong interest of the regulators.

19.3 Related Issues

19.3.1 Emerging Environmental Concerns

19.3.1.1 Global climate change

"The Influence of Global Climate Change (GCC) on the Scientific Foundations and Applications of Environmental Toxicology and Chemistry" was addressed in the framework of a Society of Environmental Toxicology and Chemistry (SETAC) Pellston workshop resulting in seven papers published in 2013 in *Environmental Toxicology and Chemistry* (Vol. 32, No. 1) and summarized in Stahl et al. (2013). It is expected that GCC will influence the fate and bioavailability of chemicals. In addition, climate can influence the sensitivity of organisms to contaminants, whereas contaminants could influence the sensitivity of organisms to climate. For instance, the fitness cost of resistance to cadmium in the least killifish (*Heterandria formosa*) included a greater susceptibility to increased temperature (Xie and Klerks, 2003), one of the most important parameter affected by GCC. The complex interactions between climate change and pollutants may be particularly problematic for species living at the edge of their physiological tolerance range where acclimation capacity may be limited (Noyes et al., 2009; Hummel et al., in Amiard-Triquet et al., 2011). The influence of temperature on pollutant toxicity has been addressed in Chapter 7 (Figure 7.3), underlying particularly the effects on contaminant kinetics and enzyme activities, many of which are used as biomarkers. In the case of pesticides, Köhler and Triebskorn (2013) have examined the potential interactions of global change and pesticide effects on the physiology and ecology of wildlife species. Kimberly and Salice (2013) have shown that combined stressors (high temperature and cadmium contamination in freshwater gastropods) may produce greater-than-additive effects, challenging predictive power. So they concluded that more studies are needed to better characterize the interactive effects of chemical contaminants and stressors related to climate change. However, is it possible to carry out such a detailed study for the thousands of contaminants entering the aquatic environment?

19.3.1.2 Plastic debris

The widespread use of plastics in every aspect of our everyday life (the vast majority being for disposable use) has led to their annual world production from 1.7 million tons in the 1950s (when mass plastic production started) to 288 million tons in 2012 generating large amounts of plastic wastes (PlasticsEurope, 2013; Koelmans, 2014). A large part of plastic wastes still ends up in landfill or accumulates in a wide environment and constitutes an unlimited source of plastic debris in oceans and seas (Eriksen et al., 2014). Although there is a degree of awareness for the potential risks, the impact of plastic particles on ecosystems is far from being understood. On average, three quarters of all marine litter consists of plastic items, well-known to be persistent and the majority of reported litter-related incidents of marine organisms such as birds, cetaceans, and fish are due to plastics. The entanglement in and ingestion of macroplastic items is well known in vertebrates (Laist, 1997). Nevertheless, information on microplastics regarding their fate, behavior, and biological impacts on marine organisms is only just emerging.

The Marine Strategy Framework Directive has defined marine litter as a full descriptor of the marine environment with a focus on plastics and degradation products as a main issue. Once present in the marine environment, plastic debris are exposed to degrading forces such as ultraviolet-B radiation and physical abrasion (Barnes et al., 2009) leading progressively to small and smaller fragments until they become microplastics usually defined as particles smaller than 5 mm (e.g., Arthur et al., 2009), but some authors set up the upper size limit at 1 mm (e.g., Dekiff et al., 2014). The polymers most widely produced as plastics are polypropylene, polyethylene, polystyrene, polyethylene terephthalate, and polyvinylchloride (Andrady et al., 2011). An emerging issue of concern is the accumulation of microplastics in the water column and in sediments. The presence of microplastics has been reported in different marine compartments (subsurface and surface waters, intertidal and subtidal sediments) worldwide (e.g., Claessens et al., 2011; Collignon et al., 2012), but the main difficulties are due to the lack of standardization of sampling methodologies and separation of plastics microparticles. Further research therefore needs to consider sampling design—namely, the number and the size of replicates, the spatial area and the sampling frequency, the methodology employed for sampling (i.e., type of net for aquatic samples or core for sediment samples), and, finally, methods used for identification of microplastics.

Plastics are biologically inert but a key factor explaining the bioavailability of microplastics is their small size, making them available to a wide range of organisms including filter-feeders, deposit-feeders, and scavengers (Thompson et al., 2004). For example, histological and fluorescence microscopy studies have shown the presence of 2-mm and 4- to 16-mm microspheres in the gut cavity and digestive tubules of mussels (*M. edulis*) (Browne et al., 2008). Biological impacts of ingested microplastics have been suggested such as blockage of enzyme production, diminished feeding stimulus, nutrient dilution, reduced growth rates, decreased steroid hormone levels, delayed ovulation, reproductive failure, and absorption of toxins (Galgani et al., 2010). Moreover, once ingested by the organism, plastics can release chemicals (nonylphenols, polybrominated diphenyl ethers, phthalates, or bisphenol A (BPA)) but also sorb hydrophobic pollutants (e.g., DDT, PCBs, dioxins), which can induce toxic responses (Teuten et al., 2009; Engler, 2012). Because lower trophic level organisms such as invertebrates are able to ingest microplastics, their introduction to the food web transfer is questionable (Wright et al., 2013). Furthermore, the increasing accumulation of microplastics into marine environment generates a new marine habitat where biological interactions are taking place. This "new" habitat and its environmental impacts represents a new research avenue.

19.3.2 Alternatives to Animal Testing

The number of experimental animals used for regulatory ecotoxicological evaluation in the European Union reported for 2008 reached 90,583 fish, 179 amphibians, and 50,603 birds (reported in Scholz et al., 2013). A measure of common sense consists in toxicological data sharing to avoid unnecessary animal testing. In the case of REACH, the legislation provides several mechanisms to ensure the sharing of data: companies may need to inquire if data have already been submitted on their substance; share information from animal studies and, with some exceptions, submit joint registration dossiers with other registrants for the same substance (ECHA, 2011). This report was also to describe the extent to which registrants used non-animal test methods within their registration dossiers to fulfill the information requirements of REACH. The 3R principles—replacement, reduction, and refinement of animal tests—were formulated as soon as 1959 by Russell and Burch (in Scholz et al., 2013). In line with the principle of reduction, Scholz et al. (2013) proposed a revision of test designs and risk assessment schemes as also the read across or category approach (i.e., a study on one substance to cover information requirements for another, or multiple substances). However, in the framework of REACH, the ECHA (2011) reports that the use of the latter is often not well-justified. In addition, the experimental data provided in the dossiers are in some cases also not sufficient to meet information requirements under REACH. Currently it is expected that more animal testing will need to be requested by the European Chemicals Association to ensure the safe use of chemical substances.

The scientific community has considerably assisted in promoting non-animal test methods to fulfill one of the objectives associated to environmental assessment worldwide (e.g., Ritter et al., 2012; Hartung et al., 2013; Scholz et al., 2013; Jang et al., 2014). Animal alternatives research has historically focused on human safety assessments and has only recently been extended to environmental testing. This is particularly for those assays that involve the use of fish (Embry et al., 2010). These authors and others (e.g., Fetter et al. (2014)) proposed the use of fish embryos as an alternative to experiments with adult animals considering that the use of fish embryos combines the complexity of a full organism with the simplicity and reproducibility of cellular assays. Another step forward was taken by other authors investigating alternatives to the fish early life-stage test (Sogorb et al., 2014; Villeneuve et al., 2014). Non-animal test methods have been reviewed by Scholz et al. (2013). They mainly include in silico, ex vivo, and in vitro methods already described for prioritization (Section 19.2.1) and modeling (Section 19.2.4). For a number of toxicological endpoints required by REACH, in vitro tests are accepted instead of the corresponding animal study (ECHA, 2011). It is generally accepted that a single experimental or nontesting approach is generally insufficient to replace an animal test. Therefore, integrated testing strategies are recommended to combine the different promising alternative methods in a powerful and predictive approach that allows significant reduction of animal testing (Hartung et al., 2013; Scholz et al., 2013).

Priority substances are designated as such on three criteria: persistent, bioaccumulative, and toxic. Thus in a regulatory context, bioaccumulation is a prominent endpoint and standardized bioaccumulation tests are available. During the past decade, several approaches have been developed to estimate bioconcentration factors using reduced sampling. Internal benchmarking improves precision and reduces animal requirements for determination of fish bioconcentration factors (Adolfsson-Erici et al., 2012). Compared with the initial OECD test guideline for bioconcentration factor testing, it has been proposed to reduce the number of sampling time during depuration (Springer et al., 2008) or the number of test concentrations (Creton et al., 2013). Recently, OECD (2012) has launched a revised version of test no. 305 that provides the option to use a minimized procedure. The minimized test design yielded similar bioconcentration results as those of the standard test for ibuprofen (Nallali et al., 2011), and these authors conclude that it has potential for use in screening approaches for pharmaceuticals and other classes of chemicals. Recently, Carter et al. (2014) have proposed to expend the use of minimized bioconcentration tests to aquatic invertebrates, assuming that this approach uses up to 70% fewer organisms.

19.4 Which Strategies Should be Adopted to Promote Environmental Protection?

19.4.1 Remediation

An extreme case provides a striking illustration of the magnitude of remediation problems: the experiences of the Minamata Bay project in Japan (Hosokawa, 1993). The remediation project commenced in 1977 was completed in 1990 after 1.5 million m^3 of Hg-contaminated sediment had been treated by careful dredging and confined reclamation at a total cost of 48,500 million yen (equivalent to US $650 million).

Remediation is costly and it is questionable if impacts from habitat changes are not more disturbing than contamination (Maughan, 2014). According to the USEPA (2005), it must be insured that long-term risk reduction provided by dredging and excavation or capping of contaminated sediments outweigh sediment disturbance and habitat disruption.

In this context, the validity of environmental quality standards is crucial: they must not be overprotective (e.g., Conder et al., 2015) or underprotective (e.g., Kattwinkel and Liess, 2014). However, uncertainty is the rule rather than the exception even for contaminants studied by these authors (mercury, pesticides) that gave rise to so many studies over decades. It is expected that uncertainty for emerging contaminants is more important, as exemplified even in the field of human toxicology despite much more funding opportunities and scientific advances compared to ecotoxicology. In the case of BPA, a ubiquitous endocrine disruptor present in many household products, the European Food Security Authority (EFSA, 2015) has recently established a temporary tolerable daily intake of 4 $\mu g\,kg^{-1}$ body weight per day, replacing the former EFSA standard (2006), which was set at 50 $\mu g\,kg^{-1}$ body weight per day, a change of more than one order of magnitude within less than a decade.

The poverty of environmental quality standards is a blight for risk management, and indirectly has a societal impact because communication about such uncertainties generates a lack of public confidence in regulatory decision-making.

19.4.2 Reducing or Replacing Contaminants at Risk

Because risk assessment and risk management are so problematic, reducing toxicant concentrations in the aquatic environment is a consummation devoutly to be wished. It requires a combination of complementary measures such as source controls, optimal usage, storage and disposal of chemicals, and wastewater management and treatment.

Source control includes reduction of emissions, targeted restrictions or finally ban. Examples include nonylphenols (EU, 2003), tributyltin-based in antifouling paints (Gubbins et al., in Amiard-Triquet et al., 2013), or polybrominated diphenyl ethers used as brominated flame retardants (De Jourdan et al., 2014). For example, take again the case of BPA. In recent years, BPA regulation has been tightened, particularly to protect against exposure during fetal and neonatal life (for details, see Eladak et al., 2015). Given these restrictions and societal pressure, bisphenol S and bisphenol F are already or are planned to be used as BPA substitutes. Unhappily, recent reports have shown the adverse effects of bisphenol S and bisphenol F on physiological functions in fish (Kinch et al., 2015), humans, and rodents (Eladak et al., 2015), similar to those already known with BPA. Similarly, alternative brominated flame retardants share structural and chemical properties comparable to the polybrominated diphenyl ethers. From a mesocosm study with fish, De Jourdan et al. (2014) concluded new brominated flame retardants have the potential to be bioaccumulative and persistent in vivo and, therefore, warrant further study of physiological effects linked to chronic, sublethal responses.

On the other hand, QSAR models or sustainable molecular design guidelines (Hernández-Altamirano et al., 2010; Connors et al., 2014) represent sound bases for "green chemistry" that aims at developing chemical products with reduced unintended biological activity while effecting their desired function. In the case of "green pharmacy," Bebianno and Gonzalez-Rey (Chapter 16), underline that not only drug design but also drug delivery, packaging, and finally disposal contribute as a whole to reduce the impact of active pharmaceutical ingredients in the environment. In addition, wastewater streams generated by hospitals or industry that contain elevated micropollutant concentrations can be pretreated before being discharged to sewers, or specific strategies (e.g., separate urine collection in hospitals) can be adopted to reduce the pollutant load entering wastewater treatment plants. For PCPPs (Chapter 16) and EDCs (Chapter 15), effluents of wastewater treatment plants are major sources of contamination and upgrading the removal of micropollutants is technically feasible but at a higher cost than in conventional wastewater treatment. Recently, the Swiss authorities decided to implement additional wastewater treatment steps and based upon this experience, Eggen et al. (2014) plead that taking action on micropollutants must not remained a luxury for rich countries only.

Reducing discharge of pollutants from diffuse sources such as agriculture is also of major concern and various measures can help improve the situation such as banning or limiting the use of the most toxic pesticides as decided for instance under the authority of the USEPA or in the European regulations; designing less toxic substances according to the principles of green chemistry (Hernández-Altamirano et al., 2010); or developing a more sustainable use of pesticides (EU, 2009) as already accepted by farmers involved in sustainable agriculture or organic farming. However, conflicts of interest may be intense, the most traditional part of the profession claiming that negative impacts of reduced uses of pesticides will result in crop yield losses, a drop in farming profits, more imports from more permissive countries and higher food prices (NFU, CPA, and AIC, 2014).

For nanomaterials, the "safer-by-design" concept is largely encouraged to reduce the toxicity of these contaminants rather than controlling hazards.

19.5 Conclusion

The importance of the acquis in the field of science and research is well illustrated for many organic pollutants, including emerging pollutants, in freshwater ecosystems (Sibley and Hanson, 2011). Information is abundant at the infraindividual and individual levels, whereas it is more and more scarce when progressing in biological levels of organization (Table 19.2).

For decades, risk assessment in a regulatory framework remained in its infancy, being mainly based on strategies whose priority was simpleness and low cost. To answer increased

Table 19.2: Degree of Current Understanding of Noxious Effects of Legacy and Emerging Organic Chemicals at Different Levels of Biological Organization in Freshwater Ecosystems

Chemical	Infraindividual	Individual	Population	Community	Ecosystem
PCBss	XXX	XXX	XX	X	X
TCDD/TCDEs	XXX	XXX	XX	X	X
PBDEs	XXX	XX	X	NE	NE
PAHs	XXX	XXX	XX	X	X
Pharmaceuticals	XX	XXX	X[a]	X[ab]	NE
Bisphenol A	XXX	XXX	X	NE	NE
Nonylphenol	XXX	XXX	XX	X	X

The higher is the number of Xs, the better is the scientific information.
NE: no evidence of impacts; PAH, polyaromatic hydrocarbon; PCB, polychlorinated biphenyl; TCDD, 2,3,7,8-tetrachlorodibenzo-p-dioxin; TCDE, Tetrachlorinated diphenyl ether; PBDE, polybrominated diphenyl ether.
[a]Sanchez et al. (2011).
[b]community level effects could occur in microbial communities (Schmieder and Edwards, 2012).
Modified after Sibley and Hanson (2011).

awareness and concern in society at large about environmental pollution, many laws and regulations were adopted and regulatory bodies are deeply involved in ensuring their implementation. For instance, all the legislation about the assessment of the ecological status of water bodies promotes the assessment of pollutants and other human pressures on these ecosystems in an integrative way, under an ecosystem approach management view (Chapter 14). We are currently witnessing a strong investment of researchers toward the applicability of academic findings, for instance under the auspices of the SETAC, such as the utility of the passive sampling methods for contaminated sediments (for details, see Chapter 3), the use of field-based biological effects and exposure techniques (Baird et al., 2007), the relevance of the tissue residue approach for toxicity assessment (for details, see Chapter 5), traits-based ecological risk assessment (*Integrated Environmental Assessment and Management*, Volume 7, Number 2), the relevance of ecological models in ecotoxicology and ecological risk assessment (Galic and Forbes, 2014), and the interaction of global climate change and contaminants (Stahl et al., 2013). During the past few years, regulatory bodies have integrated rapidly the more recent concepts and methodologies that have been developed in innovative scientific studies, producing key reports on advanced tools (use of passive samplers (USEPA, 2012b), bioavailability assessment and integration in environmental assessment (ITRC, 2011; ESTCP, 2012), "omics" (SCHER, SCENIHR, SCCS, 2013), mesocosms (OECD, 2004a; USEPA, 2011)), new concepts (AOPs (OECD, 2013) and integrated marine environmental monitoring (ICES, 2012)), emerging contaminants (EDCs (Kortenkamp et al., 2011; UNEP and WHO, 2013), nanomaterials (OECD, 2014a)), and the assessment of chemical mixtures (SCCS, SCHER, SCENIHR, 2011).

Developing and assessing adverse outcome pathways and establishing mechanistic effect models are important strategies for ensuring a robust coherent environmental risk assessment. In this aim, an effective integration of data from in silico, in vitro, and in vivo studies is needed. To address this problem, the availability of public toxicity data resources is crucial and database covering at least a part of the spectrum already exist (Dix et al., 2007; Goldsmith et al., 2014; Zhu et al., 2014). Predictive risk assessment would clearly benefit from the building of unifying databases to standardize the way data are collected, recorded, and presented (Artigas et al., 2012; Madden et al., 2014). The resulting big and multidimensional data sets provides an informatics challenge, requiring appropriate computational methods including modeling and mining tools (Dix et al., 2007; Zhu et al., 2014).

Retrospective risk assessment is indispensable to verify if the requirements laid down in predictive risk assessment are honored or if it is necessary to adopt decisions so that the environment and human health may be adequately protected. Very simplistically, Figure 19.4 questions the way in which scientific information is used in risk management decision, taking into account conventional toxicological parameters despite harsh and justified criticisms they have generated (Landis and Chapman, 2011; Jager, 2012). At very low and very high environmental concentrations, the decisions are not problematical. At intermediate levels, making the right decision (consolidate decision-making elements,

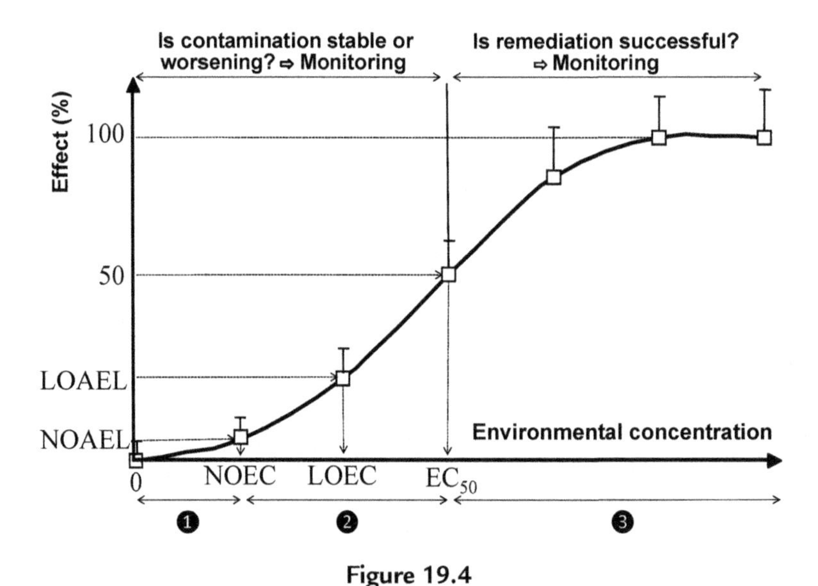

Figure 19.4

How scientific information is used in risk management decision. A simplified approach based on the concentration–effect relationship. E(L)C50: mean effective (lethal) concentration; LOAEL, low observed adverse effect level; LOEC, low observed effect concentration; NOAEL, no observed adverse effect level; NOEC, no observed effect concentration. ❶ concentration range at which no action is needed; ❷ concentration range at which is it questionable if the risk is "acceptable"; and ❸ concentration range at which remediation is needed. *(After Amiard and Amiard-Triquet, 2008 with permission)*.

establish a remediation plan) remains challenging. Anyway, on a wide range of environmental contamination, monitoring is indispensable and to date, the most traditional tools of chemical, biological and ecological monitoring have been consistently renewed (Chapters 3, 4, 7, 8 and 14).

In their safety evaluations of chemicals, public health agencies grant special prominence to studies based on standard bioassays carried out by using good laboratory practices (GLP). The GLP principles thereby help ensure that studies submitted to regulatory authorities, to notify or register chemicals, are of sufficient quality and rigor and are verifiable (e.g., OECD, 2004b about in vitro studies; OECD, 2014b for histopathology). Much less weight in risk assessments is given to a large number of non-GLP studies conducted in academic research units by the leading experts in various fields of science from around the world, despite they are crucial for emerging contaminants the study of which needs innovative practices (e.g., characterization of exposure to nanomaterials (Chapter 17), contaminants that exhibit nonlinear dose–response relationships (Chapter 15)). Although there is general agreement that regulatory submissions should be based on the best available science, concern is expressed that some regulations are framed in a way which made the integration of nonstandard methods challenging (ECETOC, 2013).

We have already mentioned (Chapter 2) than when establishing hazardous concentrations, it is frequently accepted that 5% of sensitive species may be neglected (for an example, see Figure 2.6 in this book). Thus the final objective is questionable: recovery of the Garden of Eden or sustainability of ecosystems for the benefits of nature to human beings? The notions of biological or ecological integrity and the good chemical and ecological status which are respectively at the basis of the US Clean Water Act and the EU Water Framework Directive are in agreement with the former, but in the real world, even the latter is far from being reached (EEA, 2013).

The content of this book about tools, methodologies, and concepts, coauthored by scientific experts from many different institutes worldwide, highlights how assessing environmental risks of chemicals entering the aquatic environment as a consequence of human activities is a complicated task. In only a few cases may the link between the presence of a chemical in the medium and deleterious effects be established indisputably (e.g., tributyltin in antifouling paints), leading to the ban of the incriminated substance. However, decades may be needed after the proof of noxiousness has been established and then the recovery of biota could take 15–30 years (Borja et al., 2010; Hering et al., 2010).

Recently, the European Environment Agency (EEA, 2013) has highlighted the dramatic extend of the problem, far beyond the aquatic environment. Entitled "Late lessons from early warnings: science, precaution, innovation," this report of more than 700 pages documents the major gaps of the current health and environmental regulation in Europe and elsewhere. In our domain of interest in this book, it focuses on emerging lessons from the degradation of natural systems and their wider implications for society (the pill and the feminization of fish, Jobling and Owen, in EEA, 2013); newly emerging and largescale products, technologies, and trends, which potentially offer many benefits but also potentially much harm to people and ecosystems and thereby ultimately economic development (nanotechnologies, Hansen et al., in EEA, 2013).

Considering such situations, the precautionary principle is the conceptualization of the common saying "an ounce of prevention is worth a pound of cure." Theoretically, in the absence of formal scientific proof but having an array of presumptions, a political authority can and must regulate the use of products that can pose a risk. In fact, it is needed to examine the pros and cons of decisions aiming at the protection of environmental and human health to avoid negative economic and social impacts. Unhappily, environmental (and even human) risk assessment of chemicals is not straightforward. The conventional risk assessment generates tremendous uncertainties, favoring the lobbying of opponents to protective measures. Even when more innovative and relevant approaches are adopted, it generally remains a certain amount of unavoidable uncertainties. As already mentioned (Section 19.4.2), the EU regulation of pesticides is considered as overly precautionary by farmers, and their representatives claim that it is because policy and regulatory decisions are not based on "sound science." Similarly, Dietrich et al. (2013) have written that "scientifically unfounded precaution drives European Commission's recommendations on EDC regulation, while defying common sense, well-established science and risk assessment principles," in complete contradiction with the best experts in the field (Bergman et al., 2013a,b). However, false positives are few and far between as compared with false negatives and carefully designed precautionary actions can stimulate innovation, even if the risk turns out not to be real or as serious as initially feared (Hansen and Tickner, in EEA, 2013).

Claims that environmental laws will destroy the economy have been regularly made and are consistently false. A graph prepared by Peter Gleick, Pacific Institute, shows that major environmental laws (Clean Air and Clean Water Acts, Endangered Species Act, Montreal protocol to protect the ozone layer) were passed without any negative impact on the US gross domestic product from 1929 to 2013 in real 2009 dollars (corrected for inflation) (http://www.huffingtonpost.com/peter-h-gleick/will-new-climate-regulati_b_5432090.html). It has been explained by the fact that if a peculiar sector of activity may be more or less impacted, the society at large benefits from these decisions through a reduction of health expenditures and the conservation of ecosystem services including provisioning, regulating, and cultural services that directly affect people and supporting services needed to maintain the other services (MEA, 2005).

According to the EEA (2013), societies are not making the most use of the costly lessons that can be gleaned from their histories. This regulatory body has analyzed why taking actions have been delayed. They pinpoint not only the novel and challenging nature of the issues themselves, the limited production of scientific information yet acquired but also the strong opposition by the corporate and scientific establishments of the day, and the tendency by the decision-making institutions, practices, and cultures to favor the status quo and the short-term perspective. For instance, the dossiers submitted in the framework of REACH demonstrate that companies prefer conventional procedures rather than innovative ones (ECHA, 2011; Cesnaitis et al., 2014; Tarazona et al., 2014). In addition, the registrants have widely used the read across or category approach (i.e., a study on one substance to cover information requirements for another) or multiple substances.

The use of these methods are often not well-justified, and neither are the experimental data provided in the dossiers that in some cases are not sufficient to meet information requirements under REACH (ECHA, 2011). These limitations are based on an economic rationale at the scale of companies because the chosen strategies require less labor and have lower analytical costs. However, pursuing an aim of general interest, it is absolutely needed to go beyond this corporative standpoint.

Risk assessment necessarily engages complementary expertise in different domains of "hard" sciences but in the light of the foregoing, the contribution of social and economic sciences appears clearly as crucial for better risk management and regulatory decision-making (Figure 19.5).

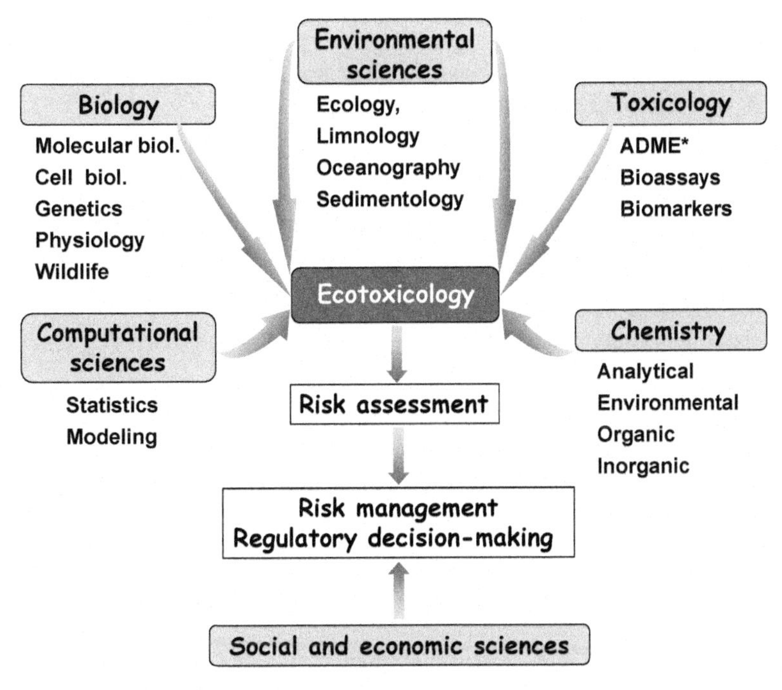

Figure 19.5
The necessary interdisciplinarity in risk assessment, risk management and regulatory decisions.
*Absorption, distribution, metabolism, excretion.

References

Adolfsson-Erici, M., Åkerman, G., McLachlan, M.S., 2012. Internal benchmarking improves precision and reduces animal requirements for determination of fish bioconcentration factors. Environ. Sci. Technol. 46, 8205–8211.

Amiard, J.C., Amiard-Triquet, C., 2008. Les biomarqueurs dans l'évaluation de l'état écologique des milieux aquatiques. Lavoisier, Tec&Doc, Paris. 372 p.

Amiard, J.C., Meunier, T., Babut, M., 2015. Les PCB. Environnement et santé. Lavoisier, Tech&Doc, Paris.

Amiard-Triquet, C., Amiard, J.C., Rainbow, P.S., 2013. Ecological Biomarkers: Indicators of Ecotoxicological Effects. CRC Press, Boca Raton.

Amiard-Triquet, C., Rainbow, P.S., 2009. Environmental Assessment of Estuarine Ecosystems. A Case Study. CRC Press, Boca Raton. 355 p.

Amiard-Triquet, C., Rainbow, P.S., Roméo, M., 2011. Tolerance to Environmental Contaminants. CRC Press, Boca Raton.

Andrady, A.L., 2011. Microplastics in the marine environment. Mar. Pollut. Bull. 62, 1596–1605.

Ankley, G.A., Bennett, R.E., Erickson, R.J., et al., 2010. Adverse outcome pathways: a conceptual framework to support ecotoxicology research and risk assessment. Environ. Toxicol. Chem. 29, 730–741.

Arthur, C., Baker, J., Bamford, H., 2009. In: Proceedings of the International Research. Workshop on the Occurence, Effects and Fate of Microplastic Marine Debris NOAA Technical Memorandum NOS-OR&R-30, p. 49.

Artigas, J., Arts, G., Babut, M., et al., 2012. Towards a renewed research agenda in ecotoxicology. Environ. Pollut. 160, 201–206.

Ashauer, R., Brown, C.D., 2013. Highly time–variable exposure to chemicals—Toward an assessment strategy. Integr. Environ. Assess. Manag. 9, e27–e33.

Ashauer, R., Escher, B.I., 2010. Advantages of toxicokinetic and toxicodynamic modelling in aquatic ecotoxicology and risk assessment. J. Environ. Monit. 12, 2056–2061.

ASTM, 2012. Standard Guide for Behavioral Testing in Aquatic Toxicology, E1604 – 12. http://www.astm.org/Standards/E1604.htm (accessed 18.02.15).

Bae, M.J., Park, Y.S., 2014. Biological early warning system based on the responses of aquatic organisms to disturbances: a review. Sci. Total Environ. 466–467, 635–649.

Baird, D.J., Burton, G.A., Culp, J.M., et al., 2007. Summary and recommendations from a SETAC Pellston workshop on in situ measures of ecological effects. Integr. Environ. Assess. Manag. 3, 275–278.

Balmer, N.V., Dao, T., Leist, M., et al., 2014. Application of "Omics" technologies to in vitro toxicology. In: Bal-Price, A., Jennings, P. (Eds.), In Vitro Toxicology Systems, Methods in Pharmacology and Toxicology. Springer Science + Business Media, New-York.

Barnes, D.K.A., Galgani, F., Thompson, R.C., et al., 2009. Accumulation and fragmentation of plastic debris in global environments. Philos. Trans. R. Soc. 364B, 1985–1998.

Basu, N., 2015. Applications and implications of neurochemical biomarkers in environmental toxicology. Environ. Toxicol. Chem. 34, 22–29.

Bejarano, A.C., Clark, J.R., Coelho, G.M., 2014. Issues and challenges with oil toxicity data and implications for their use in decision making: a quantitative review. Environ. Toxicol. Chem. 33, 732–742.

Benedetti, M., Ciaprini, F., Piva, F., et al., 2012. A multidisciplinary weight of evidence approach for classifying polluted sediments: Integrating sediment chemistry, bioavailability, biomarkers responses and bioassays. Environ. Int. 38, 17–28.

Bergman, Å., Heindel, J.J., Jobling, S., et al., 2013a. State of the Science of Endocrine Disrupting Chemicals – 2012. UNEP and WHO. 260 p.

Bergman, A., Heindel, J.J., Kasten, T., et al., 2013b. The impact of endocrine disruption: a consensus statement on the state of the science. Environ. Health Perspect. 121, 104–106.

Billoir, E., Delhaye, H., Forfait, C., et al., 2012. Comparison of bioassays with different exposure time patterns: the added value of dynamic modelling in predictive ecotoxicology. Ecotoxicol. Environ. Saf. 75, 80–86.

Bocquené, G., Abarnou, A., Boulon, A.I., et al., 2012. RISKENSEINE – Risque sanitaire et environnemental d'origine chimique dans l'estuaire de la Seine. http://seine-aval.crihan.fr/web/attached_file/componentId/kmelia63/attachmentId/32520/lang/fr/name/RISKENSEINE_mars_2012_red_pro.pdf.

Borja, A., Dauer, D.M., Elliott, M., et al., 2010. Medium and long-term recovery of estuarine and coastal ecosystems: patterns, rates and restoration effectiveness. Estuar. Coasts 33, 1250–1260.

Browne, M.A., Dissanayake, A., Galloway, T.S., et al., 2008. Ingested microscopic plastic translocates to the circulatory system of the mussel, *Mytilus edulis* (L.). Environ. Sci. Technol. 42, 5026–5031.

Burton, G.A., Johnston, E.L., 2010. Assessing contaminated sediments in the context of multiple stressors. Environ. Toxicol. Chem. 29, 2625–2643.

Calabrese, E.J., Blain, R.B., 2011. The hormesis database: the occurrence of hormetic dose responses in the toxicological literature. Regul. Toxicol. Pharmacol. 61, 73–81.

Carson, R., 1962. Silent Spring. Houghton Mifflin. Boston re-published by Mariner Books (2002).

Carter, L.J., Ashauer, R., Ryan, J.J., et al., 2014. Minimised bioconcentration tests: a useful tool for assessing chemical uptake into terrestrial and aquatic invertebrates? Environ. Sci. Technol. 48, 13497–13503.

Cesnaitis, R., Sobanska, M.A., Versonnen, B., et al., 2014. Analysis of the ecotoxicity data submitted within the framework of the REACH Regulation. Part 3. Experimental sediment toxicity assays. Sci. Total Environ. 475, 116–122.

Chapman, P.M., Wang, F., Caeiro, S.S., 2013. Assessing and managing sediment contamination in transitional waters. Environ. Int. 55, 71–91.

Claessens, M., De Meester, S., Van Landuyt, L., et al., 2011. Occurrence and distribution of microplastics in marine sediments along the Belgian coast. Mar. Pollut. Bull. 62, 2199–2204.

Coelho, G., Clark, J., Aurand, D., 2013. Toxicity testing of dispersed oil requires adherence to standardized protocols to assess potential real world effects. Environ. Pollut. 177, 185–188.

Collignon, A., Hecq, J.-H., Galgani, F., et al., 2012. Neustonic microplastic and zooplankton in the North Western Mediterranean Sea. Mar. Pollut. Bull. 64, 861–864.

Conder, J.M., Fuchsman, P.C., Grover, M.M., et al., 2015. Critical review of mercury sediment quality values for the protection of benthic invertebrates. Environ. Toxicol. Chem. 34, 6–21.

Connors, K.A., Voutchkova-Kostal, A.M., Kostal, K.J., et al., 2014. Reducing aquatic hazards of industrial chemicals: probabilistic assessment of sustainable molecular design guidelines. Environ. Toxicol. Chem. 33, 1894–1902.

Creton, S., Weltje, L., Hobson, H., 2013. Reducing the number of fish in bioconcentration studies for plant protection products by reducing the number of test concentrations. Chemosphere 90, 1300–1304.

Dagnino, A., Viarengo, A., 2014. Development of a decision support system to manage contamination in marine ecosystems. Sci. Total Environ. 466–467, 119–126.

De Hoop, L., Huijbregts, M.A.J., Schipper, A.M., et al., 2013. Modelling bioaccumulation of oil constituents in aquatic species. Mar. Pollut. Bull. 76, 178–186.

De Jourdan, B.P., Hanson, M.L., Muir, D.C.G., et al., 2014. Fathead minnow (*Pimephales promelas* Rafinesque) exposure to three novel brominated flame retardants in outdoor mesocosms: bioaccumulation and biotransformation. Environ. Toxicol. Chem. 33, 1148–1155.

Dekiff, J.H., Remy, D., Klasmeier, J., et al., 2014. Occurrence and spatial distribution of microplastics in sediments from Norderney. Environ. Pollut. 186, 248–256.

De Laender, F., De Schamphelaere, K.A.C., Vanrolleghem, P.A., et al., 2009. Comparing ecotoxicological effect concentrations of chemicals established in multi-species vs. single-species toxicity test systems. Ecotoxicol. Environ. Saf. 72, 310–315.

De Laender, F., VanOevelen, D., Middelburg, J.J., et al., 2010. Uncertainties in ecological, chemical and physiological parameters of a bioaccumulation model: implications for internal concentrations and tissue based risk quotients. Ecotox. Environ. Saf. 73, 240–246.

Depledge, M.H., 1994. The rational basis for the use of biomarkers as ecotoxicological tools. In: Fossi, M.C., Leonzio, C. (Eds.), Nondestructive Biomarkers in Vertebrates. Lewis Publishers, Boca Raton, pp. 261–285.

Devillers, J. (Ed.), 2009. Ecotoxicology Modeling. Springer, Dordrecht, p. 399.

Dietrich, D.R., von Aulock, S., Marquardt, H., et al., 2013. Scientifically unfounded precaution drives European Commission's recommendations on EDC regulation, while defying common sense, well-established science and risk assessment principles. Chem. Biol. Interact. 205, A1–A5.

Ding, G., Peijnenburg, W.J.G.M., 2013. Physicochemical properties and aquatic toxicity of poly- and perfluorinated compounds. Crit. Rev. Environ. Sci. Technol. 43, 598–678.

Dix, D.J., Houck, K.A., Martin, M.T., et al., 2007. The ToxCast program for prioritizing toxicity testing of environmental chemicals. Toxicol. Sci. 95, 5–12.

Dutton, J., Fisher, N.S., 2014. Modeling metal bioaccumulation and tissue distribution in killifish (*Fundulus heteroclitus*) in three contaminated estuaries. Environ. Toxicol. Chem. 33, 89–101.

ECETOC, 2013. Mode of Action: Recent Developments, Regulatory Application and Future Work. Workshop Report, No. 26.

ECHA (European Chemicals Agency), 2011. The Use of Alternatives to Testing on Animals for the REACH Regulation. ECHA-11-R-004.2-EN.

EEA, 2013. Late Lessons from Early Warnings: Science, Precaution, Innovation. EEA Report No 1/2013, 750 p. http://dx.doi.org/10.2800/70069.

EFSA (European Food Safety Authority), 2006. Opinion of the scientific panel on food additives, flavourings, processing aids and materials in contact with food on a request from the commission related to 2,2-bis(4-hydroxyphenyl)propane (Bisphenol A). The EFSA Journal 2006, 428, 1–75.

EFSA (European Food Security Authority), 2009. Use of the benchmark dose approach in risk assessment1 Guidance of the Scientific Committee (Question No EFSA-Q-2005-232). EFSA J. 1150, 1–72.

EFSA, 2015. Scientific opinion on the risks to public health related to the presence of bisphenol A (BPA) in foodstuffs. EFSA J. 13 (1), 3978. http://www.efsa.europa.eu/en/efsajournal/pub/3978.htm (accessed 10.02.15).

Eggen, R.I.L., Hollender, J., Joss, A., et al., 2014. Reducing the discharge of micropollutants in the aquatic environment: the benefits of upgrading wastewater treatment plants. Environ. Sci. Technol. 48, 7683–7689.

Eladak, S., Grisin, T., Moison, D., et al., 2015. A new chapter in the bisphenol A story: bisphenol S and bisphenol F are not safe alternatives to this compound. Fertil. Steril. 103, 11–21.

Embry, M.R., Belanger, S.E., Braunbeck, T.A., et al., 2010. The fish embryo toxicity test as an animal alternative method in hazard and risk assessment and scientific research. Aquat. Toxicol. 97, 79–87.

Engler, R.E., 2012. The complex interaction between marine debris and toxic chemicals in the ocean. Environ. Sci. Technol. 46, 12302–12315.

Eriksen, M., Lebreton, L.C.M., Carson, H.S., et al., 2014. Plastic pollution in the world's oceans: more than 5 trillion plastic pieces weighing over 250,000 tons afloat at sea. PLoS ONE 9 (12), e111913.

Escher, B.I., Ashauer, R., Dyer, S., et al., 2011. Crucial role of mechanisms and modes of toxic action for understanding tissue residue toxicity and internal effect concentrations of organic chemicals. Integr. Environ. Assess. Manag. 7, 28–49.

ESTCP, 2012. Demonstration and Evaluation of Solid Phase Microextraction for the Assessment of Bioavailability and Contaminant Mobility Cost and Performance Report. Environmental security technology certification program U.S. Department of Defense (ER-200624).

EU, 2003. Directive 2003/53/EC of the European Parliament and of the Council Amending Council Directive 76/769/EEC Relating to Restrictions on the Marketing and Use of Certain Dangerous Substances and Preparations (Nonylphenol, Nonylphenol Ethoxylate and Cement). http://eur-lex.europa.eu/legal-content/EN/TXT/PDF/?uri=CELEX:32003L0053&rid=1 (accessed 10.02.15).

EU, 2009. Directive 2009/128/EC of the European Parliament and of the Council Establishing a Framework for Community Action to Achieve the Sustainable Use of Pesticides. http://eur-lex.europa.eu/legal-content/EN/TXT/PDF/?uri=CELEX:02009L0128-20091125&from=EN (accessed 10.02.15).

Fetter, E., Krauss, M., Brion, F., et al., 2014. Effect-directed analysis for estrogenic compounds in a fluvial sediment sample using transgenic cyp19a1b-GFP zebrafish embryos. Aquat. Toxicol. 154, 221–229.

Forbes, V.E., Calow, P., Grimm, V., et al., 2011. Adding value to ecological risk assessement with population modeling. Hum. Ecol. Risk Assess. 17, 287–299.

Fulton, M.H., Key, P.B., 2001. Acetylcholinesterase inhibition in estuarine fish and invertebrates as an indicator of organophosphorous insecticide exposure and effects. Environ. Toxicol. Chem. 20, 37–45.

Gabsi, F., Schäffer, A., Preuss, T.G., 2014. Predicting the sensitivity of populations from individual exposure to chemicals: the role of ecological interactions. Environ. Toxicol. Chem. 33, 1449–1457.

Gaillard, J., Banas, D., Thomas, M., et al., 2014. Bioavailability and bioaccumulation of sediment-bound polychlorinated biphenyls to carp. Environ. Toxicol. Chem. 33, 1324–1330.

Galgani, F., Fleet, D., Van Franeker, J., et al., 2010. In: Zampoukas, N. (Ed.), Marine Strategy Framework Directive Task Group 10 Report, Marine Litter. EUR 24340 EN e 2010.

Galic, N., Forbes, V., 2014. Ecological models in ecotoxicology and ecological risk assessment: an introduction to the special section. Environ. Toxicol. Chem. 33, 1446–1448.

Galic, N., Hommen, U., Baveco, J.H., et al., 2010. Potential application of population models in the European ecological risk assessment of chemicals II: review of models and their potential to address environmental protection aims. Integr. Environ. Assess. Manag. 6, 338–360.

Gerhardt, A., Ingram, M.K., Kang, I.J., et al., 2006. In situ on-line toxicity biomonitoring in water: recent developments. Environ. Toxicol. Chem. 25, 2263–2271.

Giesy, J.P., Naile, J.E., Khim, J.S., et al., 2010. Aquatic toxicology of perfluorinated chemicals. In: Whitacre, D.M. (Ed.), Rev. Environ. Contam. Toxicol., 202, pp. 1–52.

Goldsmith, M.R., Grulke, C.M., Chang, D.T., et al., 2014. DockScreen: a database of in silico biomolecular interactions to support computational toxicology. Dataset Pap. Sci. Article ID 421693, 6 p.

Grimm, V., Ashauer, R., Forbes, V., et al., 2009. CREAM: a European project on mechanistic effect models for ecological risk assessment of chemicals. Environ. Sci. Pollut. Res. 16, 614–617.

Hartung, T., Luechtefeld, T., Maertens, A., et al., 2013. Food for thought …Integrated testing strategies for safety assessments. ALTEX 30, 3–18.

Hering, D., Borja, A., Carstensen, J., et al., 2010. The European Water Framework Directive at the age of 10: a critical review of the achievements with recommendations for the future. Sci. Total Environ. 408, 4007–4019.

Hernández-Altamirano, R., Mena-Cervantes, V.Y., Perez-Miranda, S., et al., 2010. Molecular design and QSAR study of low acute toxicity biocides with 4,4′-dimorpholyl-methane core obtained by microwave-assisted synthesis. Green Chem. 12, 1036–1048.

Hosokawa, Y., 1993. Remediation work for mercury contaminated bay – experiences of Minamata bay project. Jpn. Water Sci. Technol. 28, 339–348.

Hritonenko, N., Yatsenko, Y., 2013. Models of water pollution propagation. In: Hritonenko, N., Yatsenko, Y. (Eds.), Mathematical Modeling in Economics, Ecology and the Environment. Optimization and Its Applications, 88. Springer, pp. 197–217.

ICES, 2012. Integrated marine environmental monitoring of chemicals and their effects. In: Davies, I.M., Vethaak, A.D. (Eds.), ICES Cooperative Research Report No. 315, p. 277.

ITRC (Interstate Technology & Regulatory Council), 2011. Incorporating Bioavailability Considerations into the Evaluation of Contaminated Sediment Sites. CS-1. Interstate Technology & Regulatory Council, Contaminated Sediments Team, Washington, D.C. http://www.itrcweb.org/contseds-bioavailability/cs_1.pdf (accessed 31.07.14).

Jager, T., 2012. Bad habits die hard: the NOEC's persistence reflects poorly on ecotoxicology. Environ. Toxicol. Chem. 31, 228–229.

James, I.D., 2002. Modelling pollution dispersion, the ecosystem and water quality in coastal waters: a review. Environ. Model. Software 17, 363–385.

Jang, Y., Kim, J.E., Jeong, S.H., et al., 2014. Towards a strategic approaches in alternative tests for pesticide safety. Toxicol. Res. 30, 159–168.

Kapo, K.E., Holmes, C.M., Dyer, S.D., et al., 2014. Developing a foundation for eco-epidemiological assessment of aquatic ecological status over large geographic regions utilizing existing data resources and models. Environ. Toxicol. Chem. 33, 1665–1677.

Kattwinkel, M., Liess, M., 2014. Competition matters: species interactions prolong the long-term effects of pulsed toxicant stress on populations. Environ. Toxicol. Chem. 33, 1458–1465.

Kimberly, D.A., Salice, C.J., 2013. Interactive effects of contaminants and climate-related stressors. Environ. Toxicol. Chem. 32, 1337–1343.

Kinch, C.D., Ibhazehiebob, K., Jeong, J.H., et al., 2015. Low-dose exposure to bisphenol A and replacement bisphenol S induces precocious hypothalamic neurogenesis in embryonic zebrafish. PNAS 112, 1475–1480.

Koelmans, A.A., Gouin, T., Thompson, R., et al., 2014. Plastics in the marine environment. Environ. Toxicol. Chem. 33, 5–10.

Köhler, H.R., Triebskorn, R., 2013. Wildlife ecotoxicology of pesticides: can we track effects to the population level and beyond? Science 341, 759–765.

Kortenkamp, A., Evans, R., Martin, O., et al., 2011. State of the Art Assessment of Endocrine Disrupters. Final Report. Project Contract Number 070307/2009/550687/SER/D, 135 p.

Kramer, V.J., Etterson, M.A., Hecker, M., et al., 2011. Adverse outcome pathways and ecological risk assessment: bridging to population-level effects. Environ. Toxicol. Chem. 30, 64–76.

Kretschmann, A., Ashauer, R., Hollender, J., et al., 2012. Toxicokinetic and toxicodynamic model for diazinon toxicity – Mechanistic explanation of differences in the sensitivity of *Daphnia magna* and *Gammarus pulex*. Environ. Toxicol. Chem. 31, 2014–2022.

Laist, D., 1997. Impacts of marine debris: entanglement of marine life in marine debris including a comprehensive list of species with entanglement and ingestion records. In: Coe, J., Rogers, D. (Eds.), Marine Debris: Sources, Impact and Solutions. Springer Verlag, New York, pp. 99–141.

Landis, W.G., Chapman, P.M., 2011. Well past time to stop using NOELs and LOELs. Integr. Environ. Assess. Manag. 7, vi–viii.

Lu, X.X., Hong, Y., Reible, D.D., 2014. Assessing bioavailability of hydrophobic organic compounds and metals in sediments using freely available porewater concentrations. In: Reible, D.D. (Ed.), Processes, Assessment and Remediation of Contaminated Sediments, SERDP ESTCP Environmental Remediation Technology, 6, pp. 177–196.

MacDonald, D.D., Carr, R.S., Calder, F.D., et al., 1996. Development and evaluation of sediment quality guidelines for Florida coastal waters. Ecotoxicology 5, 253–278.

Madden, J.C., Rogiers, V., Vinken, M., 2014. Application of in silico and in vitro methods in the development of adverse outcome pathway constructs in wildlife. Philos. Trans. R. Soc. 369B 20130584.

Martin, J.D., Adams, J., Hollebone, B., et al., 2014. Chronic toxicity of heavy fuel oils to fish embryos using multiple exposure scenarios. Environ. Toxicol. Chem. 33, 677–687.

Maughan, J.T., 2014. Environmental Impact Analysis: Process and Methods. Chap. 7: Environmental Analysis Tools. CRC Press. pp. 261–312.

MEA (Millenium Ecosystem Assessment), 2005. Ecosystems and Human Well-Being: Synthesis. Island Press, Washington, DC. World Resources Institute. http://pdf.wri.org/ecosystems_human_wellbeing.pdf.

Nallani, G.C., Paulos, P.M., Constantine, L.A., et al., 2011. Bioconcentration of ibuprofen in fathead minnow (*Pimephales promelas*) and channel catfish (*Ictalurus punctatus*). Chemosphere 84, 1371–1377.

NAS, 1980. The International Mussel Watch. Report of a Workshop Sponsored by the Environment Studies Board. Natural Resources Commission, National Research Council, National Academy of Sciences, Washington, DC.

NFU (National Farmers' Union), CPA (Crop Protection Association), AIC (Agricultural Industries Confederation), 2014. Healthy Harvest – the Impact of Losing Plant Protection Products on UK Food Production. http://www.nfuonline.com/healthyharvest_final_digital/.

Noyes, P.D., McElwee, M.K., Miller, H.D., et al., 2009. The toxicology of climate change: environmental contaminants in a warming world. Environ. Int. 35, 971–986.

Nyman, A.M., Schirmer, K., Ashauer, R., 2014. Importance of toxicokinetics for interspecies variation in sensitivity to chemicals. Environ. Sci. Technol. 48, 5946–5954.

OECD, 2004a. Draft Guidance Document on Simulated Freshwater Lentic Field Tests (Outdoor Microcosms and Mesocosms). http://www.oecd.org/fr/securitechimique/essais/32612239.pdf (accessed 13.01.14).

OECD, 2004b. The Application of the Principles of GLP to in Vitro Studies. Series on Principles of Good Laboratory Practice and Compliance Monitoring No. 14, ENV/JM/MONO(2004)26.

OECD, 2012. Test No. 305: Bioaccumulation in Fish: Aqueous and Dietary Exposure. http://www.oecd-ilibrary.org/content/book/9789264185296-en.

OECD, 2013. Guidance Document on Developing and Assessing Adverse Outcome Pathways. Series on Testing and Assessment No. 184, ENV/JM/MONO(2013)6.

OECD, 2014a. Ecotoxicology and Environmental Fate of Manufactured Nanomaterials: Test Guidelines. Series on the Safety of Manufactured Nanomaterials No. 40, ENV/JM/MONO(2014)1.

OECD, 2014b. Guidance on the GLP Requirements for Peer Review of Histopathology. Series on Principles of Good Laboratory Practice and Compliance Monitoring No. 16, ENV/JM/MONO(2014)30.

Papa, E., Gramatica, P., 2010. QSPR as a support for the EU REACH regulation and rational design of environmentally safer chemicals: PBT identification from molecular structure. Green Chem. 12, 836–843.

Payne, J.F., Mathieu, A., Melvin, W., et al., 1996. Acetylcholinesterase, an old biomarker with a new future? Field trials in association with two urban rivers and a paper mill in Newfoundland. Mar. Pollut. Bull. 32, 225–231.

PlasticsEurope, 2013. Plastics in the Facts 2013. An Analysis of European Latest Plastics Production, Demand and Waste Data. Plastics Europe: Association of Plastic Manufacturers, Brussels. p. 40.

Powers, C.M., Grieger, K.D., Hendren, C.O., et al., 2014. A web-based tool to engage stakeholders in informing research planning for future decisions on emerging materials. Sci. Total Environ. 470–471, 660–668.

Qin, L.T., Liu, S.S., Liu, H.L., et al., 2010. Support vector regression and least squares support vector regression for hormetic dose–response curves fitting. Chemosphere 78, 327–334.

Rico-Martínez, R., Snell, T.W., Shearer, T.L., 2013. Synergistic toxicity of Macondo crude oil and dispersant Corexit 9500A_ to the *Brachionus plicatilis* species complex (Rotifera). Environ. Pollut. 173, 5–10.

Ritter, L., Austin, C.P., Bend, J.R., et al., 2012. Integrating Emerging Technologies into Chemical Safety Assessment. The Expert Panel on the Integrated Testing of Pesticides. Council of Canadian Academies, Ottawa (Ontario), Canada. 291 p.

Rubach, M., Baird, D., Boerwinkel, M.C., et al., 2012. Species traits as predictors for intrinsic sensitivity of aquatic invertebrates to the insecticide chlorpyrifos. Ecotoxicology 21, 2088–2101.

Russom, C.L., LaLone, C.A., Villeneuve, D.L., et al., 2014. Development of an adverse outcome pathway for acetylcholinesterase inhibition leading to acute mortality. Environ. Toxicol. Chem. 33, 2157–2169.

Sanchez, W., Sremski, W., Piccini, B., et al., 2011. Adverse effects in wild fish living downstream from pharmaceutical manufacture discharges. Environ. Int. 37, 1342–1348.

SCCS, SCHER, SCENIHR, 2011. Toxicity and Assessment of Chemical Mixtures. http://ec.europa.eu/health/scientific_committees/environmental_risks/docs/scher_o_150.pdf (accessed 11.12.13).

SCHER, SCENIHR, SCCS, 2013. Addressing the New Challenges for Risk Assessment. http://ec.europa.eu/health/scientific_committees/emerging/docs/scenihr_o_037.pdf.

Schmieder, R., Edwards, R., 2012. Insights into antibiotic resistance through metagenomic approaches. Future Microbiol. 7, 73–89.

Schmolke, A., Thorbek, P., Chapman, P., et al., 2010. Ecological models and pesticide risk assessment: current modelling practice. Environ. Toxicol. Chem. 29, 1006–1012.

Scholz, S., Sela, E., Blaha, L., et al., 2013. A European perspective on alternatives to animal testing for environmental hazard identification and risk assessment. Regul. Toxicol. Pharmacol. 67, 506–530.

Short, S., Yang, G., Guler, Y., et al., 2014. Crustacean intersexuality is feminization without demasculinization: implications for environmental toxicology. Environ. Sci. Technol. 48, 13520–13529.

Sibley, P.K., Hanson, M.L., 2011. Ecological impacts of organic chemicals on freshwater ecosystems. In: Sánchez-Bayo, F., van den Brink, P.J., Mann, R.M. (Eds.), Ecological Impacts of Toxic Chemicals. Bentham Science Publishers Ltd, Oak Park, IL, pp. 138–164.

Sogorb, M.A., Pamies, D., de Lapuentec, J., et al., 2014. An integrated approach for detecting embryotoxicity and developmental toxicity of environmental contaminants using in vitro alternative methods. Toxicol. Lett. 230, 356–367.

Springer, T.A., Guiney, P.D., Krueger, H.O., et al., 2008. Assessment of an approach to estimating aquatic bioconcentration factors using reduced sampling. Environ. Toxicol. Chem. 27, 2271–2280.

Stahl Jr., R.G., Hooper, M.J., Balbus, J.M., et al., 2013. The influence of global climate change on the scientific foundations and applications of environmental toxicology and chemistry: Introduction to a SETAC international workshop. Environ. Toxicol. Chem. 32, 13–19.

Tarazona, J.V., Sobanska, M.A., Cesnaitis, R., et al., 2014. Analysis of the ecotoxicity data submitted within the framework of the REACH Regulation. Part 2. Experimental aquatic toxicity assays. Sci. Total Environ. 472, 137–145.

Teuten, E.L., Saquing, J.M., Knappe, D.R.U., et al., 2009. Transport and release of chemicals from plastics to the environment and to wildlife. Philos. Trans. R. Soc. 364B, 2027–2045.

Thompson, R.C., Olsen, Y., Mitchell, R.P., et al., 2004. Lost at sea: where is all the plastic? Science 304, 838.

Truhaut, R., 1977. Eco-toxicology - objectives, principles and perspectives. Ecotoxicol. Environ. Saf. 1, 151–173.

UNEP, WHO, 2013. In: Bergman, Å., Heindel, J.J., Jobling, S., et al. (Eds.), State of the Science of Endocrine Disrupting Chemicals – 2012, p. 260.

USEPA, 2005. Contaminated Sediment Remediation Guidance for Hazardous Waste Sites. http://www.epa.gov/superfund/health/conmedia/sediment/pdfs/guidance.pdfhttp://www.epa.gov/superfund/health/conmedia/sediment/pdfs/ch6.pdf.

USEPA, 2011. Experimental Stream Facility: Design and Research. EPA 600/F-11/004. Technical Report. Cincinnati, OH. http://nepis.epa.gov/Adobe/PDF/P100E4QH.pdf (accessed 23.01.14).

USEPA, 2012a. Benchmark Dose Technical Guidance and Other Relevant Risk Assessment Documents. http://www.epa.gov/raf/publications/benchmarkdose.htm.

USEPA, 2012b. Guidelines for Using Passive Samplers to Monitor Organic Contaminants at Superfund Sediment Sites - Sediment Assessment and Monitoring Sheet (SAMS) #3. http://www.epa.gov/superfund/health/conmedia/sediment/pdfs/Passive_Sampler_SAMS_Final_Camera_Ready_-_Jan_2013.pdf (accessed 16.02.15).

USGAO (United States Government Accountability Office), 2007. Chemical Regulation – Comparison of U.S. and Recently Enacted European Union Approaches to Protect Against the Risks of Toxic Chemicals. Rep. GAO-07-825. www.gao.gov/new.items/d07825.pdf (accessed 11.02.15).

Villeneuve, D., Volz, D.C., Embry, M.R., et al., 2014. Investigating alternatives to the fish early-life stage test: a strategy for discovering and annotating adverse outcome pathways for early fish development. Environ. Toxicol. Chem. 33, 158–169.

Wang, M., Riffel, M., 2011. Making the right conclusions based on wrong results and small sample sizes: Interpretation of statistical tests in ecotoxicology. Ecotoxicol. Environ. Saf. 74, 684–692.

Wright, S.L., Thompson, R.C., Galloway, T.S., 2013. The physical impacts of microplastics on marine organisms: a review. Environ. Pollut. 178, 483–492.

Xie, L., Klerks, P.L., 2003. Responses to selection for cadmium resistance in the least killifish, *Heterandria formosa*. Environ. Toxicol. Chem. 22, 313–320.

Zhu, H., Zhang, J., Kim, M.T., et al., 2014. Big data in chemical toxicity research: the use of high-throughput screening assays to identify potential toxicants. Chem. Res. Toxicol. 27, 1643–1651.

Zhu, X.W., Liu, S.S., Qin, L.T., et al., 2013. Modeling non-monotonic dose-response relationships: model evaluation and hormetic quantities exploration. Ecotoxicol. Environ. Saf. 89, 130–136.

Zhuang, W., Gao, X., 2014. Integrated assessment of heavy metal pollution in the surface sediments of the Laizhou Bay and the coastal waters of the Zhangzi Island, China: comparison among typical marine sediment quality indices. PLoS ONE 9 (4), e94145.

Index

Note: Page numbers followed by "f" and "t" indicate figures and tables respectively.

I

L

M

N

O

P

Q

R

S

T

Printed in the United States
By Bookmasters